T0325510

IONIC LIQUIDS COMPLETELY UnCOILed

IONIC LIQUIDS COMPLETELY UNCOILED

Critical Expert Overviews

Edited by

Natalia V. Plechkova
The Queen's University of Belfast

Kenneth R. Seddon
The Queen's University of Belfast

Published by John Wiley & Sons, Inc., Hoboken, New Jersey
Published simultaneously in Canada

Library of Congress Cataloging-in-Publication Data:

Plechkova, Natalia V., editor.
Ionic liquids completely uncoiled : critical expert overviews / edited by Natalia V. Plechkova, Kenneth R. Seddon.
pages cm
Includes bibliographical references and index.
ISBN 978-1-118-43906-7 (cloth)
1. Ionic solutions. 2. Ionic structure. I. Seddon, Kenneth R., 1950– editor. II. Title.
QD561.P56 2015
541′.3723–dc23

2015025270

Cover image courtesy of Kenneth Seddon, Natalia Plechkova, and Martyn Earle

Set in 10/12pt Times by SPi Global, Pondicherry, India

Printed in the United States of America

10 9 8 7 6 5 4 3 2 1

1 2015

CONTENTS

The colored version of the figures are available in the e-versions of the books.

CONTENTS OF "IONIC LIQUIDS UNCOILED"

CONTENTS OF "IONIC LIQUIDS FURTHER UNCOILED"

COIL CONFERENCES

COIL-1	Salzburg	Austria	2005
COIL-2	Yokohama	Japan	2007
COIL-3	Cairns	Australia	2009
COIL-4	Washington	USA	2011
COIL-5	Algarve	Portugal	2013
COIL-6	Jeju Island	South Korea	2015
COIL-7	Ottawa	Canada	2017
COIL-8	Belfast[a]	UK	2019

[a] Precise location still to be confirmed.

PREFACE

This is the third and final book of three volumes of critical overviews of the key areas of ionic liquid chemistry. The first volume was entitled *Ionic Liquids UnCOILed*; the second was *Ionic Liquids Further UnCOILed*. The history and rationale behind this trilogy were explained in the preface to Volume 1 and so will not be repeated here. But we did instruct the authors as follows: 'It is important to emphasise that these are meant to be critical reviews. We are not looking for comprehensive coverage, but insight, appreciation and prospect. We want the type of review which can be read to give a sense of importance and scope of the area, highlighting this by the best published work and looking for the direction in which the field is moving. We would also like the problems with the area highlighting, *e.g.* poor experimental technique, poor selection of liquids, and variability of data'. Looking back over all three books, we are amazed at the quality of reviews produced and their 'timeless' nature – they are fresh and inciteful.

This final book includes eleven critical expert overviews of differing aspects of ionic liquids – the final chapter could almost be a stand-alone book. It is our continuing view that, in the second decade of the twenty-first century, reviews that merely regurgitate a list of all papers on a topic, giving a few lines or a paragraph (often the abstract!) to each one, have had their day – 5 min with an online search engine will provide that information. But we are sure that the growth of open-access journals and books from predatory online publishers will guarantee their prolonged existence. Such reviews belong with cassette tapes, typewriters and the printed journal – valuable in their day, but of little value now. The value of a review lies in the expertise and insight of the reviewer and their willingness to share it with the reader. It takes moral courage to say 'the work of [...] is irreproducible, or of poor quality, or that the conclusions are not valid' – but in a field expanding at the prestigious rate of ionic liquids, it is essential to have this honest feedback. Otherwise, errors are propagated. Papers still, in 2015, appear using hexafluorophosphate or tetrafluoroborate ionic liquids for synthetic or catalytic chemistry, and calculations on 'ion pairs' are still being used to rationalise liquid state properties! We trust this volume, containing eleven excellently perceptive reviews, will help guide and secure the future of ionic liquids. We believe the reviews in our volumes should be compulsory reading for all research workers in the field.

Natalia V. Plechkova
Kenneth R. Seddon

ACKNOWLEDGEMENTS

This volume is a collaborative effort. We, the editors, have our names emblazoned on the cover, but the book would not exist in its present form without the support from many people. Firstly, we thank our authors for producing such splendid, critical chapters and for their open responses to the reviewers' comments and to editorial suggestions. We are also indebted to our team of expert reviewers, whose comments on the individual chapters were challenging and thought provoking, and to Martyn J. Earle for his photographic assistance. The backing from the team at Wiley, led by Dr. Arza Seidel, has been fully appreciated – it is always a pleasure to work with such a professional group of people. Finally, this book would never have been published without the unfailing, enthusiastic support from Deborah Poland and Sinead McCullough, whose patience and endurance continue to make the impossible happen. So we thank again everyone involved in the project – we are proud to have been associated with them.

Natalia V. Plechkova
Kenneth R. Seddon

CONTRIBUTORS

ANDREW P. ABBOTT, Chemistry Department, University of Leicester, Leicester, UK

JÜRGEN ARNING, Department 10: Theoretical Ecology, UFT-Centre for Environmental Research and Sustainable Technology, University of Bremen, Bremen, Germany

STEVEN BALDELLI, Department of Chemistry, University of Houston, Houston, Texas, USA

PAUL M. BAYLEY, Institute for Frontier Materials, Geelong Technology Precinct, Deakin University, Victoria, Australia

JOSÉ N. CANONGIA LOPES, Centro de Química Estrutural, Instituto Superior Técnico, Universidade de Lisboa, Lisboa, Portugal and Instituto de Tecnologia Química e Biológica, Universidade Nova de Lisboa, Oeiras, Portugal

MARGARIDA COSTA GOMES, Institut de Chimie de Clermont-Ferrand, Université Blaise Pascal and CNRS, Aubière, France

PHILIPP EIDEN, Department for Inorganic and Analytic Chemistry, Freiburger Materialforschungszentrum (FMF) and Freiburg Institute of Advanced Studies (FRIAS: Soft Matter Science), Albert-Ludwigs-Universität Freiburg, Freiburg, Germany

MARIA FORSYTH, Institute for Frontier Materials, Geelong Technology Precinct, Deakin University, Victoria, Australia

HIRO-O HAMAGUCHI, Institute of Molecular Science and Department of Applied Chemistry, National Chiao Tung University, Hsinchu, Taiwan, and Department of Chemistry, School of Science, The University of Tokyo, Tokyo, Japan

CHRISTOPHER HARDACRE, School of Chemistry and Chemical Engineering, Queen's University of Belfast, Belfast, UK

MARCO HAUMANN, Lehrstuhl für Chemische Reaktionstechnik, Universität Erlangen-Nürnberg, Erlangen, Germany

TAKASHI HIROI, Department of Chemistry, School of Science, The University of Tokyo, Tokyo, Japan

KOICHI IWATA, Department of Chemistry, Gakushuin University, Tokyo, Japan

Ingo Krossing, Department for Inorganic and Analytic Chemistry, Freiburger Materialforschungszentrum (FMF) and Freiburg Institute of Advanced Studies (FRIAS: Soft Matter Science), Albert-Ludwigs-Universität Freiburg, Freiburg, Germany

Peter Licence, School of Chemistry, University of Nottingham, Nottingham, UK

Marianne Matzke, NERC Centre for Ecology & Hydrology Molecular Ecotoxicology, Acremann Section Maclean Building, Benson Lane Crowmarsh Gifford, Wallingford Oxfordshire, UK

Carolin Meyer, Lehrstuhl für Chemische Reaktionstechnik, Universität Erlangen-Nürnberg, Erlangen, Germany

Claire Mullan, School of Chemistry and Chemical Engineering, Queen's University of Belfast, Belfast, UK

Jan Novak, Institute for Frontier Materials, Geelong Technology Precinct, Deakin University, Victoria, Australia

Agilío A. H. Pádua, Institut de Chimie de Clermont-Ferrand, Université Blaise Pascal and CNRS, Aubière, France

Chariz Peñalber-Johnstone, Department of Chemistry, University of Houston, Houston, Texas, USA

Karl Ryder, Chemistry Department, University of Leicester, Leicester, UK

Satyen Saha, Department of Chemistry, Banaras Hindu University, Varanasi, India

Stefan Stolte, Department 3: Sustainability in Chemistry, UFT-Centre for Environmental Research and Sustainable Technology, University of Bremen, Bremen, Germany

Alasdair W. Taylor, School of Chemistry, University of Nottingham, Nottingham, UK

Peter Wasserscheid, Lehrstuhl für Chemische Reaktionstechnik, Universität Erlangen-Nürnberg, Erlangen, Germany

Tom Welton, Department of Chemistry, Imperial College London, London, UK

Sebastian Werner, Lehrstuhl für Chemische Reaktionstechnik, Universität Erlangen-Nürnberg, Erlangen, Germany

Neil Winterton, Department of Chemistry, University of Liverpool, Liverpool, UK

Tristan G. A. Youngs, ISIS Facility, Rutherford Appleton Laboratory, Chilton, UK

ABBREVIATIONS

Ionic Liquids

GNCS	guanidinium thiocyanate
GRTIL	gemini room-temperature ionic liquid
[HI-AA]	hydrophobic derivatised amino acid
IL	ionic liquid
poly(GRTIL)	polymerised gemini room-temperature ionic liquid
poly(RTIL)	polymerised room-temperature ionic liquid
RTIL	room-temperature ionic liquid
$[PSpy]_3[PW]$	[1-(3-sulfonic acid)propylpyridinium]$_3$ $[PW_{12}O_{40}]\cdot 2H_2O$

Cations

$[bm(2\text{-}Me)im]^+$	1-butyl-2,3-dimethylimidazolium
$[(bz)C_1C_5im]^+$	1-benzyl-2-methyl-3-pentylimidazolium
$[C_{10}btz]^+$	3-decylbenzothiazolium
$[C_{11}btz]^+$	3-undecylbenzothiazolium
$[C_{12}btz]^+$	3-dodecylbenzothiazolium
$[1\text{-}C_m\text{-}3\text{-}C_nim]^+$	1,3-dialkylimidazolium
$[C_nmim]^+$	1-alkyl-3-methylimidazolium
$[C_nC_mim]^+$	1-alkyl-3-alkylimidazolium
$[Hmim]^+$	1-methylimidazolium
$[C_1mim]^+$	1,3-dimethylimidazolium
$[C_1C_1im]^+$	1,3-dimethylimidazolium
$[C_1C_1C_1im]^+$	1,2,3-trimethylimidazolium
$[C_1C_1mim]^+$	1,2,3-trimethylimidazolium
$[C_2im]^+$	1-ethylimidazolium
$[C_2C_1im]^+$	1-ethyl-3-methylimidazolium
$[C_2C_1C_1im]^+$	1-ethyl-2,3-dimethylimidazolium
$[C_2C_1C_1im]^+$	1-ethyl-2,3-dimethylimidazolium
$[C_2mim]^+$	1-ethyl-3-methylimidazolium
$[C_1C_1(2\text{-}NO_2)im]^+$	1,3-dimethyl-2-nitroimidazolium
$[C_1C_1(2\text{-}Me\text{-}4\text{-}NO_2)im]^+$	1,3-dimethyl-2-methyl-4-nitroimidazolium

$[C_1C_1(4\text{-}NO_2)im]^+$ 1,3-dimethyl-4-nitroimidazolium
$[C_2C_1(4\text{-}NO_2)im]^+$ 1-ethyl-3-methyl-4-nitroimidazolium
$[C_3mim]^+$ 1-propyl-3-methylimidazolium
$[C_3C_3im]^+$ 1,3-dipropylimidazolium
$[^iC_3mim]^+$ 1-*iso*-propyl-3-methylimidazolium
$[^iC_3\ ^iC_3im]^+$ 1,3-di-*iso*-propylimidazolium
$[(^iC_3)_2im]^+$ 1,3-di-*iso*-propylimidazolium
$[C_4C_1im]^+$ 1-butyl-3-methylimidazolium
$[C_4mim]^+$ 1-butyl-3-methylimidazolium
$[^iC_4mim]^+$ 1-*iso*-butyl-3-methylimidazolium
$[^sC_4mim]^+$ 1-*sec*-butyl-3-methylimidazolium
$[^tC_4mim]^+$ 1-*tert*-butyl-3-methylimidazolium
$[(^iC_4)_2im]^+$ 1,3-di-*iso*-butylimidazolium
$[C_4C_4im]^+$ 1,3-dibutylimidazolium
$[^tC_4{}^tC_4im]$ 1,3-di-*tert*-butylimidazolium
$[C_4C_1(4,5\text{-}Br_2)im]^+$ 1-butyl-3-methyl-4,5-bromoimidazolium
$[C_5mim]^+$ 1-pentyl-3-methylimidazolium
$[C_6mim]^+$ 1-hexyl-3-methylimidazolium
$[C_6C_1im]^+$ 1-hexyl-3-methylimidazolium
$[C_6C_6im]^+$ 1,3-dihexylimidazolium
$[C_7mim]^+$ 1-heptyl-3-methylimidazolium
$[C_8mim]^+$ 1-octyl-3-methylimidazolium
$[C_9mim]^+$ 1-nonyl-3-methylimidazolium
$[C_{10}mim]^+$ 1-decyl-3-methylimidazolium
$[(C_{10})_2im]^+$ 1,3-didecylimidazolium
$[C_{11}mim]^+$ 1-undecyl-3-methylimidazolium
$[C_{12}mim]^+$ 1-dodecyl-3-methylimidazolium
$[(C_{12})_2im]^+$ 1,3-didodecylimidazolium
$[C_{13}mim]^+$ 1-tridecyl-3-methylimidazolium
$[C_{14}mim]^+$ 1-tetradecyl-3-methylimidazolium
$[C_{15}mim]^+$ 1-pentadecyl-3-methylimidazolium
$[C_{16}mim]^+$ 1-hexadecyl-3-methylimidazolium
$[C_{17}mim]^+$ 1-heptadecyl-3-methylimidazolium
$[C_{18}mim]^+$ 1-octadecyl-3-methylimidazolium
$[C_2C_1mim]^+$ 1-ethyl-2,3-dimethylimidazolium
$[C_3C_1mim]^+$ 1-propyl-2,3-dimethylimidazolium
$[C_8C_3im]^+$ 1-octyl-3-propylimidazolium
$[C_{12}C_{12}im]^+$ 1,3-bis(dodecyl)imidazolium
$[C_1OC_2mim]^+$ 1-(2-methoxyethyl)-3-methylimidazolium
$[C_4dmim]^+$ 1-butyl-2,3-dimethylimidazolium
$[C_4C_1C_1im]^+$ 1-butyl-2,3-dimethylimidazolium
$[C_4C_1mim]^+$ 1-butyl-2,3-dimethylimidazolium
$[C_6C_{7O1}im]^+$ 1-hexyl-3-(heptyloxymethyl)imidazolium
$[C_2F_3mim]^+$ 1-trifluoroethyl-3-methylimidazolium
$[C_4vim]^+$ 3-butyl-1-vinylimidazolium

$[(^{i}C_3)_2(4,5\text{-}Me_2)im]^+$	1-*iso*-propyl-3,4,5-trimethylimidazolium
$[^{i}C_3C_1(4,5\text{-}Me_2)im]^+$	1,3-di-*iso*-propyl-4,5-dimethylimidazolium
$[(^{t}C_4)_2(4\text{-}SiMe_3)im]^+$	1,3-di-*tert*-butyl-4-trimethylsilylimidazolium
$[(allyl)mim]^+$	1-allyl-3-methylimidazolium
$[P_{n}mim]^+$	polymerisable 1-methylimidazolium
$[C_2mmor]^+$	1-ethyl-1-methylmorpholinium
$[C_2py]^+$	1-ethylpyridinium
$[C_4py]^+$	1-butylpyridinium
$[C_6py]^+$	1-hexylpyridinium
$[C_8py]^+$	1-octylpyridinium
$[C_{14}py]^+$	1-tetradecylpyridinium
$[C_4m_{\beta}py]^+$	1-butyl-3-methylpyridinium
$[C_4m_{\gamma}py]^+$	1-butyl-4-methylpyridinium
$[C_6(dma)_{\gamma}py]^+$	1-hexyl-4-dimethylaminopyridinium
$[C_nC_1pyr]^+$	1-alkyl-1-methylpyrrolidinium
$[C_1C_1pyr]^+$	1,1-dimethylpyrrolidinium
$[C_1C_2pyr]^+$	1-ethyl-1-methylpyrrolidinium
$[C_2C_1pyr]^+$	1-ethyl-1-methylpyrrolidinium
$[C_2mpyr]^+$	1-ethyl-1-methylpyrrolidinium
$[C_3C_1pyr]^+$	1-propyl-1-methylpyrrolidinium
$[C_3mpyr]^+$	1-propyl-1-methylpyrrolidinium
$[C_4mpyr]^+$	1-butyl-1-methylpyrrolidinium
$[C_4C_1pyr]^+$	1-butyl-1-methylpyrrolidinium
$[C_5C_1pyr]^+$	1-pentyl-1-methylpyrrolidinium
$[C_6mpyr]^+$	1-hexyl-1-methylpyrrolidinium
$[C_nC_1pyr]^+$	1-alkyl-1-methylpyrrolidinium
$[C_1C_3pip]^+$	1-methyl-1-propylpiperidinium
$[C_2C_1pip]^+$	1-ethyl-1-methylpiperidinium
$[C_2C_6pip]^+$	1-ethyl-1-hexylpiperidinium
$[C_3C_1pip]^+$	1-methyl-1-propylpiperidinium
$[C_8quin]^+$	1-octylquinolinium
$[dabcoH]^+$	1,4-diazabicyclo[2.2.2]octan-1-ium(1+)
$[dabcoH_2]^{2+}$	1,4-diazabicyclo[2.2.2]octan-1-ium(2+)
$[dmPhim]^+$	1,3-dimethyl-2-phenylimidazolium
$[FcC_1mim]^+$	1-ferrocenyl-3-methylimidazolium
$[H_2NC_2H_4py]^+$	1-(1-aminoethyl)pyridinium
$[H_2NC_3H_6mim]^+$	1-(3-aminopropyl)-3-methylimidazolium
$[HN_{2\,2\,2}]^+$	triethylammonium
$[H_2mor]^+$	morpholinium
$[H_2pip]^+$	piperidinium
$[Hpy]^+$	pyridinium
$[H_2pyr]^+$	pyrrolidinium
$[N_{0\,1\,1\,1}]^+$	trimethylammonium
$[N_{0\,0\,1\,1}]^+$	dimethylammonium
$[N_{0\,0\,0\,1}]^+$	methylammonium

$[N_{1111}]^+$	tetramethylammonium
$[N_{1112OH}]^+$	cholinium
$[N_{1122OH}]^+$	ethyl(2-hydroxyethyl)dimethylammonium
$[N_{1114}]^+$	trimethylbutylammonium
$[N_{1444}]^+$	methyltributylammonium
$[N_{1888}]^+$	methyltrioctylammonium
$[N_{2222}]^+$	tetraethylammonium
$[N_{3333}]^+$	tetrapropylammonium
$[N_{33311}]^+$	tripropylundecylammonium
$[N_{3368}]^+$	dipropylhexyloctylammonium
$[N_{4444}]^+$	tetrabutylammonium
$[N_{5555}]^+$	tetrapentylammonium
$[N_{6666}]^+$	tetrahexylammonium
$[N_{66614}]^+$	trihexyl(tetradecyl)ammonium
$[N_{10\,10\,10\,10}]^+$	tetradecylimidazolium
$[N_{12\,12\,12\,12}]^+$	tetradodecylammonium
$[NR_3H]^+$	trialkylammonium
$[P_{222(1O1)}]^+$	triethyl(methoxymethyl)phosphonium
$[P_{444\,3a}]^+$	(3-aminopropyl)tributylphosphonium
$[P_{4444}]^+$	tetrabutylphosphonium
$[P_{5555}]^+$	tetrapentylphosphonium
$[P_{66614}]^+$	trihexyl(tetradecyl)phosphonium
$[P_{88814}]^+$	tetradecyl(trioctyl)phosphonium
$[P_{10\,10\,10\,10}]^+$	tetradecylphosphonium
$[P_{10\,10\,10}CH_2C(O)NH_2]^+$	amidomethyl-tritetradecylphosphonium
$[P_{10\,10\,10}CH_2CO_2]^+$	carboxymethyl-tritetradecylphosphonium
$[P_{18\,18\,18\,18}]^+$	tetraoctadecylphosphonium
$[PhCH_2eim]^+$	1-benzyl-2-ethylimidazolium
$[pyH]^+$	pyridinium
$[RC_nim]^+$	1,3-dialkylimidazolium
$[Rmim]^+$	1-alkyl-3-methylimidazolium
$[S_{222}]^+$	triethylsulfonium
$[S_{22\,16}]^+$	diethylhexadecylsulfonium
$[(vinyl)mim]^+$	1-vinyl-3-methylimidazolium

Anions

$[Ace]^-$	acetate
$[Ala]^-$	alaninate
$[\beta Ala]^-$	β-alaninate
$[Al(hfip)_4]^-$	tetra(hexafluoro-*iso*-propoxy)aluminate(III)
$[Arg]^-$	arginate
$[Asn]^-$	asparaginate
$[Asp]^-$	asparatinate

$[B_{4444}]^-$ tetrabutylborate
$[BBB]^-$ bis[1,2-benzenediolato(2-)-*O,O′*]borate
$[C_1CO_2]^-$ ethanoate
$[C_1SO_4]^-$, $[O_3SOC_1]^-$ methyl sulfate
$[C_8SO_4]^-$, $[O_3SOC_8]^-$ octyl sulfate
$[C_nSO_4]^-$ alkyl sulfate
$[(C_n)(C_m)SO_4]^-$ asymmetrical dialkyl sulfate
$[(C_n)_2SO_4]^-$ symmetrical dialkyl sulfate
$[CTf_3]^-$ tris{(trifluoromethyl)sulfonyl}methanide
$[Cys]^-$ cysteinate
$[dbsa]^-$ dodecylbenzenesulfonate
$[dca]^-$ dicyanamide
$[FAP]^-$ tris(perfluoroalkyl)trifluorophosphate
$[Gln]^-$ glutaminate
$[Glu]^-$ glutamate
$[Gly]^-$ glycinate anion
$[His]^-$ histidinate
$[Ile]^-$ isoleucinate
$[lac]^-$ lactate
$[Leu]^-$ leucinate
$[Lys]^-$ lysinate
$[Met]^-$ methionate
$[Nle]^-$ norleucinate
$[NDf_2]^-$ bis{bis(pentafluoroethyl)phosphinyl}amide
$[NMes_2]^-$ bis(methanesulfonyl)amide
$[NPf_2]^-$, $[BETI]^-$ bis{(pentafluoroethyl)sulfonyl}amide
$[NTf_2]^-$, $[TFSI]^-$ bis{(trifluoromethyl)sulfonyl}amide
$[O_2CC_1]^-$ ethanoate
$[O_3SOC_2]^-$, $[O_3SOC_2]^-$ ethyl sulfate
$[OMs]^-$ methanesulfonate (mesylate)
$[ONf]^-$ perfluorobutylsulfonate
$[OTf]^-$ trifluoromethanesulfonate
$[OTs]^-$ 4-toluenesulfonate, $[4\text{-}CH_3C_6H_4SO_3]^-$ (tosylate)
$[Phe]^-$ phenylalaninate
$[Pro]^-$ prolinate
$[Sacc]^-$ saccharinate
$[Ser]^-$ serinate
$[Suc]^-$ succinate
$[tfpb]^-$ tetrakis(3,5-bis(trifluoromethyl)phenyl)borate
$[Thr]^-$ threoninate
$[Tos]^-$ tosylate
$[Trp]^-$ tryphtophanate
$[Tyr]^-$ tyrosinate
$[Val]^-$ valinate

Techniques

AA	all-atom parameterisation
AES	Auger electron spectroscopy
AFM	atomic force microscopy
AMBER	Assisted Model Building with Energy Refinement
ANN	associative neural network
APPLE&P	Atomistic Polarisable Potential for Liquids, Electrolytes and Polymers
ARXPS	angle-resolved X-ray photoelectron spectroscopy
ATR-IR	attenuated total reflectance infrared spectroscopy
BPNN	back-propagation neural network
BPP	Bloembergen–Purcell–Pound theory
CADM	computer-aided design modelling
CC	Cole–Cole model
CCC	countercurrent chromatography
CD	Cole–Davidson model
CE	capillary electrophoresis
CEC	capillary electrochromatography
CHARMM	Chemistry at HARvard Molecular Mechanics
COSMO-RS	**Co**nductor-like **S**creening **Mo**del for Real Solvents
COSY	**Co**rrelation Spectroscop**y**
CPCM	conductor-like polarisable continuum model
CPMD	Car–Parrinello molecular dynamics
DFT	density functional theory
DLVO	Derjaguin, Landau, Verwey and Overbeek theory
DRS	dielectric relaxation spectroscopy
DSC	differential scanning calorimetry
ECSEM	electrochemical scanning electron microscopy
EC-XPS	electrochemical X-ray photoelectron spectroscopy
EF-CG	effective force coarse-graining method
EFM	effective fragment potential method
EI	electron ionisation
EIS	electrochemical impedance spectroscopy
EMD	equilibrium molecular dynamics
EOF	electro-osmotic flow
EPSR	empirical potential structure refinement
ES	electrospray mass spectrometry
ESI–MS	electrospray ionisation mass spectrometry
EXAFS	extended X-ray absorption fine structure
FAB	fast atom bombardment
FMO	fragment molecular orbital method
FIR	far-infrared spectroscopy
FTIR	Fourier transform infrared spectroscopy
GAMESS	general atomic and molecular electronic structure system

GC	gas chromatography
GGA	generalised gradient approximations
GLC	gas–liquid chromatography
GSC	gas–solid chromatography
HM	heuristic method
HOESY	heteronuclear Overhauser effect spectroscopy
HPLC	high-performance liquid chromatography
HREELS	high-resolution electron energy loss spectroscopy
IGC	inverse gas chromatography
IPES	inverse photoelectron spectroscopy
IR	infrared spectroscopy
IRAS	infrared reflection–absorption spectroscopy
IR-VIS SFG	infrared visible sum-frequency generation
ISS	ion scattering spectroscopy
LEIS	low-energy ion scattering
L-SIMS	liquid secondary ion mass spectrometry
MAES	metastable atom electron spectroscopy
MALDI	matrix-assisted laser desorption
MBSS	molecular beam surface scattering
MC	Monte Carlo
MD	molecular dynamics
MIES	metastable impact electron spectroscopy
MLP	multilayer perceptron
MLR	multi-linear regression
MM	molecular mechanics
MR	magnetic resonance
MRI	magnetic resonance imaging
MS	mass spectrometry
NEMD	non-equilibrium molecular dynamics
NEXAFS	near-edge absorption fine structure
NIR	near-infrared spectroscopy
NMR	nuclear magnetic resonance
NR	neutron reflectivity
NRTL	non-random two liquid
OPLS	Optimised Potentials for Liquid Simulations
PCM	polarisable continuum model
PDA	photodiode array detection
PES	photoelectron spectroscopy
PFG-NMR	pulsed field-gradient nuclear magnetic resonance
PGSE-NMR	pulsed-gradient spin-echo nuclear magnetic resonance
PPR	projection pursuit regression
QM	quantum mechanics
QSAR	quantitative structure–activity relationship
QSPR	quantitative structure–property relationship
RAIRS	reflection–absorption infrared spectroscopy

RI	refractive index
RMC	reverse Monte Carlo
RNEMD	reverse non-equilibrium molecular dynamics
RNN	recursive neural network
ROESY	rotating-frame Overhauser effect spectroscopy
RP-HPLC	reverse-phase high-performance liquid chromatography
RST	regular solution theory
SANS	small-angle neutron scattering
SCMFT	self-consistent mean field theory
SEM	scanning electron microscopy
SEM-EDX	scanning electron microscopy with energy-dispersive X-ray
SFA	surface forces apparatus
SFC	supercritical fluid chromatography
SFG	sum-frequency generation
SFM	systematic fragmentation method
SIMS	secondary ion mass spectrometry
soft-SAFT	soft statistical associating fluid theory
STM	scanning tunnelling microscopy
SVN	support vector network
TEM	tunnelling electron microscopy
TGA	thermogravimetric analysis
THz-TDS	terahertz time-domain spectroscopy
TLC	thin-layer chromatography
tPC-PSAFT	truncated perturbed chain-polar statistical associating fluid theory
TPD	temperature programmed desorption
UA	united-atom parameterisation
UHV	ultra-high vacuum
UNIFAC	*UNI*QUAC *F*unctional-group *A*ctivity *C*oefficients
UNIQUAC	*UNI*versal *QUA*si*C*hemical
UPLC	ultra-pressure liquid chromatography
UPS	ultraviolet photoelectron spectroscopy
UV	ultraviolet
UV-Vis	ultraviolet-visible
VBT	volume-based thermodynamics
XPS	X-ray photoelectron spectroscopy
XRD	X-ray powder diffraction
XRR	X-ray reflectivity

Miscellaneous

Å	1 ångstrom = 10^{-10} m
ACS	American Chemical Society
ANQ	1-amino-3-nitroguanidine
API	active pharmaceutical ingredient

ATMS	acetyltrimethylsilane
ATPS	aqueous two-phase system
a.u.	atomic units
BASF™	Badische Anilin- und Sodafabrik
BASIL	Biphasic Acid Scavenging Utilising Ionic Liquids
BATIL	Biodegradability and Toxicity of Ionic Liquids
BE	binding energy
BILM	bulk ionic liquid membrane
BNL	Brookhaven National Laboratory
BOD	biochemical oxygen demand
BP	British Petroleum
b.pt.	boiling point
BSA	bovine serum albumin
BT	benzothiophene
BTAH	benzotriazole
BTX	benzene–toluene–xylene mixture
calc.	calculated
CB	Cibacron Blue 3GA
CCDC	Cambridge Crystallographic Data Centre
CE	crown ether
CEES	2-chloroethyl ethyl sulphide
CFC MC	'continuous fractional component' Monte Carlo
cif	crystallographic information file
CL&P	Canongia Lopes and Pádua
CLM	charge lever momentum
CMC	critical micelle concentration
CMPO	octyl(phenyl)-*N,N*-diisobutylcarbamoylmethylphosphine oxide
$[C_nMeSO_4]$	alkyl methyl sulfate
CNTs	carbon nanotubes
CNRS	Centre National de la Recherche Scientifique
COIL	Congress on Ionic Liquids
CPU	central processing unit
CSA	chemical shielding anisotropy
CSD	Cambridge Structural Database
CWAs	chemical warfare agents
d	doublet (NMR)
D°_{298}	bond energy at 298 K
1D	one-dimensional
2D	two-dimensional
3D	three-dimensional
DABCO	1,4-diazabicyclo[2.2.2]octane
DBT	dibenzothiophene
DC	direct current
DC18C6	dicyclohexyl-18-crown-6
DF	Debye and Falkenhagen

DH	Debye–Hückel
DIIPA	diisopropylamine
4,6-DMDBT	4,6-dimethyldibenzothiophene
DNA	deoxyribonucleic acid
DMF	dimethylmethanamide (dimethylformamide)
DMH	dimethylhexene
2DOM	two-dimensional ordered macroporous
3DOM	three-dimensional ordered macroporous
DOS	density of states
DPC	diphenyl carbonate
DRA	drag-reducing agent
DSSC	dye-sensitised solar cell
DSTE	double stimulated-echo
E	enrichment
EDC	extractive distillation column
EE	expanded ensemble approach
EoS	equation of state
EOR	enhanced oil recovery
EPA	Environmental Protection Agency
eq.	equivalent
FCC	fluid catalytic cracking
FFT	fast Fourier transform
FIB	focussed ion beam
FMF	Freiburger Materialforschungszentrum
FRIAS	Freiburg Institute for Advanced Studies
FSE	full-scale error
ft	foot
GDDI	generalised distributed data interface
GEMC	Gibbs ensemble Monte Carlo
GSSG	glutathione disulfide
GSH	glutathione
GT	gauche-trans
HDS	hydrodesulfurisation
HEMA	2-(hydroxyethyl) methacrylate
HOMO	highest occupied molecular orbital
HOPG	highly oriented pyrolytic graphite
HV	high vacuum
i.d.	inner diameter
IFP	Institut Français du Pétrole
IgG	immunoglobulin G
IPBE	ion-pair binding energy
IPE	Institute of Process Engineering, Chinese Academy of Sciences, Beijing
ITO	indium tin oxide
IUPAC	International Union of Pure and Applied Chemistry

IVR	intramolecular vibrational energy redistribution
J	coupling constant (NMR)
KWW	Kohlrausch–Williams–Watts
LCEP	lower critical endpoint
LCST	lower critical separation temperature
LEAF	Laser-Electron Accelerator Facility
LF-EoS	lattice-fluid model equation of state
LLE	liquid–liquid equilibria
LMOG	low molecular weight gelator
LSERs	linear solvation energy relationships
LUMO	lowest unoccupied molecular orbital
m	multiplet (NMR)
M	molar concentration
MBI	1-methylbenzimidazole
MCH	methylcyclohexane
MDEA	methyl diethanolamine; bis(2-hydroxyethyl)methylamine
MEA	monoethanolamine; 2-aminoethanol
MFC	minimum fungicidal concentrations
MIC	minimum inhibitory concentrations
MMM	mixed matrix membrane
MNDO	modified neglect of differential overlap
m.pt.	melting point
MSD	mean square displacement
3-MT	3-methylthiophene
MW	molecular weight
MWCNTs	multi-walled carbon nanotubes
m/z	mass-to-charge ratio
NBB	1-butylbenzimidazole
NCA	*N*-carboxyamino acid anhydride
NE equation	Nernst–Einstein equation
NES	New Entrepreneur Scholarship
NFM	*N*-formylmorpholine
NIP	neutral ion pair
NIT	neutral ion triplet
NMP	*N*-methylpyrrolidone
NOE	nuclear Overhauser effect
NP	nanoparticle
NRTL-SAC	non-random two liquid segmented activity coefficients
o.d.	outer diameter
OECD	Organisation for Economic Co-operation and Development
OKE	optical Kerr effect
p	pressure
PAO	polyalphaolefin
PBT	persistent, bioaccumulative and toxic
PDMS	polydimethoxysilane

PEDOT	poly(3,4-ethylenedioxythiophene)
PEG	poly(ethylene glycol)
PEM	polymer electrolyte membrane
PEN	poly(ethylene-2,6-naphthalene dicarboxylate)
PES	polyethersulfone
pH	$-\log_{10}([H^+])$; a measure of the acidity of a solution
PIB	polyisobutene
pK_a	$-\log_{10}(K_a)$
pK_b	$-\log_{10}(K_b)$
PPDD	polypyridylpendant poly(amidoamine) dendritic derivative
ppm	parts per million
PQRE	platinum *quasi*-reference electrode
(PR)-EoS	Peng–Robinson equation of state
PS	polystyrene
PSE	process systems engineering
psi	1 pound per square inch = 6894.75729 Pa
PTC	phase-transfer catalyst
PTFE	poly(tetrafluoroethylene)
PTx	pressure–temperature–composition
PZC	potential of zero charge
r	bond length
RDC	rotating disc contactor
RDF	radial distribution function
REACH	*R*egistration, *E*valuation, *A*uthorisation and Restriction of ***Ch***emical Substances
RF	radio frequency
(RK) EoS	Redlich–Kwong equation of state
RMSD	root-mean-square deviation
RT	room temperature
s	singlet (NMR)
S	entropy
scCO_2	supercritical carbon dioxide
SCILL	solid catalyst with *ionic liquid* layer
SDS	sodium dodecyl sulphate
SE	Spin-echo
SED	Stokes–Einstein–Debye equation
S/F	solvent-to-feed ratio
SHOP	Shell higher olefin process
SILM	supported ionic liquid membrane
SILP	supported ionic liquid phase
SLE	solid–liquid equilibrium
SLM	supported liquid membrane
STE	stimulated spin-echo
SVHC	substance of very high concern
t	triplet (NMR)

TBP	4-(*t*-butyl)pyridine
TCEP	1,2,3-tris(2-cyanoethoxy)propane
TEA	triethylamine
TEGDA	tetra(ethylene glycol) diacrylate
THF	tetrahydrofuran
TIC	toxic industrial chemical
TMB	trimethyl borate
TMP	trimethylpentene
TMPD	*N,N,N′N′*-tetramethyl-*p*-phenylenediamine
TOF	time-of-flight
TT	*trans–trans*
UCEP	upper critical endpoint
UCST	upper critical solution temperature
VFT	Vogel–Fulcher–Tammann equations
VLE	vapour–liquid equilibria
VLLE	vapour–liquid–liquid equilibria
VMP	variable multichannel potentiostat
VOCs	volatile organic compounds
v/v	volume for volume
w/w	weight for weight
wt%	weight percent
γ	surface tension
δ	chemical shift in NMR
X	molar fraction

1 What Is an Ionic Liquid?

ANDREW P. ABBOTT and KARL RYDER
Chemistry Department, University of Leicester, Leicester, UK
PETER LICENCE and ALASDAIR W. TAYLOR
School of Chemistry, University of Nottingham, Nottingham, UK

1.1 INTRODUCTION

The history of ionic liquids is well documented, and it is widely recognised that the work of Walden produced the first recorded materials that were deliberately ionic and molten at ambient temperature [1]. The classification of ionic liquids as salts that are liquid below 100°C is arbitrary and non-satisfactory because it does not answer the philosophical question 'When is an ionic liquid not an ionic liquid?' This seemingly pointless pedantry is important, as almost all uses of ionic fluids involve the dissolution of molecular components. The issue is therefore how much solute can be added before the molecular character dominates the ionic character. It has been shown by numerous authors that the inclusion of a small amount of certain impurities can have a profound effect upon the physical and chemical properties of an ionic liquid [2–4].

Arbitrary descriptions are unsatisfactory as it is difficult to predict what materials will fall into this classification. It would be simple to view ionic liquids as an extension of molten salts or, alternatively, they could be viewed as a fundamentally different type of material. The most suitable method will probably be to describe them as extreme versions of molten salts. The main issue associated with the differentiation of high-temperature (mainly inorganic) and lower-temperature (predominantly organic) salts is accounting for the differences between the transport properties of the fluids. Clearly, the higher-temperature systems are relatively fluid because they generally have smaller ions and larger void volumes, allowing greater mobility of the ions and resulting in high ionic conductivities. What is less evident is why the properties of the higher-temperature systems vary with chemical compositions, whereas the lower-temperature systems share significant similarities.

Ionic Liquids Completely UnCOILed: Critical Expert Overviews, First Edition.
Edited by Natalia V. Plechkova and Kenneth R. Seddon.

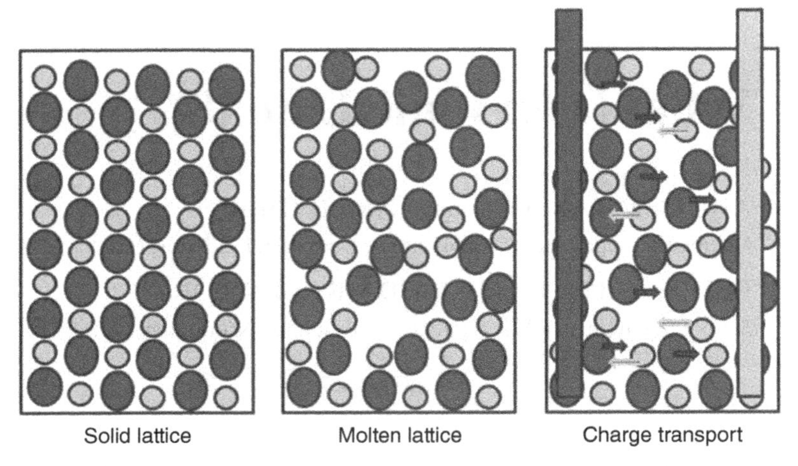

Figure 1.1 Schematic of the melting and charge transport processes in an ionic liquid.

In a seminal discussion by Bockris and Reddy [5], the issue of charge transfer in ionic liquids was discussed even before ionic liquids became a distinct research theme. The term ionic liquid is used extensively in their work: 'Pure liquid electrolytes therefore are liquids containing only ions, the ions being free or associated'. They identify the key issues of the difficulty of describing charge transfer in ionic liquids. When an electric field is applied, why do ions move with such extensive inter-ionic forces? An ionic liquid is a molten lattice, which means that empty space is poured into the fused salt. It seems logical therefore that it is this space that allows the ions in the *quasi*-lattice to move past each other. The process of melting and charge transport is shown schematically in Figure 1.1.

1.2 DILUTE AQUEOUS SOLUTIONS

To understand ionic liquids, it is informative to first review the mechanism by which charge is transported in dilute ionic solutions in molecular solvents, and how this changes with increasing salt concentration. It is also necessary to review the work on concentrated electrolyte solutions and see at which point the theories deviate from observed values, and why these deviations occur.

The conductivity of a strong 1:1 electrolyte in an aqueous solution at infinite dilution is well known [6]. The Nernst–Einstein relation shows that the limiting molar conductivity, Λ_{m}^{0}, of an electrolyte is related to the diffusion coefficient, D^0, of the ions by Equation 1.1:

$$\Lambda_{\mathrm{m}}^{0} = \left(\frac{z^2F^2}{RT}\right)\left(D_{+}^{0} + D_{-}^{0}\right) \tag{1.1}$$

where z is the charge of the ion, F is the Faraday constant, R is the gas constant, and T is the absolute temperature. The Stokes–Einstein relation, Equation 1.2, is generally used to relate the diffusion coefficient to the radius of the species, R, and the viscosity of the medium, η:

$$D^0 = \frac{kT}{6\pi\eta R} \tag{1.2}$$

It is very important to note that both Equations 1.1 and 1.2 are derived for spherical species at infinite dilution in a structureless dielectric or in a medium where the size of the diffusing species is much larger than the solvent molecules. Combining Equations 1.1 and 1.2 yields Equation 1.3:

$$\Lambda_{\mathrm{m}}^0 = \frac{z^2 Fe}{6\pi\eta}\left(\frac{1}{R_+} + \frac{1}{R_-}\right) \tag{1.3}$$

As the ionic concentration increases, the conductivity, σ, is given by Equation 1.4:

$$\sigma = \left(u_+ z_+ + u_- \,|\, z_- \,|\right) cF \tag{1.4}$$

where u_+ and u_- are the mobilities of the ions and c is the concentration of the charge carriers. The decrease in molar conductivity with increasing electrolyte concentration is due initially to ionic atmosphere effects, and then the formation of ion pairs. Once ion pairing occurs, none of the above equations are strictly valid because the concentration of charge carriers is not equal to the concentration of ions in solution. The effect of ion pairing increases as the electrolyte concentration increases, as the electrolyte becomes weaker, or as the polarity of the solvent decreases. Triple ions form as the concentration increases further, and a concomitant increase in molar conductivity is observed. The mobility of the triple ions is clearly different to that of the single ions, and Equation 1.4 has to be modified. This approach has been extensively applied to quaternary ammonium electrolytes in non-aqueous solvents, and numerous ion pairs and triple ion formation constants have been determined for many systems [7–9]. The Fuoss–Kraus equation, Equation 1.5, can be used to obtain the equilibrium constants for the ion pair dissociation, K_{p}, and triple ion dissociation, K_{T}:

$$\frac{\Lambda\sqrt{c}}{\Lambda_{\mathrm{m}}^0} = \sqrt{K_{\mathrm{p}}} + \frac{2c\sqrt{K_{\mathrm{p}}}}{3K_{\mathrm{T}}} \tag{1.5}$$

The difficulty with using this analysis was in determining Λ_{m}^0 with non-aqueous electrolyte solutions. To solve this issue, Walden derived an empirical rule where for a given electrolyte [6],

$$\Lambda_{\mathrm{m}}^0 \eta^0 = \text{constant} \tag{1.6}$$

It is however only valid at infinite dilution for a given electrolyte. The rule is clearly a consequence of Equation 1.3.

At extremely high electrolyte concentrations, the molar conductivity decreases down to a limiting value, which is that of the pure salt. This decrease has previously been ascribed to viscosity effects [6].

1.3 IONIC LIQUIDS

Clearly, if the salt is liquid under the measurement conditions, then its behaviour can be determined across the complete concentration range with a molecular solvent. This trend is shown in Figure 1.2 for various electrolyte systems. It can clearly be seen that the polarity of the molecular solvent affects the mole fraction at which the maximum in molar conductivity occurs. It could be argued that the maximum in Figure 1.2 is due to the competition between the number of charge carriers and the mobility of the ionic species. The analysis of concentrated electrolyte solutions is extremely complex.

Stokes noted that charge mobility becomes the dominant factor at high electrolyte concentrations and termed the observation the '*viscosity effect*'.

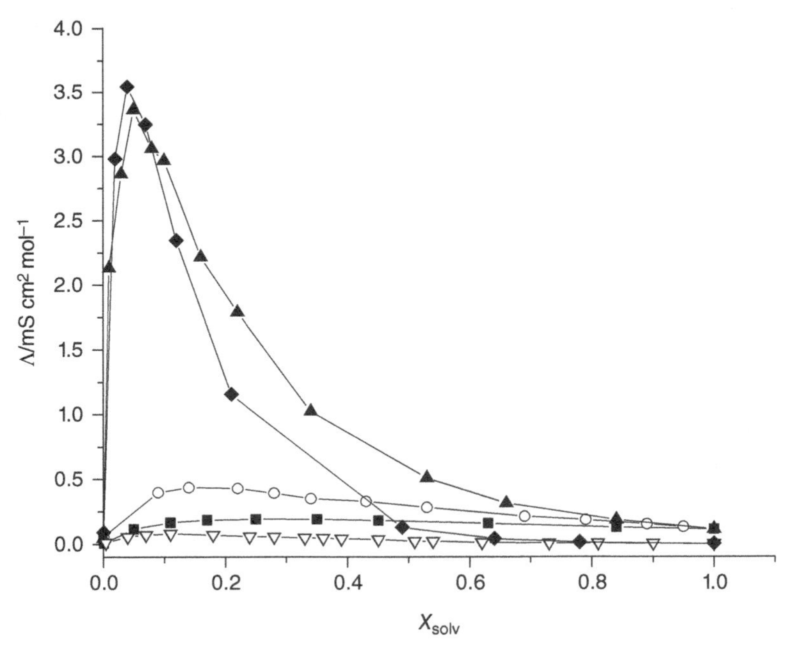

Figure 1.2 Molar conductivity of ionic liquid mixtures with various molecular solvents as a function of their mole fraction. The curves represent: ■ $[C_2mim][BF_4]$-propylene carbonate; o $[C_2mim][BF_4]$-*N*-methylformamide; ▲ $[C_2mim][BF_4]$-water; ∇ $[C_3mim]$Br-propylene carbonate; ◆ $[C_3mim]$Br-water. The data were taken from Ref. [4].

Stokes originally dealt with the molar conductivity of electrolytes in viscous solutions by empirically applying a fractional Walden rule, Equation 1.7:

$$(\Lambda_{m}\eta)^{\alpha} = \text{constant } (C) \tag{1.7}$$

where α is a decoupling parameter that depends on the ionic size. Values of α have been found to vary between 0.61 and 0.74 [10]. This was first applied to molten salts [11, 12]. Numerous groups have found empirically that the molar conductivity of an ionic liquid is inversely proportional to the viscosity of the liquid and have erroneously invoked the 'Walden rule', Equation 1.6, as a method of describing what an ionic liquid is, without noting the two critical factors upon which it is based: namely, it is defined for one given ion, and the ions are at infinite dilution. Ionic liquids that do not obey the Walden rule have been classified as super-ionic or sub-ionic [13–15]. Figure 1.3 shows the so-called Walden plots for a variety of ionic liquid systems. This is a totally empirical relationship that is observed for a group of electrolytes that have similar ion sizes, and the validity of this can be seen from Equation 1.3. It can therefore be questioned why this works if the ions are not at infinite dilution.

Schreiner et al. recently published a critique of the use of the Walden rule with ionic liquids [17]. They suggested that use of the empirical approach from Stokes, Equation 1.7, explains the observed deviations from the so-called ideal Walden plot, since each salt will have a different value of C and α:

$$\log\left(\Lambda_{m}^{0}\right) = \log(C) + \alpha\log\left(\eta^{-1}\right) \tag{1.8}$$

One approach to explain this observation has been to invoke a model of charge mobility occurring via hole transfer, which was developed by Fürth [18, 19] and applied to molten salts by Bockris [5]. The model did not fit experimental data for high-temperature systems due to the large number of suitably sized holes at high temperatures, which were able to take the small charge carriers. Ionic association decreased the number of charge carriers and hence the model broke down. The same basic model was applied to ionic liquids and found to fit extremely accurately [20, 21].

It was proposed that the reason that ionic liquids fit this model so accurately is due to the small number of suitably sized holes available at ambient temperatures that are able to accommodate the exceptionally large ions [21]. Under these conditions, the holes are effectively at infinite dilution, and this is the reason that Equation 1.3 becomes valid. It was shown that the viscosity of an ionic liquid could be accurately modelled using a gas-like model where mobility was hindered by the probability of finding a hole large enough for an ion to move into. The liquids that do not fit hole theory tend to be those whose ions are less spherical. Modifications to this theory have been made by Zhao et al., who took into account the asymmetric nature of the cation, and this significantly improved the prediction of the conductivity of long chain salts [22].

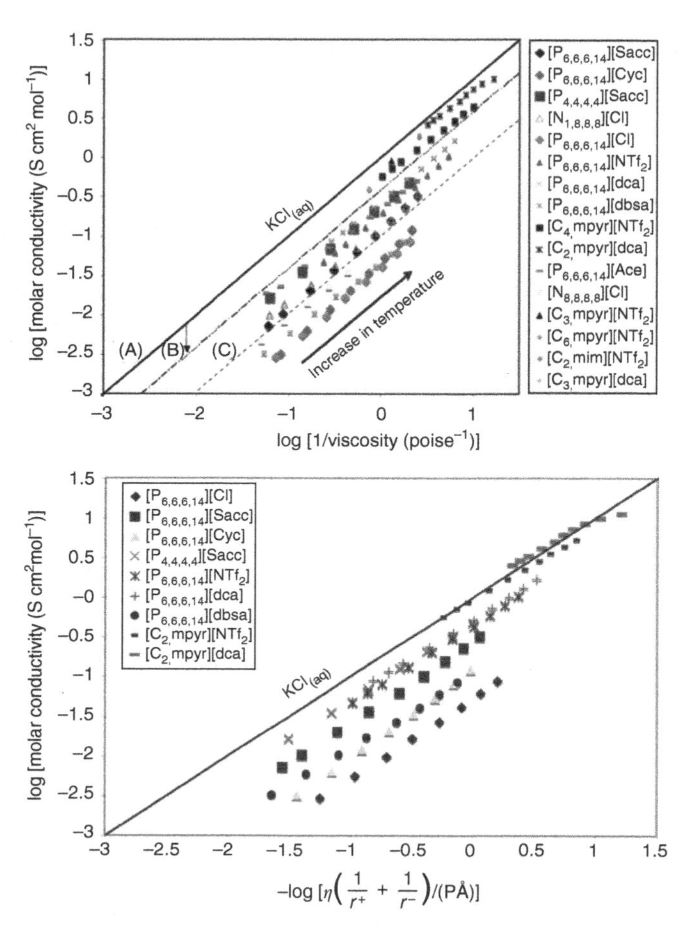

Figure 1.3 Walden plot (upper) and modified Walden plot (lower) for a variety of ionic liquids. For the upper curve, (A) is the ideal Walden plot line for KCl, (B) is the Walden plat taking account of ionic size, and (C) is the Walden plot if the conductivity was only 10% of the ideal value. Source: MacFarlane et al. 2009. Reproduced by permission of the PCCP Owner Societies.

Bockris and Reddy [5] proposed that the diffusion coefficient of species in an ionic liquid could be modelled by assuming that it is the hole itself that is diffusing. Modifying Equation 1.2 such that the term R is replaced by the radius of the void leads to theoretical diffusion coefficients that are closer to the measured values [23]. This simplification would however imply that all species in a given liquid would diffuse at the same rate in a given liquid, which is not born out in experiment. Bockris, moreover, combined the hole theory with the Stokes law to define a term that incorporated the mass of the diffusing species [5]. The difficulty of applying this theory is the requirement for a term that defines the energy required to remove a mole of material from the surface to infinity. This is naturally difficult to determine for an ionic liquid and hence this theory has never been applied.

An alternative approach has been to explain deviations from the Walden plot as arising from ion pairing, or systems that are not true ionic liquids [16]. This naturally ignores the invalidity of the Walden rule, and the correlation between Λ_m and η must be viewed as purely an empirical coincidence. The fact that molar conductivity is inversely proportional to viscosity is due to the fact that the ionic size is large and so changes in R^{-1} are relatively small. Application of Equation 1.3 to the data in Figure 1.3 (upper) brings more of the data onto the ideal slope, and this is shown in Figure 1.3 (lower), disproving that deviations from the Walden Rule are due to ion pairing. Not all data lie on the line, but those that deviate tend to correspond to large and non-spherical ions.

Watanabe and co-workers [24–26] measured the diffusion coefficient of ions in an ionic liquid as a function of the anion and cation and noted a discrepancy from the values calculated using Equation 1.2 when using the hard sphere radius of the ion. It was suggested that Equation 1.2 could be written as in Equation 1.9:

$$D = \frac{kT}{c\pi\eta R_s} \tag{1.9}$$

where c is a variable constant for each electrolyte and R_s is the hydrodynamic or Stokes' radius. Values of c were calculated between 4.1 and 4.4. However, c is only a function of whether the species diffuses in a lateral or rotational motion, and so this approach is not strictly valid [24]. The group however took the diffusion coefficient and calculated a theoretical molar conductivity, Λ_{NMR}, according to Equation 1.10:

$$\Lambda_{NMR} = \frac{N_A e^2 (D_+ + D_-)}{kT} \tag{1.10}$$

which was compared with the measured conductivity. The ratio of the molar conductivity calculated from the impedance measurement (Λ_{imp}) to that obtained from the ionic diffusivity (Λ_{NMR}) was determined, and the ratio $\Lambda_{imp}/\Lambda_{NMR}$ was found to be less than unity. This observation was ascribed to 'only part of the diffusive species in the ionic liquid contributes to the ionic conduction due to the presence of ionic association' [24]. The ratio $\Lambda_{imp}/\Lambda_{NMR}$ was defined by the term ionicity and was found to be in the range 0.5–0.8 for many ionic liquids. This is relatively high, however, and does not really correlate with the chemical differences in the liquids, that is, as the ions become larger and the charge is more delocalised, the ion pairing should decrease. While there is undoubtedly an effect of ionic interaction, it is proposed that this does not directly limit the molar conductivity.

This work, along with all other diffusion studies in ionic liquids, depends on the validity of the Stokes–Einstein equation and here lies the main discrepancy with the analysis of most diffusion coefficient data in ionic liquids. To be strictly accurate, the distance term, R, in Equation 1.3 should be replaced by the

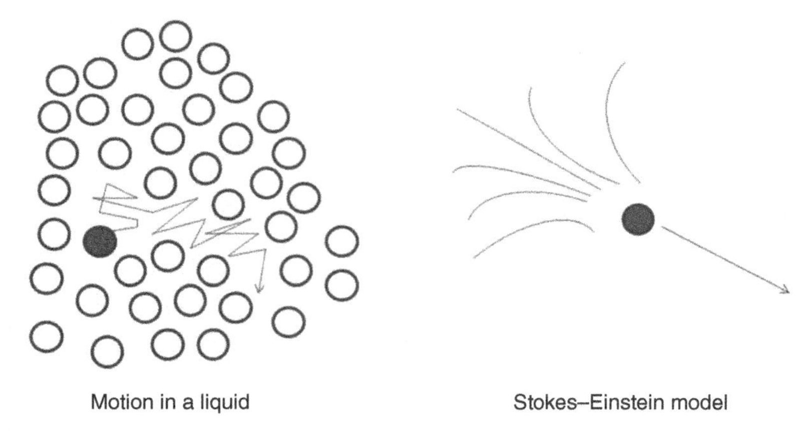

Figure 1.4 Comparison of the actual case of diffusion in an ionic liquid and the model from which the Stokes–Einstein equation is derived. Adapted from Ref. [29].

correlation length, ξ, which is only really equivalent to the radius of the diffusing species when the size of the diffusing particle is large compared with the solvent particles [27]. Equation 1.3 was initially derived to describe the random movement of microscopic particles in a liquid where the structureless dielectric assumption is valid [28]. The motion particles in a liquid and in the Stokes–Einstein model are shown schematically in Figure 1.4.

The observation that the Stokes–Einstein equation holds for ions in molecular solvents is a convenient coincidence that ignores the limitations of the model. The fact that the Stokes–Einstein equation fits the diffusion of molecules in a molecular solvent at all is remarkable, given the complexity of solvent–solute interactions. It has been noted that, in glass-forming mixtures, significant deviations from the Stokes–Einstein equation occur at temperatures below about 1.3 times the glass transition temperature, T_g [30]. This has been explained in terms of 'spatial dynamical heterogeneities', that is, there are tightly structured domains and regions with larger free volume where movement is possible.

Affouard et al. [31] used molecular dynamics simulations to model the mobility of species in simple mixtures of Lennard-Jones A–B particles. The breakdown of the Stokes–Einstein equation occurs at different temperatures depending on the particle decoupling. Simulations also show that faster moving particles show jump-like motion between vacancies. Ionic liquids are well known to be glass formers and have three-dimensional local structure [32]. Assuming that most liquids have a T_g lying between −40 and +20°C [33], that would make the temperature range where deviations from Equation 1.3 occur as anywhere from the melting point to between 303 and 380 K, which is the typical range over which diffusion coefficients are measured.

Classical diffusion can be described by Equation 1.3 when the radius of the sphere is small compared with the mean free path. With ionic liquids, the mean free path can be less than the radius of the ion, and hence the ion can be considered as moving via a series of discrete jumps where the correlation length is a measure of the size of the hole into which the ion can jump. Appreciating why deviations from the Stokes–Einstein equation occur shows why a model based on holes becomes appropriate. The approximate nature of the Stokes–Einstein equation is often overlooked and is discussed in detail by Bockris and Reddy [5, p. 379]. There are numerous aspects that need to be taken into account, including that it is derived for non-charged particles, it is the local viscosity rather than the bulk that is required, and the ordering effect of the ions exhibits an additional frictional force that needs to be explained.

The diffusion coefficients for 1-ferrocenylmethylimidazolium bis{(trifluoromethyl)sulfonyl}amide, [FcC_1mim][NTf_2], in a variety of imidazolium-based ionic liquids were measured recently by Taylor [34, 35]. Using these diffusion coefficient data in Equation 1.3, the size parameter was found to vary from 3×10^{-10} m to 9×10^{-10} m, depending on the temperature and ionic liquid used as solvent, as shown in Figure 1.5. The distance parameter should be independent of temperature and should be the size of the probe molecule (~5×10^{-10} m in radius). The observation that there is a significant change with temperature demonstrates the breakdown of the assumption of a structureless dielectric. The disparity between the radius calculated using Equation 1.3 and the hard sphere radius arises because the correlation length is now comparable with the size of the vacancy into which the ion has to jump. It can therefore be concluded that

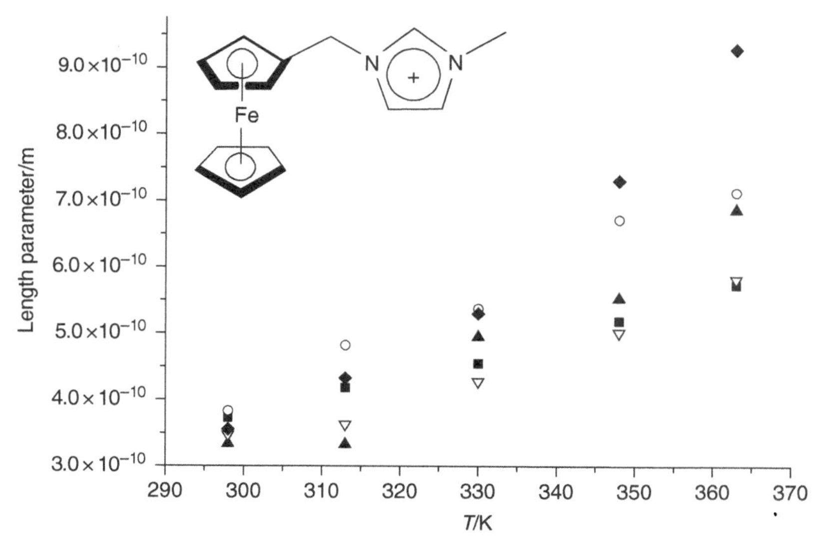

Figure 1.5 Plot of distance parameter, calculated from Equation 1.3 as a function of temperature, for 1-ferrocenylmethylimidazolium bis{(trifluoromethyl)sulfonyl}amide, [FcC_1mim][NTf_2], in a variety of ionic liquids. The points represent solutions in: ■ [C_2mim][NTf_2]; ○ [C_4mim][NTf_2]; ▲ [C_8mim][NTf_2]; ∇ [C_4mim][BF_4]; ♦ [C_8mim][BF_4]. The data were taken from Ref. [34].

diffusion coefficient data, far from giving information about the extent of ionic association, actually reveal information about the structure of the ionic liquid.

Most important, and probably overlooked until now, is the fact that the adherence of so much data to Equation 1.3 signifies that the systems are limited by the mobility of charge rather than the number of charge carriers. This further supports the idea that ion pairing is unimportant in controlling conductivity in ionic liquids.

It is proposed that an ionic liquid is the mirror image of an ideal electrolyte solution but is governed by the same rules, that is, the charge-carrying species are independent of each other. Thus, for a large proportion of ionic liquids, we can assume that conductivity occurs by a hole-hopping mechanism [21]. If these holes are at infinite dilution, classical dilute electrolyte solution thermodynamics are obeyed and ionic association can be ignored. It must however be stressed that ionic interaction clearly occurs, but this does not appear to affect the number of charge carriers since the hole concentration is much lower than the number of free ions. The concept of ionic association is probably not helpful in an ionic liquid until a significant amount of a diluent is added to the system.

1.4 CONCLUSION

It is proposed that an ionic liquid is different to a molecular solution or a molten salt in the way that it transports charge. In ionic liquids, it appears to be the mobility of the voids rather than the number of charge carriers that governs charge transport. Classical diffusion is not obeyed due to the invalidity of the central assumption of the Stokes–Einstein equation, namely, the correlation length is insignificant compared to the size of the diffusing species. It is therefore appropriate to consider a species moving in an ionic liquid by a series of discrete jumps. It is this mechanism that will be unique to ionic liquids, and a definition based on this principle will allow for the incorporation of different amounts of molecular components of different polarity. While this may not appear to advance our understanding of what an ionic liquid is, it could be concluded that an ionic liquid becomes a solution or a molten salt under the conditions when the Stokes–Einstein equation becomes valid.

REFERENCES

1 Walden, P., Über die molekulargrösse und elektrische Leitfähigkeit einiger gesehmolzenen Salze, *Bull. Acad. Sci. St. Petersburg* **8**, 405–422 (1914).

2 Seddon, K.R., Stark, A. and Torres, M.-J., The influence of chloride, water and organic solvents on the physical properties of ionic liquids, *Pure Appl. Chem.* **72**, 2275–2287 (2000).

3 Zhang, J., Wu, W., Jiang, T., Gao, H., Liu, Z., He, J. and Han B., Conductivities and viscosities of the ionic liquid [bmim][PF_6] plus water plus ethanol and [bmim][PF_6] plus water plus acetone ternary mixtures, *J. Chem. Eng. Data* **48**, 1315–1317 (2003).

4 Jarosik, A., Krajewski, S.R., Lewandowski, A. and Radzimski, P., Conductivity of ionic liquids in mixtures, *J. Mol. Liq.* **123**, 43–50 (2006).

5 Bockris, J.O'M. and Reddy, A.K.N., *Modern Electrochemistry*, Vol.**1**, Plenum Press, New York (1970).

6 Robinson, R.A. and Stokes, R.H., *Electrolyte Solutions*, Butterworths, London (1959).

7 Fuoss, R.M., Properties of electrolytic solutions, *Chem. Rev.* **17**, 27–42 (1935).

8 (a)Fuoss, R.M. and Kraus, C.A., Ionic association. II. Several salts in dioxane-water mixtures, *J. Am. Chem. Soc.* **79**, 3304–3310 (1957). (b)Fuoss, R.M. and Kraus, C.A., Properties of electrolytic solutions. II. The evaluations of Λ_0 and of K for incompletely dissociated electrolytes, *J. Am. Chem. Soc.* **55**, 476–488 (1933).

9 Abbott, A.P. and Schiffrin, D.J., Conductivity of tetra-alkylammonium salts in polyaromatic solvents, *J. Chem. Soc. Faraday Trans.* **86**, 1453–1459 (1990).

10 Longinotti, M.P. and Corti, H.R., Fractional Walden rule for electrolytes in supercooled disaccharide aqueous solutions, *J. Phys. Chem. B* **113**, 5500–5507 (2009).

11 Biltz, W. and Klemm, W., Über das elektrische Leitvermögen und den Molekularzustand geschmolzener Salze, *Z. Anorg. Allg. Chem.* **152**, 267–294 (1926).

12 Pugsley, F.A. and Wetmore, F.E.W., Molten salts: Viscosity of silver nitrate, *Can. J. Chem.* **32** (9), 839–841 (1954).

13 Angell, C.A., Byrne, N. and Belieres, J.-P., Parallel developments in aprotic and protic ionic liquids: Physical chemistry and applications, *Acc. Chem. Res.* **40**, 1228–1236 (2007).

14 Xu, W., Cooper, E.I. and Angell, C.A., Ionic liquids: Ion mobilities, glass temperatures, and fragilities, *J. Phys. Chem. B* **107**, 6170–6178 (2003).

15 Yoshizawa, M., Xu, W. and Angell, C.A., Ionic liquids by proton transfer: Vapor pressure, conductivity, and the relevance of ΔpK_a from aqueous solutions, *J. Am. Chem. Soc.* **125**, 15411–15419 (2003).

16 MacFarlane, D.R., Forsyth, M., Izgorodina, E.I., Abbott, A.P., Annat, G. and Fraser, K., On the concept of ionicity in ionic liquids, *Phys. Chem. Chem. Phys.* **11**, 4962–4967 (2009).

17 Schreiner, C., Zugmann, S., Hartl, R. and Gores, H.J., Fractional Walden rule for ionic liquids: Examples from recent measurements and a critique of the so-called ideal KCl line for the Walden plot, *J. Chem. Eng. Data* **55**, 1784–1788 (2010).

18 Fürth, R., On the theory of the liquid state. III. The hole theory of the viscous flow of liquids, *Math. Proc. Cambridge Philos. Soc.* **37**, 281–290 (1941).

19 Fürth, R., On the theory of the liquid state. II. The hole theory of the viscous flow of liquids, *Math. Proc. Cambridge Philos. Soc.* **37**, 281–290 (1941).

20 Abbott, A.P., Application of hole theory to the viscosity of ionic and molecular liquids, *ChemPhysChem.* **5**, 1242–1245 (2004).

21 Abbott, A.P., Model for the conductivity of ionic liquids based on an infinite dilution of holes, *ChemPhysChem.* **6**, 2502–2505 (2005).

22 Zhao, H., Liang, Z.-C. and Li, F., An improved model for the conductivity of room-temperature ionic liquids based on hole theory, *J. Mol. Liq.* **149**, 55–59 (2009).

23 Abbott, A.P., Harris, R.C., Ryder, K.S., D'Agostino, C., Gladden, L.F. and Mantle, M.D., Glycerol eutectics as sustainable solvent systems, *Green Chem.* **13**, 82–90 (2011).

24 Tokuda, H., Hayamizu, K., Ishii, K., Susan, M.A.B.H. and Watanabe, M., Physicochemical properties and structures of room temperature ionic liquids. 1. Variation of anionic species, *J. Phys. Chem. B* **108**, 16593–16600 (2004).

25 Tokuda, H., Hayamizu, K., Ishii, K., Susan, M.A.B.H. and Watanabe, M., Physicochemical properties and structures of room temperature ionic liquids. 2. Variation of alkyl chain length in imidazolium cation, *J. Phys. Chem. B* **109**, 6103–6110 (2005).

26 Tokuda, H., Ishii, K., Susan, M.A.B.H., Tsuzuki, S., Hayamizu, K. and Watanabe, M., Physicochemical properties and structures of room-temperature ionic liquids. 3. Variation of cationic structures, *J. Phys. Chem. B* **110**, 2833–2839 (2006).

27 Stillinger, F.H. and Hongdon, J.A., Translation-rotation paradox for diffusion in fragile glass-forming liquids, *Phys. Rev. E* **50**, 2064–2068 (1994).

28 Einstein, A. and Fürth, R., *Investigation on the Theory of Brownian Movement*, Dover, New York (1956).

29 Cussler, E.L., *Diffusion*, 3rd edition, Cambridge University Press, Cambridge (2009).

30 Kholodenko, A.L. and Douglas, J.F., Generalized Stokes–Einstein equation for spherical-particle suspensions, *Phys. Rev. E* **51**, 1081–1090 (1995).

31 Affouard, F., Descamps, M., Valdes, L.-C., Habasaki, J., Bordat, P. and Ngai, K.L., Breakdown of the Stokes–Einstein relation in Lennard-Jones glassforming mixtures with different interaction potential, *J. Chem. Phys.* **131**, 104510 (2009).

32 Del Popolo, M.G. and Voth, G.A., On the structure and dynamics of ionic liquids, *J. Phys. Chem. B* **108**, 1744–1752 (2004).

33 Zhou, Q., Lu, X., Zhang, S. and Guo, L., Physicochemical properties of ionic liquids, in *Ionic Liquids Further UnCOILed: Critical Expert Overviews*, eds. N.V. Plechkova and K.R. Seddon, John Wiley & Sons, Inc., Hoboken, NJ (2013).

34 Taylor, A.W., Electrochemistry *in vacuo*, PhD Thesis, University of Nottingham, 2010.

35 Taylor, A.W., Licence, P. and Abbott, A.P., Non-classical diffusion in ionic liquids, *Phys. Chem. Chem. Phys.* **13**, 10147–10154 (2011).

2 NMR Studies of Ionic Liquids

PAUL M. BAYLEY, JAN NOVAK, and MARIA FORSYTH
Institute for Frontier Materials, Geelong Technology Precinct, Deakin University, Victoria, Australia

2.1 INTRODUCTION

Nuclear magnetic resonance (NMR) spectroscopy is an incredibly versatile technique that has been available for about half a century, forming an integral part of the structure determination of almost any organic or biological molecule. Perhaps the most important aspect of NMR is that it is an entirely quantitative spectroscopy [1]. Revolutionary development of various experimental techniques has seen widespread application of this incredibly complex and effective spectroscopy [2, 3]. Although valuable information on chemical structure is obtained, another important function of NMR spectroscopy is in providing highly accurate kinetic data on compositionally complex liquid systems [4]. The application of NMR spectroscopy to ionic liquids has seen significant growth in recent years, as more researchers apply advanced experiments to gain valuable insights into the dynamics of these materials.

This chapter will not be an exhaustive review of the vast amount of literature available but more of a guide for the non-expert reader to gain an understanding of what information NMR spectroscopy can provide, in addition to an understanding of the recent highlights and future directions of the field. The discussion will primarily be focused on aprotonic ionic liquids, rather than on protonic ionic liquids or the older chloroaluminate systems and newer chiral ionic liquids. For further detailed information on the NMR spectroscopy of ionic liquids, the reader is referred to the excellent review by Bankmann and Giernoth of all literature up to 2007 [5], with a later extension by Giernoth including more modern ionic liquids [6] and a further review by Ananikov [7].

NMR spectroscopy is a unique structural determination technique because the observed frequency that the magnetic moment of a spin-bearing nucleus precesses depends on the molecular environment of that nucleus. During an NMR experiment, a magnetic field is applied to the sample, splitting the energy

---

Ionic Liquids Completely UnCOILed: Critical Expert Overviews, First Edition.
Edited by Natalia V. Plechkova and Kenneth R. Seddon.

levels of the spin states, to enable nuclei-specific manipulation of the populations of these states in a non-destructive manner [2, 3]. If a proton (^{1}H) is placed in an external magnetic field (B_0), the magnetic moments of the nuclei (μ) adopt either a parallel or an anti-parallel orientation to the field lines, known as the α (low energy) and β (high energy) states, respectively. The energy level for each state is defined by the magnitude of the interacting fields ($-\mu B_0$ for α and μB_0 for β), with the energy difference given by $2\mu B_0$. The system achieves the lowest energy when the population of each state (N) is given by the Boltzmann factor [8]:

$$\frac{N_\beta}{N_\alpha} = \exp\left(-\frac{2\mu B_0}{kT}\right) \tag{2.1}$$

where k is the Boltzmann's constant and T is the absolute temperature. The application of a radio frequency (RF) pulse that corresponds to the Larmor frequency, given in Equation 2.2, causes absorption of energy that equalises the spin states. The energy difference between spin states determines the intensity of the absorption [3]:

$$\omega_0 = \gamma B_0 \tag{2.2}$$

where ω_0 is the Larmor frequency, γ is the gyromagnetic ratio (a unique intrinsic property of the atom) and B_0 is the magnetic field strength.

The detected resonance frequency depends not only on the field and gyromagnetic ratio but also on the environment of the electrons and other nuclei surrounding the observed nucleus that causes modulation of the magnetic field, in a process termed shielding. Variation of the observed frequency with shielding has been termed the chemical shift. With reference to some known solvent, one can obtain an understanding of the local environment or chemical structure of the probed nuclei [3].

2.2 RELAXATION

When a nuclear spin is perturbed from its equilibrium state, it will eventually relax back to this state through two first-order processes that provide two distinct relaxation times: longitudinal or spin–lattice (T_1) and transverse or spin–spin (T_2) relaxation times. A basic understanding of relaxation phenomena is a prerequisite to understanding and performing NMR experiments, in addition to providing large amounts of detailed information on molecular and segmental motions through the analysis of relaxation data. Running basic experiments requires an idea of the T_1 time to set appropriate delays and accurately calibrate pulses, with an estimate of T_2 needed for more complex experiments such as diffusion [8, 9].

Spin–lattice (T_1) relaxation involves the dissipation of energy into the lattice of the material, the local environment of the nuclei, and causes the magnetisation vector to decrease in one, longitudinal, direction parallel to the magnetic field lines (z-direction) and grow in the other following a π-RF pulse. Although the functional form of T_2 is similar to T_1, spin–spin relaxation involves the dephasing of the precessing nuclei in the transverse (x–y) plane and, in contrast, does not include the exchange of energy from the spin to the lattice. The mechanisms of T_2 relaxation depend on lower frequency motions that generally do not provide the detail of T_1 relaxation measurements. Hence, the following discussion will mainly involve T_1 relaxation and the mechanisms that cause this phenomenon [9]. The general form of T_1 and T_2 relaxation with respect to the correlation time (τ_c) is given in Figure 2.1. While there are few examples of this plot for ionic liquids, an excellent example can be found in the work by Imanari et al. [10]. As a decrease in temperature reflects slower molecular motion, the relaxation time is typically measured with respect to this parameter to gain valuable insight into the type of motion and relevant mechanisms.

There are four main types of interaction that can provide a pathway for energy transfer, and eventual relaxation, between a pair of spins or a spin and its environment: dipole interactions, chemical shielding anisotropy (CSA), quadrupole relaxation (only present for spin ≥ 1) and scalar relaxation. Furthermore, spin–rotation interactions can provide an avenue for the relaxation of gases or highly mobile liquids, while paramagnetic relaxation is caused by nuclei with unpaired electron spin. In a 7 Tesla (300 MHz) field with a typical organic molecule, dipole and spin–rotation are often considered as the only effective mechanisms [8]. These relaxation mechanisms will be explored through the use of the Bloembergen, Purcell and Pound (BPP) theory that was developed in the early stages of NMR spectroscopy [9, 11, 12]. Although the

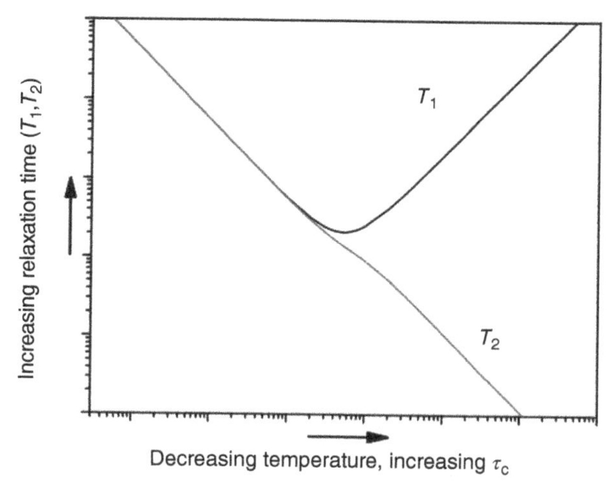

Figure 2.1 A plot of the general form of T_1 and T_2 relaxation times versus correlation time (τ_c).

theory is extensive, a brief review within the context of the ionic liquid literature will be explored, with more details given in many texts [1, 2, 8, 13].

To effectively interact with a nuclear spin, an oscillating magnetic field (or electric in the case of quadrupole nuclei) at the Larmor frequency of the nucleus is required. The rapid Brownian motion that occurs in liquids can cause these fluctuations; the Fourier analysis of these motions is a broad and continuous distribution of frequencies described by the spectral density function. The BPP theory defines the correlation time (τ_c) as the length of time a molecule or ion in a liquid remains in a given position before a collision with another molecule or ion changes its orientation [9]. Although the τ_c in solution is often described by an Arrhenius expression [9] and has been used by some researchers in the analysis of ionic liquid relaxation data [14], others have used a form of the Vogel–Fulcher and Tammann (VFT) equation [15]. The VFT equation is an empirical model first used to describe the viscosity of supercooled liquids and is applied to the physical properties of mainly viscous liquids and amorphous solids when there is an apparent increase in activation energy as the temperature is decreased [16–18]. While taking a slightly different form depending on the nucleus, the BPP equation for homonuclear T_1 relaxation of a two spin ½ system can be defined as [13]

$$\frac{1}{T_1} = C\left(\frac{\tau_c}{1+\omega_0^2\tau_c^2} + \frac{4\tau_c}{1+4\omega_0^2\tau_c^2}\right) \tag{2.3}$$

where, ω_0 is the frequency and C is a term evaluated separately for each nucleus. In the case of intramolecular dipolar relaxation between protons, C can be defined as

$$C = \frac{3}{10}\gamma_H^4\hbar^2\sum\frac{1}{r^6} \tag{2.4}$$

where γ_H is the gyromagnetic ratio of protons, ħ is the reduced Planck's constant and r the distance between interacting dipoles summed over all interacting dipoles.

As a result of this equation, several generalisations can be made, namely, the gyromagnetic ratio and the number and distance of the dipoles are significant factors in determining the relaxation rate. The gyromagnetic ratio, to the fourth power, dominates the equation to dictate the strength of the interactions causing relaxation. Since protons have the highest gyromagnetic ratio, they are typically the dominant pathway for dipolar relaxation. The inverse sixth power dependence of the distance between interacting dipoles is an indication that only the nearest intramolecular neighbours have an influence on the relaxation. Additionally, the sum of all dipoles requires an understanding of the structure of the material if one is to thoroughly analyse or predict relaxation data.

Regardless of the inherent difficulty in interpreting relaxation data, several researchers have published results on ionic liquids with varying degrees of analysis. Importantly, relaxation measurements should be made at numerous temperatures to determine the minimum relaxation time to enable detailed analysis. In the simplest terms, the minimum on a variable temperature (inverse temperature) versus T_1 relaxation time (logarithmic scale) plot will shift to lower temperatures (retaining the same relaxation time) if there is only a shift in mobility of the nucleus, while a shift on the ordinate axis is an indication of a change in environment [14]. Some authors report T_1 measurements without further analysis than this [19, 20]. The work of Hayamizu et al. involves studies of the T_1 relaxation behaviour in ionic liquids [21] and ionic liquids doped with lithium salts [22–24]. These studies use the measured minimum to numerically solve the BPP equation, based on the relationship of $\omega_0\tau_c = 1$ when a molecule or ion is tumbling at its Larmor frequency [8]. From such data, the calculation of a flip angle or a single jump distance (for lithium) is possible, allowing correlation with other techniques such as neutron spectroscopy, viscosity, conductivity and computer simulations. The correlation time for reorientational motion can also be deduced without reaching a T_1 minimum by using a non-linear least squares method [14]. By approximating the function of the correlation time as an Arrhenius expression and substituting it into the BPP equation, and assuming that the constant (C in Eq. 2.3) is temperature insensitive, the equation can be solved with the experimental T_1 versus temperature data. This method compares very closely to the results of numerically solving the BPP equation when an experimental minimum is observed [14, 24].

Carper and co-workers have performed a detailed analysis of the ^{13}C relaxation times of both $[C_4mim]^+$ [25] and $[C_9mim]^+$ [26] with $[PF_6]^-$, which was later extended with more detail to deconvolute the relative contributions of the various relaxation mechanisms [27]. They found that the contribution of CSA to the experimentally observed relaxation time was about half of the contribution from dipolar relaxation. This work raises doubts about the applicability of isotropic relaxation models to ionic liquids. It is important to note that the ^1H and ^{13}C T_1 measurements of ionic liquids in the literature show different behaviour when attached to the same ion. The random Brownian motion that occurs in most liquids leads to rapid spin diffusion between nuclei bonded to a common ion or molecule, causing them to all exhibit the same T_1. The lack of such behaviour is a clear indication that the dynamics of ionic liquids are not isotropic, with the CSA values determined for ^{13}C on the $[C_9mim]^+$ cation reminiscent of liquid crystals [26].

2.3 NUCLEAR OVERHAUSER EFFECT

As a result of relaxation, the nuclear Overhauser effect (NOE) is a phenomenon predicted by Albert Overhauser in 1953, which is the fractional change in intensity of one NMR resonance when another resonance is irradiated. It is the transfer of nuclear spin polarisation between nuclei by cross-relaxation and has become indispensable for the determination of the liquid structure of macromolecules, particularly biomolecules, since the first 2D methods were developed by K. Wüthrich, who was awarded the Nobel Prize in Chemistry in 2001 for his work [28]. It was first shown, theoretically, that saturating the electron magnetic resonance in a metal would cause the nuclear resonance intensity to increase by three orders of magnitude ($\gamma_{electron}/\gamma_{nuclei}$), and that similar, albeit much less, enhancement was caused between two nuclei [9]. Since the NOE manifests the distance between different parts of an ion or molecule, they can be directly correlated with the conformation of that ion or molecule [28, 29].

Practically quantifying the results from these experiments can be challenging, with many researchers using them as a qualitative measure of the proximity of the spin to other nuclei. There are four steps to any 2D NMR experiment that involve (i) preparing the system by exciting a nucleus with a $\pi/2$ RF pulse, (ii) the evolution step measuring the chemical shift of that nucleus before it passes its magnetisation to another, (iii) a mixing time that allows the transfer of magnetisation to other nuclei and (iv) detecting the chemical shift of the nuclei that have received any magnetisation [30]. When running an NOE experiment, a mixing time on the order of the T_1 relaxation time will provide the greatest cross-peak intensity. As a result of such a long mixing time, however, the cross peak intensities for ionic liquids may also contain a contribution from spin diffusion to convolute any quantitative analysis [31]. To remove this effect, the experiments are typically repeated with incremental changes to the mixing time to identify the linear region of NOE build-up [30, 32].

A variety of NOE experiments have been applied to ionic liquids in an effort to elucidate any existing structure of these materials in the liquid state. Although the results of such experiments and the quantification of internuclear distances can be challenging, there has been increasing interest in these experiments from the ionic liquid community. Some of the initial work by Mantz et al. [33], involving chloroaluminate ionic liquids, determined that the distance between neighbouring imidazolium chloride molecules is less than 4Å (the distance at which NOE becomes negligible and no cross-peaks are observed), which correlates well with IR measurements and theoretical calculations on the same system. By deuteriating 95% of the cations and repeating the experiment, they determined the existence of intermolecular NOE transfer. In this work, the rotating-frame Overhauser effect spectroscopy (ROESY) experiment was performed, as it is more suitable for very viscous materials and requires shorter mixing time (reducing the overall experiment time). For a full description of the ROESY experiment, the reader is referred to the excellent text by Jacobsen et al. [34].

The work of Nama et al. [35] investigated a series of $[C_4mim]^+$ ionic liquids with different anions, dissolving these liquids at varying concentrations in deuteriated dichloromethane or methanol. Application of the heteronuclear Overhauser effect spectroscopy (HOESY) experiment allows the determination of ^1H–^{19}F contacts to gain insight into potential ion pairing. The neat $[BF_4]^-$ ionic liquid shows that the ^{19}F nuclei on the anion are in close proximity to all the protons of the cation, consistent with intermolecular contacts and modelling on a similar system [36] to indicate the presence of several anions around each cation. Upon dilution, the HOESY spectra show increased selectivity with methanol, while dichloromethane retains the same contacts even at the lowest concentration of ionic liquid. These results imply that methanol is strongly solvating the ions, while strong ion pairing is retained with dichloromethane, demonstrating that HOESY experiments are a powerful technique to understand ion–solvent interactions and solvation.

Castiglione and co-workers have extensively utilised HOESY and NOESY experiments for investigating the interactions between ionic liquid cation and anion (^{1}H–^{19}F) [32, 37] and, more recently, between the ionic liquid and the lithium cation (^{1}H–^{7}Li) primarily in the [C_4mpyr][NTf_2] system [38]. They report the likely presence of fluorinated domains in both the pure and blended ionic liquids due to the high selectivity of the ^{1}H–^{19}F contacts, despite an unobservable phase separation in these materials [32, 37]. It is intriguing that similar to the anion–pyrrolidinium cation interactions between all protons, except those in the centre of the butyl chain, the pyrrolidinium–lithium interactions show the same selectivity. Even though there will be Coulomb repulsion between the two cations, the authors suggest that the strong coordination between the Li^+ and $[NTf_2]^-$ is responsible for the close proximity of the Li^+ and the pyrrolidinium cations and postulate that the result is a likely consequence of the local structure within the material [38].

Antony et al. [26] combined the equations for relaxation and NOE to form a new method of obtaining molecular reorientational dynamics from the ^{13}C relaxation data of [C_9mim][PF_6]. In this study, they calculated a pseudorotational correlation time from the experimental T_1 times, which were then used to calculate the corrected maximum NOE factors. Although the theory is complicated, the method aids in determining the relative contribution of the various relaxation mechanisms to the observed relaxation time. Interestingly, the average CSA parameters for the imidazolium ring carbons is similar to those found for aromatic rings in liquid crystal solutions, indicating a certain degree of order existing in the liquid phase.

2.4 PULSED FIELD-GRADIENT NMR SPECTROSCOPY

Particles diffusing through a medium make up the general model describing numerous physical, chemical, biological and industrial processes. NMR spectroscopy is an important technique as it can label or encode the Brownian trajectories of spin-bearing particles using very strong, ideally linear, magnetic field gradient pulses [39]. Such experiments are collectively described as pulsed field-gradient nuclear magnetic resonance (PFG-NMR), a field that has seen great expansion in recent decades. With significant developments in probe design affording increased stability of the gradient amplifiers and greater availability of actively shielded gradient coils, PFG-NMR diffusion measurements have become ubiquitous in the study of complex liquid systems [4]. PFG-NMR is not limited to the measurement of diffusion coefficients, with an enormous number of applications developed for coherence transfer and more advanced 3D and 4D experiments. These applications are, however, beyond the scope of this text and the reader is referred to the numerous reviews available in the literature [8, 40].

The most commonly applied PFG-NMR diffusion experiment is known as a spin echo (SE), illustrated in Figure 2.2: it is one of the simplest and the most easily understood. Initially, a $\pi/2$-RF pulse excites a uniform cone of magnetisation, aligning all the moments within the sample. This magnetisation is not diffusion sensitive, as the detected signal does not depend where in the sample the spins came from, until a field-gradient pulse introduces the diffusion sensitivity, winding the magnetisation into a helix of phase. This helix produces no net NMR signal because the phases are opposed; a π-RF pulse is required to refocus the magnetisation. Once refocused, the second field-gradient pulse is applied to transform the helix back into a cone with acquisition of the resulting signal. The interval between the two field-gradient pulses (denoted Δ) is known as the diffusion time, because this is where the diffusion effects manifest themselves. The amplitude of the echo is given by the Stejskal–Tanner equation for SE diffusion attenuation (Eq. 2.5) [41]. In other words, the first RF pulse prepares the spins, while the first gradient pulse encodes their position. The second gradient pulse cancels the effects of the first by being of the opposite sign, so that if the spins remained stationary throughout the experiment, a standard NMR signal is obtained. Through the rapid Brownian motion occurring in liquid materials, however, the spins move in the time between the gradient pulses (Δ) and experience a phase shift that reduces the detected signal depending on Equation 2.5. Successive experiments varying the Δ, or more commonly the gradient strength, provide the data points for a non-linear least

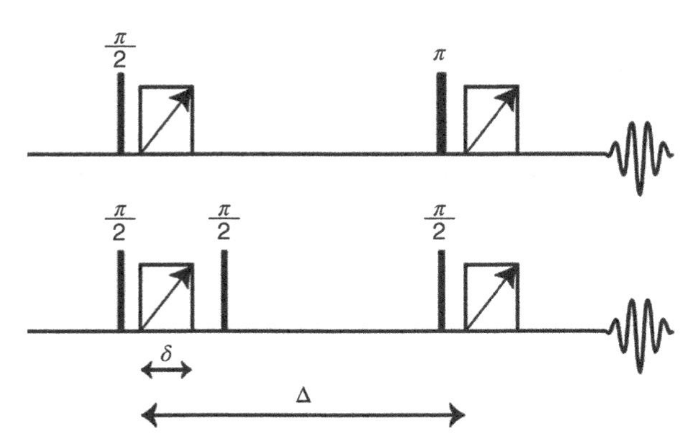

Figure 2.2 Common PFG-NMR diffusion experiment pulse sequences: above spin-echo (SE) and below stimulated spin-echo (STE). Gradient pulses, of length δ, are separated by the diffusion time Δ.

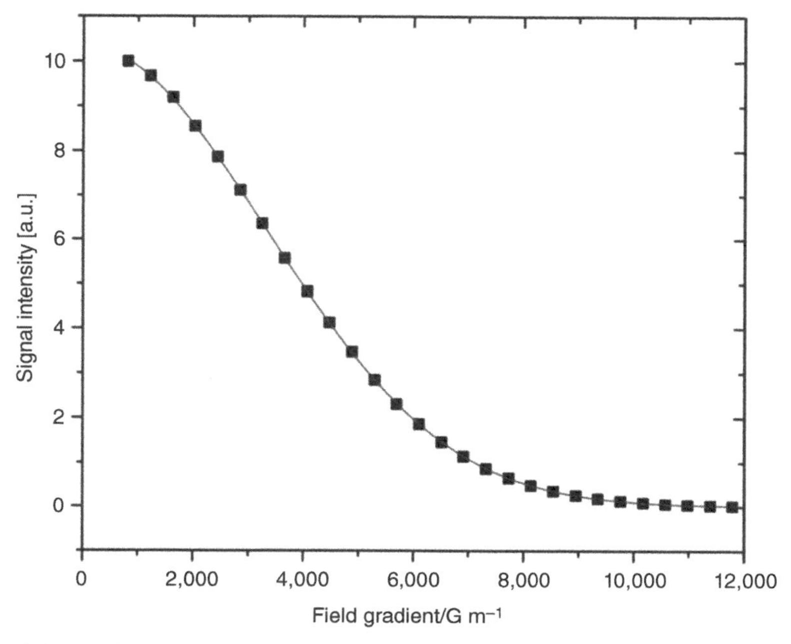

Figure 2.3 Typical experimental data from a PFG-NMR diffusion experiment with the fitted Stejskal–Tanner equation (Eq. 2.5 shown as the line).

squares fit of Equation 2.5 to solve for the diffusion coefficient. Typical experimental data with the fitted Stejskal–Tanner equation (Eq. 2.5) is shown in Figure 2.3. To produce a gradient pulse of opposite sign to the first, the polarity of the gradient coil can be physically reversed. However, this does not produce the desired level of reproducibility of the pulses, and modern NMR systems use the π-RF pulse to cause the same effect.

$$I = I_0 \exp\left[-D(\gamma G\delta)^2\left(\Delta - \frac{\delta}{3}\right)\right] \tag{2.5}$$

where D is the diffusion coefficient, G is the strength of the magnetic field gradient, Δ is the diffusion time and δ is the gradient pulse length. The immediate consequence of this equation is that there are three main experimental parameters that can be modified to acquire diffusion data: G, Δ and δ. These parameters are principally dependent on the hardware of the PFG-NMR system.

Equation 2.5 is also valid for the STE experiment. STE is a slightly different pulse sequence that involves three $\pi/2$-RF pulses, with the extra one after the first gradient pulse, as illustrated in Figure 2.2 [4, 42]. This pulse sequence has been designed for more viscous systems where there is commonly a greater discrepancy between the T_1 and T_2 relaxation times that can affect the signal intensity [42]. The primary limitation of the SE pulse sequence is that the spins must remain in the transverse (*x–y*) plane during the diffusion time period (Δ), which is limited to at least 10–20 ms on modern hardware, where they undergo T_2 relaxation processes. If the T_2 relaxation time of the material being measured is less than Δ, which can be the case in certain viscous systems and quadrupolar nuclei (such as ^{7}Li), the second gradient pulse will not properly unwind the magnetisation helix and lead to little (or no) signal intensity and greater measurement error. By applying a second $\pi/2$-RF pulse as soon as the first gradient pulse has finished, the magnetisation is flipped back into the longitudinal plane where slower relaxation processes occur ($T_1 > T_2$), retaining greater signal intensity when the experiment finishes. Since the final $\pi/2$-RF pulse has flipped the magnetisation π from the first gradient pulse, the experiment is very similar to SE; however, materials with a T_2 down to *ca.* 2 ms are still measurable.

2.4.1 PFG-NMR Diffusion Measurement Issues

The greatest problems with PFG-NMR diffusion measurements are introduced when performing variable temperature measurements; convection and RF pulses are the main issues. Large changes in the conductivity of ionic liquids with temperature will cause not only the tuning and matching to change (beyond the effect of probe temperature) but also the length of the RF pulse. To ensure that phase cycling does not cause cumulative error in the experiment, the probe should be tuned and matched at each temperature with the RF pulses checked and calibrated. A convective flow is set up in the NMR sample tube when the experiment temperature is far from ambient due to the existence of a thermal gradient along the length of the tube. In a diffusion measurement, this thermally induced flow causes an acceleration of the echo decay to cause an overestimation of the true self-diffusion coefficient of the species in question [43]. Convection is an issue for all diffusion measurements, but is particularly accentuated by viscous fluids [44] and the design of specialised diffusion probes that provide very large pulsed field gradients. Because such a large gradient coil is required around the sample, there is little room for thermal insulation and the heater coil (that is further from the sample than most probes). Schematically illustrated in Figure 2.4, the thermal gradient causes a driving force for the diffusion of material at the bottom of the sample tube to the cooler top of the tube.

There are numerous techniques that can be used to minimise the effect of convection: utilising specialised sample tubes [45], employing pulse sequences designed to accommodate the presence of flow [43, 46] and minimising sample volume or diffusion time (Δ) [47]. Figure 2.5 shows the dependence of the measured diffusion coefficient on the diffusion time (Δ) and sample volume for the common ionic liquid [C_3mpyr][NTf_2]. The diffusion time is the period in which the diffusion effects manifest, and it is expected that the longer this delay is set to, the higher the apparent diffusion coefficient (D_{app}) measured

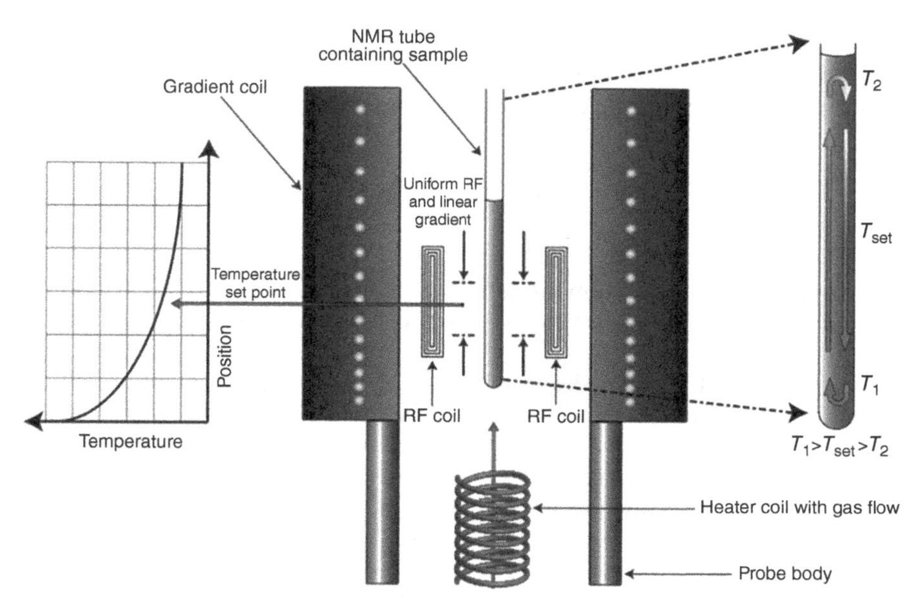

Figure 2.4 Basic schematic of a typical diffusion probe illustrating the origin of convection.

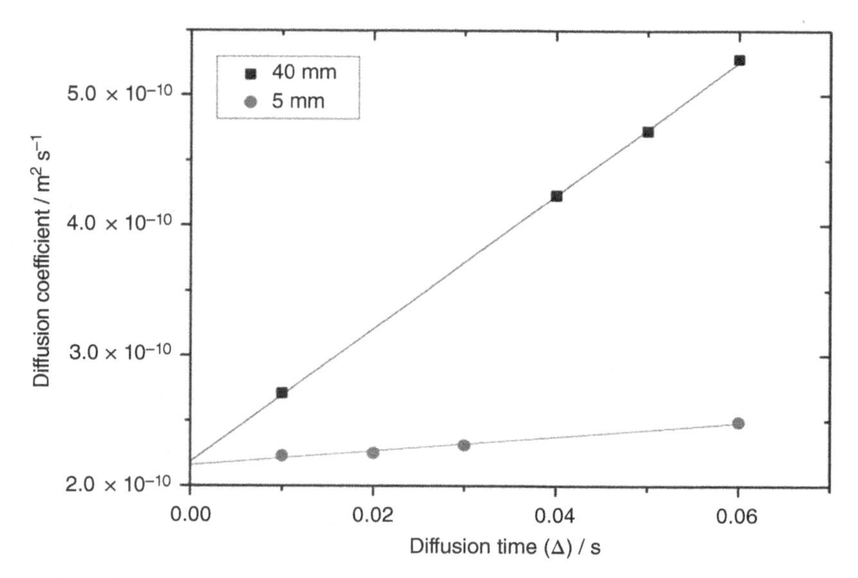

Figure 2.5 The dependence of ^{1}H diffusion coefficient on diffusion time (Δ) and sample volume at 373 K. NMR tubes of 5 mm (o.d.) were filled with [C_3mpyr][NTf_2] to a height of either 5 or 40 mm – the intercept provides the true diffusivity. The dry (<100 ppm H_2O) ionic liquid was packed and sealed under an N_2 atmosphere (<10 ppm H_2O). Measurements were performed on a 300 MHz Bruker Avance NMR spectrometer with a Diff30 diffusion probe [48].

will be. While a clear sign of convection is the failure of the Stejskal–Tanner equation to describe the data, in the event of good quality fits the apparent diffusion coefficient can be defined by Equation 2.6:

$$D_{app} = D_{true} + \frac{\Delta v^2}{2} \tag{2.6}$$

This equation is derived from modelling the convective flow as equal amounts of liquid travelling at equal velocity (v) in opposite directions in a cylindrical sample tube [47].

The gradient of the lines shown in Figure 2.5 are proportional to the velocity of convection, clearly showing the effect of sample volume and minimising the diffusion time. These lines intercept the ordinate axis at the true diffusion coefficient and show that the combination of short diffusion time and small sample volume is an effective means of reducing the measured convection to the limitations of the technique's accuracy. Unless the probe has been shimmed to accommodate small volumes, however, spectral resolution and intensity suffer, as well as additional field gradients potentially existing due to differences in magnetic susceptibility above and below the sample. An effective means of employing small sample volumes and limiting these disadvantages is the use of specialised sample tubes or capillaries [45].

Figure 2.6 is an example of how convective flow increases as the temperature is raised or lowered from the ambient. Comparison of the diffusion measured using a standard sample volume (40 mm) with the small sample volume (5 mm), which Figure 2.5 indicates is practically convection free, allows the quantification of the convective flow. Pulse sequences designed to limit the effect of convection on diffusion measurements have also been designed [43, 46]. In particular, the double-stimulated-echo (DSTE) pulse sequence has been shown to effectively remove convection effects and has also been extended to include bipolar gradients to remove eddy current effects (mainly an issue without gradient shielding). The presence of convective flow causes a flow-velocity-phase factor during the first STE period that is effectively unwound during the second, removing the effect of the velocity to give the true diffusion coefficient. Under the condition of constant laminar flow, the DSTE method is very effective, but cannot compensate for turbulent convection. The reader is referred to the work of Jerschow et al. for a complete description and the mathematics behind this experiment [46]. The limitation of this method is the length of the pulse sequence causing lower signal intensity at the end of the experiment, particularly when the material under investigation has a relatively short transverse relaxation time. Further development, however, has led to optimisations that include preparatory gradient pulses that minimise the length of the sequence [49].

Radiation damping is another issue that must be considered when performing experiments on pure samples, which can cause phase cycling errors and

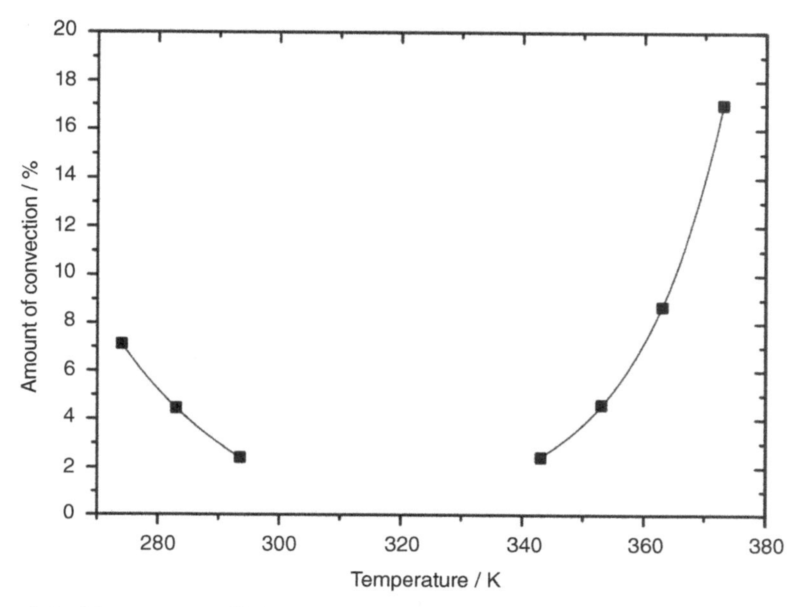

Figure 2.6 The amount of ^{1}H convective flow included in the diffusion measurements with temperature using 40 mm sample height in a 5 mm (o.d) NMR tube. Taken with neat $[C_3mpyr][NTf_2]$ using $\Delta = 10\,ms$. The dry (<100 ppm H_2O) ionic liquid was packed and sealed in an N_2 atmosphere (<10 ppm H_2O). Lines are an exponential fit to guide the eye. Measurements were performed on a 300 MHz Bruker Avance NMR spectrometer with a Diff30 diffusion probe [48].

signal broadening. When the NMR signal is extremely intense, the signal generated in the probe's RF coil can be strong enough to act as a further RF pulse on the sample. This effect is minimised by either setting the transmitter frequency off-resonance or slightly detuning the probe [2].

2.4.2 Theories of Diffusion

Throughout the development of the field of ionic liquids, several parameters have become essential in characterising and understanding the nature of the physicochemical properties of these materials. Molar conductivity is one of the most important characteristics to consider, providing an insight into both the degree of anion–cation aggregation and the electrostatic forces of the ionic liquid [50]. Several other physical parameters can be combined into the ionicity ratio to form a description of the ionic state and ion dynamics that have a significant influence on the physicochemical properties of the ionic liquid [51]. It is important to also note that a molecular solvent electrolyte only contains two species of ion, the cation and the anion of the added salt, while, in contrast, ionic liquids are entirely composed of ions that make the selective transport of either the anion or the cation impossible [52].

Diffusion in liquids is initially approached from the Stokes–Einstein equation (Eq. 2.7) [53, 54].

$$D = \frac{kT}{c\pi\eta r_s} \tag{2.7}$$

where k is the Boltzmann constant, T is the absolute temperature, c is a constant, η is the viscosity and r_s is the hydrodynamic (Stokes) radius.

This model is essentially derived from classical thermodynamics and fluid mechanics, treating the diffusing species as a spherical particle in a continuum of fluid. Further development of this model has also seen it modified to involve other geometries; however, its main purpose is to highlight the relationship between diffusion and viscosity. The main disadvantage in applying this model to ionic liquids is that it assumes that there are no interactions between the ions/molecules, but it has been found to show some correlation [51, 55].

Combining the measurements of conductivity and self-diffusion can provide a greater understanding of the ion dynamics within the framework of the Nernst–Einstein equation. Self-diffusion of ionic liquids measured by the PFG-NMR technique will give an average of the ionic species in question, regardless of their state of charge (in the case of clusters) or association. Conversely, conductivity measurements from impedance spectroscopy measure the ion mobility under the influence of an applied field. It has been found by Watanabe [50, 51, 56, 57] and co-workers that the ratio of molar conductivity directly measured to that calculated by the Nernst–Einstein equation using the diffusion coefficient ($\Lambda_{imp}/\Lambda_{NMR}$) is invariably less than unity for ionic liquids. This discrepancy has been rationalised as the fraction of ions available to participate in ionic conductivity and has been termed ionicity [50]. The ionicity, however, is a quantification of how the sample conforms to the Nernst–Einstein equation and cannot be directly assumed to be a measure of the ion availability in the chemical sense [55]. The Walden plot is a more traditional approach to assessing the ionicity of ions in solution and was developed for use with dilute aqueous solution with its applicability to a wide range of ionic liquids recently demonstrated [55]. Although a Walden plot gives a reasonable quantification of ionicity, it is more appropriate to consider it as a qualitative approach that complements the ionicity ratio as calculated from the Nernst–Einstein equation. Nernst's work in the late 1880s was primarily focused on the fundamentals of physical chemistry, as we know them today. An understanding of the origins of the Nernst–Einstein equation is imperative if it is to be used in a meaningful manner to further understand the physical chemistry of ionic liquids. For a fully dissociated, strong electrolyte, the ion mobility is related to its molar conductivity by Equation 2.8 [53].

$$\lambda = zuF \tag{2.8}$$

where λ is the molar conductivity of the ion, z is the charge, u is the mobility and F is Faraday's constant.

Since Equation 2.8 defines both cations and anions, for an ionic solution at the limit of zero concentration, where there is an absence of any interionic interactions, the molar conductivity can be written as [53]

$$\Lambda = (z_+ u_+ v_+ + z_- u_- v_-) F \tag{2.9}$$

In a typical ionic liquid, the number of cations (v_+) is equal to the number of anions (v_-) with charge neutrality retained ($z_+ = z_- = z$) to allow Equation 2.9 to be expressed in the form [53]

$$\Lambda = z(u_+ + u_-) F \tag{2.10}$$

It can be shown that an ion in solution experiences a force due to the application of an electric field, following on from Fick's first law of diffusion, and that the mobility of that ion can be expressed as a function of the diffusion coefficient (D) to give the Einstein relation [53]

$$D = \frac{uRT}{zF} \tag{2.11}$$

Through a combination of Equation 2.10 and the Einstein relation defined in Equation 2.11, the effect of the electric field cancels out and the Nernst–Einstein equation can be derived as [53]

$$\Lambda = \frac{z^2 F^2}{RT}(D_+ + D_-) \tag{2.12}$$

Primarily, the most important aspect of these equations is that the ions are considered infinitely dilute, completely dissociated and not interacting with each other. This is clearly an oversimplification for concentrated ionic materials, which are well known to form associated clusters and strongly interact with each other. Within an ionic liquid, there exists long-range Coulombic forces, hydrogen bonding and often π–π stacking that will cause the application of the Nernst–Einstein equation (Eq. 2.12) to give an overestimation of the molar conductivity. The deviation of the actual molar conductivity measured through electrochemical techniques to that obtained from the diffusion coefficients (PFG-NMR measurements) and Nernst–Einstein equation quantifies the extent of interactions between the ions and is defined as the ionicity ratio ($\Lambda_{imp}/\Lambda_{NMR}$). It must also be noted that the theory does not implicitly define the origins of any deviation or allow for the treatment of any concentration dependence.

The variation of the ionicity ratio with temperature is much harder to use in reaching any conclusions because convective flow will cause an overestimation of the PFG-NMR diffusion measurements the further away the measurement is from the ambient temperature. While convection can be minimised through the techniques previously outlined, care must still be taken to ensure that the difference between the observed and the actual diffusion coefficients is minimal. The effect of convection on the ionicity ratio of [C_3mpyr][NTf_2] can potentially make the trend with temperature contain large errors [20].

An NMR investigation of the melting behaviour of several pyrrolidinium ionic liquids has been published by Kunze et al. [58], with the ionicity ratio calculated at various stages of the melting process. The ratio is very close to zero when in the solid phase, and rises steadily throughout the melting process for all of the ionic liquids investigated. The increase in ionicity during melting is an indication that the ions in the system remove themselves from the comparatively well-ordered crystals in which they form almost entirely uncharged clusters. Once the ions dissociate and form a liquid, there is an apparent re-association as the temperature is increased further. The ionicity ratio for several imidazolium and pyridinium ionic liquids, calculated over a wide temperature range by Noda et al. [59], does not show much change with temperature in the liquid state. The work by Tokuda et al. [51, 56, 57] includes calculating the ionicity ratio over a large temperature range for a wide variety of ionic liquids. Overall, the change in ionicity with temperature in these studies does not display any clear trends. A viscous ionic liquid based on a morpholinium cation, however, appears to exhibit a decrease in the ionicity ratio when heated above room temperature [60]. Indeed, Watanabe and co-workers have reviewed the literature on the ionicity ratio of a wide range of ionic liquids, and related it to various polarity scales and physicochemical properties to show some interesting correlations [61].

2.5 RECENT DEVELOPMENTS AND FUTURE DIRECTIONS

2.5.1 Recent Developments

NMR spectroscopy, and its application to ionic liquids, is far from a stagnant field, with continual development ensuring the value of this technique in future. Researchers are beginning to study more exotic NMR-active nuclei, such as ^{81}Br [62], ^{79}Br and ^{127}I [63], in both the solid and the liquid states. PFG-NMR diffusion measurements at variable pressure [64] will aid in the design of ionic liquids for certain industrial and engineering applications. Recent work has involved the use of ionic liquids as solvent probes for cryoporometry measurements, with the melting temperature monitored by NMR spectroscopy to determine the change in melting temperature of the ionic liquid, which is then related to the pore size [65].

Recent developments also include *in situ* NMR spectroscopy of lithium battery cycling [66] and electrophoretic PFG-NMR [67, 68]. In the lithium battery study, a static solids probe contained a small lithium battery pouch cell that was connected to an external potentiostat by shielded wires. Monitoring the ^{7}Li spectrum while cycling the battery sees a shoulder on the lithium metal resonance appear when dendritic lithium deposits form.

This unique approach allows the real-time, non-invasive monitoring of the performance of lithium metal batteries at a very detailed and quantitative level. Electrophoretic PFG-NMR is the measurement of the ionic mobility under an applied electric field and has the advantage of providing translational diffusion information in conditions close to those of a real electrochemical device. Although early measurements indicate over an order of magnitude increase over the self-diffusion measurements with reasonably large electric fields [68], more recent work suggests that there might be other issues associated with this type of measurement [67].

As the use of ionic liquids as solvents for natural polymers increases, the use of NMR spectroscopy in quantifying and understanding the dissolution process is becoming more important. Some researchers are investigating the dissolution of cellulose with a combination of ^{1}H and ^{13}C chemical shift data [69, 70], while others are using ^{31}P chemical shifts to measure lignin dissolution *in situ* [71]. Due to the complex spectrum of plant cell walls, Jiang et al. used deuteriated ionic liquids and co-solvents with 2D correlation spectroscopy and chemical shift data to study the lignin structure [72].

Among the many other recent developments in applying NMR spectroscopy to the study of ionic liquids, a recent relaxometry study on fuel cell membranes has provided detailed information on both ion mobility and structure [73]. The correlation of NMR data and computer simulations on the dynamics of ionic liquids can help gain detailed insight into these materials and is sure to become more prevalent in the literature [36, 74, 75]. The use of NMR imaging techniques is likely to become popular as the applications available to the field of chemistry and engineering develop. With such potential importance, Section 2.5.2 is devoted to describing the main potential uses for the ionic liquid community.

2.5.2 Magnetic Resonance Imaging (MRI)

NMR imaging, more commonly referred to as magnetic resonance imaging (MRI), is a technique usually associated with medical diagnostics. However, it is also widely used in the physical sciences including chemistry [76], chemical engineering [77] and physics. MRI uses magnetic field gradients to give nuclear spins spatial dependence allowing regions of differing chemical composition or physical properties to be resolved. The Larmor frequency at a given position r produced under the influence of magnetic field gradients is shown in Equation 2.13, where $\boldsymbol{G}$ is the pulsed field gradient component parallel to $\boldsymbol{B}_0$.

$$\omega(r) = \gamma \boldsymbol{B}_0 + \gamma \boldsymbol{G} \cdot r \tag{2.13}$$

Gradients are used in conjunction with frequency-selective soft (also referred to as narrowband) RF pulses. These types of experiments in the physical sciences are usually referred to as MR microimaging [78] and can be conducted on a standard wide or narrow bore NMR spectrometer using specialised probes and triple-axis gradients. A resolution of tens of microns (similar to optical microscopy) is achievable with MR microimaging. A more comprehensive explanation of the MRI techniques in the physical sciences can be found in one of the excellent referenced texts [77, 79].

For MRI to be useful, contrast between neighbouring regions is required. Contrast in MR images can be obtained *via* four different methods: spin density, relaxation times, chemical shift and motion contrast. Spin density contrast is obtained where one region has a different number of nuclear spins than a neighbouring region, resulting in different signal intensity. This could be used for the observation of precipitation or gelation during chemical processes. Both T_1 and T_2 relaxation times may not be homogeneous for a sample. MRI pulse sequences can be tailored to produce T_1 and T_2 contrast so that changes can be visualised over space. As explained (see Section 2.2), both T_1 and T_2 measurements can be used to describe physicochemical changes in ionic liquids and ionic liquid–diluent mixtures. An appropriate example of this was the study of corrosion processes by Britton et al. [80] in a corrosion cell with an aqueous electrolyte. This work enabled the visualisation of electrochemical processes, which could potentially be extended to corrosion of metals in the presence of ionic liquids. This would represent the first steps towards possibly imaging working cells and batteries for which ionic liquids have shown promise as electrolytes.

Chemical shift contrast works on the principle of selectively exciting a single peak in a spectrum (again using soft RF pulses), so that only pixels with the peak or spectral region of interest will show signal in the image. This could potentially be used for *in situ* imaging of chemical reactions where species with different spectral properties are formed. One example of this would be the dissolution of water into an ionic liquid that can cause a chemical shift of the water peak. This could possibly be utilised for chemical shift imaging in metal–air batteries or fuel cells to visualise uptake and production of water as a function of position.

Motional contrast can be obtained *via* two mechanisms: coherent or incoherent motion. Incoherent motion refers to diffusion that, as explained in the preceding section, can be measured using a PFG-NMR experiment. This can be combined with an imaging pulse sequence so that diffusion coefficients can be calculated for individual pixels. Diffusion in batteries is incredibly important and any inhomogenieties in diffusion could possibly affect cell performance. Quantification of this *via* NMR diffusion imaging could be invaluable. Coherent motion (velocity) can also be measured using a PFG experiment. Magnitude calculation is not performed so phase information is retained. Coherent motion causes a phase shift that is related to the velocity by Equation 2.14 where ν is the velocity and φ is the phase shift in degrees.

$$\varphi = \frac{\Delta G \gamma \nu \delta}{2\pi} \times 360 \qquad (2.14)$$

52 Ohno, H. H., Functional design of ionic liquids, *Bull. Chem. Soc. Jpn* **79**, 1665–1680 (2006).

53 Atkins, P., *Physical Chemistry*, 8th Ed., W.H. Freeman and Co.: New York (2005).

54 Helfferich, F. G., *Ion Exchange*, Dover Publications: New York (1995).

55 MacFarlane, D. R., Forsyth, M., Izgorodina, E. I., Abbott, A. P., Annat, G. and Fraser, K., On the concept of ionicity in ionic liquids, *Phys. Chem. Chem. Phys.* **11**, 4962–4967 (2009).

56 Tokuda, H., Hayamizu, K., Ishii, K., Susan, M. A. B. H. and Watanabe, M., Physicochemical properties and structures of room temperature ionic liquids. 2. Variation of alkyl chain length in imidazolium cation, *J. Phys. Chem. B* **109**, 6103–6110 (2005).

57 Tokuda, H., Ishii, K., Susan, M. A. B. H., Tsuzuki, S., Hayamizu, K. and Watanabe, M., Physicochemical properties and structures of room-temperature ionic liquids. 3. Variation of cationic structures, *J. Phys. Chem. B* **110**, 2833–2839 (2006).

58 Kunze, M., Montanino, M., Appetecchi, G. B., Jeong, S., Schönhoff, M., Winter, M. and Passerini, S., Melting behavior and ionic conductivity in hydrophobic ionic liquids, *J. Phys. Chem. A* **114**, 1776–1782 (2010).

59 Noda, A., Hayamizu, K. and Watanabe, M., Pulsed-gradient spin-echo ^{1}H and ^{19}F NMR ionic diffusion coefficient, viscosity, and ionic conductivity of non-chloroaluminate room-temperature ionic liquids, *J. Phys. Chem. B* **105**, 4603–4610 (2001).

60 Lane, G. H., Bayley, P. M., Clare, B. R., Best, A. S., MacFarlane, D. R., Forsyth, M. and Hollenkamp, A. F., Ionic liquid electrolyte for lithium metal batteries: physical, electrochemical, and interfacial studies of *N*-methyl-*N*-butylmorpholinium bis(fluorosulfonyl)imide, *J. Phys. Chem. C* **114**, 21775–21785 (2010).

61 (a) Ueno, K., Tokuda, H. and Watanabe, M., Ionicity in ionic liquids: correlation with ionic structure and physicochemical properties, *Phys. Chem. Chem. Phys.* **12**, 1649–1658 (2010); (b) Watanabe, M. and Tokuda, H., Ionicity in Ionic Liquids: Origin of Characteristic Properties of Ionic Liquids, in *Ionic Liquids Further UnCOILed: Critical Expert Overviews*, N.V. Plechkova and K.R. Seddon, eds., John Wiley & Sons, Inc.: Hoboken, NJ, pp. 217–234 (2014).

62 Balevicius, V., Gdaniec, Z., Aidas, K. and Tamuliene, J., NMR and quantum chemistry study of mesoscopic effects in ionic liquids, *J. Phys. Chem. A* **114**, 5365–5371 (2010).

63 Gordon, P. G., Brouwer, D. H. and Ripmeester, J. A., Probing the local structure of pure ionic liquid salts with solid- and liquid-state NMR, *Chem. Phys. Chem.* **11**, 260–268 (2010).

64 Harris, K. R., Kanakubo, M., Tsuchihashi, N., Ibuki, K. and Ueno, M., Effect of pressure on the transport properties of ionic liquids: 1-alkyl-3-methylimidazolium salts, *J. Phys. Chem. B* **112**, 9830–9840 (2008).

65 Schulz, P. S., Ionic liquids as solvent probes for NMR cryoporometry, *Chem. Phys. Chem.* **11**, 87–89 (2010).

66 Bhattacharyya, R., Key, B., Chen, H., Best, A. S., Hollenkamp, A. F. and Grey, C. P., In situ NMR observation of the formation of metallic lithium microstructures in lithium batteries, *Nat. Mater.* **9**, 504–510 (2010).

67 Hayamizu, K. and Aihara, Y., Correlating the ionic drift under Pt/Pt electrodes for ionic liquids measured by low-voltage electrophoretic NMR with chronoamperometry, *J. Phys. Chem. Lett.* **1**, 2055–2058 (2010).

68 Umecky, T., Saito, Y. and Matsumoto, H., Direct measurements of ionic mobility of ionic liquids using the electric field applying pulsed gradient spin-echo NMR, *J. Phys. Chem. B* **113**, 8466–8468 (2009).

69 Hesse-Ertelt, S., Heinze, T., Kosan, B., Schwikal, K. and Meister, F., Solvent effects on the NMR chemical shifts of imidazolium-based ionic liquids and cellulose therein, *Macromol. Symp.* **294**, 75–89 (2010).

70 Zhang, J., Zhang, H., Wu, J., Zhang, J., He, J. and Xiang, J., NMR spectroscopic studies of cellobiose solvation in EmimAc aimed to understand the dissolution mechanism of cellulose in ionic liquids, *Phys. Chem. Chem. Phys.* **12**, 1941–1947 (2010).

71 King, A. W. T., Zoia, L., Filpponen, I., Olszewska, A., Xie, H., Kilpeläinen, I. and Argyropoulos, D. S., In Situ determination of Lignin Phenolics and Wood Solubility in imidazolium chlorides using ^{31}P NMR, *J. Agric. Food Chem.* **57**, 8236–8243 (2009).

72 Jiang, N., Pu, Y., Samuel, R. and Ragauskas, A. J., Perdeuterated pyridinium molten salt (ionic liquid) for direct dissolution and NMR analysis of plant cell walls, *Green Chem.* **11**, 1762–1766 (2009).

73 Neves, L. A., Sebastião, P. J., Coelhoso, I. M. and Crespo, J. G., Proton NMR relaxometry study of nafion membranes modified with ionic liquid cations, *J. Phys. Chem. B* **115**, 8713–8723 (2011).

74 Borodin, O. and Smith, G. D., Structure and dynamics of *N*-methyl-*N*-propylpyrrolidinium bis(trifluoromethanesulfonyl)imide ionic liquid from molecular dynamics simulations, *J. Phys. Chem. B* **110**, 11481–11490 (2006).

75 Borodin, O., Smith, G. D. and Henderson, W., Li^+ cation environment, transport, and mechanical properties of the LiTFSI doped *N*-methyl-*N*-alkylpyrrolidinium$^+$TFSI$^-$ ionic liquids, *J. Phys. Chem. B* **110**, 16879–16886 (2006).

76 Britton, M. M., Magnetic resonance imaging of chemistry, *Chem. Soc. Rev.* **39**, 4036–4043 (2010).

77 Gladden, L. F., Nuclear-magnetic-resonance in chemical-engineering – principles and applications, *Chem. Eng. Sci.* **49**, 3339–3408 (1994).

78 Callaghan, P. T., *Principles of Nuclear Magnetic Resonance Microscopy*, Clarendon Press/Oxford University Press: Oxford, 1991.

79 Blümich, B. and Kuhn, W., *Magnetic Resonance Microscopy: Methods and Application in Materials Science, Agriculture, and Biomedicine*, VCH: Weinheim, 1992.

80 Britton, M. M., Davenport, A. J. and Forsyth, M., Visualisation of chemical processes during corrosion of zinc using magnetic resonance imaging, *Electrochem. Commun.* **12**, 44–47 (2010).

81 Madsen, L. A., Hou, J. B. and Zhang, Z. Y., Cation/anion associations in ionic liquids modulated by hydration and ionic medium, *J. Phys. Chem. B* **115**, 4576–4582 (2011).

82 Callaghan, P. T., Kilfoil, M. L. and Samulski, E. T., Chain deformation for a polymer melt under shear, *Phys. Rev. Lett.* **81**, 4524–4527 (1998).

83 Xia, Y. and Callaghan, P. T., Study of shear thinning in high polymer-solution using dynamic NMR microscopy, *Macromolecules* **24**, 4777–4786 (1991).

84 Wassenius, H. and Callaghan, P. T., NMR velocimetry studies of the steady-shear rheology of a concentrated hard-sphere colloidal system, *Eur. Phys. J. E* **18**, 69–84 (2005).

85 Britton, M. M. and Callaghan, P. T., Two-phase shear band structures at uniform stress, *Phys. Rev. Lett.* **78**, 4930–4933 (1997).

86 Somers, A., Howlett, P., Sun, J., MacFarlane, D. and Forsyth, M., Transition in wear performance for ionic liquid lubricants under increasing load, *Tribol. Lett.* **40**, 279–284 (2010).

87 Watanabe, M., Ueno, K., Hata, K., Katakabe, T. and Kondoh, M., Nanocomposite ion gels based on silica nanoparticles and an ionic liquid: ionic transport, viscoelastic properties, and microstructure, *J. Phys. Chem. B* **112**, 9013–9019 (2008).

3 'Unusual Anions' as Ionic Liquid Constituents

PHILIPP EIDEN and INGO KROSSING
Department for Inorganic and Analytic Chemistry, Freiburger Materialforschungszentrum (FMF) and Freiburg Institute of Advanced Studies (FRIAS: Soft Matter Science), Albert-Ludwigs-Universität Freiburg, Freiburg, Germany

3.1 INTRODUCTION

The first ionic liquids were formed by simple anions such as $[NO_3]^-$ [1] or tetrahaloaluminates(III) [2–6]. In particular, the latter ionic liquids became quite famous in electrochemical processes [7]. Due to the instability of the aluminates against hydrolysis, the expansion to other fields stagnated. The next step in the evolution of ionic liquids was the introduction of the more stable, fluorinated species such as $[BF_4]^-$ and $[PF_6]^-$, which usually form ionic liquids with a moderate viscosity and rather good electrical conductivity [8]. Nevertheless, it turned out that these ionic liquids also exhibit slow hydrolysis, forming HF [8]. By using complex anions such as bis{(trifluoromethyl)sulfonyl}amide ($[NTf_2]^-$) and triflate ($[OTf]^-$), it was possible to generate stable ionic liquids with low viscosity and a high electrical conductivity. These two anion types are today among the most popular anions in application and research [7–9].

The correct choice of the anion usually determines many of the fundamental properties of an ionic liquid (e.g. hydrophobic/hydrophilic nature, good/bad conduction, low/high viscosity, glass forming, gas solubility and many others), while the choice of the cation allows these properties to be fine-tuned. Hence, it is of special importance to find new anions to introduce new or improved properties or special combinations of properties. Over the last few years, several interesting developments have emerged in this field. This chapter gives an overview of, in our personal opinion, the most exciting relevant examples, and reports many of their principal physical properties.

Ionic Liquids Completely UnCOILed: Critical Expert Overviews, First Edition.
Edited by Natalia V. Plechkova and Kenneth R. Seddon.

3.2 VARIABLES RESPONSIBLE FOR THE CHANGE OF IONIC LIQUID PROPERTIES

With the following selected examples, it will be shown how a change in entropy or size (volume, radius) influences key ionic liquid properties, for example, melting point or viscosity. We will return to these relationships when describing the new anions subsequently. The melting point, T_{fus}, is the ratio of the melting enthalpy to the melting entropy, according to Equation 3.1,

$$T_{fus} = \frac{\Delta H_{fus}}{\Delta S_{fus}} \tag{3.1}$$

This holds at the experimental melting temperature, when the liquid and solid phases have the same chemical potential: the entropy of fusion, ΔS_{fus}, almost exclusively governs the melting point. It has been shown that the enthalpy of fusion ΔH_{fus} is almost a constant, and usually amounts to *ca.* 20 kJ mol^{-1} for ionic liquids [10]. The fusion entropy, that is, the entropic freedom gained upon relaxing the restricted frozen conformation present in the solid state for the flexible mixture of conformers in the liquid state, is determined mainly by the symmetry number, σ, of the compound, and by its torsional flexibility, τ – the number of torsional angles that lead to new conformations (Eq. 3.2):

$$\Delta S_{fus} = a\log(\sigma) + b\tau + c \tag{3.2}$$

The symmetry also has a certain influence on the viscosity. However, in contrast to the entropy of fusion, where a higher τ normally indicates a lower σ, there is – in the case of viscosity – a more complicated interplay of variables: namely, the symmetry number, σ, the ion size expressed by r_m^* and the Gibbs energy of solvation of the isolated ionic liquid ions [11]:

$$\ln\frac{\eta(T)}{\eta_0} = d + e\,\ln\left(r_m^*\right) + f\ln(\sigma) + g\frac{\Delta G_{solv}^{*,\infty}}{RT} \tag{3.3}$$

r_m^* is the molecular radius and a measure of the size of the ions constituting the ionic liquid. In turn, the conductivity also depends on the same variables, as conductivity and viscosity are related by the Walden product, which was shown to hold for most ionic liquids [12–15].

3.3 PHOSPHATES AND AMIDES

Classical ionic liquids quite often use the hexafluorophosphate anion. It introduced high electrical conductivity at an acceptable level of viscosity. In the form of its lithium salt (which is not an ionic liquid), this ion has found its way to industrial application (lithium batteries). Nevertheless, the use of $[PF_6]^-$-based ionic liquids is not recommended due to hydrolysis, with formation of highly toxic and corrosive HF. To find a new anion with comparable properties, but without the disadvantages of $[PF_6]^-$, Ignatiev et al. substituted three fluoride ions by perfluoroethyl groups, giving the tris(pentafluoroethyl)trifluorophosphate anion, $[FAP]^-$ [16]. This modification made the anion stable against hydrolysis, while keeping good electrical conductivity and high hydrophobicity. The viscosity of the ionic liquids formed by this anion is decreased by about one order of magnitude in comparison to $[PF_6]^-$-based ionic liquids, likely due to a combination of less interaction potential and larger volume, which weakens the electrostatic and dispersive interactions with the counter ions [16]. However, the lithium salt of $[FAP]^-$ does not perform as well as $Li[PF_6]$ does as an electrolyte. In a recent contribution [17, 18], Ignatiev et al. substituted the fluorine atoms of $[PF_6]^-$ by oxalato ligands (or ethanedioate, see, e.g., Fig. 3.1c). These anions were successfully tested by them, and other groups, for electrochemical applications. In particular, the lithium salt of the tetrafluoro(oxalato)phosphate anion, $[PF_4(C_2O_4)]^-$, shows a similar behaviour to $Li[PF_6]$ in carbonate mixtures. So far, these new anions could solve the problem of hydrolysis, and can be used in ionic liquids for electrochemical processes. However, the chemical stability, for example, against fluorine abstraction by strong Lewis acids, limits their usage as reaction media in chemical reactions [19].

The use of the $[NTf_2]^-$ anion in ionic liquids brought a wide range of new, low viscosity and water stable (in fact, highly hydrophobic) liquid salts [7, 8]. In principle, the properties of these ionic liquids are quite satisfying, but the potential to fine-tune the viscosity, conductivity, hydrophobicity and so forth, not just by modifying the cation side, was desirable. The $[NTf_2]^-$ anion offers several options to do this – so, it is possible to introduce additional rotational freedom in the anion by elongating the perfluorinated alkyl residue. MacFarlane et al. showed that by changing $-CF_3$ to $-C_2F_5$, salts with quite a large range of plastic crystalline behaviour [20] in (for large anions) an extended sense ($\Delta S_f < 40\,J\,K^{-1}\,mol^{-1}$) of the Timmermans definition ($\Delta S_f < 20\,J\,K^{-1}\,mol^{-1}$) were obtained [21]. Apart from these solids, ionic liquids were also reported with this anion [20, 22]. The quite low entropy of fusion of these compounds is explained by a high degree of freedom in the pre-melted state. It would be

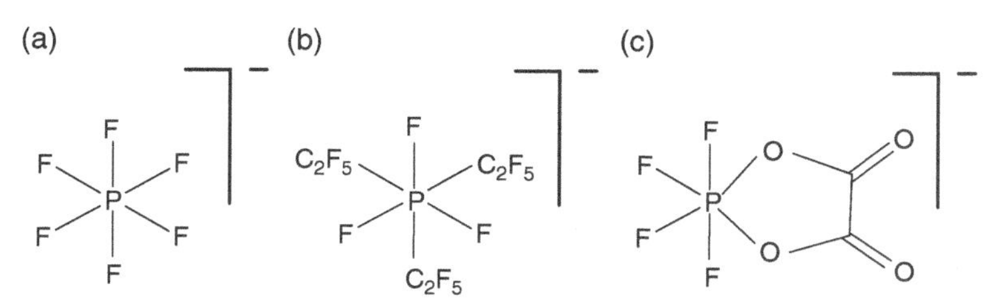

Figure 3.1 Chemical structures of some (modified) phosphate anions: (a) hexafluorophosphate, (b) trifluorotris(pentafluoroethyl)phosphate, $[FAP]^-$ and (c) tetrafluoro (oxalato)phosphate.

interesting to investigate compounds with even longer, partially fluorinated alkyl residues, because those flexible chains should significantly increase the entropy of fusion, and, therefore, a decrease in the melting temperature should be observed (see Eq. 3.1). However, perfluorination of the alkyl chain is not advisable, due to the high rigidity that would induce the inverse effect on the desired increase of the entropy of fusion. Also, perfluorinated alkyl chains normally cause high (eco-)toxicity.

Willner and co-workers recently showed an interesting access to new amide-based anions. They synthesised the bis{bis(pentafluoroethyl)phosphinyl}amide ($[NDf_2]^-$ (see Fig. 3.2a), which shows some structural similarity to the $[NTf_2]^-$ anion [19]. The $[NDf_2]^-$-ionic liquids possess a higher viscosity and slightly lower conductivity than $[NTf_2]^-$-based ionic liquids (see Table 3.1), which can be easily explained by the more than double size of $[NDf_2]^-$ (0.451 nm^3 compared to 0.223 nm^3). It is likely that due to the high rotational freedom of the $[NDf_2]^-$ anion, the ionic liquids formed with it do not show a well-defined crystallisation temperature [19]. The thermal stability and the electrical stability

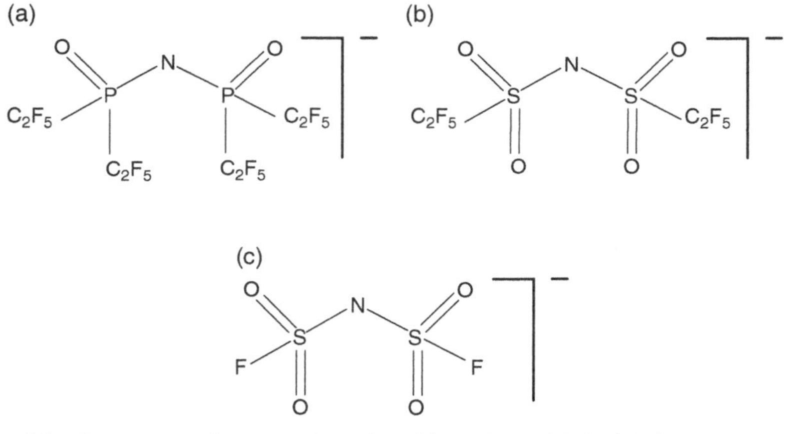

Figure 3.2 Structures of some selected amide anions: (a) bis{bis(pentafluoroethyl) phosphinoyl}amide, $[NDf_2]^-$, (b) bis{(pentafluoroethyl)sulfonyl}amide, $[NPf_2]^-$ and (c) bis(fluorosulfonyl)amide.

TABLE 3.1 Selected Physical Properties of Some Amide-Based Ionic Liquids

Ionic Liquid	T_m/°C	T_d/°C	η/mPa s^a
$[C_4mim][NTf_2]$	−2 [23]	422 [23]	66 [24]
$[C_4mim][NDf_2]$	7	320	151
$[C_{18}mim][NDf_2]$	14	300	376
$[C_4mpyr][NDf_2]$	19	300	325
$[P_{6\,6\,6\,14}][NDf_2]$	−2	300	536
$[N_{4\,4\,4\,4}][NDf_2]$	150	Not reported	Not reported

^{a} At 25°C.

in particular are rather high. Especially the large electrochemical window of the $[NDf_2]^-$ compounds and the absence of sulphur, which sometimes leads to incompatibility with some electrode materials, makes them very interesting for electrochemical applications.

3.4 BORATES

Like $[PF_6]^-$, the tetrafluoroborate anion $[BF_4]^-$ is used in a wide range of ionic liquids. Their viscosities are in general lower than those of $[PF_6]^-$, but they are still relatively high, with $[C_2mim][BF_4]$ being a remarkable exception ($\eta(25°C) = 25.2\,mPa\,s$). Nevertheless, ionic liquids based on $[BF_4]^-$ suffer from insufficient possibilities for separation from halide impurities. Due to the higher hydrophilicity of these ionic liquids (in contrast to $[PF_6]^-$), their drying process is more difficult than that of the highly hydrophobic $[NTf_2]^-$, $[FAP]^-$ and $[NDf_2]^-$ ionic liquids. In addition, they suffer hydrolysis with formation of toxic and corrosive HF [25–29]. To overcome this problem, Zhou et al. substituted one fluorine atom by a perfluorinated alkyl group (see Fig. 3.3b) [25–29]. The ionic liquids obtained by this route turned out to be both highly hydrophobic (thus overcoming the problems during the purification process) and water stable (making them more suitable for industrial applications). Due to the higher flexibility introduced by the perfluoroalkyl chains in the anion, the entropy of fusion of these compounds is much higher than those of symmetric anions such as $[BF_4]^-$ or $[PF_6]^-$. Nevertheless, some of the compounds reported by Zhou et al. have entropies of fusion less than $40\,kJ\,mol^{-1}$ ($\Delta S([N_{1\,2\,2\,3}][BF_3CF_3]) = 7.4\,J\,K^{-1}\,mol^{-1}$). Hence, they form plastic crystals, which make them suitable for electrochemical application over a wide temperature range. Besides this feature, the $[BF_3(C_nF_{2n+1})]^-$ salts show in general a lower

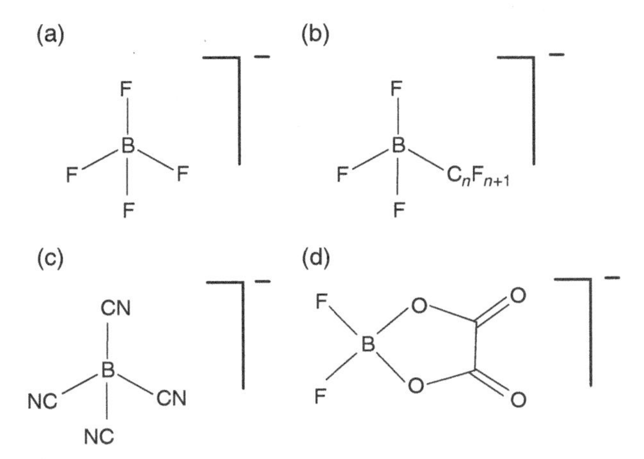

Figure 3.3 Structures of some selected borate anions: (a) tetrafluoroborate, (b) trifluoroperfluoroalkylborate, (c) tetracyanoborate and (d) difluoro(oxalato)borate.

TABLE 3.2 Melting Points, Viscosity and Specific Conductivity of Perfluoralkyltrifluoroborates[a]

Ionic Liquid[b]	T_m/°C	η/mPa s	κ/mS cm^{-1}	Ionic Liquid[b]	T_m/°C	η/mPa s	κ/mS cm^{-1}
$[N_{1224}][BF_4]$	165	—		$[COC_2OC_2mim][BF_3CF_3]$	−87	62	3.4
$[N_{1224}][BF_3CF_3]$	−3	210		$[C_2mim][BF_3CF_3]$	−20	26	14.8
$[N_{1224}][BF_3C_2F_5]$	15	104		$[C_2mim][BF_3C_2F_5]$	1	27	12
$[N_{1224}][BF_3C_3F_7]$	50	—		$[C_2mim][BF_3C_3F_7]$	8	32	8.6
$[N_{1224}][BF_3C_4F_9]$	60	—		$[C_2mim][BF_3C_4F_9]$	−4	38	5.2
$[N_{112COC2}][BF_4]$	4	335		$[C_3mim][BF_4]$	−17	103	5.9
$[N_{112COC2}][BF_3CF_3]$	−22	97		$[C_3mim][BF_3CF_3]$	−21	43	8.5
$[N_{112COC2}][BF_3C_2F_5]$	−33	58		$[C_3mim][BF_3C_2F_5]$	−42	35	7.5
$[N_{112COC2}][BF_3C_3F_7]$	−113[c]	70		$[C_3mim][BF_3C_3F_7]$	−5	44	5.3
$[N_{112COC2}][BF_3C_4F_9]$	−28	102		$[C_3mim][BF_3C_4F_9]$	−12	59	3.5
$[C_2mim][BF_4]$	15	42	13.6				

[a] From Refs. 27, 28.
[b] The indices of N and C indicate the length of an alkyl chain residue (1 = methyl, 2 = ethyl and so on). COC_n indicates an ether residue where a number indicates the length of the aliphatic chain (C = CH_2 or CH_3 if terminal, C_2 = C_2H_4 or C_2H_5 if terminal, and so on); $[C_n mim]$ = 1-alkyl-3-methylimidazolium.
[c] Glass transition: no melting was observed.

melting point than their $[BF_4]^-$ analogues. This can be understood again by the lowered symmetry, and hence by a lowered tendency of packing in a lattice. Also, the higher torsional and rotational freedom of the alkyl chains influences the melting point directly (see Eq. 3.2) [10]. Another advantage of $[BF_3(C_nF_{2n+1})]^-$-based ionic liquids is their remarkably low viscosity and high conductivity compared to their $[BF_4]^-$ analogues (see Table 3.2). Similarly to the melting points, the viscosity is lowered by the diminution of symmetry. Also, the perfluoroalkyl chains introduce a larger charge delocalisation, and thereby decrease the interactions with the environment. These ionic liquids with rather long alkyl chains and even polyether residues show a rather low viscosity and comparably high specific conductivity.

3.5 WEAKLY COORDINATING ANIONS – CARBORATES, FLUOROALKOXYALUMINATES AND FLUOROALKOXYBORATES

Many of the anions used for ionic liquids belong to the group of weakly coordinating anions as they show a minimal interaction with the cation [30]. The most weakly coordinating anions such as carborates, dodecaborates or (fluoroalkoxy)aluminates are usually used to stabilise reactive cations [31–33]. Taking this into consideration, it appeared logical to test these types of anions for ionic liquids. Reed et al. showed that it is possible to use carborane anions for the formation of ionic liquids [34]. In contrast to the expectation derived from the minimal Coulombic and dispersive interactions, most of these compounds do not melt below 100°C, as the carborane-based anions are rather rigid and do not promote entropy upon melting.

It followed nicely from this work that the most symmetric anions form the salts with the highest melting points. An introduction of symmetry-breaking groups to the anion – such as substitution of the (cage)C–H for (cage)C-alkyl – lowers the melting point considerably.[1] Willner et al. found similar effects in their work with the trimethylsilylium cation stabilised with various carborane anions [35]. By introducing an ethyl group to the perfluorocarborate $[H\text{-}CB_{11}F_{11}]^-$, the melting point of the $[SiMe_3][Et\text{-}CB_{11}F_{11}]$ salt drops to 75°C, and it therefore can be classified as an ionic liquid. Its conductivity of 2.5 mS cm^{-1} (in the melt) is not remarkably high. Nevertheless, this compound is very interesting, as it shows not only that stable ions are able to form an ionic liquid but also that it is possible to synthesise ionic liquids from very reactive ions, such as $[SiMe_3]^+$, which could be used for further reactions. A further series of ionic liquids containing boron clusters was reported by Viñas and co-workers [36] and Dymon et al. [37]. Also in these studies, a lowering of the melting points was achieved by symmetry breaking; for example, $[nido\text{-}C_2B_9H_{11}]^-$ ionic liquids melt at lower temperatures than $[closo\text{-}CB_{11}H_{12}]^-$ or $[closo\text{-}B_{12}Cl_{12}]^-$ ionic liquids (see Fig. 3.4 for structure).

Due to the large size of some anions such as the (fluoroalkoxy)aluminates ($\gg$0.5 nm^3), it was expected that ionic liquids formed with these anions should

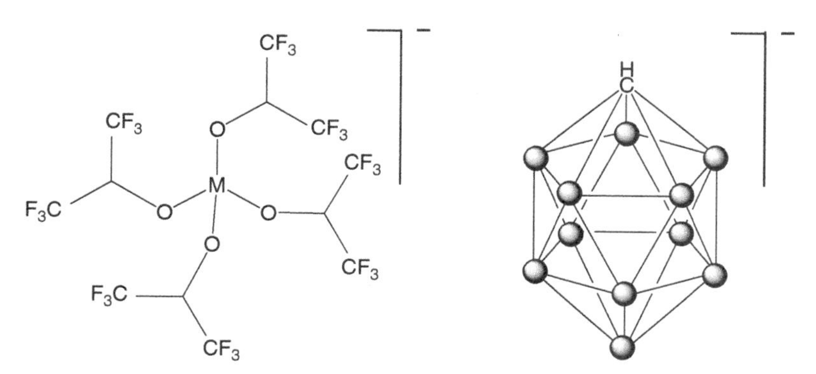

Figure 3.4 Chemical structures of two weakly coordinating anions. Left: Tetrakis-(1,1,1,3,3,3-hexafluoro-2-propanolato)aluminium(III)/boron(III) (M = Al or B). Right: 1-Carba-*closo*-dodecaborate. The spheres can be BH or BX (X = F, Cl, Br or I) units.

[1] This finding is in agreement with the effects of substitution at $[BF_4]^-$ and $[PF_6]^-$ (see Sections 2.3 and 2.4), which in principle lowers melting points and viscosities.

show some remarkable physical properties, such as very low static dielectric constants. Also (fluoroalkoxy)aluminates should melt at lower temperatures than carborates and dodecaborates, as they have a lower symmetry and a much higher rotational and torsional degree of freedom. These suggestions were confirmed in part in a recent contribution about tetrakis(hexafluoroisopropoxy) aluminate(III), $[Al(hfip)_4]^-$, ionic liquids (see Fig. 3.4 for structure) [24, 38]. The melting points, viscosities and conductivities were found in the expected favourable range, but the static dielectric constants ε_r were found to be rather high ($\varepsilon_r = 12–19$). Ionic liquids with much smaller anions, $[NTf_2]^-$, $[BF_4]^-$ or $[N(CN)_2]^-$, commonly viewed to be more polar, show similar or even lower dielectric constants [39], so it was expected to find ε_r values of less than 10. This finding is remarkable, since the ε_r-measurements on the $[Al(hfip)_4]^-$ salts were also performed at higher temperatures (70°C).

Another amazing property of $[Al(hfip)_4]^-$ ionic liquids is their high solubility for gases such as dihydrogen. 'Normal' ionic liquids show a low solubility compared to regular solvents [40] of *ca.* 0.8 mmol l^{-1}, but $[Al(hfip)_4]^-$ ionic liquids dissolve up to 6.5 mmol l^{-1}, approximately twice the amount dissolved in molecular solvents such as alcohols or toluene (see Table 3.3) [41].

The disadvantage of the $[Al(hfip)_4]^-$ anion is its sensitivity towards hydrolysis. However, Bulut showed that using boron instead of aluminium as the central atom leads to $[B(hfip)_4]^-$ ionic liquids, which are compatible with the usual laboratory atmosphere for days [42]. This gain in stability is traded in by about 30–50°C higher melting points and higher viscosities. Nevertheless, other properties such as the dielectric behaviour or the solubility of gases should show a similarly exciting trend.

TABLE 3.3 Selected Physical Properties of $[Al(hfip)_4]^-$ Ionic Liquids[a]

Ionic Liquid[#]	T_m/°C	T_c/°C	η(25°C)/ mPa s	κ(25°C)/ mS cm^{-1}	Solubility of H_2/mmol l^{-1}
$[C_2mim][Al(hfip)_4]$	31	<25	24.4[b]	6.22[b]	—
$[(allyl)mim][Al(hfip)_4]$	12	−24	37.7	4.24	5.23
$[C_2dmim][Al(hfip)_4]$	39	34	21.5[c]	4.68[c]	—
$[C_4py][Al(hfip)_4]$	36	<25	37.9[d]	10.22[e]	—
$[C_4mim][Al(hfip)_4]$	40	<25	21.8[b]	5.71[c]	—
$[C_3mpip][Al(hfip)_4]$	69	58	—	—	—
$[C_4mpyr][Al(hfip)_4]$	50	39	16.7[f]	7.15[d]	—
$[C_4dmim][Al(hfip)_4]$	0	−35	42.0	2.53	6.50
$[C_4mmorph][Al(hfip)_4]$	31	6	86.6	1.99[b]	—
$[C_6mim][Al(hfip)_4]$	5	−5	43.5	2.36	5.53

[a] From Ref. [24].
[b] 40°C.
[c] 50°C.
[d] 30°C.
[e] 70°C.
[f] 60°C.

3.6 MAGNETIC IONIC LIQUIDS

3.6.1 Transition Metal-Based Anions

Several publications deal with transition metals dissolved in ionic liquids. Most of them intend to generate already known complexes in a 'solvent-free' environment: for example, by utilising the Lewis acidity of halogenoaluminate(III) mixtures, or to synthesise new complexes [43–45], but this is not the focus here. Another set of publications showed that it is not only possible to dissolve transition metals in ionic liquids but also to form pure ionic liquids with transition metal-based complex anions [46, 47]. Yoshida and Saito reported a series of imidazolium salts with iron(III) chlorides and bromides, showing paramagnetic properties combined with low viscosity (see Table 3.4) [47]. The magnetic moment of the ionic liquids ranged from 5.80–5.85μ_B ($[FeCl_4]^-$) and 5.73–5.78μ_B ($[FeBr_4]^-$). Wilkes and co-workers reported other magnetic ionic liquids formed with transition metal complex anions [48]. They synthesised tetraalkylphosphonium and 1,3-dialkylimidazolium salts of chloro- and bromo-complexes with iron(III), cobalt(II) and manganese(II), as well as tetrathiocyanatocobalt(II) room-temperature ionic liquids.

3.6.2 Multiply Charged Anions

Wilkes and co-workers [49] presented, in 2005, a series of room temperature liquids formed with remarkable anions such as the $[Co(C_2B_9H_{11})]^-$ anion. They also reported other room-temperature ionic liquids formed by cyano, thiocyanato and selenothiocyanato complexes of the transition metals iron(II), cobalt(II) and nickel(II). These anions are quite unusual with respect to their higher negative charge. Normally, ionic liquids are formed by a singly charged cation and a singly charged anion. Higher charges imply higher Coulombic interactions between cation and anion, which results in

TABLE 3.4 Selected Physical Properties of Tetrahaloferrate Ionic Liquids[a]

Ionic Liquid	T_g/°C	T_m/°C	η [b]/mPa s	σ [b]/mS cm^{-1}	χ_{mol} [b]/emu mol^{-1}
$[C_2mim][FeCl_4]$	Not reported	18	18	20	1.42
$[C_4mim][FeCl_4]$	−88	Not reported	36	8.9	1.41
$[C_6mim][FeCl_4]$	−86	<rt	45	4.7	1.44
$[C_8mim][FeCl_4]$	−84	<rt	77	2.2	1.44
$[C_4mim][FeBr_4]$	−83	−2	62	5.5	1.38
$[C_6mim][FeBr_4]$	−82	<rt	95	2.8	1.39
$[C_8mim][FeBr_4]$	−81	<rt	121	1.4	1.40

[a] From Ref. [47].
[b] Measured at 25°C.

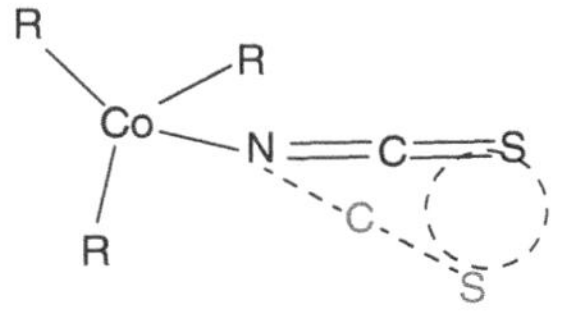

Figure 3.5 Sketch of the possible rotation around the axis along the M–N bond, leading to torsional freedom and thus a high entropy in the liquid state [50].

much higher lattice energies (*cf.* Kapustinskii equation). Therefore, the Gibbs energy of fusion should be high (implying high melting points). By analogy, the viscosity of molten salts of those complexes should be rather high. The latter assumption was found to be true by Wilkes and co-workers, but since they synthesised phosphonium salts with quite long alkyl chains, this finding is not astonishing. Recently, Köckerling and co-workers presented another series of $[Co(NCS)_4]^{2-}$ ionic liquids. They isolated dark blue room-temperature liquids with the $[C_2mim]^+$ and $[C_4mim]^+$ cation, which showed a startling low viscosity: $\eta = 145$ mPa s for $[C_2mim]_2[Co(NCS)_4]$ at 25°C [50] (*cf.* $\eta = 2436$ mPa s for $[P_{6\,6\,6\,14}]_2[Co(NCS)_4]$ at 25°C [49]. The low melting points of these ionic liquids might be accounted for by the larger torsional freedom of the thiocyanato- *vs.* the halo-complexes: because the angle $\measuredangle$(M–N–C) $\neq$ 180°, a variation in the torsion angle τ(M–N–C–S) results in a rotation cone along the M–N axis. This greatly increases the torsional degrees of freedom to be populated upon melting, leading to a large gain in fusion entropy when going from the solid to the liquid state. This increased flexibility is likely responsible for the superior melting behaviour of these thiocyanate ionic liquids (Fig. 3.5).

Moreover, all cation–anion interactions are mismatching according to the HSAB-concept; that is, 'soft' sulfur only interacts weakly with the 'hard' hydrogen in the cation [50]. The combination of both effects appears to lead to low melting and low viscosity tetrathiocyanatocobaltate(II) ionic liquids.

3.6.3 Lanthanides in Ionic Liquids

Most publications in this area treat the dissolution of rare earth salts in ionic liquids to form new complexes that may be used as catalysts or to induce useful properties such as luminescence, rather than using rare earth ions as complex anions for ionic liquids [51–60]. Nockemann et al. reported a series of lanthanide-based ionic liquids with thiocyanate complexes as the anion (see Table 3.5) [61]. The composition of the anion was determined by X-ray diffraction of single crystals obtained from the liquid phase. The identity of the structure found in the solid phase with the one in the liquid phase was proven by electronic absorption spectroscopy. The anions, depending on how much water is involved, exhibit a negative charge of 3–5,

TABLE 3.5 Selected Physical Properties of Lanthanide-based Ionic Liquids[a]

Ionic Liquid	T_m/°C	T_{dec}/°C	ρ/g cm^{-3}	Refractive Index
$[C_4mim]_4[La(NCS)_7(H_2O)_2]$	38	308	1.32	1.564
$[C_4mim]_4[Pr(NCS)_7(H_2O)]$	33	340	1.34	1.568
$[C_4mim]_4[Nd(NCS)_7(H_2O)]$	28	294	1.35	1.565
$[C_4mim]_4[Sm(NCS)_7(H_2O)]$	<rt	312	1.37	1.572
$[C_4mim]_4[Eu(NCS)_7(H_2O)]$	<rt	343	1.38	1.579
$[C_4mim]_4[Gd(NCS)_7(H_2O)]$	<rt	332	1.44	1.580
$[C_4mim]_4[Tb(NCS)_7(H_2O)]$	<rt	297	1.45	1.580
$[C_4mim]_4[Ho(NCS)_7(H_2O)]$	<rt	289	1.48	1.579
$[C_4mim]_4[Er(NCS)_7(H_2O)]$	<rt	302	1.49	1.581
$[C_4mim]_4[Yb(NCS)_7(H_2O)]$	<rt	296	1.53	1.582
$[C_4mim]_5[La(NCS)_8]$	<rt	349	1.35	1.563

[a] From Ref. [61].

TABLE 3.6 Magnetic Properties of Some Dysprosium-Based Ionic Liquids[a]

Ionic Liquid	T_m	μ_{eff}/μ_B	χ_g/emu g^{-1}	χ_{mol}/emu mol^{-1}
$[C_6mim]_3[Dy(NCS)_6(H_2O)_2]$	<rt	10.4	43×10^{-6}	0.045
$[C_6mim]_4[Dy(NCS)_7(H_2O)]$	<rt	10.6	38×10^{-6}	0.047
$[C_6mim]_5[Dy(NCS)_8]$	<rt	10.4	32×10^{-6}	0.047

[a] From Ref. [62].

but still most of them are classified as room-temperature ionic liquids. Mudring and co-workers prepared room-temperature ionic liquids with thiocyanate complexes of dysprosium (see Table 3.6) [62]. Comparing to Nockemann et al. they found varying amounts of water coordinated to the metal centre ($[Dy(SCN)_{8-x}(H_2O)_x]^{(5-x)-}$, $x=0$–2). The ionic liquids were found to be paramagnetic with a high magnetic moment (>10μ_B), as well as to exhibit excellent luminescence properties with rather long lifetimes (between 24 and 30 μs).

3.7 INORGANIC IONIC LIQUIDS

Almost all reported ionic liquids consist of 'organic' cations and 'inorganic' anions. Only a few publications deal with ionic liquids completely formed by inorganic compounds and most of them treat eutectic mixtures. The work of Shan and co-workers is an exception in this field, as it describes ionic liquids formed by just inorganic cations and anions [63]. In their work, they describe a class of anions formed by polyoxometallates with the Keggin structures bound to rare earth or transition metals following the formula $[M^{III}(TiW_{11}O_{39})_2]^{13-}$ for trivalent rare earths, M, and $[M^{n+}(TiW_{11}O_{39})]^{(8-n)-}$ for transition metals M^{n+}. Sodium is the counter ion in all cases. The reported inorganic ionic liquids are insoluble in common organic solvents and only soluble in water. They show a quite low viscosity (42–82 mPa s) and melting points below 273 K in general. The electrical conductivities are about 20 mS cm^{-1}. The problem with this category of ionic liquids is the necessity of water to form a liquid (27–40 molecules per formula unit). Here, the water is likely bound to Na^+ to form a hydrated, but defined $[Na(OH_2)_x]^+$ cation. This weakly complexed water results in a quite low thermal stability, as the water is lost during heating. Since these compounds have tended to be sidelined, it would be interesting to synthesise other analogues, with stronger bound solvates than sodium hydrates, to increase thermal and chemical stability. In this manner, it should be possible to increase the range of these ionic liquids substantially and maybe to bring them to application.

3.8 SUMMARY

Anions are the main determinant of the physical properties of ionic liquids: quite amazing properties, such as magnetism or fluorescence, can be induced if the ion is chosen properly. In addition, fine-tuning of viscosity and related properties can be achieved. A quite new, and as yet not intensely investigated, category of materials are inorganic ionic liquids. Here, it seems to be necessary to include solvent complexes of the cation in our definition of ionic liquid cations. Using suitable solvates with high entropic flexibility will clearly allow lowering the melting point even further. An extreme example for this notion are likely the hydrates of the proton with mineral acid counter ion, for example, $[H_3O]^+[HSO_4]^-$ (melting point 8.59°C) or $[H_3O]^+[NO_3]^-$ (melting point −37.63°C) [64]. They exhibit very low melting points, consist only of ions and might justly also be described as inorganic ionic liquids.

REFERENCES

1 Walden, P., Über die Molekulargrösse und elektrische Leitfähigkeit einiger gesehmolzenen Salze, *Bull. Acad. Imper. Sci.* **8**, 405–422 (1914).

2 Robinson, J. and Osteryoung, R.A., Electrochemical and spectroscopic study of some aromatic-hydrocarbons in the room-temperature molten-salt system aluminum chloride-N-butylpyridinium chloride, *J. Am. Chem. Soc.* **101**, 323–327 (1979).

3 Wilkes, J.S., Levisky, J.A., Wilson, R.A. and Hussey, C.L., Dialkylimidazolium chloroaluminate melts – a new class of room-temperature ionic liquids for electrochemistry, spectroscopy, and synthesis, *Inorg. Chem.* **21**, 1263–1264 (1982).

4 Fannin, A.A., Floreani, D.A., King, L.A., Landers, J.S., Piersma, B.J., Stech, D.J., Vaughn, R.L., Wilkes, J.S. and Williams, J.L., Properties of 1,3-dialkylimidazolium chloride aluminum-chloride ionic liquids. 2. Phase-transitions, densities, electrical conductivities, and viscosities, *J. Phys. Chem.* **88**, 2614–2621 (1984).

5 Fannin, A.A., King, L.A., Levisky, J.A. and Wilkes, J.S., Properties of 1,3-dialkylimidazolium chloride aluminum-chloride ionic liquids. 1. Ion interactions by nuclear magnetic-resonance spectroscopy, *J. Phys. Chem.* **88**, 2609–2614 (1984).

6 Sanders, J.R., Ward, E.H. and Hussey, C.L., Aluminum bromide-1-methyl-3-ethylimidazolium bromide ionic liquids, *J. Electrochem. Soc.* **133**, 325–330 (1986).

7 Endres, F., Abbott, A. and MacFarlane, D., eds., *Electrodeposition from Ionic Liquids*, Wiley-VCH, Weinheim (2008).

8 Wasserscheid, P. and Welton, T., eds., *Ionic Liquids in Synthesis*, 2nd Ed., Wiley-VCH, Weinheim (2008).

9 Bonhôte, P., Dias, A.-P., Papageorgiou, N., Kalyanasundaram, K. and Grätzel, M., Hydrophobic, highly conductive ambient-temperature molten salts, *Inorg. Chem.* **35**, 1168–1178 (1996).

10 Preiss, U., Bulut, S. and Krossing, I., In silico prediction of the melting points of ionic liquids from thermodynamic considerations: a case study on 67 salts with a melting point range of 337°C, *J. Phys. Chem. B* **114**, 11133–11140 (2010).

11 Eiden, P., Bulut, S., Köchner, T., Friedrich, C., Schubert, T. and Krossing, I., In Silico predictions of the temperature-dependent viscosities and electrical conductivities of functionalized and nonfunctionalized ionic liquids, *J. Phys. Chem. B* **115**, 300–309 (2010).

12 Xu, W. and Angell, C.A., Solvent-free electrolytes with aqueous solution-like conductivities, *Science* **302** (5644), 422–425 (2003).

13 Fraser, K.J., Izgorodina, E.I., Forsyth, M., Scott, J.L. and MacFarlane, D.R., Liquids intermediate between 'molecular' and 'ionic' liquids: liquid ion pairs? *Chem. Commun.* **37**, 3817–3819 (2007).

14 Johansson, K.M., Izgorodina, E.I., Forsyth, M., MacFarlane, D.R. and Seddon, K.R., Protic ionic liquids based on the dimeric and oligomeric anions: $[(AcO)xHx_{-1}]^-$, *Phys. Chem. Chem. Phys.* **10**, 2972–2978 (2008).

15 MacFarlane, D.R., Forsyth, M., Izgorodina, E.I., Abbott, A.P., Annat, G. and Fraser, K., On the concept of ionicity in ionic liquids, *Phys. Chem. Chem. Phys.* **11**, 4962–4967 (2009).

16 Ignatiev, N.V., Welz-Biermann, U., Kucheryna, A., Bissky, G. and Willner, H., New ionic liquids with tris(perfluoroalkyl)trifluorophosphate (FAP) anions, *J. Fluor. Chem.* **126**, 1150–1159 (2005).

17 Schmidt, M., Ignatiev, N., Pitner, W.R. and Eichhorn, J., *Reactive ionic liquids*, DE 102008021271 (2010).

18 Schmidt, M., Ignatiev, N. and Pitner, W.R., *Reactive ionic liquids*, WO 2009132740 (2009).

19 Bejan, D., Ignatiev, N. and Willner, H., New ionic liquids with the bis[bis(pentafluoroethyl)phosphinyl]imide anion, $[(C_2F_5)_2P(O)]_2N^-$ Synthesis and characterization, *J. Fluor. Chem.* **131**, 325–332 (2010).

20 Johansson, K.M., Adebahr, J., Howlett, P. C., Forsyth, M. and MacFarlane, D.R., N-methyl-N-alkylpyrrolidinium bis(perfluoroethylsulfonyl) amide ($[NPf_2]^-$) and tris(trifluoromethanesulfonyl) methide ($[CTf_3]^-$) salts: synthesis and characterization, *Aust. J. Chem.* **60**, 57–63 (2007).

21 Timmermans, J., Plastic crystals – a historical review, *J. Phys. Chem. Solids* **18**, 1–8 (1961).

22 Tokuda, H., Tsuzuki, S., Susan, M., Hayamizu, K. and Watanabe, M., How ionic are room-temperature ionic liquids? An indicator of the physicochemical properties, *J. Phys. Chem. B* **110**, 19593–19600 (2006).

23 Fredlake, C.P., Crosthwaite, J.M., Hert, D.G., Aki, S. and Brennecke, J.F., Thermophysical properties of imidazolium-based ionic liquids, *J. Chem. Eng. Data* **49**, 954–964 (2004).

24 Bulut, S., Klose, P., Huang, M.-M., Weingärtner, H., Dyson, P.J., Laurenczy, G., Friedrich, C., Menz, J., Kümmerer, K. and Krossing, I., Synthesis of room-temperature ionic liquids with the weakly coordinating $[Al(ORF)_4]^-$ anion (RF=C(H)$(CF_3)_2$) and the determination of their principal physical properties, *Chem. Eur. J.* **16**, 13139–13154 (2010).

25 Zhou, Z.B., Matsumoto, H. and Tatsumi, K., Low-viscous, low-melting, hydrophobic ionic liquids: 1-alkyl-3-methylimidazolium trifluoromethyltrifluoroborate, *Chem. Lett.* **33**, 680–681 (2004).

26 Zhou, Z.B., Matsumoto, H. and Tatsumi, K., Low-melting, low-viscous, hydrophobic ionic liquids: N-alkyl(alkyl ether)-N-methylpyrrolidinium perfluoroethyltrifluoroborate, *Chem. Lett.* **33**, 1636–1637 (2004).

27 Zhou, Z.B., Matsumoto, H. and Tatsumi, K., Low-melting, low-viscous, hydrophobic ionic liquids: 1-alkyl(alkyl ether)-3-methylimidazolium perfluoroalkyltrifluoroborate, *Chem. Eur. J.* **10**, 6581–6591 (2004).

28 Zhou, Z.B., Matsumoto, H. and Tatsumi, K., Low-melting, low-viscous, hydrophobic ionic liquids: aliphatic quaternary ammonium salts with perfluoroalkyltrifluoroborates, *Chem. Eur. J.* **11**, 752–766 (2005).

29 Zhou, Z.B., Takeda, M. and Ue, M., New hydrophobic ionic liquids based on perfluoroalkyltrifluoroborate anions, *J. Fluor. Chem.* **125**, 471–476 (2004).

30 Krossing, I. and Raabe, I., Noncoordinating anions – fact or fiction? A survey of likely candidates, *Angew. Chem. Int. Ed.* **43**, 2066–2090 (2004).

31 Krossing, I. and Reisinger, A., Chemistry with weakly-coordinating fluorinated alkoxyaluminate anions: gas phase cations in condensed phases? *Coord. Chem. Rev.* **250**, 2721–2744 (2006).

32 Bolli, C., Derendorf, J., Kessler, M., Knapp, C., Scherer, H., Schulz, C. and Warneke, J., Synthesis, crystal structure, and reactivity of the strong methylating agent $Me_2B_{12}Cl_{12}$, *Angew. Chem. Int. Ed.* **49**, 3536–3538 (2010).

33 Kessler, M., Knapp, C., Sagawe, V., Scherer, H. and Uzun, R., Synthesis, characterization, and crystal structures of silylium compounds of the weakly coordinating dianion $[B_{12}Cl_{12}]^{2-}$, *Inorg. Chem.* **49**, 5223–5230 (2010).

34 Larsen, A.S., Holbrey, J.D., Tham, F.S. and Reed, C.A., Designing ionic liquids: imidazolium melts with inert carborane anions, *J. Am. Chem. Soc.* **122**, 7264–7272 (2000).

35 Kuppers, T., Bernhardt, E., Eujen, R., Willner, H. and Lehmann, C.W., $[Me_3Si][R\text{-}CB_{11}F_{11}]$-synthesis and properties, *Angew. Chem. Int. Ed.* **46**, 6346–6349 (2007).

36 Nieuwenhuyzen, M., Seddon, K.R., Teixidor, F., Puga, A.V. and Viñas, C., Ionic liquids containing boron cluster anions, *Inorg. Chem.* **48**, 889–901 (2009).

37 Dymon, J., Wibby, R., Kleingardner, J., Tanski, J.M., Guzei, I.A., Holbrey, J.D. and Larsen, A.S., Designing ionic liquids with boron cluster anions: alkylpyridinium and imidazolium [*nido*-$C_2B_9H_{11}$] and [*closo*-$CB_{11}H_{12}$] carborane salts, *Dalton Trans.* **22**, 2999–3006 (2008).

38 Timofte, T., Pitula, S. and Mudring, A.V., Ionic liquids with perfluorinated alkoxyaluminates, *Inorg. Chem.* **46**, 10938–10940 (2007).

39 Hunger, J., Stoppa, A., Schrödle, S., Hefter, G. and Buchner, R., Temperature dependence of the dielectric properties and dynamics of ionic liquids, *Chem. Phys. Chem.* **10**, 723–733 (2009).

40 Dyson, P.J., Laurenczy, G., Ohlin, C.A., Vallance, J. and Welton, T., Determination of hydrogen concentration in ionic liquids and the effect (or lack of) on rates of hydrogenation, *Chem. Commun.* **19**, 2418–2419 (2003).

41 Linke, W.F., Seidell, A., *Solubilities of Inorganic and Metal–Organic Compounds*, American Chemical Society, Washington, DC (1958), p. 1075.

42 Bulut, S., Novel ionic liquids with weakly coordinating anions: synthesis, characterization, principal physical properties and their predictions, Ph. D. Thesis, Albert-Ludwigs-Universität Freiburg (Freiburg, 2010).

43 Dyson, P.J., Grossel, M.C., Srinivasan, N., Vine, T., Welton, T., Williams, D.J., White, A.J.P. and Zigras, T., Organometallic synthesis in ambient temperature chloroaluminate (III) ionic liquids. Ligand exchange reactions of ferrocene, Dalton Trans. 3465–3469 (1997).

44 Hasan, M., Kozhevnikov, I.V., Siddiqui, M.R.H., Steiner, A. and Winterton, N., A novel N-heterocyclic carbene of platinum(II): synthesis in ionic liquids and crystal structure, *J. Chem. Res.* **8**, 392–393 (2000).

45 Stark, A., MacLean, B.L. and Singer, R.D., 1-Ethyl-3-methylimidazolium halogenoaluminate ionic liquids as solvents for Friedel-Crafts acylation reactions of ferrocene, Dalton Trans. 63–66 (1999).

46 Sitze, M.S., Schreiter, E.R., Patterson, E.V. and Freeman, R.G., Ionic liquids based on $FeCl_3$ and $FeCl_2$. Raman scattering and ab initio calculations, *Inorg. Chem.* **40**, 2298–2304 (2001).

47 Yoshida, Y. and Saito, G.Z., Influence of structural variations in 1-alkyl-3-methylimidazolium cation and tetrahalogenoferrate(III) anion on the physical properties of the paramagnetic ionic liquids, *J. Mater. Chem.* **16**, 1254–1262 (2006).

48 Del Sesto, R.E., McCleskey, T.M., Burrell, A.K., Baker, G.A., Thompson, J.D., Scott, B.L., Wilkes, J.S. and Williams, P., Structure and magnetic behavior of transition metal based ionic liquids, *Chem. Commun.* **4**, 447–449 (2008).

49 Del Sesto, R.E., Corley, C., Robertson, A. and Wilkes, J.S., Tetraalkylphosphonium-based ionic liquids, *J. Organomet. Chem.* **690**, 2536–2542 (2005).

50 Peppel, T., Köckerling, M., Geppert-Rybczykńsa, M., Ralys, R.V., Lehmann, J.K., Verevkin, S.P. and Heintz, A., Low-viscosity paramagnetic ionic liquids with doubly charged $[Co(NCS)_4]^{2-}$ ions, *Angew. Chem. Int. Ed.* **49**, 7116–7119 (2010).

51 Baba, T. and Watanuki, Y., Nonaqueous quaternary ammonium salt electrolytic solutions containing cyclic boron or phosphorus compounds for electrolytic capacitors, JP 2007123631 (2007).

52 Babai, A. and Mudring, A.-V., Anhydrous praseodymium salts in the ionic liquid $[bmpyr][Tf_2N]$: structural and optical properties of $[bmpyr]_4[PrI_6][Tf_2N]$ and $[bmyr]_2[Pr(Tf_2N)_5]$, *Chem. Mater.* **17**, 6230–6238 (2005).

53 Babai, A. and Mudring, A.-V., Rare-earth iodides in ionic liquids: the crystal structure of $[SEt_3]_3[LnI_6]$ (Ln = Nd, Sm), *Inorg. Chem.* **44**, 8168–8169 (2005).

54 Babai, A. and Mudring, A.-V., Rare-earth iodides in ionic liquids: crystal structures of $[bmpyr]_4[LnI_6][Tf_2N]$ (Ln = La, Er), *J. Alloys Compd.* **418**, 122–127 (2006).

55 Bhatt, A.I., May, I., Volkovich, V.A., Collison, D., Helliwell, M., Polovov, I.B. and Lewin, R.G., Structural characterization of a lanthanum bistriflimide complex, La$(N(SO_2CF_3)_2)_3(H_2O)_3$, and an investigation of La, Sm, and Eu electrochemistry in a room-temperature ionic liquid, $[Me_3NnBu][N(SO_2CF_3)_2]$, *Inorg. Chem.* **44**, 4934–4940 (2005).

56 Billard, I. and Gaillard, C., Actinide and lanthanide speciation in imidazolium-based ionic liquids, *Radiochim. Acta* **97**, 355–359 (2009).

57 Dzudza, A. and Marks, T.J., Efficient intramolecular hydroalkoxylation of unactivated alkenols mediated by recyclable lanthanide triflate ionic liquids: scope and mechanism, *Chem. Eur. J.* **16**, 3403–3422 (2010).

58 Hines, C.C., Cordes, D.B., Griffin, S.T., Watts, S.I., Cocalia, V.A. and Rogers, R.D., Flexible coordination environments of lanthanide complexes grown from chloride-based ionic liquids, *New J. Chem.* **32**, 872–877 (2008).

59 Mudring, A.-V., Babai, A., Arenz, S., Giernoth, R., Binnemans, K., Driesen, K. and Nockemann, P., Strong luminescence of rare earth compounds in ionic liquids: luminescent properties of lanthanide(III) iodides in the ionic liquid 1-dodecyl-3-methylimidazolium bis(trifluoromethanesulfonyl)imide, *J. Alloys Compd.* **418**, 204–208 (2006).

60 Nockemann, P., Thijs, B., Lunstroot, K., Parac-Vogt, T.N., Gorller-Walrand, C., Binnemans, K., Van Hecke, K., Van Meervelt, L., Nikitenko, S., Daniels, J., Hennig, C. and Van Deun, R., Speciation of rare-earth metal complexes in ionic liquids: a multiple-technique approach, *Chem. Eur. J.* **15**, 1449–1461 (2009).

61 Nockemann, P., Thijs, B., Postelmans, N., Van Hecke, K., Van Meervelt, L. and Binnemans, K., Anionic rare-earth thiocyanate complexes as building blocks for low-melting metal-containing ionic liquids, *J. Am. Chem. Soc.* **128**, 13658–13659 (2006).

62 Mallick, B., Balke, B., Felser, C. and Mudring, A.V., Dysprosium room-temperature ionic liquids with strong luminescence and response to magnetic fields, *Angew. Chem. Int. Ed.* **47**, 7635–7638 (2008).

63 Dai, L., Yu, S., Shan, Y. and He, M., Novel room temperature inorganic ionic liquids, *Eur. J. Inorg. Chem.* **2**, 237–241 (2004).

64 Wiberg, N., ed., *Holleman-Wiberg's Lehrbuch der Anorganischen Chemie*, **101** Ed., Walter de Gruyter, Berlin (1995).

4 Investigating the Structure of Ionic Liquids and Ionic Liquid: Molecular Solute Interactions

CHRISTOPHER HARDACRE and CLAIRE MULLAN
School of Chemistry and Chemical Engineering, Queen's University of Belfast, Belfast, UK

TRISTAN G. A. YOUNGS
ISIS Facility, Rutherford Appleton Laboratory, Chilton, UK

4.1 INTRODUCTION

Due to the potential number of ionic liquid systems possible, selecting candidates for specific applications is particularly difficult and means that quantifying the interactions in the liquid state is of enormous benefit. However, understanding the macroscopic properties of a system requires an understanding of the microscopic structure in the system. The complexity of interactions between species in the liquid state presents a significant challenge, even for simple molecular solvents. In the case of ionic liquids, the introduction of charged species provides even more variety and directionality to the interactions and deepens the problem. Even relatively simple (and useful) concepts such as polarity and degree of hydrogen bond-accepting/hydrogen bond-donating ability remain largely unprobed in these media. From a structural perspective, however, much more headway has been made, and several experimental techniques have begun to provide answers to these questions, including neutron diffraction, X-ray scattering, extended X-ray absorption fine structure and NMR spectroscopy [1]. However, due to the difficulties associated with studying the essential interactions on an ionic (or molecular) level, a key part of these investigations is the parallel use of theoretical methods such as classical or quantum molecular dynamics (MD). In every sense, the application of combined experimental and theoretical approaches allows one to validate the other, and a much clearer picture on the liquid state may be formed.

Ionic Liquids Completely UnCOILed: Critical Expert Overviews, First Edition.
Edited by Natalia V. Plechkova and Kenneth R. Seddon.

4.2 STRUCTURE OF SIMPLE MOLTEN SALTS

Atomic molten salts such as the alkali halides have been studied extensively using experimental methods such as neutron diffraction and extended X-ray absorption fine structure, enabling their structures in the liquid melt to be quantified. Investigations were pioneered by Enderby and co-workers, who began with molten NaCl [2], perhaps now the best known example of such structural research. Their work demonstrated clearly the charge ordering within the system and has become the standard template for the liquid structure of purely ionic binary melts. The radial distribution functions (RDFs) obtained from the study for all pairs of ions in molten NaCl are shown in Figure 4.1, in which the prominent feature is the maximum in the unlike-ion (Na—Cl) curve at approximately 2.7 Å. Thereafter, a minimum occurs at approximately 4.3 Å corresponding to the maximum in the like-ion curves (Na—Na and Cl—Cl). A key observation to make is the lack of overlap above $g = 1$ between the Na—Cl RDF and either of the like-ion traces, indicating that concentric shells of anions and cations exist without penetration of one into the other. Such structuring persists out to at least the third coordination shell. Integration of the RDF curves up to some distance (usually a minimum in the curve) provides the number of ions within this range, that is, the coordination number. For the prominent maximum in the Na—Cl curve, the value is 5.8 ± 0.1, meaning each ion is surrounded by approximately six ions of opposite charge in the first shell, and strongly suggests a loosely octahedral arrangement of ions. Integration of the second Na—Cl RDF peak gives a second shell coordination number of 13.0 ± 0.5. Despite the fluidity of the system at the high temperatures of the melt, these numbers may be directly compared to similar numbers

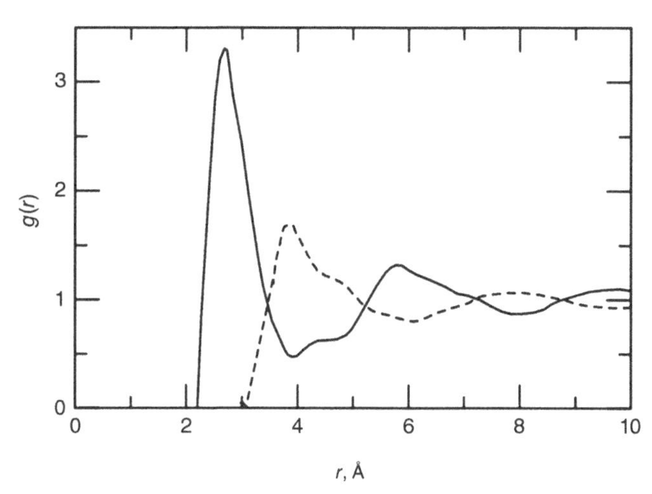

Figure 4.1 Cation–anion (solid line) and cation–cation (dashed line) RDFs of molten NaCl (data taken from Ref. [2]) determined from neutron diffraction data.

obtained from the crystalline FCC structure of the salt. Subsequent to this landmark study, a number of molten salts have been studied, including mixtures and eutectics [3]. However, even the simplest ionic liquids are much more complex than an atomic molten salt, with both cation and anion often being polyatomic and asymmetric.

4.3 STRUCTURE OF PURE IONIC LIQUIDS

Despite the vast number of potential systems and extensive simulation studies, significantly less direct and experimentally derived structural data for ionic liquids is currently available. The 1,3-dialkylimidazolium ionic liquids are perhaps the most widely investigated and utilised ionic liquid systems, since they were among the first ionic liquids to be synthesised and utilised as molten salt solvents. Consequently, they have received the most attention both experimentally and theoretically, with neutron diffraction, in particular, playing a critical role in elucidating structural information. From a theoretical perspective, MD and Monte Carlo techniques have been applied to the determination of many bulk properties, not least of which the liquid structure, and in combination with experimental neutron studies have proven to be valuable tools. For atomic molten salts, the combination of theory and experiment has been widely applied since the original work on NaCl [4, 5]. The past decade has seen a large increase in the number of computational studies focussing on ionic liquids, due in part to the variety of potential systems and, therefore, the strong need to provide structure–property/activity relationships in order to formulate design criteria for specific applications. Experimental results have assisted in the evaluation of force fields and thus the formation of more accurate potentials with which to model and compare with the experimental data [6].

4.3.1 1,3-Dimethylimidazolium Chloride

The first ionic liquid to be studied by neutron diffraction was 1,3-dimethylimidazolium chloride, [C_1mim]Cl, which possesses the simplest of the possible alkylimidazolium cations. In order to extract microscopic quantities such as spatial probabilities and RDFs from collected differential cross scattering data, analysis is typically performed following reverse Monte Carlo (RMC) or empirical potential structure refinement (EPSR) procedures [7]. The latter procedure can be viewed as a Monte Carlo simulation of system utilising a model potential similar to a classical molecular mechanics force field. This model potential is modified in order to bring the total structure factor calculated from the model system as close as possible to the experimental data. From configurations generated with this refined potential, standard quantities (such as RDFs) may be calculated.

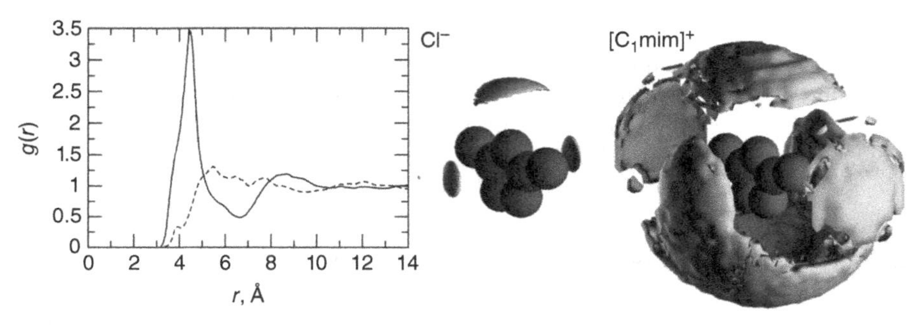

Figure 4.2 Cation–anion (solid line) and cation–cation (dashed line) RDFs and spatial probabilities for $[C_1mim]Cl$ determined from neutron diffraction data [8].

Hardacre et al. [8] were able to show that the general structure of the liquid mirrors that of NaCl quite closely, despite the obvious differences in size and chemistry between Na^+ and $[C_1mim]^+$. For example, ion–ion RDFs (Figure 4.2) show a clear oscillatory structure, with the cation–anion and cation–cation minima and maxima occupying mutually exclusive positions, strongly inferring the presence of ionic shells containing only one ionic species. The predominant peak in the cation–anion curve at 5.5 Å indicates that this is the most likely distance at which to find an anion from the centre of mass of the cation. Furthermore, integration of the cation–anion RDF gives a coordination number of approximately 6, again comparable with the atomic salt.

However, the lack of spherical symmetry of the cation necessitates that a simple octahedral arrangement of anions around a central cation is unlikely, and this is confirmed by the extracted spatial probabilities. While the RDF is a two-dimensional depiction of the likelihood of finding species at a given distance from a central point, spatial probabilities represent exactly the same quantity taking into account its angular dependence, that is, the relative spatial position around the central ion. Such quantities are especially important when species are not spherically symmetric and/or contain strongly directional interaction sites.

The spatial probabilities shown in Figure 4.2 confirm the strong anisotropy of anions around the $[C_1mim]^+$ cation, with a clear preference for the Cl^- to be positioned along the vectors of the C—H bonds of the ring. Also evident are the possibility for anions to lie above and below the plane of the aromatic ring and the lack of significant density located around the methyl groups. In the crystal, the dominating interactions are C—H···Cl contacts between both imidazolium ring hydrogen atoms and methyl hydrogen atoms [9], but in the liquid state, only the former interaction is strong enough to be sufficiently long-lived to be found. Nevertheless, the liquid and solid states of $[C_1mim]Cl$ bear a great deal of similarity.

Classical simulation studies of $[C_1mim]Cl$ [10–14] present stimulated RDFs that are consistent with those shown in Figure 4.2; however, none have shown a strong probability of finding anions above and below the imidazolium ring, a feature clearly evident from the experiments. Largely, such disagreements lay with the choice of charges placed on the five imidazolium ring atoms, and it is possible to show that reproduction of this particular feature of the experimental data can be achieved by employing a cation model in which the atomic charges in the system give an overall slight positive charge on the aromatic ring, rather than a slightly negative charge as is the case in the models of reference [11]. However, care should be taken regarding such changes in the potentials since the finer details of structural investigations, such as sub-peaks in RDFs, are sensitive to the representation of charges on the ions and the distribution of total charge [15–17]. Typically, the regions above and below the ring are predominantly populated by other cations, albeit at such a distance that they do not penetrate into the first anion shell to any significant degree. Several *ab initio* MD simulations have also been performed [18–20], which are also consistent with the neutron diffraction data except for, as in most classical simulations, the presence of anions above and below the imidazolium ring.

4.3.2 1,3-Dimethylimidazolium Hexafluorophosphate

Retaining the same simple imidazolium cation and replacing the chloride anion for hexafluorophosphate complete the move away from a partially atomic room-temperature molten salt to a true ionic liquid (melting point 362 K *cf.* 450 K for [C_1mim]Cl). Replacing the atomic cation with a large organic molecule has comparatively little effect on the typical characteristics of the true atomic molten salt, and here the inclusion of a large, weakly coordinating anion further hinders strong interaction between ions. Neutron diffraction of 1,3-dimethylimidazolium hexafluorophosphate, [C_1mim][PF_6], shows that the alternating cation/anion shell structure still pervades, accompanied by a slight shift in the positions of the RDF maxima to longer distances as a result of accommodating the larger anion [21]. However, a more complex cation–anion interaction is alluded to by the presence of three distinct maxima, suggesting three separate types of interaction between the two or, alternatively, three distinct (favourable) interaction sites on the cation. Spatial probabilities of anions around a central cation (Figure 4.3)

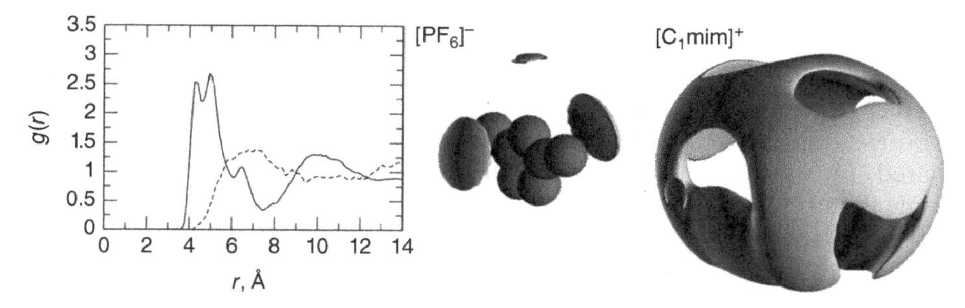

Figure 4.3 Cation–anion (solid line) and cation–cation (dashed line) RDFs and spatial probabilities for [C_1mim][PF_6], determined from neutron diffraction data [21].

show anion density along the imidazolium ring hydrogen vectors and also above and below the plane of the ring. However, in contrast to [C_1mim]Cl, it is the latter interaction that is the strongest. Anions will always occupy positions around the cation, which present the most favourable electrostatic interactions, and since the C—H···F hydrogen bond is very weak in this system, the anions localise above and below the ring. There is a small density associated with H(2), since it is the most acidic (in terms of charge the most positive) hydrogen in the system, but the H(4) and H(5) positions show little associated anion interaction.

Integration of the first peak in the cation–anion RDF gives a coordination number of 6.8, which is greater than that observed for [C_1mim]Cl. The spatial distributions of ions around a central $[PF_6]^-$ anion reveal a preference for cations to be positioned at the faces of the PF_6 octahedron, while the $[PF_6]^-$ anions are positioned directly along the P–F vectors. Such an arrangement is presumably dictated more by cation positioning rather than favourable anion–anion interactions and leads to the cations and anions occupying mutually exclusive positions.

Classical simulation studies of [C_1mim][PF_6] (as well as the related [C_4mim][PF_6]) have also been performed and again show a generally good agreement with experimental data. A variety of models including all-atom (i.e. explicitly including all hydrogen atoms) [22], partially united-atom [10], and fully united-atom [23] have been used, and each displays the same essential features seen from neutron diffraction, although the finer details of the main cation–anion RDF peak do differ between models. Other trends are reproduced well, for example, the preference of $[PF_6]^-$ to coordinate near to the H(2) position over the H(4,5) positions [22] and the location of anions above and below the ring [10]. *Ab initio* simulations carried out by Bhargava and Balasubramanian [24] also agree well with experiment although, in this case (and also with Hanke's partially united-atom classical model of Ref. 10), no significant density along the C(2)—H(2) bond vector was found, while such contacts are seen in both the liquid and solid states [25].

4.3.3 1,3-Dimethylimidazolium Bis{(trifluoromethyl)sulfonyl}amide

The bis{(trifluoromethyl)sulfonyl}amide anion ($[NTf_2]^-$) is considerably different compared with anions such as hexafluorophosphate, tetrafluoroborate and chloride; it is significantly larger and more weakly coordinating, as well as possessing an inherent degree of flexibility. The resulting weakening of the cation–anion interaction corresponds to a reduction in viscosity compared, for example, to the analogous chloride and hexafluorophosphate ionic liquids [26].

A combined neutron diffraction and MD study of [C_1mim][NTf_2] made by Deetlefs et al. [27] reveals the salt-like oscillatory structure in the RDFs, but it was noted that the oscillations themselves are significantly weaker than observed previously owing to the weaker attraction between ions. The first peak in the cation–anion RDF appears at approximately 7.0 Å, which reflects

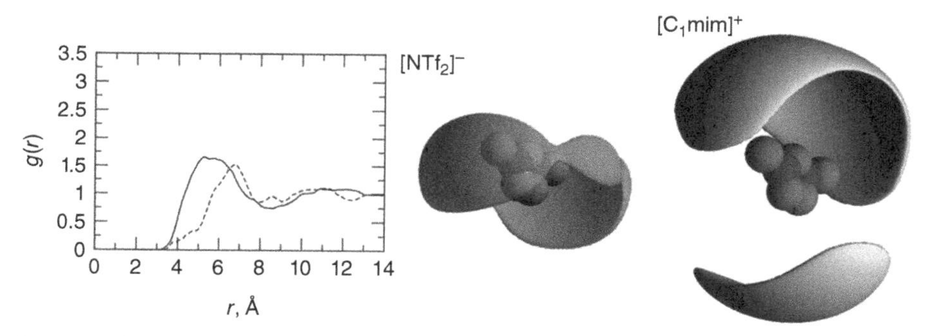

Figure 4.4 Cation–anion (solid line) and cation–cation (dashed line) RDFs and spatial probabilities for $[C_1mim][NTf_2]$, determined from neutron diffraction data [27].

an expansion of the liquid due to the increased size of the anion. Notably, in this case, there is overlap between the cation–anion and cation–cation RDFs, which suggests that the first shell around a cation contains both types of ion. This is a result of the softer ionic binding between the cations and $[NTf_2]^-$ anion, which permits more close contacts between cations than might be expected when taking into account the associated charge–charge repulsion.

The spatial probability distributions derived from EPSR (Figure 4.4) indicate that the regions of highest probability of locating anions are above and below the plane of the ring and also associated with the methyl groups, creating a band of anions around the cation. It has been observed that as the hydrogen-bonding ability of the anion decreases, the probability density of anions above and below the ring increases, which suggests the liquid structure of $[C_1mim][NTf_2]$ is driven more by packing than by one or two specific favourable cation–anion interactions. The cation density is located around the hydrogen atoms of the aromatic ring; however, this may also be due to packing rather than strong specific intermolecular contacts.

In the cases of $[C_1mim]Cl$ and $[C_1mim][PF_6]$, the solid- and liquid-state structures were quite similar. In contrast, the crystal structure of $[C_1mim][NTf_2]$ [28] is not reflected in the liquid-state structure. The results above indicate anion density above and below the ring, whereas the crystal structure presents stacks of cations resulting from π–π interactions. Furthermore, in the solid state, the $[NTf_2]^-$ anion adopts the *cis* conformation, but the liquid structure demonstrates a mixture of *cis* and *trans* conformers in the ratio 1:4. Force field simulations performed as part of the study predicted a similar ratio of conformers.

4.3.4 1-Ethyl-3-methylimidazolium Ethanoate

The liquid structure of 1-ethyl-3-methylimidazolium ethanoate, $[C_2mim][O_2CCH_3]$, has been probed by neutron diffraction experiments and MD simulations [29]. The ethanoate anion is small and polar, possessing a hydrophobic methyl group and a hydrophilic carboxylate group: it presents an interesting study since the polar, hydrogen-bonding, bidentate ethanoate anions have a clear directional interaction to the cation. Where the spherical shape of the chloride and hexafluorophosphate anions facilitated hydrogen-bonding interaction to multiple cations, the ethanoate anion no longer has this ability, owing to the presence of the methyl group.

The ion–ion RDFs calculated from the EPSR analysis of neutron diffraction data are shown in Figure 4.5. The average cation–cation and cation–anion separations at approximately 7 and 5 Å, respectively, are comparable with those found for other ionic liquids. A degree of periodicity is observed over the first two shells about the cation centre of mass, with anion shells found at approximately 5 and 10 Å, and a cation shell at approximately 7 Å, which is again similar to the RDF observed for [C_1mim]Cl. The typical features expected for these systems are found; however, the first peak in the cation–anion distribution is noticeably broader than that observed for [C_1mim]Cl and bears more similarity to [C_1mim][NTf_2] [8, 27], as is also the case for 1-butyl-3-methylimidazolium trifluoroethanoate, [C_4mim][CF_3COO] [30]. This is despite the [NTf_2]$^-$ anion being much larger and possessing more diffuse charge than the ethanoate, indicating that the shell of anions around each cation is significantly broader than the length of the ethanoate anion. This implies that the anions form an approximate bilayer structure with charged carboxylate groups oriented towards cations in each adjacent shell and hydrophobic methyl groups in a central domain. Integration of the cation–anion RDF up to the first minimum at 7 Å gives a coordination number of 6.8, which is comparable to those found for other ionic liquid systems: 6 for [C_1mim]Cl and 6.8 for [C_1mim][PF_6].

The spatially resolved probability distributions of anions and cations around a central cation from the EPSR model are shown in Figure 4.5. As found previously in other ionic liquid systems, the most probable positions at which to find anions are along the C—H bond vectors of the imidazolium ring. The distribution of anions about a central cation is not symmetric about the central axis of the imidazolium ring since anisotropy exists between the regions associated with

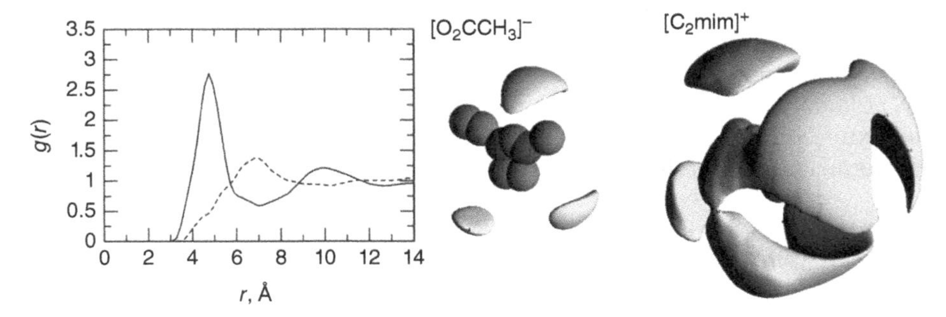

Figure 4.5 Cation–anion (solid line) and cation–cation (dashed line) RDFs and spatial probabilities for [C_2mim][O_2CCH_3] determined from neutron diffraction data [29].

the H(4) and H(5) carbon atoms, owing to the presence of the ethyl side chain. Free rotation of the ethyl group hinders access of anions over the H(5) position and the density is skewed towards the H(4) hydrogen of the cation. The anion density associated with the H(2) position is also offset towards the methyl group of the imidazolium cation for the same reason. It is evident that the region of closest cation–cation interaction in the first shell is directly above and below the cation ring plane, but there also exists a regular placement of other cations around the central cation with distinct lobes present in Figure 4.5. The two regions located near the H(2) position are commonly observed, for example, in $[C_1mim][NTf_2]$ or $[C_1mim]Cl$, as is the region associated with the H(4) and H(5) atoms of the cation. Lobes to the left and right of the alkyl chains are similar to those observed in $[C_1mim]Cl$ but in contrast with those in $[C_1mim][NTf_2]$, which are split into two, located in the same positions in the *xy* plane but situated above and below the ring plane.

Examination of the distributions of anions and cations around a central anion shows firstly that cations are predominantly found below the bidentate jaws of the anion. Only at lower probabilities (*ca.* top 20% of molecules) do regions appear elsewhere towards the sides of the anion. A structural feature not observed in previous studies with ionic liquids containing Cl^-, $[PF_6]^-$ or $[NTf_2]^-$ is evident in the distribution of the ethanoate anions around a central anion: a strong correlation of anion CH_3 groups is found, indicating significant self-structuring.

MD simulations of the system have also been performed for comparison with experimental data. The RDFs derived from EPSR analysis of the neutron diffraction data and those of the MD show good agreement, except for the pre-peak at 4 Å in the cation–cation correlation, which is less pronounced in the EPSR data. Possible explanations for this may lie either with the MD, for which the cation–cation interaction may be overly favoured, or in the EPSR simulation, in which the increased flexibility of the molecules may prevent such geometries from persisting. The peak at approximately 4 Å (MD) and small shoulder at approximately 5 Å (EPSR) in the cation–cation curve are indicative of ring stacking and depends strongly on the nature of the anion. Such features may also be found in simulations of, for example, $[C_1mim]Cl$, and in the solution ^1H NMR spectra of 1-ethyl-3-methylimidazolium chloride, $[C_2mim]Cl$ [31]. Increasing the size and reducing the coordinating ability of the anion reduce the probability of cation–cation stacking – for example, some evidence of weak, low-probability stacking is observed for $[C_1mim][PF_6]$, but not in $[C_1mim][NTf_2]$. Delocalisation of the charge on the anion leads to weaker directionality of cation–anion interactions and consequently reduces the probability of cation stacking, a structural feature that requires strong 'bridging' anions to exist [32].

A consequence of the high probability of ethanoate anions forming close contacts with the cation through ring C—H···O hydrogen bonding is that anion domains leading to methyl-to-methyl interactions are formed. These anion domains bridge the cations with three ethanoate carboxylate groups pointing towards one cation and three pointing away towards another cation in the adjacent shell. This alternating orientation leads to an expansion of the width of the ethanoate coordination shell, as observed in the broadening of the first peak in the anion RDF, and results in the formation of a pocket of hydrophobicity in the centre of the anion shell, produced from the increased probability of anion methyl–methyl contacts over carboxylate–methyl contacts.

4.3.5 1-Methyl-4-cyanopyridinium Bis{(trifluoromethyl)sulfonyl}amide

Ionic liquids with 1-alkyl-cyanopyridinium cations have recently been synthesised to study the influence of core substitution and modification on the characteristics of ionic liquid-forming salts [33]. In the case of these ionic liquids, the influence of an electron-withdrawing group on the cation was examined in detail in the liquid structure of 1-methyl-4-cyanopyridinium bis{(trifluoromethyl)sulfonyl} amide, $[(NC)_{\gamma}C_1py][NTf_2]$. This ionic liquid incorporates a nitrile functionality on the pyridinium ring of the cation, which polarises the charge distribution in the cation, with greater charge localisation in the heteroatom end of the ring compared with the analogous 1-alkylpyridinium systems.

Ion–ion partial radial distributions based on the centre of mass of the cations and anions extracted from EPSR are shown in Figure 4.6. The average cation–anion and cation–cation separations at approximately 5 and 8Å, respectively, are comparable with those found for the liquid structure of $[C_1mim][NTf_2]$ measured at 303K [27], and this is consistent with the small changes in molar volume and, therefore, atomic density of the two cations. However, as expected from the larger molar volume of the $[NTf_2]^-$ anion compared with Cl^- and $[PF_6]^-$, a significant increase in the cation–cation separation is found in contrast with those of 1,3-dimethylimidazolium chloride and hexafluorophosphate [8, 21]. Although a degree of periodicity is observed over the first three shells about the cation centre of mass, with anion shells at approximately 5 and 12Å and a cation shell at approximately 8Å, significant overlap of the cation–cation/anion–anion RDFs with the cation–anion curve is observed. This is similar to,

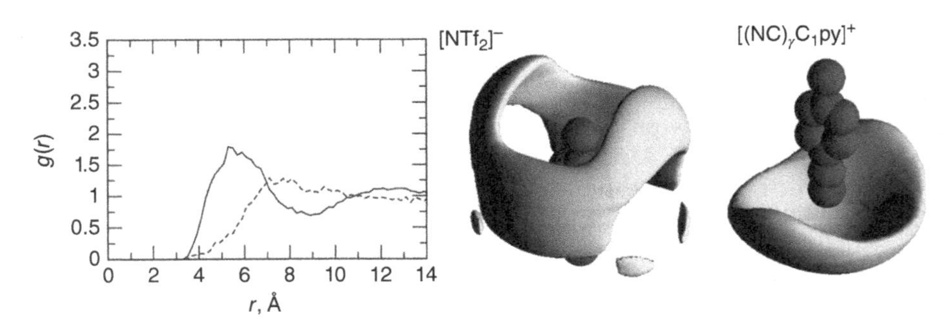

Figure 4.6 Cation–anion (solid line) and cation–cation (dashed line) RDFs and spatial probabilities for $[(NC)_{\gamma}C_1py][NTf_2]$ determined from neutron diffraction data [33].

although more pronounced than, observations made in [C_1mim][NTf_2] and is a result of the softer ionic bonding of the [NTf_2]$^-$ anion coupled with a cation somewhat larger than [C_1mim]$^+$.

The spatially resolved probability distributions of the cation and anion around a central 1-methyl-4-cyanopyridinium cation obtained from EPSR are also shown in Figure 4.6. At high probabilities, the anions sit above and below the pyridinium ring, slightly offset towards the ring nitrogen atom. With decreasing probability, the anion distribution spreads and forms four lobes in the plane of the pyridinium ring associated with the four aromatic hydrogen atoms, and no significant interactions are found with the $-CH_3$ or $-CN$ groups. In contrast, the cation–cation spatial distribution shows localisation about the nitrile group, which increases in size as the probability decreases. At high probabilities, the anion and cation spatial distributions form mutually exclusive regions. Similarities exist with the distributions found for [C_1mim][NTf_2]. For example, in both cases, the anions and cations occupy different regions of the cation, with the anion mostly associated with the π-density on the cation. However, in the case of [C_1mim][NTf_2], a significant anion interaction with the methyl groups is found at high probability. This may be associated with the increased hydrogen bond-donating ability of the aromatic hydrogen atoms in the case of the cyanopyridinium cation due to the electron-withdrawing cyano group compared with the weakly acidic hydrogen atoms at the C(4,5) positions on the imidazolium cation. The effect of the stronger hydrogen bonding with the anion, together with the reduction in electron cloud–anion Coulombic repulsion, increases the interaction with the cation as shown by the anion interaction axial to the pyridinium-ring hydrogen bonds. This interaction dominates over any methyl hydrogen-bonding interaction that may exist.

Investigating the atom-specific partial RDFs relating to cation-to-cation association reveals the presence of first shell contacts between cations caused by the penetration of nitrile groups into the first coordination sphere of adjacent cations with a distance of approximately 3.5 Å to the nitrile, ring and methyl groups, compared with the average centre-of-mass separation of approximately 8 Å. This apparent association of cations with cations could be simply steric in nature, since the nitrile group projects out from the cation centre of mass. However, the methyl group that protrudes from the opposite pole shows no close contacts with adjacent cations, which suggests that some preferential local ordering of the first shell around individual cations exists and includes interactions of the nitrile groups in adjacent cations.

Both EPSR fitting of experimentally obtained neutron diffraction data and MD simulations give remarkably consistent results for the liquid structure of [$(NC)_\gamma C_1$py][NTf_2]. The region of highest probability of finding cations is close to the cyano group of the central cation and orientation is governed by interactions between cyano groups. Analysis of the MD simulations reveals a fine structure of ordering of cyano groups around the central cation. The anions occupy regions directly above and below the pyridinium ring and around the acidic protons of the central cation at high probabilities. This structuring of the ionic liquid is observed in both models, despite the different methodology used in each case, with each cation surrounded by anions in the first shell at approximately 5 Å. Careful screening of group–group interactions shows that the nitrile groups of cations project into the first anion-rich shell of the next neighbour cations. This is a consequence of the non-spheroidal shape of the ions and may structurally relate to the clustering of cations observed in the crystal structure. The anion shows both *cis* and *trans* conformations, with the latter dominating the structure, a distribution that evolves naturally from MD simulations and EPSR refinement of the neutron diffraction data.

4.3.6 Summary

The very nature of ionic liquid systems means that the single largest influence on liquid structure is charge–charge interaction between cation and anion. In this sense, it is to be expected that the concentric shell structure of the liquid, so commonly observed for atomic molten salts, should be observed frequently in the present systems. Where the ions are of relatively small size and relatively moderate hardness, for example, as for $[C_1mim]Cl$, $[C_1mim][O_2CCH_3]$ and $[C_1mim][PF_6]$, the molten salt-like structural picture remains relatively unmodified. Any differences that arise can be directly related to either the presence of specific directional interactions (e.g. the aromatic C—H···X interaction) or, conversely, the lack thereof (e.g. regions above and below the plane of the imidazolium ring, which present an electrostatic field in the form of a quadrupole but which offer no 'strong' structure-making interactions) occurring between the cation and anion. From the anion's perspective, these serve to create regions of high and low interaction strength around the cation, resulting in an anisotropic anion distribution both spatially and in terms of density around the cation. Where one or both of the constituent ions become larger and softer in character, the packing forces become more dominant. For example, in the case of $[C_1mim][NTf_2]$, the cation still presents the same electrostatic field as in $[C_1mim]Cl$, but the softness and flexibility of the anion dictates that more spatially accessible regions must be populated, even if the associated electrostatic field is less favourable. In such cases, other cations begin to occupy regions in the primary coordination shell, despite the clear repulsive interactions that this would seemingly introduce.

4.4 STRUCTURE OF IONIC LIQUIDS AND SOLUTES

One of the important features of ionic liquids is their ability to dissolve a wide range of chemically different solutes. This may be driven by specific interaction with either the cation or the anion, as the Lewis donor/acceptor ability of the solute is often an important feature defining the dissolution process. Since the nature of the cation and anion may be varied significantly, the result is ionic liquids, which can dissolve and solvate both organic and inorganic species.

An understanding of the interactions within the solvent and between the solvent and solute is important if a degree of predictability for processes within the family of ionic liquids is to be achieved. The miscibility of aromatic molecules in ionic liquids, forming a biphase between the ionic liquid/aromatic phase and the excess of aromatic, was recognised by Atwood and co-workers in the 1970s [34]. The term 'liquid clathrate' was introduced to describe these ionic liquid/aromatic liquid phases, which, in many cases, contain a huge excess of aromatic and display ionic liquid-induced structural ordering. Despite the number of studies confirming the phenomena, it is still remarkable that the solubility of simple aromatic molecules in ionic liquids is so high, especially if one considers the disparity in molecular types and interactions. The primary characteristic of an ionic liquid is the Coulombic charge lattice between ions, whereas in liquid aromatics the molecules exhibit π–π and alkyl–π interactions. It is also worth noting that, in general, the solubility of aliphatic hydrocarbons in ionic liquids is exceptionally low. The high solubility of aromatics in these types of salts, both in solid and liquid phases, has been attributed to the formation of clathrate structures, and both the liquid and solid states of ionic liquid/aromatic mixtures have been investigated. Association of cations and aromatic guests though π-stacking has been observed by neutron scattering in the liquid state [35] and by X-ray crystallography in the solid state [25, 36]. The formation and control of this interaction has implications with respect to the use of ionic liquids as entrainers and extraction solvents for the separation and removal of aromatic compounds, for example, the separation of benzene–toluene–xylene (BTX) mixtures [37] and the deep desulfurisation of fuels [38].

Although the interactions and structures found in molecular liquids have been examined extensively, the examination of ionic liquid–solute interactions has been much more limited. Such studies provide an insight into which interactions dominate the liquid phase and, consequently, which determine the chemical and physical properties of the liquid, such as density, viscosity, polarity and power of solvation. A wide range of experimental techniques have been employed to this end, including extended X-ray absorption fine structure to examine to the structure of metal complexes [39], IR spectroscopy to examine the interactions of H_2O and CO_2 [40], line broadening in NMR spectroscopy to probe the chloride/glucose interaction [41] and NOE effects to understand the interaction of water with ionic liquids [42]. A number of theoretical studies have highlighted the contrasting interactions between various solutes and ionic liquids. For example, Hanke and co-workers [43] showed that, in $[C_1mim]Cl$, polar, hydrogen-bonding molecules such as water and methanol associate strongly with the anion, while dimethyl ether and propane favour interactions with the cation. Ammonia is found to interact predominantly with the imidazolium cations through hydrogen bonding to the ring hydrogens [44], while the driving forces behind carbon dioxide solubility are predominantly associated with the anion [45]. In contrast, relatively few experimentally determined ionic liquid–solute structural investigations have been performed. In order to gain a detailed understanding of ionic liquid–solute interactions, neutron scattering methods have been used to probe the liquid structure of a number of systems and have been compared with MD simulations of the same systems as cross validation.

4.4.1 Benzene:1,3-Dimethylimidazolium Hexafluorophosphate

To probe solvation interactions, Deetlefs et al. performed neutron diffraction experiments on $[C_1mim][PF_6]$ containing 33 and 67 mol% concentrations of benzene [33] and observed a reduction in the anion coordination number around the cations, as the benzene molecules take up positions around the imidazolium ring. In the 33 mol% case, around 1.3 anions are replaced by benzene, while for 67 mol%, this increases to 2.7. The primary coordination shell of the cation is still anion rich and, as in the case of the pure liquid, relatively free of other cations. In addition, the high-probability regions of anions located around the cation (Figure 4.7) are relatively unchanged from the pure liquid, and this suggests that the anions that have been replaced by the benzene were only weakly interacting with the cation. Nevertheless, spatial probability distributions extracted from neutron data, while confirming that the high-probability anion distribution is almost identical to that of the pure liquid, show the existence of similar regions of benzene molecules, which occupy the same region spatial domains. This indicates that there is at least some dynamic competition between solute and anion for these positions around the cation. In terms of the cation distribution, the most significant change from the pure liquid is the disappearance of density associated with the methyl groups, having been replaced by benzene molecules. Despite the obvious perturbations made to the cation–anion interaction, the aromatic is nevertheless soluble to a high degree in the ionic liquid, with only relatively modest effects on the overall liquid structure.

Equimolar mixtures of $[C_1mim][PF_6]$ with benzene, 1,3,5-trifluorobenzene and hexafluorobenzene have also been investigated by MD simulations [46]. It was observed that, depending on the nature of the quadrupole present on

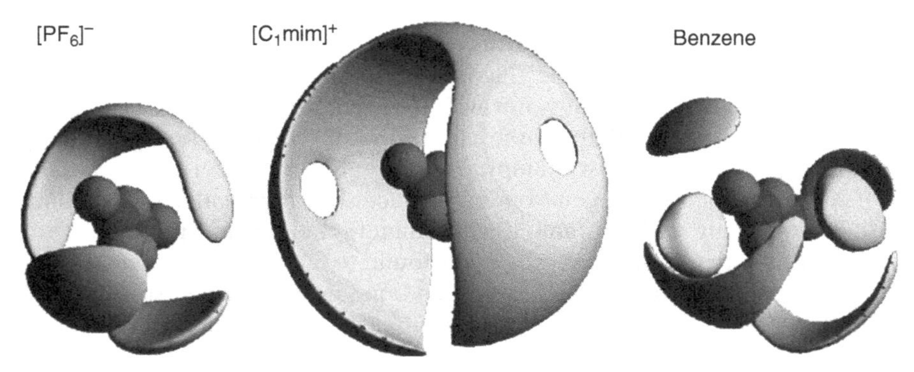

Figure 4.7 Spatial probabilities for $[C_1mim][PF_6]$:benzene determined from neutron diffraction data [33].

the aromatic, the solvation environments can be markedly different. For instance, benzene associates strongly with the cation, as was found experimentally, while hexafluorobenzene prefers to associate with anions. Normal benzene possesses a quadrupole moment, which presents a negative electrostatic field above and below the aromatic plane, and hence provides a favourable environment in which cations may sit. In contrast, hexafluorobenzene is almost exactly opposite, presenting a positive electrostatic field, making anions more likely to approach. While this is a convincing argument, careful analysis therein revealed that in both cases, the overall interaction energy between the aromatic and ions above and below the plane is positive (i.e. repulsive). Of more importance is the favourable interaction between aromatic and cation/anion in the equatorial region – for both benzene (with anion) and hexafluorobenzene (with cation), this is the dominant interaction, and the presence of the counter ion above and below the plane in the quadrupolar field is a result of either ionic association or simple packing effects.

4.4.2 Glucose:1-Alkyl-3-methylimidazolium Ionic Liquids

Swatloski et al. showed previously that [C_4mim]Cl is an excellent solvent in which to solubilise cellulose [47], and recently, it has been demonstrated that ionic liquids based on ethanoate and formate anions are also efficient solvents for this purpose [48, 49]. In particular, [C_2mim][O_2CCH_3] has been shown to be a potential solvent for a new commercial process for the preparation of cellulose fibres. Neutron diffraction and complementary MD simulations have been performed on both the [C_1mim]Cl and [C_2mim][O_2CCH_3]/glucose system in order to increase our understanding of ionic liquid properties relevant to the biomass dissolution process [29, 50, 51].

It is immediately apparent that the general ionic liquid structure is perturbed very little by the dissolved solute, despite its size, as has been observed previously for other systems with equally large solutes, for example, benzene [33]. A slight expansion (*ca.* 0.4Å) of the coordination shells in the 6:1 system can be observed with a shift in the maxima of the RDFs to longer distances. Glucose ion RDFs show that both cations and anions approach the sugar fairly closely and indicate that the first solvation shell will be a mixture of cations and anions. The small shoulder at *ca.* 4Å in the cation–cation curve is much less well defined in the 6:1 [C_2mim][O_2CCH_3]/glucose system in comparison with the neat ionic liquid system, which may suggest the cation–cation stacking interaction is disrupted on the addition of glucose to the system. Similar observations were also made in 5:1 [C_1mim]Cl/glucose systems [47, 48]. Some glucose–glucose contact is present, suggesting that occasional hydrogen bonds between neighbouring sugars occur, but the predominant glucose–glucose RDF peak is further out at 10Å. A structural feature compatible with this length scale is the positioning of two sugar molecules hydrogen bonding to the same ethanoate anion, but on opposite oxygen centres.

Integration of the cation–anion RDF up to 7.4 Å, that is, the first minimum in the cation–anion RDF, gives a coordination number of 6.3, which is reduced from 6.8 anions in the first shell around the cation compared with that found in the neat ionic liquid system, as is to be expected owing to the presence of the solute, and is consistent with the idea that the ethanoate anions now also form close contacts with the dissolved sugar as well as with the ionic liquid cations. The total coordination of ethanoate anions around glucose is 6.2, which is somewhat larger than the 3.9 chloride anions, which coordinate to glucose in 1,3-dimethylimidazolium chloride [47, 48].

The spatial probability distributions of anions and cations around a central cation are largely unaffected by the presence of glucose in the system. The most probable positions at which to find anions remain along the C–H bond vectors of the cation ring, with retention of cation density directly above and below the plane of the central cation. The cation–anion interactions remain well defined, despite the presence of glucose in the ionic liquid system, which indicates that any interaction glucose has with the ions must compete with the strong hydrogen bonding present in the parent ionic liquid system. Albeit small, it is observed that the regions of cation density around a central cation are less well defined in the spatial distributions of the 6:1 $[C_2mim][O_2CCH_3]$/glucose system in comparison with the neat ionic liquid system. This was also evident in the cation–cation RDF and may be a result of slightly weaker cation–cation interactions on addition of glucose. Spatial probability distributions of species around a central glucose (Figure 4.8) reveal the likely positions of the species around the sugar. With respect to the ethanoate anion, the positioning is guided almost entirely by hydrogen bonding to the hydroxyl groups, and the predominant regions of density around the sugar are close to the OH sites around the ring. In contrast, high-probability regions for the cation are generally found

Figure 4.8 Spatial probabilities of anions (red) and cations (blue) around a central glucose molecule, calculated from molecular dynamics simulations [47, 48]. The distribution of cation H(2), H(4) and H(5) protons is shown as a light blue surface.

outside the anion density around the equator of the sugar and tend to suggest that the cation is positioned as a result of interaction with the local anions, rather than as a result of direct interaction with the sugar. This trend appears not to be followed in the vicinity of the ring oxygen, around which a clear band of cations exists and which does not immediately appear to be the result of neighbouring anions since there are no high-probability regions in the same locality.

Addition of glucose to [C_2mim][O_2CCH_3] results in only subtle differences in the RDFs and three-dimensional probability distributions. This is not dissimilar to the findings in the case of a 5:1 system of [C_1mim]Cl with glucose [47, 48]. In this system, there were no significant changes in the RDFs of the ions, and in terms of probability distributions, the cation–cation high-probability regions were displaced on addition of glucose, but no change was observable in the cation–anion interaction. The interaction between anions and glucose molecules is through hydrogen-bonding interactions between ethanoate oxygen atoms and glucose hydroxyl hydrogen atoms; however, further investigation into the individual regions of anion density around the glucose molecule is required to determine which regions of anion density are associated with specific hydroxyl groups.

The data associated with the cation indicates little interaction between the imidazolium ring and the glucose. Recently, there has been significant debate [52] on the importance of interactions with the cation for the solubilisation of the cellulose. Moyna and co-workers maintain that no significant interaction exists, whereas Zhang et al. have proposed a model whereby hydrogen bonding between the cation and the solute plays an important role [53]. Evidence from MD and neutron diffraction measurements, coupled with the fact that a wide range of cations have been utilised for the solubilisation of cellulose, suggests it is likely that the cation only plays a minor role in the solvation process. Clearly, for charge neutrality, the cation has to be in the vicinity of the anion and, therefore, its size and strength of interaction with the anion will have a bearing on the ability for the anion to induce dissolution of the cellulose. However, it is also likely that some of the observations between the solubilities of cellulose as a function of the cation are associated with kinetics rather than thermodynamics. Mass transport, for example, the viscosity of the ionic liquid, changes significantly as a function of the cation and the systems become increasingly more viscous as the cellulose is solubilised making it increasingly difficult to stir.

4.4.3 PCl_3/$POCl_3$:1,3-Dimethylimidazolium Bis{(trifluoromethyl)sulfonyl}amide

Since ionic liquids can act as excellent solvents for many different chemical materials, their use as storage and/or reaction media is attracting some attention. Examples are the highly moisture-sensitive PCl_3 and $POCl_3$, reagents used heavily in many industries. Certain hydrophobic ionic liquids, such as those based on the [NTf_2]$^-$ anion, are able to dissolve and stabilise these materials with respect to hydrolysis, even if the ionic liquid itself has a non-negligible

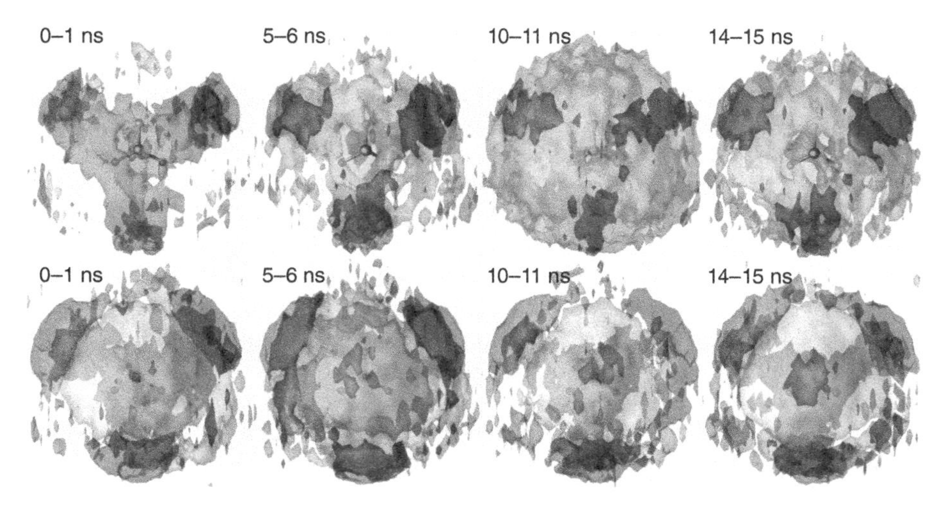

Figure 4.9 Spatial distributions of cations (blue), anions (red) and solute (green) around a central solute molecule for 14 mol% PCl_3 (top) and $POCl_3$ (bottom) in $[C_1mim][NTf_2]$, calculated from MD simulations [57].

water content [54]. Subsequent studies have shown that these systems may prove useful as storage and/or reaction media for these materials [55, 56].

Experimental neutron scattering data on the $[C_1mim][NTf_2]$ ionic liquid containing up to 14 mol% of the reactive compound [57] did not reveal any significant differences to the scattering traces obtained for the pure liquid. This perhaps suggests that the ionic liquid can readily accommodate the solute owing to its small size, making detection of structural changes difficult, even up to the solubility limit of the material. Nevertheless, MD of the 14 mol% PCl_3 and $POCl_3$ systems reveals some significant differences between the solvation of the two compounds. Over long simulation times, it was observed that PCl_3 tends to form high-concentration microdomains within the ionic liquid, whereas the $POCl_3$ tends to be more diffuse throughout the system (Figure 4.9). While the P—Cl bonds in both reagents are relatively non-polar, the oxygen atom present in $POCl_3$ permits a favourable interaction with the cation and enables the molecules to spread more easily into the ionic liquid (or, conversely, means that the reagent molecules are not clustered together by the ionic liquid).

4.4.4 1-Methylnaphthalene:1-Methyl-4-cyanopyridinium Bis{(trifluoromethyl)-sulfonyl}amide

It was observed that contacting cyanopyridinium ionic liquids with hydrocarbons containing electron-rich aromatics resulted in a distinct colour change in the ionic liquid phase, from colourless to an intense bright yellow, indicative of the formation of a charge-transfer complex. In addition, on solidification of the liquid phase, in the case of 4-cyanopyridinium ionic liquids, the colour was maintained. Cyanopyridinium cations have previously been ion exchanged into zeolite pores and have been shown to form 1:1 charge-transfer complexes with aromatic donors within the zeolite cavities and demonstrated as shape-selective visual probes to monitor the physisorption of water into zeolites [58]. 1-Methylnaphthalene and ionic liquids containing relatively electron-deficient 1-methyl-4-cyanopyridinium cations [59] were examined by neutron scattering and MD simulations [60]. In addition, the structure in the solid state determined using X-ray diffraction has been compared with the liquid structure.

The RDFs derived from the neutron diffraction data (Figure 4.10) showed that the strong ordering of anions and cations into alternating shells is maintained in the presence of the solute, as expected from other reports of ionic liquid–solute interactions. Little change in the cation–anion distance is observed on addition of the aromatic compared with the pure ionic liquid, with the peak occurring at approximately 5 Å in both cases, as found for other solutes in ionic liquids. A small increase in the cation–cation spacing was found, increasing from 7.5 to 8.5 Å as expected. Although the aromatic was found in the same coordination shell around the cation as the anion, little contact between the anion and the aromatic was observed. Therefore, the aromatic–cation interaction was postulated as being the determining factor in the formation of the charge-transfer complex. On addition of the solute to the ionic liquid, an increased probability of finding the cation above the ring of another cation was observed, which mimics the spatial distribution found for the aromatic. The latter shows high density above and below the cation ring, providing evidence that short-range aromatic–cation π–π stacking dominates in the liquid state. MD simulations provided more detailed conformation information. Analysis of the cation/aromatic pairs showed that most of the rings were tilted less than 25° from coplanarity. This substructure is also found in the solid state; however, in this case, infinite π–π stacks were found, whereas this is not likely in the liquid state, which may be best viewed as a transient sandwich.

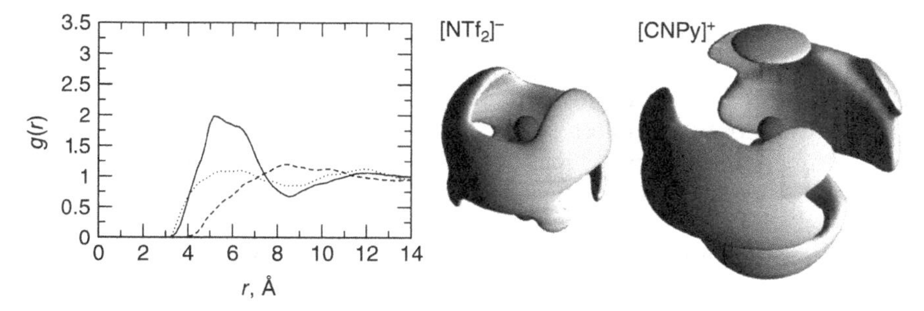

Figure 4.10 Cation–anion (solid line) and cation–cation (dashed line) RDFs and spatial probabilities for 1:1 $[(NC)_\gamma C_1py][NTf_2]$:methylnaphthalene, determined from neutron diffraction data [60]. The cation–aromatic RDF is shown as a dotted line.

4.4.5 Summary

The predominant solubilising properties of ionic liquids are largely governed by the hardness of the constituent ions. For ionic liquids based on small anions, such as Cl^- or $[O_2CCH_3]^-$, the primary solvation processes observed in the literature rely on strong hydrogen-bonding interactions with the target solute. The cation in these systems plays a relatively minor role in the actual solvation process but, unavoidably, is nevertheless always nearby. For the dissolution of glucose, it has been suggested that the cation plays a measurable role, hydrogen bonding to glucose hydroxyl oxygen atoms via the imidazolium H(2) position [51]; however, it remains to be seen whether this is a truly measurable effect or simply a secondary interaction brought about by the close proximity of the anions (to which the cations are strongly associated). As the size of the anion is increased, spreading its charge over a larger area, there is more likelihood of the same cation playing a more direct role in solvation, for example, as for $POCl_3$ and hexafluorobenzene. Solutes that possess only a weak external electrostatic potential, for instance, PCl_3, are still soluble within certain ionic liquids, but in such cases, the solubility itself is passive and the strong cation–anion interaction forces the solute to segregate or self-associate into microdomains. Nevertheless, such heterogeneous solutions still provide a protective influence and may certainly be exploited.

The Coulombic cation–anion interaction is the prevailing cohesive force in all ionic liquid systems of modest chain length, yet it is still remarkable that even the presence of reasonable concentrations of large solutes, such as glucose or benzene, has little effect on the liquid from a structural perspective, despite the need for solute–anion interactions to facilitate dissolution. Of course, such strong association of the solute with the ionic liquid is not necessarily a useful property – for reactions conducted in ionic liquid systems, the strong association of reactants with the ionic liquid may be beneficial from some perspectives (e.g. stability of sensitive reagents) but may ultimately be detrimental to the performance of the process if the interactions cannot be overcome.

4.5 HETEROGENEITY IN IONIC LIQUID STRUCTURE

A significant feature of room-temperature ionic liquids, distinct from higher-temperature inorganic molten salts, is the presence of non-polar, non-charge-bearing functional groups, which increase the size/volume of the ions, thereby reducing the effective ion–ion interactions. In many cases, these non-charge-bearing groups are anisotropically distributed on the cation (or anion) of the ionic liquid, such as in 1,3-dialkylimidazolium compared with tetraalkylammonium cations, and this can distort the local packing of ions in the first and second shells around each ion, potentially leading to local or even longer-scale structural anisotropy. The presence of local mesoscopic inhomogeneity, with nanoscale ionic and non-polar domains, in ionic liquids was proposed on the basis of MD simulations [13, 61–63]. A picture of well-defined microphases has not been demonstrated explicitly by experiment, at least not for the short-chain length materials. A range of experimental techniques have indicated that these domains do exist. For example, Triolo et al. showed the presence of scattering from a low-Q peak using X-ray scattering of medium-chain length imidazolium ionic liquids [64]. This correlation length is found to be between one and two times the long axis of the cation, that is, similar to the structural relationship found in longer-chain imidazolium ionic liquid crystals, which do show separate polar and non-polar domains [65]. Therein, a strong correlation was observed between the correlation length and the alkyl chain length. Spectroscopic data have also been used to probe the existence of the domain structure; however, it has been demonstrated that nanoscale segregation is only one of a number of possible interpretations of the data [66]. Furthermore, Hu and Margulis [67] have recently shown that the simulations of these ionic liquids may only represent transient structures, rather than a true time-dependent representation of the ionic liquid structure. In these calculations, the ionic liquids were described as glasses rather than liquids on the time frame of electronic probe relaxation processes. Therefore, the correlation between the optical Kerr effect and dielectric relaxation spectroscopy measurements and the MD simulation derived structures may not be robust as the domains may not persist at time scales beyond the order of pico- to nanoseconds [68].

Neutron scattering data for the series of 1-alkyl-3-methylimidazolium hexafluorophosphate ionic liquids, $[C_n mim][PF_6]$ ($n=4$, 6, or 8), collected over a wide Q range (0.05–50 Å^{-1}) in the liquid state using H/D isotopic substitution has recently been used to further investigate the contributions of the various components within the system to the total scattering factor [69]. In good agreement with previous studies, the neutron diffraction data showed a low-Q feature less than 0.8 Å^{-1}, which could be correlated with the chain length on the imidazolium cation. However, difference scattering patterns from the range of isotopically substituted cations clearly illustrate that the origin of this peak is not associated with the alkyl chain–alkyl chain van der Waals interactions (Fig. 4.11). For the ionic liquid based on the butyl chain length, no contribution is observed and low-Q feature at 0.44 Å^{-1} is dominated by cation–anion correlations. As the chain length increases, a greater contribution from the alkyl groups was seen; however, the contribution from the Coulombic interaction is still found to be high.

While these data do not preclude the presence of alkyl–alkyl chain interactions, it is unlikely that these are structure-determining interactions for the short-chain length materials. In fact, the trend of the correlation length with alkyl chain length [13, 62] may be extended to include data from the $[C_1mim]^+$ cation, which will have very few van der Waals interactions, and is completely dominated by Coulombic and hydrogen-bonding interactions [63]. It is likely that some of the differences between the experimental and theoretical data may be associated with the time scales probed in each case. Further studies need to be undertaken to examine whether such domains can be induced at the short chain lengths by reducing the Coulombic contributions or with the application of pressure, for example.

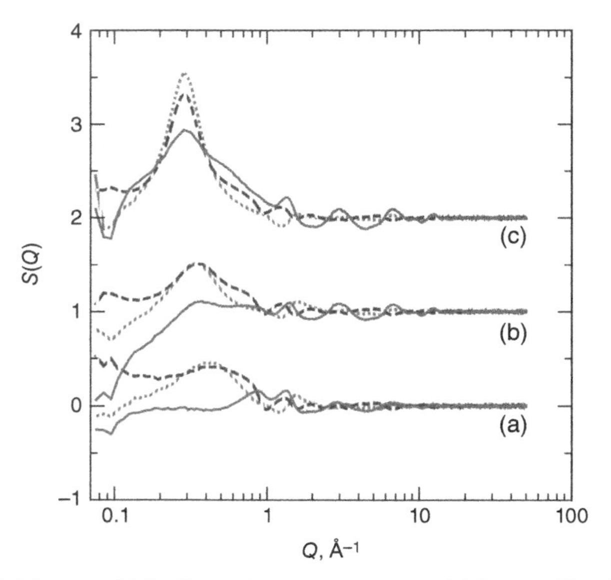

Figure 4.11 Difference S(Q) diffraction spectra for the (a) [C_4mim][PF_6], (b) [C_6mim][PF_6] and (c) [C_8mim][PF_6] systems, showing subtraction of data for the fully protiated minus methyl deuterated (dots), perdeuterated minus alkyl-deuterated (dashed) and alkyldeuterated minus fully protiated (solid) systems, corresponding to the pair contributions of the methyl hydrogens, imidazolium ring+methyl hydrogens and alkyl chains, respectively. The significant contribution of the imidazolium head groups and minimal contribution from the alkyl chains to the low-Q scattering region is clearly visible with a shift to low Q (longer correlation length) and increase in resolution and intensity as the alkyl chain length on the cation increases [69] .

REFERENCES

1 Wasserscheid, P. and Welton, T., *Ionic Liquids in Synthesis*, Wiley-VCH, Weinheim (2007).

2 Edwards, F.G., Enderby, J.E., Howe, R.A. and Page, D.I., The structure of molten sodium chloride, *J. Phys. C Solid State Phys.* **8**, 3483–3490 (1975).

3 (a) Blander, M., Bierwagen, E., Calkins, K.G., Curtiss, L.A., Price, D.L. and Saboungi, M.-L., Structure of acidic haloaluminate melts: Neutron diffraction and quantum chemical calculations, *J. Chem. Phys.* **97**, 2733–2741 (1992); (b) Lee, Y.-C., Price, D.L., Curtiss, L.A., Ratner, M.A. and Shriver, D.F., Structure of the ambient temperature alkali metal molten salt $AlCl_3$/LiSCN, *J. Chem. Phys.* **114**, 4591–4594 (2001); (c) Takahashi, S., Suzuya, K., Kohara, S., Koura, N., Curtiss, L.A. and Saboungi, M.-L., Structure of 1-ethyl-3-methylimidazolium chloroaluminates: Neutron diffraction measurements and ab initio calculations, *Z. Phys. Chem.* **209**, 209–221 (1999); (d) Trouw, F.R. and Price, D.L., Chemical applications of neutron scattering, *Annu. Rev. Phys. Chem.* **50**, 571–601 (1999).

4 Lantelme, F., Turq, P., Quentrec, B. and Lewis, J.W.E., Application of the molecular dynamics method to a liquid system with long range forces (Molten NaCl), *Mol. Phys.* **28**, 1537–1549 (1974).

5 (a) Tosi, M.P., Pastore, G., Saboungi, M.L. and Price, D.L., Liquid structure and melting of trivalent metal, *Phys. Scr. T* **39**, 367–371 (1991); (b) Lai, S.K., Li. W. and Tosi, M.P., Evaluation of liquid structure for potassium, zinc, and cadmium, *Phys. Rev. A* **42**, 7289–7303 (1990); (c) Mason, P.E., Neilson, G.W., Enderby, J.E., Saboungi, M.-L. and Brady, J. W., Structure of aqueous glucose solutions as determined by neutron diffraction with isotopic substitution experiments and molecular dynamics calculations, *J. Phys. Chem. B* **109**, 13104–13111 (2005).

6 Canongia Lopes, J.N. and Padua, A.A.H., Molecular force field for ionic liquids composed of triflate or bistriflylimide anions, *J. Phys. Chem. B* **108**, 16893–16898 (2004).

7 (a) Soper, A.K., Empirical potential Monte Carlo simulation of fluid structure, *Chem. Phys.* **202**, 295–306 (1996); (b) Soper, A.K., The radial distribution functions of water and ice from 220 to 673 K and at pressures up to 400 MPa, *Chem. Phys.* **258**, 121–137 (2000); (c) Soper, A.K., Tests of the empirical potential structure refinement method and a new method of application to neutron diffraction data on water, *Mol. Phys.* **99**, 1503–1516 (2001).

8 Hardacre, C., Holbrey, J.D., McMath, S.E.J., Bowron, D.T. and Soper, A.K., Structure of molten 1,3-dimethylimidazolium chloride using neutron diffraction, *J. Chem. Phys.* **118**, 273–278 (2003).

9 Arduengo, I.M., Dias, H.V.R., Harlow, R.L. and Kline, M., Electronic stabilisation of nucleophilic carbenes, *J. Am. Chem. Soc.* **114**, 5530–5534 (1992).

10 Hanke, C.G., Price, S.L. and Lynden-Bell, R.M., Intermolecular potentials for simulations of liquid imidazolium salts, *Mol. Phys.* **99**, 801–809 (2001).

11 Canongia Lopes, J.N., Deschamps, J. and Padua, A.A.H., Modeling ionic liquids using a systematic all-atom force field, *J. Phys. Chem. B* **108**, 2038–2047 (2004).

12 Liu, Z.P., Huang, S.P. and Wang, W.C., A refined force field for molecular simulation of imidazolium-based ionic liquids, *J. Phys. Chem. B* **108**, 12978–12989 (2004).

13 Urahata, S.M. and Ribeiro, M.C.C., Structure of ionic liquids of 1-alkyl-3-methylimidazolium cations: A systematic computer simulation study, *J. Chem. Phys.* **120**, 1855–1863 (2004).

14 Youngs, T.G.A., Del Pópolo, M.G. and Kohanoff, J., Development of complex classical force fields through force matching to ab initio data: Application to a room-temperature ionic liquid, *J. Phys. Chem. B* **110**, 5697–5707 (2006).

15 Lynden-Bell, R.M. and Youngs, T.G.A., Using DL_POLY to study the sensitivity of liquid structure to potential parameters, *Mol. Simul.* **32**, 1025–1033 (2006).

16 Youngs, T.G.A. and Hardacre, C., Application of static charge transfer within an ionic liquid force field and its effect on structure and dynamics, *ChemPhysChem.* **9**, 1548–1558 (2008).

17 Bhargava, B.L. and Balasubramanian, S., Refined potential model for atomistic simulations of ionic liquid, *J. Chem. Phys.* **127**, 114510 (2007).

18 Del Pópolo, M.G., Lynden-Bell, R.M. and Kohanoff, J., Ab initio molecular dynamics simulation of a room temperature ionic liquid, *J. Phys. Chem. B* **109**, 5895–5902 (2005).

19 Bühl, M., Chaumont, A., Schurhammer, R. and Wipff, G., Ab initio molecular dynamics of liquid 1,3-dimethylimidazolium chloride, *J. Phys. Chem. B* **109**, 18591–18599 (2005).

20 Bhargava, B.L. and Balasubramanian, S., Intermolecular structure and dynamics in an ionic liquid: A Car-Parrinello molecular dynamics simulation study of 1,3-dimethylimidazolium chloride, *Chem. Phys. Lett.* **417**, 486–491 (2006).

21 Hardacre, C., McMath, S.E.J., Nieuwenhuyzen, M., Bowron, D.T. and Soper, A. K., Liquid structure of 1,3-dimethylimidazolium salts, *J. Phys. Condens. Matter* **15**, S159–S166 (2003).

22 Morrow, T.I. and Maginn, E.J., Molecular dynamics study of the ionic liquid 1-*n*-butyl-3-methylimidazolium hexafluorophosphate, *J. Phys. Chem. B* **106**, 12807–12813 (2002).

23 Shah, J.K., Brennecke, J.F. and Maginn, E.J., Thermodynamic properties of the ionic liquid 1-*n*-butyl-3-methylimidazolium hexafluorophosphate from Monte Carlo simulations, *Green Chem.* **4**, 112–118 (2002).

24 Bhargava, B.L. and Balasubramanian, S., Insights into the structure and dynamics of a room-temperature ionic liquid: Ab initio molecular dynamics simulation studies of 1-*n*-butyl-3-methylimidazolium hexafluorophosphate ([bmim][PF_6]) and the [bmim][PF_6]-CO_2 mixture, *J. Phys. Chem. B* **111**, 4477–4487 (2007).

25 Holbrey, J.D., Reichert, W.M., Nieuwenhuyzen, M., Sheppard, O., Hardacre, C. and Rogers, R.D., Liquid clathrate formation in ionic liquid-aromatic mixtures, *Chem. Commun.* 476–477 (2003).

26 Matsumoto, H., Matsuda, T. and Miyazaki, Y., Room temperature molten salts based on trialkylsulfonium cations and bis(trifluoromethylsulfonyl)imide, *Chem. Lett.* **12**, 1430–1431 (2000).

27 Deetlefs, M., Hardacre, C., Nieuwenhuyzen, M., Padua, A.A.H., Sheppard, O. and Soper, A.K., Liquid structure of the ionic liquid 1,3-dimethylimidazolium bis(trifluoromethyl) sulfonylamide, *J. Phys. Chem. B* **110**, 12055–12061 (2006).

28 Holbrey, J.D., Reichert, W.M. and Rogers, R.D., Crystal structures of imidazolium bis(trifluoromethanesulfonyl)imide ionic liquid salts: The first organic salt with a *cis*-TFSI anion conformation, *Dalton Trans.* 2267–2271 (2004).

29 Bowron, D.T., D'Agostino, C., Gladden, L.F., Hardacre, C., Holbrey, J.D., Lagunas, C., McGregor, J., Mantle, M.D., Mullan, C.L. and Youngs, T.G.A., Structure and dynamics of 1-Ethyl-3-methylimidazolium ethanoate via molecular dynamics and neutron diffraction, *J. Phys. Chem. B* **114**, 7760–7768 (2010).

30 Schroder, C., Rudas, T., Neumayr, G., Gansterer, W. and Steinhauser, O., Impact of anisotropy on the structure and dynamics of ionic liquids: A computational study of 1-butyl-3-methyl-imidazolium trifluoroethanoate *J. Chem. Phys.*, **127**, 044505 (2007).

31 Avent, A.G., Chaloner, P.A., Day, M.P., Seddon, K.R. and Welton, T., Evidence for hydrogen bonding in solutions of 1-ethyl-3-methylimidazolium halides, and its implications for room-temperature halogenoaluminate(III) ionic liquids, *J. Chem. Soc. Dalton Trans.*, 3405–3413 (1994).

32 Hardacre, C., Holbrey, J.D., Nieuwenhuyzen, M. and Youngs, T.G.A., Structure and solvation in ionic liquids, *Acc. Chem. Res.* **40**, 1146–1155 (2007).

33 Hardacre, C., Holbrey, J.D., Mullan, C.L., Nieuwenhuyzen, M., Youngs, T.G.A. and Bowron, D.T., Liquid structure of the ionic liquid, 1-methyl-4-cyanopyridinium bis{(trifluoromethyl) sulfonyl}imide determined from neutron scattering and molecular dynamics simulations, *J. Phys. Chem. B* **112**, 8049–8056 (2008).

34 (a) Atwood, J.L., Liquid clathrates, in *Inclusion Compounds*, Vol. **1**, eds. J.L. Atwood, J.E.D. Davies and D.D. MacNicol, Academic Press, London (1984), pp. 375–406; (b) Atwood, J.L. and Atwood, J.D., Non-stoichiometric liquid enclosure compounds ('liquid clathrates'), in *Inorganic Compounds with Unusual Properties*, Vol. **150**, ed. R.B. King, American Chemical Society, Washington, DC (1976), pp. 112–127; (c)Atwood, J.L., Liquid clathrates, in *Recent Developments in Separation Science*, Vol. **3**, ed. N.N. Li, CRC Press, Cleveland (1977), pp. 195–209.

35 Deetlefs, M., Hardacre, C., Nieuwenhuyzen, M., Sheppard, O. and Soper, A.K., Structure of ionic liquid-benzene mixtures, *J. Phys. Chem. B* **109**, 1593–1598 (2005).

36 (a) Holbrey, J.D., Reichert, W.M., Nieuwenhuyzen, M., Sheppard, O., Hardacre, C. and Rogers, R.D., Liquid clathrate formation in ionic liquid-aromatic mixtures, *Chem. Commun.* 476–477 (2003); (b) Lachwa, J., Bento, I., Duarte, M.T., Lopes, J.N.C. and Rebelo, L.P.N., Condensed phase behaviour of ionic liquid-benzene mixtures: Congruent melting of a [emim][NTf_2]···C_6H_6 inclusion crystal, *Chem. Commun.* 2445–2447 (2006).

37 Meindersma, G.W., Hansmeier, A.R. and de Haan, A.B., Ionic liquids for aromatics extraction present status and future outlook, *Ind. Eng. Chem. Res.* **49**, 7530–7540 (2010).

38 (a) Holbrey, J.D., Lopez-Martin, I., Rothenberg, G., Seddon, K.R., Silvero, G. and Zheng, X., Desulfurisation of oils using ionic liquids: Selection of cationic and anionic components to enhance extraction efficiency, *Green Chem.* **10**, 87–92 (2008); (b) Bosmann, A., Datsevich, L., Jess, A., Lauter, A., Schmitz, C. and Wasserscheid, P., Deep desulfurization of diesel fuel by extraction with ionic liquids, *Chem. Commun.* 2494–2495 (2001).

39 Hardacre, C., Application of EXAFS to molten slats and ionic liquid technology, *Annu. Rev. Mater. Res.* **35**, 29–49 (2005).

40 (a) Cammarata, L., Kazarian, S.G., Salter, P.A. and Welton, T., Molecular states of water in room temperature ionic liquids, *Phys. Chem. Chem. Phys.* **3**, 5192–5200 (2001); (b) Kazarian, S.G. Briscoe, B.J. and Welton, T., Combining ionic liquids and supercritical fluids: In situ ATR-IR study of CO_2 dissolved in two ionic liquids at high pressures, *Chem. Commun.* 2047–2048 (2000).

41 Remsing, R.C., Swatloski, R.P., Rogers, R.D. and Moyna, G., Mechanism of cellulose dissolution in the ionic liquid 1-*n*-butyl-3-methylimidazolium chloride: A ^{13}C and $^{35/37}Cl$ NMR relaxation study on model systems, *Chem. Commun.* 1271–1273 (2006).

42 (a) Mele, A., Tran, C.D. and Lacerda, S.H.D., The structure of a room-temperature ionic liquid with and without trace amounts of water: The role of C–H···O and C–H···F interactions in 1-*n*-butyl-3-methylimidazolium tetrafluoroborate, *Angew. Chem. Int. Ed.* **42**, 4364–4366 (2003); (b) Mele, A., NOE experiments for ionic liquids: Tools and strategies, *Chim. Oggi* **28**, 48–55 (2010).

43 Hanke, C.G., Atamas, N.A. and Lynden-Bell, R.M., Solvation of small molecules in imidazolium ionic liquids: A simulation study, *Green Chem.* **4**, 107–111 (2002).

44 Shi, W. and Maginn, E.J., Molecular simulation of ammonia absorption in the ionic liquid 1-ethyl-3-methylimidazolium bis(trifluoromethylsulfonyl)imide ([emim][Tf_2N]), *AICHE J.*, **55**, 2414–2421 (2009).

45 Cadena, C., Anthony, J.L., Shah, J.K., Morrow, T.I., Brennecke, J.F. and Maginn, E.J., Why is CO_2 so soluble in imidazolium-based ionic liquids? *J. Am. Chem. Soc.* **126**, 5300–5308 (2004).

46 Harper, J.B. and Lynden-Bell, R.M., Macroscopic and microscopic properties of solutions of aromatic compounds in an ionic liquid, *Mol. Phys.* **102**, 85–94 (2004).

47 (a)Swatloski, R.P., Spear, S.K., Holbrey, J.D. and Rogers, R.D., Dissolution of cellulose with ionic liquids, *J. Am. Chem. Soc.* **124**, 4974–4975 (2002); (b)Zhang, H., Wu, J., Zhang, J. and He, J., 1-Allyl-3-methylimidazolium chloride room temperature ionic liquid: A new and powerful nonderivatizing solvent for cellulose, *Macromolecules* **38**, 8272–8277 (2005); (c)Schlufter, K., Schmauder, H.-P., Dorn, S. and Heinze, T., Bacterial cellulose in the ionic liquid 1-*n*-butyl-3-methylimidazolium chloride, *Macromol. Rapid Commun.* **27**, 1670–1676 (2006).

48 Sun, N., Rahman, M., Qin, Y., Maxim, M.L., Rodriguez, H. and Rogers, R.D., Complete dissolution and partial delignification of wood in the ionic liquid 1-ethyl-3-methylimidazolium ethanoate, *Green Chem.* **11**, 646–655 (2009).

49 Fukaya, Y., Sugimoto, A. and Ohno, H., Superior solubility of polysaccharides in low viscosity, polar, and halogen-free 1,3-dialkylimidazolium formates, *Biomacromolecules* **7**, 3295–3297 (2006).

50 Youngs, T.G.A., Holbrey, J.D., Deetlefs, M., Nieuwenhuyzen, M., Gomes, M.F.C. and Hardacre, C., A molecular dynamics study of glucose solvation in the ionic liquid 1,3-dimethylimidazolium chloride, *ChemPhysChem.* **7**, 2279–2281 (2006).

51 Youngs, T.G.A., Holbrey, J.D. and Hardacre, C., A molecular dynamics study of glucose solvation in the ionic liquid 1,3-dimethylimidazolium chloride, *J. Phys. Chem. B* **111**, 13765–13774 (2007).

52 Remsing, R.C., Petrik, I.D., Liu, Z. and Moyna, G., Comment on 'NMR spectroscopic studies of cellobiose solvation in EmimAc aimed to understand the dissolution mechanism of cellulose in ionic liquids' by Zhang, J., Zhang, H., Wu, J., Zhang, J., He, J. and Xiang, J. Phys. Chem. Chem. Phys. 2010, 12, 1941, *Phys. Chem. Chem. Phys.* **12**, 14827–14828 (2010), plus references therein.

53 Zhang, J., Zhang, H., Wu, J., Zhang, J., He, J. and Xiang, J., NMR spectroscopic studies of cellobiose salvation in EmimAc aimed to understand the dissolution mechanism of cellulose in ionic liquids, *Phys. Chem. Chem. Phys.* **12**, 1941–1947 (2010).

54 Amigues, E., Hardacre, C., Keane, G., Migaud, M. and O'Neill, M., Ionic liquids – media for unique phosphorus chemistry, *Chem. Commun.* 72–74 (2006).

55 Amigues, E.J., Hardacre, C., Keane, G. and Migaud, M.E., Solvent-modulated reactivity of PCl_3 with amines, *Green Chem.* **10**, 660–669 (2008).

56 Amigues, E.J., Hardacre, C., Keane, G., Migaud, M.E., Norman, S.E. and Pitner, W.R., Selective synthesis of chlorophosphoramidites using ionic liquids, *Green Chem.* **11**, 1391–1396 (2009).

57 Holbrey, J.D., Hughes, K., Youngs, T.G.A. and Hardacre, C., Unpublished results.

58 (a) Yoon, K.B. and Kochi, J.K., Shape-selective access to zeolite supercages – arene charge-transfer complexes with viologens as visible probes, *J. Am. Chem. Soc.* **111**, 1128–1130 (1989); (b) Yoon, K.B. and Kochi, J.K., Shape-selective modulation of ion-pair and triple-ion equilibria in zeolites – charge-transfer salts of methylviologen iodides, *J. Phys. Chem.* **95**, 1348–1356 (1991).

59 Hardacre, C., Holbrey, J.D., Mullan, C.L., Nieuwenhuyzen, M., Reichert, W.M., Seddon, K.R. and Teat, S.J., Ionic liquid characteristics of 1-alkyl-*n*-cyanopyridinium and 1-alkyl-*n*-(trifluoromethyl)pyridinium salts, *New J. Chem.* **32**, 1953–1967 (2008).

60 Hardacre, C., Holbrey, J.D., Mullan, C.L., Nieuwenhuyzen, M., Youngs, T.G.A., Bowron, D.T. and Teat, S.J., Solid and liquid charge transfer complex formation between 1 methylnaphthalene and 1 alkyl cyanopyridinium bis{(trifluoromethyl) sulfonyl}imide ionic liquids, *Phys. Chem. Chem. Phys.* **12**, 1842–1853 (2010).

61 Canongia Lopes, J.N.A. and Pádua, A.A.H., Nanostructural organization in ionic liquids, *J. Phys. Chem. B* **110**, 3330–3335 (2006).

62 (a) Urahata, S.M. and Ribeiro, M.C.C., Structure of ionic liquids of 1-alkyl-3-methylimidazolium cations: A systematic computer simulation study, *J. Chem. Phys.* **120**, 1855–1863 (2004); (b) Wang, Y.T. and Voth, G.A., Unique spatial heterogeneity in ionic liquids, *J. Am. Chem. Soc.*, **127**, 12192–12193 (2005); (c) Wang, Y.T. and Voth, G.A.,Tail aggregation and domain diffusion in ionic liquids, *J. Phys. Chem. B* **110**, 18601–18608 (2006); (d) Schroeder, C., Rudas, T. and Steinhauser, O., Simulation studies of ionic liquids: Orientational correlations and static dielectric properties, *J. Chem. Phys.* **125**, 244506 (2006); (e) de Andrade, J., Boes, E.S. and Stassen, H., Liquid-phase structure of dialkylimidazolium ionic liquids from computer simulations, *J. Phys. Chem. B* **112**, 8966–8974 (2008); (f) Habasaki, J. and Ngai, K.L., Heterogeneous dynamics of ionic liquids from molecular dynamics simulations, *J. Chem. Phys.* **129**, 194501 (2008); (g) Bhargava, B.L., Devane, R., Klein, M.L. and Balasubramanian, S., Nanoscale organization in room temperature ionic liquids: A coarse grained molecular dynamics simulation study, *Soft Matter* **3**, 1395–1400 (2007).

63 Bhargava, B.L., Balasubramanian, S. and Klein, M.L., Modelling room temperature ionic liquids, *Chem. Commun.* 3339–3351 (2008).

64 Triolo, A., Russina, O., Bleif, H.-J. and Di Cola, E., Nanoscale segregation in room temperature ionic liquids, *J. Phys. Chem. B* **111**, 4641–4644 (2007).

65 Bradley, A.E., Hardacre, C., Holbrey, J.D., Johnston, S., McMath, S.E.J. and Nieuwenhuyzen, M., Small-angle X-ray scattering studies of liquid crystalline 1-alkyl-3-methylimidazolium salts, *Chem. Mater.* **14**, 629–635 (2002).

66 Kuhrmi, C. and Berg, M.A., Dispersed kinetics without rate heterogeneity in an ionic liquid measured with multiple population-period transient spectroscopy, *J. Phys. Chem. Lett.* **1**, 161–164 (2010).

67 Hu, Z. and Margulis, C. J., Heterogeneity in a room-temperature ionic liquid: Persistent local environments and the red-edge effect, *Proc. Natl. Acad. Sci. U. S. A.* **103**, 831–836 (2006).

68 (a) Russina, O., Triolo, A., Gontrani, L., Caminiti, R., Xiao, D., Hines Jr., L.G., Bartsch, R.A., Quitevis, E.L., Plechkova, N. and Seddon, K.R., Morphology and intermolecular dynamics of 1-alkyl-3-methylimidazolium bis{(trifluoromethane)

sulfonyl}amide ionic liquids: Structural and dynamic evidence of nanoscale segregation, *J. Phys. Condens. Matter* **21**, 424121 (2009); (b)Turton, D.A., Hunger, J., Stoppa, A., Hefter, G., Thoman, A., Walther, M., Buchner, R. and Wynne, K., Dynamics of imidazolium ionic liquids from a combined dielectric relaxation and optical Kerr effect study: Evidence for mesoscopic aggregation, *J. Am. Chem. Soc.* **131**, 11140–11146 (2009).

69 Hardacre, C., Holbrey, J.D., Mullan, C.L., Youngs, T.G.A. and Bowron, D.T., Small angle neutron scattering from 1-alkyl-3-methylimidazolium hexafluorophosphate ionic liquids ($[C_n mim][PF_6]$, $n = 4, 6$, and 8), *J. Chem. Phys.* **133**, 074510 (2010).

5 Molecular Modelling of Ionic Liquids

JOSÉ N. CANONGIA LOPES

Centro de Química Estrutural, Instituto Superior Técnico, Universidade de Lisboa, Lisboa, Portugal and Instituto de Tecnologia Química e Biológica, Universidade Nova de Lisboa, Oeiras, Portugal

MARGARIDA COSTA GOMES and AGILÍO A. H. PÁDUA

Institut de Chimie de Clermont-Ferrand, Université Blaise Pascal and CNRS, Aubière, France

5.1 INTRODUCTION

Molecular modelling and simulation include computational techniques developed within the frameworks of quantum, molecular or statistical mechanics; they are able to analyse the links between the properties and macroscopic behaviour of matter and its characteristics at a molecular level. Modelling and simulation studies are traditionally used either as predictive or interpretative research instruments, but in the case of ionic liquids – a relatively recent research area – they have also assumed the role of exploratory tools leading to some discoveries that were only later corroborated by experimental evidence. In other words, the knowledge concerning the physical chemistry of ionic liquids is rapidly advancing through the interplay between experiments, theory and modelling, each providing challenges, guidelines and checks to the others. In the following sections, we will analyse the evolution of this fruitful interplay and illustrate how many of the breakthroughs in this area have been revealed to the scientific community at the first three COIL fora.

5.2 FORCE FIELD DEVELOPMENT: FELDER/畑 (HATAKE)/PADDOCKS

Molecular modelling starts with the definition of a suitable force field capable of describing the intra- and intermolecular interactions taking place between the molecular components that constitute the systems to be studied.

Ionic Liquids Completely UnCOILed: Critical Expert Overviews, First Edition.
Edited by Natalia V. Plechkova and Kenneth R. Seddon.

At the beginning of the century, the number of models capable of describing the molecular characteristics of ionic liquids was very limited and fragmented – only a few ionic liquids had been studied, on an almost case-by-case basis [1–4]. When Canongia Lopes and Pádua (two of the authors of the present contribution) developed and introduced their own force field (CL&P) [5], their main goal was to provide a systematic model that could be generalised to describe entire families of ionic liquids. In order to meet that objective and to take into account the 'modular' nature of ionic liquids (where one can recombine different anions and cations to yield new ionic liquids), three basic specifications were built into the model: internal consistency, transferability and compatibility (Fig. 5.1). The underlying rationale was that, since ionic liquids were a new field of study and these compounds can exist in enormous variety, it made more sense to have a more general, though less precise model, instead of a more meticulous model that would represent just one specific ionic liquid. This somewhat flexible approach proved to be quite successful – nowadays, the CL&P force field is one of the most widely used parameterisations for the molecular simulation of

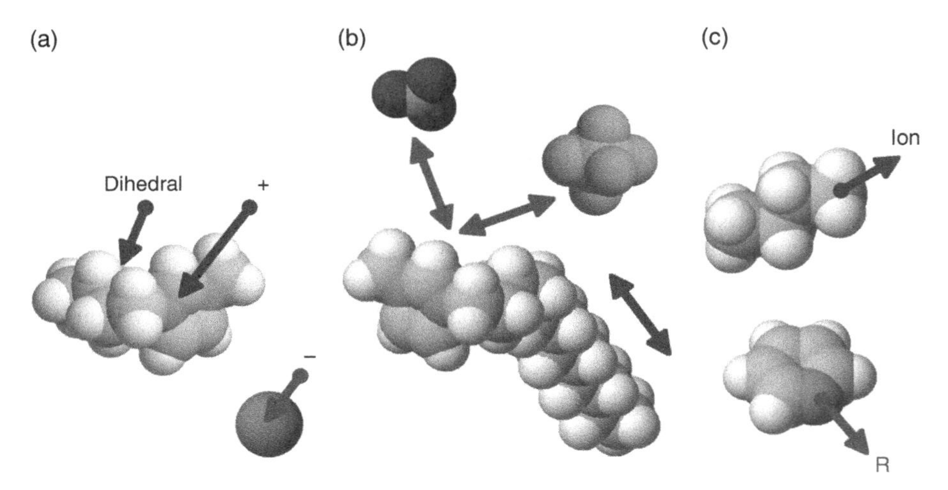

Figure 5.1 Basic specifications of a general, systematic force field for ionic liquids. (a) Internal consistency: anions and cations are parameterised with the same force field functional, with special attention given to the parameterisation of atomic partial point charges and the flexibility (dihedral angles) of the ions. The transfer of parameters from different sources is also checked for adequacy (for instance, the parameters for the chloride ion are taken from molten salt data, not from aqueous solution results). (b) Transferability: force field parameters are valid within the same homologous family (for instance, when modelling ions with different alkyl side-chain lengths – bottom double arrow) and allow the possibility of ion interchange to yield different ionic liquids (top double arrows). (c) Compatibility: molecular residues and moieties are taken directly from well-established force fields like OPLS [6]. Simple rules are established to join seamlessly neutral molecules to an existing ion (top arrow) or an ion to a neutral residue (bottom arrow).

ionic liquids. The five articles [5, 7–10] that currently describe the force field (two in 2004, 2006, 2008 and 2010) have been cited more than 450 times and, curiously, the history of COIL is intertwined with that of the CL&P force field: the latter was presented for the first time as an invited oral presentation in a major conference at COIL-1 (Salzburg, 2005), and since then, every 2 years, an extension of the force field has been released matching in opposition of phase – even *versus* odd years – the biennial character of COIL.

Other important contributions in the area of force field parameterisation for ionic liquids have also appeared at different editions of COIL.

Some of the contributions that were presented at COIL-1 represented force field developments similar to the CL&P force field applied to other sets of ions. These included either united-atom (UA) or all-atom (AA) parameterisations based on extensions of well-established force fields like CHARMM [11], AMBER [12] or OPLS [6], with extra parameters obtained ab initio. As examples, we can cite the work based on the UA force field of Maginn and co-workers [3] used by Maurer and co-workers to predict gas solubilities [13], the structure and dynamics studies by Mauro Ribeiro using his own UA force field [14] or the introduction by Maginn and co-workers of a new AA version of their own force field, used to predict the solubility of CO_2 in imidazolium ionic liquids [15]. These examples reflect a shift from UA models to AA models, attempts to encompass whole families of ionic liquids (breaking with the past tradition of modelling ionic liquids on a one-to-one basis) and some extra care in incorporating into the model some degree of flexibility (the description of the torsional movements of the ions is attempted in some cases). These advances show, in fact, a general convergence between the different approaches to force field development, in agreement with the basic ideas behind the development of CL&P (Fig. 5.1).

In contrast, del Popolo et al. presented at COIL-1 a model based on ab initio molecular dynamics (MD) simulation of ionic liquids [16]. This is a departure from the more traditional approaches mentioned in the previous paragraph, with enormous potential for development both in terms of accuracy and applicability of the model. However, only a handful of rather simple ionic liquids have been modelled so far, due to the higher computational costs and parameterisation complexity associated to the model.

A special mention must also be made at this point to empirical potential structure refinement (EPSR) models that are usually used to interpret X-ray diffraction and extended X-ray absorption fine structure (EXAFS) spectroscopy data. MD simulations of ionic liquids using these models were presented at COIL-1 by Hardacre and co-workers [17]. These are not traditional force fields, in the sense that they are developed to match structural data and thus generally lack information concerning intramolecular parameterisation (which is already provided by the structural data itself). Their contribution to the modelling of ionic liquids can be very important as far as structural properties are concerned but are limited otherwise.

COIL-2 witnessed the emergence of the modelling of ionic liquids plus molecular solvent systems. In terms of force field development, this means that the premises of transferability/compatibility claimed by most models have to be valid not only between the ions of the ionic liquid but also between those ions and molecular (neutral) species. AA models based on traditional force fields (CHARMM [11], AMBER [12] or OPLS [6]) have a clear advantage in addressing this problem, since most of the neutral species are already parameterised within those force fields, which means that most of the modelling studies presented at COIL-2 were based on these types of force field. This includes presentations by Rebelo and Canongia Lopes [18] based on work using the CL&P force field [19], the discussion of solvation effects in ionic liquids by Song [20] using MD results obtained by Shim et al. using a CHARMM-based rigid force field [21], the analysis of ionic liquid–solute interactions by Hardacre and the analysis of the interplay between solvation and nanostructure by Padua, both based on the CL&P force field [5, 7, 8, 22].

The modelling of pure ionic liquids was also pursued in various other areas using different force fields: transport properties were analysed by Maginn et al. [23] using their own AA CHARMM-based force field (with some input information from the CL&P force field as far as the parameterisation of bistriflamide anions was concerned [7]), the nano-heterogeneity in pure ionic liquids was addressed by Hu and Margulis [24] (see also Section 5.3) using an OPLS-based force field developed by Berne [4], and Umebayashi et al. [25] addressed the issue of molecular conformations in ionic liquids using the CL&P force field [7].

Finally, important contributions concerning the development of new or refined force fields were given by Ludwig and co-workers [26] and Siehl [27]. The latter author introduced the possibility of modelling charge transfer between ions as a way to account for polarisability effects in ionic liquids and their solutions; the former authors produced a refined version of the CL&P force field [5, 7] capable of taking into account the specific interactions between the acidic hydrogen atoms of 1,3-dialkylimidazolium cations and various anions. These contributions represent valid routes to the development of force fields that are fine-tuned to certain classes of ionic liquids, where problems such as polarisability, charge transfer, hydrogen bonding or the occurrence of specific interactions may hinder the use of more general models.

In COIL-3, ionic liquid modelling witnessed a shift from bulk (albeit non-isotropic) conditions to non-equilibrium or interfacial settings: Leuw [28] discussed the interfacial structure of ionic liquids using X-ray reflectometry studies aided by a self-consistent mean field theory (SCMFT) model and MD simulations using a polarisable force field based on AMBER parameterisations; Grant [29] presented a polarisable, transferable and quantum-chemistry-based force field denominated as Atomistic Polarisable Potential for Liquids, Electrolytes and Polymers (APPLE&P) and used it to characterise ionic liquids that can be used as electrolytes for batteries and supercapacitors; Voth [30] has shown how different properties of ionic liquids (such as interfacial structure, self-diffusion and viscosity) can be simulated using coarse-grained models, namely, the effective force coarse-graining (EF-CG) method; Daily and Micci [31] used an AMBER-based AA [2] and the EF-CG [30] force fields to study ionic liquids under non-equilibrium conditions (high electric fields); Kislenko et al. [32] studied the structure of the electrochemical interface between a graphite surface and an ionic liquid using an AA AMBER-based force field; Watanabe and collaborators [33] studied the self-diffusion coefficients in a series of ionic liquids using an adapted and extended version of the CL&P force field; and finally, Zhang and co-workers [34] extended their own AA AMBER-based force field to encompass amino-functionalised ionic liquids and used it to model different properties of these ionic liquids, including self-diffusion coefficient.

Simulation work on the structure of bulk ionic liquids and their interactions with molecular solvents was also present at COIL-3: Umebayashi and co-workers [35] used extended versions of the CL&P force field [5] to interpret large-angle X-ray scattering experimental data, Maginn et al. [36] used Monte Carlo simulations to compute pure and binary gas isotherms in ionic liquids using parameters from their own force field [23] to model the cations and the CL&P force field to model the bistriflamide anion [7], and Santini et al. [37] used this last force field [7] to study the solvation of toluene in different ionic liquids in order to interpret their specific catalytic capabilities.

Finally, Izgorodina [38] revisited the issue of ab initio methods as an alternative and effective way to perform simulations in ionic liquids. Other empirical models based on group contribution, parachor or neural network methods were also presented at COIL-3 [39–41]. These are not models to be used within the context of computer simulation and the frameworks of quantum, molecular or statistical mechanics. Albeit very important and useful within their areas of application (property prediction, system tailoring), they fall outside the scope of the present chapter and will not be further discussed.

The schematic representation of Figure 5.2 summarises the development, and use, of different types of force field for ionic liquids, presented and discussed at the different COIL meetings. Panel A shows force fields with different degrees of resolution, from coarse-grained to atomistic models. AA models run through the complete timeline due to their general character and the straightforward way that they can be integrated with vast force field parameterisations for common molecules. UA models (halfway between AA and coarse-grained models) are losing popularity, with most simulations being run at the atomistic or coarse-grained levels. The latter types of models are specially suitable for non-equilibrium conditions, but sometimes lack the detail necessary to the correct description of specific interactions, a situation that is particularly relevant in the case of mixtures of ionic liquids with molecular solvents. Ab initio simulation methods represent the ultimate models in terms of 'resolution' but are costly in terms of computing time and were only applied to a few systems. In fact, this type of model represents a sort of transition between panel A, where the degree of atomic description of the models is schematically presented, and panel B, where different types of correction/approximations are applied to those descriptions.

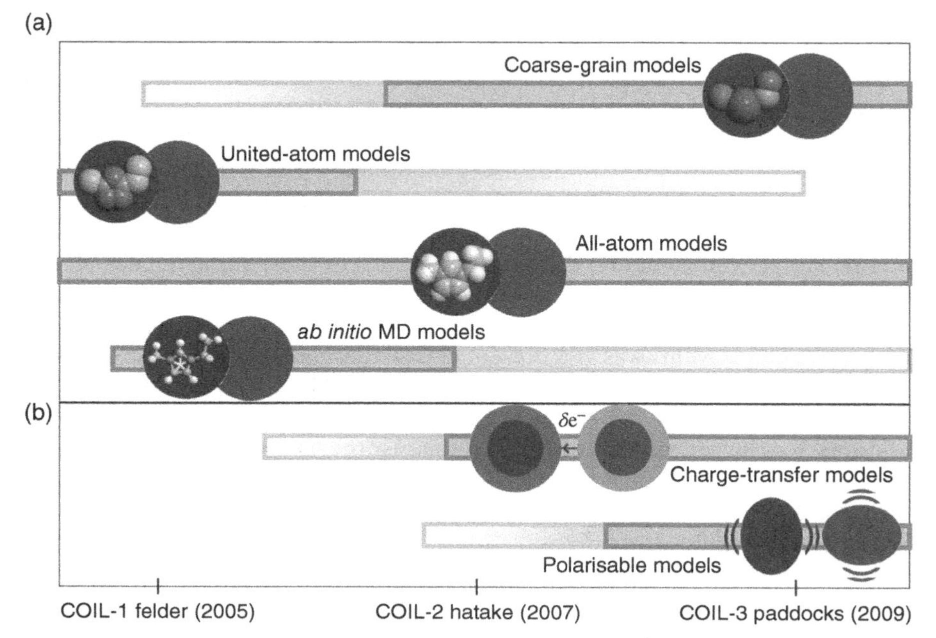

Figure 5.2 Schematic chronology of the main different types of force field models used to simulate ionic liquid systems. Panel (a) describes models with different degrees of resolution, whereas panel (b) shows two of the most common methods used to refine the models of panel (a). All models are assumed to take into account the flexible nature of most ionic liquid ions.

In order to improve transferability (see Fig. 5.1), the parameterisation of ionic liquids in systematic atomistic force fields was not meant to be too specific because the objective was to model families of homologous ions that could be combined with different counter-ions. This meant that, on one hand, the force field parameterisation was concentrated on parts of the ions that defined entire homologous series and, on the other hand, that the anions and cations were parameterised independently. This sort of transferability implies different types of approximation. For example, it does not account for the possibility of charge-transfer effects between different ions (the charges are usually obtained *ab initio* from isolated ions). In the condensed phase, where each ion is surrounded by neighbours of opposite charge, such an effect may not have a profound effect in terms of structure, but will certainly play a role in terms of the intensity of possible specific interactions between ions and the formation of (transient) ion pairs [26]. Also, polarisation of electron clouds is not taken into account explicitly, although was included a posteriori through a number of refinement schemes [28, 29]. Inclusion of explicit polarisation accelerates the microscopic dynamics obtained with the models but has little effect on the estimation of equilibrium or structural properties [42]. The introduction of these two types of correction/approximation (charge-transfer and polarisation methods) represents two major routes to the refinement of systematic (either atomistic or coarse-grained) models.

As a corollary to the present chapter, it must be stressed that Barbara Kirchner presented a contribution at COIL-3 summing up most of the theoretical work performed on ionic liquid systems over the last decade: this work has been recently published in a volume of the series *Topics in Current Chemistry* [43].

5.3 FLUID-PHASE ORGANISATION: STRUKTUR/KŌZŌ/STRUCTURE

Ionic liquids differ from molecular fluids in their microscopic liquid-state structure: in molecular fluids, ordering is primarily determined by packing considerations or by specific interactions such as multipolar forces or hydrogen bonds. But in ionic systems, electroneutrality is an overwhelming driving force that imposes charge ordering [44]. When looking at the distribution of ions surrounding a specific cation or anion, one finds successive, alternating layers of counter-ions and co-ions, which become less ordered as the distance from the central ion increases. This charge ordering is sometimes described as a remnant of the crystal structure, but it would be more logical to perceive both the solid and liquid structures as resulting from the same underlying electrostatic interactions.

Although the basic charge ordering exists both in inorganic, high-temperature molten salts and in room-temperature ionic liquids, the latter have much richer structures (see Chapter 3). The additional structural features of ionic liquids, which nonetheless must comply with the unavoidable charge ordering, are due to the more complex organic ions present: in order to obtain a salt that does not crystallise easily, one or both the composing ions have to be large, asymmetric and flexible, have delocalised electrostatic charge and possess significant non-polar moieties, or combinations of these factors. Specific interactions such as hydrogen-bond or π-interactions also give rise to positional and orientational correlations that mark the short-range structure of many ionic liquids, just as they do so in molecular fluids. The presence of covalently bonded ionic and non-polar chemical groups in the same ion, such as an imidazolium ring with an alkyl side chain, also leads to remarkable structures in the liquid state, since these antagonistic moieties have to coexist in a highly cohesive ionic medium while always respecting the underlying charge ordering. The COIL congresses have spanned the crucial period of discovery and investigation by physical chemists of many of the interesting and unique structural features of ionic liquid phases and were the venue for the presentation and discussion of a large number of important findings.

Subsequent to the prevalent charge ordering, the structure of room-temperature ionic liquids may be the result of specific associating interactions. Hydrogen bonding between cations and anions (and also with solutes) has been identified as an important structuring mechanism in ionic liquids [45] – and also in their crystalline states – although many of the main families of these compounds do not form hydrogen bonds, for example, quaternary ammonium or phosphonium salts, to name just a few. Even in 1,3-dialkylimidazolium cations, hydrogen bonding between the anions and the hydrogen atoms of the imidazolium ring becomes less significant as the size of the anion increases, and thus, its density of negative charge diminishes [46].

Certain common families of ionic liquids are based on aromatic cations, namely, imidazolium or pyridinium, and their spatial ordering is affected by interactions involving π-systems. The structure of solutions of benzene in imidazolium ionic liquids has been studied both experimentally by neutron diffraction and through simulation [47]. Inclusion compounds have been identified in the solid–liquid phase diagrams [48] and the role of cation–π-interactions confirmed, also in the structure of solutions of toluene in ionic liquids [37].

Specific interactions involving hydrogen bonds and π-electrons induce orientational and spatial ordering among neighbouring ions that propagate throughout the crystal structures, but in the liquid phases, these are essentially short-range correlations, that is, up to a few ångström. These features are not fundamentally different from what is observed in molecular fluids possessing the same kinds of interaction.

The complexity of most ionic liquids goes beyond this scale. One of the remarkable structural features of room-temperature ionic liquids occurs when non-polar side chains or moieties of significant, but not too large, size are present in one or both the ions. If these moieties are small, they will have little impact on the liquid structure [17, 46], which will resemble very much that of a simple molten salt – dominated by charge ordering – with added short-range correlations due to specific interactions. At the other extreme, if alkyl side chains are sufficiently long, then anisotropy will appear in the form of mesophases and the ionic fluid will be a liquid crystal [49]. But side chains of intermediate length will have to be accommodated in the structure of a highly cohesive ionic fluid. The non-polar moieties will tend to segregate from the cohesive ionic head groups that hold most of the electrostatic charge density. The liquid will still be isotropic but heterogeneous at length scales of the order of a nanometre [50]. The structure is characterised by spatial domains resulting from tail aggregation and by a network of cation–anion contacts, the disruption of which is energetically unfavourable. If the side chains are long enough, eventually, the non-polar domains may coalesce. The overall picture is very similar to that of melts of diblock copolymers with antagonistic groups: two 'immiscible' moieties are covalently bonded, leading to microphase separation and thus to a nanostructured liquid phase. Different topologies are possible, with a variety of either dispersed of percolating microphases (Figure 5.3) [51].

Of course, preferential interactions between side chains on one side and ionic head groups on the other will be seen in the crystal structures [52], since these lead to the minimum of free energy in the solid, as well as in the liquid. Once again, instead of considering the heterogeneous structure of the liquid

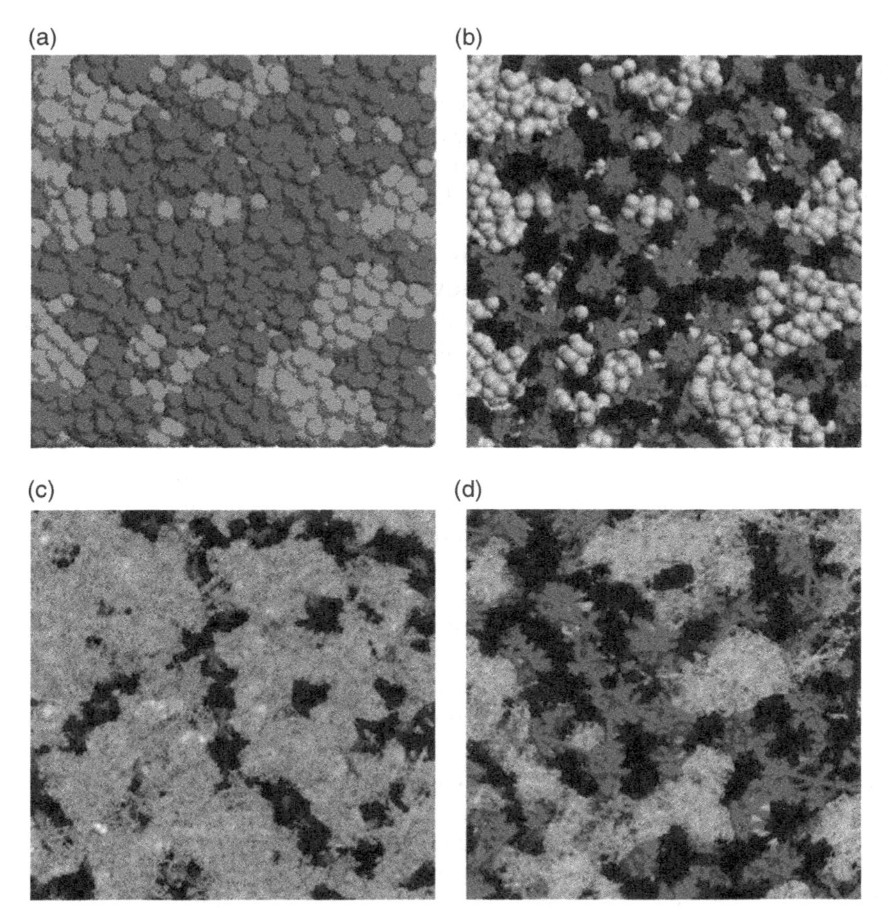

Figure 5.3 Molecular dynamics simulation snapshots showing the nanostructured nature of different ionic liquids: (a) the 'red-and-green' colour coding showing the segregation between polar (dark grey) and non-polar (light grey) domains in 1-hexyl-3-methylimidazolium hexafluorophosphate; (b) an alternative rendering of the same snapshot showing the charge ordering in the polar domain, with the anions represented in black and the polar parts of the cations in dark grey. The non-polar domains are depicted in light grey; the same rendering is used in (c) and (d) to depict charge ordering and the different segregation patterns in *N*-(tetradecyl)-*N*,*N*,*N*-trihexylammonium chloride, $[N_{66614}]Cl$ (filamentous-like morphology of the polar part), and 1-tetradecyl-3-methylimidazolium bistriflamide, $[C_{14}mim]Cl$ (globular-like morphology), respectively.

simply as a remnant of that of the crystal, it is more logical to understand the structures of both phases as an outcome of the same underlying balance between electrostatic and van der Waals forces.

It is important to realise that the system is fluid, and therefore, the structural heterogeneities give also rise to dynamic ones. The interplay between a structural view and a dynamic one is recurrent in complex fluids, relating the behaviour of transport properties or relaxation phenomena with 'static' concepts such as free volume, molecular association or heterogeneous structures. A more complete understanding of the physics of these systems requires a combination of both the structural and dynamic information, and in the ionic liquids, community significant effort has been undertaken on both sides of the question.

In COIL-1, all of the main aspects of the liquid-state structure of room-temperature ionic liquids were already laid out in contributions by several groups, although sometimes still in an incipient way. Hardacre and co-workers reported neutron diffraction studies of short-chain dialkylimidazolium ionic liquids, perfectly illustrating the charge ordering and π-interactions [17]. They also used different spectroscopic and simulation methods to study how the solvation of aromatic and polar molecules (benzene and ethanenitrile, respectively) in the ionic liquids modifies the structure of the media [46] and the balance between the different terms in the interactions: coulombic, van der Waals (dispersive), hydrogen bonds and multipolar.

2005 was the year in which the nanostructuring of ionic liquids into non-polar and ionic domains was postulated by several groups using molecular simulation. Ribeiro and co-workers identified pre-peaks in the structure factors of alkylimidazolium ionic liquids with varying side-chain lengths [14], from methyl to octyl. These pre-peaks indicate intermediate-range ordering related to the increasing length of the alkyl chains. The focus of Ribeiro's work is two-fold, with analyses of both structure and dynamics of the systems, namely, the anisotropic displacement of the cations parallel or perpendicular to the imidazolium rings. Canongia Lopes and Padua presented the now famous 'red and green' picture of the nano-segregated structure of ionic liquids, in which a colour code allows visual recognition of the ionic and non-polar domains [50]. The highly cohesive nature of ionic liquids is translated into the very small free volume in these systems, probed by the population distribution of spontaneous cavities, obtained from molecular simulation. The solvation of CO_2 and other gases is decomposed into contributions from the different terms in the interactions [53].

Watanabe reported on the concept of ionicity, which is a ratio between the ionic conductivity measured and that deduced from NMR diffusion coefficient data (that accounts for total mobility of ions, even in the form of associated 'pairs' or clusters that do not contribute to charge transport). Ionicity is less than unity if potential charge carriers are not available for transport, and thus reflects the degree of ionic association in the liquid. Transport properties such as viscosity, diffusion coefficient and ionicity do not vary monotonically as the alkyl side chain increases, which is consistent with the appearance of nano-segregated structures at intermediate chain lengths [54].

With a larger scope, turned towards nanostructured materials, Antonietti and co-workers make use of the strong structure of ionic liquids, here ionic liquid crystals, as templates for the synthesis of mesoporous nanomaterials.

The link between the length of the alkyl side chains, the presence of hydrogen bonds and other factors and the nanostructured features of the ionic liquids were established, though in a semi-empirical way, but the vision is clearly there that ionic liquids as nanostructured solvents can have a major role in the synthesis of nanoparticles and other nanomaterials [55].

In COIL-2, the contributions of most research groups had integrated the structural and dynamic complexity of ionic liquids, and many results were interpreted in the context of nanostructured, heterogeneous liquids. Triolo and co-workers presented the first direct experimental evidence of the structural heterogeneities in pure ionic liquids: the pre-peaks in their X-ray structure factors indicate that the heterogeneities increase in size linearly with the length of the side chains in imidazolium ionic liquids [56], results that corroborate the MD simulations of Canongia Lopes and Padua [50]. Deki and co-workers also reported small-angle X-ray scattering results with clear pre-peaks in alkyltrimethylammonium ionic liquids having alkyl chain lengths from 2 to 12 carbon atoms. The pre-peaks become more intense as the chain length increases (indicating a higher degree of structuring) and are also displaced to lower wave vectors, therefore longer characteristic lengths [57].

Margulis and co-workers presented a perspective on solvation dynamics in ionic liquids based on the existence of persistent local environments. These heterogeneous structures manifest on the spectra of probe molecules, which show a 'red-edge' effect [58] similar to that observed in glassy media. The spectroscopic studies had been performed by Samanta and co-workers and were elegantly interpreted by Margulis using MD simulation [24]. Quitevis and co-workers presented a spectroscopic study using the optical Kerr effect that probes local environments and dynamics on mixtures of ionic liquids, showing the presence of heterogeneities also in ionic liquid mixtures [59].

Protic ionic liquids were the object of a structural X-ray study by Ishiguro and co-workers, but the compound chosen, ethylammonium nitrate, has small, simple ions, and the structure is determined by charge ordering and hydrogen bonding [60].

Hardacre and co-workers focussed on the interactions and solvation of D-glucose, as a model to understand the dissolution of biomacromolecules such as cellulose, for which certain ionic liquids are excellent solvents [22, 61]. Since the ionic liquid chosen, 1,3-dimethylimidazolium chloride, does not have non-polar groups, the structure of the solution is dominated by charge ordering and hydrogen bonds.

Padua and co-workers showed how the nanoscale segregation of ionic liquids into polar and non-polar domains defines the solvation of various kinds of molecular compound. The structural heterogeneities mean that the ionic liquid can offer different solvation environments to solutes that, according to their polarity or ability to form specific interactions such as hydrogen bonds, will interact preferentially with certain domains in the ionic liquid [19]. Non-polar solutes, such as hexane, will be solvated preferentially in the non-polar domains formed by aggregation of the alkyl side chains. Dipolar solutes, such as ethanenitrile, will be solvated near the ionic domain but also in contact with the non-polar ones (depending on their ability to intercalate between cation and anion head groups). Solutes such as methanol or water will form strong hydrogen bonds with the ionic head groups, and longer-chain alcohols are expected to have a dual behaviour. This analysis of solute–solvent interactions in terms of the dominant terms in each of the heterogeneous domains brought a novel way to understand solvation in ionic liquids.

Spanning the composition range in a mixture of an ionic liquid with a molecular compound, different regimes of association will be found, from the concentrated ionic liquid containing diluted molecules of solute to systems in which the ionic liquid becomes a dilute electrolyte. The miscibility limit of a molecular compound in an ionic liquid may be determined by how much 'swelling' of the ionic network is allowed to happen before the system separates into two phases [62].

Structure of ionic liquids at interfaces, another fascinating and important subject, was studied by Baldelli and co-workers [63] and by Ouchi [64] using surface spectroscopic techniques. Non-polar alkyl chains are segregated towards vacuum at the free surface of ionic liquids, as a result of the balance between coulombic and van der Waals forces [63, 64].

The link between the self-organisation of ionic liquids and the synthesis of nanostructured materials was established in a series of contributions. Dupont discussed the effect of the three-dimensional nanoscale arrangement of ionic liquids on the production and stabilisation of transition metal nanoparticles [65]. As shown also by Santini and co-workers, metal nanoparticles of several metals can be synthesised *in situ* in ionic liquids, leading to narrow size distributions and spontaneously stable suspensions without the need for additional surface-active agents [66].

Firestone and co-workers showed the effect of self-assembly in mixtures of ionic liquids with water, in which persistent structures serve as a template for the synthesis of colloidal particles and more sophisticated structures exploiting the architecture of the mesophases [67]. Other outstanding contributions using the same templating strategy relied on ionic liquid crystals, but here the ionic liquid nature of the systems is not the main point, but it is rather that they form liquid crystal phases [68].

By COIL-3, the segregated structure of ionic liquids in nanometre-size polar and non-polar domains had become so mainstream that it was the focus of specific research aimed either to better understand the structural features – by proving or disproving them – or to use such self-organisation in order to perform something new. A great many contributions related findings and results to the heterogeneous structure of the ionic liquid phases.

The relations between liquid structure and solvation of molecular compounds were studied from a variety of angles. For instance, the oral presentation by Hardacre at COIL-3 described charge-transfer complexes between functionalised ionic liquids and aromatic solutes, whereas the one presented by Hamaguchi showed three different solvation modes for I_2, indicating the presence of three different solvation sites, or local environments in $[C_4mim][NTf_2]$ as a result of nano-heterogeneities. Other research efforts, making use of probe molecules to investigate the nanostructures, are those of Diamond and co-workers, who used photochromic dyes [69], and Iwata, who showed that the vibrational relaxation of probe molecules is not correlated with the macroscopic properties of ionic liquids such as thermal diffusivity, a result compatible with heterogeneous local structures [70]. Quitevis and co-workers compared the optical Kerr effect spectrum of CS_2 dissolved in pentane and in $[C_5mim][NTf_2]$, showing that the solvation environments of this small molecule are similar in the two fluids: the CS_2 solute molecules in the ionic liquid are isolated from each other in the non-polar domains [71].

Yao and co-workers used ultrasonic and microwave spectroscopy to relate non-Newtonian behaviour of ionic liquids with longer side chains with the existence of non-polar domain structures [72], while Triolo and co-workers presented further scattering data on the nanometre-scale heterogeneities of ionic liquids with alkyl chains of varying length [56].

Notable novelties concerning the structure of ionic liquids came from the contribution of Atkin and co-workers, who showed the existence of pre-peaks in the X-ray structure factors of protic ionic liquids (alkylammonium nitrates) with small side chains [73]. This intermediate-range ordering is likely to be due to different kinds of structural features than those observed in ionic liquids with longer side chains, but demonstrates once more how ionic liquids are full of surprises. Strong pre-peaks in other protic ionic liquids were also described in a number of contributions by Umebayashi and co-workers [60].

In another quite original contribution, Mele and co-workers showed evidence of fluorous domain segregation in ionic liquids, obtained in ionic liquids with pyrrolidinium cations and perfluorinated anions of the family of fluorosulfonylamides [74].

Concerning more complex systems such as colloidal suspensions of nanoparticles, Santini and co-workers reported studies on the influence of the size of the nanostructures in ionic liquids and the size distribution of ruthenium nanoparticles synthesised therein by reduction of a non-polar organometallic complex. The stabilisation mechanism proposed is based on a template effect linked to the structure of the ionic liquids [66, 75, 76], not to electrostatic stabilisation (due to DLVO-type forces as observed in colloidal suspensions in aqueous electrolytes). Watanabe and co-workers also observed that silica nanoparticle colloids couldn't be stabilised in ionic liquids without surface-grafted polymer chains, again indicating that DLVO-type forces are insufficient for effective stabilisation, as expected from the short screening length in these media composed mainly of ions [77].

The main directions of research uncovered in COIL-3 lead us to think that the concept of heterogeneous liquid-phase structures is being considered in the design and study of more diverse and complex systems, including protic and aprotic ionic liquids with more creative chemical structures, and in truly heterogeneous systems such as nanoparticle suspensions or those involving interfaces. As an example, Endres related the sophisticated nature of ionic liquids with the different deposit morphologies obtained by electrodeposition [78], pointing to the enormous potential of ionic liquids for the fabrication of nanostructured materials.

In this section, we have examined the structure of ionic liquid phases, including short-, medium- and long-range features. As usual when dealing with ionic liquids, the overall picture is not a simple one. These are truly complex fluids, and several structural features coexist, at coincident or distinctive length scales, to compose a multi-scaled picture. First and foremost, the liquid structure is defined by charge ordering. This is inescapable and imprints a 'strong', robust structure to the fluids. This robust structure is, arguably, one of the main reasons for the success of ionic liquids as designer solvents: the ions can be functionalised in view of specific applications, and the inclusion of different chemical groups does not easily destroy the defining physical properties of an ionic liquid, such as low volatility and wide liquid range. The same would not be possible with molecular liquids, which have a much 'weaker' structure, and in these compounds, the inclusion of essentially any functional group leads to major changes in physical properties.

Secondly, the heterogeneous medium-range structure of ionic liquids, which presents segregation between ionic and non-polar domains, appears as an interesting and original feature of ionic liquids. These structural heterogeneities are present in ionic liquids with non-polar side chains or moieties of intermediate length: too short and the structure is essentially determined by the charge ordering; too long and mesophases appear and we are in the realm of liquid crystals. But for intermediate lengths, segregation produces spatial heterogeneities of nanometre length. The heterogeneous domains mean that solutes of different polarity can find sites of preferential solvation in the ionic liquid. Also, a templating effect of the self-organised, nanostructured ionic medium can be exploited for the production of nanostructured materials, including synthesis and stabilisation of nanoparticle suspensions, mesoporous solids or gels.

Research on the heterogeneous, self-organised structure of ionic liquids, arguably one of several unique and fascinating properties of these materials, is a very active field, with detailed studies being undertaken on the nature of the structural (and dynamic) heterogeneities [52, 79–81]. A clearer picture emerges in which ionic liquids are not organised like micellar systems, are not like liquid crystals (provided the side chains are not too long), but are not simple molten salts either. This complexity, resulting from subtle balances in the molecular interactions, is the source of their appeal to physical chemists.

5.4 IONIC LIQUID MODELLING: EIGENSCHAFTEN/SEISHITSU/PROPERTIES

Because ionic liquids are a relatively new field of research, a comprehensive body of experimental information to which theoretical studies can be anchored is still rather limited. From COIL-1 to COIL-3, substantial data were gathered on the macroscopic properties of ionic liquids and their mixtures. As in other fields, molecular simulation studies can provide reasonable predictions of properties, but especially they can offer insights on how the fluid properties depend on their behaviour at the molecular level. In the case of ionic liquids, molecular simulation also takes the role of an exploratory tool, leading to new discoveries and stimulating experimentalists to obtain new data. Knowledge about the physical chemistry of ionic liquids has advanced rapidly, as illustrated by the contributions of the international scientific community to the different COIL congresses.

At the time of COIL-1, several authors had confirmed the peculiar behaviour of some of the physico-chemical properties of ionic liquids, but the nanostructured nature of these fluids underlying that behaviour had not yet been evidenced, either by molecular simulation or experimentally. However, the existing atomistic force field [5] already contained a detailed description of the interactions and conformations needed to estimate properties such as viscosity, electrical conductivity or diffusion coefficients [82] and surface behaviour. Maginn and collaborators [83] and independently Maurer and co-workers [13, 84] proposed the first molecular simulations to predict gas solubility, phase behaviour (including melting points) and thermophysical properties of ionic liquids and their mixtures, still using simplified UA models. The results obtained were satisfactory, but the molecular mechanism underlying the macroscopic behaviours found could not yet be fully understood.

By the time COIL-2 took place in 2007, the nanostructured nature of the ionic liquids had been postulated using molecular simulation [50] and evidenced by indirect experimental data [54, 85] or by direct X-ray or neutron diffraction studies [56]. This microscopic vision of these fluids changed the way their physico-chemical properties could be explained. The concept of ionicity was supported by this microscopic vision, and indirect experimental evidence came from viscosity and conductivity measurements, as presented by Watanabe et al. [54, 86]. This molecular approach pointed towards alternative ways to probe the structure of ionic liquids, not by considering only the structure of the component ions but also by using external probes (e.g. neutral molecular species). Solubility experiments with selected solute molecules proved to be the most obvious experimental route: different molecular solutes, according to their polarity or tendency to form associative interactions, would not only interact selectively with certain parts of the individual ions but might also be solvated in distinct local environments in the ionic liquid.

Two distinct groups of experimental techniques were presented in COIL-2 to study the behaviour of different solutes in ionic liquids, spectroscopic and thermodynamic, giving access to different scales of the properties studied – one microscopic and the other macroscopic. It was possible to explain microscopically the phase behaviour of ionic liquid solutions by balancing the effects of the solute–solvent interactions and the dynamics of the solutions. Bases were established to assess the microscopic mechanisms responsible by the properties observed and so to open the way to the rapid advancement of the field contributing to the development of novel applications in a growing variety of disciplines including catalysis, synthesis, nanomaterial synthesis or pharmaceutics.

Maginn [87] and Margulis [24, 88] presented their studies on the transport properties of ionic liquids and explained why these media have to be regarded as a different class of solvents. Several other groups associated spectroscopic techniques and molecular simulation calculations to assess the interfacial behaviour of the solutions containing ionic liquids. The use of AA force fields allows an accurate balance between the specific intermolecular interactions and the structural/conformational effects responsible for the properties determined experimentally, whether they are thermodynamic or spectroscopic.

The solvation of neutral species in ionic liquids is distinct from those of charged ions because the solute–solvent interactions can be complex and the presence of neutral solutes can affect the structure of the solution. In COIL-2, several groups addressed solvation and its dynamics in ionic liquid media. Rebelo and collaborators [18] associated molecular simulation studies and phase equilibria measurements to explain the behaviour of pure ionic liquids and their mixtures, and Canongia Lopes et al. [19] explained how molecular solutes, depending on their molecular nature, are solvated in different molecular regions of the ionic liquids. Modelling the interactions of small ions with ionic liquids is very important for several applications, namely, for electrochemistry or energy storage, and Tsuzuki et al. addressed this issue from first principles calculations [89].

Several studies were presented concerning the solvation structure [22] and solvation dynamics [90] in ionic liquids, associating spectroscopic measurements with first principles calculations [91]. It was possible to identify different molecular mechanisms of solvation that confirm the concept that the molecular structure of the ionic solvent determines the way the solutes are dissolved in this media, the solvation sites and solute–solvent interactions depending on the molecular nature of the solute [92].

Equipped with a molecular view of ionic liquids, liquids as complex self-organised media, many different research groups presented in COIL-3 tentative explanations for their properties aimed at different applications, ranging from chemistry to nanomaterials. Other properties, not only phase equilibria, could now be predicted using mainly AA force fields. Important advances were made concerning the explanation of the mechanisms of solvation of various groups of solutes. For example, as shown by Canongia Lopes et al., non-polar molecules tend to be solvated in or near the non-polar domains as, due to the large cohesive energy between the charged groups, they are not able to integrate into the polar network. Dipolar solutes will in turn interact closely with the charged head groups of the ions, also showing affinity for the non-polar domains through their aliphatic moieties. Associating solutes will be able to form hydrogen bonds with the ions and become solvated in the ionic domain [19]. Aromatic compounds constitute a particular group [93] as they have significant quadrupole moments due to the aromatic electron cloud above and below the plane of the rings [94]. In benzene, negatively charged regions exist above and below the ring plane, and a rim of positive charges is found in the ring plane due to the hydrogen atoms. The mechanism of solvation of toluene in ionic liquids has been presented by Santini and co-workers [37], showing that when the solute interacts preferentially only with the cation or the anion (and is excluded from the non-polar domains), at a given concentration, the integrity of the polar network can be disrupted, and the system will phase-separate in order to avoid the energetic cost of separating the ions.

These general mechanisms allowed Maginn to predict the solubility of gaseous solutes and their mixtures in different ionic liquids using MD simulation with AA force fields [95]. This has shown that the prediction of the gas solubility in ionic liquids requires the use of more physically sound models based on good molecular descriptions rather than empirical theories of solvation, for example, regular solution theory. These are only useful for the correlation of gas solubility, and not for the prediction of this property. As stressed by Rebelo and collaborators [18], it is now possible to assess the phase diagrams of ionic liquids and their mixtures and to understand them from a molecular perspective. Different solvation mechanisms also lead to solutions with different microscopic structures. The microscopic mechanisms are confirmed by different experimental studies using Raman spectroscopy or neutron diffraction [46].

The microscopic structure of ionic liquids also determines the way mass transfer occurs in these media, as shown by Tsuzuki et al., who have shown by MD simulation that the translational dynamics of atoms in the non-polar domains is significantly different from that of atoms in the polar domains. The different motions of atoms in the two types of domain will be important for predicting the transport properties of ions and solute molecules in ionic liquids [33].

In order to fully explain the way in which molecular level structures and interactions bridge to macroscopic liquid-state properties, Voth stressed that a wide range of length and timescale calculations are required to predict different properties, like the interfacial tension, self-diffusion and viscosity [30]. Both Voth and Borodin demonstrated the necessity of using polarisable force fields to accurately predict several properties of ionic liquids and their mixtures, namely, the transport of ionic species in the bulk or at metallic interfaces [96].

The prediction of volatilities and melting points of pure ionic liquids was possible by associating first principles calculations and calorimetric measurements, as presented, for example, by Verevkin and co-workers. They showed that the knowledge of gas-phase thermodynamic properties is important for the modelling of vapour–liquid equilibrium on pure ionic fluids [97].

5.5 CONCLUSION

This chapter was mainly written between the COIL-3 (Cairns, Australia) and COIL-4 (Washington, DC, United States) conferences, whereas the later stages of editing and proofing were done already after the COIL-4 meeting. As a befitting conclusion to the present chapter, many of the issues explored in the last three sections have been readdressed and discussed in Washington, DC. Topics such as force field refinements (Section 5.2) and the complex structure of ionic liquids at a nanoscale level (Section 5.3) occupied centre stage in some of the most successful parallel sessions of the conference. Such facts confirm ionic liquids as a vibrant research topic and molecular modelling as one of its most valuable tools.

REFERENCES

1 Hanke, C.G., Price, S.L. and Linden-Bell, R., Intermolecular potentials for simulations of liquid imidazolium salts, *Mol. Phys.* **99**, 801–809 (2001).

2 de Andrade, J., Boes, E.S. and Stassen, H., A force field for liquid state simulations on room temperature molten salts: 1-ethyl-3-methylimidazolium tetrachloroaluminate, *J. Phys. Chem. B* **106**, 3546–3548 (2002).

3 Shah, J.K., Brennecke, J.F. and Maginn, E.J., Thermodynamic properties of the ionic liquid 1-*n*-butyl-3-methylimidazolium hexafluorophosphate from Monte Carlo simulations, *Green Chem.* **4**, 112–118 (2002).

4 Margulis, C.J., Stern, H.A. and Berne, B.J., Computer simulation of a 'green chemistry' room-temperature ionic solvent, *J. Phys. Chem. B* **106**, 12017–12021 (2002).

5 Canongia Lopes, J.N., Deschamps, J. and Pádua, A.A.H., Modelling ionic liquids using a systematic all-atom force field, *J. Phys. Chem. B* **108**, 2038–2047 (2004).

6 Jorgensen, W.L., Maxwell, D.S. and Tirado-Rives, J., Development and testing of the OPLS all-atom force field on conformational energetics and properties of organic liquids, *J. Am. Chem. Soc.* **118**, 11225–11236 (1996).

7 Canongia Lopes, J.N. and Pádua, A.A.H., Molecular force field for ionic liquids composed of triflate or bistriflylimide anions, *J. Phys. Chem. B* **108**, 16893–16898 (2004).

8 Canongia Lopes, J.N. and Pádua, A.A.H., Molecular force field for ionic liquids III: Imidazolium, pyridinium, and phosphonium cations; chloride, bromide, and dicyanamide anions, *J. Phys. Chem. B* **110**, 19586–19592 (2006).

9 Canongia Lopes, J.N., Pádua, A.A.H. and Shimizu, K., Molecular force field for ionic liquids IV: Trialkylimidazolium and alkoxycarbonyl-imidazolium cations; alkylsulfonate and alkylsulfate anions, *J. Phys. Chem. B* **112**, 5039–5046 (2008).

10 Shimizu, K., Almantariotis, D., Costa Gomes, M.F., Pádua, A.A.H. and Canongia Lopes, J.N., Molecular force field for ionic liquids V: Hydroxyethylimidazolium, dimethoxy-2-methylimidazolium, and fluoroalkylimidazolium cations and bis(Fluorosulfonyl) Amide, perfluoroalkanesulfonylamide, and fluoroalkylfluorophosphate anions, *J. Phys. Chem. B* **114**, 3592–3600 (2010).

11 Brooks, B.R., Bruccoleri, R.E., Olafson, B.D., States, D.J., Swaminathan, S. and Karplus. M., CHARMM – a program for macromolecular energy, minimization, and dynamics calculations, *J. Comput. Chem.* **4**, 187–217 (1983).

12 Case, D.A., Cheatham, T.E., Darden, T., Gohlke, H., Luo, R., Merz Jr., K.M., Onufriev, A., Simmerling, C., Wang, B. and Woods, R., The Amber biomolecular simulation programs, *J. Comput. Chem.* **26**, 1668–1688 (2005).

13 Kumelan, J., Kamps, A.P.S., Urukova, I., Tuma, D. and Maurer, G., Solubility of oxygen in the ionic liquid [bmim][PF_6]: Experimental and molecular simulation results, *J. Chem. Thermodyn.* **37**, 595 (2007).

14 Urahata, S.M. and Ribeiro, M.C.C., Structure of ionic liquids of 1-alkyl-3-methylimidazolium cations: A systematic computer simulation study, *J. Chem. Phys.* **120**, 1855–1863 (2004).

15 Morrow, T.I. and Maginn, E.J., Molecular dynamics study of the ionic liquid 1-*n*-butyl-3-methylimidazolium hexafluorophosphate, *J. Phys. Chem. B* **106**, 12807–12813 (2002).

16 Del Popolo, M.G., Lynden-Bell, R.M. and Kohanoff, J., Ab initio molecular dynamics simulation of a room temperature ionic liquid, *J. Phys. Chem. B* **109**, 5895 (2005).

17 Deetlefs, M., Hardacre, C., Nieuwenhuyzen, M., Pádua, A.A.H., Sheppard, O. and Soper, A.K., Liquid structure of the ionic liquid 1,3-dimethylimidazolium bis{(trifluoromethyl)sulfonyl}amide, *J. Phys. Chem. B* **110**, 12055–12061 (2006).

18 Rebelo, L.P.N., Lopes, J.N.C., Esperança, J.M.S.S., Guedes, H.J.R., Lachwa, J., Visak, V.N. and Visak, Z.P., Accounting for the unique, doubly dual nature of ionic liquids from a molecular thermodynamic, and modelling standpoint, *Acc. Chem. Res.* **40**, 1114–1121 (2007).

19 Canongia Lopes, J.N., Costa Gomes, M.F. and Pádua, A.A.H., Nonpolar, polar, and associating solutes in ionic liquids, *J. Phys. Chem. B* **110**, 16816–16818 (2006).

20 Chowdhury, P.K., Halder, M., Sanders, L., Calhoun, T., Anderson, J.L., Armstrong, D.W., Song, X. and Petrich, J.W., Dynamic solvation in room-temperature ionic liquids, *J. Phys. Chem. B* **108**, 10245–10255 (2004).

21 Shim, Y., Duan, J., Choi, M.Y. and Kim, H.J., Solvation in molecular ionic liquids, *J. Chem. Phys.* **119**, 6411–6414 (2003).

22 Youngs, T.G.A., Holbrey, J.D., Deetlefs, M., Nieuwenhuyzen, M., Gomes, M.F.C. and Hardacre, C., A molecular dynamics study of glucose solvation in the ionic liquid 1,3-dimethylimidazolium chloride, *ChemPhysChem.* **7**, 2279–2281 (2006).

23 Cadena, C., Zhao, Q., Snurr, R.Q. and Maginn, E.J., Molecular modelling and experimental studies of the thermodynamic and transport properties of pyridinium-based ionic liquids, *J. Phys. Chem. B* **110**, 2821–2832 (2006).

24 Hu, Z.H. and Margulis, C.J., Heterogeneity in a room-temperature ionic liquid: Persistent local environments and the red-edge effect, *Proc. Natl. Acad. Sci. U. S. A.* **103**, 831–836 (2006).

25 Canongia Lopes, J.N., Shimizu, K., Pádua, A.A.H., Umebayashi, Y., Fukuda, S., Fujii, K. and Ishiguro, S., A tale of two ions: The conformational landscapes of bis(trifluoromethanesulfonyl) amide and *N,N*-dialkylpyrrolidinium, *J. Phys. Chem. B* **112**, 1465–1472 (2008).

26 Köddermann, T., Paschek, D. and Ludwig, R., Molecular dynamic simulations of ionic liquids: A reliable description of structure, thermodynamics and dynamics, *ChemPhysChem.* **8**, 2464–2470 (2007).

27 Tanaka, M. and Siehl, H.-U., An application of the consistent charge equilibration (CQEq) method to guanidinium ionic liquid systems, *Chem. Phys. Lett.* **457**, 263–266 (2008).

28 Lauw, Y., Horne, M.D., Rodopoulos, T., Webster, N.A.S., Minofar, B. and Nelson, A., X-Ray reflectometry studies on the effect of water on the surface structure of $[C_4mpyr][NTf_2]$ ionic liquid, *Phys. Chem. Chem. Phys.* **11**, 11507–11514 (2009).

29 Bedrov, D., Borodin, O., Li, Z. and Grant, S.D., Influence of polarization on structural, thermodynamic, and dynamic properties of ionic liquids obtained from molecular dynamics simulations, *J. Phys. Chem. B* **114**, 4984–4997 (2010).

30 Wang, Y., Feng, S. and Voth, G.A., Transferable coarse-grained models for ionic liquids, *J. Chem. Theory Comput.* **5**, 1091–1098 (2009).

31 Daily, J.W. and Micci, M.M., Ionic velocities in an ionic liquid under high electric fields using all-atom and coarse-grained force field molecular dynamics, *J. Chem. Phys.* **131**, 094501 (2009).

32 Kislenko, S.A., Samoylov, I.S. and Amirov, R.H., Molecular dynamics simulation of the electrochemical interface between a graphite surface and the ionic liquid $[BMIM][PF_6]$, *Phys. Chem. Chem. Phys.* **11**, 5584–5590 (2009).

33 Tsuzuki, S., Shinoda, W., Saito, H., Mikami, M., Tokuda, H. and Watanabe, M., Molecular dynamics simulations of ionic liquids: Cation and anion dependence of self-diffusion coefficients of ions, *J. Phys. Chem. B* **113**, 10641–10649 (2009).

34 Liu, X., Zhou, G., Zhang, S. and Yao, X., Molecular dynamics simulation of dual amino-functionalized imidazolium-based ionic liquids, *Fluid Phase Equilib.* **284**, 44–49 (2009).

35 Fujii, K., Mitsugi, T., Takamuku, T., Yamaguchi, T., Umebayashi, Y. and Ishiguro, S., Effect of methylation at the C2 position of imidazolium on the structure of ionic liquids revealed by large angle X-ray scattering experiments and MD simulations, *Chem. Lett.* **38**, 340–341 (2009).

36 Shi, W. and Maginn, E.J., Atomistic simulation of the absorption of carbon dioxide and water in the ionic liquid 1-*n*-hexyl-3-methylimidazolium bis(trifluoromethylsulfonyl) imide ($[hmim][Tf_2N]$, *J. Phys. Chem. B* **112**, 2045–2055 (2008).

37 Gutel, T., Santini, C.C., Pádua, A.A.H., Fenet, B., Chauvin, Y., Canongia Lopes, J.N., Bayard, F., Gomes, M.F.C. and Pensado, A.S., Interaction between the pi-system of toluene and the imidazolium ring of ionic liquids: A combined NMR and molecular simulation study, *J. Phys. Chem. B* **113**, 170–177 (2009).

38 Izgorodina, E.I., Bernard, U.L. and MacFarlane, D.R., Ion-pair binding energies of ionic liquids: Can DFT compete with ab initio-based methods? *J. Phys. Chem. A* **113**, 7064–7072 (2009).

39 Deetlefs, M. and Seddon, K.R., Assessing the greenness of some typical laboratory ionic liquid preparations, *Green Chem.* **12**, 17–30 (2010).

40 Torrecilla, J.S., Deetlefs, M., Seddon, K.R. and Rodriguez, F., Estimation of ternary liquid-liquid equilibria for arene/alkane/ionic liquid mixtures using neural networks, *Phys. Chem. Chem. Phys.* **10**, 5114–5120 (2008).

41 Preiss, U., Bulut, S. and Krossing, I., In silico prediction of the melting points of ionic liquids from thermodynamic considerations: A case study on 67 salts with a melting point range of 337 degrees C, *J. Phys. Chem. B* **114**, 11133–11140 (2010).

42 Bhargava, B.L., Balasubramanian, S. and Klein, M., Modelling room temperature ionic liquids, *Chem. Commun.* 3339–3351 (2008).

43 Kirchner, B., Ionic liquids from theoretical investigations, *Top. Curr. Chem.* **290**, 213–262 (2009).

44 Hansen, J.P. and McDonald, I.R., *Theory of Simple Liquids*, 2nd edition, Academic Press, London (1986).

45 Dupont, J., On the solid, liquid and solution structural organization of imidazolium ionic liquids, *J. Braz. Chem. Soc.* **15**, 341–350 (2004).

46 Hardacre, C., Holbrey, J.D., Nieuwenhuyzen, M. and Youngs, T.G.A., Structure and solvation in ionic liquids, *Acc. Chem. Res.* **40**, 1146–1155 (2007).

47 Deetlefs, M., Hardacre, C., Nieuwenhuyzen, M., Sheppard, O. and Soper, A., Structure of ionic liquid-benzene mixtures, *J. Phys. Chem. B* **109**, 1593–1598 (2005).

48 Łachwa, J., Bento, I., Duarte, M., Lopes, J. and Rebelo, L., Condensed phase behaviour of ionic liquid–benzene mixtures: Congruent melting of a [emim][NTf_2]·C_6H_6 inclusion crystal, *Chem. Commun.* 2445–2447 (2006).

49 Gordon, C., Holbrey, J.D., Kennedy, A. and Seddon, K.R., Ionic liquid crystals: Hexafluorophosphate salts, *J. Mater. Chem.* **8**, 2627–2636 (1998).

50 Canongia Lopes, J.N.A. and Pádua, A.A.H., Nanostructural organization in ionic liquids, *J. Phys. Chem. B* **110**, 3330–3335 (2006).

51 Canongia Lopes, J.N., Microphase separation in mixtures of Lennard-Jones particles, *Phys. Chem. Chem. Phys.* **4**, 949–954 (2002).

52 Annapureddy, H.V.R., Kashyap, H.K., de Biase, P.M. and Margulis, C.J., What is the origin of the prepeak in the X-ray scattering of imidazolium-based room-temperature ionic liquids? *J. Phys. Chem. B* **114**, 16838–16846 (2010).

53 Deschamps, J. and Pádua, A.A.H., Interactions of gases with ionic liquids: Molecular simulation, in *Ionic Liquids IIIa: Fundamentals, Progress, Challenges, and Opportunities, Properties and Structure*, ACS Symposium Series, Vol. **901**, eds. R.D. Rogers and K.R. Seddon, American Chemical Society, Washington, DC (2005), pp. 150–158.

54 Tokuda, H., Hayamizu, K., Ishii, K., Susan, M. and Watanabe, M., Physicochemical properties and structures of room temperature ionic liquids. 2. Variation of alkyl chain length in imidazolium cation, *J. Phys. Chem. B* **109**, 6103–6110 (2005).

55 Antonietti, M., Kuang, D., Smarsly, B. and Yong, Z., Ionic liquids for the convenient synthesis of functional nanoparticles and other inorganic nanostructures, *Angew. Chem. Int. Ed.* **43**, 4988–4992 (2004).

56 Triolo, A., Russina, O., Bleif, H.-J. and Di Cola, E., Nanoscale segregation in room temperature ionic liquids, *J. Phys. Chem. B* **111**, 4641–4644 (2007).

57 Mizuhata, M., Maekawa, M. and Deki, S., Ordered structure in room temperature molten salts containing aliphatic quaternary ammonium ions, *ECS Trans.* **3**, 89–95 (2007).

58 Mandal, P., Sarkar, M. and Samanta, A., Excitation-wavelength-dependent fluorescence behavior of some dipolar molecules in room-temperature ionic liquids, *J. Phys. Chem. A* **108**, 9048–9053 (2004).

59 Xiao, D., Rajian, J.R., Cady, A., Li, S., Bartsch, R.A. and Quitevis, E.L., Nanostructural organization and anion effects on the temperature dependence of the optical Kerr effect spectra of ionic liquids, *J. Phys. Chem. B* **111**, 4669–4677 (2007).

60 Ishiguro, S.-I., Umebayashi, Y., Kanzaki, R. and Fujii, K., Structure, solvation, and acid-base property in ionic liquids, *Pure Appl. Chem.* **82**, 1927–1941 (2010).

61 Youngs, T.G.A., Hardacre, C. and Holbrey, J.D., Glucose solvation by the ionic liquid 1,3-dimethylimidazolium chloride: A simulation study, *J. Phys. Chem. B* **111**, 13765–13774 (2007).

62 Del Popolo, M.G., Mullan, C.L., Holbrey, J.D., Hardacre, C. and Ballone, P., Ion association in [bmim][PF_6]/naphthalene mixtures: An experimental and computational study, *J. Am. Chem. Soc.* **130**, 7032–7041 (2008).

63 Baldelli, S., Surface structure at the ionic liquid-electrified metal interface, *Acc. Chem. Res.* **41**, 421–431 (2008).

64 Iwahashi, T., Nishi, T., Yamane, H., Miyamae, T., Kanai, K., Seki, K., Kim, D. and Ouchi, Y., Surface structural study on ionic liquids using metastable atom electron spectroscopy, *J. Phys. Chem. C* **113**, 19237–19243 (2009).

65 Dupont, J. and Scholten, J.D., On the structural and surface properties of transition-metal nanoparticles in ionic liquids, *Chem. Soc. Rev.* **39**, 1780–1804 (2010).

66 Gutel, T., Garcia-Anton, J., Pelzer, K., Philippot, K., Santini, C.C., Chauvin, Y., Chaudret, B. and Basset, J.-M., Influence of the self-organization of ionic liquids on the size of ruthenium nanoparticles: Effect of the temperature and stirring, *J. Mater. Chem.* **17**, 3290–3292 (2007).

67 Batra, D., Seifert, S. and Firestone, M.A., The effect of cation structure on the mesophase architecture of self-assembled and polymerized imidazolium-based ionic liquids, *Macromol. Chem. Phys.* **208**, 1416–1427 (2007).

68 Ichikawa, T., Yoshio, M., Hamasaki, A., Mukai, T., Ohno, H. and Kato, T., Self-organization of room-temperature ionic liquids exhibiting liquid-crystalline bicontinuous cubic phases: Formation of nano-ion channel networks, *J. Am. Chem. Soc.* **129**, 10662–10663 (2007).

69 Coleman, S., Byrne, R., Minkovska, S. and Diamond, D., Investigating nanostructuring within imidazolium ionic liquids: A thermodynamic study using photochromic molecular probes, *J. Phys. Chem. B* **113**, 15589–15596 (2009).

70 Iwata, K., Okajima, H., Saha, S. and Hamaguchi, H.-O., Local structure formation in alkyl-imidazolium-based ionic liquids as revealed by linear and nonlinear Raman spectroscopy, *Acc. Chem. Res.* **40**, 1174–1181 (2007).

71 Xiao, D., Hines, L.G.J., Bartsch, R.A. and Quitevis, E.L., Intermolecular vibrational motions of solute molecules confined in nonpolar domains of ionic liquids, *J. Phys. Chem. B* **113**, 4544–4548 (2009).

72 Mizoshiri, M., Nagao, T., Mizoguchi, Y. and Yao, M., Dielectric permittivity of room temperature ionic liquids: A relation to the polar and nonpolar domain structures, *J. Chem. Phys.* **132**, 164510 (2010).

73 Atkin, R. and Warr, G.G., The smallest amphiphiles: Nanostructure in protic room-temperature ionic liquids with short alkyl groups, *J. Phys. Chem. B* **112**, 4164–4166 (2008).

74 Castiglione, F., Moreno, M., Raos, G., Famulari, A., Mele, A., Appetecchi, G.B. and Passerini, S., Structural organization and transport properties of novel pyrrolidinium-based ionic liquids with perfluoroalkyl sulfonylimide anions, *J. Phys. Chem. B* **113**, 10750–10759 (2009).

75 Gutel, T., Santini, C.C., Philippot, K., Pádua, A., Pelzer, K., Chaudret, B., Chauvin, Y. and Basset, J.-M., Organized 3D-alkyl imidazolium ionic liquids could be used to control the size of in situ generated ruthenium nanoparticles? *J. Mater. Chem.* **19**, 3624–3631 (2009).

76 Campbell, P.S., Podgorsek, A., Gutel, T., Santini, C.C., Pádua, A.A.H., Costa Gomes, M.F., Bayard, F., Fenet, B. and Chauvin, Y., How do physical-chemical parameters influence the catalytic hydrogenation of 1,3-cyclohexadiene in ionic liquids? *J. Phys. Chem. B* **114**, 8156–8165 (2010).

77 Ueno, K., Imaizumi, S., Hata, K. and Watanabe, M., Colloidal interaction in ionic liquids: Effects of ionic structures and surface chemistry on rheology of silica colloidal dispersions, *Langmuir* **25**, 825–831 (2009).

78 Armand, M., Endres, F., Macfarlane, D.R., Ohno, H. and Scrosati, B., Ionic–liquid materials for the electrochemical challenges of the future, *Nat. Mater.* **8**, 621–629 (2009).

79 Hardacre, C., Holbrey, J.D., Mullan, C.L., Youngs, T.G.A. and Bowron, D.T., Small angle neutron scattering from 1-alkyl-3-methylimidazolium hexafluorophosphate ionic liquids ([$C_n$mim][PF_6], $n = 4$, 6, and 8), *J. Chem. Phys.* **133**, 074510 (2010).

80 Russina, O., Triolo, A., Gontrani, L., Caminiti, R., Xiao, D., Hines Jr., L., Bartsch, R.A., Quitevis, E., Plechkova, N. and Seddon, K., Morphology and intermolecular dynamics of 1-alkyl-3-methylimidazolium bis{(trifluoromethane)sulfonyl}amide ionic liquids: structural and dynamic evidence of nanoscale segregation, *J. Phys. Condens. Matter* **21**, 424121 (2009).

81 Xiao, D., Hines, L.G.J., Li, S., Bartsch, R.A., Quitevis, E.L., Russina, O. and Triolo, A., Effect of cation symmetry and alkyl chain length on the structure and intermolecular dynamics of 1,3-dialkylimidazolium bis(trifluoromethanesulfonyl)amide ionic liquids, *J. Phys. Chem. B* **113**, 6426–6433 (2009).

82 Xu, W. and Angell, C.A., Solvent-free electrolytes with aqueous solution – like conductivities, *Science* **302**, 422–425 (2003).

83 Shah, J.K. and Maginn, E.J., Monte Carlo simulations of gas solubility in the ionic liquid 1-*n*-butyl-3-methylimidazolium hexafluorophosphate, *J. Phys. Chem. B* **109**, 10395–10405 (2005).

84 Urukova, I., Vorholz, J. and Maurer, G., Solubility of CO_2, CO, and H-2 in the ionic liquid [bmim][PF_6] from Monte Carlo simulations, *J. Phys. Chem. B* **109**, 12154–12159 (2005).

85 Tokuda, H., Hayamizu, K., Ishii, K., Susan, M.A.B.H. and Watanabe, M., Physicochemical properties and structures of room temperature ionic liquids. 1. Variation of anionic species, *J. Phys. Chem. B* **108**, 16593–16600 (2004).

86 Tokuda, H., Tsuzuki, S., Susan, M.A.B.H., Hayamizu, M. and Watanabe, M., How ionic are room-temperature ionic liquids? An indicator of the physicochemical properties, *J. Phys. Chem. B* **110**, 19593–19600 (2006).

87 Maginn, E.J., Atomistic simulation of the thermodynamic and transport properties of ionic liquids, *Acc. Chem. Res.* **40**, 1200–1207 (2007).

88 Margulis, C.J., Room-temperature ionic liquids: Slow dynamics, viscosity, and the red edge effect, *Acc. Chem. Res.* **40**, 1097–1105 (2007).

89 Tsuzuki, S., Tokuda, H., Hayamizu, K. and Watanabe, M., Magnitude and directionality of interaction in ion pairs of ionic liquids: Relationship with ionic conductivity, *J. Phys. Chem. B* **109**, 16474–16481 (2005).

90 Arzhantsev, S., Jin, H., Baker, G.A. and Maroncelli, M., Measurements of the complete solvation response in ionic liquids, *J. Phys. Chem. B* **111**, 4978–4989 (2007).

91 Umebayashi, Y., Mitsugi, T., Fukuda, S., Fujimori, T., Fujii, K., Kanzaki, R., Takeuchi, M. and Ishiguro, S.-I., Lithium ion solvation in room-temperature ionic liquids involving bis(trifluoromethanesulfonyl) imide anion studied by Raman Spectroscopy and DFT calculations, *J. Phys. Chem. B* **111**, 13028–13032 (2007).

92 Pádua, A.A.H., Costa Gomes, M.F. and Canongia Lopes, J.N.A., Molecular solutes in ionic liquids: A structural, perspective, *Acc. Chem. Res.* **40**, 1087–1096 (2007).

93 Blesic, M., Canongia Lopes, J.N., Pádua, A.A.H., Shimizu, K., Costa Gomes, M.F. and Rebelo, L.P.N., Phase equilibria in ionic liquid–aromatic compound mixtures, including Benzene Fluorination effects, *J. Phys. Chem. B* **113**, 7631–7636 (2009).

94 Shimizu, K., Canongia Lopes, J.N., Costa Gomes, M.F., Pádua, A.A.H. and Rebelo, L.P.N., On the role of the dipole and quadrupole moments of aromatic compounds in the solvation by ionic liquids, *J. Phys. Chem. B* **113**, 9894–9900 (2009).

95 Shi, W. and Maginn, E.J., Molecular simulation and regular solution theory modelling of pure and mixed gas absorption in the ionic liquid 1-*n*-hexyl-3-methylimidazolium bis(Trifluoromethylsulfonyl)amide ([hmim][Tf_2N]), *J. Phys. Chem. B* **112**, 16710–16720 (2008).

96 Borodin, O., Smith, G.D. and Henderson, W., Li+ cation environment, transport, and mechanical properties of the LiTFSI doped *N*-methyl-*N*-alkylpyrrolidinium + TFSI- ionic liquids, *J. Phys. Chem. B* **110**, 16879–16886 (2006).

97 Emel'yanenko, V.N., Verevkin, S.P. and Heintz, A., The gaseous enthalpy of formation of the ionic liquid 1-butyl-3-methylimidazolium dicyanamide from combustion calorimetry, vapor pressure measurements, and ab initio calculations, *J. Am. Chem. Soc.* **129**, 3930–3937 (2007).

6 Chemical Engineering of Ionic Liquid Processes

CAROLIN MEYER, SEBASTIAN WERNER,
MARCO HAUMANN, and PETER WASSERSCHEID
Lehrstuhl für Chemische Reaktionstechnik, Universität Erlangen-Nürnberg,
Erlangen, Germany

6.1 GENERAL ASPECTS OF CHEMICAL ENGINEERING WITH IONIC LIQUIDS

The development of engineering applications with ionic liquids started in the mid-1990s, when the first examples of continuous catalytic processes using ionic liquids (e.g. Difasol by the Institut Français du Pétrole (IFP) [1]) and the first examples of technically relevant ionic liquid-based extraction studies [2] were published. Quickly, it was realised that some of the characteristics of ionic liquids enable more efficient and greener process concepts. In most cases, these advantages were realised by clever multiphase engineering enabling ionic liquid recycling within the process. The extremely low vapour pressures of typical ionic liquids were quickly found to be an important advantage in some applications (e.g. if an ionic liquid catalyst solution is contacted with a gas); however, it is also fair to say that in many engineering applications, process schemes were found to need a fundamental redesign due to the ionic liquid involatility.

The unique property profile of ionic liquids originates from a complex interplay of coulombic, hydrogen-bonding and van der Waals interactions among the ionic liquid's ions and between the ionic liquid and dissolved substances and surfaces. Understanding and utilising these complex interactions to operate a chemical process in the most efficient manner is the central goal in developing ionic liquid applications in chemical engineering. The sometimes hesitative attitude of the engineering community towards ionic liquids often stems from the fact that their enormous structural variability (often praised in the literature as a key advantage of ionic liquids [3]) can be pretty scary for somebody who wants to identify the best ionic liquid for a given application and – at the

Ionic Liquids Completely UnCOILed: Critical Expert Overviews, First Edition.
Edited by Natalia V. Plechkova and Kenneth R. Seddon.

same time – is interested in knowing all relevant physico-chemical properties of this 'best' ionic liquid to design the process.

In this context, it is very encouraging that our general knowledge about ionic liquid properties has increased a lot in the recent years [4]. Moreover, theoretical predictability of some ionic liquid properties by means of theoretical tools (MD simulations) has been dramatically improved in the recent years [5]. With the much better knowledge on ionic liquid properties developed over the last 5 years, it has also been found that the concept of using ionic liquids comes along with a number of general features that are – more or less – intrinsic to the approach. Table 6.1 shows an overview of typical ionic liquid properties that are relevant for engineering applications. The table gives typical values for selected properties, as well as known upper and lower limits (for more detailed information about specific properties, see Ref. 6).

Engineering is always about applying scientific principles in some sort of technical and real-life context. As a consequence, 'engineering' starts where a first rough cost/benefit analysis of a scientific finding or principle has already been performed, and the owner of the respective knowledge is convinced that further development towards application is reasonable. From all scientific findings around ionic liquids made in the last 30 years, there is therefore only a relatively small percentage that has made the transition into the world of engineering. To evaluate whether a certain finding should be 'engineered', the following five issues/questions have proven to provide a helpful guideline in the experience of the authors.

6.1.1 Technical Performance: Unique Selling Points

Ionic liquid technology is still quite new for engineers and, like every new technology, it comes with considerable technical risk. In general terms, this 'risk' just reflects the lack of operational experience on a technical scale for most potential ionic liquid applications. Of specific interest are questions regarding the long-term stability of the ionic liquid inventory or the long-term effects of the ionic liquid material on process equipment. Corrosion, swelling and embrittlement of seals or other sensitive plant components are relevant aspects in this context. Other typical discussion points relate to the handling, storage and disposal of ionic liquids on a larger scale. Overcoming such justified concerns for a decision in favour of an ionic liquid process requires strong arguments. The strongest of these arguments is a unique technical performance, some technical feature that has not been possible so far. Such a unique technical selling point inspires the creative thinking of engineers and attracts them to the new technology. More important, if after careful evaluation the unique ionic liquid performance is considered to be of economic interest (and the main arguments to evaluate this will follow in Sections 6.1.2–6.1.5), there is no alternative way to circumvent 'ionic liquid technology'. With the solution to all 'risky' points during a certain development process, the technical risk of applying ionic liquids is reduced for all successive applications of a similar nature.

TABLE 6.1 Selected Properties of Ionic Liquids of Relevance for Applications in Chemical Engineering[a]

Property	Lower Limit Example	Typical Range of Most Ionic Liquids	Upper Limit Example
Melting point/ glass transition	$[C_2mim]Cl-AlCl_3$ (1:2) −96°C (glass transition)	0–60°C	100°C by definition
Density	$[C_6mpyr][N(CN)_2]$ 0.92 g l^{-1b}	1.1–1.6 g l^{-1b}	$[C_2mim]Br-AlBr_3$ (1:2) 2.2 g l^{-1b}
Viscosity	$[C_2mim]Cl-AlCl_3$ (1:2) = 14 mPa s[b]	40–800 mPa s[b]	$[C_4mim]Cl$ (super-cooled) = 40.89 Pa s[b]
Thermal stability	$[C_2mim][O_2CMe]$ *ca.* 120°C[c]	230–300°C[c]	$[C_2mim][NTf_2]$ 400°C[c]
Surface tension	$[C_{12}mim][PF_6]$ 23.6 mN m^{-1}	30–50 mN m^{-1}	$[C_1mim][MeSO_4]$ 59.8 mN m^{-1}
Heat capacity	$[C_4mim][MeSO_4]$ 247 J mol^{-1} K^{-1}	300–400 J mol^{-1} K^{-1}	$[C_8mim][NTf_2]$ 654 J mol^{-1} K^{-1}
Water miscibility	$[NTf_2]^-$, $[P(C_2F_5)_3F_3]^-$	Many ionic liquids do mix with water but can be extracted from water	$[RSO_3]^-$, $[RSO_4]^-$, $[R_2PO_4]^-$
Hydrolytic stability	$[BF_4]^-$, $[PF_6]^-$	Also heterocyclic cations can hydrolyse under extreme conditions	$[NTf_2]^-$, $[OTf]^-$, $[CH_3SO_3]^-$
Corrosion	$[NTf_2]^-$, $[OTf]^-$	Most ionic liquids are corrosive versus Cu; protective additives are known	Cl^-, HF formed from $[MF_x]^-$ hydrolysis
Toxicity	[Cholinium] $[O_2CMe]$	Often: increasing toxicity for aquatic systems with increasing lipophilicity	$[C_2mim][CN]$
Price	$[HNR_3][HSO_4]$ *ca.* 3 € kg^{-1d}	25–250 € kg^{-1}	$[C_4dmim][NTf_2]$ *ca.* 1.000 € kg^{-1d}

[a] From Ref. 6 and references cited therein.
[b] At room temperature.
[c] Long-term TGA experiments.
[d] Estimation made for a production scale of 1 kg and for a purity less than 98%.

6.1.2 Added Value

It is not a trivial point that a technically unique property adds value to a process or product. A particular striking example is the unique finding that some ionic liquids can dissolve wood [7]. While this effect is scientifically highly interesting, it is not evident in which way this finding can lead to a process or product of high added value. The main problem is obviously the fact that technical ways of effective fractionation of wood components from ionic liquid solutions, and high price utilisation of wood components, are actually not at hand. Luckily, in many other cases, the situation is more favourable, and the unique technical ionic liquid performance often allows for significantly better product qualities, shortcuts in processing sequences or more attractive operational windows. It is obvious that a unique performance of an ionic liquid that results in a very high added value is the ultimate driver to engineer such application.

6.1.3 Amount of Ionic Liquid Needed to Realise the Effect

It is a matter of fact that in the last 20 years of ionic liquid research, many technically interesting properties of these new materials have been discovered. However, the number of absolutely unique ionic liquid features – performance aspects that cannot be achieved by any other material or technology – is comparatively small to date. This has much to do with the fact that engineers in the earlier years of ionic liquid research tried to improve existing processes by using ionic liquids to achieve drop-in solutions. While such an optimisation strategy certainly offered the potential to save investment cost, it led inevitably to the effect that fully engineered and optimised processes using large amount of a cheap conventional solvents were tried to be replaced by alternatives requiring large amounts of ionic liquids. Despite many interesting new technical features identified for the ionic liquid version of these processes, such comparison was rarely in favour of the ionic liquid process when weighing benefits, risk and cost aspects. Obviously, cost and risk are a strong function of the amount of ionic liquid that is required to achieve the special 'ionic liquid effect' in a given application. A smaller amount of ionic liquid reduces the specific ionic liquid cost and with that the technical risk of ionic liquid degradation or loss.

It is a striking fact that many engineering applications that have proved to be successful, or of high potential, in the past 5 years are in fact 'ionic liquid thin film applications'. Typically, relatively small amounts of ionic liquid are used as a kind of 'magic ingredient' in the process or product. Successful and technically very promising thin film applications in catalysis (such as 'supported ionic liquid phase' (SILP) catalysis and 'solid catalyst with ionic liquid layer' (SCILL)), in separation technologies, in sensing and in mechanical engineering will be highlighted in detail later.

6.1.4 Ionic Liquid Stability under Process Conditions

In every engineering application, the ionic liquid has to fulfil certain requirements regarding its chemical, thermal, electrochemical or hydrolytic stability. For the development process of an engineering application, it is crucial to define at a very early stage these stability criteria and also the experimental methods to verify them. Stability issues count among the most important aspects that have ruled out ionic liquids from many otherwise promising development projects in the past. Moreover, it is very painful and costly to realise in a late stage of development that a specific ionic liquid that has been selected (e.g. for cost reasons) does not fulfil some of the stability criteria. The authors of this chapter recommend taking stability issues very seriously right from the beginning of any ionic liquid process development.

6.1.5 Ionic Liquid Cost per Kilogram

Despite the fact that a larger number of ionic liquids are commercially available today (a few of them even on the ton scale), their cost per kilogram remains typically by a factor of 10–50 higher than that of ordinary organic solvents. This price difference quite simply reflects the higher number of synthetic steps in ionic liquid production, as well as scale effects with regard to the production quantity. Again, this price difference favours applications where a small amount of ionic liquid has a unique, significant and economically relevant effect. In addition, it draws a lot of attention to aspects of ionic liquid recycling and lifetime in those processes where a large amount of ionic liquid is required technically.

In the following, we will restrict ourselves mainly to the description of engineering applications for which – in our view – the above-named five issues have been answered in a positive manner. This does not mean that all other applications are not worth consideration by engineers, but it certainly provides a reasonable selection to describe the most relevant aspects of ionic liquids in engineering applications from the authors' points of view.

6.2 CATALYTIC APPLICATIONS INVOLVING STRONGLY ACIDIC IONIC LIQUIDS

6.2.1 Characteristic Properties of Acidic Halometallate Ionic Liquids

Liquids of high acidity that can be tuned in their solubility, miscibility and acid strength are highly interesting for many acid-catalysed technical reactions. Acidic halometallate ionic liquids provide unique properties for such applications that have led to a great academic and industrial interest in such liquids and the related processes [8]. The best investigated and most established acidic halometallate ionic liquids are formed by mixing a molar excess of aluminium(III) halide with an organic halide salt, [cation]X (X = Cl or Br).

The first systems of this kind were discovered in the 1940s, when mixtures of 1-ethylpyridinium bromide and aluminium(III) chloride were used for the electrodeposition of aluminium [9–11]. Halometallate ionic liquids are at all compositions (except at the exact 1:1 stoichiometric mixture of aluminium(III) halide and the halide salt, forming the defined salt [cation][AlX_4]) characterised by the presence of more than one anionic species. These complex anions are in chemical equilibrium, with the composition of this equilibrium being a strong function of the molar ratio of aluminium(III) halide and the halide salt. Table 6.2 displays the most relevant species present as complex anions, for example, chloroaluminate(III) ionic liquids. Figure 6.1 illustrates the composition of these anionic species as a function of the mole fraction $X(AlCl_3)$ for the ionic liquid 1-ethyl-3-methylimidazolium chloride ([C_2mim]Cl—$AlCl_3$) [12].

TABLE 6.2 Thermodynamic Equilibrium Parameters of Reactions Occurring in Chloroaluminate Ionic Liquid [12]

	Equilibrium	$\log_{10}(K)$ at 473 K
I	$[AlCl_4]^- \rightleftharpoons Cl^- + \text{'}AlCl_3\text{'}$	−16.67
II	$[Al_2Cl_7]^- \rightleftharpoons [AlCl_4]^- + \text{'}AlCl_3\text{'}$	−2.54
III	$[Al_3Cl_{10}]^- \rightleftharpoons [Al_2Cl_7]^- + \text{'}AlCl_3\text{'}$	−0.90
IV	$[Al_4Cl_{13}]^- \rightleftharpoons [Al_3Cl_{10}]^- + \text{'}AlCl_3\text{'}$	−0.34

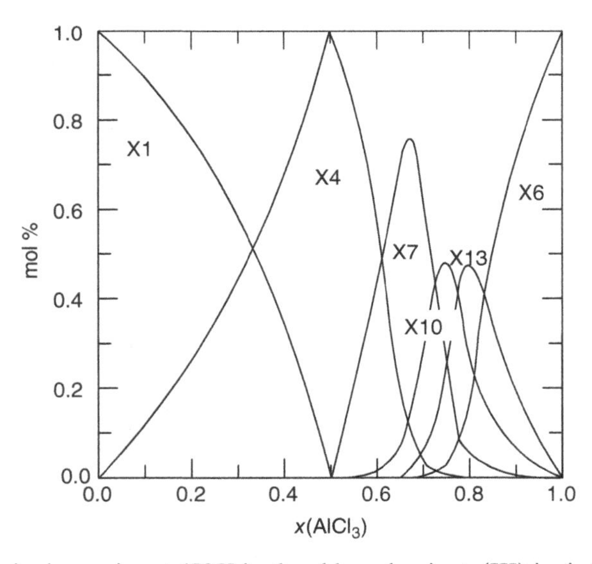

Figure 6.1 Anionic species at 473 K in the chloroaluminate(III) ionic liquid [C_2mim] Cl—$AlCl_3$ as a function of the mole fraction $X(AlCl_3)$, calculated from a thermodynamic model. X1 = Cl^-, X4 = $[AlCl_4]^-$, X6 = $[Al_2Cl_6]$, X7 = $[Al_2Cl_7]^-$, X10 = $[Al_3Cl_{10}]^-$, X13 = $[Al_4Cl_{13}]^-$ [12].

Chloroaluminate melts are extremely hygroscopic and in all practically relevant applications more or less contaminated with water [13]. Accordingly, chloroaluminate(III) ionic liquids contain at least traces of oxo- and/or hydroxo-chloroaluminate species and molecular HCl. It has been known since the early 1980s that Brønsted superacidity results if gaseous HCl is added to acidic chloroaluminate melts [14], and quantitative studies of the acidity of HCl in $[C_2mim]Cl-AlCl_3$ as a function of HCl pressure and ionic liquid composition (51.0–66.4 mol % $AlCl_3$) have been published [15].

6.2.2 Acidic Halometallate Ionic Liquids: Unique Selling Points and Typical Applications in Catalysis

A number of key industrial processes in oil refinery, petrochemistry and chemistry are acid catalysed. The industrial use of cheap and strong acids, such as aluminium(III) chloride or sulfuric acid, in typical workup protocols involves complete catalyst hydrolysis for product isolation. This results in high levels of corrosive waste water, which has to be treated in sophisticated downstream processes due to the presence of organic product residues. For obvious reasons, product isolation involving hydrolysis leads to a complete loss of the catalyst acidity in these conventional technologies.

In sharp contrast, acidic ionic liquids, and in particular chloroaluminate(III) ionic liquids, offer the unique advantage of tuneable acid strengths (by adjustment of the $AlCl_3$ mole fraction) combined with some important physicochemical properties that open alternative product isolation strategies. The latter involve the extremely low ionic liquid vapour pressure (note that hydrogen chloride and/or aluminium(III) chloride may still evaporate/sublime from acidic ionic liquids) and the tuneable miscibility of acid catalyst phase with the products [16]. By using acidic ionic liquids as catalyst phase, it is possible to separate the products from the catalyst phase either by liquid–liquid phase separation (requiring a miscibility gap between the ionic liquid and the product phase) or by product evaporation (requiring an appreciable vapour pressure of the product). In both cases, the ionic liquid catalyst remains in the reaction system and can be reused. In this way, no neutralisation of the ionic liquid catalyst is required, a fact that results directly in a much better E-factor [17] compared to the same reaction with traditional systems, such as aluminium(III) chloride or sulfuric acid. Instructive examples of acidic chloroaluminate(III) ionic liquids acting as catalyst and solvent in petrochemical processes are described in the following sections.

6.2.2.1 *Friedel–Crafts Reactions*

Friedel–Crafts reactions are conversions that form one additional carbon–carbon bond with an aromatic compound, catalysed by strong Lewis acids. Important reactions are Friedel–Crafts alkylation, Friedel–Crafts acylation and carbonylation.

6.2.2.1.1 Friedel–Crafts Alkylation Friedel–Crafts alkylations (Scheme 6.1) of benzene using haloalkanes were among the first reactions to be investigated in chloroaluminate(III) ionic liquids using the acidic ionic liquid $[C_2mim]Cl{-}AlCl_3$ [18]. Primary and secondary haloalkanes were applied, yielding polyalkylated products. An excess of aromatic substrate favours the formation of monoalkylated products.

BP patented the synthesis of alkylated aromatics by the reaction of an olefin with benzene in acidic ionic liquids [19]. Advantages of this process compared to the reaction with aluminium(III) chloride in organic solvents are the simple product separation, catalyst recycling and higher selectivities to alkylation products. The alkylation of benzene with linear olefins (C_{10}—C_{14}) is largely used industrially to produce alkylbenzenes. They serve as precursors for alkylbenzene sulfonates, which are employed as surfactants and detergents intermediates [20]. AkzoNobel developed the ionic liquid catalyst and solvent trimethylammonium chloride–aluminium(III) chloride, $[N_{1\,1\,1\,0}]Cl{-}AlCl_3$ ($X(AlCl_3) = 0.67$), for the alkylation of benzene with 1-dodecene, a cheaper alternative to imidazolium-based ionic liquids [21]. This reaction proceeds in a biphasic mode, implying simple catalyst recovery and recycling.

Varying mole fractions of aluminium(III) chloride in the acidic ionic liquid lead to different ionic liquid acidities. The effect of ionic liquid acidity on the yield and selectivity was tested in the alkylation of diphenyl oxide with 1-dodecene using $[C_4mim]Cl{-}AlCl_3$ [22]. Ionic liquids with $X(AlCl_3) \leq 0.55$ were not catalytically active. A maximum yield (about 90%) of monoalkylated product was achieved with $X(AlCl_3) = 0.6$ ($T_{reaction} = 353\,K$). Ranking of ionic liquids with mixed halogenoaluminate(III) anions according to their Lewis acidity gives the following order [23]:

$$[C_{12}mim][Al_2Cl_6Br] < [C_8mim][Al_2Cl_6Br] \approx [C_4mim][Al_2Cl_7] < [C_4mim][Al_2Cl_6Br] < [C_4mim][Al_2Cl_6I]$$

These different ionic liquid acidities were compared using the results of the alkylation of benzene with 1-dodecene. However, note that ionic liquids with more Lewis acidic anions are less hydrolytically stable.

Detailed kinetic investigations of the reaction of cumene with propene in $[C_2mim]Cl{-}AlCl_3$ ($X(AlCl_3) = 0.67$) were conducted by Joni et al. in a liquid–liquid biphasic reaction mode [24]. Various products (di-, tri- and tetraisopropylbenzene) result from a series of consecutive alkylation reactions. It is

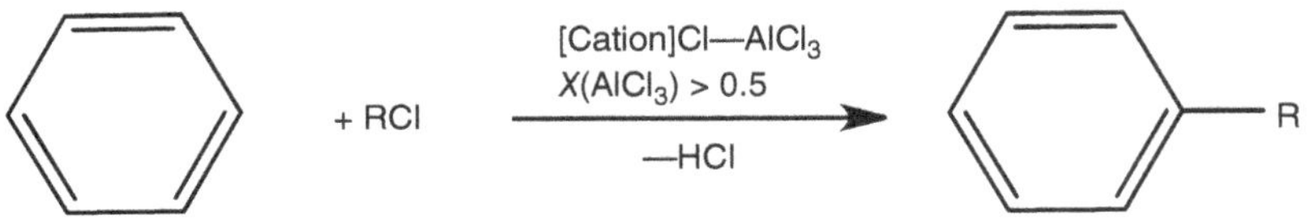

$$\text{C}_6\text{H}_6 + \text{RCl} \xrightarrow[\text{—HCl}]{[\text{Cation}]\text{Cl—AlCl}_3,\ X(\text{AlCl}_3) > 0.5} \text{C}_6\text{H}_5\text{—R}$$

Scheme 6.1 Friedel–Crafts alkylation.

necessary to take the solubility of these products into account to fit kinetic models to the data. A conductor-like screening model for real solvent (COSMO-RS) method was used to predict the relative solubilities of the products. Higher alkylated products are less soluble in the reactive ionic liquid phase, leading to an improved selectivity for the monoalkylated product.

6.2.2.1.2 Friedel-Crafts Acylation Friedel–Crafts acylation (Scheme 6.2) is, in contrast to Friedel–Crafts alkylation, selective for the formation of monosubstituted ketones, which are deactivated for further electrophilic substitutions.

Aromatic ketones are important fine chemicals and intermediates. A Lewis acid, for example, aluminium(III) chloride, is required to generate the active acylium ions $[RCO]^+$ from the acylating agent. Usually, acid chlorides or acid anhydrides are used as acylating agents. The carbonyl oxygen of the product forms a stable adduct with the highly oxophilic aluminium(III) chloride. Thus, stoichiometric amounts of aluminium(III) chloride relative to the acylating agent are consumed (hence the reaction is not truly catalytic), and the product must be liberated by hydrolysis.

The first aromatic acylations in ionic liquids were performed in acidic chloroaluminates [18], and the same systems were later extensively studied as solvent and catalyst to perform acylation reactions [25]. Reactions of ethanoyl chloride with aromatic compounds (including naphthalene and anthracene) in the Lewis acidic ionic liquid $[C_2mim]Cl–AlCl_3$ ($X(AlCl_3) = 0.67$) were conducted and compared favourably with the results of conventional molecular solvents [26], and the product of the acylation of ferrocene also formed an adduct with aluminium(III) chloride [27]. Hence, the focus of this research area lies on the development of less oxophilic, Lewis acidic ionic liquids to circumvent the hydrolysis step for product recovery. Recently, promising results using acidic chloroindate(III) ionic liquids have been reported [28].

6.2.2.1.3 Aromatic Carbonylation Aromatic carbonylation (Scheme 6.3) represents another important type of Friedel–Crafts reaction. In 1985, Texaco patented the reaction of toluene with carbon monoxide to 4-tolualdehyde in a Lewis acidic ionic liquid [29]. Brausch et al. used a new type of highly acidic ionic liquid with the formula $[cation][NTf_2]–AlCl_3$ for the carbonylation of toluene, which resulted in selectivities of about 85% to 4-tolualdehyde [30].

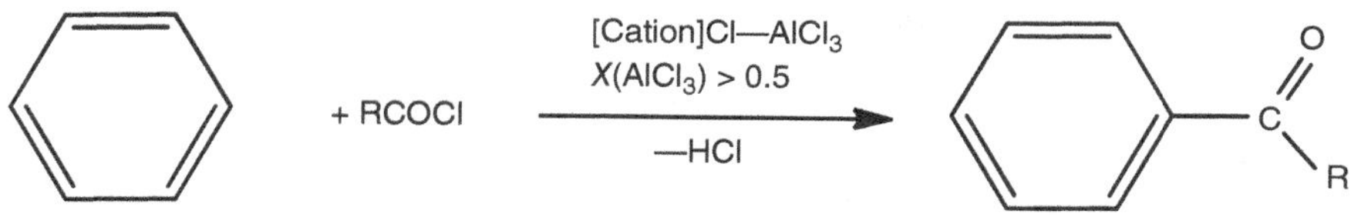

+ RCOCl; [Cation]Cl—$AlCl_3$, $X(AlCl_3) > 0.5$; —HCl

Scheme 6.2 Friedel–Crafts acylation.

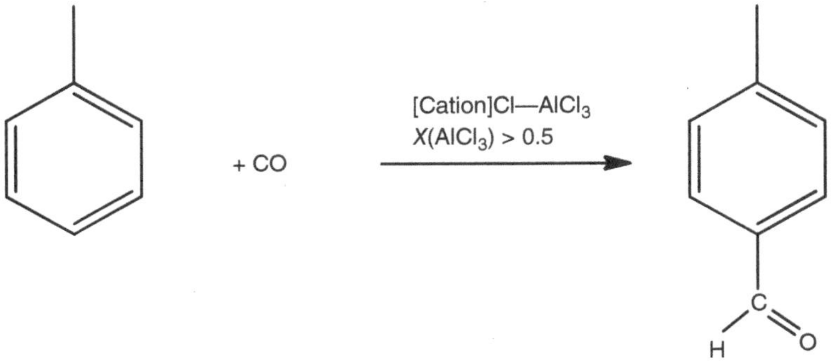

[Cation]Cl—AlCl$_3$
X(AlCl$_3$) > 0.5
+ CO

Scheme 6.3 Carbonylation of toluene.

Product isolation was carried out by a hydrolysis step, but the hydrophobic ionic liquid [cation][NTf_2] could be recycled after the hydrolysis step and then reloaded with aluminium(III) chloride. The group of White investigated toluene carbonylation at room temperature in the acidic ionic liquids [C_2mim]Cl—$AlCl_3$ or [C_4mim]Cl—$AlCl_3$ combined with HCl gas [31]. High selectivities to 4-tolualdehyde could be achieved, even at very high conversions. However, solid formation was found at high conversion: complex formation between the acidic chloroaluminate anions and the product aldehyde (cf. Section 6.4) was suggested as reason for this observation.

Carbonylation of benzene to benzaldehyde has been studied in different acidic ionic liquids of the type [C_4mim]X—$AlCl_3$ (X = Cl or Br), resulting in high selectivities (96%) towards benzaldehyde at relatively high yields (91%) [32]. After the reaction, benzaldehyde was present in a complex with the acidic aluminate ionic liquid: dichloromethane and diethyl ether were added to the reaction mixture for product extraction. 10 wt. % of the initial aluminium(III) chloride had to be added after each recycle to compensate the loss of aluminium(III) chloride in the further workup process. During recycle, a decreasing yield of benzaldehyde was observed. After the fifth recycling, product yield went down to 72% of the initial yield of the first run.

6.2.2.2 Refinery Alkylation Alkylation (Scheme 6.4) or 'refinery alkylation' is the reaction of isobutane with short-chain olefins (C_3—C_5) in the presence of highly acidic catalysts. Alkylates are particularly suitable gasoline-blending components due to their non-aromatic, high octane and paraffinic nature. Catalysts based on acidic ionic liquids have been demonstrated as greener alternatives to the actual processes based on sulfuric and hydrofluoric acids. The latter processes show severe drawbacks concerning handling issues, corrosion and toxic plant inventory.

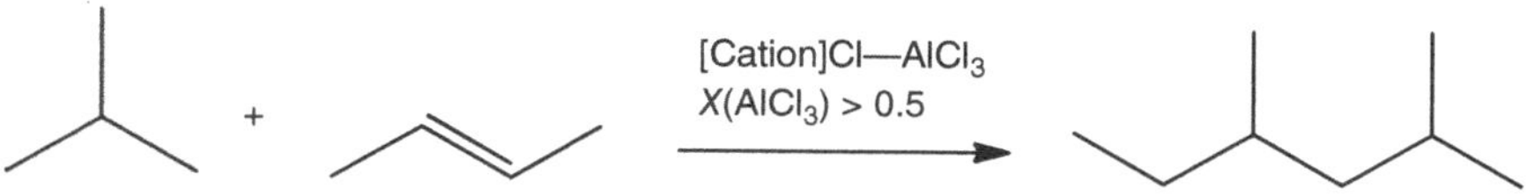

[Cation]Cl—$AlCl_3$
$X(AlCl_3) > 0.5$

Scheme 6.4 Refinery alkylation.

The group of Chauvin from the IFP were the first to publish the alkylation of isobutane with 2-butene in the acidic ionic liquid $[C_4mim]Cl–AlCl_3$ [33]. Acidic chloroaluminate(III) ionic liquids were found to be highly suitable catalysts for alkylation due to the fact that proper alkylation selectivities require a fine-tuning of the liquid acidity [34]. Cracking reactions are favoured at too high ionic liquid acidities; heavy by-product formation occurs at too low acidities. Continuous butene alkylation was performed by the group of Chauvin for 500 h, without any loss of catalytic activity and selectivity [35]. A sufficient level of mixing was found to be essential to reach high selectivities and a good quality alkylate. Moreover, these authors realised that the addition of copper(I) chloride to the acidic chloroaluminate ionic liquid is beneficial for the reaction performance.

A detailed study of the alkylation of isobutane with 2-butene in acidic ionic liquids using $[C_nmim]X–AlCl_3$ ($n = 4$, 6 or 8; X = Cl, Br or I) was conducted by Yoo et al. [36]. The $[C_8mim]Br–AlCl_3$ system showed the highest catalytic activity. The catalytic results obtained with $[C_8mim]Br–AlCl_3$ were compared to those achieved with a sulfuric acid catalyst. The authors stated that the ionic liquid catalyst showed a very high activity but exhibited slower deactivation, due to a lower level of by-product formation. However, lower concentrations of trimethylpentanes were produced with the ionic liquid catalyst. Both results can be explained by the higher acidity of $[C_8mim]Br–AlCl_3$ compared to sulfuric acid. The moisture sensitivity of these ionic liquids led to a gradual deactivation of the ionic liquid catalyst and solvent over time.

PetroChina has introduced a commercial alkylation process based on a composite catalyst system consisting of an acidic chloroaluminate(III) ionic liquid and copper(I) chloride. The addition of copper(I) chloride led to higher selectivities to octanes and a higher trimethylpentane/dimethylhexane ratio [37, 38]. In 2006, this process replaced an existing 65,000 $t\,a^{-1}$ sulfuric acid alkylation plant in China [39].

The group of Jess investigated the isobutene/2-butene alkylation with the Lewis acidic ionic liquid catalyst $[C_4mim]Cl–AlCl_3$ ($X(AlCl_3) = 0.64$) and two different promoters [40]. *tert*-Bromobutane and *tert*-chlorobutane were added to the Lewis acidic ionic liquid: the results were compared to those obtained with the neat Lewis acidic ionic liquid catalyst, sulfuric acid, and to catalytic runs in which water was added to the Lewis acidic ionic liquid. The reactions with *tert*-halobutanes as promoter were accelerated by up to two orders of magnitude. Pure Lewis acidic ionic liquid catalysts and $[C_4mim]Cl–AlCl_3$ with the *tert*-halobutane promoter were found to be more active than $[C_4mim]$

Cl—$AlCl_3$ with added water or sulfuric acid. Another result of these experiments was the gradual formation of a Brønsted acidic ionic liquid catalyst system with increasing catalyst deactivation over time, due to an increasing decomposition of the *tert*-butyl cation into H^+ and isobutene. This catalyst deactivation resulted in inconstant product selectivities.

6.2.2.3 Oligomerisation and Polymerisation Lewis acidic chloroaluminate(III) ionic liquids are known to catalyse the oligomerisation/polymerisation (Scheme 6.5) of olefins via a cationic reaction mechanism, even in the presence of the proton scavenger dichloroethylaluminium(III) [41,42].

Two mechanisms of the carbocation formation from olefins are possible in acidic chloroaluminate ionic liquids (Scheme 6.6).

In 1995, BP patented the polyisobutene (PIB) synthesis using isobutene-rich C_4 cuts as feedstock [43]. This technology offers many advantages compared to aluminium(III) chloride or boron(III) fluoride catalysts: simple product separation due to biphasic reaction mode, increased rate of polymerisation, minimisation of consecutive reactions of the olefinic end group in the polymer (e.g. isomerisation) as the polymer product forms a separate phase, higher molecular weight of the polymer (even at higher reaction temperatures) and incorporation of butene in the polymer product. Thus, the direct use of raffinate is possible, without prior isolation of isobutene.

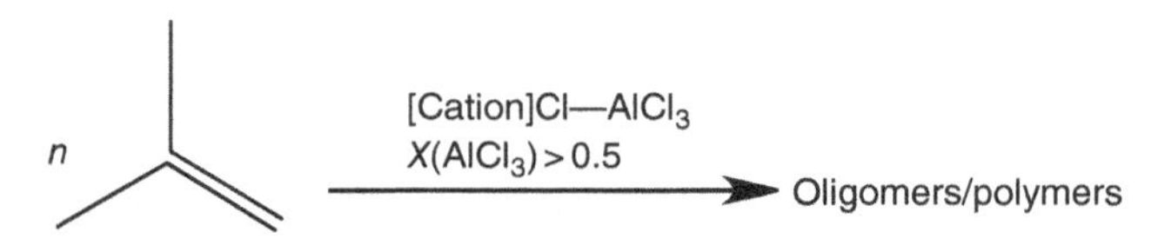

Scheme 6.5 Olefin oligomerisation/polymerisation.

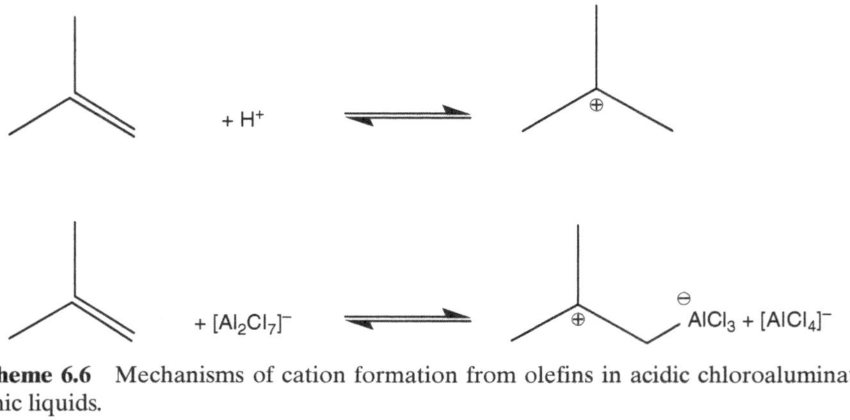

Scheme 6.6 Mechanisms of cation formation from olefins in acidic chloroaluminate ionic liquids.

In 2000, Symyx and Bayer patented the polymerisation of isobutene into very high molecular weight polymers (average molecular weight of $M > 100{,}000\,g\,mol^{-1}$), which possess low oxygen permeability and mechanical resilience and are therefore used in the automobile industry as rubber products [44]. Chloroaluminate(III) ionic liquid catalysts are advantageous because of the need of extremely low temperatures for the production of very high molecular weight polymers. The possibility of PIB production was also shown by BASF in 2008 [45]. The use of alkyl-substituted pyrazolium cations in combination with heptachlorodialuminate anions distinguishes this invention: the average molecular weight was more than four times higher ($M = 700{,}000\,g\,mol^{-1}$) than the product obtained from the $[C_2mim]Cl{-}AlCl_3$ system ($M = 150{,}000\,g\,mol^{-1}$).

Synthetic lubricating oils with low pour points and high viscosity indices are typically synthesised by oligomerisation of α-olefins, like 1-decene or 1-dodecene. In 1997, BP showed that it is possible to oligomerise a mixture of $C_6{-}C_{12}$ α-olefins to high-viscosity polyalphaolefins (PAOs) using Lewis acidic ionic liquid catalysts [46]. Chevron described later a process for the production of very-high-viscosity PAOs using the acidic ionic liquid $[N_{1\,1\,1\,0}]Cl{-}AlCl_3$ [47]. The reaction was carried out in the absence of organic solvents, which hitherto have been used as a diluent for the feed that consists of 1-decene or 1-dodecene. The continuous process for the production of PAOs using the same catalysts was patented by Chevron in 2003 [48].

6.2.2.4 Acidic Ionic Liquid-Catalysed Cracking of Hydrocarbons Catalytic cracking is a very important refinery process to convert long-chain hydrocarbons into shorter-chain lengths of higher market demand. One example is the catalytic cracking of polyethylene, an important reaction to recycle polyethylene waste by conversion into valuable short-chain hydrocarbons. Acidic chloroaluminate(III) ionic liquids ($X(AlCl_3) = 0.67$) were found to crack polyethylene in the presence of protons $[C_2mim][HCl_2]$ (1 mol %) or sulfuric acid (2 mol %) ($T = 393–473\,K$) while producing gaseous alkanes and cyclic alkanes [49]. No significant amount of aromatic compounds or olefins was formed.

Cracking of linear or branched alkanes (Scheme 6.7) was possible below 343 K with the acidic ionic liquid 1-hexylpyridinium chloride–aluminium(III) chloride, $[C_6py]Cl{-}AlCl_3$, system [50]. The addition of a proton scavenger, like dichloroethylaluminium(III) or calcium hydride, reduced the rate of cracking, while the addition of protons in the form of hydrogen chloride increased the catalytic cracking activity.

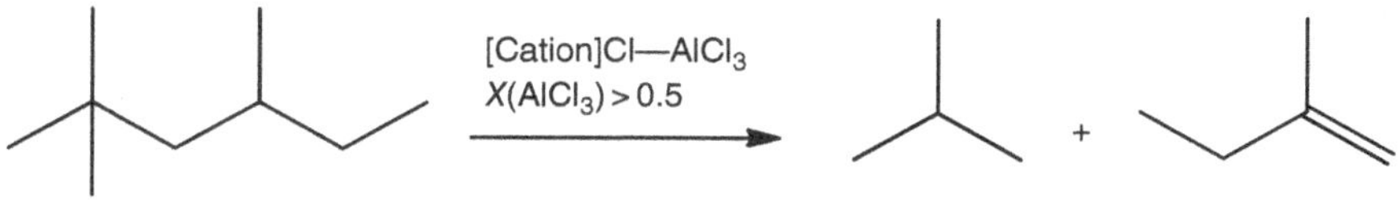

$[Cation]Cl{-}AlCl_3$
$X(AlCl_3) > 0.5$
+

Scheme 6.7 Alkane cracking using acidic ionic liquid systems.

6.3 IMMOBILISATION CONCEPTS FOR TRANSITION METAL CATALYSTS USING IONIC LIQUIDS

Apart from acidic catalysis, ionic liquids have been intensively tested in the last two decades for the immobilisation of homogeneously dissolved transition metal catalysts. Successful catalyst immobilisation techniques are essential for industrial homogeneous catalysis to solve the problem of catalyst/product separation and to recover and recycle the often very expensive dissolved transition metal complexes. Different immobilisation concepts applying ionic liquids have been developed, including the use of organic–ionic liquid multiphase reaction systems and the use of SILP catalysis. These concepts will be described in the following sections.

6.3.1 Organic–Ionic Liquid Biphasic Catalysis

Multiphase reaction systems for homogeneous transition metal catalysis consist of one – often polar – phase that contains the dissolved catalyst (the so-called immobilising phase or catalyst phase) and at least one second phase that preferably contains the reaction products. Generally, a suitable catalyst phase must exhibit the following properties:

- Sufficient solubility for the reactants and rapid mass transfer of the feedstock into the catalyst phase
- A pronounced miscibility gap with the desired product and by-products
- Excellent solvation of the catalyst to ensure full catalyst immobilisation and no leaching
- No deactivation of the immobilised catalyst by the catalyst solvent

Realistic options for using multiphase catalysis with dissolved transition metal catalysts have been described in the literature for organic–organic, organic–aqueous, organic–fluorous phase, organic–supercritical fluids, organic–liquid polymer, organic–ionic liquid and supercritical fluid–ionic liquid solvent pairs [51]. Prominent examples of successfully industrially applied multiphase reaction systems are the Shell higher olefin process (SHOP) (organic–organic) [52a] and the Ruhrchemie/Rhône-Poulenc's oxo process (organic–water) [52b].

A schematic representation of a continuous liquid–liquid biphasic catalytic process is shown for the case of an organic–ionic liquid system in Figure 6.2. Over the last decade, hundreds of successful examples ionic liquid-based liquid–liquid biphasic catalysis have been reported, far too numerous to mention here. The reader more interested in a complete picture of this huge research activity is referred to the large number of excellent reviews on the topic, with recent examples originating from Wu et al. [53]; Geldbach [54], Parvulescu and Hardacre [55]; Haumann and Riisager [56]; Wasserscheid and Kuhlmann [57]; Olivier-Bourbigou et al. [58]; and Jutz et al. [59].

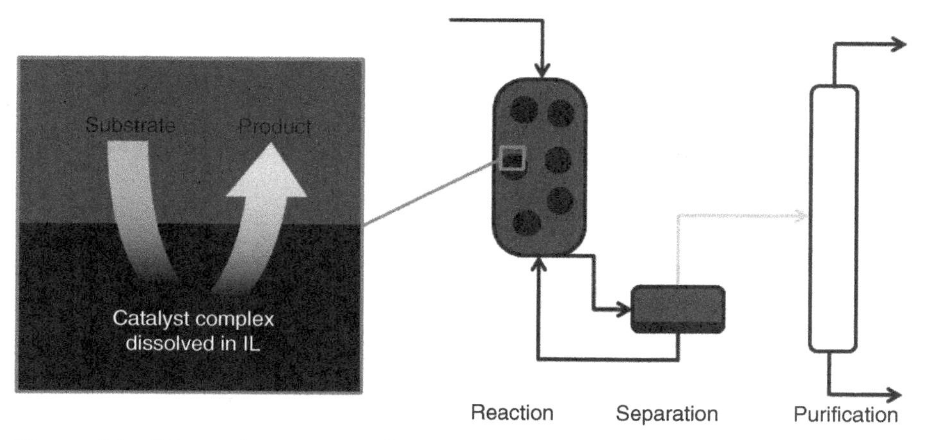

Figure 6.2 Schematic representation of continuous organic–ionic liquid biphasic catalysis – general principle (left) and process flow scheme (right).

When applying ionic liquids as the catalyst solvent in liquid–liquid biphasic catalysis, additional challenges arise from mass transfer limitations between the organic substrate/product phase and the ionic liquid catalyst phase. Note that, in general, the relatively high viscosity of ionic liquids compared to classical organic solvents [60] slows mass transfer processes down. To a rough approximation, diffusion rate is a function of the size of the diffusing molecule and the viscosity of the medium. High stirring rates or special mixing devices (e.g. static mixers) have to be thus applied to increase the mass transfer rate between organic and ionic liquid phases and to allow full utilisation of the ionic liquid catalyst solvent in the case of a fast reaction. The interplay between the mass transfer rate and the rate of reaction is illustrated in Figure 6.3, for the case of a fast catalytic reaction in a highly viscous ionic liquid. The organic substrate **1** from the low-viscosity organic phase reaches the phase boundary with little mass transfer resistance caused by the diffusion layer on the organic side. At the phase boundary, a step change in concentration occurs that reflects the equilibrium solubility of substrate **1** in the respective ionic liquid, according to the two-film model. Finally, in the diffusion layer on the ionic liquid side, slow mass transfer and fast reaction result in complete consumption of substrate **1**. In this case, no reaction takes place in the bulk of the ionic liquid simply because there is no substrate **1** left to react (reaction engineering details on fluid–fluid reaction systems can be found in chemical engineering textbooks, such as [20]).

As illustrated in Figure 6.3, many liquid–liquid biphasic reactions using active catalysts dissolved in ionic liquids suffer from severe mass transport limitation issues. Thus, only a small part of the ionic liquid takes effectively part in the reaction, a fact that the reaction engineer calls 'low ionic liquid utilisation'.

One very attractive way to circumvent this problem is to apply the catalytic ionic liquid in the form of a supported thin film. Such supported thin films

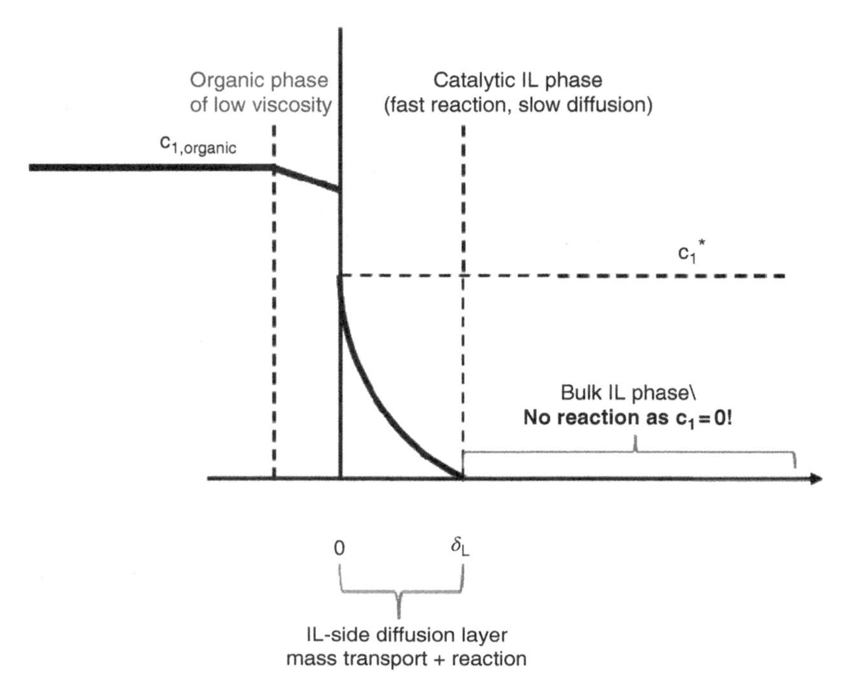

Figure 6.3 Concentration profile of a substrate **1** in liquid–liquid biphasic catalysis, assuming a fast catalytic reaction and slow mass transport – all catalytic reaction takes place in the diffusion film on the ionic liquid side, close to the phase boundary.

provide the whole volume of the ionic catalyst solution close to the fluid–fluid phase boundary, thus within the diffusion layer. The realisation of this concept has been introduced into the literature under the term 'SILP' catalysis.

6.3.2 SILP Catalysis

A SILP catalyst comprises a thin film of ionic catalyst solution immobilised on a highly porous support material. Outstanding advantages of the SILP system compared to a classical biphasic system are the very thin film of ionic liquid and the high specific surface area. Thus, mass transfer influences can be excluded as long as there is no other liquid in the pores of the support. The absence of mass transfer limitation leads to an extremely efficient use of the ionic liquid catalyst volume and, therefore, to a higher catalytic productivity with respect to the applied amount of ionic liquid and catalytic metal. Besides, the SILP concept combines in a very attractive manner the advantages of classical homogeneous catalysis (high activity under mild condition and high, ligand-modified selectivity) and heterogeneous catalysis (simple product separation, catalyst recycling and continuous processing). Typically, the ionic liquid is held on the porous support by physisorption. Capillary forces and coulombic interactions (e.g. between the ionic liquid ions and the surface —OH groups of the support) play an important role in immobilising the ionic liquid on the porous support.

SILP catalysts are highly suitable for continuous gas-phase processes in a fixed-bed reactor due to the extremely low volatility of typical ionic catalyst solutions. A slurry-phase reaction mode, in contrast, is only reasonable if the cross-solubility of the physisorbed ionic liquid with the reactant/product/solvent phase is extremely low, as in the case of an acidic chloroaluminate(III) ionic liquid and an alkane.

SILP catalysis was introduced by Mehnert, in 2002, for slurry-phase hydroformylation and hydrogenation reactions [61, 62]. Shortly later, Riisager et al. published the first successful example of continuous gas-phase SILP catalysis [63]. Recently, many technically relevant examples of SILP catalysis have been published, including examples of hydroformylation [64, 65], hydrogenation [66, 67], enantioselective hydrogenation [68, 69], water–gas shift reaction [70, 71], alkene metathesis [72, 73], hydroamination [74] and carbonylation of methanol [75]. Additionally, valuable information about SILP catalysis has been collected in reviews by Riisager et al. [76, 77], Gu and Li [78], van Doorslaer et al. [79] and Virtanen et al. [80].

The ionic liquid film distribution in SILP catalysts was studied by solid-state NMR spectroscopy [81]. In these studies, no homogeneous ionic liquid film was observed, but island formation was proposed for ionic liquid loadings less than 10 vol. % (referred to the total pore volume of the porous support). Lemus et al. characterised SILP materials that were prepared with different supports [82]: pore size distributions with varying amounts of $[C_8mim][PF_6]$ on activated carbon were measured and analysed with a combination of adsorption/desorption isotherms of N_2 at 77 K and mercury porosimetry. They suggested hierarchical pore filling: firstly the micropores, followed by the mesopores and finally the macropores. Lercher and co-workers reported on the complete coverage of the porous silica surface with ionic liquid analysed by infrared (IR) absorption spectroscopy [83]. Jess et al. investigated $[C_4mim][C_8SO_4]$ loadings on porous solid catalyst Ni/SiO_2 [84]: N_2 adsorption measurements indicated that the surface area decreases with an increasing degree of ionic liquid pore filling. The micro- and mesopores were partially or completely filled for a pore filling degree of 20 vol. %. These authors concluded and calculated that a monolayer with a thickness of 0.5 nm was formed at an ionic liquid pore filling degree of around 10 vol. %. The micro- and mesopores were successively blocked with increasing ionic liquid loadings.

It should also be mentioned that SILP catalysis has not only been successfully applied for the immobilisation of homogeneously dissolved transition metal complexes but also for acidic catalysis. The first example of an immobilised acidic chloroaluminate(III) on a support was reported, as early as 2000, by Hölderich and co-workers for the alkylation of different aromatics (benzene, toluene, naphthalene and phenol) [85]. The acidic ionic liquid $[C_4mim]Cl{-}AlCl_3$ was added to dried and calcined supports (SiO_2, Al_2O_3, TiO_2 or

ZrO_2), and the excess of ionic liquid was extracted with dichloromethane in a Soxhlet extraction system. Formation of HCl was observed during immobilisation, indicating reaction of the chloroaluminate(III) anion with the metal-OH groups of the support. Only the silica-based supports were active in the alkylation reaction. In 2002, the group of Hölderich published different covalent immobilisation methods for acidic ionic liquids on porous supports [86]. Systems with either anion or cation covalently bonded to silica have been reported.

Joni et al. developed a well-defined and optimised pretreatment of silica support materials prior to impregnation with acidic chloroaluminate(III) ionic liquid [87]. A solution of $[C_2mim]Cl{-}AlCl_3$ ($X = 0.67$) and dichloromethane was contacted with the calcined silica. After this chemical pretreatment step, surface —OH groups of the silica support were completely removed as the anion was covalently bound to the basic surface sites of the silica support. The pretreated supports themselves showed no catalytic activity in the investigated reaction, diisopropylbenzene isomerisation. The surface-modified support was shown to give reproducible results after a further immobilisation of a defined amount of acidic chloroaluminate ionic liquid. Pictures of this acidic SILP catalyst and the Lewis acidic ionic liquid $[C_2mim]Cl{-}AlCl_3$ ($X = 0.67$) are shown in Figure 6.4. In contrast to the work reported by Hölderich, this additional ionic liquid film remains free-flowing on the support's surface while being fixed to the support by capillary forces and physisorption. These SILP catalysts were successfully tested in the slurry-phase Friedel–Crafts alkylation of cumene and in the continuous gas-phase isopropylation of cumene and toluene. The catalysts operated with constant high selectivity and unchanged catalytic activity for 210 h time on stream [88].

Figure 6.4 Lewis acidic ionic liquid $[C_2mim]Cl{-}AlCl_3$ ($X = 0.67$) (left), acidic SILP catalyst $[C_2mim]Cl{-}AlCl_3$ ($X = 0.67$) immobilised on pretreated silica, according to Joni [88] (right).

6.4 SCILL

The main purpose of a SCILL system is not to immobilise an ionic liquid, but to modify the catalytic reactivity of a solid surface. For this purpose, the internal surface of a classical heterogeneous catalyst is coated with a thin film of ionic liquid. A schematic picture of a SCILL catalyst is shown in Figure 6.5.

The following effects on the activity and selectivity of the solid catalyst have been observed through such ionic liquid coatings:

- The ionic liquid coating may change the effective concentration of reactant(s) and intermediate(s) at the solid catalytic site compared to the uncoated solid catalyst, according to the solubility of the gaseous reactants in the ionic liquid and related mass transfer processes [89].
- The ionic liquid may interact with the solid catalytic sites and thereby modify its adsorption and reaction properties [90, 91]; this includes the possibility that the ionic liquid may poison specific catalytic sites of the catalytic material, thus suppressing unwanted side reactions [92].

Jess and co-workers introduced the SCILL concept using a consecutive reaction as model transformation, the hydrogenation of cyclooctadiene to cyclooctene and cyclooctane [90, 93]. A commercial Ni/SiO_2 catalyst was coated with the ionic liquid [C_4mim][C_8SO_4]. The effect of ionic liquid coating on the selectivity for the intermediate cyclooctene was very pronounced. One explanation for the much higher yield (70%) of this intermediate by applying the SCILL catalyst compared to the uncoated Ni/SiO_2 catalyst (40%) is the lower

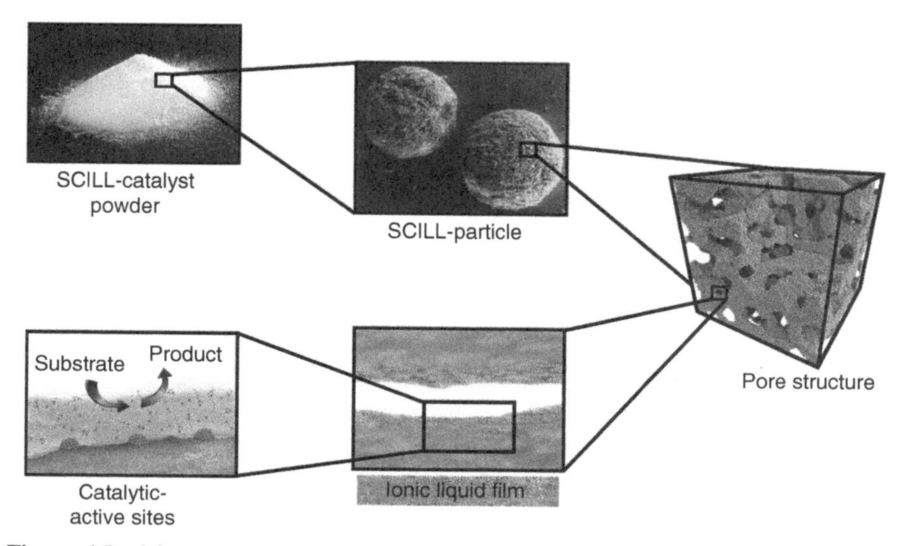

Figure 6.5 SCILL catalyst materials – schematic view on a classical heterogeneous catalyst coated with a thin layer of ionic liquid.

solubility of the intermediate product, which decreases the reaction rate for the consecutive hydrogenation. However, the influence of the ionic liquid layer on the effective concentrations of cyclooctadiene and cyclooctene alone could not explain the total change in selectivity. In fact, the ionic liquid acted probably as a competing adsorbent and inhibited the (re)adsorption of cyclooctene on the nickel sites. The ionic liquid coating led also to a decrease of specific surface area of the porous catalyst and to lower concentrations of cyclooctadiene in the ionic liquid than in the organic phase. Both factors led to a decline in activity. Although the SCILL catalyst was used in liquid-phase catalysis, no leaching of the ionic liquid into the organic phase was detectable. The Jess group extended their SCILL research to further selective liquid- and gas-phase hydrogenations [85]. For those systems in which the feedstock is less soluble in the ionic liquid compared to the organic phase, the reaction rate was lowered (cyclooctadiene, octyne, naphthalene). For a higher solubility, the effect was reversed, as for cinnamaldehyde. The same correlation is valid for the selectivity of intermediates. For intermediates with a lower solubility in the ionic liquid than in the feed (octene compared to octyne, hydrocinnamaldehyde compared to cinnamaldehyde and tetraline compared to naphthalene), the maximum yield of the intermediate increased compared to the uncoated catalyst. The yield for the intermediate cinnamyl alcohol, which has a higher solubility than the feedstock cinnamaldehyde, decreased when the SCILL catalyst was applied.

The group of Claus investigated SCILL catalysts in the regioselective hydrogenation of citral [91, 94]. Supported palladium catalysts were coated with $[N(CN)_2]^-$-, $[NTf_2]^-$- or $[PF_6]^-$-based ionic liquids, respectively. The modification of the applied palladium catalyst on silica by ionic liquid coating led to very high selectivities for citronellal as consecutive and side reactions were inhibited effectively. The best results were achieved with the dicyanamide ionic liquids (90% selectivity to citronellal at 80% conversion). Except for $[C_4mim][PF_6]$, the citral conversions were in a similar range compared to the uncoated catalyst. The lower dihydrogen solubility in ionic liquids compared to organic solvents [95] did not influence the observed reaction rates in a negative manner.

Further studies were conducted to investigate the interaction of the applied ionic liquid with the active palladium. Such investigations form the basis for a better molecular understanding of promoting effects in the citral hydrogenation [91] and other applications of SCILL catalysis. The coating of Pd/SiO_2 with ionic liquids was shown to lower the dihydrogen uptake and enthalpies of adsorption remarkably. The reduction of H_2 uptake was much higher for $[N(CN)_2]^-$-based ionic liquids compared to $[NTf_2]^-$-based ionic liquids. X-ray photoelectron spectroscopy (XPS) measurements revealed that ionic liquids deposited on Pd/SiO_2 modify the chemical state of surface palladium, which is partially transformed to Pd^{2+} and undergoes complexation with the $[N(CN)_2]^-$: the ionic liquid anion acts as ligand to the metal. This finding was also underlined by IR spectroscopy. Besides lower dihydrogen solubility in ionic liquids, the coordination between ionic liquid and palladium may also contribute to a lower accessibility of dihydrogen to Pd surface sites. X-ray absorption spectroscopy (XAS) characterisation excluded an influence of the ionic liquid on the metal dispersion, which may affect dihydrogen uptake and adsorption strength. Selective citral hydrogenation with SCILL catalysts based on palladium and the ionic liquid $[C_4mim][N(CN)_2]$ was also applied continuously in a trickle-bed reactor [96].

The SCILL catalyst system $Ru/Al_2O_3/[cation][NTf_2]$ also influenced the conversion and selectivity of the citral hydrogenation [97]. The conversions decreased for all SCILL catalysts in comparison to the Ru/Al_2O_3 system. Lower dihydrogen solubility in ionic liquids compared to organic solvents accounts reasonably for this observation. Bistriflamide-based ionic liquids inhibit primarily the C=C hydrogenation, which leads to different product selectivities.

Silica-supported platinum catalysts with a thin film of $[C_4dmim][OTf]$ (1-butyl-2,3-dimethylimidazolium trifluoromethanesulfonate) were investigated by the Lercher group, with a focus on the interactions between ionic liquid, support and platinum particles [83]. The interaction of silica and the ionic liquid occurs via hydrogen bonds. Platinum clusters modify the electron density of the ionic liquid, which changes the polarity of the ionic liquid within certain levels. The ionic liquid protects the platinum cluster from oxidation, as the platinum particles in ionic liquid-coated systems are in the zero oxidation state. SCILL-catalysed ethene hydrogenation showed that the interaction of the ionic liquid with the platinum surface is weaker than the interaction of ethene and the platinum surface.

The groups of Steinrück, Libuda and Wasserscheid studied the interaction of the ionic liquid $[C_4mim][NTf_2]$ with platinum and palladium nanoparticles supported on thin and ordered alumina films under UHV conditions [91, 98]. The interaction of $[C_4mim][NTf_2]$ with the nanoparticles is strong enough to partially replace strongly chemisorbed adsorbates, like carbon monoxide. The ionic liquid showed a ligand-type effect that is specific for the applied ionic liquid–metal combination and selective to particular sites.

To summarise SILP and SCILL catalysis, a comparison of the characteristic features of these two ionic liquid thin film technologies is given in Table 6.3. This table is intended to highlight important differences and some similarities.

Since the first reports on SILP and SCILL catalysis, the industrial interest in these two technologies has been very high. Both technologies aim to realise beneficial ionic liquid features in catalysis with the smallest possible amount of ionic liquid in a most efficient manner. This greatly reduces the ionic liquid specific investment and all potential ionic liquid risk that scales with the applied quantity of ionic liquid. In process technologies, the immobilisation of the ionic liquid limits the ionic liquid-related risk to a small section of the overall plant. The presence of ionic liquid in pumps, storage tanks or recycling loops can be avoided if the entire ionic liquid remains located in one particular unit operation. Moreover, the remaining risk is further reduced if the supported

TABLE 6.3 Comparison of SILP and SCILL Catalysts

	SILP	SCILL
Catalyst nature	Homogeneous metal catalyst dissolved in an ionic liquid and immobilised on porous support	Heterogeneous catalyst coated with ionic liquid
Purpose	Immobilisation of homogeneous catalyst	Modification of heterogeneous catalyst, activity and/or selectivity improvement
Reactant/product solubility	Optimal: high for reactants and low for products	Optimal: high for reactants and low for products
Ionic liquid loading	Low for high specific surface area but enough for homogeneous dissolution of metal (complex)	Low for high specific surface area but enough to cover active centres
Ionic liquid–support interaction	Good wettability required	Good wettability required
Metal and metal complex solubility	As high as possible	As low as possible
Ionic liquid–metal interaction	Desired to stabilise catalyst in solution	Desired to enhance activity and/or selectivity
Maximum reaction temperature	Limited by stability of metal (complexes) and ligand	Limited by ionic liquid stability

ionic liquid material can be tested in the form of a drop-in solution in existing fixed-bed reactors of miniplants or pilot plants. In catalyst development, nothing is more convincing than a proven superior performance in a 10,000 h long-term test run under commercially relevant or even fast-ageing conditions!

Some successful developments into this level of commercial relevance have also triggered the question of reproducible and efficient large-scale production of SILP and SCILL materials. The state-of-the-art preparation method on the laboratory scale is an incipient wetness-type impregnation of the support material (SILP) or heterogeneous catalyst (SCILL) by a solution of the ionic liquid in an auxiliary solvent. This is followed by subsequent removal of that auxiliary solvent under vacuo. Typically, the solvent removal is carried out in an ordinary rotary evaporator, which limits this kind of preparation to batches of a few hundred grams of material. Recently, a new and scalable preparation method has been published using fluidised-bed spray coating to disperse the

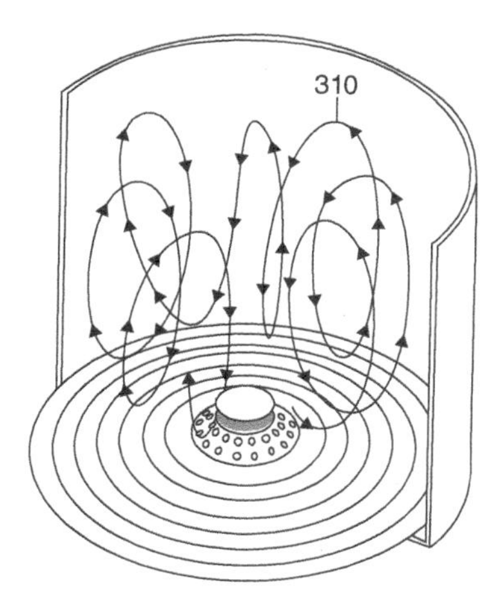

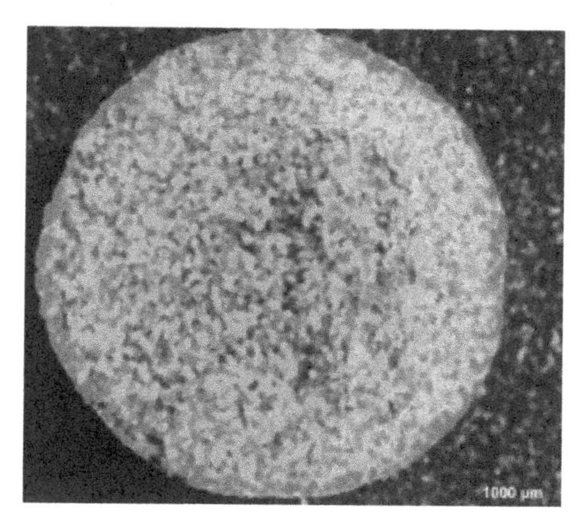

Figure 6.6 Large-scale production and characterisation of SILP and SCILL catalyst materials. Left: a schematic drawing of the toroidal movement of the support particles in an air coater during impregnation with the solvent-diluted ionic liquid catalyst solution. Right: an SEM–EDX picture for material characterisation, exemplified for a spherical γ-alumina coated with [{$Ru(CO)_3Cl_2$}$_2$] in [C_4dmim][OTf] to check the elemental distribution: ruthenium (dark grey area at the external surface of the catalyst pellet); sulfur (light grey spots distributed over the catalyst pellet) [100].

ionic liquid film onto the support's surface [99]. It involves the fluidisation of the uncoated support or catalyst material by means of a temperature-controlled gas flow, such as air, dinitrogen or argon. Once the support material is fluidised, a solution of the ionic liquid in an auxiliary solvent is sprayed onto the material through a nozzle. The auxiliary solvent is rapidly evaporated due to the gas flow, resulting in a good dispersion of the ionic liquid onto the support. The fluidised bed, with its inherent mixing pattern, ensures that the support material is well mixed and evenly coated by the ionic liquid film (Figure 6.6).

The process described has been already applied to produce SILP and SCILL materials on the 100 kg scale. In general, the production of SILP and SCILL materials by fluidised-bed coating is easily scalable from grams to tons and allows catalyst production with high reproducibility and at low costs. It is anticipated that these recent advances in SILP and SCILL material production will significantly support the successful further implementation of these materials into more and more industrial catalytic processes.

6.5 STABILISATION AND IMMOBILISATION OF NANOPARTICLES IN IONIC LIQUIDS

Ionic liquids are also valuable media to prepare and stabilise catalytic nanoparticles [100]. In most cases, the reactions are multiphase systems. The nanoparticles dispersed in the ionic liquid form the dense phase, and the substrate and product remain in the upper organic phase. Hence, the ionic liquid–nanoparticle phase is easily recovered by simple decantation. Metal nanoparticles of 1–10nm diameters have unique properties that result from their large surface-to-volume ratio and quantum size effects [101]. Nanoparticles are only kinetically stable because the formation of bulk metal is thermodynamically favoured. Thus, nanoparticles must be stabilised in order to prevent agglomeration by electrostatic and/or steric protection. Ionic liquids have been shown to be a promising medium for the synthesis and stabilisation of nanoparticles [102, 103]. Transition metal nanoparticles are probably stabilised by protective layers of supramolecular structures $\{[(C_n mim)_x(X)_{x-n})]^{n+}[(C_n mim)_{x-n}(X)_x)]^{n-}\}_m$ (X = anion) through the loosely bound anionic moieties and/or *N*-heterocyclic carbenes together with an oxide layer, if present, on the metal surface. However, the surface-bound protective species are easily displaced by other substances in the media. This fact explains on the one hand to some extent their catalytic activity and on the other hand their relatively low stability against aggregation/agglomeration. More stable catalytic systems in ionic liquids can be obtained by the addition of ligands or polymeric stabilisers.

Transition metal nanoparticles dispersed in ionic liquids are active catalysts for various reactions, such as the hydrogenation of alkenes, arenes or ketones [102]. Further examples of ionic liquid-based catalysis with metal nanoparticles are described in the reviews by Gu [78] and Dupont [101].

Interestingly, first examples can also be found in the literature describing the immobilisation of an ionic liquid in which metal nanoparticles are immobilised on solid supports. This concept follows in close analogy the idea of SILP catalysis but applies ionic liquid-stabilised nanoparticles on a support instead of in an ionic catalyst solution [104].

6.6 IONIC LIQUIDS IN SEPARATION PROCESSES

In most cases, chemical reactions in industry are neither 100% selective nor operated at full conversion. This makes efficient separation processes a crucial part of chemical technology. Separation units are costly both in investment and operation. 60–80% of the total cost of a chemical plant is typically attributed to its separation units [20]. In industrial separation processes, the rules of chemical thermodynamics are applied to separate the different components of a reaction mixture and to obtain the product in the desired purity.

Distillation/rectification is by far the most common separation process in the chemical industry. Here, the difference in vapour pressure/fugacity of the different components of a reaction mixture is used as the driving force for separation. Extractive distillation, extraction and absorption processes gain importance, if distillation proves unfeasible, for example, due to azeotropic points or very small separation factors. Extraction uses the relative difference in activity coefficients to separate components in a liquid–liquid biphasic system, whereas absorption processes are employed to remove specific compounds from a gas stream. All three processes utilise specific solvents to alter the partition coefficients in a favourable way. Ionic liquids, with their tuneable miscibility and solubility properties, are very attractive candidates for this task and have been intensively studied in the last decade, both by academic groups and industrial companies.

6.6.1 Extractive Distillation

Extractive distillation is based on addition of a solvent to the product mixture that selectively interacts with one component of the mixture. By doing so, the non-ideality of the resulting mixture enables effective separation of components that would be otherwise difficult to separate. By addition of a volatile solvent, often called the 'entrainer' in the engineering literature, an azeotropic distillation process is established. The addition of a non-volatile solvent results in an extractive distillation process [105]. The recovery and reuse of the selective solvent after separation are crucial for the process economics. Here, the application of an almost non-volatile entrainer, as in the case of ionic liquids, causes a redesign of existing separation units and restricts the opportunities for simple drop-in solutions. Despite this complication, the use of ionic liquids as selective entrainers is highly interesting due to their tuneable solvent nature. The technology is covered by a patent by Arlt and BASF AG [106]. Ionic liquid entrainers are usually applied in amounts of 30–50 wt. % [107]. This amount is smaller than required in the case of classic organic entrainers, such as dimethylmethanamide. One additional advantage of using an ionic liquid entrainer arises from its extremely low vapour pressure, saving the solvent regeneration column and thus a lot of energy. A benchmark calculation carried by BASF for a technically relevant example has revealed a saving potential of *ca.* 37% for energy cost and 22% for the investment by replacing a classical entrainer by an ionic liquid [108]. BASF has operated an extractive distillation process with an ionic liquid entrainer in a pilot plant continuously for 3 months. The applied ionic liquid maintained its full performance over this testing time. Relevant application scenarios for extractive distillation with an ionic liquid entrainer are organic–water separations (e.g. tetrahydrofuran/water [109]), alkene–alkane separations (e.g. 1-hexene/hexane [110, 111]) and aromatic–aliphatic separations [112].

6.6.2 Extraction

Extraction requires a miscibility gap between two liquids, a pre-requisite that is fulfilled by many ionic liquids with respect to unipolar organic solvents. In addition, highly fluorinated ionic liquids also offer typically a miscibility gap with water. For all extractions with ionic liquids, the knowledge of the distribution coefficients and the ionic liquid selectivity is crucial. From the engineering point of view, it is important to note that the higher viscosity of ionic liquids compared to organic solvents changes the hydrodynamic design of extraction columns drastically [113]. More knowledge about coalescence phenomena in biphasic mixtures containing ionic liquids is required for a more efficient design of industrial extraction equipment. Relevant examples of ionic liquid extraction processes include the removal of heavy metal traces from water (either using complexing agents dissolved in the respective ionic liquid [114, 115] or functionalised ionic liquids carrying the complexing group in one of their ions) have been successfully applied in metal ion extraction. These systems work both as extractants and as solvents [116, 117] and for the removal of aromatic from aliphatic hydrocarbon compounds [118, 119], fuel desulfurisation [120, 121] and protein extraction [122, 123].

6.6.3 Absorption

Absorption requires a detailed knowledge of the solubility of gases in ionic liquids. Many groups have worked in these important areas over the last decade (see, e.g. Refs. 124, 125). In general, one can state that ionic liquids show low gas solubilities for dihydrogen, dioxygen, dinitrogen, argon and carbon monoxide, while carbon dioxide, ammonia, nitrogen oxides and sulfur dioxide show generally much higher solubilities in ionic liquids. Apart from selective solubility of a gas component in the ionic liquid, the capacity of a given ionic liquid for the respective gas component is also important. Low capacity leads, in a technical realisation, to a very large ionic liquid volume and to large and costly equipment. Regeneration of the absorber solvent is pretty straightforward in the case of ionic liquids, as flash distillation or stripping can be performed without the need to re-condensate the absorbent. Relevant examples of absorption processes including ionic liquid absorbents are CO_2 absorption from power plant exhaust gas streams [126–128] and the removal of S-containing gases from natural gas [129]. Supported ionic liquids have been also applied very successfully in absorption processes. Just very recently, PETRONAS, a Malaysian petrochemical company, disclosed at the EuCheMS Molten Salt and Ionic Liquids Conference 2012 in the Celtic Manor, Wales, the commercial operation of a SILP material for mercury removal from natural gas on a technical refinery scale (60 tons of SILP-type material) [130]. To the best of our knowledge, this marks the first publication of a commercial SILP application in the field of separation technologies. A very similar approach of SILP absorption had been reported before for the desulfurisation of natural gas using supported stannate(II) and zincate(II) ionic liquids [129].

6.6.4 Membrane Processes

Membrane processes offer another attractive option to apply ionic liquids in separation processes. Two general approaches have been reported: the use of ionic liquids as bulk ionic liquid membranes (BILM) [131, 132] or as supported ionic liquid membranes (SILM) [133, 134]. In the former case, the ionic liquid acts as selective separation layer; in the latter case, the membrane supports the liquid separation layer and separates two compartments, resulting in a very efficient ionic liquid use and separation unit. Reported examples for SILMs include reactions with *in situ* product removal [135], extraction of bioorganic substances [136, 137] and removal of CO_2 and SO_2 from gas streams [138, 139]. SILMs have also been successfully tested in the separation of alkenes and alkanes [140] and in the removal of aromatic hydrocarbons [141, 142].

6.7 CONCLUSIONS

There is absolutely no doubt that the unique properties of ionic liquids offer a great potential to improve existing engineering applications or to develop very attractive new ones. However, the largest part of the reported examples in literature still represents 'proof-of-concept' research, and important practical questions, such as recycling efficiency, ionic liquid recovery or the degree of ionic liquid degradation over time, have not yet been answered completely.

One important question that still largely determines the efficiency of the process development process is the rational selection of the best ionic liquid for a given engineering problem. This question cannot be discussed in all detail here, but it should be shortly addressed for the particular question of optimising the solubility of a specific molecular organic compound in ionic liquids, a property of utmost importance for almost all engineering application of ionic liquids.

While in recent years a steeply growing number of experimental solubility data have been published [143], in addition the first theoretical predictions have been described using the COSMO-RS approach [144]. The COSMO-RS method can be applied to predict the solubility of a compound in an ionic liquid in thermodynamic equilibrium. It starts from separate DFT calculations of the ionic liquid ions and the solute molecules. Interestingly, data files can be combined so that – for example – a file generated for the $[C_2mim]^+$ ion can be used in combination with different anion files and substrate files to calculate solubilities in different systems. This makes the method particularly efficient for ionic liquids. The DFT results for each moiety are converted into a statistics of surface elements of the same charge density. This distribution – the so-called σ-profile – is used to calculate the activity coefficient of each component by pairwise interacting the surface elements of the different components and minimising the system energy in this process. As one result of this exercise, the activity coefficient of the solute in the cation–anion mixture forming the ionic liquid is obtained, which allows calculating properties, such as solubility data, distribution coefficients or vapour pressures of volatile substances in the respective ionic liquid.

So far, this chapter has strongly focussed on the application of ionic liquids in catalytic applications and separation technologies. It should be noted, however, that ionic liquids also started to make, in the last decade, a significant impact in other areas of engineering, most importantly in the areas of lubrication, process machinery and absorption cooling.

Following the first pioneering studies in 2001 [145], the application of ionic liquids as lubricants and working fluids in process machines has developed dynamically. The first commercial application of this approach has already been realised, the so-called Linde ionic compressor [146]. In this type of compressor, an ionic liquid substitutes for the metallic piston for compressing the gas. The concept builds not only on the extremely low vapour pressure of the ionic liquid but also on its excellent tribologic behaviour and its very low solubility for the gases to be compressed. Most relevant, from a practical point of view, is the compression of dihydrogen to pressures up to 1000 bar using this technology. Advantages against the traditional metallic piston compressor are a drastic reduction in the number of movable parts, a very high volumetric efficiency and a *quasi*-isothermal compression (note that the ionic liquid piston can be cooled using a classical heat exchanger). Very recently, the same concept has also been applied to the compression of pure dioxygen, a gas that would undergo heavy oxidation reaction with normal organic lubricants [147]. Due to the high oxidation stability and strong flame resistance of the applied ionic liquid, even very pure dioxygen (quality 6.0) could be compressed with the ionic compressor to 16 bar for 3000 h service time, without any technical problem. Apart from the ionic compressor concept, ionic liquids have been proposed as lubricants in combustion engines, as hydraulic fluids, as working fluids in liquid ring pumps and as hydraulic oil in hydraulically driven diaphragm pumps [148].

Another very interesting engineering application of ionic liquids is their use in absorption chillers. It has been demonstrated that some of the major actual restrictions for a broader application of heat transformation technologies can be effectively overcome by introducing ionic liquid working pairs. Novel working pairs based on ionic liquids (e.g. $[C_2mim][O_2CMe]$/water) enable elimination of crystallisation and corrosion problems while meeting other crucial chiller requirements, including heat and mass transfer, as well as efficiency [149].

We are convinced that ionic liquids offer a huge potential for better, more efficient, safer or totally new, engineering applications, but there is still much to do. Alternative ionic liquid structures, more physico-chemical and engineering data, large-scale ionic liquid production, registration of new ionic liquids, improved theoretical prediction tools and the development of dedicated process units are on the way. The outcome of these efforts will certainly facilitate in the future the application of these fascinating liquids in more and more technical applications.

REFERENCES

1 Chauvin, Y. and Olivier-Bourbigou, H., Nonaqueous ionic liquids as reaction solvents, *Chem. Tech.* **25**(9), 26–30 (1995).

2 Huddleston, J.G., Willauer, H.D., Swatloski, R.P., Visser, A.E. and Rogers, R.D., Room temperature ionic liquids as novel media for 'clean' liquid-liquid extraction, *Chem. Commun.* 1765–1766 (1998).

3 Seddon, K.R., Ionic liquids: Designer solvents for green synthesis, *Chem. Eng.* **730**, 33–35 (2002).

4 (a) Hallett, J.P. and Welton, T., Room-temperature ionic liquids: Solvents for synthesis and catalysis. 2, *Chem. Rev.* **111**, 3508–3576 (2011); (b) Zhou, Q., Lu, X., Zhang, S. and Guo, L., 'Physicochemical Properties of Ionic Liquids', in *Ionic Liquids Further UnCOILed: Critical Expert Overviews*, Eds. N.V. Plechkova and K.R. Seddon, John Wiley & Sons, Inc., Hoboken, NJ (2013), pp. 275–307.

5 (a) Preiss, U.P., Beichel, W., Erle, A.M.T., Paulechka, Y.U. and Krossing, I., Is universal, simple melting point prediction possible? *Chem. Phys. Chem.* **12**, 2959–2972 (2011); (b) Shah, J.K. and Maginn, E.J., 'Molecular Simulation of Ionic Liquids: Where We Are and the Path Forward', in *Ionic Liquids Further UnCOILed: Critical Expert Overviews*, Eds. N.V. Plechkova and K.R. Seddon, John Wiley & Sons, Inc, Hoboken, NJ (2013), pp. 149–192.

6 National Institute of Standards and Technology, 'IUPAC ionic liquids database, ILThermo (NIST standard reference database #147)', http://ilthermo.boulder.nist.gov/ILThermo/mainmenu.uix, 2006 (accessed from 11 July 2015).

7 Sun, N., Rahman, M., Qin, Y., Maxim, M.L., Rodriguez, H. and Rogers, R.D., Complete dissolution and partial delignification of wood in the ionic liquid 1-ethyl-3-methylimidazolium acetate, *Green Chem.* **11**(5), 646–655 (2009).

8 Earle, M., 'Stoichiometric Organic Reactions and Acid-catalyzed Reactions in Ionic Liquids', in *Ionic Liquids in Synthesis*, 2nd Ed., Eds. P. Wasserscheid and T. Welton, Wiley-VCH, Weinheim (2008), pp. 292–367.

9 Wier Jr., T.P. and Hurley, F.H., Electrodeposition of Aluminum, U.S. Pat. 2,446,349 (1948).

10 Hurley, F.H. and Wier, T.P., Electrodeposition of metals from fused quaternary ammonium salts, *J. Electrochem. Soc.* **98**, 203–206 (1951).

11 Hurley, F.H. and Wier, T.P., The electrodeposition of aluminium from nonaqueous solutions at room temperature, *J. Electrochem. Soc.* **98**, 207–212 (1951).

12 Øye, H.A., Jagtoyen, M., Oksefjell, T. and Wilkes, J.S., 'Vapor Pressure and Thermodynamics of the System 1-Methyl-3-Ethylimidazolium Chloride – Aluminium Chloride', in *Molten Salt Chemistry and Technology*, Eds. M. Chemla and D. Devilliers, Materials Science Forum, Vol. **73**, Trans Tech Publications Ltd, Zurich (1991), pp. 183–189.

13 Noel, M.A.M., Trulove, P.C. and Osteryoung, R.A., Removal of protons from ambient-temperature chloroaluminate ionic liquids, *Anal. Chem.* **63**, 2892–2896 (1991).

14 Smith, G.P., Dworkin, A.S., Pagni, R.M. and Zingg, S.P., Brønsted superacidity of HCl in a liquid chloroaluminate – $AlCl_3$-1-ethyl-3-methyl-1H-imidazolium chloride, *J. Am. Chem. Soc.* **111**, 525–530 (1989).

15 Smith, G.P., Dworkin, A.S., Pagni, R.M. and Zingg, S.P., Quantitative study of the acidity of HCl in a molten chloroaluminate system ($AlCl_3$/1-ethyl-3-methyl-1H-imidazolium chloride) as a function of HCl pressure and melt composition (51.0–66.4 mol% $AlCl_3$), *J. Am. Chem. Soc.* **111**, 5075–5077 (1989).

16 Wang, Y., Gong, X., Wang, Z. and Dai, L., SO_3H-functionalized ionic liquids as efficient and recyclable catalysts for the synthesis of pentaerythritol diacetals and diketals, *J. Mol. Catal. A Chem.* **322**, 7–16 (2010).

17 (a) Sheldon, R.A., Consider the environment quotient, *Chem. Tech.* **24**(3), 38–47 (1994); (b) Sheldon, R.A., The E factor: Fifteen years on, *Green Chem.* **9**, 1273–1283 (2007).

18 Boon, J.A., Levisky, J.A., Pflug, J.L. and Wilkes, J.S., Friedel-Crafts reactions in ambient-temperature molten salts, *J. Org. Chem.* **51**, 480–483 (1986).

19 Abdul-Sada, A.K., Atkins, M.P., Ellis, B., Hodgson, P.K.G., Morgan, M.L.M. and Seddon, K.R., Process and catalysts for the alkylation of aromatic hydrocarbons, World Pat. 95 21806 (1995).

20 Jess, A. and Wasserscheid, P., *Chemical Technology*, Wiley-VCH, Weinheim (2013).

21 Sherif, F.G., Shyu, L.J., Greco, C.G., Talma, A.G. and Lacroix, C.P.M., Linear alkylbenzene formation using low-temperature ionic liquid and long-chain alkylating agent, WO Pat. 9803454 (1998).

22 Piao, L., Fu, X., Yang, Y., Tao, G.H. and Kou, Y., Alkylation of diphenyl oxide with α-dodecene catalyzed by ionic liquids, *Catal. Today* **93–95**, 301–305 (2004).

23 Xin, H.L., Wu, Q., Han, M.H., Wang, D.Z. and Jin, Y., Alkylation of benzene with 1-dodecene in ionic liquids Rmim + $Al_2Cl_6X^-$ (R = butyl, octyl and dodecyl; X = chlorine, bromine and iodine), *Appl. Catal. A Gen.* **292**, 354–361 (2005).

24 Joni, J., Schmitt, D., Schulz, P.S., Lotz, T.J. and Wasserscheid, P., Detailed kinetic study of cumene isopropylation in a liquid-liquid biphasic system using acidic chloroaluminate ionic liquids, *J. Catal.* **258**, 401–409 (2008).

25 Liu, Z.C., Meng, X.H., Zhang, R. and Xu, C.M., Friedel-Crafts acylation of aromatic compounds in ionic liquids, *Pet. Sci. Technol.* **27**, 226–237 (2009).

26 Adams, C.J., Earle, M.J., Roberts, G. and Seddon, K.R., Friedel-Crafts reactions in room temperature ionic liquids, *Chem. Commun.* 2097–2098 (1998).

27 Surette, J.K.D., Green, L. and Singer, R.D., 1-Ethyl-3-methylimidazolium halogenoaluminate melts as reaction media for the Friedel-Crafts acylation of ferrocene, *Chem. Commun.* 2753–2754 (1996).

28 Earle, M.J., Hakala, U., Hardacre, C., Karkkainen, J., McAuley, B.J., Rooney, D.W., Seddon, K.R., Thompson, J.M. and Wähälä, K., Chloroindate(III) ionic liquids: Recyclable media for Friedel-Crafts acylation reactions, *Chem. Commun.* 903–905 (2005).

29 Knifton, J.F., Process for producing *p*-tolualdehyde from toluene using an aluminum halide alkyl pyridinium halide 'melt' catalyst, US Pat. 4554383A (1985).

30 Brausch, N., Metlen, A. and Wasserscheid, P., New, highly acidic ionic liquid systems and their application in the carbonylation of toluene, *Chem. Commun.* 1552–1553 (2004).

31 Angueira, E.J. and White, M.G., Arene carbonylation in acidic, chloroaluminate ionic liquids, *J. Mol. Catal. A Chem.* **227**, 51–58 (2005).

32 Zhao, W.J. and Jiang, X.Z., Efficient synthesis of benzaldehyde by direct carbonylation of benzene in ionic liquids, *Catal. Lett.* **107**, 123–125 (2006).

33 Olivier, H., Chauvin, Y. and Hirschauer, A., Room-temperature molten-salts – both novel catalyst for isobutane alkylation and solvent for catalytic dimerization of alkenes, *Abstr. Pap. Am. Chem. Soc.* **203**(Pt 2), 780–785 (1992).

34 Chauvin, Y., Hirschauer, A. and Olivier, H., Alkylation of isobutane with 2-butene using 1-butyl-3-methylimidazolium chloride aluminum-chloride molten-salts as catalysts, *J. Mol. Catal.* **92**, 155–165 (1994).

35 Chauvin, Y., Hirschauer, A. and Olivier, H., Catalytic composition and process for the alkylation of aliphatic hydrocarbons, US Pat. 5750455 (1996).

36 Ya, K.S., Namboodiri, V.V., Varma, R.S. and Smirniotis, P.G., Ionic liquid-catalyzed alkylation of isobutane with 2-butene, *J. Catal.* **222**, 511–519 (2004).

37 Huang, C.P., Liu, Z.C., Xu, C.M., Chen, B.H. and Liu, Y.F., Effects of additives on the properties of chloroaluminate ionic liquids catalyst for alkylation of isobutane and butene, *Appl. Catal. A Gen.* **277**, 41–43 (2004).

38 Liu, Y., Hu, R., Xu, C. and Su, H., Alkylation of isobutene with 2-butene using composite ionic liquid catalysts, *Appl. Catal. A Gen.* **346**, 189–193 (2008).

39 Liu, Z.C., Zhang, R., Xu, C.M. and Xia, R.G., Ionic liquid alkylation process produces high-quality gasoline, *Oil Gas J.* **104**(40), 52–56 (2006).

40 Aschauer, S., Schilder, L., Korth, W., Fritschi, S. and Jess, A., Liquid-phase isobutane/ butene-alkylation using promoted lewis-acidic IL-catalysts, *Catal. Lett.* **141**, 1405–1419 (2011).

41 Goledzinowski, M., Birss, V.I. and Galuszka, J., Oligomerization of low-molecular-weight olefins in ambient-temperature molten salts, *Ind. Eng. Chem. Res.* **32**(8), 1795–1797 (1993).

42 Stenzel, O., Brull, R., Wahner, U.M., Sanderson, R.D. and Raubenheimer, H.G., Oligomerization of olefins in a chloroaluminate ionic liquid, *J. Mol. Catal. A Chem.* **192**, 217–222 (2003).

43 Abdul-Sada, A.K., Ambler, P.W., Hodgson, P.K.G., Seddon, K.R. and Stewart, N.J., Ionic liquids of imidazolium halide for oligomerization or polymerization of olefins, World Pat. 9521871 (1995).

44 Murphy, V., Ionic liquids and processes for production of high molecular weight polyisoolefins, WO Pat. 0032658 (2000).

45 Wissel, K., Halbritter, K., Massonne, K. and Stegmann, V., Verfahren zur Herstellung von Homo- oder Copolymeren aus olefinischen Monomeren mittels Pyrazoliumsalzen, DE Pat. 102007040919A1 (2008).

46 Atkins, M.P., Smith, M.R. and Ellis, B., Lubricating oils, EP Pat. 791643A1 (1997).

47 Hope, K.D., Driver, M.S. and Harris, T.V., High-viscosity poly(alpha-olefins) prepared with ionic liquid catalysts, US Pat. 6395948 (2002).

48 Hope, K.D., Twomey, D.W., Stern, D.A. and Collins, J.B., Method for manufacturing high viscosity poly(a-olefins) using ionic liquid catalysts, World Pat. 2003089390 (2003).

49 Adams, C.J., Earle, M.J. and Seddon, K.R., Catalytic cracking reactions of polyethylene to light alkanes, *Green Chem.* **2**, 21–24 (2000).

50 Li, X.A., Johnson, K.E. and Treble, R.G., Alkane cracking, alkene polymerization, and Friedel-Crafts alkylation in liquids containing the acidic anions HX_2^-, $XH(AlX_4)^-$, $XH(Al_2X_7)^-$, and $Al_2X_7^-$ (X = chlorine, bromine), *J. Mol. Catal. A Chem.* **214**, 121–127 (2004).

51 Cornils, B., Herrmann, W.A., Horváth, I.T., Leitner, W., Mecking, S., Olivier-Bourbigou, H. and Vogt, D. (eds.), *Multiphase Homogeneous Catalysis*, Vol. **1**, Wiley-VCH, Weinheim (2005).

52 (a) Keim, W., Nickel: An element with wide application in industrial homogeneous catalysis, *Angew. Chem. Int. Ed. Engl.* **29** (3), 235–244 (1990). (b) Kohlpaintner, C.W., Fischer, R.W. and Cornils, B., Aqueous biphasic catalysis: Ruhrchemie/Rhône-Poulenc oxo process, *Appl. Catal. A Gen.* **221**, 219–225 (2001).

53 Wu, B., Liu, W., Zhang, Y. and Wang, H., Do we understand the recyclability of ionic liquids? *Chem. Eur. J.* 1804–1810 (2009).

54 Geldbach, T.J., 'Organometallics in Ionic Liquids-Catalysis and Coordination Chemistry', in *Organometallic Chemistry, Vol. 34: A Review of the Literature Published between January 2004 and December 2005*, Eds. I.J.S. Fairlamb and J. Lynam, Royal Society of Chemistry, Cambridge (2008), pp. 58–73.

55 Parvulescu, V.I. and Hardacre, C., Catalysis in ionic liquids, *Chem. Rev.* **107**, 2615–2665 (2007).

56 Haumann, M. and Riisager, A., Hydroformylation in room temperature ionic liquids (RTILs): Catalyst and process developments, *Chem. Rev.* **108**, 1474–1497 (2008).

57 Wasserscheid, P. and Kuhlmann, S., 'Multiphasic Catalysis Using Ionic Liquids in Combination with Compressed CO_2', in *Ionic Liquids in Synthesis*, 2nd Ed., Eds. P. Wasserscheid and T. Welton, Wiley-VCH, Weinheim (2008), pp. 558–569.

58 Olivier-Bourbigou, H., Magna, L. and Morvan, D., Ionic liquids and catalysis: Recent progress from knowledge to applications, *Appl. Catal. A Gen.* **373**, 1–56 (2010).

59 Jutz, F., Andanson, J.-M. and Baiker, A., Ionic liquids and dense carbon dioxide: A beneficial biphasic system for catalysis, *Chem. Rev.* **111**, 322–353 (2011).

60 Camper, D., Becker, C., Koval, C. and Noble, R., Diffusion and solubility measurements in room temperature ionic liquids, *Ind. Eng. Chem. Res.* **45**, 445–450 (2006).

61 (a) Mehnert, C.P., Cook, R.A., Dispenziere, N.C. and Afeworki, M., Supported ionic liquid catalysis – A new concept for homogeneous hydroformylation catalysis, *J. Am. Chem. Soc.* **124**, 12932–12933 (2002); (b) Mehnert, C.P., Mozeleski, E.J. and Cook, R.A., Supported ionic liquid catalysis investigated for hydrogenation reactions, *Chem. Commun.* 3010–3011 (2002).

62 Mehnert, C.P., Supported ionic liquid phases, *Chem. Eur. J.* **11**, 50–56 (2005).

63 Riisager, A., Wasserscheid, P., van Hal, R. and Fehrmann, R., Continuous fixed-bed gas-phase hydroformylation using supported ionic liquid-phase (SILP) Rh catalysts, *J. Catal.* **219**, 452–455 (2003).

64 Jakuttis, M., Schönweiz, A., Werner, S., Franke, R., Wiese, K.-D., Haumann, M. and Wasserscheid, P., Rhodium-phosphite SILP catalysis for the highly selective hydroformylation of mixed C-4 feedstocks, *Angew. Chem. Int. Ed.* **50**, 4492–4495 (2011).

65 Haumann, M., Jakuttis, M., Franke, R., Schönweiz, A. and Wasserscheid, P., Continuous gas-phase hydroformylation of a highly diluted technical C4 feed using supported ionic liquid phase catalysts, *ChemCatChem.* **3**, 1822–1827 (2011).

66 Mikkola, J.-P.T., Virtanen, P.P., Kordas, K., Karhu, H. and Salmi, T.O., SILCA-supported ionic liquid catalysts for fine chemicals, *Appl. Catal. A Gen.* **328**, 68–76 (2007).

67 Virtanen, P., Karhu, H., Kordas, K. and Mikkola, J.-P., The effect of ionic liquid in supported ionic liquid catalysts (SILCA) in the hydrogenation of α,β-unsaturated aldehydes, *Chem. Eng. Sci.* **62**, 3660–3671 (2007).

68 Öchsner, E., Schneider, M.J., Meyer, C., Haumann, M. and Wasserscheid, P., Challenging the scope of continuous, gas-phase reactions with supported ionic liquid phase (SILP) catalysts-Asymmetric hydrogenation of methyl acetoacetate, *Appl. Catal. A Gen.* **399**, 35–41 (2011).

69 Hintermair, U., Höfener, T., Pullmann, T., Francio, G. and Leitner, W., Continuous enantioselective hydrogenation with a molecular catalyst in supported ionic liquid phase under supercritical CO_2 flow, *ChemCatChem.* **2**, 150–154 (2010).

70 Werner, S., Szesni, N., Bittermann, A., Schneider, M.J., Haerter, P., Haumann, M. and Wasserscheid, P., Screening of supported ionic liquid phase (SILP) catalysts for the very low temperature water-gas-shift reaction, *Appl. Catal. A Gen.* **377**, 70–75 (2010).

71 Werner, S., Szesni, N., Kaiser, M., Fischer, R.W., Haumann, M. and Wasserscheid, P., Ultra-low-temperature water-gas shift catalysis using supported ionic liquid phase (SILP) materials, *ChemCatChem.* **2**, 1399–1402 (2010).

72 Duque, R., Öchsner, E., Clavier, H., Caijo, F., Nolan, S.P., Mauduit, M. and Cole-Hamilton, D.J., Continuous flow homogeneous alkene metathesis with built-in catalyst separation, *Green Chem.* **13**, 1187–1195 (2011).

73 Scholz, J., Loekman, S., Szesni, N., Hieringer, W., Görling, A., Haumann, M. and Wasserscheid, P., Ethene-induced temporary inhibition of grubbs metathesis catalysts, *Adv. Synth. Catal.* **353**, 2701–2707 (2011).

74 Breitenlechner, S., Fleck, M., Müller, T.E. and Suppan, A., Solid catalysts on the basis of supported ionic liquids and their use in hydroamination reactions, *J. Mol. Catal. A Chem.* **214**, 175–179 (2004).

75 Riisager, A., Jørgensen, B., Wasserscheid, P. and Fehrmann, R., First application of supported ionic liquid phase (SILP) catalysis for continuous methanol carbonylation, *Chem. Commun.* 994–996 (2006).

76 Riisager, A., Fehrmann, R., Haumann, M. and Wasserscheid, P., Supported ionic liquids: Versatile reaction and separation media, *Top. Catal.* **40**, 91–102 (2006).

77 Riisager, A., Fehrmann, R., Haumann, M. and Wasserscheid, P., Supported ionic liquid phase (SILP) catalysis: An innovative concept for homogeneous catalysis in continuous fixed-bed reactors, *Eur. J. Inorg. Chem.* 695–706 (2006).

78 Gu, Y. and Li, G., Ionic liquids-based catalysis with solids: State of the art, *Adv. Synth. Catal.* **315**, 817–847 (2009).

79 van Doorslaer, C., Wahlen, J., Mertens, P., Binnemans, K. and De Vos, D., Immobilization of molecular catalysts in supported ionic liquid phases, *Dalton Trans.* **39**, 8377–8390 (2010).

80 Virtanen, P., Salmi, T.O. and Mikkola, J.-P., Supported ionic liquid catalysts (SILCA) for preparation of organic chemicals, *Top. Catal.* **53**, 1096–1103 (2010).

81 Haumann, M., Schönweiz, A., Breitzke, H., Buntkowsky, G., Werner, S. and Szesni, N., Solid-state NMR investigations of supported ionic liquid phase water-gas shift catalysts: Ionic liquid film distribution vs. catalyst performance, *Chem. Eng. Technol.* **35**, 1421–1426 (2012).

82 Lemus, J., Palomar, J., Gilarranz, M.A. and Rodriguez, J.J., Characterization of Supported Ionic Liquid Phase (SILP) materials prepared from different supports, *Adsorption* **17**, 561–571 (2011).

83 Knapp, R., Jentys, A. and Lercher, J.A., Impact of supported ionic liquids on supported Pt catalysts, *Green Chem.* **11**, 656–661 (2009).

84 Jess, A., Kern, C. and Korth, W., Solid catalyst with ionic liquid layer (SCILL) – A concept to improve the selectivity of selective liquid and gas phase hydrogenations, *Oil Gas Eur. Mag.* **38**(1), 38–45 (2012).

85 DeCastro, C., Sauvage, E., Valkenberg, M.H. and Hölderich, W.F., Immobilised ionic liquids as lewis acid catalysts for the alkylation of aromatic compounds with dodecene, *J. Catal.* **196**, 86–94 (2000).

86 Valkenberg, M.H., deCastro, C. and Holderich, W.F., Immobilisation of ionic liquids on solid supports, *Green Chem.* **4**, 88–93 (2002).

87 Joni, J., Haumann, M. and Wasserscheid, P., Development of a supported ionic liquid phase (SILP) catalyst for slurry-phase Friedel-Crafts alkylations of cumene, *Adv. Synth. Catal.* **351**, 423–431 (2009).

88 Joni, J., Haumann, M. and Wasserscheid, P., Continuous gas-phase isopropylation of toluene and cumene using highly acidic Supported Ionic Liquid Phase (SILP) catalysts, *Appl. Catal. A Gen.* **372**, 8–15 (2010).

89 Kernchen, U., Etzold, B., Korth, W. and Jess, A., Solid catalyst with ionic liquid layer (SCILL) – A new concept to improve selectivity illustrated by hydrogenation of cyclooctadiene, *Chem. Eng. Technol.* **30**, 985–994 (2007).

90 Arras, J., Paki, E., Roth, C., Radnik, J., Lucas, M. and Claus, P., How a supported metal is influenced by an ionic liquid: In-depth characterization of SCILL-type palladium catalysts and their hydrogen adsorption, *J. Phys. Chem. C* **114**, 10520–10526 (2010).

91 Steinrück, H.P., Libuda, J., Wasserscheid, P., Cremer, T., Kolbeck, C., Laurin, M., Maier, F., Sobota, M., Schulz, P.S. and Stark, M., Surface science and model catalysis with ionic liquid-modified materials, *Adv. Mater.* **23**, 2571–2587 (2011).

92 Arras, J., Steffan, M., Shayeghi, Y., Ruppert, D. and Claus, P., Regioselective catalytic hydrogenation of citral with ionic liquids as reaction modifiers, *Green Chem.* **11**, 716–723 (2009).

93 Jess, A., Korth, W. and Etzold, B., Mit einer ionischen Flüssigkeit beschichteter poröser heterogener Katalysator, DE Pat. 102006019460A1 (2007).

94 Arras, J., Steffan, M., Shayeghi, Y. and Claus, P., The promoting effect of a dicyanamide based ionic liquid in the selective hydrogenation of citral, *Chem. Commun.* 4058–4060 (2008).

95 Dyson, P.J., Laurenczy, G., Ohlin, C.A., Vallance, J. and Welton, T., Determination of hydrogen concentration in ionic liquids and the effect (or lack of) on rates of hydrogenation, *Chem. Commun.* 2418–2419 (2003).

96 Wörz, N., Arras, J. and Claus, P., Continuous selective hydrogenation of citral in a trickle-bed reactor using ionic liquid modified catalysts, *Appl. Catal. A Gen.* **391**, 319–324 (2011).

97 Arras, J., Ruppert, D. and Claus, P., Supported ruthenium catalysed selective hydrogenation of citral in presence of $[NTf_2]^-$ based ionic liquids, *Appl. Catal. A Gen.* **371**, 73–77 (2009).

98 Sobota, M., Happel, M., Amende, M., Paape, N., Wasserscheid, P., Laurin, M. and Libuda, J., Ligand effects in SCILL model systems: Site-specific interactions with Pt and Pd nanoparticles, *Adv. Mater.* **23**, 2617–2621 (2011).

99 Werner, S., Szesni, N., Kaiser, M., Haumann, M. and Wasserscheid, P., A scalable preparation method for SILP and SCILL ionic liquid thin-film materials, *Chem. Eng. Technol.* **35**, 1962–1967 (2012).

100 Scholten, J.D., Leal, B.C. and Dupont, J., Transition metal nanoparticle catalysis in ionic liquids, *ACS Catal* **2**, 184–200 (2012).

101 Migowski, P. and Dupont, J., Catalytic applications of metal nanoparticles in imidazolium ionic liquids, *Chem. Eur. J.* **13**, 32–39 (2007).

102 Dupont, J. and Silva D.D.O., 'Transition-metal Nanoparticle Catalysis in Imidazolium Ionic Liquids', in *Nanoparticles and Catalysis*, Ed. D. Astruc, Wiley-VCH, Weinheim (2008), pp. 195–218.

103 Antonietti, M., Kuang, D.B., Smarsly, B. and Yong, Z., Ionic liquids for the convenient synthesis of functional nanoparticles and other inorganic nanostructures, *Angew. Chem. Int. Ed.* **43**, 4988–4992 (2004).

104 Hagiwara, H., Sasaki, H., Tsubokawa, N., Hoshi, T., Suzuki, T., Tsuda, T. and Kuwabata, S., Immobilization of Pd on nanosilica dendrimer as SILC: Highly active and sustainable cluster catalyst for Suzuki-Miyaura reaction, *Syn. Lett.* 1990–1996 (2010).

105 Lei, Z.G., Li, C.Y. and Chen, B.H., Extractive distillation: A review, *Sep. Purif. Rev.* **32**, 121-213 (2003).

106 Arlt, W., Seiler, M., Jork, C. and Schneider, T., Ionic liquids as selective additives for the separation of close-boiling or azeotropic mixtures, WO Pat. 2002074718 (2002).

107 Beste, Y., Schoenmakers, H., Arlt, W., Seiler, M. and Jork, C., Recycling of ionic liquids with extractive distillation, WO Pat. 2005016484 (2005).

108 Maase, M., 'Industrial Applications of Ionic Liquids', in *Ionic Liquids in Synthesis*, 2nd Ed., Eds. P. Wasserscheid and T. Welton, Wiley-VCH, Weinheim (2008), pp. 663–687.

109 Jork, C., Seiler, M., Beste, Y.A. and Arlt, W., Influence of ionic liquids on the phase behavior of aqueous azeotropic systems, *J. Chem. Eng. Data* **49**, 852–857 (2004).

110 Lei, Z., Arlt, W. and Wasserscheid, P., Selection of entrainers in the 1-hexene/n-hexane system with a limited solubility, *Fluid Phase Equilib.* **260**, 29–35 (2007).

111 Westerholt, A., Liebert, V. and Gmehling, J., Influence of ionic liquids on the separation factor of three standard separation problems, *Fluid Phase Equilib.* **280**, 56–60 (2009).

112 Liebert, V., Nebig, S. and Gmehling, J., Experimental and predicted phase equilibria and excess properties for systems with ionic liquids, *Fluid Phase Equilib.* **268**, 14–20 (2008).

113 Meindersma, G.W., Sanchez, L.M.G., Hansmeier, A.R. and de Haan, A.B., Application of task-specific ionic liquids for intensified separations, *Mon. Chem.* **138**, 1125–1136 (2007).

114 Visser, A.E., Swatloski, R.P., Reichert, W.M., Griffin, S.T. and Rogers, R.D., Traditional extractants in nontraditional solvents: Groups 1 and 2 extraction by crown ethers in room-temperature ionic liquids, *Ind. Eng. Chem. Res.* **39**, 3596–3604 (2000).

115 Sayar, N.A., Filiz, M. and Sayar, A.A., Extraction of Zn(II) from aqueous hydrochloric acid solutions into Alamine 336-*m*-xylene systems. Modeling considerations to predict optimum operational conditions, *Hydrometallurgy* **86**, 27–36 (2007).

116 Visser, A.E., Swatloski, R.P., Reichert, W.M., Mayton, R., Sheff, S., Wierzbicki, A., Davis, J.H. and Rogers, R.D., Task-specific ionic liquids incorporating novel cations for the coordination and extraction of Hg^{2+} and Cd^{2+}: Synthesis, characterization, and extraction studies, *Environ. Sci. Technol.* **36**, 2523–2529 (2002).

117 Nockemann, P., Thijs, B., Pittois, S., Thoen, J., Glorieux, C., Van Hecke, K., Van Meervelt, L., Kirchner, B. and Binnemans, K., Task-specific ionic liquid for solubilizing metal oxides, *J. Phys. Chem. B* **110**, 20978–20992 (2006).

118 Domańska, U., Pobudkowska, A. and Królikowski, M., Separation of aromatic hydrocarbons from alkanes using ammonium ionic liquid C_2NTf_2 at T = 298.15 K, *Fluid Phase Equilib.* **259**, 173–179 (2007).

119 Meindersma, G.W., Podt, A.J.G. and de Haan, A.B., Selection of ionic liquids for the extraction of aromatic hydrocarbons from aromatic/aliphatic mixtures, *Fuel Process. Technol.* **87**, 59–70 (2005).

120 Esser, J., Wasserscheid, P. and Jess, A., Deep desulfurization of oil refinery streams by extraction with ionic liquids, *Green Chem.* **6**, 316–322 (2004).

121 Liu, D., Gui, J., Song, L., Zhang, X. and Sun, Z., Deep desulfurization of diesel fuel by extraction with task-specific ionic liquids, *Pet. Sci. Technol.* **26**, 973–982 (2008).

122 Tae, E.L., Lee, S.H., Lee, J.K., Yoo, S.S., Kang, E.J. and Yoon, K.B., A strategy to increase the efficiency of the dye-sensitized TiO2 solar cells operated by photoexcitation of Dye-to-TiO_2 charge-transfer bands, *J. Phys. Chem. B* **109**, 22513–22522 (2005).

123 Jiang, Y., Xia, H., Yu, J., Guo, C. and Liu, H., Hydrophobic ionic liquids-assisted polymer recovery during penicillin extraction in aqueous two-phase system, *Chem. Eng. J.* **147**, 22–26 (2009).

124 Anthony, J.L., Anderson, J.L., Maginn, E.J. and Brennecke, J.F., Anion effects on gas solubility in ionic liquids, *J. Phys. Chem. B* **109**, 6366–6374 (2005).

125 Scovazzo, P., Camper, D., Kieft, J., Poshusta, J., Koval, C. and Noble, R., Regular solution theory and CO_2 gas solubility in room-temperature ionic liquids, *Ind. Eng. Chem. Res.* **43**, 6855–6860 (2004).

126 Anderson, J.L., Dixon, J.K. and Brennecke, J.F., Solubility of CO_2, CH_4, C_2H_6, C_2H_4, O_2, and N_2 in 1-hexyl-3-methylpyridinium bis(trifluoromethylsulfonyl) imide: Comparison to other ionic liquids, *Acc. Chem. Res.* **40**, 1208–1216 (2007).

127 Sanchez, L.M.G., Meindersma, G.W. and de Haan, A.B., Solvent properties of functionalized ionic liquids for CO_2 absorption, *Chem. Eng. Res. Des.* **85**(A1), 31–39 (2007).

128 Huang, J. and Rüther, T., Why are ionic liquids attractive for CO_2 absorption? An overview, *Aust. J. Chem.* **62**, 298–308 (2009).

129 Kohler, F., Roth, D., Kuhlmann, E., Wasserscheid, P. and Haumann, M., Continuous gas-phase desulfurisation using supported ionic liquid phase (SILP) materials, *Green Chem.* **12**, 979–984 (2010).

130 Atkins, M.P., Keynote Lecture: 'Ionic liquids for Hg removal', EuChem 2012, 6 August 2012, Celtic Manor, Wales.

131 Chakraborty, M. and Bart, H.-J., Highly selective and efficient transport of toluene in bulk ionic liquid membranes containing Ag^+ as carrier, *Fuel Process. Technol.* **88**, 43–49 (2007).

132 Branco, L.C., Crespo, J.G. and Afonso, C.A.M., Ionic liquids as an efficient bulk membrane for the selective transport of organic compounds, *J. Phys. Org. Chem.* **21**, 718–723 (2008).

133 Branco, L.C., Crespo, J.G. and Afonso, C.A.M., Highly selective transport of organic compounds by using supported liquid membranes based on ionic liquids, *Angew. Chem. Int. Ed.* **41**, 2771–2773 (2002).

134 Miyako, E., Maruyama, T., Kamiya, N. and Goto, M., Use of ionic liquids in a lipase-facilitated supported liquid membrane, *Biotechnol. Lett.* **25**, 805–808 (2003).

135 Hernandez-Fernandez, F.J.,de los Rios, A.P., Rubio, M., Tomas-Alonso, F., Gomez, D. and Villora, G., A novel application of supported liquid membranes based on ionic liquids to the selective simultaneous separation of the substrates and products of a transesterification reaction, *J. Membr. Sci.* **293**, 73–80 (2007).

136 Fortunato, R., González-Muñoz, M.J., Kubasiewicz, M., Luque, S., Alvarez, J.R., Afonso, C.A.M., Coelhoso, I.M. and Crespo, J.G., Liquid membranes using ionic liquids: The influence of water on solute transport, *J. Membr. Sci.* **249**, 153–162 (2005).

137 Matsumoto, M., Ohtani, T. and Kondo, K., Comparison of solvent extraction and supported liquid membrane permeation using an ionic liquid for concentrating penicillin G, *J. Membr. Sci.* **289**, 92–96 (2007).

138 Luis, P., Neves, L.A., Afonso, C.A.M., Coelhoso, I.M., Crespo, J.G., Garea, A. and Irabien, A., Facilitated transport of CO_2 and SO_2 through supported ionic liquid membranes (SILMs), *Desalination* **245**(1–3), 485–493 (2009).

139 Cserjési, P., Nemestóthy, N., Vass, A., Csanádi, Z. and Bélafi-Bakó, K., Study on gas separation by supported liquid membranes applying novel ionic liquids, *Desalination* **245**, 743–747 (2009).

140 DeJong, F. and DeWith, J.F., Process for the separation of olefins and paraffins, WO Pat. 2005061422 (2005).

141 Matsumoto, M., Inomoto, Y. and Kondo, K., Selective separation of aromatic hydrocarbons through supported liquid membranes based on ionic liquids, *J. Membr. Sci.* **246**, 77–81 (2005).

142 Wang, B., Lin, J., Wu, F. and Peng, Y., Stability and selectivity of supported liquid membranes with ionic liquids for the separation of organic liquids by vapor permeation, *Ind. Eng. Chem. Res.* **47**, 8355–8360 (2008).

143 Kakiuchi, T., Mutual solubility of hydrophobic ionic liquids and water in liquid-liquid two-phase systems for analytical chemistry, *Anal. Sci.* **24**, 1221–1230 (2008).

144 Jork, C., Kristen, C., Pieraccini, D., Stark, A., Chiappe, C., Beste, Y.A. and Arlt, W., Tailor-made ionic liquids, *J. Chem. Thermodyn.* **37**, 537–558 (2005).

145 Ye, C.F., Liu, W.M., Chen, Y.X. and Yu, L.G., Room-temperature ionic liquids: A novel versatile lubricant, *Chem. Commun.* 2244–2245 (2001).

146 Kömpf, M., Linde Technology Report (2006), pp. 24–29.

147 Predel, T. and Schlücker, E., Ionic liquids in oxygen compression, *Chem. Eng. Technol.* **32**, 1183–1188 (2009).

148 Hilgers, C., Uerdingen, M., Wagner, M., Wasserscheid, P. and Schlücker, E., Processing or working machine comprising an ionic liquid as the service fluid, WO Pat. 2006087333 (2006).

149 Wasserscheid, P. and Seiler, M., Leveraging gigawatt potentials by smart heat-pump technologies using ionic liquids, *Chem. Sustain. Chem.* **4**, 459–463 (2011).

7 Vibrational Spectroscopy of Ionic Liquid Surfaces

CHARIZ PEÑALBER-JOHNSTONE and STEVEN BALDELLI
Department of Chemistry, University of Houston, Houston, Texas, USA

7.1 INTRODUCTION

Room-temperature ionic liquids have drawn significant attention in the scientific community since their first discovery in 1914 [1]. Indeed, even after all these years, these remarkable salts, with melting points below 100°C, still present a plethora of cation–anion combinations that remains to be explored, offering new and promising solutions to current scientific and technological demands. Most notable among the properties of room-temperature ionic liquids are their conductivity, extremely low volatility and exceptional miscibility behaviour, making them ideal for several applications, such as in electrochemistry [2, 3], catalysis [4–6] and fluids engineering [7–9].

Owing to the extremely low vapour pressures of room-temperature ionic liquids [10], surface science techniques have been used to characterise them, as they are able to withstand the demanding conditions required by such methods, namely, the ultra-high-vacuum (UHV) pressures in the 10^{-10}Torr range [11]. Because the chemical nature of the liquid surface is completely unique relative to that of the bulk, the use of surface-sensitive techniques is critical to probe such surfaces with atomic level accuracy. A list of UHV-based methods applied to room-temperature ionic liquids would include, inter alia, X-ray photoelectron spectroscopy (XPS) [12–16], UV photoelectron spectroscopy (UPS) [17, 18], inverse photoelectron spectroscopy (IPES) [18, 19], metastable ion spectroscopy (MIES) [17], direct recoil spectroscopy (DRS) [20, 21], near-edge absorption fine structure (NEXAFS) [18] and low-energy ion scattering (LEIS) [12]. Comprehensive reviews of these UHV-based methods of surface science applied to ionic liquids have recently been reported in the literature [6, 11]. However, despite the wealth of information provided by these techniques,

Ionic Liquids Completely UnCOILed: Critical Expert Overviews, First Edition.
Edited by Natalia V. Plechkova and Kenneth R. Seddon.

vibrational spectroscopy provides the most detailed model for the molecular structure and properties of ionic liquids on surfaces. More precise information at the molecular level may be derived from quantitative analysis of vibrational spectra. This, in tandem with the atomic level accuracy of surface-sensitive techniques, will serve to further the rapidly evolving field of ionic liquid surface studies.

Presented here is a brief review of the vibrational spectroscopic techniques used to characterise ionic liquids on surfaces. Specifically, this review includes results obtained from sum-frequency generation spectroscopy (SFG), infrared reflection–absorption spectroscopy (IRAS), surface-enhanced Raman scattering (SERS) and high-resolution electron energy loss spectroscopy (HREELS).

Up until the 1990s, the last three techniques have been most useful in studies involving adsorption of molecules on metal single crystals at UHV pressures [22] and so are typically not available to most neat liquid or solid–liquid interfaces. More recently, SFG has been widely used in investigating the molecular structure and orientation of several ionic liquids at various interfaces, as this method is inherently surface sensitive [23]. As a non-linear, vibrational spectroscopic technique, SFG selectively detects molecules in a non-centrosymmetric environment. IRAS, on the other hand, lacks intrinsic surface sensitivity, since it equally probes both the surface and bulk regions of the material. Moreover, its high chemical resolution [22] allows detection of even the smallest changes in molecular structure. It must be noted, however, that a limitation for HREELS is the requirement for ultra-high-vacuum pressures. Other vibrational spectroscopies mentioned herein, that is, infrared (IR), Raman and SFG, have virtually no pressure restrictions as they are designed to work at ambient and more realistic pressure conditions, with a variety of surfaces.

7.2 VIBRATIONAL SPECTROSCOPY

Vibrational spectroscopy is well suited to elucidating the details on the surface of ionic liquids. This is due to the high degree of chemical information that is provided in the vibrational spectrum. The various techniques mentioned earlier each have advantages and limitations, but the level of information is clearly unique and highly informative. From the vibrational spectrum, information as to the identity of the molecules is available. However, more specifically, the chemical functional groups are identified as they each have unique vibrational resonances. Thus, for example, it is easy to distinguish CH_2, CH_3, CH and X–CH_n (n=1, 2 or 3) from each other. Other functional groups are readily identified also [24].

Another advantage of vibrational spectroscopy is the possibility to determine the orientation of the various functional groups at the surface. Both orientation and identity contribute to the surface properties. This is most commonly achieved using polarised IR light at the surface. The maximum interaction (absorption) occurs when the electric vector of light is aligned to the transition

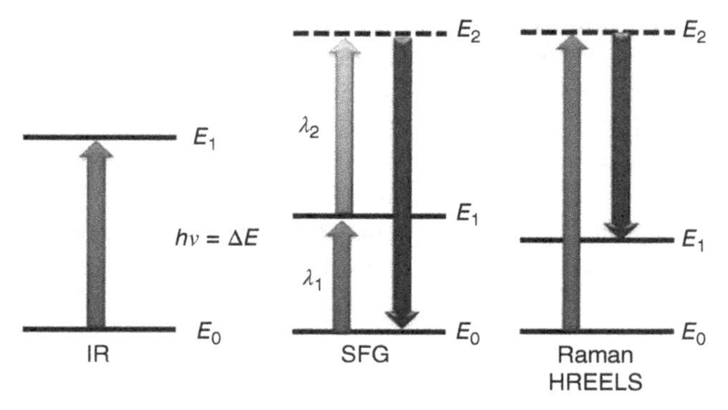

Figure 7.1 Energy level transitions for different vibrational spectroscopic techniques: infrared absorption, SFG nonlinear spectroscopy, Raman and HREELS (where an electron is inelastically scattered).

dipole of the functional group. By varying the polarisation of the light field and measuring the response, the orientation can be estimated. This can be performed on several vibrational modes to provide an overall structure of the ionic liquid ions at the surface [24].

In some situations, the vibrational peaks can be integrated, and the concentration at the surface is determined. However, since the peak intensities also depend on the orientation, concentration is usually best determined by other techniques.

Of the four most common vibrational spectroscopies used, each has value depending on the precise interface or information required. The general schemes are outlined in Figure 7.1.

Photon-based techniques such as Raman [25], IR [26] and SFG [27] are beneficial as they are able to probe any interface accessible by light. This means that systems such as gas–liquid, solid–liquid and liquid–liquid are possible. However, HREELS, which requires UHV conditions, is not possible for the last two interfaces [28]. Similarly, Raman and IR absorption are not only sensitive to the molecules at the surface but also detect the bulk-phase species. When probing the interface of bulk liquids, the Raman and IR signals originate over a distance from approximately 10^{-7} to 10^{-6} m. This makes analysis or experimental design complicated if explicit surface detection is desired. IR and Raman spectroscopy typically are able to cover from 400 to 4000 cm^{-1} but could be limited by materials (optical window) constraints. HREELS is an exclusively surface-sensitive technique, due to the limited mean free path of electrons (few nanometres), but of course limited to the UHV vacuum–liquid interface. HREELS has the ability to collect the vibrational spectrum from 200 to 4000 cm^{-1}. Although the resolution is very low, greater than 50 cm^{-1}, the large spectral range is useful in interpreting the structure of the ions at the surface. SFG spectroscopy has the unique ability to detect molecules at the liquid surface in contact with any medium. The signal typically originates in the first nanometre at the interface, where the isotropic bulk environment no longer exists. SFG is best used when the surface must be distinguished from the bulk phase and thus is ideal in the case of the neat gas–liquid interface, as well as solid–liquid and liquid–liquid boundaries. The SFG spectra are usually limited to a spectral range above $1000\,cm^{-1}$ because of the IR laser source needed. This sometimes limits the species observable by SFG [29].

Overall, vibrational spectroscopy provides very useful chemical information on the surface of ionic liquids, and some specific examples are given in the following.

7.3 AIR–LIQUID INTERFACES

There are numerous publications focussed on the use of SFG spectroscopy to model the gas–liquid interface of several ionic liquids. Investigations have covered, to a great extent, 1,3-alkylimidazolium ionic liquids with various anions [30–34]. Other ionic liquid classes explored include tetraalkylammonium [35, 36] and 1,1-dialkylpyrrolidinium [36] ionic liquids. Recently, efforts have been extended towards investigating the surface orientation of tetraalkylphosphonium ionic liquids, which are currently under study [37].

Figure 7.2a and b depicts the numbering scheme and the model for the molecular orientation of the 1-butyl-3-methylimidazolium cation, $[C_4mim]^+$, in ionic liquids at the air–liquid interface, respectively. SFG results show that the butyl chain of the cation extends towards the gas phase with the imidazolium ring lying parallel to the surface plane. This configuration has been found to be consistent, regardless of the anion [32], for 1-butyl-3-methylimidazolium room-temperature ionic liquids, particularly where the anions are composed of Br^-, I^-, $[PF_6]^-$, $[BF_4]^-$, $[N(CF_3SO_2)_2]^-$, $[SCN]^-$, $[CH_3SO_3]^-$, $[CH_3SO_4]^-$ or $[N(CN)_2]^-$. The SFG results reported were supported by FT-IR and polarised Raman spectroscopy with isotopic labelling. Studies by Ouchi et al. have reached similar conclusions [38, 39].

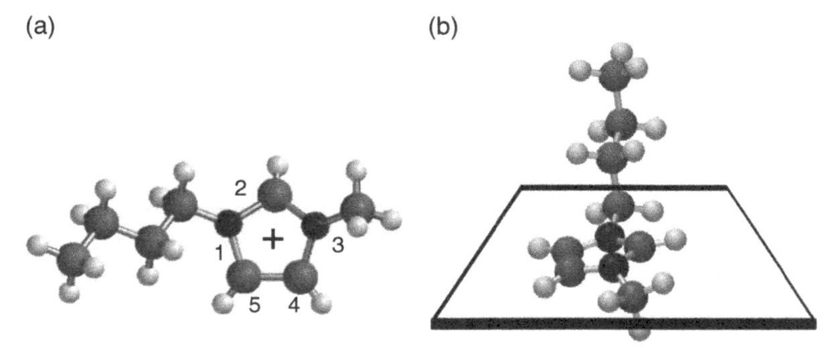

Figure 7.2 (a) Numbering scheme of $[C_4mim]^+$. (b) Molecular orientation of $[C_4mim]^+$ at the gas–liquid interface.

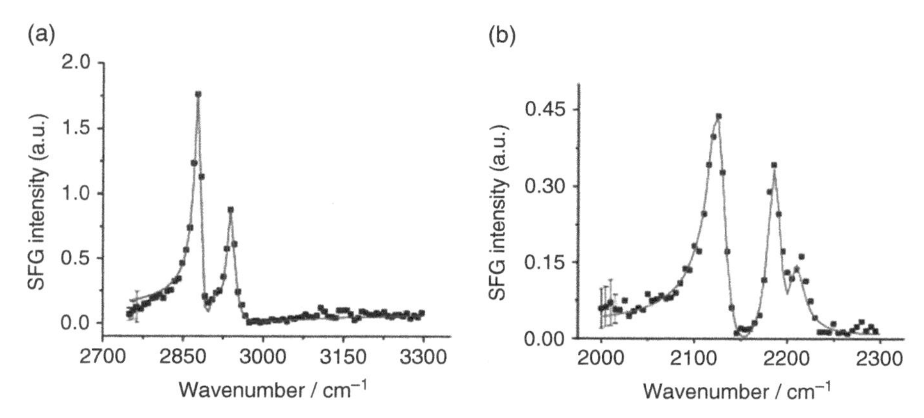

Figure 7.3 Sum-frequency spectra of $[C_4mim][N(CN)_2]$ at the gas–liquid interface at *ssp* polarisation: (a) C—H stretching region and (b) C—N stretching region [35].

For $[C_4mim][CH_3SO_4]$ and $[C_4mim][CH_3SO_3]$, both anion and cation are found to occupy the first layer of the gas–liquid interface [31]. The systematic variation of the alkyl chain on both cation and anion for 1,3-dialkylimidazolium alkylsulfates, to investigate the driving force for orientation and partitioning of the ions at the air–liquid interface, shows findings that are consistent with the model described earlier [33, 34]. Coulombic (as seen from ring orientation) and chain–chain interactions were considered in the study, which suggested that both ions are found at the interface, as evidenced by the presence of the methyl group vibration from the shortest-chain cation and anion. The absence of ring modes from the imidazolium cation suggests that it lies parallel to the surface plane.

Figure 7.3 shows SFG spectra taken for $[C_4mim][N(CN)_2]$ at the gas–liquid interface, using an *ssp* polarisation (*s*-polarised SFG beam, *s*-polarised visible beam and *p*-polarised IR beam) combination in the C—H and C—N stretching regions [35].

The C—H stretching region contains the symmetric methyl stretch and its Fermi resonance at ~2879 and ~2940 cm^{-1}, respectively. However, in the C—N stretching region, the spectra are dominated by peaks at ~2130 and ~2180 cm^{-1}, which correspond to the antisymmetric and the symmetric C≡N stretches, respectively. The peak at approximately 2210 cm^{-1} is a combination band of the symmetric and antisymmetric C—N stretching modes. This clearly shows that both cations and anions are present at the interface.

Figure 7.4 presents the sum-frequency spectra for $[C_4mim][BF_4]$ taken at different polarisation combinations. Vibrational assignments of the peaks observed in these spectra are presented in Table 7.1 [40]. The polar orientation of the molecule on the surface is what dictates the magnitude of the sum-frequency signal. By varying the polarisations of the input and output beams, the Cartesian components of the susceptibility tensor are determined, which allows for the determination of the molecular orientation relating to the surface normal [23].

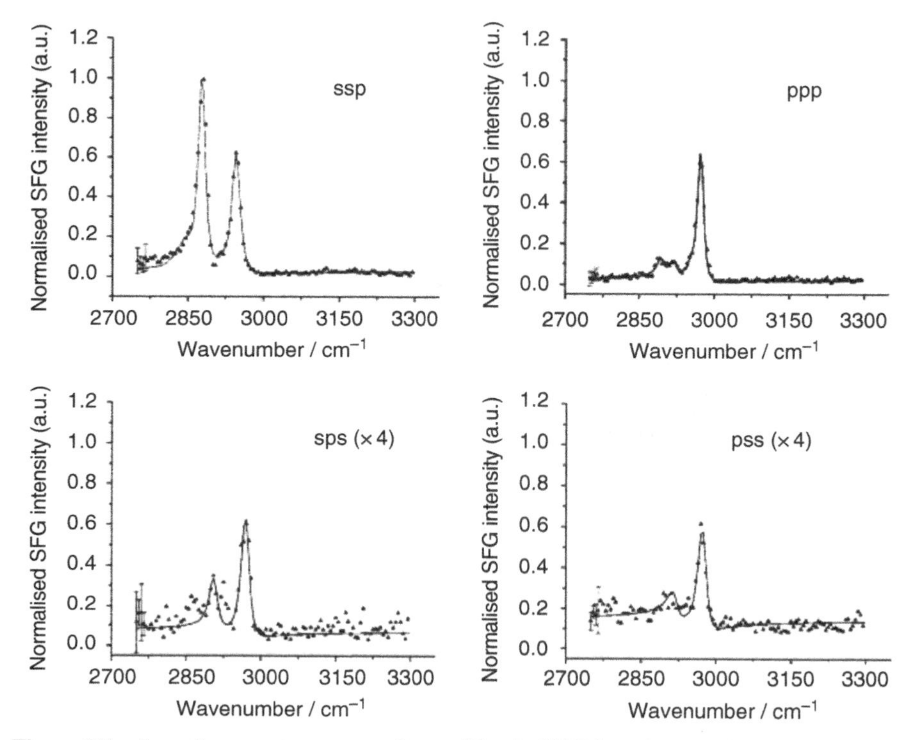

Figure 7.4 Sum-frequency spectra of neat $[C_4mim][BF_4]$ at the gas–liquid interface at four polarisation combinations [40].

TABLE 7.1 Vibrational Peak Assignments for $[C_4mim][BF_4]$[a]

Peak	Assignment	Frequency/cm^{-1}
1	CH_2 (ss)	~2850
2	CH_3 (ss)	~2880
3	CH_2 (as)	~2911
4	CH_3 (FR)	~2945
5	CH_3 (as)	~2965
6	C–H (ss)	~3070 (C_6H_6)
7	H–C(4)–C(5)–H (ss)	~3175

[a] From Ref. 40.

Using surface potential and surface tension measurements along with SFG, the results support that the relative position of the anion is in a slightly lower plane compared to the cation for 1-butyl-3-methylimidazolium ionic liquids with anions of varying shapes and sizes. Surface potential, in addition to SFG vibrational spectra, helps determine the positions of the ions by measuring its contribution to the total potential. Figure 7.5 shows an overall representation of the arrangement of

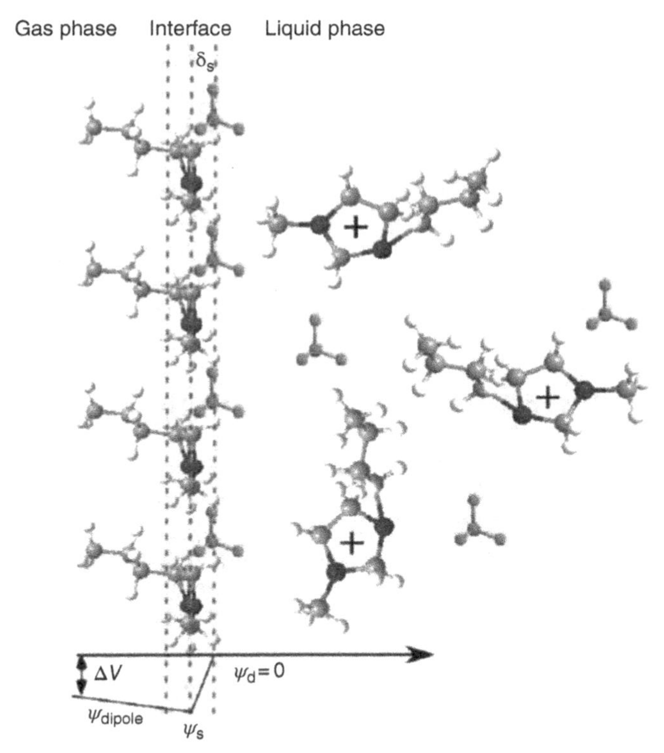

Figure 7.5 Overall arrangement of $[C_4mim]^+$ ions on the surface and their contribution to the surface potential [30].

$[C_4mim]^+$ ions with anions of varying shapes and sizes at the gas–liquid surface, as concluded from the three combined techniques [30]. Furthermore, it has been reported that, at this interface, there is a decreasing trend in surface tension values as the alkyl chain length on the cation is increased [34].

To further illustrate the arrangement of anions at the air–liquid interface, 1-butyl-3-methylimidazolium trifluoromethanesulfonate, $[C_4mim][OTf]$, has been studied using SFG. CF_3 symmetric and SO_3 symmetric stretching modes pertaining to the $[OTf]^-$ anion in the low-frequency region were observed, implying the presence of anions at the interface. Moreover, it was found that the amplitude of the CF_3 symmetric stretch peak of the triflate anion has an opposite sign with respect to that of the SO_3 symmetric peak, indicating that $[OTf]^-$ anions have polar ordering at the surface. The non-polar CF_3 group points towards the gas phase, whereas the SO_3 group points towards the bulk liquid [41]. As for HREELS backed up by density functional theory (DFT) calculations on 1-ethyl-3-methylimidazolium bis{(trifluoromethyl)sulfonyl}amide, $[C_2mim][NTf_2]$, surface–vacuum interface [42], it was found that the most pronounced structures in the spectra were related to coupled antisymmetric O=S=O and CF_3 stretching modes of the anion.

In SFG studies involving tetraalkylammonium and 1,1-dialkylpyrrolidinium ionic liquids, structures were systematically varied by attaching alkyl substituents of different lengths and functional groups. Ordering of the chains at the topmost layer was found to be strongly dependent on alkyl chain length with *gauche* defects observed for those with longer alkyl chains. Moreover, interfacial structure changed upon the addition of an oxygen atom to the aliphatic chain, as it introduced additional interactions due to the presence of a heteroatom, that is, hydrogen bonding, ion–dipole and dipole–dipole. A similar structure for pyrrolidinium bistriflamides was observed as for their imidazolium analogues, in that the ring appears to be lying parallel to the liquid surface plane [36].

7.4 ROOM-TEMPERATURE IONIC LIQUID MIXTURES

SFG studies on mixtures of ionic liquids have so far been confined to looking at effects of a molecular solvent such as benzene [40] on the interfacial ordering of the ions at the surface, as well as effects of water on the orientation of both hydrophilic and hydrophobic room-temperature ionic liquids [43–45]. Upon addition of benzene to hydrophobic 1-butyl-3-methylimidazolium hexafluorophosphate, $[C_4mim][PF_6]$, or hydrophilic 1-butyl-3-methylimidazolium tetrafluoroborate, $[C_4mim][BF_4]$, it was found that at concentrations greater than 0.01 mol fraction, benzene is present at the surface affecting mainly the conformation of the butyl chain and not significantly the imidazolium ring. This was evidenced by the presence of methylene symmetric and antisymmetric modes along with the absence of aromatic ring modes [40].

The first demonstration of the formation of a non-polar alkyl chain layer between a room-temperature ionic liquid and alcohol interface was reported by Iwahashi et al. [46]. In their study, the room-temperature ionic liquid $[C_4mim][PF_6]$ in contact with deuterated linear alcohols, namely, ethanol and butanol, was probed using SFG. SFG spectral features were found to change when the room-temperature ionic liquid surface was in contact with alcohol, in which a broadening effect was observed attributed to a range of intermolecular interactions available to the butyl chain of the cation, which was found to be oriented towards the alcohol phase. The authors further proposed that the alcohol molecules orient with their methyl groups facing each other at the surface, such that a non-polar interfacial layer model is suggested for the alcohol/ionic liquid interface. A molecular picture, proposed based on these results, is depicted in Figure 7.6.

In order to look at the effect of water on the structure of both hydrophobic and hydrophilic room-temperature ionic liquids, SFG measurements were taken at water partial pressures of 5×10^{-5} Torr and 20 Torr. Results showed that ionic liquid behaviour at the surface differed depending on whether the ionic liquid was hydrophobic or hydrophilic. For hydrophobic ionic liquids, the imidazolium ring reorients towards the surface normal upon addition of water, while for hydrophilic ionic liquids, the ring remains flat on the surface. The process was found to be reversible, with the tilting of the cation attributed to the interaction of water with C(2)—H. Moreover, for water-miscible ionic liquids, water molecules were said to be stabilised by favourable intermolecular interactions, that is, hydrogen bonding and dipole–dipole forces, such that it was less likely to partition to the surface [43]. In a similar study involving the effect of water on $[C_4mim][BF_4]$ [44], no obvious changes in orientation were found, especially at ionic liquid mol fractions greater than or equal to 0.05. The butyl chain of the cation was found pointing towards the gas phase, with the CH_3 tilt angle slightly higher for the dry ionic liquid compared to the mixture with water. A molecular model showing the effect of water on the orientation of the cation at the gas–liquid interface is illustrated in Figure 7.7.

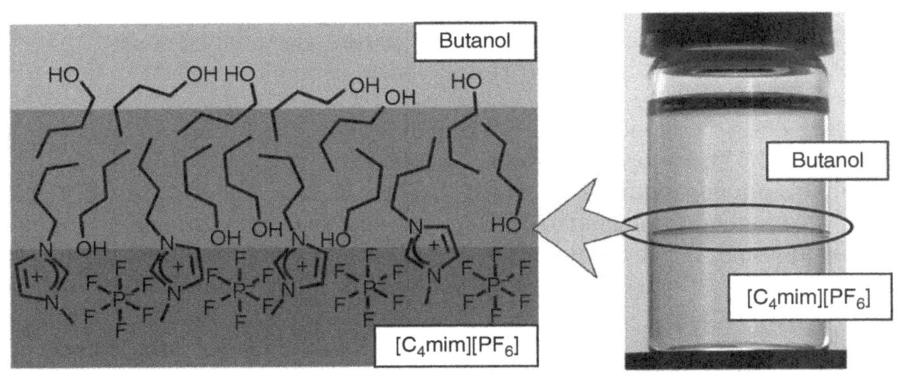

Figure 7.6 Non-polar interfacial layer model proposed for the alcohol/room-temperature ionic liquid interface [46].

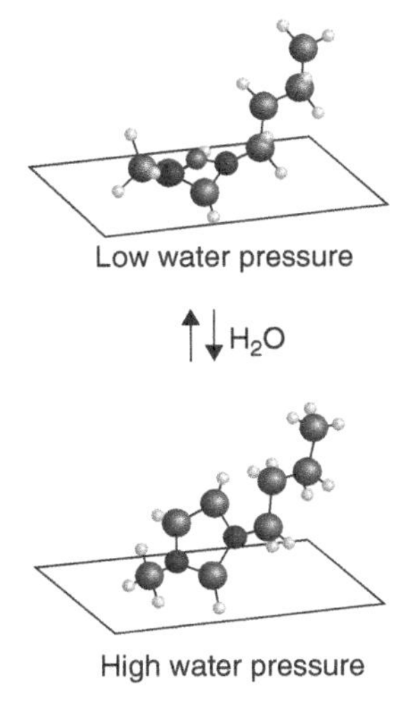

Figure 7.7 A model of the $[C_4mim]^+$ cation reorientation at the gas–liquid interface upon addition of water [43].

The same system involving [C_4mim][BF_4] and water mixtures with varying concentrations has been studied by attenuated total reflection infrared (ATR-IR) absorption and Raman spectroscopy [47]. Upon addition of water, vibrational modes of the imidazolium ring remain unaffected, while those of the terminal methyl group of the butyl chain were blue-shifted by approximately 10 cm^{-1}. Additionally, the hydrogen-bonding network of water molecules was found to break down rapidly upon addition of room-temperature ionic liquids, as seen in the change in spectral shape in the O—H stretching region.

Cation orientation effects upon water addition were further demonstrated by taking SFG spectra of 1-butyl-3-methylimidazolium bis{(trifluoromethyl) sulfonyl}amide, [C_4mim][NTf_2], at water partial pressures of 5×10^{-5} Torr and 200 Torr. It was found that below 10^{-4} Torr, the imidazolium ring is oriented parallel to the surface plane. At water partial pressures above 10^{-4} Torr, however, the ring tips from the surface with an angle of 40–55° along the surface normal [45].

Other efforts by Ouchi et al. [48] to study ionic liquid mixtures involved using ATR-IR absorption and Raman spectroscopy. [C_4mim]I or [C_4mim][BF_4] and their aqueous mixtures were investigated. Results showed that the ATR spectrum scanned from 2800 to 3200 cm^{-1} were very different for the two ionic liquids, although the spectral features in this region were from the butyl chain and imidazolium ring of the same cation. The presence of water was found to have no appreciable effect on the spectrum of [C_4mim][BF_4]; in contrast, a pronounced effect was observed for [C_4mim]I, in that as the water concentration increased, vibrational modes from the imidazolium ring appeared. For very dilute solutions of aqueous mixtures of both room-temperature ionic liquids, the spectra proved to be very similar. Moreover, Raman spectra around 600 cm^{-1} indicative of the butyl chain conformation supported conclusions that the I^- anion must be closer to the C(2)—H of the imidazolium cation, and interaction more specific, as compared to [BF_4]$^-$.

7.5 LIQUID–SOLID SURFACES

Several publications found in the literature involving ionic liquid studies on liquid–solid surfaces include the ionic liquid/quartz [49, 50], ionic liquid/TiO_2 [51] and ionic liquid/Pt interfaces [52–55].

Data has shown that for [C_4mim][BF_4] and [C_4mim][PF_6] on the surface of an equilateral quartz prism, spectra for both ionic liquids display similar vibrations from both alkyl chain and aromatic ring. The anion size influenced the orientation of the molecule such that tilt angles from the surface normal to the aromatic ring differed for both, with [C_4mim][BF_4] higher than that of [C_4mim][PF_6]. Also, the study suggests that the anion is positioned next to the cation adsorbed on the quartz surface. The methyl group of the butyl chain lies nearly parallel to the surface for [C_4mim][BF_4] [49]. Comparing these observations with similar studies done by Conboy and co-workers [56, 57] on the same interface, they postulated that, as the size of the anion is increased, the orientation of the cation is tilted upward towards surface normal. Since the $[PF_6]^-$ anion is 10% larger than $[BF_4]^-$, it causes the ring of the cation to tilt more towards the surface normal [49]. Furthermore, the authors concluded that, independent of the anion composition, the imidazolium ring orients more parallel to the silica surface as the alkyl chain length decreases [57].

The structure of ionic liquids on fused IR quartz modified by deuterated dodecyltrichlorosilane to give a hydrophobic surface has been investigated [50]. Using $[C_4mim][BF_4]$ and $[C_4mim][PF_6]$, it was found that the terminal methyl group was arranged towards the alkylsilane monolayer, with the imidazolium ring lying in the plane of the interface. Orientation of the methyl group was found to be only slightly affected for both ionic liquids since the anion was found close to the cation ring due to coulombic interactions. Comparing results on the three different interfaces, the data suggest the following:

1. At the ionic liquid/quartz interface, the methyl group orients towards the surface normal as the surface becomes more hydrophobic.
2. At the gas–liquid interface, the anion does not affect the cation orientation [32].
3. For the hydrophobic quartz interface, the orientation of the cation is influenced by the type of anion, such that the more hydrophobic anion, $[PF_6]^-$, has greater attraction to the non-polar silane monolayer, while the hydrophilic anion, $[BF_4]^-$, is more attracted to the polar C(2)—H group on the imidazolium ring.

On hydrophilic surfaces, ionic liquids form weak hydrogen bonds between the nitrogen atoms of the ring and anions adsorb to the surface causing the cation to tilt towards the surface normal. Figure 7.8 shows a comparison of SFG spectra (in *ssp* polarisation) of $[C_4mim][PF_6]$ at the three different interfaces.

In contrast to the mirror-like, polished quartz surface, ionic liquids on a relatively 'rougher' surface of TiO_2 have been studied using ionic liquids $[C_4mim][N(CN)_2]$ and $[C_4mim][CH_3SO_4]$ probed by SFG [51]. Both cation and anion were present at the interface for $[C_4mim][N(CN)_2]$, while only the cation

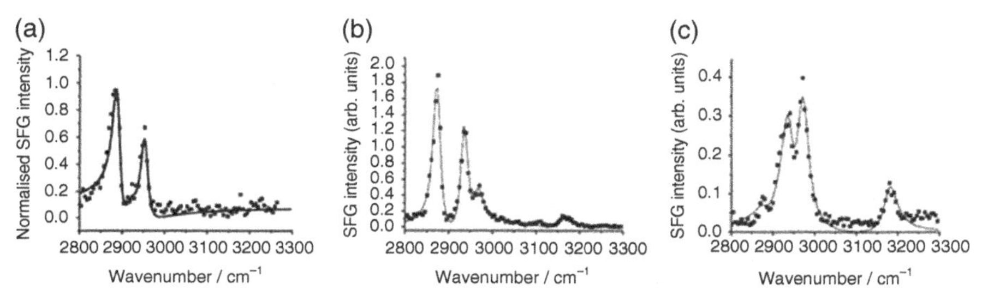

Figure 7.8 SFG spectra (*ssp*) of $[C_4mim][PF_6]$ at three different interfaces (a) air/liquid, (b) hydrophobic quartz/liquid and (c) hydrophilic quartz/liquid [50].

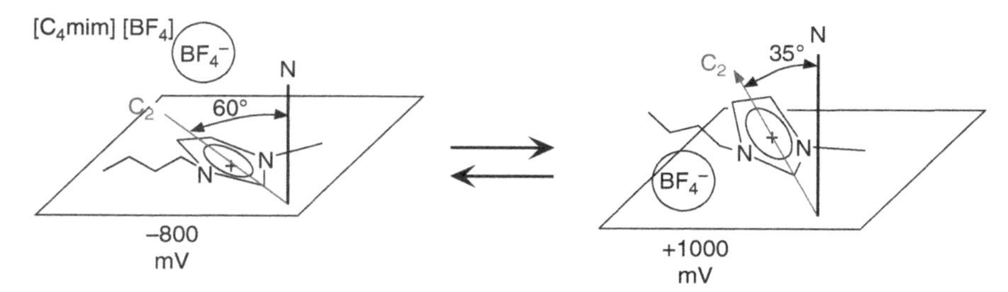

Figure 7.9 Changes in surface orientation of the cation and anion in $[C_4mim][BF_4]$ at the platinum electrode [54, 55].

was detected for $[C_4mim][CH_3SO_4]$. The imidazolium ring lies nearly parallel to the TiO_2 surface, as further supported by contact angle and surface charge density calculations. Specific adsorption of $[N(CN)_2]^-$ was found, in contrast to a weak adsorption of $[CH_3SO_4]^-$, as $[N(CN)_2]^-$ tends to bind to titanium sites at the surface of TiO_2. This was seen in the unusually strong SFG signal in the C—N stretching region. Moreover, strong methylene contributions were attributed to the surface roughness of the TiO_2 as it is composed of particles of approximately 2.5 nm in diameter as opposed to a polished quartz surface.

In an effort to provide a model for ionic liquids similar to the Gouy–Chapman–Stern model, which appropriately describes the organisation of solvent and ions in normal electrolytic solutions, a few studies have probed various ionic liquid/Pt electrode interfaces using SFG. To date, there is still no adequate description to explain the arrangement of the ions in pure ionic compounds, such as ionic liquids adjacent to electrode surfaces. Initial studies using $[C_4mim][BF_4]$ and $[C_4mim][PF_6]$ ionic liquids showed that the imidazolium ring conformation changes from 35° from the surface normal at positive surface charge to 60° at negative surface charge. A molecular representation is found in Figure 7.9. The methyl group for $[C_4mim][BF_4]$ shows a more perpendicular orientation at positive potentials, while for $[C_4mim][PF_6]$, methyl group analysis suggests that the butyl chain is more parallel to the surface [54]. Moreover, the ions were found to orient at the interface depending on electrode charge with a layer structure at the electrode surface composed of a single-ion layer of the Helmholtz type and ruling out a multilayer structure [53].

Using electrochemical impedance spectroscopy (EIS) with SFG, a description of the $[C_4mim][BF_4]$/Pt electrode interface was further investigated [55]. Vibrational Stark shift served as an independent measure of the thickness of the 'double layer', which suggested that the ions organised in a Helmholtz-like layer in agreement with initial studies. SFG results showed that the structure at the interface was potential-dependent with anions adsorbed to the surface at positive potentials relative to the potential of zero charge (PZC). This causes the imidazolium ring to be repelled and thereby orient more along the surface normal. At negative potentials, however, the cation is oriented more parallel to the surface plane with the anions repelled from the surface. A very thin 'double-layer' structure at the surface with ions forming a single layer (as seen from interfacial capacitance and vibrational Stark shift of CO) at the interface and screening the electrode charge was proposed. Similarly, *in situ* SFG results from Zhou et al. [58] proved the existence of a single-ion layer and a diffuse layer species. Their findings suggested that the ions on the $[C_4mim][OTf]/Pt$ interface are present as a double-layer structure. It was further postulated that the adsorption/desorption hysteresis of ionic liquids at electrode surfaces can be changed by varying the structure of the anions and the cations, as it is most possibly an intrinsic property of the room-temperature ionic liquids.

In contrast to the behaviour observed for $[C_4mim][BF_4]$ and $[C_4mim][PF_6]$, potential-dependent spectra for $[C_4mim][N(CN)_2]$ at the liquid/Pt interface suggest a multilayer structure of the Helmholtz type. Strong C—N vibrations observed for the anion suggest increased ordering as the $[N(CN)_2]^-$ anion tends to align itself to the surface normal with increasing potential. Signal strength may also come from increased $[N(CN)_2]$ concentration at the electrode surface for positive surface charge excesses. The appearance of symmetric ring vibrations from the cation at positive potentials denotes that the ring is repelled from the electrode and tilting towards the surface [52].

Model studies under UHV conditions, aimed to better understand the interaction of ionic liquid films with oxide supports, have been performed using thin films of $[C_4mim][NTf_2]$ [59]. With the use of IRAS in combination with DFT, thin films of the ionic liquid were grown on an atomically flat, well-ordered alumina film on NiAl(110) by evaporation. Results from the time-resolved IRAS measured during the growth and desorption of the films indicated no decomposition. Strong changes in intensity for individual anion bands were observed in the monolayer region, indicating orientation effects. The results further suggested that $[NTf_2]^-$ anions adopted a cis-conformation in the submonolayer region, adsorbing in a slightly tilted orientation with respect to the surface and mainly interacting with the oxide support through the sulfonyl groups. Figure 7.10 illustrates a schematic model of the most probable orientation of the $[NTf_2]^-$ anion on Al_2O_3/NiAl(110) in the submonolayer region, resulting from fitting DFT calculations to IRAS measurements.

To obtain more insight into the surface chemistry of supported ionic liquid-based catalysts, combined methods of time-resolved IRAS and XPS were used to investigate thin films of the ionic liquid $[C_4mim][NTf_2]$ deposited by physical vapour deposition (PVD) onto a model surface consisting of palladium nanoparticles (NPs) grown in UHV on an ordered alumina film on NiAl(110) [60]. Results showed that the ionic liquid molecularly adsorbs onto both the palladium particles and alumina at 300 K. IR spectra reveal that the anions interact with the palladium sites preferentially through the sulfonyl groups. Furthermore, upon heating to temperatures higher than 400 K, which is the desorption temperature of the ionic liquid, molecular desorption competes with decomposition. On the other hand, CO pre-adsorbed on the palladium particles is partially displaced by the ionic liquid, even at 300 K.

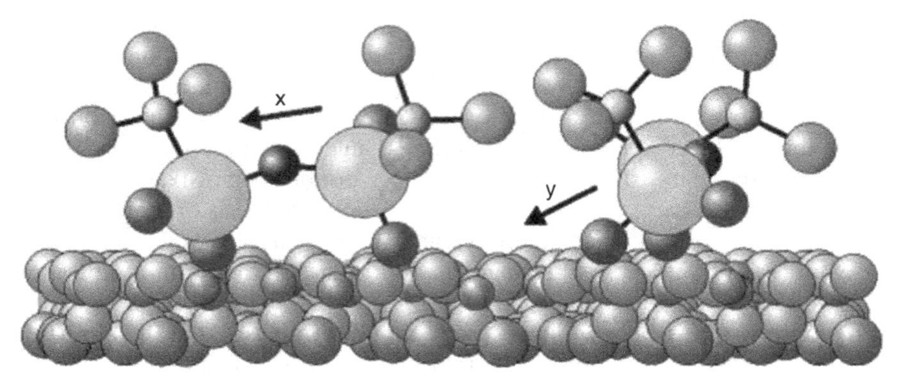

Figure 7.10 Proposed model of the orientation of the $[NTf_2]^-$ anion on Al_2O_3/NiAl(110) in the submonolayer region, from DFT calculations and IRAS measurements [59].

SERS spectra at ionic liquid/silver [61, 62] and ionic liquid/copper [63, 64] surfaces have been reported in the literature, where the first report of *in situ* SERS was for $[C_4mim][PF_6]$ adsorbed on a silver electrode. Results showed that the cation adsorbs on the silver electrode for potentials more negative than –0.4V versus a platinum *quasi*-reference electrode (PQRE). Moreover, at potentials less negative than the PZC (*ca.* –1V vs. PQRE), the cation adsorbs on the silver surface with the imidazolium ring almost perpendicular to the surface. At negative potentials relative to the PZC, however, the imidazolium ring stays almost parallel to the surface and is reduced to a $[C_4mim]^+$ carbene at –3.0V [61].

SERS spectra taken using silver NPs dispersed in the ionic liquid $[C_4mim][BF_4]$ and, from an island film spin coated with $[C_4mim][BF_4]$ [62], showed enhancement of the $[BF_4]^-$ symmetric stretching signal for silver island films. No enhancement, however, was observed for the silver NP/ionic liquid interface. Results were explained from the point of view of the effect of water molecules on the structure of the ionic liquid on the different substrates used, supporting that the presence of water in the silver NP/ionic liquid promoted aggregation of the silver NPs. Figure 7.11 shows a schematic picture for the interaction of $[C_4mim][BF_4]$ on two different silver surfaces.

Electrochemical measurements in tandem with SERS at the $[C_4mim][BF_4]$/copper electrode interface [63] reveal that the copper electrode has an electrochemical window of approximately 2.5V in the ionic liquid and that the Cu_2O film is present at the copper electrode surface prior to the SERS activation procedure. No considerable interaction between the anions and the metal surface was found; however, some cation vibrational modes show significant changes in the applied potentials as shifts in the wavenumber position and intensities. At potentials negative to the PZC, the cation interacts with the surface through a partial charge transfer, culminating in the reduction of the cation at approximately –2.6V.

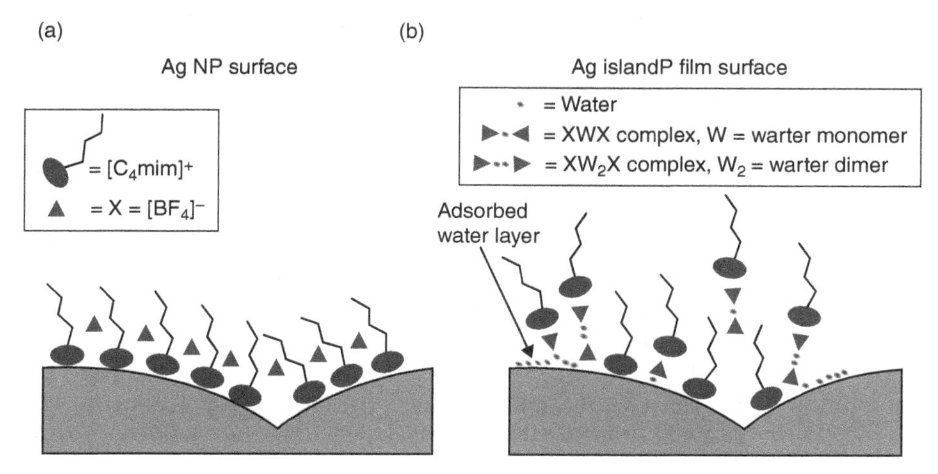

Figure 7.11 Model for the interaction of $[C_4mim][BF_4]$ with (a) an Ag NP aggregate in the Ag NPs/$[C_4mim][BF_4]$ colloidal solution and (b) with an Ag aggregated nanostructure on Ag island film deduced from SERS spectra [62].

A related study on the spectroelectrochemical behaviour of a copper electrode in $[C_4mim][BF_4]$ containing benzotriazole (BTAH) has been investigated by cyclic voltammetry and SERS [64]. A considerable decrease in the anodic currents in the presence of BTAH, as seen in the cyclic voltammograms, suggests that BTAH inhibits the oxidation of copper in the ionic liquid. The SERS results have shown that at potentials positive to the PZC for copper in the ionic liquid, a $[Cu(I)(BTA)]_n$ polymeric film is formed as BTAH interacts with the copper(I). The polymeric film is responsible for the corrosion inhibition process. However, at potentials negative to the PZC, BTAH is found to adsorb chemically on the copper surface. This phenomenon is determined by the observation of the ν(Cu—N) stretching mode and by the δ(N—H) bending mode.

REFERENCES

1 Welton, T., Room-temperature ionic liquids. Solvents for synthesis and catalysis, *Chem. Rev.* **99**(8), 2071–2084 (1999).

2 Armand, M., Endres, F., MacFarlane, D.R., Ohno, H. and Scrosati, B., Ionic-liquid materials for the electrochemical challenges of the future, *Nat. Mater.* **8**(8), 621–629 (2009).

3 Su, Y.-Z., Fu, Y.-C., Wei, Y.-M., Yan, J.-W. and Mao, B.-W., The electrode/ionic liquid interface: Electric double layer and metal electrodeposition, *Chem. Phys. Chem.* **11**(13), 2764–2778 (2010).

4 Dyson, P.J. and Geldbach, T.J., Applications of ionic liquids in synthesis and catalysis, *Electrochem. Soc. Interface* **16**, 50–53 (2007).

5 Parvulescu, V.I. and Hardacre, C., Catalysis in ionic liquids, *Chem. Rev.* **107**(6), 2615–2665 (2007).

6 Steinrück, H.P., Libuda, J., Wasserscheid, P., Cremer, T., Kolbeck, C., Laurin, M., Maier, F., Sobota, M., Schulz, P.S. and Stark, M., Surface science and model catalysis with ionic liquid-modified materials, *Adv. Mater.* **23**, 2571–2587 (2011).

7 Zhao, H., Innovative applications of ionic liquids as 'green' engineering liquids, *Chem. Eng. Commun.* **193**, 1660–1677 (2006).

8 Mokrushin, V., Assenbaum, D., Paape, N., Gerhard, D., Mokrushina, L., Wasserscheid, P., Arlt, W., Kistenmacher, H., Neuendorf, S. and Göke, V., Ionic liquids for Propene-Propane separation, *Chem. Eng. Technol.* **33**, 63–73 (2010).

9 Han, X. and Armstrong, D.W., Ionic liquids in separations, *Acc. Chem. Res.* **40**, 1079–1086 (2007).

10 Armstrong, J.P., Hurst, C., Jones, R.G., Licence, P., Lovelock, K.R.J., Satterley, C.J. and Villar-Garcia, I.J., Vapourisation of ionic liquids, *Phys. Chem. Chem. Phys.* **9**, 982–990 (2007).

11 Steinrück, H.-P., Surface science goes liquid! *Surf. Sci.* **604**, 481–484 (2010).

12 Caporali, S., Bardi, U. and Lavacchi, A., X-ray photoelectron spectroscopy and low energy ion scattering studies on 1-buthyl-3-methyl-imidazolium bis(trifluoromethane) sulfonimide, *J. Electron. Spectros. Relat. Phenomena* **151**, 4–8 (2006).

13 Lockett, V., Sedev, R., Bassell, C. and Ralston, J., Angle-resolved X-ray photoelectron spectroscopy of the surface of imidazolium ionic liquids, *Phys. Chem. Chem. Phys.* **10**, 1330–1335 (2008).

14 Maier, F., Cremer, T., Kolbeck, C., Lovelock, K.R.J., Paape, N., Schulz, P.S., Wasserscheid, P. and Steinruck, H.P., Insights into the surface composition and enrichment effects of ionic liquids and ionic liquid mixtures, *Phys. Chem. Chem. Phys.* **12**, 1905–1915 (2010).

15 Smith, E.F., Garcia, I.J.V., Briggs, D. and Licence, P., Ionic liquids in vacuo; solution-phase X-ray photoelectron spectroscopy, *Chem. Commun.*, 5633–5635 (2005).

16 Smith, E.F., Rutten, F.J.M., Villar-Garcia, I.J., Briggs, D. and Licence, P., Ionic liquids in vacuo: Analysis of liquid surfaces using ultra-high-vacuum techniques, *Langmuir* **22**, 9386–9392 (2006).

17 Hofft, O., Bahr, S., Himmerlich, M., Krischok, S., Schaefer, J.A. and Kempter, V., Electronic structure of the surface of the ionic liquid [EMIM][Tf_2N] studied by metastable impact electron spectroscopy (MIES), UPS and XPS, *Langmuir* **22**, 7120–7123 (2006).

18 Nishi, T., Iwahashi, T., Yamane, H., Ouchi, Y., Kanai, K. and Seki, K., Electronic structures of ionic liquids and studied by ultraviolet photoemission, inverse photoemission, and near-edge X-ray absorption fine structure spectroscopies, *Chem. Phys. Lett.* **455**, 213–217 (2008).

19 Kanai, K., Nishi, T., Iwahashi, T., Ouchi, Y., Seki, K., Harada, Y. and Shin, S., Electronic structures of imidazolium-based ionic liquids, *J. Electron Spectros. Relat. Phenomena* **174**, 110–115 (2009).

20 Gannon, T.J., Law, G., Watson, P.R., Carmichael, A.J. and Seddon, K.R., First observation of molecular composition and orientation at the surface of a room-temperature ionic liquid, *Langmuir* **15**, 8429–8434 (1999).

21 Law, G., Watson, P.R., Carmichael, A.J. and Seddon, K.R., Molecular composition and orientation at the surface of room-temperature ionic liquids: Effect of molecular structure, *Phys. Chem. Chem. Phys.* **3**, 2879–2885 (2001).

22 Hoffmann, F.M., Infrared reflection-absorption spectroscopy of adsorbed molecules, *Surf. Sci. Rep.* **3**, 107–192 (1983).

23 Wang, H.-F., Gan, W., Lu, R., Rao, Y. and Wu, B.-H., Quantitative spectral and orientational analysis in surface sum frequency generation vibrational spectroscopy (SFG-VS), *Int. Rev. Phys. Chem.* **24**, 191–256 (2005).

24 Colthup, N.B., Daly, L.H. and Wilberley, S.E., *Introduction to Infrared and Raman Spectroscopy*, 3rd Ed., Academic Press, San Diego (1990).

25 Diem, M., *Introduction to Modern Vibrational Spectroscopy*, John Wiley & Sons, Inc., New York (1993).

26 Greenler, R.G., Infrared study of adsorbed molecules on metal surfaces by reflection technique, *J. Chem. Phys.* **44**, 310–315 (1966).

27 Shen, Y.R., Surface properties probed by second-harmonic and sum-frequency generation, *Nature* **337**, 519–525 (1989).

28 Vickerman, J.C., *Surface Analysis: The Principal Techniques*, John Wiley & Sons, Inc., New York (1998).

29 Bain, C.D., Sum-frequency vibrational spectroscopy of the solid/liquid interface, *J. Chem. Soc. Faraday Trans.* **91**(9), 1281–1296 (1995).

30 Martinez, I.S. and Baldelli, S., On the arrangement of ions in imidazolium-based room temperature ionic liquids at the gas-liquid interface, using sum frequency generation, surface potential, and surface tension measurements, *J. Phys. Chem. C* **114**, 11564–11575 (2010).

31 Santos, C.S., Rivera-Rubero, S., Dibrov, S. and Baldelli, S., Ions at the surface of a room-temperature ionic liquid, *J. Phys. Chem. C* **111**, 7682–7691 (2007).

32 Rivera-Rubero, S. and Baldelli, S., Surface characterization of 1-butyl-3-methylimidazolium Br^-, I^-, PF_6^-, BF_4^-, $(CF_3SO_2)_2N^-$, SCN^-, $CH_3SO_3^-$, $CH_3SO_4^-$, and $(CN)_2N^-$ ionic liquids by sum frequency generation, *J. Phys. Chem. B* **110**, 4756–4765 (2006).

33 Santos, C.S. and Baldelli, S., Surface orientation of 1-methyl-, 1-ethyl-, and 1-butyl-3-methylimidazolium methyl sulfate as probed by sum-frequency generation vibrational spectroscopy, *J. Phys. Chem. B* **111**, 4715–4723 (2007).

34 Santos, C.S. and Baldelli, S., Alkyl chain interaction at the surface of room temperature ionic liquids: Systematic variation of alkyl chain length (R = C1–C4, C8) in both Cation and Anion of [RMIM][R-SO_3] by sum frequency generation and surface tension, *J. Phys. Chem. B* **113**, 923–933 (2009).

35 Aliaga, C. and Baldelli, S., Sum frequency generation spectroscopy of dicyanamide based room-temperature ionic liquids. Orientation of the cation and the anion at the gas-liquid interface, *J. Phys. Chem. B* **111**, 9733–9740 (2007).

36 Aliaga, C., Baker, G.A. and Baldelli, S., Sum frequency generation studies of ammonium and pyrrolidinium ionic liquids based on the bis-trifluoromethanesulfonimide anion, *J. Phys. Chem. B* **112**, 1676–1684 (2008).

37 Peñalber, C., Adamová, G., Plechkova, N.V., Baldelli, S. and Seddon, K.R., Sum-frequency generation spectroscopy of tetraalkylphosphonium-based ionic liquids at the air-liquid interface (to be published).

38 Iimori, T., Iwahashi, T., Ishii, H., Seki, K., Ouchi, Y., Ozawa, R., Hamaguchi, H.-o. and Kim, D., Orientational ordering of alkyl chain at the air/liquid interface of ionic liquids studied by sum frequency vibrational spectroscopy, *Chem. Phys. Lett.* **389**, 321–326 (2004).

39 Jeon, Y., Sung, J., Bu, W., Vaknin, D., Ouchi, Y. and Kim, D., Interfacial restructuring of ionic liquids determined by sum-frequency generation spectroscopy and X-ray reflectivity, *J. Phys. Chem. C* **112**, 19649–19654 (2008).

40 Santos, C.S. and Baldelli, S., Gas-liquid interface of hydrophobic and hydrophilic room-temperature ionic liquids and benzene: Sum frequency generation and surface tension studies, *J. Phys. Chem. C* **112**, 11459–11467 (2008).

41 Iwahashi, T., Miyamae, T., Kanai, K., Seki, K., Kim, D. and Ouchi, Y., Anion configuration at the air/liquid interface of ionic liquid [bmim]OTf studied by sum-frequency generation spectroscopy, *J. Phys. Chem. B* **112**, 11936–11941 (2008).

42 Krischok, S., Eremtchenko, M., Himmerlich, M., Lorenz, P., Uhlig, J., Neumann, A., Ottking, R., Beenken, W.J.D., Hofft, O., Bahr, S., Kempter, V. and Schaefer, J.A., Temperature-dependent electronic and vibrational structure of the 1-ethyl-3-methylimidazolium bis(trifluoromethylsulfonyl)amide room-temperature ionic liquid surface: A study with XPS, UPS, MIES, and HREELS, *J. Phys. Chem. B* **111**, 4801–4806 (2007).

43 Rivera-Rubero, S. and Baldelli, S., Influence of water on the surface of hydrophilic and hydrophobic room-temperature ionic liquids, *J. Am. Chem. Soc.* **126**, 11788–11789 (2004).

44 Rivera-Rubero, S. and Baldelli, S., Influence of water on the surface of the water-miscible ionic liquid 1-butyl-3-methylimidazolium tetrafluoroborate: A sum frequency generation analysis, *J. Phys. Chem. B* **110**, 15499–15505 (2006).

45 Baldelli, S., Influence of water on the orientation of cations at the surface of a room-temperature ionic liquid: A sum frequency generation vibrational spectroscopic study, *J. Phys. Chem. B* **107**, 6148–6152 (2003).

46 Iwahashi, T., Sakai, Y., Kanai, K., Kim, D. and Ouchi, Y., Alkyl-chain dividing layer at an alcohol/ionic liquid buried interface studied by sum-frequency generation vibrational spectroscopy, *Phys. Chem. Chem. Phys.* **12**, 12943–12946 (2010).

47 Jeon, Y., Sung, J., Kim, D., Seo, C., Cheong, H., Ouchi, Y., Ozawa, R. and Hamaguchi, H.-o., Structural change of 1-butyl-3-methylimidazolium tetrafluoroborate + water mixtures studied by infrared vibrational spectroscopy, *J. Phys. Chem. B* **112**, 923–928 (2008).

48 Jeon, Y., Sung, J., Seo, C., Lim, H., Cheong, H., Kang, M., Moon, B., Ouchi, Y. and Kim, D., Structures of ionic liquids with different anions studied by infrared vibration spectroscopy, *J. Phys. Chem. B* **112**, 4735–4740 (2008).

49 Romero, C. and Baldelli, S., Sum frequency generation study of the room-temperature ionic liquids/quartz interface, *J. Phys. Chem. B* **110**, 6213–6223 (2006).

50 Romero, C., Moore, H.J., Lee, T.R. and Baldelli, S., Orientation of 1-butyl-3-methylimidazolium based ionic liquids at a hydrophobic quartz interface using sum frequency generation spectroscopy, *J. Phys. Chem. C* **111**, 240–247 (2007).

51 Aliaga, C. and Baldelli, S., A sum frequency generation study of the room-temperature ionic liquid-titanium dioxide interface, *J. Phys. Chem. C* **112**(8), 3064–3072 (2008).

52 Aliaga, C. and Baldelli, S., Sum frequency generation spectroscopy and double-layer capacitance studies of the 1-butyl-3-methylimidazolium dicyanamide-platinum interface, *J. Phys. Chem. B* **110**, 18481–18491 (2006).

53 Baldelli, S., Probing electric fields at the ionic liquid-electrode interface using sum frequency generation spectroscopy and electrochemistry, *J. Phys. Chem. B* **109**, 13049–13051 (2005).

54 Rivera-Rubero, S. and Baldelli, S., Surface spectroscopy of room-temperature ionic liquids on a platinum electrode: A sum frequency generation study, *J. Phys. Chem. B* **108**, 15133–15140 (2004).

55 Baldelli, S., Surface structure at the ionic liquid-electrified metal interface, *Acc. Chem. Res.* **41**, 421–431 (2008).

56 Fitchett, B.D. and Conboy, J.C., Structure of the room-temperature ionic liquid/SiO_2 interface studied by sum-frequency vibrational spectroscopy, *J. Phys. Chem. B* **108**, 20255–20262 (2004).

57 Rollins, J.B., Fitchett, B.D. and Conboy, J.C., Structure and orientation of the imidazolium cation at the room-temperature ionic liquid/SiO_2 interface measured by sum-frequency vibrational spectroscopy, *J. Phys. Chem. B* **111**, 4990–4999 (2007).

58 Zhou, W., Inoue, S., Iwahashi, T., Kanai, K., Seki, K., Miyamae, T., Kim, D., Katayama, Y. and Ouchi, Y., Double layer structure and adsorption/desorption hysteresis of neat ionic liquid on Pt electrode surface – An in-situ IR-visible sum-frequency generation spectroscopic study, *Electrochem. Commun.* **12**, 672–675 (2010).

59 Sobota, M., Nikiforidis, I., Hieringer, W., Paape, N., Happel, M., Steinrück, H.-P., Görling, A., Wasserscheid, P., Laurin, M. and Libuda, J.r., Toward ionic-liquid-based model catalysis: Growth, orientation, conformation, and interaction mechanism of the [Tf_2N] anion in [BMIM][Tf_2N] thin films on a well-ordered alumina surface, *Langmuir* **26**, 7199–7207 (2010).

60 Sobota, M., Schmid, M., Happel, M., Amende, M., Maier, F., Steinruck, H.-P., Paape, N., Wasserscheid, P., Laurin, M., Gottfried, J.M. and Libuda, J., Ionic liquid based model catalysis: Interaction of [BMIM][Tf_2N] with Pd nanoparticles supported on an ordered alumina film, *Phys. Chem. Chem. Phys.* **12**, 10610–10621 (2010).

61 Santos, V.O., Alves, M.B., Carvalho, M.S., Suarez, P.A.Z. and Rubim, J.C., Surface-enhanced Raman scattering at the silver electrode/ionic liquid ($BMIPF_6$) interface, *J. Phys. Chem. B* **110**, 20379–20385 (2006).

62 Rubim, J.C., Trindade, F.A., Gelesky, M.A., Aroca, R.F. and Dupont, J., Surface-enhanced vibrational spectroscopy of tetrafluoroborate 1-*n*-butyl-3-methylimidazolium ($BMIBF_4$) ionic liquid on silver surfaces, *J. Phys. Chem. C* **112**, 19670–19675 (2008).

63 Brandão, C.R.R., Costa, L.A.F., Breyer, H.S. and Rubim, J.C., Surface-enhanced Raman scattering (SERS) of a copper electrode in 1-*n*-butyl-3-methylimidazolium tetrafluoroborate ionic liquid, *Electrochem. Commun.* **11**, 1846–1848 (2009).

64 Costa, L.A.F., Breyer, H.S. and Rubim, J.C., Surface-enhanced Raman scattering (SERS) on copper electrodes in 1-*n*-butyl-3-methylimidazoliun tetrafluorbarate (BMI.BF4): The adsorption of benzotriazole (BTAH), *Vib. Spectrosc.* **54**, 103–106 (2010).

8 Raman Spectroscopy and the Heterogeneous Liquid Structure in Ionic Liquids

SATYEN SAHA
Department of Chemistry, Banaras Hindu University, Varanasi, India

TAKASHI HIROI
Department of Chemistry, School of Science, The University of Tokyo, Tokyo, Japan

KOICHI IWATA
Department of Chemistry, Gakushuin University, Tokyo, Japan

HIRO-O HAMAGUCHI
Institute of Molecular Science and Department of Applied Chemistry, National Chiao Tung University, Hsinchu, Taiwan, and Department of Chemistry, School of Science, The University of Tokyo, Tokyo, Japan

8.1 INTRODUCTION

Intensive physicochemical studies in the past decade have shed a good deal of light on the multifaceted nature of ionic liquids, both from basic and applied science viewpoints [1]. The liquid structure in ionic liquids is one of the foci in these studies, which has raised more questions than it has answered. Since ionic liquids consist solely of ions, cations and anions, the long-range Columbic interaction must play a role in their structure formation. This fact contrasts sharply with the case of molecular liquids, in which only short-range multipole interactions operate. Thus, ionic liquids may well have characteristic liquid structures that cannot be associated with molecular liquids. The many distinctive properties of ionic liquids, namely, low melting points, high viscosities, marked amphiphilicities, and so on, are likely to originate from these characteristic liquid structures. Elucidating these structures is, therefore, crucial for a thorough understanding as well as for the full utilisation of ionic liquids in the future.

Ionic Liquids Completely UnCOILed: Critical Expert Overviews, First Edition.
Edited by Natalia V. Plechkova and Kenneth R. Seddon.

Raman spectroscopy is based on Raman scattering that was discovered by Sir C. V. Raman in 1928 [2]. Raman scattering is inelastic light scattering in which an incident photon gives/receives an energy quantum to/from the scatterer to create a new scattered photon having less/more energy. The spectrum of the scattered photons is called the Raman spectrum. Raman scattering is most often observed for vibrational transitions and 'Raman spectra' usually mean 'vibrational Raman spectra'. Once Raman spectra are obtained, we are able to extract otherwise unobtainable information on the structure: the dynamics and environments of molecules/ions in the solid (crystalline), liquid and gaseous states of materials of any kind. Ionic liquids are no exception. Thus, Raman spectra are sometimes called 'molecular fingerprints'. As reviewed in the following Sections, Raman spectra of $[C_4mim]X$ (X = Cl, Br, and I) in the crystalline and liquid states have proven that the known two polymorphs of $[C_4mim]Cl$ are due to the *trans* and *gauche* rotational isomers in the butyl chain of the imidazolium cation, and that the two rotational isomers do co-exist in the liquid phase [3–6]. This structural complexity has been discussed in relation to the low melting points of $[C_4mim]X$, in which the co-existence of the two different butyl structures may well hinder crystallisation, which requires structural unification in the crystallising liquid. The thermal diffusion dynamics in ionic liquids has also been studied in relation to the possible heterogeneous liquid structure of ionic liquids. The 'picosecond Raman thermometer' of S_1 (the lowest excited singlet state) of *trans*-stilbene, which shows peak shifts linearly proportional to the local temperature [7], has been used to monitor the picosecond cooling process of the molecule after photoexcitation in various ionic liquids. Marked deviation of the cooling rates in ionic liquids from those in molecular liquids has been discussed in terms of the local structure present in ionic liquids [8]. Low-frequency Stokes/anti-Stokes Raman spectroscopy has also been used to trace the melting process of ionic liquids. The disappearance of low-frequency lattice vibration bands has been used as an indicator of melting of a crystal to discuss unusually slow *trans/gauche* thermal equilibration just after the melting of a small crystal of $[C_4mim]Cl$ [9]. Raman spectroscopy has very good affinity with ionic liquid studies.

Vibrational spectroscopy excels in detecting and analysing water contained in ionic liquids. Infrared spectroscopy is more sensitive in detecting water, but Raman can also do it very well. Existence of water molecules, though in a very small amount, can affect the physicochemical properties of ionic liquids such as viscosity [10], electrical conductivity [11], in addition to polarity [12]. The reactivity of ionic liquid changes with water content [13], sometimes so markedly that it changes the course of the reaction altogether [14]. In fact, the existence of a water molecule in the unit cell can change the conformation of the alkyl group of the cation [15]. Assessment of a trace amount of water in ionic liquids is, therefore, very important. At the same time, the effect of contained water in a larger scale should also be of great importance in practical applications. Water containing ionic liquids may be regarded as a new class of ionic liquids whose properties can be changed continuously with careful regulation of the water content.

This chapter is not meant to provide a comprehensive review of the many ionic liquid studies using Raman spectroscopy just as a routine analytical tool. Rather, it focuses on structural chemistry approaches taken in our Tokyo group to elucidate the possible heterogeneous liquid structure in ionic liquids. Only those studies in which Raman spectroscopy plays crucial and irreplaceable roles are included. Readers are recommended to refer also to two earlier reviews, 'Structure of ionic liquids and ionic liquid compounds: are ionic liquids genuine liquids in the conventional sense?' [16], and 'Local structure formation in alkyl-imidazolium-based ionic liquids as revealed by linear and nonlinear Raman spectroscopy' [17].

8.2 RAMAN SPECTROSCOPY FOR IONIC LIQUID STUDIES

The methodology of Raman spectroscopy is rapidly becoming matured. Here, we describe two classes of new Raman spectroscopies, Raman microspectroscopy and low-frequency Raman spectroscopy, that are considered to be important for future applications to ionic liquid studies.

Raman microspectroscopy is a hybrid of Raman spectroscopy and confocal optical microscopy. It enables Raman spectroscopy of materials under a microscope, with typical spatial resolution of 0.5 μm (lateral) and 2 μm (depth), to be determined. A very small crystal of micrometer size can be easily measured. It also enables space-resolved measurements of heterogeneous samples for which distributions of different molecules/ions are of primary interest. This capability of Raman microspectroscopy is well demonstrated in Figure 8.1, in which the deliquescent process of a small [C_4mim]Cl crystal is studied. The point (1) in the inset of Figure 8.1 corresponds to the remaining crystalline state of the original [C_4mim]Cl crystal. It generates the Raman spectrum of a *trans* crystal that shows a strong *trans* marker band at 625 cm^{-1}, with no *gauche* marker at 603 cm^{-1}. The point (3) marks the area of aqueous solution of [C_4mim]Cl. This liquid gives a spectrum that very much resembles that of super-cooled liquid [C_4mim]Cl (see following discussion). It shows a clear *gauche* marker band. The point (2) is between (1) and (3) and corresponds to deliquescing [C_4mim]Cl. The spectrum here is close to that at (3), except for the stronger *trans* marker at 625 cm^{-1}, indicating that a small amount of absorbed water can change the bulk structure of [C_4mim]Cl in a drastic way. This example clearly shows the power of Raman microspectroscopy in studying spatially heterogeneous systems composed of ionic liquids.

The low-frequency region of a Raman spectrum contains valuable information on intermolecular/interionic interactions that play roles in the structure formation in liquids and crystals. Our group in Tokyo has developed a high-speed low-frequency Raman spectrometer with the use of an iodine vapour filter (Fig. 8.2) [18]. The iodine vapour filter eliminates Rayleigh scattering by virtue of a very narrow (<0.01 cm^{-1}) and intense absorption line of iodine. Thanks to the high Rayleigh scattering elimination efficiency of the filter, the developed Raman spectrometer is capable of recording down to ±5 cm^{-1} in

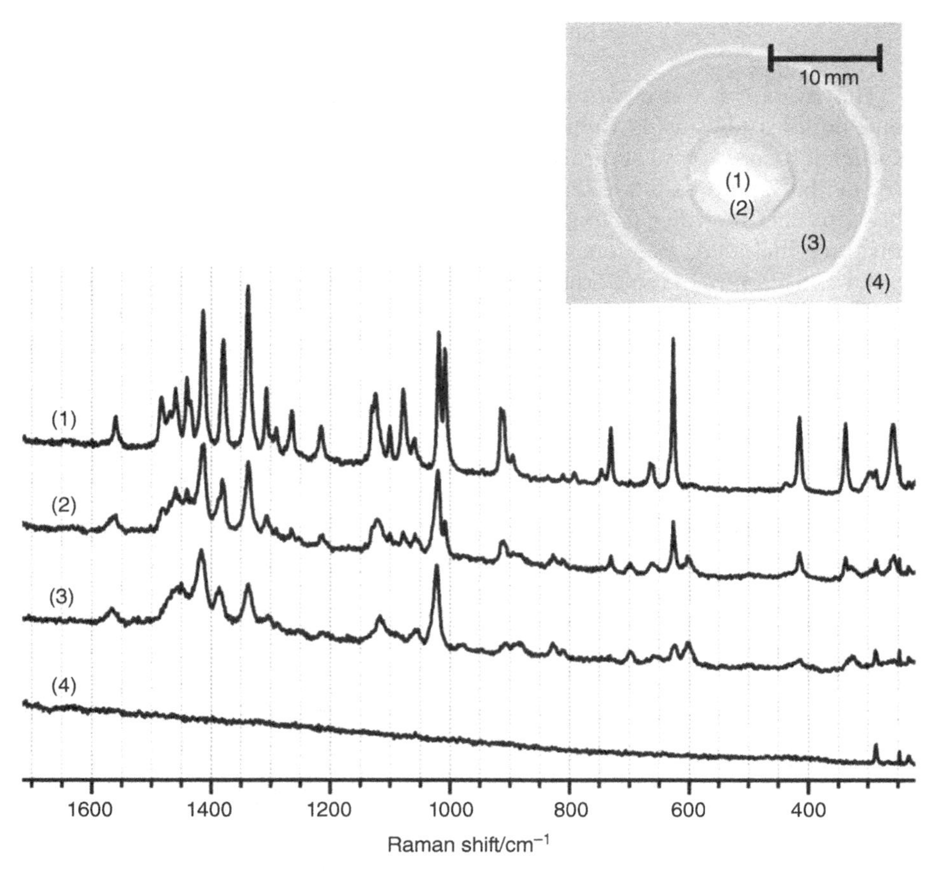

Figure 8.1 Space-resolved Raman spectra of deliquescing [C_4mim]Cl. (1) [C_4mim]Cl crystal, (2) deliquescing [C_4mim]Cl crystal, (3) aqueous solution of [C_4mim]Cl, and (4) air. Inset shows a microscope image.

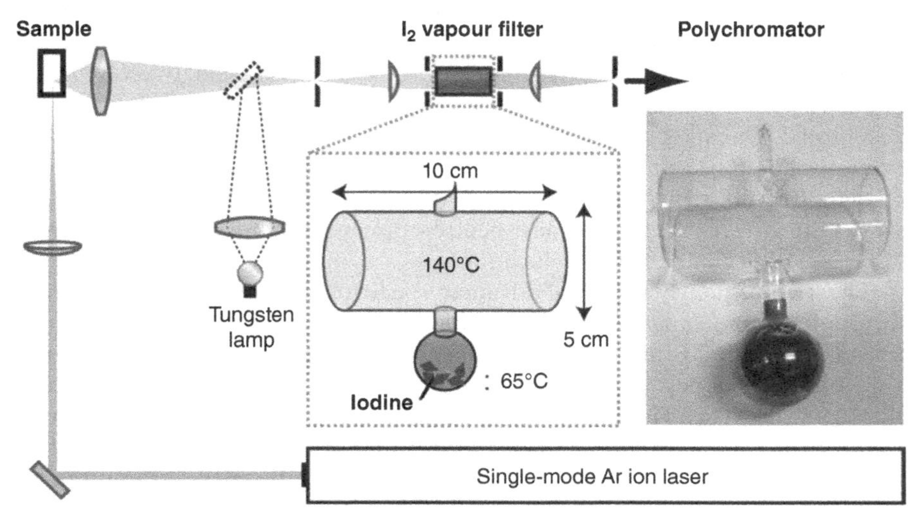

Figure 8.2 Fast low-frequency Raman spectrometer with an iodine vapour filter.

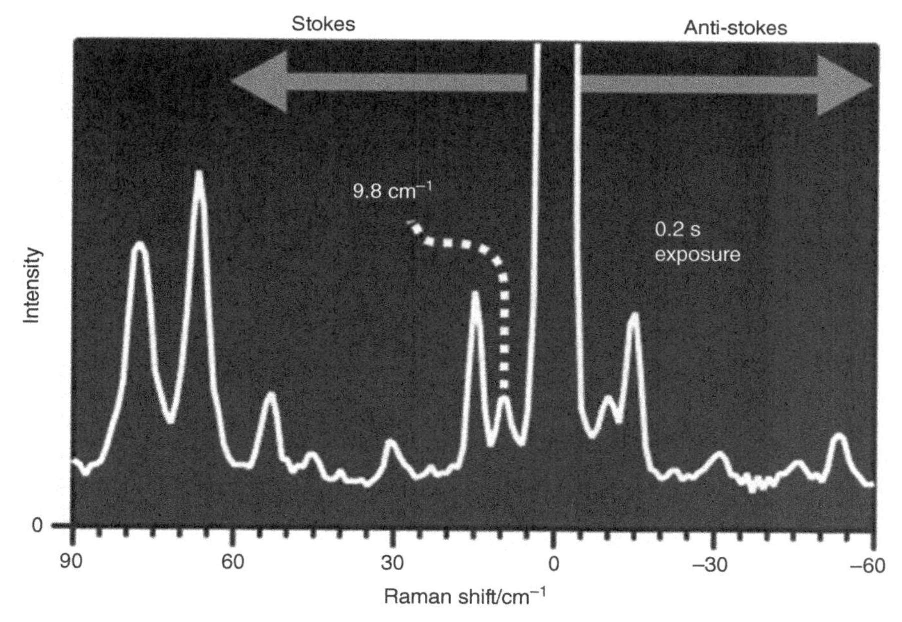

Figure 8.3 Low-frequency Raman spectrum of L-cystine.

both the Stokes and anti-Stokes sides simultaneously. The measurement time of the apparatus is determined only by the exposure time of the multichannel detector. It is now possible to investigate the fast changes of lattice vibrations during melting or crystallisation: this is extremely useful for understanding the phase transition mechanism.

As an example, the Stokes/anti-Stokes low-frequency Raman spectrum of L-cystine is shown in Figure 8.3. L-Cystine is commonly used for testing the low-frequency capability of a Raman spectrometer. The $\pm 9.8\,cm^{-1}$ Raman bands are measurable in as short as 0.2 s (Fig. 8.3). If the laser power is raised, faster measurements with 10 ms exposure time are possible.

8.3 ROTATIONAL ISOMERISM IN IONIC LIQUID

Rotational isomerism and Raman spectroscopy are historically related. The *trans/gauche* rotational isomerism with respect to a C–C single bond was discovered by Mizushima et al. by Raman spectroscopy for 1,2-dichloroethane [19]. The discovery was accidentally made by unexpected freezing of a 1,2-dichloroethane liquid sample during a low temperature measurement. Some of the Raman bands observed in the liquid were found to be missing in the crystalline state. Note that the *trans* and *gauche* isomers do co-exist in the liquid state of 1,2-dichloroethane, but that the crystalline state contains only the *trans* form. Therefore, the number of Raman bands from the liquid is much larger than those found in the crystal. It is beyond doubt that the versatile nature of Raman spectroscopy, which is capable of measuring both liquid and crystalline samples *in situ*, has helped in this serendipitous discovery of the rotational isomerism. After more than 60 years since the discovery, similar rotational isomerism was found in ionic liquids. This time, the discovery was made with the synergy between Raman spectroscopy with X-ray crystallography. The details of this synergy have been reviewed [16], with the crystal structures of the two polymorphs of $[C_4mim]Cl$ and $[C_4mim]Br$, the Raman spectra of crystalline and liquid $[C_4mim]X$ (X = Cl, Br, or I), and the vibrational analysis of the $[C_4mim]^+$ cation. Here, we briefly summarise the results.

Two crystal polymorphs, referred to as Crystal (1) and Crystal (2), were found for $[C_4mim]Cl$. They were characterised by their distinctive X-ray powder patterns and Raman spectra [3]. Although $[C_4mim]Cl$ is solid at ambient temperatures, it can also exist as a supercooled liquid under the same experimental conditions. Thanks to this peculiar property of $[C_4mim]Cl$, the Raman spectra of the two crystal polymorphs as well as that of supercooled liquid are obtainable at ambient temperatures. The Raman spectra of the two polymorphs show notable differences in the 500–800 cm^{-1} wavenumber region (Fig. 8.4). The bands at 625, 730, and 790 cm^{-1} are observed only for Crystal (1), while those at 500, 603, and 701 cm^{-1} are observed only for Crystal (2). Interestingly, both sets of these characteristic Raman bands are found simultaneously in the supercooled liquid state. This result indicates that two distinct structures of the $[C_4mim]^+$ cation co-exist in the supercooled liquid state. Note that the chlorine atom does not give any Raman band.

Subsequently, the crystal structures of $[C_4mim]Cl$ Crystal (1), $[C_4mim]Cl$ Crystal (2) and $[C_4mim]Br$ were determined by X-ray structural analysis [4–6]. The structural difference between the two crystal polymorphs was found to arise from the rotational isomerism with regard to the C7—C8 bond of the butyl chain. Crystal (1) has the *trans* conformation both around the C7—C8 bond and the C8—C9 bond (*TT* form, Fig. 8.5a). In contrast, Crystal (2) has the *gauche* conformation around the C7—C8 bond and the *trans* conformation around the C8—C9 bond (*GT* form; Fig. 8.5b).

In order to confirm the relationship between the observed Raman spectral differences and the structural variation of the $[C_4mim]^+$ cation, a DFT calculation of the $[C_4mim]Cl$ with energy optimisation at the *TT* and *GT* conformations was performed [6]. The calculation reproduces successfully the characteristic bands in the 500–800 cm^{-1} wavenumber region, both for the *TT* form and the *GT* form. The 625 and 730 cm^{-1} bands of the *TT* form arise from the deformation vibrations of the imidazolium ring. They correspond to the 603 and 700 cm^{-1} bands of the *GT* form, respectively. The frequency differences in the two bands result from the different coupling scheme with the CH_2 rocking vibration of the C8 methylene group. In addition to these bands, the 500 cm^{-1} band can be used as a marker band for the *GT* form because this band arises from the C7—C8—C9 deformation that reflects the gauche conformation around the C7—C8 bond. With the use of these marker bands, the co-existence of at least two rotational isomers of $[C_4mim][BF_4]$, one of the most commonly

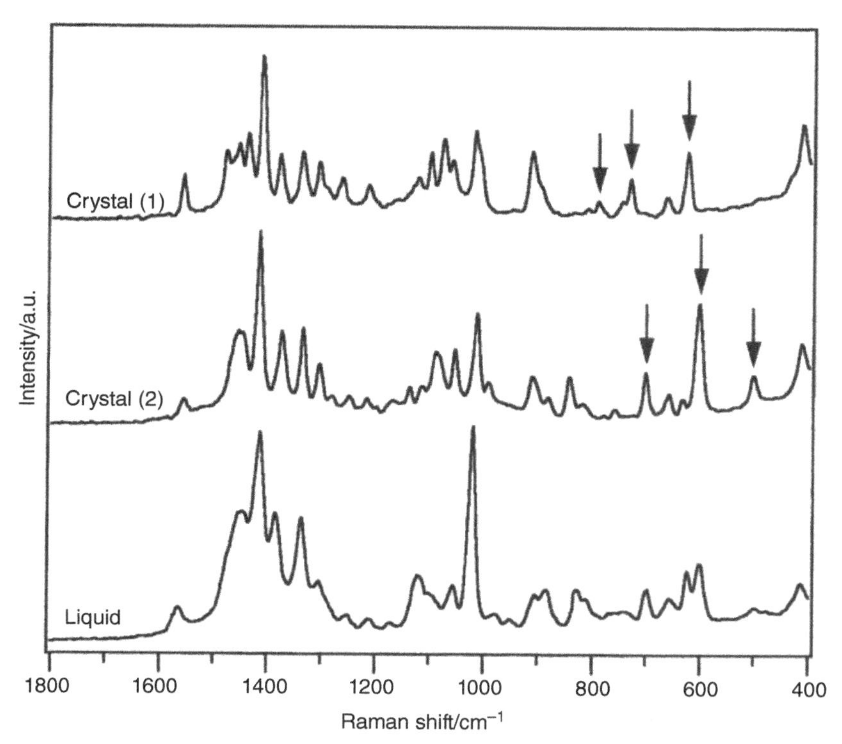

Figure 8.4 Raman spectra of Crystal (1) (top), Crystal (2) (middle) and liquid, $[C_4mim]Cl$ in a supercooled liquid state (bottom) [3]. The arrows indicate characteristic Raman bands for Crystals (1) and (2).

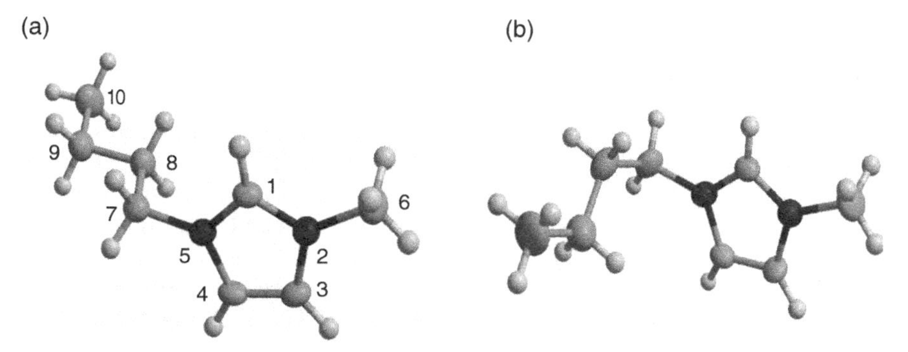

Figure 8.5 Two rotational isomers, (a) *trans-trans* and (b) *gauche-trans*, of the $[C_4mim]^+$ cation.

studied room-temperature ionic liquids, was confirmed (Fig. 8.6). As in the case of the supercooled liquid state of $[C_4mim]Cl$, the marker bands of both the *TT* and *GT* forms are clearly observed in the spectrum of $[C_4mim][BF_4]$ (Fig. 8.6a).

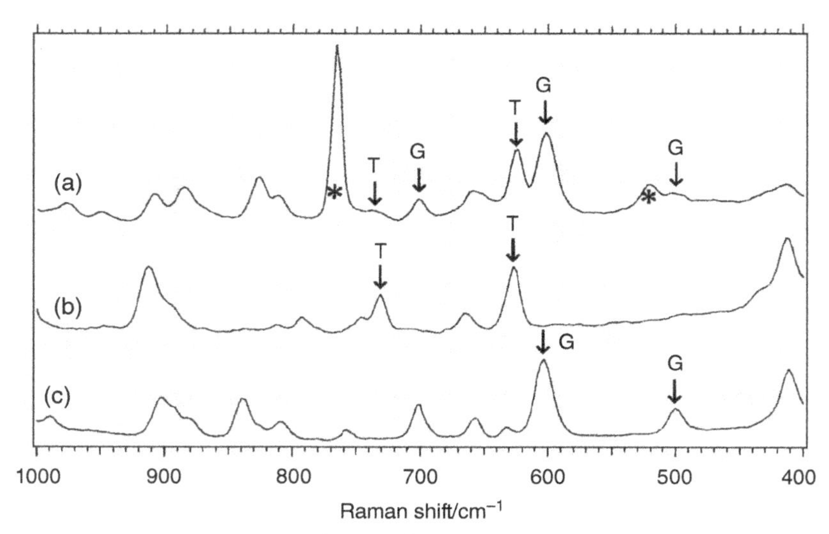

Figure 8.6 Raman spectra of (a) $[C_4mim][BF_4]$ (liquid), (b) $[C_4mim]Cl$ Crystal (1) (*TT* form) and (c) $[C_4mim]Br$ (*GT* form). *Indicates anion bands.

Strong aliphatic interactions between the two butyl chains (molecular zippers) of $[C_4mim]^+$ have been found for both of the two crystal polymorphs of $[C_4mim]Cl$ [4,5]. The enthalpy differences between the *trans* and *gauche* forms in several ionic liquids, determined from the Arrhenius plot of the temperature dependent relative Raman intensity changes, have suggested that similar strong aliphatic interactions may also exist in the liquid state [16]. During a rapid melting of crystalline $[C_4mim]Cl$, the conformational change between the *trans* and *gauche* conformations has been found to need an unusually long time, suggesting a strong interaction between the butyl chains [16].

In order to elucidate further these intriguing findings, sub-second time-resolved Raman tracing of the melting process of $[C_4mim]Cl$ Crystal (2) have been performed [9], with the use of the low-frequency Raman spectrometer described earlier [18]. In this experiment, a small piece (*ca.* 1 mm) of single crystalline $[C_4mim]Cl$ was rapidly heated by using a heat gun from 20 to 90°C, which is 30°C higher than the melting point of $[C_4mim]Cl$. The sample melted within 1 min of the start of the heating. Raman spectral changes during the melting process were measured continuously with a 0.5 s recording time. The observed Raman spectral change is shown in Figure 8.7. Before the heating (−10 s), a few sharp bands assigned to lattice vibrations exist below 200 cm^{-1}. As the heating proceeds, these bands gradually become weaker and eventually disappear. After 54 s, when all the lattice vibration bands totally disappear, a broad and intense spectral feature remains in the low-frequency region. The spectral shape looks quite similar to those of $[C_4mim]^+$-based

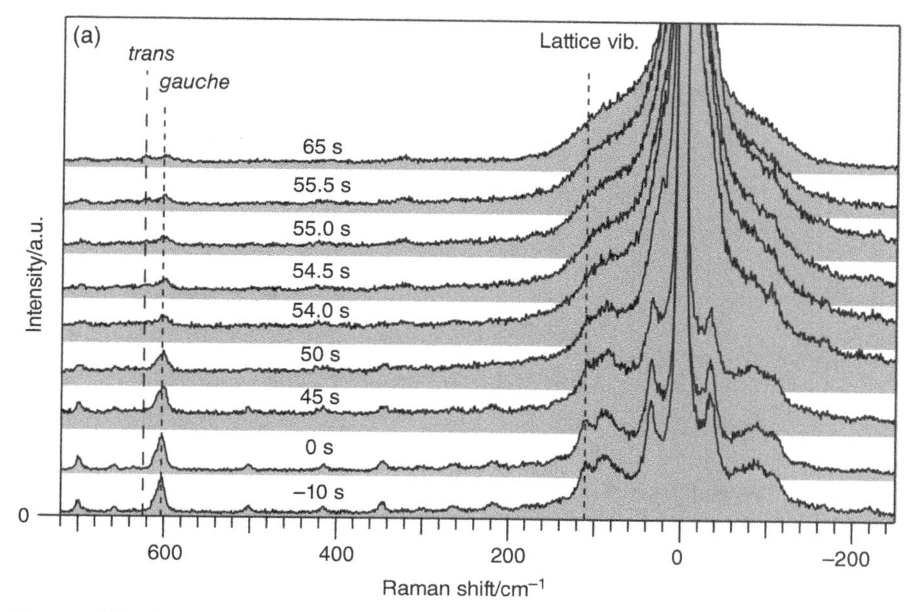

Figure 8.7 Raman spectral changes during the melting process of $[C_4mim]Cl$ Crystal (2).

room-temperature ionic liquids, such as $[C_4mim][PF_6]$, or $[C_4mim][BF_4]$ [16]. This fact indicates that the $[C_4mim]Cl$ becomes liquid after 54 s.

The higher frequency region of the spectral change in Figure 8.7 shows the conformational change of the $[C_4mim]^+$ cation. In the 0-s spectrum, the *gauche* marker band exists at 603 cm^{-1}, while there is no *trans* marker band at 625 cm^{-1}. After the heating, the intensity of the *gauche* marker decreases and the *trans* marker gradually appears. The intensity ratio of the *trans* and *gauche* markers becomes constant after 65 s. Interestingly, a time lag is observed between the disappearance of the lattice vibrations and the appearance of the *trans* marker. They do not happen simultaneously. The lattice vibration bands are almost missing at 54 s, while the *trans* marker is not obviously seen. The *trans* marker appears a few seconds after the loss of the lattice vibrational bands. In order to make this discussion clear, the intensity of the 111 cm^{-1} band and the intensity ratio (625 cm^{-1}/603 cm^{-1}) of the two markers are plotted in Figure 8.8. The intensity of the 111 cm^{-1} band gradually decreases as the heating proceeds. The decreasing rate becomes large after 48 s, showing the start of melting. After 57 s, the intensity of the lattice vibration band is smaller than the noise level. The *trans/gauche* intensity ratio starts to increase at 54 s and becomes constant after 57 s. Thus, the conformational change occurs during 54 and 57 s. There is a clear time lag of a few seconds between the loss of the crystal structure (disappearance of the 111 cm^{-1} band) and the conformational change from the *gauche*

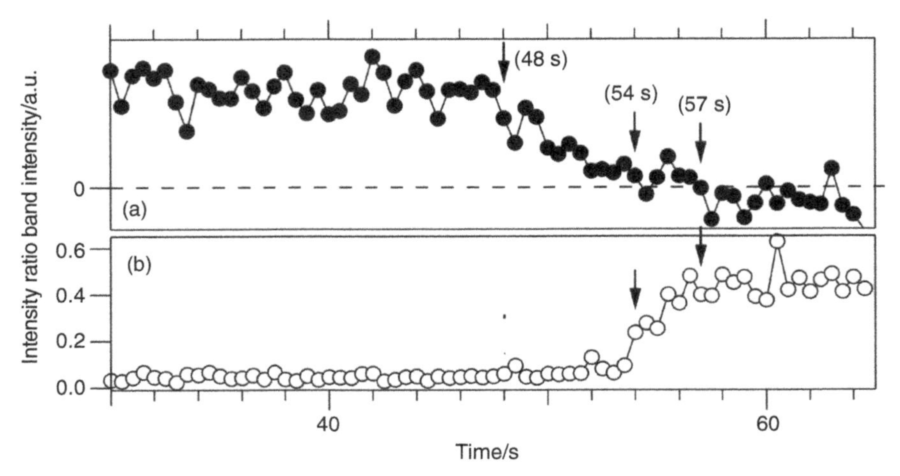

Figure 8.8 Plot with time of (a) the intensity of the 111-cm^{-1} lattice vibration band and (b) the intensity ratio (625-cm^{-1}/603-cm^{-1}) of the *trans*/*gauche* marker bands.

form to the *trans* (the 625 cm^{-1}/603 cm^{-1} ratio). The result again indicates that the alkyl chains of $[C_4mim]^+$ are strongly interacting with each other. This strong interaction remains even after the loss of the crystal structure, causing the long equilibration time of the butyl conformations. The thermal equilibration dynamics between the two rotational isomers provides valuable information on the interaction between the alkyl groups of the $[C_4mim]^+$, which is likely to play a key role in the formation of local structures in imidazolium-based ionic liquids.

There have been more studies of the rotational isomerism in ionic liquids using Raman spectroscopy and quantum chemical calculations. The *TT* form of the $[C_4mim]^+$ is found to be more abundant in liquid $[C_4mim]I$ than in molten $[C_4mim]Cl$ and $[C_4mim]Br$ [20]. Two isomers of $[C_2mim]^+$, planar and nonplanar forms of the ethyl group relative to the imidazolium ring plane, are found and reproduced by a quantum chemical calculation [21]. Rotational isomers of the bis{(trifluoromethane)sulfonyl}amide ($[NTf_2]^-$) anion have also been found. The anion has two isomers of C_2 and C_1 symmetry in $[C_4mim][NTf_2]$ [22]. Two isomers were found for the 1-hexyl-3methylimidazolium cation, $[C_6mim]^+$. They have the *TTTT* form (all-*trans* conformation) and the *GTTT* form (*gauche* conformation around the C7—C8 bond and *trans* conformation around the others) [23]. A distorted GT form of the $[C_4mim]^+$ has been reported for $[C_4mim]Cl$ under high pressure [24]. A water molecule in the unit cell changes the conformation of the alkyl group of the cation [15], as discussed in more detail in the following text. All these studies suggest that rotational isomerism is universally occurring in ionic liquids, and that elucidation of the structure and dynamics related to rotational isomerism helps to unveil the many characteristic properties of ionic liquids.

8.4 THERMAL DIFFUSION DYNAMICS AND HETEROGENEOUS LIQUID STRUCTURE IN IONIC LIQUIDS

Thermal diffusion dynamics in liquids reflects intermolecular interactions that transfer energy from one molecule to the other. The energy transfer takes place in the time regime of picosecond and nanosecond and, therefore, we need a high speed temperature measuring system for tracing the thermal diffusion dynamics in liquids. Our Tokyo group has previously reported that the C=C stretch peak position in the Raman spectrum of S_1 *trans*-stilbene changes linearly with temperature (Fig. 8.9) and that this linear relationship can be used to measure the local temperature of the micro-environment where the probe S_1 *trans*-stilbene molecule is situated [7]. Since the Raman spectrum of S_1 *trans*-stilbene can be measured with picosecond time-resolution, we have a 'picosecond thermometer' that can trace picosecond temperature changes in the micro-environment of interest.

When *trans*-stilbene is photoexcited to the S_1 state, some excess vibrational energy (the difference between the exciting photon energy and the 0-0 electronic transition energy) is deposited into the molecule. This excess energy is very quickly (in <1 ps) distributed among the many vibrational modes by intramolecular vibrational energy re-distribution (IVR) and hot S_1 *trans*-stilbene is formed. This hot S_1 *trans*-stilbene then cools down by dissipating the excess energy thermally to the surrounding solvent molecules (vibrational cooling).

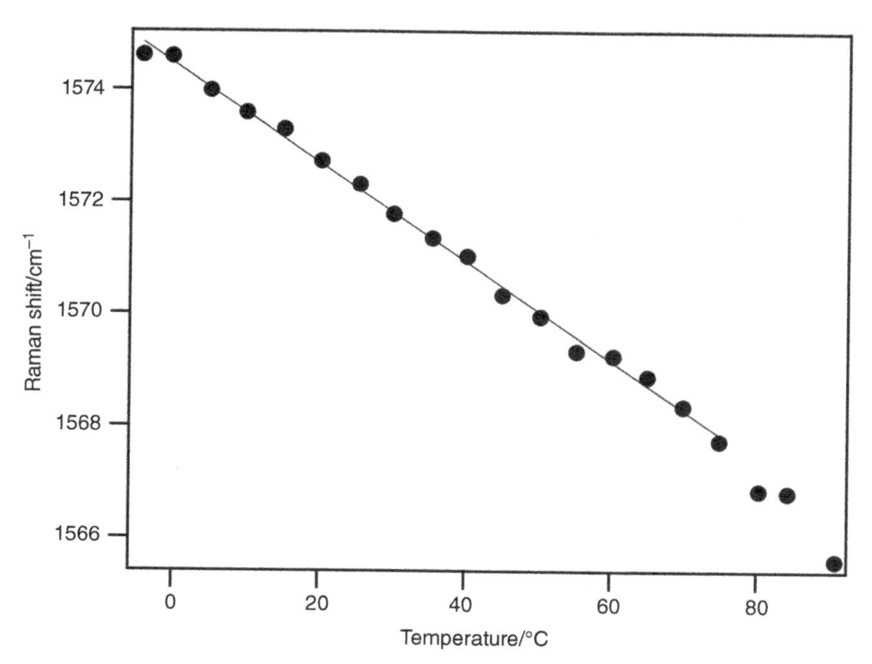

Figure 8.9 Plot of the C=C stretch peak position of S_1 *trans*-stilbene in decane vs. temperature.

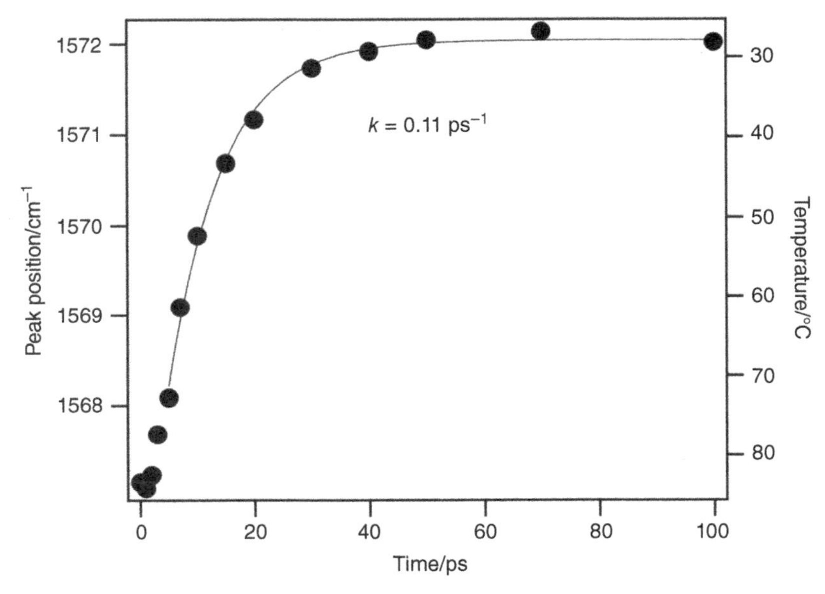

Figure 8.10 Picosecond peak position change of S_1 *trans*-stilbene representing the cooling process in decane.

The vibrational cooling process of S_1 *trans*-stilbene is very well monitored by the change of the C=C stretch peak position just after the photoexcitation. Figure 8.10 shows a representative peak position change observed in decane. The left ordinate in Figure 8.10 is the peak position in cm^{-1}, which is converted to temperature in celsius in the right ordinate by using the linear relationship in Figure 8.9. Just after the photoexcitation (0 ps), the temperature of the microenvironment around S_1 *trans*-stilbene is approximately 80°C and it quickly lowers to come back to the room temperature after 50 ps. From an exponential fitting of the curve, the cooling rate k in decane is obtained as $k = 0.11\ ps^{-1}$. The cooling rate k thus obtained reflects quantitatively the microscopic thermal diffusion dynamics in liquids.

The cooling rates of S_1 *trans*-stilbene have been measured in a number of molecular liquids as well as in ionic liquids (Fig. 8.11) [8]. The results show a marked contrast between the molecular and ionic liquids. In molecular liquids, including alkanes (heptane, hexane, octane, nonane, decane), alcohols (methanol, ethanol, ethylene glycol) and others (chloroform, toluene), a clear linear relation is observed between the cooling rate k and the bulk thermal diffusivity κ (right-hand box in Fig. 8.11). Thermal diffusivity κ is defined in the diffusion equation of heat, Equation 8.1:

$$\frac{\partial\theta}{\partial t} = \frac{\lambda}{c\rho}\Delta\theta = \kappa\Delta\theta \tag{8.1}$$

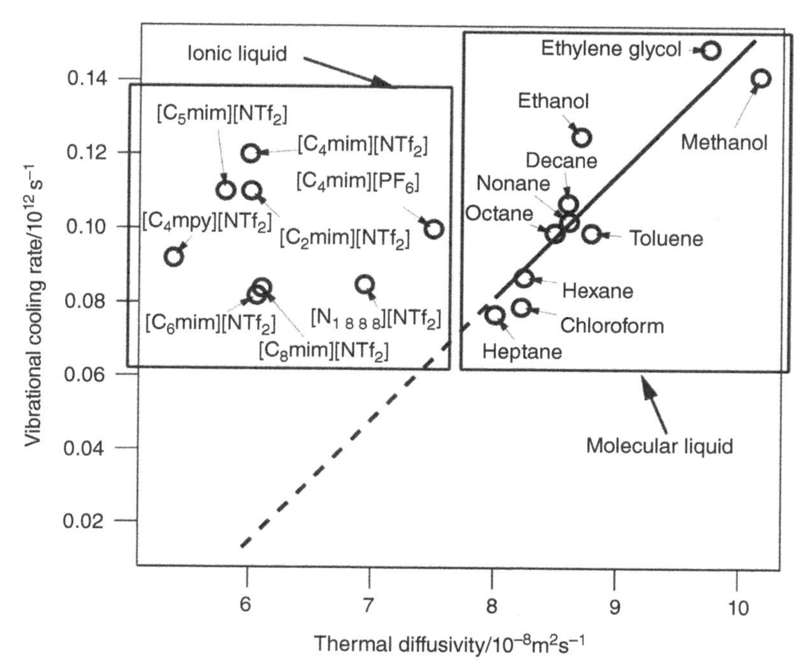

Figure 8.11 Vibrational cooling rate of S_1 *trans*-stilbene and thermal diffusivity of solvent. The solid line represents the correlation between the vibrational cooling rate and thermal diffusivity observed in molecular liquids, whereas the dotted portion indicates the extrapolated values.

where θ is the temperature, t is the time, λ is the thermal conductivity, c is the specific heat, ρ is the mass density, and Δ is the Laplacian $\left(\Delta = \frac{\partial^2}{\partial x^2} + \frac{\partial^2}{\partial y^2} + \frac{\partial^2}{\partial z^2}\right)$. The thermal diffusivity κ represents the rate of temperature change caused by the bulk heat conduction.

The linear relationship between k and κ can be explained as follows. Just after photoexcitation, hot S_1 *trans*-stilbene is formed, and it quickly shares the excess energy with the surrounding solvent molecules in the first solvation shell. In the case of decane in Figure 8.10, the temperature of this first solvation shell is approximately 80°C at 0 ps. The excess energy is then transferred from the solvent molecules in the first solvent shell to those in the outer shells and eventually to the bulk solvent molecules. These energy transfers occur on the picosecond time scale through the solvent/solvent intermolecular interactions. Thus, the cooling rate reflects the speed of the solvent/solvent energy transfer, whose rate is most likely to be correlated to the bulk thermal diffusivity κ of the solvent. It is interesting that this simple scheme holds for many molecular liquids of different kinds including alkanes, alcohols and others.

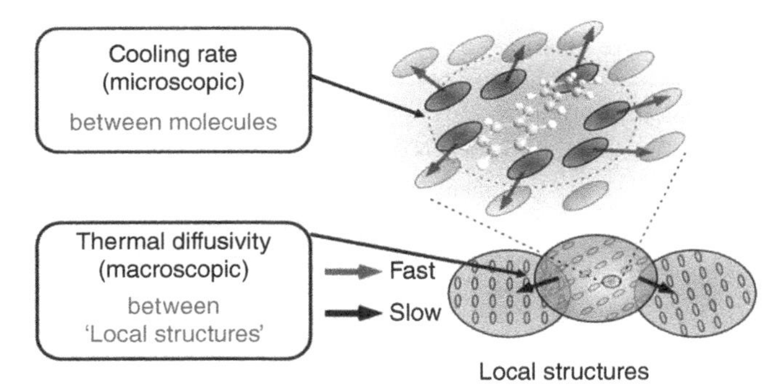

Figure 8.12 Two-step thermal diffusion in ionic liquids. Energy transfer is fast within a local structure, but is slow across the two local structures.

In contrast to the case of molecular liquids, no linear relationship is found between the cooling rate and the thermal diffusivity in ionic liquids (left-hand box in Fig. 8.11). The eight points corresponding to the eight ionic liquids (six with imidazolium cations, and two others) randomly scatter, showing no clear dependence of the cooling rate on the thermal diffusivity. Furthermore, all the points for ionic liquids are markedly deviated to the upper side from the line extrapolated from the points for molecular liquids. Ionic liquids generally have much smaller bulk thermal diffusivities than molecular liquids, indicating that bulk thermal diffusion is much slower in ionic liquids than in molecular liquids. The deviation from the extrapolated line means that, despite much slower bulk thermal diffusion, the micro-environmental cooling rates in ionic liquids are as fast as those in molecular liquids. In other words, the microscopic vibrational cooling rate is well correlated to the bulk thermal diffusivity for molecular liquids, but there is no such correlation found for ionic liquids.

The inconsistent microscopic and the macroscopic thermal properties suggest that thermal diffusion in ionic liquids is more complex than it is in molecular liquids. A structure model that accounts for this inconsistency has already been given [17]. This model assumes that ionic liquids have heterogeneous liquid structure consisting of microscopic local structures and that thermal diffusion takes place in two steps, firstly within a local structure and secondly across two adjacent local structures (Fig. 8.12). The energy transfer within a local structure is as fast as it is in molecular liquids, while that across the boundary of two local structures is likely to be slow. The former corresponds to the microscopic cooling rate measured by the 'Raman thermometer'. The latter determines the bulk thermal diffusivity that is a measure of heat transfer in a macroscopic scale larger than the size of a local structure. This structural model and the two-step thermal diffusion in ionic liquids account for the micro/macro inconsistency very well. The fast cooling rate corresponds to the heat transfer within a local structure and the slow bulk diffusion involves slow energy transfer across the boundary of the two local structures. Note that no

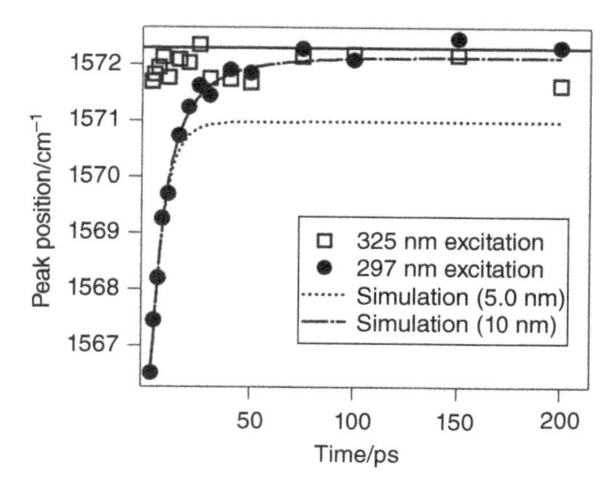

Figure 8.13 The cooling kinetics of S_1 *trans*-stilbene in [C_5mim][NTf_2] excited at 325 nm (□) and at 297 nm (●). The simulated curves are also shown for local structure size of 5 nm (dotted line) and 10 nm (broken line) [25].

local structure boundaries exist in molecular liquids, and therefore the thermal diffusivity is determined, both microscopically and macroscopically, by the solvent/solvent energy transfer rate alone.

What is the size of the local structure? In order to address this question, a quantitative analysis of the cooling curve has been carried out. Figure 8.13 shows two cooling curves of S_1 *trans*-stilbene in [C_5mim][NTf_2], with and without excess energy [25]. With the excitation at 297 nm (●), an excess energy of 2500 cm^{-1} is deposited to S_1 *trans*-stilbene and a large peak position change reflecting vibrational cooling is observed. No such change is observed for the excitation at 325 nm, which carries only a very small excess energy. Peak positions after 50 ps, with and without the excess energy, agree with each other within the experimental uncertainties, indicating that the temperature increase just after the photoexcitation recovers to room temperature after 50 ps. The excess energy is thought to be fully dissipated within a local structure without reaching the boundary. The volume of local structures in ionic liquids is large enough to make the temperature rise at 50 ps smaller than the detection limit.

The vibrational cooling kinetics of S_1 *trans*-stilbene can be expressed, Equation 8.2, with a solution of the diffusion equation of heat [26]:

$$\theta(x,y,z,t) - T_{\mathrm{r.t.}} = (4\pi\kappa t)^{-3/2} \int_{-\infty}^{\infty}\int_{-\infty}^{\infty}\int_{-\infty}^{\infty} f(x',y',z') \times \exp\left\{-\frac{\left[(x-x')^2+(y-y')^2+(z-z')^2\right]}{4\kappa t}\right\} dx'dy'dz' \tag{8.2}$$

where $\theta(x,y,z)$ is the temperature at the position x,y,z at time t, $T_{r.t.}$ is the room temperature, and $f(x,y,z)$ is the initial distribution of temperature change,

$$f(x,y,z) = \theta(x,y,z,0) - T_{r.t.} \tag{8.3}$$

The cooling curve is simulated very well for molecular liquids, with an assumption that the initial temperature distribution is represented by a box corresponding to an S_1 *trans*-stilbene molecule and the first solvation shell and that the heat is dissipated to the bulk solution with an infinitely large volume [7]. The same approach is taken for ionic liquids with a condition that heat is dissipated only within a finite volume of L^3/nm^3 of a local structure. The simulated curves for $L = 5.0$ nm (dotted line) and $L = 10.0$ nm (broken line) are shown in Figure 8.13. For $L = 5.0$ nm, the curve is saturated much earlier than that observed, showing that the deposited excess energy causes a significant temperature rise even after 50 ps. Such a temperature rise is not experimentally observed, as mentioned above. On the other hand, for $L = 10.0$ nm, the simulated curve reproduces the observed curve very well. These results indicate that L must be larger than 10 nm in order to explain the observed cooling kinetics by the simulation based on the heat diffusion theory. In other words, it is suggested that the size of the local structures formed in [C_5mim][NTf_2] is larger than 1000 nm^3.

8.5 WATER IN IONIC LIQUIDS

The interaction of dissolved water with ionic liquids has been extensively studied using IR spectroscopy and NIR spectroscopy [27–30]. These methods are sensitive to environment and intermolecular interactions, and, therefore, they have been used to study the interactions between water and many chemical substances such as ionic liquids. Strong interaction with water is found to be ubiquitous in all ionic liquids, irrespective of their hydrophilic or hydrophobic nature. This is evident from the fact that while [C_4mim][BF_4] (known to be a hydrophilic ionic liquid) absorbs 0.320 M in 24 h exposure to 40% humidity, [C_4mim][NTf_2] and [C_4mim][PF_6] (known to be hydrophobic ionic liquids) each absorb 0.097 and 0.083 M, respectively, which is far from negligible [29]. These differences in water absorption can fairly be correlated with the strength of the hydrogen bonding between water and the anions of the ionic liquids.

It has long been speculated that the water molecule can change the structure of ionic liquids. We have been able to show very clearly that there is indeed a change in the molecular conformation upon absorption of water, with the help of combined studies of NIR Raman spectroscopy and single crystal X-ray diffraction [15]. Though IR spectroscopy is more sensitive to water, pre-treatment is highly problematic for very hygroscopic ionic liquids, like [C_3mimCN]I (shown in Fig. 8.14).

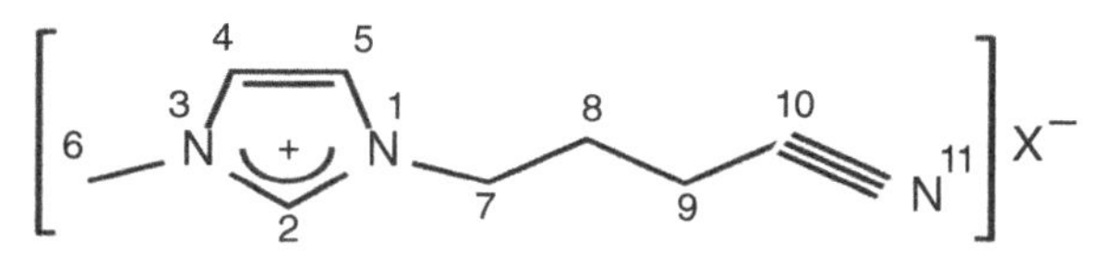

Figure 8.14 1-(Butylnitrile)-3-methylimidazolium halide, $[NCC_3mim]X$ (where X = Cl or I).

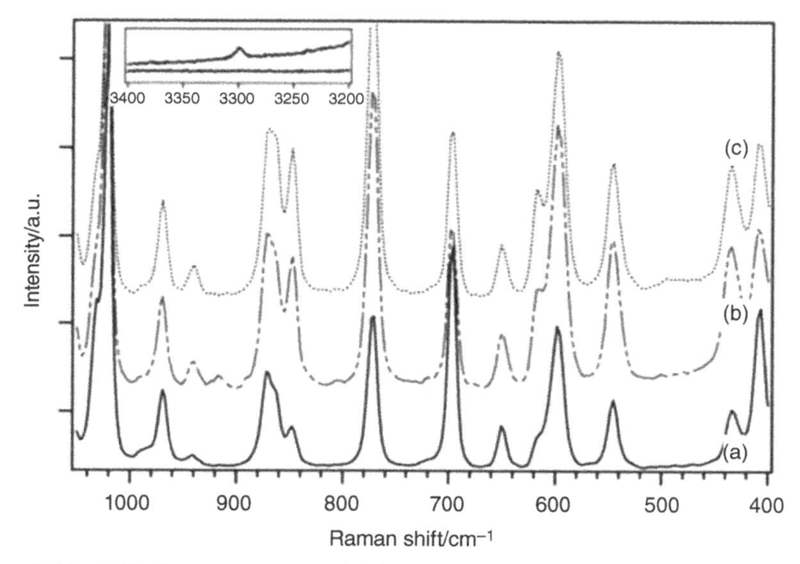

Figure 8.15 NIR Raman spectra of (a) transparent, (b) opaque and (c) moisture-included crystals of $[NCC_3mim]I$ [15]. The inset is for the opaque crystal.

The beauty of NIR Raman spectroscopy is that it is much less affected by the fluorescence that is often associated with newly synthesised ionic liquids. NIR Raman spectra can be measured directly from a sample kept in a sealed glass vial. These advantages of NIR Raman spectroscopy make it possible to record the vibrational spectra of moisture-sensitive ionic liquids. The Raman spectrum of $[NCC_3mim]I$ (a transparent crystal) along with the spectrum of a $[NCC_3mim]I$ crystal containing water (opaque) are shown in Figure 8.15. The opaque crystal was obtained from ordinary ethanenitrile (not pre-dried), and was not a single crystal.

Figure 8.15 shows that several skeletal modes of the imidazolium ring give rise to characteristic Raman bands. The bands at 600 and 695 cm^{-1} are the characteristic marker bands for the *gauche* conformation around the C7–C8 bond. These bands, similar to the case of $[C_4mim]Cl$, arise from the deformation modes of the imidazolium ring that are coupled with the CH_2 rocking vibration of the C8 methylene group. A comparison of the Raman spectrum of the transparent $[NCC_3mim]I$ with that of the opaque crystal reveals that the two spectra

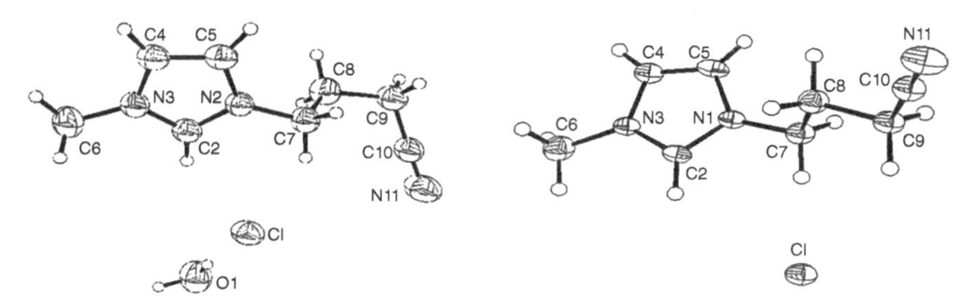

Figure 8.16 Crystal structures of $[NCC_3mim]Cl.H_2O$ (left), in which the butylnitrile chain has a *trans-gauche* (with respect to C7—C8 and C8—C9 bond, respectively) conformation, and anhydrous $[NCC_3mim]Cl$ (right) in which the butylnitrile side chain has an all *trans* conformation [15].

differ considerably with respect to the intensities of several bands. The intensity ratio of the 695 cm^{-1} band to 771 cm^{-1} differs in the transparent and opaque crystals. The same is true for the 848 and 872 cm^{-1} bands. Notably, the marker band for the *trans* conformation appears clearly at 617 cm^{-1} in the opaque crystal. The appearance of this marker band on absorption of moisture indicates the formation of the *trans* conformation around the C7—C8 bond. In order to make sure that the water is responsible for this structural change, the transparent crystal (pure $[NCC_3mim]I$) has been deliberately exposed to the open air for a few seconds followed by a Raman measurement, illustrated in Figure 8.15c. The recorded spectrum is similar to that of the opaque crystal (with a stronger appearance of the 617 cm^{-1} band), indicating that the structural change is due to the moisture. This finding also clearly signifies the role of the solvent used in the synthesis and the subsequent treatment of ionic liquids. The change in structural features studied by ATR-IR spectroscopy on the absorption of moisture from the air has also been reported earlier by Oliveira and co-workers [29]. A weak peak of the O—H stretch band at 3300 cm^{-1} is observed in the Raman spectrum of the $[NCC_3mim]I$ opaque crystal, and is shown in the inset in Figure 8.15. This peak was absent for a transparent crystal. The position of the peak (3300 cm^{-1}) suggests that H_2O is present in the system, and is strongly hydrogen-bonded. A clearer picture of the structural change at the molecular level has been obtained with the help of single crystal X-ray diffraction technique [15]. We were able to obtain single crystals of $[NCC_3mim]Cl$: having absorbed water and pure $[NCC_3mim]Cl$, that is, without a water molecule (shown in Fig. 8.16).

It is clear from the X-ray structure that the nitrile-substituted butyl chain of the cation is in a *gauche* conformation around the C8—C9 bond in the presence of water, but that the *trans* conformation is observed for pure, anhydrous $[NCC_3mim]Cl$. It was further shown that the water molecule is hydrogen-bonded to the anion and to another water molecule, and that no self-associated water clusters have been found [30]. It further revealed that two strong classical hydrogen bonds keep the water molecules together with the chloride anions, and form rhombohedral-like $\{(H_2O)_2(Cl^-)\}_2$ complexes (see Fig. 8.17) [15]. No direct

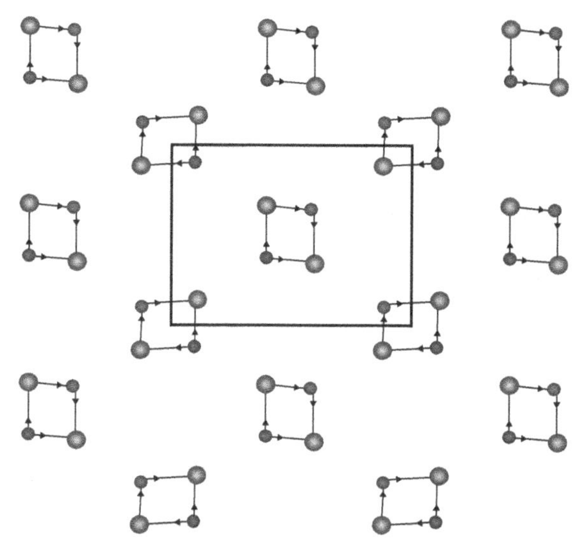

Figure 8.17 The hydrogen-bonded assembly of water molecules (smaller circle) and chloride anions (larger circle) in a $[NCC_3mim]Cl.H_2O$ crystal [15]. The cations are not shown, for clarity.

interaction between water and the cation is observed, though the presence of water is responsible for the change in the structure of the cation.

In-depth analyses of Raman spectra have been performed in subsequent work where the amount of water in $[C_4mim][BF_4]$ was varied systematically [27]. Both characteristic vibrations of CH_x ($x=1, 2,$ or 3) and OH of the added water have been measured as shown in Figure 8.18.

While peaks from CH bonds did not change their positions, those from the terminal methyl group of the butyl chain shifted by $10\,cm^{-1}$ with the addition of water. Changes in the spectral shape in the OH stretching region indicate that the hydrogen-bonded network of water molecules breaks down rapidly as the ionic liquid is added. A quantitative analysis of the Raman spectra (shown in Fig. 8.18, inset) shows that the ratio between the *gauche* and *trans* band areas increases with increasing water concentration and reaches a maximum at $45\,mol\,l^{-1}$ water concentration, and then decreases rapidly afterwards. This concentration dependence is indicative of some conformational changes of the butyl chain around this concentration. The blue shift of the stretching band of $[BF_4]^-$at $765\,cm^{-1}$ is also found to be informative, and can provide quantitative estimation of the structural changes.

Kinetics of water absorption by ionic liquids has been found to be relatively complex. They neither follow simple first-order nor second-order rate equations. At every stage of water absorption, the physical properties (e.g. polarity, viscosity, etc.) change at the molecular level, making the phenomenon of absorption highly complex in nature. This fact has recently been proved by

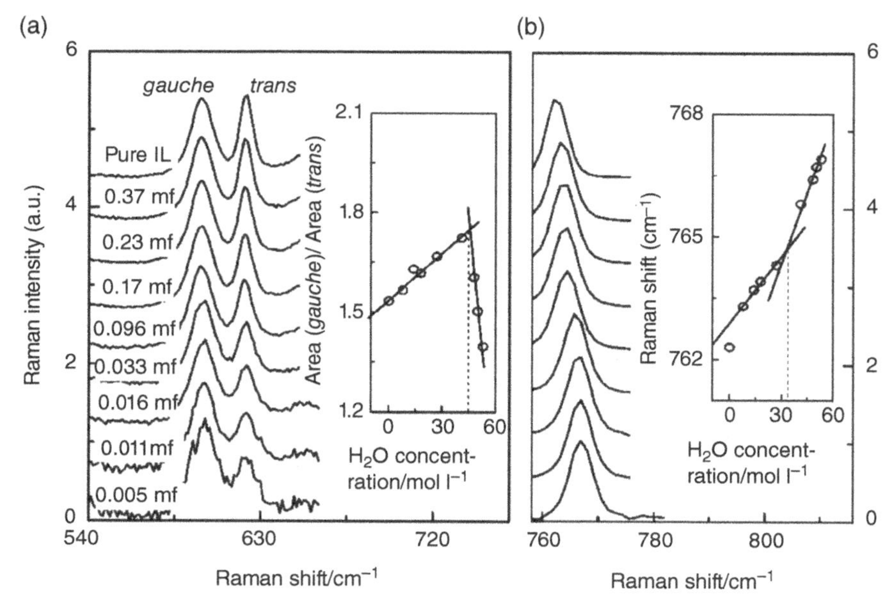

Figure 8.18 Normalised Raman spectra of $[C_4mim][BF_4]$ between 590 and 780 cm^{-1} with varying water concentrations. (a) The peaks at approximately 600 and 620 cm^{-1} are for the *gauche* and *trans* conformations (see text). The open circles are the area ratios between the *gauche* and *trans* peaks. (b) The peak at around 765 cm^{-1} is from $[BF_4]^-$. The open circles show the peak shift.

NIR spectroscopy, measuring the overtone and combination bands of the O–H stretch modes [29].

8.6 CONCLUDING REMARKS

Raman spectroscopy has played, is playing and will play a significant role in the structural studies of ionic liquids. It provides information not only on the molecular (ionic) structure of the component ions, but also on the higher order micro/macroscopic liquid structure. As summarised in this chapter, the possibility of heterogeneous liquid structure has been repeatedly discussed by our Tokyo group with a working hypothesis of local structure formation in ionic liquids. Strong aliphatic interactions between the alkyl groups, as well as the long-range Coulombic interactions, are among the likely causes of the unique local structure formation. An estimate of the size of the local structure ($L > 10$ nm) has been given recently by a simulation analysis of the thermal diffusion dynamics measured with picosecond time-resolved Raman spectroscopy. Other physicochemical methods, X-ray/neutron diffraction, and theoretical simulation in particular, will hopefully confirm this hypothesis from different bases.

ACKNOWLEDGEMENTS

The authors are grateful to all their collaborators in the work presented in this chapter.

REFERENCES

1 (a)Freemantle, M., *An Introduction to Ionic Liquids*, RSC Publications, Cambridge (2010);(b)Wasserscheid, P. and Welton, T. (eds.), *Ionic Liquids in Synthesis*, 2nd ed., Wiley-VCH, Weinheim (2008).

2 Raman, C.V. and Krishnan, K.S., A new type of secondary radiation, *Nature*, **121**, 501–502 (1928).

3 Hayashi, S., Ozawa, R. and Hamaguchi, H., Raman spectra, crystal polymorphism, and structure of a prototype ionic-liquid [bmim]Cl, *Chem. Lett.* **32**(6), 498–499 (2003).

4 Saha, S., Hayashi, S., Kobayashi, A. and Hamaguchi, H., Crystal structure of 1-butyl-3-methylimidazolium chloride. A clue to the elucidation of the ionic liquid structure, *Chem. Lett.* **32**(8), 740–741 (2003).

5 Holbrey, J.D., Reichert, W.M., Nieuwenhuyzen, M., Johnston, S., Seddon, K.R. and Rogers, R.D., Crystal polymorphism in 1-butyl-3-methylimidazolium halides: supporting ionic liquid formation by inhibition of crystallization, *Chem. Commun.*, 1636–1637 (2003).

6 Ozawa, R., Hayashi, S., Saha, S., Kobayashi, A. and Hamaguchi, H., Rotational isomerism and structure of the 1-butyl-3-methylimidazolium cation in the ionic liquid state, *Chem. Lett.* **32**(10), 948–949 (2003).

7 Iwata, K. and Hamaguchi, H., Microscopic mechanism of solute-solvent energy dissipation probed by picosecond time-resolved Raman spectroscopy, *J. Phys. Chem. A* **101**, 632–637 (1997).

8 Iwata, K., Yoshida, K., Takada, Y. and Hamaguchi, H., Vibrational cooling process of S_1 *trans*-stilbene in ionic liquids observed with picosecond time-resolved Raman spectroscopy, *Chem. Lett.* **36**, 504–505 (2007).

9 Okajima, H. and Hamaguchi, H., Unusually long *Trans/Gauche* conformational equilibration time during the melting process of BmimCl, a prototype ionic liquid, *Chem. Lett.* **40**, 1308–1309 (2011).

10 Widegren, J.A., Laesecke, A. and Magee, J.W., The effect of dissolved water on the viscosities of hydrophobic room-temperature ionic liquids, *Chem. Commun.*, 1610–1612 (2005).

11 Widegren, J.A., Saurer, E.M., Marsh, K.N. and Magee, J.W., Electrolytic conductivity of four imidazolium-based room-temperature ionic liquids and the effect of a water impurity, *J. Chem. Thermodyn.* **37**, 569–575 (2005).

12 Aki, S.N.V.K., Brennecke, J.F. and Samanta, A., How polar are room-temperature ionic liquids? *Chem. Commun.*, 413–414 (2001).

13 Brown, R.A., Mollet, P., McKoon, E., Eckert, C.A., Liotta, C.L. and Jessop, P.G., Asymmetric hydrogenation and catalyst recycling using ionic liquid and supercritical carbon dioxide, *J. Am. Chem. Soc.* **123**, 1254–1255 (2001).

14 Walker, A.J. and Bruce, N.C., Cofactor-dependent enzyme catalysis in functionalized ionic solvents, *Chem. Commun.*, 2570–2571 (2004).

15 Saha, S. and Hamaguchi, H., Effect of water on the molecular structure and arrangement of nitrile-functionalized ionic liquids, *J. Phys. Chem. B* **110**(6), 2777–2781 (2006).

16 Hamaguchi, H. and Ozawa, R., Structure of ionic liquids and ionic liquid compounds: are ionic liquids genuine liquids in the conventional sense? *Adv. Chem. Phys.* **131,** 85–104 (2005).

17 Iwata, K., Okajima, H., Saha S. and Hamaguchi, H., Local structure formation in alkyl-imidazolium-based ionic liquids as revealed by linear and nonlinear Raman spectroscopy, *Acc. Chem. Res.* **40**, 1174–1181 (2007).

18 Okajima, H. and Hamaguchi, H., Fast low frequency (down to 10 cm^{-1}) multichannel Raman spectroscopy using an iodine vapor filter, *Appl. Spectrosc.* **63**, 958–960 (2009).

19 Mizushima, S., Morino, Y. and Noziri, S., Raman effect and free rotation, *Nature* **137**, 945 (1936).

20 Katayanagi, H., Hayashi, S., Hamaguchi, H. and Nishikawa, K., Structure of an ionic liquid, 1-*n*-butyl-3-methylimidazolium iodide, studied by wide-angle X-ray scattering and Raman spectroscopy, *Chem. Phys. Lett.* **392**, 460–464 (2004).

21 Umebayashi, Y., Fujimori, T., Sukizaki, T., Asada, M., Fujii, K., Kanzaki, R. and Ishiguro, S., Evidence of conformational equilibrium of 1-Ethyl-3-methylimidazolium in its ionic liquid salts: Raman spectroscopic study and quantum chemical calculations, *J. Phys. Chem. A* **109**, 8976–8982 (2005).

22 Fujii, K., Fujimori, T., Takamuku, T., Kanzaki, R., Umebayashi, Y. and Ishiguro, S., Conformational equilibrium of bis(trifluoromethanesulfonyl) imide anion of a room-temperature ionic liquid: Raman spectroscopic study and DFT calculations, *J. Phys. Chem. B* **110,** 16, 8179–8183 (2006).

23 Berg, R.W., Deetlefs, M., Seddon, K.R., Shim, I. and Thompson, J.M., Raman and ab initio studies of simple and binary 1-alkyl-3-methylimidazolium ionic liquids, *J. Phys. Chem. B* **109**(40), 19018–19025 (2005).

24 Chang, H.-C., Chang, C.-Y., Su, J.-C., Chu, W.-C., Jiang, J.-C. and Lin, S.H., Conformations of 1-butyl-3-methylimidazolium chloride probed by high pressure Raman spectroscopy, *Int. J. Mol. Sci.* **7**, 417–424 (2006).

25 Yoshida, K., Iwata, K., Nishiyama, Y., Kimura, Y., Hamaguchi, H., Local structures in ionic liquids probed and characterized by microscopic thermal diffusion monitored with picosecond time-resolved Raman spectroscopy, *J. Chem. Phys.* **136**, 104504-1–104504-8 (2012).

26 Carslaw, H.S. and Jaeger, J.C., *Conduction of Heat in Solids*, 2nd ed., Oxford University, Oxford (1959).

27 Jeon, Y., Sung, J., Kim, D., Seo, C., Cheong, H., Ouchi, Y., Ozawa R. and Hamaguchi, H.O., Structural change of 1-butyl-3-methylimidazolium tetrafluoroborate+water mixtures studied by infrared vibrational spectroscopy, *J. Phys. Chem. B* **112**, 923–928 (2008).

28 Dominguez-Vidal, A., Kaun, N., Ayora-Canada, M.J. and Lendl, B., Probing intermolecular interactions in water/ionic liquid mixtures by far-infrared spectroscopy, *J. Phys. Chem. B* **111,** 4446–4452 (2007).

29 Tran, C.D., Lacerda, S.H.D.P. and Oliveira, D., Absorption of water by room-temperature ionic liquids: effect of anions on concentration and state of water, *Appl. Spectrosc.* **57,** 152–157 (2003).

30 Köddermann, T., Wertz, C., Heintz, A. and Ludwig, R., The association of water in ionic liquids: a reliable measure of polarity, *Angew. Chem. Int. Ed.* **45**, 3697–3702 (2006).

9 (Eco)Toxicology and Biodegradation of Ionic Liquids

STEFAN STOLTE

Department 3: Sustainability in Chemistry, UFT-Centre for Environmental Research and Sustainable Technology, University of Bremen, Bremen, Germany

MARIANNE MATZKE

NERC Centre for Ecology & Hydrology Molecular Ecotoxicology, Acremann Section Maclean Building, Benson Lane Crowmarsh Gifford, Wallingford Oxfordshire, UK

JÜRGEN ARNING

Department 10: Theoretical Ecology, UFT-Centre for Environmental Research and Sustainable Technology, University of Bremen, Bremen, Germany

9.1 INTRODUCTION

This chapter critically reviews the state of the art in the fields of ionic liquid (eco)toxicology and biodegradability up to the year 2011. The developments in both fields will be presented in a brief discussion of the most relevant literature in each. Then, the authors will point out the existing data gaps and further research needs, with the aim of providing a framework for future research initiatives leading to an effective and sustainable hazard and for the risk assessment of ionic liquids. This should help to prioritise ionic liquid investigations and future testing programs in order to generate data that will also be useful from a regulatory point of view.

The overall focus in this chapter will be on issues connected with the following three guiding questions: (i) What more needs to be done for the design and development of inherently safer ionic liquids?; (ii) What information, and how much of it, do we need to achieve a sound and legally relevant risk assessment of ionic liquids?; and (iii) Which strategies and techniques seem to be the most promising ones for achieving the goal of inherently safer ionic liquids with a

Ionic Liquids Completely UnCOILed: Critical Expert Overviews, First Edition.
Edited by Natalia V. Plechkova and Kenneth R. Seddon.

low hazard and risk potential for man and the environment, on the one hand, but with high-quality technological performance on the other?

To address these questions, issues like the data, and hence the toxicological and ecotoxicological tests that will be needed for the future design of ionic liquids as well as actual bottlenecks in hazard and risk assessment will be discussed. Among these issues, some crucial parameters relevant to the exposure, biodegradability and toxicology of ionic liquids and current knowledge of them need to be analysed. It will also be pointed out at which stage of industrial development the different hazard and risk measures can be most effectively implemented for the sustainable use of ionic liquids in products and processes.

The discussion of these issues in this chapter is subdivided into two sections and a concluding summary is provided. It begins with a presentation of the main issues dealing with ionic liquid toxicology and ecotoxicology, followed by a discussion of how existing knowledge can be integrated and of what is needed to complete the picture. The second part deals with the biodegradability of ionic liquid structures and substructures. Here, we draw attention to what we have been able to learn so far from existing data, after which we embark on a critical discussion on how this important field for a sound risk assessment of ionic liquids might be improved. Hence, while the first part covers the toxicological and ecotoxicological hazards of ionic liquids and their use, the second section describes highly important issues closely related to environmental hazards like persistency, exposure and bioaccumulation, as well as the spatiotemporal range of these substances.

The final part will summarise the key issues from each of the two subsections and will also briefly discuss strategies and techniques for the sustainable reuse and recycling of ionic liquids in industrial processes, with the aim of reducing exposure as well as making sensible economical use of them.

9.2 THE TOXICOLOGY OF IONIC LIQUIDS – BEYOND THE 'SIDE CHAIN EFFECT'

The term 'ionic liquid' subsumes all combinations of a cation and an anion that have a melting point below 100°C. Besides the fact that this 100°C threshold is an arbitrary one, this assumption involves the danger of an inadequate simplification regarding the structural variability of these substances. In fact, one cannot draw general conclusions about the toxicity of 'ionic liquids', as one similarly cannot when talking about the effects of, say, pyrethroid pesticides. A more differentiated view on ionic liquids is, therefore, necessary.

When talking about ionic liquids, two situations generally need to be differentiated: (i) ionic liquids as solvents and (ii) ionic liquids as solutes. While in the former case, ionic liquids can be treated as a distinct phase with certain solvent parameters like density, viscosity or polarity, in the latter case – and this is the most relevant one for toxicological studies – we are dealing with more or less separately dissolved cations and anions. Thus, all the following discussions are based on the situation where the cations and anions of ionic liquids are dissolved in a physiological or environmental medium of a certain ionic strength, pH, temperature, and with specific components like proteins, other biomolecules or organic substances like humic acids and soil components. Moreover, since ionic liquids are composed mainly of bulky and asymmetric organic cations bound to bulky anions with a widely distributed negative charge, ionic liquids can be referred to as 'weakly coordinating ions'. This implies that, compared to normal ions, their interactions with, for example, each other, solvent molecules and biomolecules cannot be explained and described exclusively by electrostatic properties; van der Waals interactions, other dispersive forces, and hydrogen-bond donor and acceptor potentials, among other things, need to be considered as well. Consequently, an ionic liquid in solution can be present as different molecular species, ranging from completely dissolved and hydrated ions to ion pairs and even larger clusters or aggregates. Which species are predominant in a given medium depends strongly on the ionic liquid's concentration, the chemical environment of the ions and the available molecular interaction potentials [1]. From a toxicological point of view, this means that the situation with ionic liquids becomes more complex compared to other salt solutions, since the effects, modes of action and bioavailability may change significantly when, for instance, the anion species is exchanged or the concentration is increased, resulting in ion pair formation. Hence, unexpected non-linear effects due to species and concentration-dependent synergistic and/or antagonistic effects can be found, and have to be kept in mind when interpreting toxicological studies of ionic liquids [2–4]. Apart from these more general features of ionic liquids, what makes them really challenging is their diverse structural and chemical space. Here again, it becomes obvious that talking about the toxicology of ionic liquids in general is pointless. To handle this diverse pool of substances, one has to group ionic liquids at least according to their cationic head groups and to the anionic species used as counter ions, as is done in the literature available on ionic liquid toxicology. Hence, we can discuss and compare, say, the toxicology of different *N*-alkylimidazolium chlorides, but not the toxicology of ionic liquids in general.

In this context, the literature data available today on the toxicology and ecotoxicology of ionic liquids can only shed light on certain, selected compounds out of the huge pool of available ionic liquids. The state of the art as regards the toxicology and ecotoxicology of ionic liquids has recently been summarised in some comprehensive review papers [5–9]. According to these review papers, a broad data set is available for the most common ionic liquid species, like the imidazolium-based salts. The available data cover aquatic test systems (various limnic and marine green algae, limnic and marine bacteria, crustaceans, fish embryos and whole organisms like fish and plants), terrestrial test systems (soil bacteria, invertebrates, earthworms and plants), cellular assays using human and other mammalian cell lines, and primary cells. Furthermore, there are some outcomes from animal tests in rodents, owing to regulatory data needs. Despite the fact that single studies address only a very few specific ionic liquids, some general conclusions can be drawn on the toxicological behaviour of the vast majority of the hitherto investigated head groups, side chains and anions. Looking at the data available at the meta level, the predominant paradigm describing the biological activity of ionic liquids seems to be the so-called 'side-chain effect'. Nearly all studies – regardless of whether they are dealing with the acute aquatic toxicity of ionic liquids or with cytotoxic effects on human and animal cells, or even with the organismic effects in mammals – have identified the length of the side chain attached to the cationic core structure (a measure of its hydrophobicity) as a reliable molecular descriptor for the observed effects. In this context, it has been shown that the introduction of polar functional groups into the aliphatic side chains of the cation can, in some cases, reduce the observed effects by as much as several orders of magnitude, and that polar, non-aromatic head groups exert significantly smaller effects than their aromatic, and hence more hydrophobic, analogues [10–12]. The same trend was observed in a series of different anionic species combined with the same cation [3]. In other words, increasing the hydrophobicity of the ionic liquid cation and/or anion enhances the reported biological effects, and *vice versa*. In toxicology, this is a well-known phenomenon and can be observed for all substances acting exclusively *via* non-specific interactions with lipid membranes, membrane proteins, or hydrophobic regions of proteins and cellular structures in general. This mode of action is the minimal toxic effect every chemical can exert on living organisms, and is often referred to in the literature as 'narcosis' or 'baseline toxicity' [11, 13–16]. Knowing the hydrophobicity of an ionic liquid cation and anion, one can easily set up quantitative structure activity relationships (QSAR) regressions and, therefore, predict the effect range by using well-known baseline toxicants and linking their hydrophobicity with the baseline toxicity observed in any *in vitro* or *in vivo* test system. This was shown, for example, by Ranke et al. [16], who used high performance liquid chromatography (HPLC)-derived lipophilicity parameters of ionic liquids and correlated them with the effective concentrations from cell viability assays at which half of the cells investigated were negatively affected (EC_{50}). Using the same data set of cytotoxicity data, Torrecilla et al. [17] used mathematical models (MLP, MLR and non-linear models) to predict EC_{50} values of ionic liquid cations and anions. Hence, such non-specific toxic effects are, in principle, easy to predict and thus of minor concern in contemporary regulatory hazard and risk assessments.

Thus, the question, whether ionic liquids (or at least the most common representatives among this enormous pool of compounds) can be viewed exclusively as baseline toxicants or whether there are some more specific effects and issues to look at beyond the side chain effect, needs to be answered for a prospective approach in ionic liquid toxicology. We think that the field of ionic liquid toxicology can only move forward if future research activities address more specific effects. To demonstrate the side chain effect again and again in more and more biological test systems simply consolidates the data already available, but cannot contribute substantially to the prospective design of inherently safer ionic liquids.

Hence, what toxicological information and data on more specific modes of action do we need to improve our understanding of the effects of ionic liquids in biological systems? Figure 9.1 summarises, in schematic form, some of the specific targets and modes of toxic action a chemical can generally exert at the molecular and cellular level, which can be considered in future testing schemes.

Looking at the current discussions at the regulatory level of environmental legislation, and at which endpoints and test systems should be measured and used with priority, there are some important issues that can be addressed in rapid *in vitro* screening assays – at least on the first tier of a weight-of-evidence based hazard assessment approach. We propose first to look at these parameters, which will be presented below, in order to push the development of inherently safer ionic liquids and their applications forward. For example, regarding the needs and demands for data under the European regulation for the registration, evaluation, authorisation and restriction of chemicals (REACh regulation), there are some specific toxicological effects and endpoints of most commercially available ionic liquids, for which data is almost completely non-existent. In particular, a more vigorous approach should be taken for the investigation of the properties classifying a chemical as a 'substance of very high concern (SVHC)' with regard to ionic liquids. These properties include:

- Endocrine disruption
- Mutagenicity, carcinogenicity, teratogenicity
- Long-term and chronic toxicity
- Chemosensitising potential
- Bioaccumulation
- Persistency

While the last mentioned parameter – persistency – dealing with the biodegradation of ionic liquids will be discussed in Section 9.3, the other issues are directly relevant to toxicology and ecotoxicology. The next paragraph gives some examples from recent research to show that all these properties with respect to ionic liquids have to be investigated in more detail. But before presenting this selected literature review, we would like to point out some additional specific effects to look at, which are especially relevant to ionic liquids, since they are composed mostly of weakly coordinating ions, with some special physicochemical properties pointed out in the introduction to this section.

In this context, the degree of dissolution of the cations and anions of an ionic liquid is of high relevance. This was shown, for example, by Stolte et al. [2] and Matzke et al. [3], who applied the model of concentration addition to interpret the observed synergistic effects of ionic liquid combinations in a cytotoxicity assay [2], in limnic green algae and in terrestrial plants [3]; and by Zhang et al., who demonstrated the synergistic and antagonistic effects of ionic

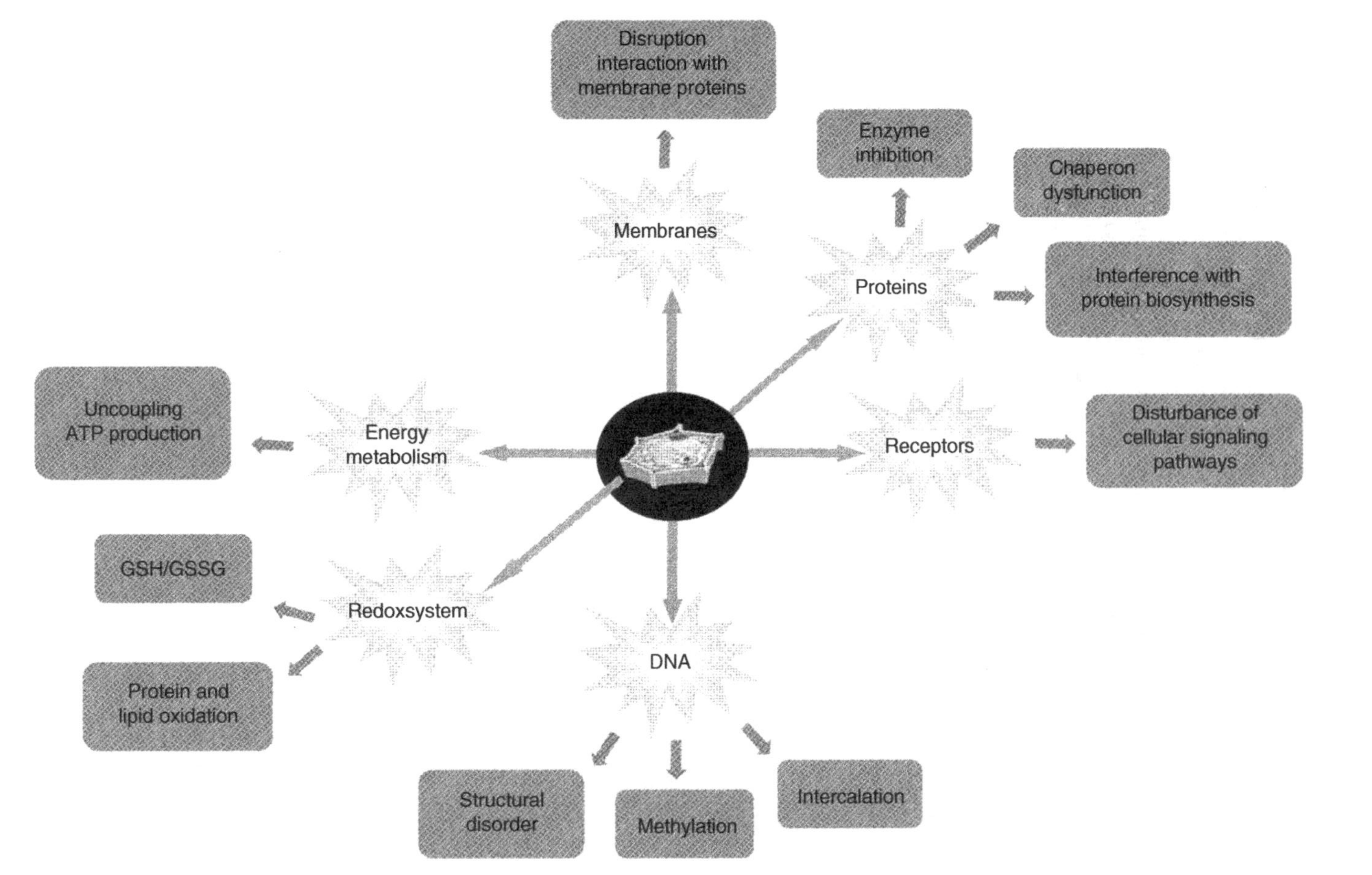

FIGURE 9.1 A schematic of the specific targets and modes of toxic action that a chemical can generally exert at the molecular and cellular level.

liquid mixtures on bacteria [4]. We, therefore, support the arguments of Kulacki et al. [18], that the mere reporting of EC_{50} values from ecotoxicological and toxicological studies is insufficient. Thus, research efforts to look at the entire concentration range in order to discover specific and non-linear effects that may arise only above a certain threshold need to be included in future testing approaches. Additionally, the influence of varying salt concentrations and pH values on the speciation of ionic liquids in biological test media needs more attention in order to complete the picture of the relevant windows for toxic effects and modes of action.

Coming back now to the specific effects mentioned above connected with SVHC parameters, the next paragraph briefly summarises the current knowledge of ionic liquids with respect to their possible specific modes of action.

Some studies at the subcellular level show that certain ionic liquid structures can interact with cellular structures and functions, with isolated proteins and DNA. In this context, Yu et al. [19] showed, for a series of imidazolium-based ionic liquids, that they can influence the antioxidative systems in *Daphnia magna*, and that oxidative stress is most likely to be involved in the observed toxic effects. Regarding the interaction with specific proteins, Arning et al. [20] showed (in an *in vitro* test system) that especially imidazolium and *N*,*N*-dimethylamino cations could exert drastic inhibitory effects on the enzyme acetylcholinesterase. Furthermore, using an *in vitro* assay as well, Składanowski et al. [21] demonstrated that some imidazolium ionic liquids are able to inhibit the enzyme AMP-deaminase, which is essential for nucleotide homeostasis in cells. Luo et al. showed that ATPase in earthworms was inhibited by dosing imidazolium-based ionic liquids into soils [22], and Xie et al. [23] demonstrated that imidazolium-based ionic liquids could interact electrostatically with isolated DNA molecules. Going a step further, Zhang et al. [24] presented the first hints of possible genotoxic effects for 1-butyl-3-methyl-imidazolium chloride, using the bacterial SOS/umu indicator assay. They found a significant increase of the activity of SOS genes (which are involved in the recovery after DNA damaging events), starting at a concentration of 670 μM in a *Salmonella thyphimurium* strain after incubating these bacteria for 2 h with 1-butyl-3-methyl-imidazolium chloride. However, since the SOS/umu assay is only an indicator assay, no conclusions can be made at this point about the resulting mutations or permanent changes of the DNA structure owing to the genotoxic events that are observed. Furthermore, Zhang and co-workers were the first to show the genotoxic effects of an ionic liquid, and hence these data should not be taken to be representative of imidazolium-based ionic liquids in general. However, their work demonstrates that especially the genotoxic, mutagenic and teratogenic effects of ionic liquids should be included in further hazard assessment studies.

An attempt to investigate the long-term effects of ionic liquids in organisms was made by Wang et al., who demonstrated the sub-chronic effects of 1-octyl-3-methylimidazolium bromide on fish embryos after a 72 h incubation [25]. Hence, some specific effects of certain ionic liquid structures are already known; but data are still missing, particularly with regard to the endocrine disruption potential of ionic liquids and possible long-term and chronic effects. Moreover, the interactions of ionic liquid cations and anions – representing charged species with a lot of other molecular interaction potentials – with ion channels and other cellular transporters leading, for example, to chemosensitising effects, still need to be investigated.

Having now identified some of the most important data gaps in the prospective design of ionic liquid structures, the question about the best strategy and the most promising test systems to be implemented for generating these data still remains. Since endocrine disruption properties, genotoxicity and a high bioaccumulation in combination with persistency and poor biodegradability (see Section 9.3) can mean the knockout of ionic liquids under regulatory frameworks like the REACh regulation, efforts are urgently needed to look at these properties in detail. However, the final classification of a substance into one of these SVHC categories, for instance, can only be made after extensive testing using animals and other organisms. But such tests are ethically questionable, expensive and time-consuming, and should, therefore, be restricted to a very limited number of substances, as pointed out by Hartung et al. [26]. Thus, especially for the infinite number of possible combinations of ionic liquid cations and anions and their possible degradation products in organisms and the environment, an intelligently tiered testing strategy is needed to screen a large pool of substances and to identify suspicious substances for higher tier testing. Such a tiered testing strategy must be based on sound *in vitro* screening assays that will be capable of generating reliable indicators of specific effects, like genotoxicity or endocrine disruptive behaviour. Looking at the available test guidelines, for example, from the OECD, one can find some *in vitro* assays for the endpoints mentioned above, but a debate is currently in progress on how to improve these assays at the international level. Hence, future research efforts in ionic liquid toxicology should first focus on the development of test systems and endpoints, in order to rapidly screen the SVHC as well as the persistent, bioaccumulative and toxic (PBT) potential of ionic liquids. Secondly, the development of sound and reliable prediction tools like QSAR methods, based on the structure–activity relationship data gained in, for example, cellular screening assays, should be extended to handle the enormous numbers of possible ionic liquid structures and to reduce their complexity for further testing.

9.3 BIODEGRADABILITY OF IONIC LIQUIDS

The extensive use and production of man-made chemicals in almost all fields of human endeavour has resulted in the continuous release of chemicals into the biosphere. Since the early 1940s, a number of serious environmental problems have been caused by especially persistent chemicals that are highly resistant to biotic and abiotic transformation. Both the biotic and the abiotic degradability of organic chemicals are influenced largely by the potential exposure of the substances to organisms and, hence, are key parameters in risk assessment for predicting spatio-temporal environmental concentrations and long-term adverse effects on biota (potentially resulting in disorders, effects on reproduction, genetic mutations, cancer, mortality and adverse effects on the immune and nervous systems); hence, the demands of environmental legislation in Europe and the United States for non-persistent chemicals. For 'sustainable' or 'green' chemicals, complete and rapid biotic and/or abiotic degradation is a crucial prerequisite in accordance with Paul T. Anastas' and John C. Warner's tenth principle of green chemistry: 'Design chemicals and products to degrade after use: Design chemical products to break down to innocuous substances after use so that they do not accumulate in the environment' [27].

Degradability, in particular, could be a bottleneck in designing inherently safer and sustainable ionic liquids, because the technologically high desirable tendency of certain ionic liquids to be very stable, both thermally and chemically, is often reflected by their recalcitrance towards biological degradation processes.

In the past 10 years, some 30 papers and a few reviews [7, 9, 16, 28–30] have reported on the biodegradability of different ionic liquid structures, based mainly on 'ready biodegradability' test procedures under aerobic conditions. Fundamental studies in the field were performed by the groups of Scammels [31–38] and Gathergood [31–34, 39]. Guided by their knowledge on the development of biodegradable surfactants, they investigated different imidazolium, pyridinium or phosphonium compounds substituted with side chains containing, for example, ester, ether, polyether and amide functional groups combined with several anions in different tests for ready biodegradability. Wells and Coombe [40] investigated the biodegradability of ammonium, imidazolium, phosphonium and pyridinium compounds by measuring the biochemical oxygen demand (BOD). 1-Methylimidazolium and 3-methylpyridinium compounds alkylated with butyl, hexyl and octyl side chains, and with bromide as the anion, were examined by Docherty et al. [41], who applied a dissolved organic carbon (DOC) die-away test. Stolte et al. [42] investigated different imidazoles, imidazolium, pyridinium and 4-(dimethylamino)pyridinium compounds substituted with various alkyl side chains as well as their analogues containing functional groups in a primary biodegradation test, and were able to identify different biodegradation products originating from the 1-methyl-3-octyl-imidazolium cation. Several biological degradation products have also been identified for alkylpyridinium compounds by the teams associated with Pham [43], Docherty [44] and Zhang [45]. Stepnowski et al. [46] used the closed bottle test to examine the biodegradability of 1-alkoxymethyl-3-hydroxypyridinium cations combined with acesulfamate, saccharinate and chloride anions. A set of oxygenated and non-oxygenated imidazolium-based ionic liquids were investigated by Modelli et al. [47] in experiments with soil microorganisms.

Markiewicz et al. [48] explored the primary biodegradation and sorption on sewage sludge flocs of the 1-methyl-3-octylimidazolium cation with respect to its concentration. The anaerobic biodegradability and metabolisation of nine different imidazolium, pyridinium and dimethylaminopyridinium-based cations were examined over a period of 11 months [49]. Krossing et al. applied a closed bottle test and a manometric respirometry test to investigate different imidazolium compounds and one morpholinium ionic liquid [50]. Zhang [45] and Abrusci et al. [51] used axenic bacterial cultures to investigate the biodegradability of different ionic liquid structures.

On the basis of the above publications, we now know that some ionic liquid cations and anions are biodegradable under certain conditions, but most of them are recalcitrant towards biodegradation when test procedures according to 'ready biodegradability' test conditions are performed. With respect to structure–biodegradability relationships, the following rules of thumb can be applied to ionic liquids:

- The pyridinium core generally exhibits a higher degree of biodegradation than imidazolium, 4-(dimethylamino)pyridinium or phosphonium head groups.
- Increased biodegradability was observed in pyridinium, imidazolium and 4-(dimethylamino)pyridinium cations with elongated alkyl side chains (C8 > C6 > C4 > C2).
- The biodegradability of cations is reduced when cations with very long alkyl side chains ≥C10 are substituted, probably based on its increased toxicity to microorganisms.
- The pyridinium head group itself seems to be ultimately biodegradable, even when it is linked to shorter side chains (C2), and when the stringent-ready biodegradation test conditions are applied. In contrast, only the side chains of imidazolium compounds decompose; the core structure remains recalcitrant to biodegradation.
- The introduction of ester functional groups into the side chains can increase the biodegradability of the side chain.
- The biodegradability of the common imidazolium ionic liquids tested seems to be even worse under anaerobic than under aerobic conditions.
- Anions showing good biodegradability are alkylsulfates (e.g. methylsulfate or octylsulfate), linear alkylsulfonates (e.g. methylsulfonate), linear alkylbenzene sulfonates (e.g. 4-toluenesulfonate) and salts of organic acids (e.g. ethanoate or lactate), whereas typical fluorine-containing ionic liquid anions (such as $[N(CF_3SO_2)_2]^-$, $[(C_2F_5)_3PF_3]^-$, or $[CF_3SO_3]^-$) have been found to be recalcitrant to biodegradation.
- For inorganic anions such as $[BF_4]^-$ or $[PF_6]^-$, biodegradation tests – that are based on the measurement of oxidisable carbon in the molecule – are not relevant. In this case, abiotic processes such as hydrolysis are more important for calculating the residence time in the environment.

These rules of thumb provide proactive starting points for the design of inherently safer ionic liquids; nonetheless, they are still generalisations with limited expressiveness, especially in view of the small experimental database. In the following sections, therefore, we indicate a number of indispensable investigations as regards biodegradability, which will have to be carried out with ionic liquids in the near future. We shall also make some general statements about the misinterpretation of experimental biodegradation data, from which misleading conclusions may be drawn in the context of the design of inherently safer ionic liquids.

9.3.1 Test Procedures

So far, most of the studies investigating the biodegradability of ionic liquids have focussed on tests carried out according to OECD protocols (301 A-F) [52] to evaluate 'ready biodegradability' under aerobic conditions. The term 'ready biodegradability' is an arbitrary classification of chemicals that have passed a screening test for ultimate biodegradation, and implies the complete mineralisation of the test chemical, *inter alia*, to CO_2, H_2O or $[NO_3]^-$. A ready biodegradation test is based on stringent conditions (high ratio between test chemical concentration and cell density of inocula, which are not pre-exposed to the test chemical, only to inorganic salts in the medium). With respect to typical environmental conditions, however, this test is not realistic; it is nevertheless of significance, especially in the context of environmental legislation. If a chemical passes this test, we can assume that it is very likely to be readily biodegradable in the environment [53]. These chemicals come with just a low risk of persistency, or even none at all. Even if a chemical fails this test, it does not necessarily follow that it will not degrade in the environment, just that further testing will be necessary to clear up this initial suspicion of persistency. Many ionic liquids have been proven to be non-biodegradable when these stringent ready biodegradability test conditions are applied, which raises suspicions of persistency. 'Unconstrained', that is, environmentally more realistic test conditions (lower ratios of test chemical concentration to cell density of inoculum; the use of an inoculum pre-exposed to the test chemical or derived from industrial wastewater treatment plants; and the addition of organic co-substrates to the medium) should be used to acquire general information about ionic liquid biodegradability in the environment. At the second Conference on Biodegradability and Toxicity of Ionic Liquids (BATIL 2) [54], it was stated that the imidazolium core structure itself could also be biodegraded over a test period of 28 days when an inoculum from an industrial wastewater treatment plant was used instead of the OECD-recommended inoculum from domestic clarification plants. Recently, Abrusci et al. investigated a set of different pyrrolidinium, pyridinium, imidazolium and phosphonium compounds using an axenic culture of *Sphingomonas paucimobilis* [51]. This isolated bacterial strain seems to be able to degrade ionic liquid cations under the applied test conditions of 30 and 45°C. However, even if a chemical is degradable under these favourable test conditions, it does not necessarily follow that it will degrade in the environment. These few examples show that also the imidazolium core structure can generally be a substrate for metabolism when conditions are favourable, and where bacteria or other organisms have a suitable enzyme system for converting them.

In general, the still limited biodegradation data need to be complemented by systematic investigations to examine the biodegradability of different head groups under different aerobic as well as anaerobic conditions, by applying different test procedures not solely focussing on 'ready biodegradability' protocols. Besides biological degradation, abiotic transformation processes have to be considered if we wish to acquire a complete description of the environmental fate and the environmental lifetime of ionic liquid structures. Generally, only very limited data are as yet available on the abiotic conversion of ionic liquid cations and anions. Abiotic mechanisms like oxidation, reduction, hydrolysis and photolysis (under environmental conditions) may be the initial conversion step for non-biodegradable cations and anions, which may then yield biodegradable conversion products. These important degradation pathways should also be investigated in more detail with respect to ionic liquids to improve the hazard and risk assessment of these chemicals.

9.3.2 Interpretation of Biodegradation Data

In general, when assessing the biodegradability of ionic liquids, it has to be borne in mind that anions (whether biodegradable or not) may influence the biodegradability of a cation. For instance, the combination of a readily biodegradable cation (measured as a halide salt) with a toxic anion may reduce the biodegradability of the cation. Besides, if the anion reduces the water solubility of the cation, the biodegradation rate may be lower because of the decreased bioavailability of the cation to the microorganisms in the surrounding medium.

Moreover, the fact that ionic liquids are formed by two separate ionic species can lead to a misinterpretation of biodegradation data, and may result in misleading or even false conclusions – as has already occurred in the case of some ionic liquids. The course of biodegradation can be followed by applying non-specific composite parameters like CO_2 production. Here, the final result of such a test is expressed as '% degradation', based on the carbon that is converted to CO_2 (measured *via* manometric measurements, for instance) in relation to the total carbon content of the molecule, which for ionic liquids means the carbon content of both the cation and anion. Some compounds have been classified as 'readily biodegradable' (showing a biodegradation rate of >60% during 28 days) according to, for example, the ISO CO_2 Headspace test procedure, but this categorisation is based on a misleading molecular design and the wrong declaration of chemicals. To give an example, Figure 9.2a: a biodegradation rate of 41% ('CO_2 Headspace' test) was found for the 1-(pentoxycarbonyl)-3-methylimidazolium cation (tested as bromide) [34]. This percentage corresponds quite well to the oxidisable C-content of the ester side chain, or rather, to the pentanol formed after the suggested enzymatic hydrolysis. This means that the core structure is neither touched nor degraded by the microorganisms, and may well be

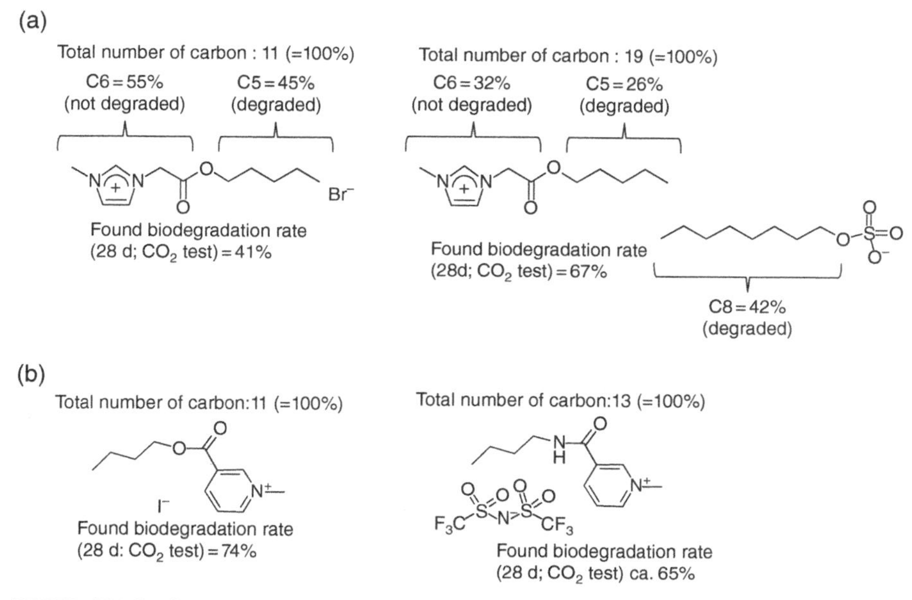

FIGURE 9.2 Some sample misleading degradation studies: (a) for imidazolium-based cations and (b) for pyridinium-based cations [34, 35].

resistant to further biodegradation. The authors of this study recommended the combination of this cation with the excellently biodegradable octylsulfate anion. Of course, this is accompanied by an increase in biodegradability, now that the C-content of the biodegradable parts of the side chain and of the anion exceeds the OECD level of 60%, see Figure 9.2a. However, this putative 'readily biodegradable' compound still contains a non-biodegradable core structure. Even more misleading is the 'ready' classification of the substance shown in Figure 9.2b. This ionic liquid is formed by a readily biodegradable cation and the non-biodegradable $[N(CF_3SO_2)_2]^-$ anion. In this case, the C-content of the cation is overbalanced and the C-content of the anion does not contribute significantly to the overall C-content of the ionic liquid's structural formula, illustrated in Figure 9.2b. Thus, the measured CO_2 evolution is more or less exclusively related to the cation. This ionic liquid, containing the $[N(CF_3SO_2)_2]^-$ anion, is thus classified as readily biodegradable [35], which is a mistake, because the anion structure is non-biodegradable and potentially persistent under normal environmental conditions. Ideally, ionic liquids (and chemicals in general) should not be classified in this arbitrary manner. The aim of biodegradation experiments should not be to pass certain threshold values, but to find out whether a chemical, or a combination of chemical entities, and their first conversion products undergo or do not undergo further and rapid metabolisation, ending in complete biodegradation within the environment. Thus, the interpretation of results should be guided by the following question [55]: 'Which test was performed under what conditions, and with which microorganisms?' If biodegradation was observable, two questions arise, namely, 'at what rate?' and 'breakdown to what?'. Without careful consideration of these questions, the results of biodegradability studies can lead to false or contradictory conclusions.

9.3.3 Transformation Products

For a sound hazard assessment of ionic liquids, it is obligatory to link biotic and abiotic degradation studies with chemical analysis and (eco)toxicity tests. The nature of the biological conversion pathways and products of ionic liquid substructures have to be known to avoid erroneous classifications of ionic liquids if a pass level is fulfiled but degradation products are recalcitrant toward biodegradation, as discussed above. Moreover, knowledge of the chemical structure of the metabolites or abiotic transformations products is crucial with respect to their hazard assessment. Transformation products or intermediates have their own characteristic (eco)toxicological profiles, and they may well be more toxic than their parent compounds. Although little information is available regarding this issue, most of the ionic liquid metabolites detected so far are the more polar ones, and a lower (eco)toxicity of these compounds compared to their parent compounds has been postulated. Nevertheless, some restrictions do have to be imposed, because aldehydes, which are intermediates in the oxidation pathway from a terminal hydroxyl group to a carboxylic group, have not yet been analysed for their (eco)toxicity. In principle, the formation of highly reactive epoxides is also conceivable, especially during the hydroxylation step of the pyridinium core structure.

Thus, analytically identified or theoretically predicted transformation products of ionic liquids should be synthesised and (eco)toxicologically characterised, and included in the evaluation of the hazard and risk potential of their parent structures.

9.4 CONCLUSION

It was the aim of this chapter to critically discuss the current state of knowledge of ionic liquid (eco)toxicology and degradability under environmental conditions. Therefore, the two main sections of this chapter briefly presented the most recent and relevant literature in both fields, the outcome is discussed critically according to the three guiding questions presented in the introductory section: (i) What more needs to be done for the design and development of inherently safer ionic liquids?; (ii) What information, and how much of it, do we need to achieve a sound and legally relevant risk assessment of ionic liquids?;

and (iii) Which strategies and techniques seem to be the most promising ones for achieving the goal of inherently safer ionic liquids with a low hazard and risk potential for man and the environment on the one hand but with high-quality technological performance on the other?

In this concluding part, we would now like to summarise our key issues and point out what, from our point of view, needs to be done in future research to take a big step towards inherently safer and hence more sustainable ionic liquids.

Generally, the environmental risk of ionic liquids can be described as the product of exposure and hazard of a given compound. Even if the main exposure pathways, the expected technological uses and the amounts and distributions of certain ionic liquid structures in different environmental compartments do still remain unclear, we have tried to address the exposure level by discussing the degradability and persistency of ionic liquids. The hazard potential of ionic liquids as the second important risk parameter can be discussed on the basis of reported (eco)toxicity data.

Looking at the (eco)toxicological issues of the ionic liquids investigated so far, it seems most likely that the predominant mechanism of their acute toxicity is unspecific, hydrophobicity-driven baseline toxicity. As such, this mode of action can easily be predicted with simple regression models for each relevant test system or organism and is thus of minor regulatory concern for further testing. Additionally, the acute toxicity originating from this mode of action can be easily controlled and reduced by decreasing the hydrophobicity of ionic liquid cations and anions; various examples are given for this strategy in the literature.

However, there are still data gaps where more specific and long-term or chronic modes of the toxic action of ionic liquids are concerned. Here, an need for screening tests has been identified, especially with respect to the highly relevant SVHC criteria. More attention needs to be given to the non-linear effects discussed above (e.g. increased bioavailability and synergistic effects in toxicological studies), resulting from the combination of certain ionic liquid cations and anions (especially the technologically relevant hydrophobic and bulky ones). The same applies to the combinatorial effects of ionic liquids and other surfactants in the aquatic environment, since the most likely exposure pathways seem to be wastewater streams.

As discussed above, many ionic liquid structures seem to be persistent, or at least are only slowly degradable in the environment. This tendency of persistency, combined with the mostly relatively high hydrophobicity of ionic liquid cations and anions, bears the hazard of bioaccumulation. Thus, we suggest that a greater research effort be made to elucidate the bioaccumulation potential of ionic liquids and the underlying mechanisms involved. Since ionic liquids can be both hydrophobic and ionic, special mechanisms and distribution effects may come into play in their bioaccumulation behaviour. And, as already mentioned above, the possibly high persistency and slow degradation rates require more and detailed chronic effect studies in order to complete the hazard picture of at least the most frequently used ionic liquid structures.

Coming now to the issue of biodegradability, it can be concluded that the presented design criteria may lead to ionic liquids with improved biodegradability and ecotoxicological properties, but that they may also limit practical applications with respect to physicochemical properties such as stability (thermal or electrochemical) or viscosity. Therefore, for a particular technological application, it may be necessary to use a non-biodegradable and/or (eco)toxic ionic liquid. The authors would like to point out that although these ionic liquids can be components of sustainable products and processes, comprehensive risk management is needed in such cases with respect to the operational safety of employees. Furthermore, it has to be ensured that environmental contamination is minimal. Thus, within such processes, strategies and techniques are needed to remove these compounds, for instance, from processing effluents *via* regenerative methods (like membrane filtration [56]) or *via* destructive advanced oxidation techniques like electrochemical treatment [42] or UV degradation [57].

Finally, economic aspects and, especially production costs, will determine whether ionic liquids will be widely used in technical applications in the near future. From a structural point of view, they certainly have a considerable potential to contribute to the development of more sustainable products and processes.

REFERENCES

1 Weingärtner, H., Understanding ionic liquids at the molecular level: Facts, problems, and controversies, *Angew. Chem. Int. Ed.* **47**, 654–670 (2008).

2 Stolte, S., Arning, J., Bottin-Weber, U., Matzke, M., Stock, F., Thiele, K., Uerdingen, M., Welz-Biermann, U., Jastorff, B. and Ranke, J., Anion effects on the cytotoxicity of ionic liquids, *Green Chem.* **8**, 621–629 (2006).

3 Matzke, M., Stolte, S., Thiele, K., Juffernholz, T., Arning, J., Ranke, J., Welz-Biermann, U. and Jastorff, B., The influence of anion species on the toxicity of 1-alkyl-3-methylimidazolium ionic liquids observed in an (eco) toxicological test battery, *Green Chem.* **9**, 1198–1207 (2007).

4 Zhang, J., Liu, S.S., Dou, R.N., Liu, H.L. and Zhang, J., Evaluation on the toxicity of ionic liquid mixture with antagonism and synergism to Vibrio qinghaiensis sp.-Q67, *Chemosphere* **82**, 1024–1029 (2011).

5 Ranke, J., Stolte, S., Störmann, R., Arning, J. and Jastorff, B., Design of sustainable chemical products – The example of ionic liquids, *Chem. Rev.* **107**, 2183–2208 (2007).

6 Frade, R.F.M. and Afonso, C.A.M., Impact of ionic liquids in environment and humans: An overview, *Hum. Exp. Toxicol.*, **29**, 1038–1054 (2010).

7 Pham, T.P., Cho, C.W. and Yun, Y.S., Environmental fate and toxicity of ionic liquids: A review, *Water Res.*, **44**, 352–372 (2010).

8 Wood, N. and Stephens, G., Accelerating the discovery of biocompatible ionic liquids, *Phys. Chem. Chem. Phys.* **12**, 1670–1674 (2010).

9 Petkovic, M., Seddon, K.R., Rebelo, L.P.N. and Pereira, C.S., Ionic liquids: A pathway to environmental acceptability, *Chem. Soc. Rev.* **40**, 1383–1403 (2011).

10 Stolte, S., Arning, J., Bottin-Weber, U., Müller, A., Pitner, W.R., Welz-Biermann, U., Jastorff, B., Ranke, J., Effects of different head groups and functionalised side chains on the cytotoxicity of ionic liquids, *Green Chem.* **9**, 760–767 (2007).

11 Stolte, S., Matzke, M., Arning, J., Böschen, A., Pitner, W.R., Welz-Biermann, U., Jastorff, B. and Ranke, J., Effects of different head groups and functionalised side chains on the aquatic toxicity of ionic liquids, *Green Chem.* **9**, 1170–1179 (2007).

12 Samori, C., Malferrari, D., Valbonesi, P., Montecavalli, A., Moretti, F., Galletti, P., Sartor, G., Tagliavini, E., Fabbri, E. and Pasteris, A., Introduction of oxygenated side chain into imidazolium ionic liquids: Evaluation of the effects at different biological organization levels, *Ecotoxicol. Environ. Saf.* **73**, 1456–1464 (2010).

13 Overton, C.E., *Studien über die Narkose, zugleich ein Beitrag zur allgemeinen Pharmakologie*, Gustav Fischer, Jena (1901).

14 Escher, B.I. and Hermens, J.L.M., Modes of action in ecotoxicology: Their role in body burdens, species sensitivity, QSARs, and mixture effects, *Environ. Sci. Pollut. Res.* **36**, 4201–4217 (2002).

15 Escher, B.I. and Schwarzenbach, R.P., Mechanistic studies on baseline toxicity and uncoupling of organic compounds as a basis for modeling effective membrane concentrations in aquatic organisms, *Aquat. Sci.* **64**, 20–35 (2002).

16 Ranke, J., Müller, A., Bottin-Weber, U., Stock, F., Stolte, S., Arning, J., Störmann, R. and Jastorff, B., Lipophilicity parameters for ionic liquid cations and their correlation to in vitro cytotoxicity, *Ecotoxicol. Environ. Saf.* **67**, 430–438 (2007).

17 Torrecilla, J.S., Garcia, J., Rojo, E. and Rodriguez, F., Estimation of toxicity of ionic liquids in Leukemia Rat Cell Line and Acetylcholinesterase enzyme by principal component analysis, neural networks and multiple lineal regressions, *J. Hazard. Mater.* **164**, 182–194 (2009).

18 Kulacki, K.J., Chaloner, D.T., Costello, D.M., Docherty, K.M., Larson, J.H., Bernot, R.J., Brueseke, M.A., Kulpa, C.F. and Lamberti, G.A., Aquatic toxicity and biodegradation of ionic liquids – A synthesis, *Chim. Oggi* **25**, 32–36 (2007).

19 Yu, M., Wang, S.H., Luo, Y.R., Han, Y.W., Li, X.Y., Zhang, B.J. and Wang, J.J., Effects of the 1-alkyl-3-methylimidazolium bromide ionic liquids on the antioxidant defense system of Daphnia magna, *Ecotoxicol. Environ. Saf.* **72**, 1798–1804 (2009).

20 Arning, J., Stolte, S., Böschen, A., Stock, F., Pitner, W.R., Welz-Biermann, U., Jastorff, B. and Ranke, J., Qualitative and quantitative structure activity relationships for the inhibitory effects of cationic head groups, functionalised side chains and anions of ionic liquids on acetylcholinesterase, *Green Chem.* **10**, 47–58 (2008).

21 Składanowski, A.C., Stepnowski, P., Kleszczyński, K. and Dmochowska, B., AMP deaminase *in vitro* inhibition by xenobiotics: A potential molecular method for risk assessment of synthetic nitro- and polycyclic musks, imidazolium ionic liquids and *N*-glucopyranosyl ammonium salts, *Environ. Toxicol. Pharmacol.* **19**, 291–296 (2005).

22 Luo, Y.R., Wang, S.H., Li, X.Y., Yun, M.X., Wang, J.J. and Sun, Z.J., Toxicity of ionic liquids on the growth, reproductive ability, and ATPase activity of earthworm, *Ecotoxicol. Environ. Saf.* **73**, 1046–1050 (2010).

23 Xie, Y.N., Wang, S.F., Zhang, Z.L. and Pang, D.W., Interaction between room temperature ionic liquid [bmim]BF_4 and DNA investigated by electrochemical micromethod, *J. Phys. Chem. B* **112**, 9864–9868 (2008).

24 Zhang, Z., Liu, J.F., Cai, X.Q., Jiang, W.W., Luo, W.R. and Jiang, G.B., Sorption to dissolved humic acid and its impacts on the toxicity of imidazolium based ionic liquids, *Environ. Sci. Technol.* **45**, 1688–1694 (2011).

25 Wang, S.H., Huang, P.P., Li, X.Y., Wang, C.Y., Zhang, W.H. and Wang, J.J., Embryonic and developmental toxicity of the ionic liquid 1-methyl-3-octylimidazolium bromide on Goldfish, *Environ. Toxicol.* **25**, 243–250 (2010).

26 Hartung, T. and Rovida, C., Chemical regulators have overreached, *Nature* **460**, 1080–1081 (2009).

27 Anastas, P.T. and Warner, J.C., *Green Chemistry: Theory and Practice*, Oxford University Press, New York (1998).

28 Boethling, R.S., Sommer, E. and DiFiore, D., Designing small molecules for biodegradability, *Chem. Rev.* **107**, 2207–2227 (2007).

29 Coleman, D. and Gathergood, N., Biodegradation studies of ionic liquids, *Chem. Soc. Rev.* **39**, 600–637 (2010).

30 Stolte, S., Steudte, S., Igartua, A. and Stepnowski, P., Biodegradation of ionic liquids – A view from a chemical structure perspective, *Curr. Org. Chem.* **15**, 1946–1973 (2011).

31 Gathergood, N. and Scammells, P.J., Design and preparation of room-temperature ionic liquids containing biodegradable side chains, *Aust. J. Chem.* **55**, 557–560 (2002).

32 Gathergood, N., Garcia, M.T. and Scammells, P.J., Biodegradable ionic liquids: Part I. Concept, preliminary targets and evaluation, *Green Chem.* **6**, 166–175 (2004).

33 Garcia, M.T., Gathergood, N. and Scammells, P.J., Biodegradable ionic liquids – Part II. Effect of the anion and toxicology, *Green Chem.* **7**, 9–14 (2005).

34 Gathergood, N., Scammells, P.J. and Garcia, M.T., Biodegradable ionic liquids – Part III. The first readily biodegradable ionic liquids, *Green Chem.* **8**, 156–160 (2006).

35 Harjani, J.R., Singer, R.D., Garcia, M.T. and Scammells, P.J., The design and synthesis of biodegradable pyridinium ionic liquids, *Green Chem.* **10**, 436–438 (2008).

36 Atefi, F., Garcia, M.T., Singer, R.D. and Scammells, P.J., Phosphonium ionic liquids: Design, synthesis and evaluation of biodegradability, *Green Chem.* **11**, 1595–1604 (2009).

37 Harjani, J.R., Farrell, J., Garcia, M.T., Singer, R.D. and Scammells, P.J., Further investigation of the biodegradability of imidazolium ionic liquids, *Green Chem.* **11**, 821–829 (2009).

38 Ford, L., Harjani, J.R., Atefi, F., Garcia, M.T., Singer, R.D. and Scammells, P.J., Further studies on the biodegradation of ionic liquids, *Green Chem.* **12**, 1783–1789 (2010).

39 Morrissey, S., Pegot, B., Coleman, D., Garcia, M.T., Ferguson, D., Quilty, B. and Gathergood, N., Biodegradable, non-bactericidal oxygen-functionalised imidazolium esters: A step towards 'greener' ionic liquids, *Green Chem.* **11**, 475–483 (2009).

40 Wells, A.S. and Coombe, V.T., On the freshwater ecotoxicity and biodegradation properties of some common ionic liquids, *Org. Process Res. Dev.* **10**, 794–798 (2006).

41 Docherty, K.M., Dixon, J.K. and Kulpa, C.F., Biodegradability of imidazolium and pyridinium ionic liquids by an activated sludge microbial community, *Biodegradation* **18**, 481–493 (2007).

42 Stolte, S., Abdulkarim, S., Arning, J., Blomeyer-Nienstedt, A., Bottin-Weber, U., Matzke, M., Jastorff, B. and Thöming, J., Primary biodegradation of ionic liquid cations, identification of degradation products of 1-methyl-3-octylimidazolium chloride and electrochemical wastewater treatment of poorly biodegradable compounds, *Green Chem.* **10**, 214–242 (2008).

43 Pham, T.P., Cho, C.W., Jeon, C.O., Chung, Y.J., Lee, M.W. and Yun, Y.S., Identification of metabolites involved in the biodegradation of the ionic liquid 1-butyl-3-methylpyridinium bromide by activated sludge microorganisms, *Environ. Sci. Pollut. Res.* **43**, 516–521 (2009).

44 Docherty, K.M., Joyce, M.V., Kulacki, K.J. and Kulpa, C.F., Microbial biodegradation and metabolite toxicity of three pyridinium-based cation ionic liquids, *Green Chem.* **12**, 701–712 (2010).

45 Zhang, C., Wang, H., Malhotra S.V., Dodge C.J. and Francis, A.J., Biodegradation of pyridinium based ionic liquids by an axenic culture of soil Corynebacteria, *Green Chem.* **12**, 851–858 (2012).

46 Stasiewicz, M., Mulkiwicz, E., Tomaczak-Wanzel, R., Kumirska, J., Siedlecka, E.M., Gołebiowski, M., Gajdus, J., Czerwicka, M. and Stepnowski, P., Assessing toxicity and biodegradation of novel, environmentally benign ionic liquids (1-alkoxymethyl-3-hydroxypyridinium chloride, saccharinate and acesulfamates) on cellular and molecular level, *Ecotoxicol. Environ. Saf.* **71**, 157–165 (2008).

47 Modelli, A., Sali, A., Galletti, P. and Samori, C., Biodegradation of oxygenated and non-oxygenated imidazolium-based ionic liquids in soil, *Chemosphere* **73**, 1322–1327 (2008).

48 Markiewicz, M., Jungnickel, C., Markowska, A., Szczepaniak, U., Paszkiewicz, M. and Hupka, J., 1-methyl-3-octylimidazolium chloride-sorption and primary biodegradation analysis in activated sewage sludge, *Molecules* **14**, 4396–4405 (2009).

49 Neumann, J., Grundmann, O., Thöming, J., Schulte, M. and Stolte, S., Anaerobic biodegradability of ionic liquid cations under denitrifying conditions, *Green Chem.* **20**, 620–627 (2010).

50 Bulut, S., Klose, P., Huang, M.M., Weingartner, H., Dyson, P.J., Laurenczy, G., Friedrich, C., Menz, J., Kummerer, K. and Krossing, I., Synthesis of room-temperature ionic liquids with the weakly coordinating $[Al(OR^F)_4]^-$ anion ($R^F = C(H)(CF_3)_2$) and the determination of their principal physical properties, *Chem. Eur. J.* **16**, 13139–13154 (2010).

51 Abrusci, C., Palomar, J., Pablos, J.L., Rodriguez, F. and Catalina, F., Efficient biodegradation of common ionic liquids by Sphingomonas paucimobilis bacterium, *Green Chem.* **13**, 709–717 (2011).

52 OECD 301. OECD guideline for testing of chemicals: 301 Ready Biodegradability. Adopted by the Council on 17th July 1992-Ready, 1-62, 1992.

53 OECD. Revised introduction to the OECD guidelines for the testing of chemicals. Section 3 Part 1: Principles and Strategies related to the testing of degradation of organic chemicals. 2006.

54 Stolte, S., (Eco)toxicity and biodegradation of ionic liquids – progress in designing inherently safer chemicals, BATIL-2 (Biodegradability and Toxicity of Ionic Liquids), 29/9/2009, Frankfurt. http://events.dechema.de/batil2.html (accessed 9 July 2015).

55 Brown, D., Introduction to surfactant biodegradation, in *Biodegradability of Surfactants*, D.R. Karsa and M.R. Porter, eds., Blackie Academic and Professional, Glasgow (1995), pp. 1–28.

56 Fernandez, J.F., Waterkamp, D. and Thöming, J., Recovery of ionic liquids from wastewater: Aggregation control for intensified membrane filtration, *Desalination* **224**, 52–56 (2008).

57 Stepnowski, P. and Zaleska, A., Comparison of different advanced oxidation processes for the degradation of room temperature ionic liquids, *J. Photochem. Photobiol. Chem.* **170**, 45–50 (2005).

10 Ionic Liquids and Organic Reaction Mechanisms

TOM WELTON

Department of Chemistry, Imperial College London, London, UK

10.1 INTRODUCTION

There are not many things that you can do to a molecule or ion: you can add something to it (*addition*), you can take something away from it (*elimination*), you can add something to it while taking something away (*substitution*), or you can rearrange the atoms within it without adding or subtracting anything (*isomerisation*, or *rearrangement*); electron transfer can occur with or without atom transfer (*reduction/oxidation* or *redox*). These are the fundamental chemical reactions.

Another way of thinking about a chemical reaction is that it is what happens when you bring (usually) two or more reactants together to form products. However, when looking at the products of these processes, it is common for these not to appear to be the simple outcome of just one of the fundamental reactions described above. That is because these fundamental steps do not have to happen in isolation. The products of a chemical reaction are often formed by several of these steps occurring in succession. This succession of events provides the route by which the reactants transform into the products. This route is the *reaction mechanism*.

For a reaction to occur, two conditions must be met. The first is that the products should be more energetically favourable than the reactants. This energy difference controls the equilibrium constant of the reaction:

$$\Delta G = -RT \ \ln(K) \tag{10.1}$$

The second condition is that there must be an available mechanism for the reaction. If there is no accessible route to get from reactants to products, no reaction will occur.

Ionic Liquids Completely UnCOILed: Critical Expert Overviews, First Edition.
Edited by Natalia V. Plechkova and Kenneth R. Seddon.

Most chemists would have witnessed at some time an organic chemistry professor beginning with sketches of commercially available reactants on a board, then with a flourish of *curly arrows* (a representation of the movement of electron pairs) generating a number of intermediates that obviously (in his/her eyes) will go on to give some desired product, and pronouncing the mechanism of a reaction that has yet to be performed. Some hapless student is then sent to perform the synthesis having had it made clear that failure can only arise due to the student's incompetence. However, it is often true that the imagined mechanism does not occur in practice, because it is energetically disfavoured and something else (or indeed nothing) happens instead. Thermodynamics determine the extent of reaction that is observed, whereas the energies of the mechanism determine the rate of a reaction. Hence, the way in which reaction mechanisms are investigated is by rate measurements. Occasionally, the selectivities of a reaction can be used as an alternative when these arise from competing rates of formation of the different products. When a reaction is thermodynamically possible, but does not occur because the mechanism is energetically disfavoured in comparison to another (or none), it is said to be under kinetic control. This is because the reaction is too slow for the expected (or any) products to be formed.

Kinetic control of reactions occurs because there is an energy barrier that needs to be overcome for a reaction to occur, even for the fundamental steps described above. This is because the arrangement of atoms that must happen for the transition from the reactant to the products to occur is energetically less favoured than those of either the reactants or the products (Fig. 10.1). The difference in energy between the starting conditions and this *transition state*, $[AB]^{\ddagger}$, is the *activation energy* ($\Delta G^{\ddagger}$) of the step. The chemical species at the transition state is the *activated complex*. The relationship

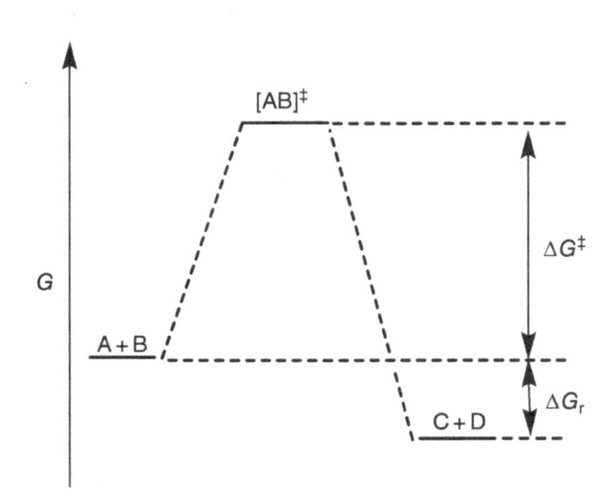

Figure 10.1 A simple reaction energy profile.

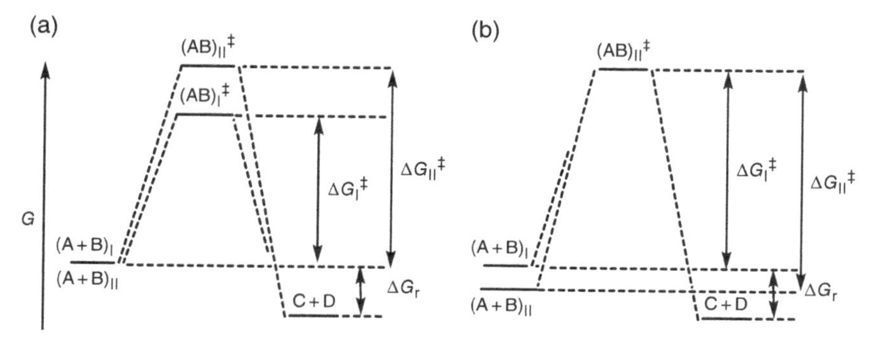

Figure 10.2 (a) Stabilisation of the transition state, and (b) stabilisation of the starting materials.

between this energy and the rate of a reaction is given by the Arrhenius equation, Equation 10.2:

$$k = \mathrm{A}e^{-\Delta G^{\ddagger}/RT} \tag{10.2}$$

So, if the activation energy is larger, the rate of the reaction is slower. The usual way in which ionic liquids, or any solvent, can affect the rate of a reaction step is to change its activation energy. It does this by interactions between it and the starting materials and activated complex. Interactions with the products of the reaction are irrelevant in this situation. The solvent could, at least hypothetically, stabilise or destabilise the transition state, that is reduce or increase its energy, without changing the energy of the starting conditions (Fig. 10.2a), or it could stabilise or destabilise the starting conditions, that is reduce or increase its energy, without changing the energy of the transition state (Fig. 10.2b).

Usually, the energies of all states change on changing the solvent, often in the same direction, and it is the relative changes in these that determine the differences in the activation energies for the reaction in the different solvents (see Fig. 10.3).

Multistep reactions have more complex reaction profiles, with each step having its own starting conditions, transition state and products, where the products of the first step are the starting materials for the second … these are called *reaction intermediates* (see Fig. 10.4).

Fortunately, when considering a reaction with a multistep mechanism, we do not have to consider every step. Anyone who has ever found themselves coming up to a lane closure on a busy motorway will recognise that the flow of traffic is determined by the rate at which cars can get through this restriction, and not by the free-flowing traffic. The same is true for reaction mechanisms. The rate of a reaction, no matter how complex the mechanism, is determined by its slowest step, known as the *rate controlling* or *rate determining step*. It is not that this is the only step that we need to consider; it is the only step that we

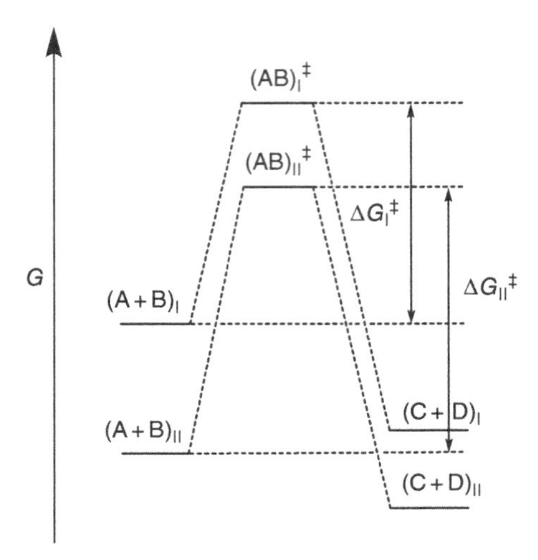

Figure 10.3 The stabilisation changes of all energy states upon transfer from solvent I to solvent II.

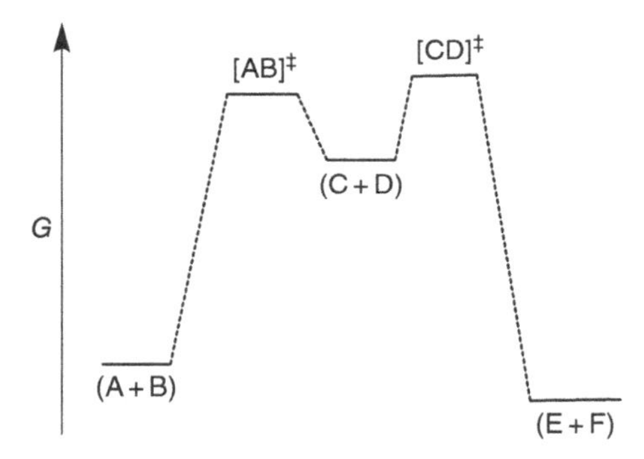

Figure 10.4 A reaction energy profile for a two-step reaction.

can easily investigate. Making changes that would affect other steps in the reaction mechanism will have no effect on the rate of the reaction. If we knew how an ionic liquid could influence steps of reactions, and we knew a particular reaction's rate determining step, we would be able to predict the effect of the ionic liquid upon the rate of that reaction.

10.2 SOLVENT EFFECTS ON THE RATES AND MECHANISMS OF CHEMICAL REACTIONS

The study of the effects of solvents on the rates of reactions, and thereby on reaction mechanisms, is not new. It can be traced back to the work of Hughes and Ingold in the 1930s [1–5]. This led to decades of work that continues to this day [6]. They studied nucleophilic substitutions in great depth in order to

TABLE 10.1 The Effect of Changing Solvent on Nucleophilic Substitution Reactions

Reaction Type	Reactants	Activated Complex	Effect of an Increased Polarity Solvent on Rate
(a) S_N1	R-X	$^{\delta+}R\cdots X^{\delta-}$	Large increase
(b) S_N1	$R\text{-}X^+$	$^{\delta+}R\cdots X^{\delta+}$	Small decrease
(c) S_N2	Y + R-X	$^{\delta-}Y\cdots R\cdots X^{\delta-}$	Large increase
(d) S_N2	Y^- + R-X	$^{\delta-}Y\cdots R\cdots X^{\delta-}$	Small decrease
(e) S_N2	Y + $R\text{-}X^+$	$^{\delta+}Y\cdots R\cdots X^{\delta+}$	Small decrease
(f) S_N2	Y^- + $R\text{-}X^+$	$^{\delta-}Y\cdots R\cdots X^{\delta+}$	Large decrease

understand how solvents can affect reaction rates (Table 10.1). They selected these reactions because of the variety that could be achieved with respect to charge formation, destruction or transfer.

Using a simple electrostatic model of solvation in nucleophilic substitution reactions, Hughes and Ingold concluded that more polar solvents will preferentially stabilise more polar solutes and proposed a set of rules to predict the outcome of changing the solvent on the rate of a reaction:

1. Increasing the polarity of the solvent increases the rates of reactions in which the charge density of the activated complex is greater than that of the reactants.
2. Increasing the polarity of the solvent decreases the rate of reactions in which the charge density of the activated complex is lower than that of the reactants.
3. Changing the polarity of the solvent has no effect on the rates of reactions in which the charge density of the activated complex is the same as that of the reactants.

This set of simple rules has stood the test of time well and are still the basis of most organic chemists' approach to solvent effects on the rates of reactions. The Hughes–Ingold rules do not consider specific interactions, such as hydrogen bonds, but it is easy to see how these can be incorporated into the analysis.

With this in mind, the Welton group reasoned that the same approach could be taken to the study of the effects of ionic liquids on the rates of reactions. They used the Kamlet-Taft polarity scales to develop Linear Solvation Energy Relationships (LSERs) to describe the effects of solvents on the reaction kinetics of various S_N2 nucleophilic substitution reactions (Schemes 10.1–10.4). S_N2 reactions occur in a concerted step in which the *nucleophile* replaces the *nucleofuge* or *leaving group* as it dissociates from the substrate. Their reaction profiles have the form of that in Figure 10.1.

The reactions of butylamine, dibutylamine and tributylamine with methyl 4-nitrobenzenesulfonate (Scheme 10.1; Table 10.2) have been studied in 1-butyl-1-methylpyrrolidinium bis{(trifluoromethyl)sulfonyl)}amide, $[C_4C_1pyrr]$

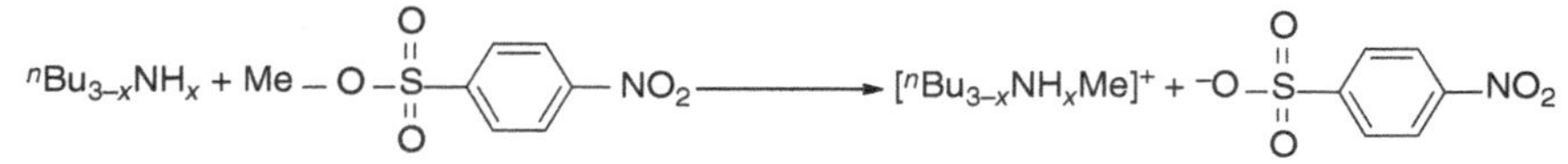

Scheme 10.1 The reactions of primary, secondary and tertiary amines with methyl 4-nitrobenzenesulfonate.

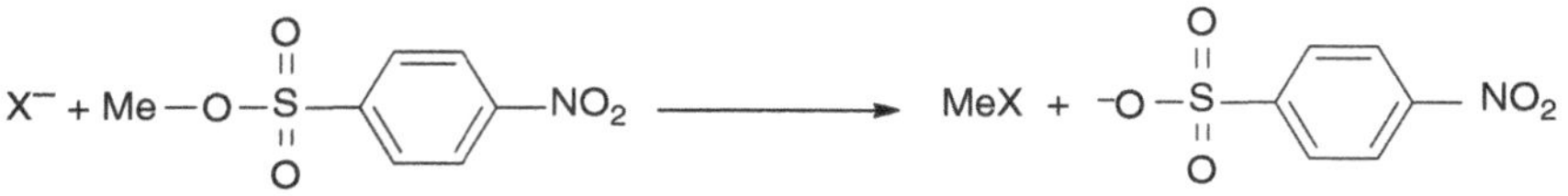

Scheme 10.2 The reactions of anionic nucleophiles with methyl 4-nitrobenzenesulfonate.

Scheme 10.3 The reactions of primary, secondary and tertiary amines with *S*,*S*-dimethyl-4-nitrophenylsulfonium salts.

Scheme 10.4 The reaction of chloride (anionic nucleophile) with $[4\text{-}NO_2C_6H_4S(CH_3)_2]^+$.

$[NTf_2]$, 1-butyl-1-methylpyrrolidinium triflate, $[C_4C_1pyrr][OTf]$, and 1-butyl-3-methylimidazolium triflate $[C_4C_1im][OTf]$, and their reactivities compared to the same reactions in the molecular solvents dichloromethane and ethanenitrile [7]. It was shown that all of the amines were more nucleophilic in the ionic liquids than in the molecular solvents used, as one would expect from the Hughes–Ingold rules if the ionic liquids are considered to be polar solvents.

All three reactions were accelerated by solvents with large values of π^* (dipolar and polarisability contributions to polarity), again as expected from the Hughes–Ingold rules. Hydrogen bonds from the starting materials to the solvent also accelerated the reactions, as shown by the positive effect of β on the reactions of $BuNH_2$ and Bu_2NH, with the effect on the reaction of $BuNH_2$ being twice that on the reaction of Bu_2NH. This is because hydrogen bond acceptor solvents will interact more with the emerging ammonium protons of the activated complex than with the reactant amine protons (see Fig. 10.5). This leads to a greater solvent-induced stabilisation of the transition state and, hence, a lower activation energy and a faster reaction. Bu_3N has no protons, and so it is not possible for its reactivity to be influenced by a hydrogen bond accepting solvent; in

TABLE 10.2 LSER Correlations for $\ln(k_2)$ Obtained for Butylamines with Methyl 4-Nitrobenzenesulfonate in $[C_4C_1pyrr][NTf_2]$, $[C_4C_1pyrr][OTf]$, $[C_4C_1im][OTf]$, Dichloromethane and Ethanenitrile [7]

Amine	LSER	R^2
$BuNH_2$	$\ln(k_2) = -8.77 + 4.57\beta + 6.32\pi^*$	0.93
Bu_2NH	$\ln(k_2) = -8.57 + 2.23\beta + 7.30\pi^*$	0.92
Bu_3N	$\ln(k_2) = 0.87 - 2.56\alpha + 12.80\pi^*$	0.70

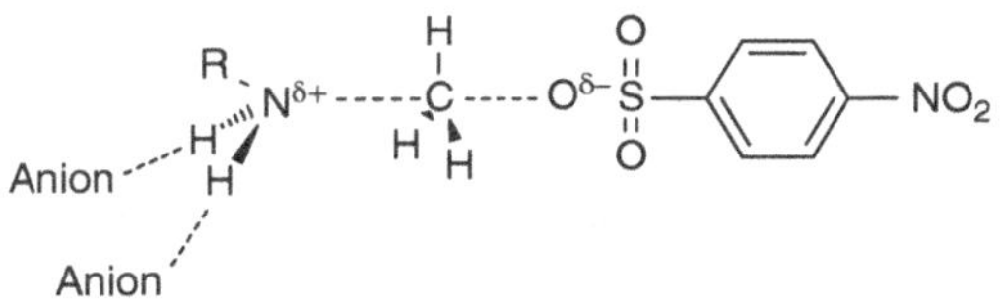

Figure 10.5 The ionic liquid anions hydrogen bonding with the activated complex for the reaction of $BuNH_2$ with methyl 4-nitrobenzenesulfonate.

its LSER, the reduction of its reactivity by hydrogen bond donor solvents is seen. Overall, the Kamlet-Taft analysis of these reactions show that the Hughes–Ingold rules can be applied to ionic liquids with the ionic liquids being polar (but not excessively so) solvents that are capable of acting as both hydrogen bond donors and acceptors. Skrzypczak and Neta found that the rates of the reactions of 1,2-dimethylimidazole with benzyl bromide in the 12 ionic liquids $[C_nC_1im]X$, $[C_ndmim]X$ {$[C_ndmim]^+$ = 1-alkyl-2,3-dimethylimidazolium, $n = 4$ or 8, X = $[NTf_2]$, $[PF_6]$ or $[BF_4]$}, $[C_4C_1pyrr][NTf_2]$ and $[N_{4\,4\,4\,6}][NTf_2]$ {$[N_{4\,4\,4\,6}]^+$ = tributylhexylammonium} had broad agreement with these results, and were similar to those found in polar aprotic molecular solvents [8]. Chiappe et al. found that the rates of the reactions of benzyl chlorides with 1-methylimidazole were consistently faster in ionic liquids {$[C_4mim][PF_6]$, $[C_4mim][NTf_2]$, $[C_4dmim][NTf_2]$ or $[C_4C_1pyrr][NTf_2]$} than molecular solvents (MeCN, or DMF), although the precise ordering changed with differently substituted benzyl chlorides [9].

The LSER for the Bu_3N reactions with methyl 4-nitrobenzenesulfonate had a considerably lower R^2 value than for the other two reactions (see Table 10.2) [7]. Close inspection of the data showed that k_2 for the reaction of Bu_3N in $[C_4C_1pyrr][OTf]$ was higher than expected. An Eyring analysis suggested that the neutral starting materials interacted with the ionic liquid ions only weakly and were not able to disrupt the anion-cation Coulombic attraction. It was suggested that this gives the starting materials a restricted volume in which to move. The developing charges on the activated complex allowed it to interact with the ionic liquid ions. In so doing, it disrupted the anion–cation interactions and caused some breakdown of the local structure of the ionic liquid. This has a dramatic effect on the $\Delta S^‡$ for the reaction, which leads to a reduction of the activation energy and can be thought of as a solvatophobic rate effect.

TABLE 10.3 LSER Correlations for ln(k_2) Obtained for Some Anionic Nucleophiles in [C_4mim][NTf_2], [C_4mim][OTf], [C_4C_1pyrr][NTf_2], DMSO, CH_2Cl_2 and MeOH

Anion	LSER	R^2
Cl^- (β=1.00)	$\ln(k_2)=0.21-7.56\alpha$	0.99
Br^- (β=0.67)	$\ln(k_2)=-0.87-5.38\alpha$	0.97
I^- (β=0.30)	$\ln(k_2)=-2.57-3.05\alpha+1.16\beta$	0.95
$[CH_3CO_2]^-$ (β=1.49)	$\ln(k_2)=-2.37-7.60\alpha+2.65\beta+1.83\pi^*$	1.00
$[CF_3CO_2]^-$	$\ln(k_2)=-9.18-4.94\alpha+5.76\pi^*$	0.97
SCN^- (β=0.33)	$\ln(k_2)=-3.87-2.46\alpha$	0.89
CN^- (β=1.37)	$\ln(k_2)=-3.16-5.07\alpha+5.78\pi^*$	0.99

TABLE 10.4 LSERs for Reactions of Butylamines with $[4\text{-}NO_2PhS(CH_3)_2]^+$

Amine	LSER	R^2
$BuNH_2$	$\ln(k_2)=-2.38-3.59\alpha-4.16\beta+2.10\pi^*$	0.96
Bu_2NH	$\ln(k_2)=-2.66-2.79\alpha-5.01\beta+2.89\pi^*$	0.99
Bu_3N	$\ln(k_2)=-5.62-6.46\beta+4.26\pi^*$	0.87

Application of the Hughes–Ingold rules to nucleophilic substitutions of anionic nucleophiles with neutral electrophiles would suggest that these reactions should be somewhat slower than in molecular solvents. This was confirmed by investigations of the reactions of a range of anionic nucleophiles {Cl^-, Br^-, I^-, $[CH_3CO_2]^-$, $[CF_3CO_2]^-$, $[CN]^-$ and $[SCN]^-$; Scheme 10.2; Table 10.3} with methyl 4-nitrobenzenesulfonate in ionic liquids {[C_4mim][NTf_2], [C_4C_1pyrr][NTf_2] or [C_4C_1pyrr][OTf]} and molecular solvents (CH_2Cl_2, DMSO or MeOH) [10–13].

Although the generality of the success of the Hughes–Ingold rules was shown, more detailed analysis showed the importance of hydrogen-bonding effects. The Kamlet-Taft LSERs always revealed a negative coefficient for α, due to the solvent hydrogen bonding to the nucleophile, stabilising it with respect to the activated complex. This effect was stronger for the better hydrogen-bond accepting nucleophiles. Two of the nucleophiles, I^- and $[CH_3CO_2]^-$, also had best fit LSERs with β in the correlations, but with a much lower contribution that was attributed to antagonistic hydrogen-bonding effects. π^* appeared in the best-fit LSERs for three of the nucleophiles, $[CN]^-$, $[CH_3CO_2]^-$ and $[CF_3CO_2]^-$.

The reactivities of butylamine, dibutylamine and tributylamine with *S*, *S*-dimethyl-4-nitrophenylsulfonium salts {$[4\text{-}NO_2C_6H_4S(CH_3)_2]X$, Scheme 10.3} in [$C_4$mim]X and [$C_4C_1$pyrr]X {X=[$NTf_2$] or [OTf]} and a range of molecular solvents (toluene, dichloromethane, tetrahydrofuran, ethanenitrile, or methanol) lead to similar general conclusions [14]. The Kamlet-Taft LSERs (Table 10.4) showed very strong negative β effects, due to interactions between the solvents and acidic protons on $[4\text{-}NO_2C_6H_4S(CH_3)_2]^+$, which reduce the electrophilicity of the reacting carbon centre. The favourable interactions with the N-H protons

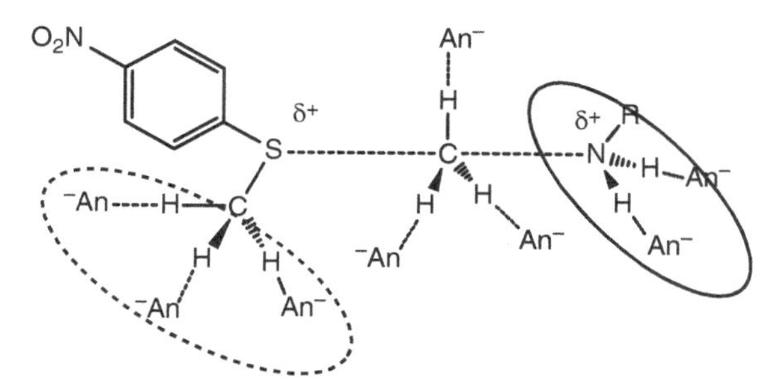

Figure 10.6 The activated complex for the reaction of $[4\text{-}NO_2C_6H_4S(CH_3)_2]^+$ with dibutylamine, showing potential interactions with anions that can accelerate (solid ellipse) and decelerate (broken ellipse) the reaction.

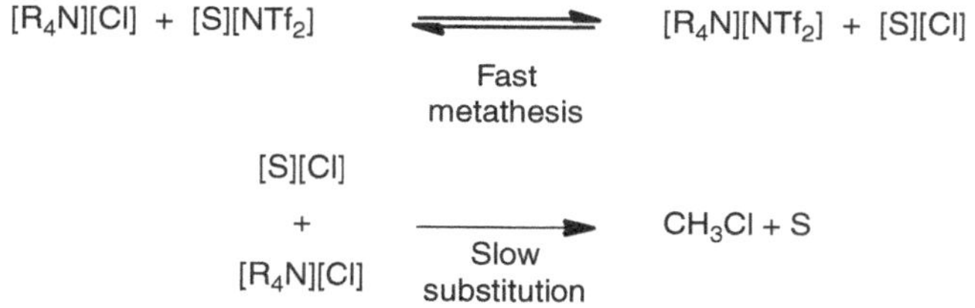

$[R_4N][Cl]$ + $[S][NTf_2]$ ⇌ (Fast metathesis) $[R_4N][NTf_2]$ + $[S][Cl]$

$[S][Cl]$ + $[R_4N][Cl]$ → (Slow substitution) CH_3Cl + S

Scheme 10.5 The reaction of chloride (anionic nucleophile) with $[4\text{-}NO_2C_6H_4S(CH_3)_2]^+$ in a molecular solvent, where $[S]^+$ represents $[4\text{-}NO_2C_6H_4S(CH_3)_2]^+$ and S represents $4\text{-}NO_2C_6H_4S(CH_3)$.

of $BuNH_2$ and Bu_2NH that were seen in the reactions of these with the neutral electrophile (see, e.g. Fig. 10.5) still take place. These are shown by a reduction in the negative β effect for Bu_2NH and even more so for $BuNH_2$. An α effect was observed for Bu_3N, with a small effect for $BuNH_2$ and a slightly greater effect for Bu_2NH. A positive dependence on π^* was observed for all of the reactions. These were again rationalised by consideration of the different strengths of solute-solvent interactions for the starting materials and activated complex of the reaction (Fig. 10.6).

As expected by the application of the Hughes–Ingold rules, the reactions of Cl^- with dimethyl-4-nitrophenylsulfonium salts $\{[4\text{-}NO_2C_6H_4S(CH_3)_2]X$, Scheme 10.4$\}$ in ionic liquids were considerably slower than the same reactions in most molecular solvents. However, these reactions also followed a different reaction mechanism in the ionic liquids to those seen in molecular solvents [15]. In molecular solvents, the reactions follow a mechanism that is initiated by a fast ion metathesis to form the reactive ion pairs, which go on to give the products *via* a nucleophilic substitution reaction (Scheme 10.5).

When the reaction was performed under attempted *pseudo*-first-order conditions, with a large excess of the nucleophile, two different kinetic behaviours were observed in molecular solvents. In the first, when the concentration of the nucleophile was increased, the observed rate constant of the reaction increased, but not in the linear way that would be expected for a simple bimolecular reaction. This is because the relationship between the concentration of the actual reacting species with the $[R_4N]Cl$ is non-linear. Hence, *pseudo*-first-order conditions are not met. In the second reaction type, the observed rate constant decreases as the concentration of the chloride salt is increased. This is because the product of the metathesis precipitates from the solution, reducing the concentration of the electrophile, thereby reducing the rate of the reaction.

In ionic liquids, the kinetics of these reactions followed the linear behaviour expected for a second-order reaction being performed under *pseudo*-first-order conditions. The authors used this as evidence to suggest that ion pairs were not formed in these reactions, and that the reaction mechanism was a simple S_N2 reaction of free solvated ions.

Unimolecular substitution reactions in ionic liquids have been far less studied than bimolecular substitutions. The hydrolysis reactions of 1-adamantyl mesylate were found to have rate constants in $[C_4mim][NTf_2]$ between those found for protic molecular solvents (CF_3CH_2OH, CH_3OH or CD_3CO_2D) and DMSO-d_6 [16]. The authors concluded that carbocation intermediates did form in these reactions, but that the rates of ionisation were not extraordinary, and that they depended upon the nature of the leaving group.

Arenediazonium salts, $[ArN_2]X$, are highly reactive intermediates that often undergo substitution *via* a S_N1 pathway. They are so reactive that they have even been shown to react with the $[NTf_2]^-$ anion [17, 18]. An unusual observation was made when $[PhN_2][BF_4]$ was dissolved in $[C_4mim]Br/[C_4mim][NTf_2]$ (1:2, 1:1 or 3:1) and only products of the reaction with $[NTf_2]^-$ were observed. While it is not uncommon that, in the presence of a suitably reactive reagent, a usually inert material can be made to react, one would still expect the relative reactivities of the nucleophiles to remain unchanged. The same reaction conducted in water with a 1:1 $[C_4mim]Br/[C_4mim][NTf_2]$ mixture yielded the expected products, bromobenzene and phenol. The authors suggested that the strong interaction of the $[C_4mim]^+$ cation of the ionic liquid with the Br^- reduced its nucleophilicity to such an extent that the strongest nucleophile remaining in the system was the $[NTf_2]^-$ ion. While this, in principle, makes sense, the generality of this result is yet to be demonstrated.

10.3 COMPETING MECHANISMS

With a general framework set for understanding how ionic liquids can affect the rates of different reactions in place, it becomes possible to make predictions of the effects of ionic liquids upon the rates of reactions that have not yet been performed in ionic liquids. To do this, it is necessary to know how charge is generated, destroyed or redistributed during the reaction. However, in many cases, more than one mechanism is energetically available to the reactants. In these cases, the mixtures of products formed depend upon the relative rates of the different reactions (kinetic control). If changing the solvent sufficiently affects the activation energies of the different reactions differently, it can lead to a change of the preferred mechanism, and thereby to reaction products. The effect upon $\Delta G^{\ddagger}$ does not have to be large for this to occur, with it requiring a change of *ca.* 12 kJ mol^{-1} to change the ratio of products from 90:10 to 10:90. There are a number of reaction types for which such mechanism changes are common.

10.3.1 Unimolecular/Bimolecular Substitutions

Unimolecular substitution reactions begin with the formation of a carbocation reaction intermediate, which then goes on to form the product in a second step. The rate determining step is the formation of the carbocation, and the activated complex for this transition has its charge distributed over just two atoms ($^{\delta+}$R···X$^{\delta-}$ for a neutral starting material, $^{\delta+}$R···X$^{\delta+}$, for a cationic starting material). Bimolecular substitutions occur in one step and have activated complexes that have their charge distributed over three atoms (e.g. $^{\delta-}$Y···R···X$^{\delta-}$ when both starting materials are neutral, see Table 10.1 for other examples). Hence, as polar solvents, the Hughes–Ingold rules tell us to expect that the transition state for the S_N1 reaction will be preferentially stabilised in comparison to that of the S_N2 reaction in ionic liquids (see Fig. 10.7). If the two activation energies are similar enough to give a mixture of products, this can be sufficient to change the selectivity of the reaction.

When the starting material is a neutral solute, it might be expected that ionic liquids will accelerate this process in comparison to molecular solvents. The hydrolysis reactions of a number of substrates in [C$_4$mim][NTf$_2$] have been shown to proceed *via* carbocations [16]. The rate constants for the ionic liquid reactions were greater than the same reactions in protic molecular solvents (CF_3CH_2OH, CH_3OH or CD_3CO_2D), but slower than those in DMSO-d_6. As with the S_N2 reac-

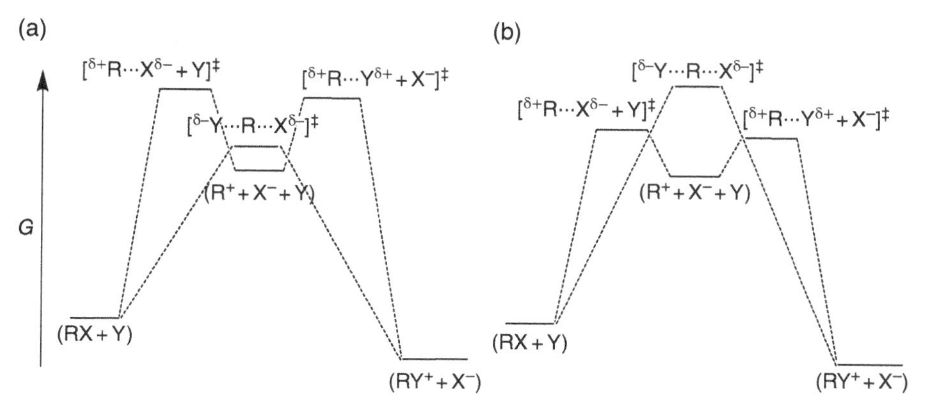

Figure 10.7 The reaction profiles of competing S_N1 and S_N2 reactions of a neutral nucleophile and neutral electrophile with (a) S_N2 favoured, and (b) S_N1 favoured.

tions above, this shows that the ionic liquids should be considered to have similar polarities to polar molecular solvents, but not as especially polar. The ionic liquids are likely only to favour unimolecular over bimolecular processes to the extent of other polar solvents and depending upon their donor/acceptor properties, rather than because of their ionic character.

10.3.2 Substitution/Elimination

The S_N2 and E2 reactions are mechanistically closely related and have similar activated complexes with delocalised charges. However, in the E2 activated complex, the charge is delocalised over more atoms than in that of the S_N2 reaction (see Fig. 10.8). The Hughes–Ingold rules would, therefore, lead us to expect a preference for S_N2 over E2 reactions. Is there any evidence for this?

β-Phenylethyl bromides, mesylates, and triflates are known to be susceptible to elimination reactions to form styrene. However, the nucleophilic substitutions of these with a range of anionic nucleophiles ($[CN]^-$, $[N_3]^-$, or $[NO_2]^-$) or $PhCH_2NH_2$ in $[P_{6\,6\,6\,14}][NTf_2]$ showed no elimination products in most of the reactions [19]. The one exception to this was the reaction of 2-phenyl-1-bromoethane with $NaNO_2$ at temperatures exceeding 85°C, but this could be prevented at lower temperatures. This suggests that at least this ionic liquid will favour S_N2 over E2 reactions.

In spite of the evidence presented above that ionic liquids favour nucleophilic substitution over base-induced elimination, it is possible to heavily favour elimination by selecting an appropriate substrate. 1,1,1-tribromo-2,2-bis(dimethoxyphenyl) ethanes have just such a strong preference for elimination, and hence their reactions with a range of amines in $[C_4mim][BF_4]$, $[C_4mim][PF_6]$ and $[C_4dmim][BF_4]$ have been investigated in

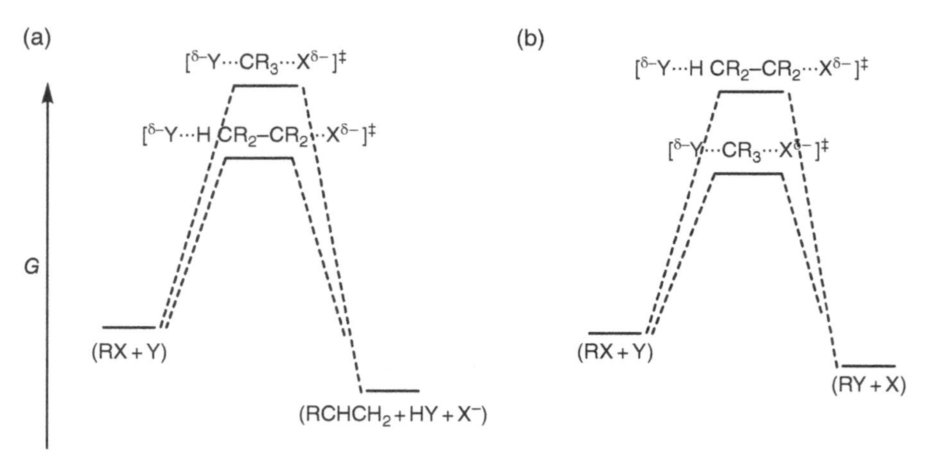

Figure 10.8 The reaction profiles of competing E2 and S_N2 reactions of a neutral nucleophile and neutral electrophile with (a) E2 favoured, and (b) S_N2 favoured.

some detail [20]. Remarkably, no elimination was observed for any of the primary or tertiary amines that were investigated, and only secondary cyclic amines were found to promote the reaction. This is in spite of the authors selecting primary amines with similar aqueous basicities (pK_{BH+}) as the secondary amine employed. Consequently, the importance of amine geometry, flexibility and steric hindrance in determining the outcomes of these reactions were emphasised. This, together with additional kinetics data, led the authors to propose a shift in mechanism from E1cB in methanol to E2 in the ionic liquids, that is, from more charge separation during the reaction to less. As above, this suggests that the ionic liquids are not especially ionising solvents under these conditions.

10.3.3 Diels-Alder Reactions

The Diels-Alder cycloaddition reaction is a concerted reaction, with no change in the charge density upon going from reactants to the activated complex and finally to the products. Therefore, from the Hughes–Ingold approach, little solvent effect is expected upon these reactions. In spite of this, dramatic solvent effects are often observed when a non-symmetric dienophile is used, and a mixture of products is formed. For instance, the reaction of methyl acrylate with cyclopentadiene gives a mixture of two products (see Scheme 10.6).

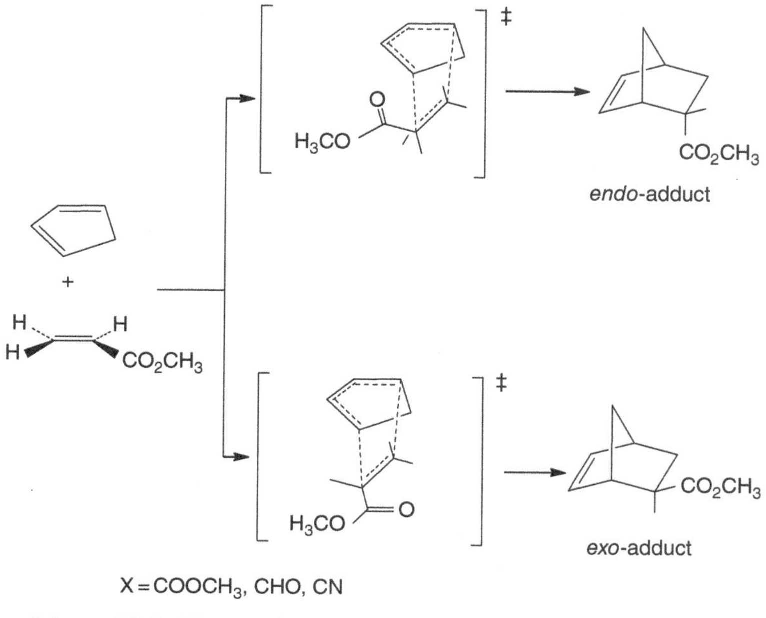

Scheme 10.6 The reaction of methyl acrylate with cyclopentadiene.

As the product ratio is under kinetic control, it is determined by the difference in the rates of the two competing reactions to give *endo* and *exo* products. The solvent effects upon this ratio are so strong that it has even been used to define a solvent polarity scale, Ω, based on the log of the ratio of the *endo* to the *exo* product, Equation 10.3 [21]:

$$\log_{10}\left(\frac{[endo]}{[exo]}\right) = \Omega \tag{10.3}$$

This reaction is also known to be greatly accelerated when using water, a solvent in which the starting materials and products are barely soluble, and the selectivities are greatly enhanced for the reaction. There is also another huge increase in rate when LiCl is added to the solution. Explanations for these observations have included the alignment of the diene and dienophile dipoles, either in parallel or antiparallel, giving greater or lower polarity-activated complexes which can be differently stabilised by solvents [20] or hydrophobic interactions [22–24].

Dyson et al. [25] conducted a very thorough study of the selectivities and rates of the reactions of cyclopentadiene with methyl acrylate in 30 different $[NTf_2]^-$ ionic liquids. They showed that the selectivities were indeed kinetically controlled. Interestingly, they demonstrated that the rate of formation of the *exo* adduct was not affected by the choice of solvent, but that the rate of formation of the *endo* adducts was sensitive to the solvent. These different solvent dependencies for the rates of formation of the two isomers led to the changing selectivities with changing the solvent. They concluded that interactions between the cation of the ionic liquid, including (but not only) hydrogen bonding, and the carbonyl oxygen of the dienophile in the activated complex stabilised this transition state with respect to that of the *exo* adduct.

Chiappe et al. [26] used fewer ionic liquids, but extended the range of dienophiles to include acrolein, methyl acrylate and acrylonitrile and used the Kamlet-Taft LSER approach. As with the examples above, the selectivities were found to be under kinetic control. Both the selectivities and rates of these reactions were found to depend upon the solvent - hydrogen bond donation ability for the reactions of acrolein and methyl acrylate, which both have strong hydrogen bond acceptor sites, but not of acrylonitrile, which does not. A theoretical study [27] of the two limiting cases with the highest *endo*/*exo* selectivities, acrolein and methyl acrylate in $[HC_4im][NTf_2]$ {$[HC_4im]^+$=1-butylimidazolium}, compared to the lowest selectivity case of acrylonitrile in $[C_4dmim][NTf_2]$ {$[C_4dmim]^+$ = 1-butyl-2,3-dimethylimidazolium} again found that hydrogen bonding from the cation to the dienophiles' hydrogen-bond acceptor sites were key to explaining the rates and selectivities of the reactions. In these studies, the hydrogen-bonding ability of the ionic liquids was derived from measurements on probe molecules. In neither case are these likely to give a 'pure' hydrogen-bonding value, and the two studies probably revealed the same results, but with slightly different emphasis.

10.3.4 Electron Transfer Reactions

Electron transfer reactions have been well studied in ionic liquids, with some very interesting results. They are such fast reactions that bimolecular electron transfers are almost always diffusion controlled. The diffusion-limited rate coefficient is related to the bulk viscosity of the pure solvent by Equation 10.4:

$$k_{\text{diff}} = \frac{8000RT}{3\eta} \tag{10.4}$$

Therefore, the comparatively high viscosities of ionic liquids should lead to these reactions being much slower than their equivalents in free-flowing molecular solvents [28].

In several cases, this has shown to be true. For example, the photoelectron transfer between the tris(2,2′-bipyridine)ruthenium(II) dication, $[Ru(bipy)_3]^{2+}$, and methyl viologen {MV^{2+}; 1,1′-dimethyl-4,4′-bipyridinium dication} has been investigated in a number of ionic liquids and is shown to occur at an approximately diffusion controlled rate [29, 30], as was the bimolecular decay of $[Br_2]^-$ and of $[I_2]^-$ radical anions in a series of $[NTf_2]^-$ ionic liquids [31]. Other electron transfer reactions, on the other hand, have had rates up to an order of magnitude greater than would be predicted from the ionic liquids' viscosities. These include reactions benzophenone with naphthalene [32], alkylpyridinyl radicals with duroquinone [33], 9,10-dicyanoanthracene with 20 different aromatic single electron donors [34], pyrene [35] or coumarin dyes with *N,N*-dimethylaniline [36].

Since the rates of these reactions are diffusion controlled, the answer to the question as to why these different types of reactions have such different rates cannot be that the electron transfer itself has become faster. This would not have any effect on the rate of the overall process. It must be that the diffusion of the starting materials is faster in these latter reactions. This is not as surprising as it might at first seem. The diffusion rates of *N,N,N′N′*-tetramethyl-*p*-phenylenediamine, TMPD, and its mono- and di-cations in the ionic liquids $[C_n mim][NTf_2]$ ($n = 4$ or 10), $[C_4pyr][NTf_2]$ or $[P_{6\,6\,6\,14}][NTf_2]$ were found to vary in the order TMPD > $[TMPD]^{\cdot+}$ > $[TMPD]^{2+}$ [37], with the relative values 1, 0.53 and 0.33; whereas in ethanenitrile the values were 1, 0.89 and 0.51.

Bringing the above facts together, it can be seen that salts dissolved in an ionic liquid interact directly with the charge-bearing part of ionic liquid's ions, and the frictional forces and range of motion experienced by these solute ions are of a similar nature to those experienced by the ions of the pure ionic liquid that give rise to its bulk viscosity. Consequently, bulk viscosities give a good estimate of the diffusion rates of solute ions in ionic liquids. However, molecular solutes do not have such strong interactions with the ions of the ionic liquids. Hence, the frictional forces experienced by some molecules moving through the ionic liquid will be very different to those that give rise to the bulk viscosities of the ionic liquids, particularly lacking the ion–ion interactions. Consequently, bulk viscosity measurements do not give a guide to the diffusion of these solute molecules in ionic liquids, which can be considerably faster. These increases in the rates of these reactions, therefore, do not arise out of any change in the mechanism of the reaction. However, there appears to be one exception to this. Pyridinium cations are capable of accepting an electron, which they can then transfer to an acceptor. This gives rise to a solvent-mediated mechanism for electron transfer reactions in pyridinium ionic liquids [38].

10.3.5 Effects of Specific Ionic Liquids

The examples above detail how ionic liquids in general might affect the mechanism of a reaction. They demonstrate that the ionic liquids have effects that are equivalent to those of polar molecular solvents, with only one example of qualitatively different behaviour [15]. Ionic liquids may also affect the mechanisms of the reactions of solutes *via* the particular ions from which they are composed.

HNO_3 has been used as a nitrating agent for electrophilic aromatic substitutions [39]. In ionic liquids with triflate or hydrogensulfate anions, the expected mononitrated arenes were produced. However, when halide ionic liquids were used, monohalogenated arenes were produced (Scheme 10.7). Since halide ions are not themselves electrophilic, a halogenating agent must first be produced. HNO_3 can act as an oxidising agent, so the most likely mechanism for the reaction is that it reacts with the halide to give the hypohalous acid halogenating agent. Clearly, this mechanism change can only occur in halide ionic liquids. It was also demonstrated that the crucial factor was the combination of ions in the solution, not the order of their addition nor the form in which they were added, so the same result was seen when HCl was the acid used in $[C_4mim][NO_3]$. In methanesulfonate ionic liquids yet another reaction was seen. Here, the nitric acid acted as an oxidising agent for toluene to yield benzoic acid. Again, using methanesulfonic acid dissolved in $[C_4mim][NO_3]$ yielded the same result.

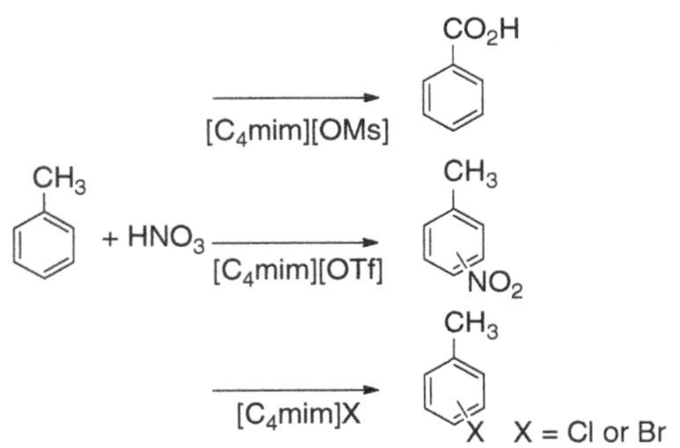

CO_2H
$[C_4mim][OMs]$
CH_3 CH_3
$+ HNO_3$ $[C_4mim][OTf]$ NO_2
CH_3
$[C_4mim]X$ X X = Cl or Br

Scheme 10.7 The reactions of toluene with nitric acid in three different ionic liquids [39].

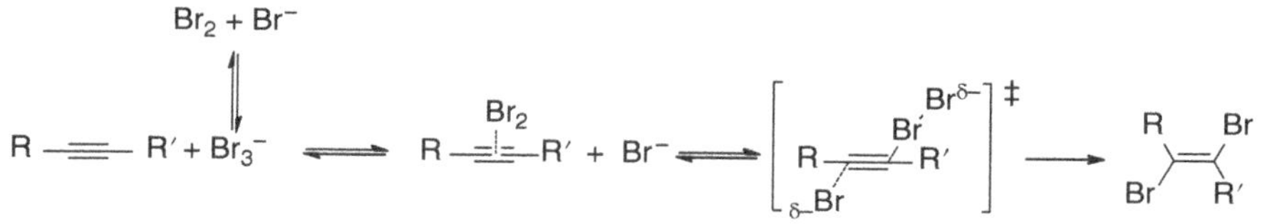

Scheme 10.8 A proposed mechanism for the bromination of alkynes in ionic liquids.

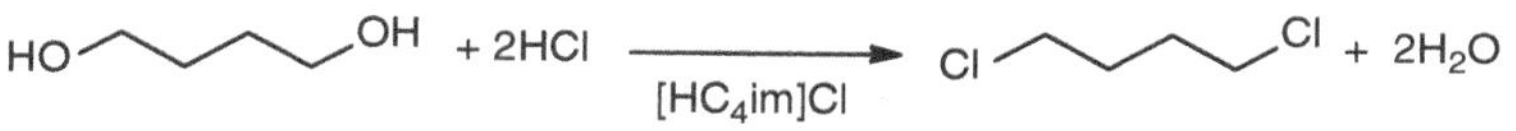

Scheme 10.9 Use of HCl for the conversion of alcohols to alkyl halides in an ionic liquid [43].

Electrophilic addition across a double or triple carbon–carbon bond is the characteristic reaction of alkenes and alkynes. Chiappe et al. [40, 41] studied the bromination of alkenes in $[C_4mim]X$ {X = Br, $[PF_6]$ or $[BF_4]$} with Br_2. In $[C_4mim][PF_6]$ and $[C_4mim][BF_4]$, the reactions gave the expected mixtures of *syn* and *anti* products of the electrophile Br_2. However, the reaction in $[C_4mim]Br$ showed a completely different mechanism, with $[Br_3]^-$ acting as the electrophile and producing *anti*-stereospecific brominations. When Cl_2 was used as a chlorinating agent in $[C_4mim]Br$, 1,2-bromochloroalkenes were produced, which was again taken as an indication of the presence of a trihalide electrophile, $[BrCl_2]^-$ in this case. The reaction intermediate was identified as being similar to that previously proposed to explain the kinetics of the bromination of alkynes in ionic liquids [42], in which early attachment by a halide nucleophile occurs on a 1:1 unsaturated alkene-halogen, or alkyne-halogen, π-complex (Scheme 10.8). This time, the change in reaction mechanism occurs because in the halide ionic liquid, no molecular halogen molecules are present, so the conventional mechanism is not possible.

Another example where the use of a halide ionic liquid gives a different starting material, and therefore leads to a different mechanism for the reaction, is in BASF's so-called 'nucleophilic HCl' conversion of alcohols to alkyl halides (Scheme 10.9) [43]. Under the right conditions, HCl can act as a chlorination agent for alcohols. However, the usefulness of this is normally hindered by the formation of ether by-products. This is usually circumvented by using chlorinating agents that can also act as a dehydrating agent, such as $COCl_2$, $SOCl_2$, PCl_3, and so on. BASF found that when used with a chloride ionic liquid, HCl itself could be used without the formation of by-products, probably due to the formation of the $[HCl_2]^-$ ion in the ionic liquid [44], which is the actual chlorinating agent.

In the presence of even moderate bases, imidazolium cations can be deprotonated to form *N*-heterocyclic carbenes [45]. These are reactive bases [46],

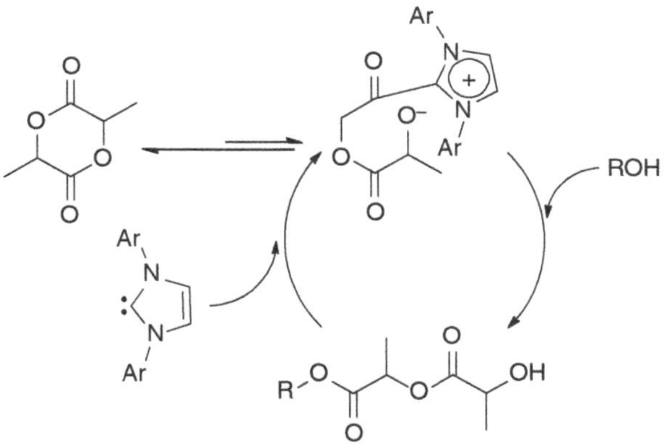

Scheme 10.10 A carbene as an active catalyst in ring-opening polymerisation [52, 53].

and can act as catalysts [47–51]. These are usually discussed in the context of facing challenges. For example, the disparity between the high benzaldehyde conversions (as determined by GC) and low product yields in the base-catalysed Baylis-Hillman reaction of methyl acrylate and benzaldehyde, catalysed by 3-hydroxyquinuclidine or 1,4-diazabicyclo[2.2.2]octane (DABCO) in the presence of [C_4mim]Cl, was explained by a reaction of the benzaldehyde with [C_4mim]$^+$, *via* the generated carbene [52]. They can be avoided by substituting the imidazolium ring at the 2-position, usually with a methyl group. However, the formation of carbenes in ionic liquids has been used to generate a catalyst in the biphasic ring-opening polymerisation of lactides and lactones [53, 54]. [C_2mim][BF_4] with tetrahydrofuran provided an excellent medium for the reaction (Scheme 10.10), which was initiated by the addition of a small amount of potassium *tert*-butoxide. The neutral carbene transfers to the tetrahydrofuran layer, where the reaction occurs. The reaction is terminated by addition of a protic ammonium salt, which regenerates the imidazolium cation by proton transfer.

10.4 CONCLUSIONS

It can be seen from the above that ionic liquids can have dramatic effects upon chemical reaction mechanisms. They can lead to changes in rates, changes in product selectivity and even changes in the reaction that occurs. They can do this through specific reactions with the ionic liquid's ions, but more generally do this *via* ionic liquid-solute interactions with the starting materials and activated complexes of the reaction. However, these effects have seldom fallen outside the range of effects that are known for molecular solvents, and can be accounted for using the same theories that are applied to these.

REFERENCES

1 Ingold, C.K., *Structure and Mechanism in Organic Chemistry*, 2nd Ed., Bell, London (1969).

2 Hughes, E.D. and Ingold, C.K., Mechanism of substitution at a saturated carbon atom. Part IV. A discussion of constitutional and solvent effects on the mechanism, kinetics, velocity, and orientation of substitution, *J. Chem. Soc.* 244–255 (1935).

3 Hughes, E.D., Mechanism and kinetics of substitution at a saturated carbon atom, *Trans. Faraday Soc.* **37**, 603–631 (1941).

4 Hughes, E.D. and Ingold, C.K., The mechanism and kinetics of elimination reactions, *Trans. Faraday Soc.* **37**, 657–685 (1941).

5 Cooper, K.A., Dhar, M.L., Hughes, E.D., Ingold, C.K., MacNulty, B.J. and Woolf, L.I., Mechanism of elimination reactions. Part VII. Solvent effects on rates and product-proportions in uni- and bi-molecular substitution and elimination reactions of alkyl halides and sulphonium salts in hydroxylic solvents, *J. Chem. Soc.* 2043–2049 (1948).

6 Reichardt, C. and Welton, T., *Solvents and Solvent Effects in Organic Chemistry*, 4th updated and enlarged Ed., Wiley-VCH, Weinheim (2011).

7 Crowhurst, L., Lancaster, N.L., Arlandis, J.M.P. and Welton, T., Manipulating solute nucleophilicity with room temperature ionic liquids, *J. Am. Chem. Soc.* **126**, 11549–11555 (2004).

8 Skrzypczak, A. and Neta, P., Rate constants for reaction of 1,2-dimethylimidazole with benzyl bromide in ionic liquids and organic solvents, *Int. J. Chem. Kinet.* **36**, 253–258 (2004).

9 Bini, R., Chiappe, C., Pomelli, C.S. and Parisi, B., Effect of ionic liquids on the Menschutkin reaction: An experimental and theoretical study, *J. Org. Chem.* **74**, 8522–8530 (2009).

10 Lancaster, N.L., Welton, T. and Young, G.B., A study of halide nucleophilicity in ionic liquids, *J. Chem. Soc. Perk. T.* **2**, 2267–2270 (2001).

11 Lancaster, N.L., Salter, P.A., Welton, T. and Young, G.B., Nucleophilicity in ionic liquids. 2. Cation effects on halide nucleophilicity in a series of bis(trifluoromethyl sulfonyl)imide ionic liquids, *J. Org. Chem.* **67**, 8855–8861 (2002).

12 Lancaster, N.L. and Welton, T., Nucleophilicity in ionic liquids. 3.1 Anion effects on halide nucleophilicity in a series of 1-butyl-3-methylimidazolium ionic liquids, *J. Org. Chem.* **69**, 5986–5992 (2004).

13 Crowhurst, L., Falcone, R., Lancaster, N.L., Llopis-Mestre, V. and Welton, T., Using Kamlet-Taft solvent descriptors to explain the reactivity of anionic nucleophiles in ionic liquids, *J. Org. Chem.* **71**, 8847–8853 (2006).

14 Ranieri, G., Hallett, J.P. and Welton, T., Nucleophilic reactions at cationic centers in ionic liquids and molecular solvents, *Ind. Eng. Chem. Res.* **47**, 638–644 (2008).

15 Hallett, J.P., Liotta, C.L., Ranieri, G. and Welton, T., Charge screening in the S_N2 reaction of charged electrophiles and charged nucleophiles: An ionic liquid effect, *J. Org. Chem.* **74**, 1864–1868 (2009).

16 Creary, X., Willis, E.D. and Gagnon, M., Carbocation-forming reactions in ionic liquids, *J. Am. Chem. Soc.* **127**, 18114–18120 (2005).

17 Bini, R., Chiappe, C., Marmugi, E. and Pieraccini, D., The 'non-nucleophilic" anion $[Tf_2N]^-$ competes with the nucleophilic Br^-: An unexpected trapping in the dediazoniation reaction in ionic liquids, *Chem. Commun.*, 897–899 (2006).

18 Laali, K.K., Okazaki, T. and Bunge, S.D., *N*-(trifluoromethylsulfonyl)aryloxy trifluoromethylsulfoximines [ArO-SO(CF_3)=NTf] and *N*-aryltriflimides $ArN(Tf)_2$ by thermal and photolytic dediazoniation of [ArN_2][BF_4] in [BMIM][Tf_2N] ionic liquid: Exploiting the ambident nucleophilic character of a 'Nonnucleophilic' anion, *J. Org. Chem.* **72**, 6758–6762 (2007).

19 McNulty, J., Nair, J.J., Cheekoori, S., Larichev, V., Capretta, A. and Robertson, A.J., Scope and mechanistic insights into the use of tetradecyl(trihexyl)phosphonium bistriflimide: A remarkably selective ionic liquid solvent for substitution reactions, *Chem. Eur. J.* **12**, 9314–9322 (2006).

20 D'Anna, F., Frenna, V., Pace, V. and Noto, R., Effect of ionic liquid organizing ability and amine structure on the rate and mechanism of base induced elimination of 1,1,1-tribromo-2, 2-bis(phenyl-substituted)ethanes, *Tetrahedron* **62**, 1690–1698 (2006).

21 Berson, J.A., Hamlet, Z. and Mueller, W.A., The correlation of solvent effects on the stereoselectivities of Diels-Alder reactions by means of linear free energy relationships. A new empirical measure of solvent polarity, *J. Am. Chem. Soc.* **84**, 297–304 (1962).

22 Rideout, D.C. and Breslow, R., Hydrophobic acceleration of Diels-Alder reactions, *J. Am. Chem. Soc.* **102**, 7816–7817 (1980).

23 Breslow, R., Hydrophobic effects on simple organic reactions in water, *Acc. Chem. Res.* **24**, 159–164 (1991).

24 Grieco, P.A., Organic chemistry in unconventional solvents, *Aldrichim. Acta* **24**, 59–66 (1991).

25 Vidis, A., Ohlin, C.A., Laurenczy, G., Kusters, E., Sedelmeier, G. and Dyson, P.J., Rationalisation of solvent effects in the Diels-Alder reaction between cyclopentadiene and methyl acrylate in room temperature ionic liquids, *Adv. Synth. Catal.* **347**, 266–274 (2005).

26 Bini, R., Chiappe, C., Mestre, V.L., Pomelli, C.S. and Welton, T., A rationalization of the solvent effect on the Diels-Alder reaction in ionic liquids using multiparameter linear solvation energy relationships, *Org. Biomol. Chem.* **6**, 2522–2529 (2008).

27 Bini, R., Chiappe, C., Mestre, V.L., Pomelli, C.S. and Welton, T., A theoretical study of the solvent effect on Diels-Alder reaction in room temperature ionic liquids using a supermolecular approach, *Theor. Chem. Acc.* **123**, 347–352 (2009).

28 Grodkowski, J. and Neta, P., Reaction kinetics in the ionic liquid methyltributyl ammonium bis(trifluoromethylsulfonyl)imide. Pulse radiolysis study of $\bullet CF_3$ radical reactions, *J. Phys. Chem. A* **106**, 5468–5473 (2002).

29 Gordon, C.M. and McLean, A.J., Photoelectron transfer from excited-state ruthenium(II) tris(bipyridyl) to methylviologen in an ionic liquid, *Chem. Commun.*, 1395–1396 (2000).

30 Álvaro, M., Carbonell, E., Ferrer, B., Garcia, H. and Herance, J.R., Ionic liquids as a novel medium for photochemical reactions. $Ru(bpy)_3^{2+}$/ viologen in imidazolium ionic liquid as a photocatalytic system mimicking the oxido-reductase enzyme, *Photochem. Photobiol.* **82**, 185–190 (2006).

31 Takahashi, K., Sakai, S., Tezuka, H., Hiejima, Y., Katsumura, Y. and Watanabe, M., Reaction between diiodide anion radicals in ionic liquids, *J. Phys. Chem. B* **111**, 4807–4811 (2007).

32 McLean, A.J., Muldoon, M.J., Gordon, C.M. and Dunkin, I.R., Bimolecular rate constants for diffusion in ionic liquids, Chem. Commun., 1880–1881 (2002).

33 Skrzypczak, A. and Neta, P., Diffusion-controlled electron-transfer reactions in ionic liquids, *J. Phys. Chem. A* **107**, 7800–7803 (2003).

34 Vieira, R.C. and Falvey, D.E., Photoinduced electron-transfer reactions in two room-temperature ionic liquids: 1-butyl-3-methylimidazolium hexafluorophosphate and 1-octyl-3-methylimidazolium hexafluorophosphate, *J. Phys. Chem. B* **111**, 5023–5029 (2007).

35 Paul, A. and Samanta, A., Photoinduced electron transfer reaction in room temperature ionic liquids: A combined laser flash photolysis and fluorescence study, *J. Phys. Chem. B* **111**, 1957–1962 (2007).

36 Sarkar, S., Pramanik, R., Seth, D., Setua, P. and Sarkar, N., Photoinduced electron transfer (PET) from N,N-dimethylanilineto 7-amino Coumarin dyes in a room temperature ionic liquid (RTIL): Slowing down of electron transfer rate compared to conventional solvent, *Chem. Phys. Lett.* **477**, 102–108 (2009).

37 Evans, R.G., Klymenko, O.V., Price, P.D., Davies, S.G., Hardacre, C. and Compton, R.G., A comparative electrochemical study of diffusion in room temperature ionic liquid solvents versus acetonitrile, *Chem. Phys. Chem.* **6**, 526–533 (2005).

38 Behar, D., Neta, P. and Schultheisz, C., Reaction kinetics in ionic liquids as studied by pulse radiolysis: Redox reactions in the solvents methyltributylammonium bis(trifluoroniethylsulfonyl)imide and N-butylpyridinium tetrafluoroborate, *J. Phys. Chem. A* **106**, 3139–3147 (2002).

39 Earle, M.J., Katdare, S.P. and Seddon, K.R., Paradigm confirmed: The first use of ionic liquids to dramatically influence the outcome of chemical reactions, *Org. Lett.* **6**, 707–710 (2004).

40 Chiappe, C., Capraro, D., Conte, V. and Pieraccini, D., Stereoselective halogenations of alkenes and alkynes in ionic liquids, *Org. Lett.* **3**, 1061–1063 (2001).

41 Bortolini, O., Bottai, M., Chiappe, C., Conte, V. and Pieraccini, D., Trihalide-based ionic liquids. Reagent-solvents for stereoselective iodination of alkenes and alkynes, *Green Chem.* **4**, 621–627 (2002).

42 Chiappe, C., Conte, V. and Pieraccini, D., Bromination of alkynes in ionic liquids – A kinetic investigation, *Eur. J. Org. Chem.* 2831–2837 (2002).

43 Stegmann, V. and Massonne, K., *Method for producing haloalkanes from alcohols*, World Patent, WO 2005 026089 (2005).

44 Trulove, P.C. and Osteryoung, R.A., Proton speciation in ambient-temperature chloroaluminate ionic liquids, *Inorg. Chem.* **31**, 3980–3985 (1992).

45 Carmichael, A.J., Deetlefs, M., Earle, M.J., Fröhlich, U. and Seddon, K.R., 'Ionic liquids: Improved syntheses and new products', in *Ionic Liquids as Green Solvents: Progress and Prospects*, eds. R.D. Rogers and K.R. Seddon, *ACS Symp. Ser.*, Vol. **856**, American Chemical Society, Washington, DC (2003), pp. 14–31; Earle, M.J. and Seddon, K.R., *Imidazole Carbenes*, WO Patent, WO 0177081 (2001).

46 Nair, V., Bindu, S. and Sreekumar, V., N-heterocyclic carbenes: Reagents, not just ligands!, *Angew. Chem. Int. Edit. Engl.* **43**, 5130–5135 (2004).

47 Enders, D., Niemeier, O. and Henseler, A., Organocatalysis by N-heterocyclic, carbenes, *Chem. Rev.* **107**, 5606–5655 (2007).

48 Connor, E.F., Nyce, G.W., Myers, M., Mock, A. and Hedrick, J.L., First example of N-heterocyclic carbenes as catalysts for living polymerization: Organocatalytic ring-opening polymerization of cyclic esters, *J. Am. Chem. Soc.* **124**, 914–915 (2002).

49 Nyce, G.W., Lamboy, J.A., Connor, E.F., Waymouth, R.M. and Hedrick, J.L., Expanding the catalytic activity of nucleophilic N-heterocyclic carbenes for transesterification reactions, *Org. Lett.* **4**, 3587–3590 (2002).

50 Grasa, G.A., Kissling, R.M. and Nolan, S.P., *N*-heterocyclic carbenes as versatile nucleophilic catalysts for transesterification/acylation reactions, *Org. Lett.* **4**, 3583–3586 (2002).

51 Grasa, G.A., Guveli, T., Singh, R. and Nolan, S.P., Efficient transesterification/acylation reactions mediated by N-heterocyclic carbene catalysts, *J. Org. Chem.* **68**, 2812–2819 (2003).

52 Aggarwal, V.K., Emme, I. and Mereu, A., Unexpected side reactions of imidazolium-based ionic liquids in the base-catalysed Baylis-Hillman reaction, *Chem. Commun.*, 1612–1613 (2002).

53 Nyce, G.W., Glauser, T., Connor, E.F., Möck, A., Waymouth, R.M. and Hedrick, J.L., In situ generation of carbenes: A general and versatile platform for organocatalytic living polymerization, *J. Am. Chem. Soc.* **125**, 3046–3056 (2003).

54 Dove, A.P., Pratt, R.C., Lohmeijer, B.G.G., Culkin, D.A., Hagberg, E.C., Nyce, G.W., Waymouth, R.M. and Hedrick, J.L., *N*-Heterocyclic carbenes: Effective organic catalysts for living polymerization, *Polymer* **47**, 4018–4025 (2006).

11 Crystallography of Ionic Liquids

NEIL WINTERTON
Department of Chemistry, University of Liverpool, Liverpool, UK

11.1 INTRODUCTION

Historically, an ionic liquid has been seen as any salt in the molten or liquid state and presumed to be homogeneous and isotropic (although a moment's thought immediately raises questions about what characterises homogeneity in the context of a substance comprising two distinct and oppositely charged entities). Studies of molten salts, until relatively recently, have largely been confined to inorganic materials [1, 2]. Indeed, it was possible for Ubbelohde [1], as recently as 1978, to write authoritatively that 'a [...] subdivision of organic salts in which both anion and cation are organic does not appear to have been investigated at all extensively' and 'up to the present comparatively little detail has been published concerning salts with organic cations'. At the time, most organic salts appeared to be unstable when melted, with much discussion as to whether this was an intrinsic property of organic salts or arose because of the presence of impurities. Replacement of the anion in tetraalkylammonium halides by some now widely used in ionic liquids, such as $[BF_4]^-$ and $[PF_6]^-$, did not necessarily lead to low-melting materials. $[N_{4444}]Br$ melts at 119.5°C, $[N_{4444}][BF_4]$ at 162°C and $[N_{4444}][PF_6]$ at 247°C, according to one 1966 study [3], which also reported $[N_{6666}][BF_4]$ to melt at 91°C, a true ionic liquid. In the 1970s, J. E. Gordon and co-workers [4] carried out an extensive study of organic molten salts, investigating the relationship between the characteristics of the component ions and the properties of the salt. For instance [4], they described a series of straight-chain quaternary ammonium salts (some of which melted below room temperature) with a total carbon number of 20. They showed that melting point could be reduced by lowering cation symmetry. For instance, while $[N_{5555}][ClO_4]$ melted at 118°C, the isomeric $[N_{3368}][ClO_4]$ was a liquid at room temperature. $[N_{33311}][ClO_4]$, however, melted at 65.5°C and other isomers showed more complex behaviour,

Ionic Liquids Completely UnCOILed: Critical Expert Overviews, First Edition.
Edited by Natalia V. Plechkova and Kenneth R. Seddon.

including polymorphism, mesophase and so-called 'pre-freezing' behaviour. Papers cited by Gordon and Subbarao [4] provide a rich source of early data on low-melting salts. Despite such significant early work, the rapid growth in interest in low-melting ionic liquids has been a relatively recent phenomenon.

Not surprisingly, most published work on ionic liquids has sought to characterise and understand the properties and behaviour of these materials in the liquid state, including the nature of ion–ion interactions and the degree of short-range order in the molten state, so-called microheterogeneity [5, 6]. The relationship between the structure of a crystalline solid and of the liquid formed on melting close to the melting point is an aspect that was the focus of much early study [7]. Understanding of ionic liquids can, as has been noted by Laus et al. [8], be enhanced by studies of the solid state, particularly the crystalline state, the subject of this chapter, and of the thermodynamics and dynamics of the processes of fusion and melting. The phase behaviour of ionic liquids has become a major topic of interest in its own right, extending the range of materials exhibiting fragility [9], the glassy state [10] and so-called 'plastic'[11, 12] phases. Studies of ionic liquids, therefore, have provided additional insights into a range of related phenomena.

Such phenomena are of both fundamental and technological relevance, as crystallisation is one of a very small number of techniques generally available as a practical means for the purification of chemical compounds (particularly involatile salts), important in both characterisation as materials for research and in their industrial use, recovery and recycle. While a classic test of purity of an organic molecular compound is the comparison of its melting point with that of an authentic sample, the use of such comparisons for ionic liquids must be approached with a degree of caution. This is so both because of the difficulties in obtaining materials of established purity and because of the nature of the phase behaviour of ionic liquids. As is now well known, published melting point data for a single ionic liquid can vary widely. For instance, SciFinder lists no less than 12 values for the melting point of 1-ethyl-3-methylimidazolium bromide that range from 50–51°C to 83°C. Such variation may arise not simply because of questions of relative purity but also because of the circumstances under which the measurement is made, the history of the sample, as well as the intrinsic characteristics of the particular material under study. Some ionic liquids melt over a wide range of temperatures and cannot be said to have a melting 'point'; others have no observable melting point at all.

Reliability of melting points takes on particular significance since ionic liquids have, by a convention [13] now widely accepted, come to encompass a narrower group of molten salts than historically envisaged, that is, those that are liquid below 100°C, with early reviews [14–17] listing the small number of materials that fell into this category. While the convenience of this definition is evident when considering the practical application of these intensely interesting materials (such as, as electrolytes, energetic materials or pharmaceuticals [18]), it is less than useful when seeking to understand their structure and properties that, clearly, are not and cannot be bound by such arbitrary distinctions. Furthermore, the precise definition, if strictly applied, may inhibit consideration of an extensive body of relevant published work, old and new, on compounds that fall on the wrong side of this artificial boundary (or that have simply been ignored by those studying ionic liquids). Perhaps more of concern is that attempts to test theory and correlate computationally derived estimates with physical observation may be compromised by data of low accuracy or precision.

For an ionic material to be included in this review, its melting temperature (generally abbreviated to m.pt., however, measured) must be known (or, at least, explicit reference made to its liquid state at ambient temperature) and be below 100°C. Unfortunately, a significant number of crystallographic studies initially surveyed for inclusion in this review either fail to report the melting points of the salts being investigated or cite an original synthesis that itself fails to provide this information. Too few studies examine phase behaviour using techniques such as differential scanning calorimetry (DSC), from which it is possible to extract more reliable and reproducible melting and crystallisation temperatures.

Bearing in mind the arbitrary nature of the definition, little is to be gained by excluding a salt with a melting point (m.pt.) of 101°C (or even 110°C). Indeed, there may well be some trend or phenomenon, relevant to our understanding of what may lead to low-melting temperatures among ionic materials, which is better identified from studies of materials that fall outside the classification. For all these reasons, while the prime focus of this survey will be 'ionic liquids' in the current conventional sense, we will not feel bound to limit our review to such materials (though to make our survey manageable, we will limit inclusion, generally, to materials melting up to about 110°C).

Studies of the phase behaviour of ionic liquids reveal that many of them do not display crystallisation behaviour at all, rather forming glasses on cooling from a more fluid state. The availability of a wide variety of such ionic liquids has allowed their characteristics to be seen in the context of other glassy materials [9, 10]. Only when we fully understand, for instance, the nature, relative importance and net consequence of the interionic forces (instantaneously and dynamically) and their manifestation on the macroscopic level that lead to one ionic liquid ([C_4mim][BF_4], 1-butyl-3-methylimidazolium tetrafluoroborate) to be (so far, see Ref. [19]) non-crystallisable and another, closely related, (e.g. [C_4mim][CF_3SO_3], 1-butyl-3-methylimidazolium trifluoromethanesulfonate) to be readily crystallised, can we be said to understand ionic liquids. This is a general problem, as a recent report [20] exemplifies that very similar compounds, such as α-pentaacetylglucose and β-pentaacetylglucose, can display quite different crystallisation mechanisms.

The purpose of this chapter (written from a chemical, as opposed to a specifically crystallographic, perspective) is, primarily, to bring together published material on the crystallography of ionic liquids from single-crystal X-ray diffraction studies, on the basis of the conventional definition. An initial assessment is made of the material assembled, though this will be cursory. Mudring has provided an excellent review [21] of ionic liquid solidification phenomena and techniques for crystallisation. A proper analysis of the information is a much longer-term activity and beyond the scope of the present chapter.

The collection includes salts that, coincidentally, melt below 100°C and may not have been considered as ionic liquids by those who studied them. Many of these may well be of limited interest in ionic liquid applications. However, relevant insights may be gained from a consideration of their crystal and molecular structures. Some such salts may also provide pointers to wholly new classes of ionic liquids beyond those that have been very extensively studied. Because of limitations of space, only an initial assessment of the current state of knowledge on the nature of interionic interactions in the solid state can be made. It is hoped that the assembly and organisation of the material collected in this chapter can stimulate more detailed studies of these phenomena, as discussed in the concluding Section 11.21. The reader is referred to some recent reviews that have begun to address this topic in more detail [21–24].

It should be said that not all ionic materials that crystallise at low temperatures (and for which single-crystal X-ray diffraction studies have been reported) are included in the practical definition of ionic liquids. We exclude materials (some of considerable scientific interest) that are unstable at ambient temperatures or that decompose before melting or at (or shortly above) their melting temperatures. We do not cover, therefore, much interesting work on the characterisation of carbenium ions [25] and of rare gas salts [26]. In addition, zwitterions are not included, though some that have been characterised by X-ray crystallography, such as (*S*)-*N*-acetyl-2-methoxycarbonyl-2-(3-methylpyridinium-1-yl)-*N*-propylsulfamate (Fig. 11.1) with a m.pt. of 56–58°C [27], are low melting.

In addition, reasons of space lead us to exclude ionic liquid crystal investigations, except where a single-crystal investigation has been reported on a material that becomes an isotropic liquid below 100°C. An excellent review of ionic liquid crystals has appeared [28].

As we are interested in materials that comprise solely of ions, we also automatically exclude (apart from a small number of illustrative examples) crystalline solvates, some co-crystals, and inclusion compounds (even if their melting points are <100°C), as these materials, self-evidently, do not give molten states that contain only ions. Reference to the contributions of Mootz, for instance, who has characterised a large number of very low-melting salt solvates, is limited to a relatively small number of single-crystal X-ray structural

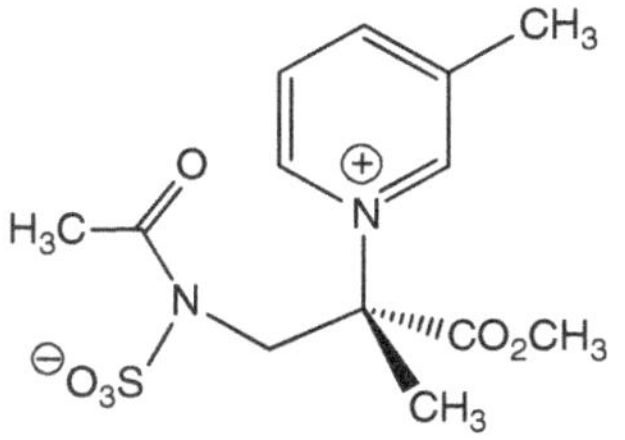

Figure 11.1 (*S*)-*N*-acetyl-2-methoxycarbonyl-2-(3-methylpyridinium-1-yl)-*N*-propylsulfamate [27].

investigations of materials in which anions (such as $[HCl_2]^-$ or $[H_2F_3]^-$) are believed to remain the dominant species on melting. Similarly, the review of the crystal structures of acid and other hydrates is limited.

More generally, crystal structure determinations are important in characterising the products formed from reactions of weak acids and bases, particularly to establish whether or not proton transfer has resulted in salt formation. This is an area of intense study in which three outcomes are noted, differentiated by X-ray crystallographic investigations, namely, salt formation, co-crystallisation without proton transfer and simple recrystallisation, governed by differences in pK_a, the formation of intermolecular and interionic hydrogen-bond networks and associated substituent effects. Some very subtle phenomena have been noted, making it sometimes difficult to predict whether salt formation will be the outcome. In addition, some low-melting coordination complexes display contacts in the solid state between a cationic metal centre and a counter-ion that suggest the absence of salt formation in the crystal but which are weak and may not survive on melting.

Such distinctions are, therefore, a matter of degree not of kind. Identification of materials that are ionic in the solid state, but which do not melt congruently, that is, may involve dissociation of either the component anion or cation (or both) to give uncharged components, requires the establishment of phase diagrams to determine congruency of melting or otherwise. However, in many instances, phase behaviour simply has not been investigated. In other instances, ionic liquid phases that may be formed on initial melting can display dissociation at higher temperatures, even to the extent of complete (and reversible) dissociation. The dissociation of some protic salts of ammonia and alkylamines may lead to sublimation, rather than melting (a characteristic that may make such materials easier to separate and recover and render them of greater technological interest).

While we have sought to make this review comprehensive using a range of databases and searching methods, our judgement is that it is likely to be incomplete. The use of the Cambridge Structural Database (CSD) [29] of the Cambridge Crystallographic Data Centre (CCDC) (now accessible via the Royal Society of Chemistry [30]) has proved invaluable. However, finding reports of melting points for structurally characterised salts, necessary for their inclusion in this review, has been problematic. We have found SciFinder of particular use in this respect, but there will be an unknown number of compounds that should be included but for which, so far, we have been unable to track down a melting point. Melting point compilations, such as that by Krossing and colleagues [31] of 538 melting points, have been useful. However, it is a measure of the large number of ionic liquids now known that, of the *ca.* 350 salts with melting points below 100°C listed by Krossing et al., only 50 or so have been subject to an X-ray crystallographic investigation.

The structures of about 750 salts are identified in this chapter. These are collected in the main Tables 11.1, 11.2 and 11.3,[1] in order of their reported melting

[1] The grey cells in the tables indicate that the salt type (p:q) ratio is different from 1:1, which is the most common ratio.

TABLE 11.1 Room-Temperature Ionic Liquids (m.pt. ≤ 30°C)

m.pt./°C	CSD Refcode	References	Cation $[cat]^{q+}$	Anion $[A]^{p-}$	Salt Type (p:q)	Data Temp/K	Space Group	$V_{cell}/10^3$ Å^3	Z	$U_L(V)$/ kJ mol^{-1}	Formula Weight	ρ(calc)/ g cm^{-3}	$U_L(D)$/ kJ mol^{-1}
Liq	NOMXAP	[52]	1-Ethyl-3-methylimidazolium	$[N(SO_2Me)_2$	1:1	123	$P2_1$	0.64268	2	446.31	283.37	1.464	446.30
Liq	WUSTOU	[53]	1,3-Di-*tert*-butylimidazolium	$[Me_3Al(\mu^3\text{-}CH_2)(AlMe_2)_2(\mu^2\text{-}CH_3)]^-$	1:1	173	$P2_1/c$	2.752315	4	369.53	396.55	0.957	369.55
Liq	EYIYOB	[54]	1-(3-Cyanopropyl)pyridinium	$[N(SO_2CF_3)_2]$	1:1	100	$P2_1/n$	1.703003	4	415.65	427.37	1.667	415.67
Liq	EYIZAO	[54]	1-(3-Cyanopropyl)-3-methylpyridinium	$[N(SO_2CF_3)_2]$	1:1	100	$P2_1/n$	1.756165	4	412.47	441.39	1.669	412.46
Liq	EYIZES	[54]	1-(Cyanomethyl)-2,5-dimethylpyridinium	$[N(SO_2CF_3)_2]$	1:1	100	$P\bar{1}$	0.823972	2	419.08	427.37	1.722	419.07
Liq	RAQGUN	[55]	Bromotriphenylphosphonium	$[Br_7]$	1:1	200	$P2_1$	1.247183	2	378.40	901.55	2.401	378.42
Liq	RAQGIB	[55]	Benzyltriphenylphosphonium	$[Br_8]$	2:1	200	$P\bar{1}$	1.208727	1	752.27	1346.07	1.849	752.27
Liq	RAQHAU	[55]	Chlorotriphenylphosphonium	$[Cl_2I_{14}]$	2:1	200	$P2_1/c$	2.996813	2	687.93	2443.05	2.702	687.38
Ref DSC		[56]	1,3-Dimethylimidazolium	$[F(HF)_3]$	1:1	173	$P\bar{1}$	0.86317	4	494.99	176.16	1.356	494.99
−27	GEQBAF	[57]	Triethylammonium	$[F(HF)_2]$	1:1	123	$Pbca$	1.957	8	478.91	161.21	1.094	478.90
−26	RENSEJ	[58]	1-Ethyl-3-methylimidazolium	$[N(SO_2CF_3)_2]$	1:1	230	$Pca2_1$	3.0726	8	426.54	391.3	1.69	426.45
−26	Cif	(UoL, unpublished work)	1-Ethyl-3-methylimidazolium	$[N(SO_2CF_3)_2]$	1:1	200	$Pna2_1$	3.25518	8	420.39	391.32	1.597	420.41
−2	RENSEJ01	[59]	1-Ethyl-3-methylimidazolium	$[N(SO_2CF_3)_2]$	1:1	120	$Pna2_1$	3.0793	8	426.31	391.32	1.688	426.31
−26	RENSIN	[58]	1-Ethyl-3-methylimidazolium	$[CF_3SO_3]$	1:1	150	$Pbca$	2.307	8	458.89	260.2	1.50	459.05
−18.5		[60]	Oxonium dihydrate	$[NO_3]$	1:1	85	$P2_12_12_1$	0.47845	4	579.94	117.06	1.625	579.96
−17	DEHSUE01	[61]	Pyridinium	$[F(HF)_2]$	1:1	173	$P\bar{1}$	0.339575	2	527.47	139.12	1.361	527.53
−11	LAZREK	[62]	1-Butyl-1-methylpyrrolidinium	$[N(SO_2CF_3)_2]$	1:1	130	$P2_12_12_1$	1.8111	4	409.32	422.4	1.55	409.39
−11	XUZWIY	[63]	*N,N,N′*-triethylethylenediammonium	$[CF_3C(O)CHC(O)CF_3]$	1:1	133	$C2/c$	3.479668	8	413.43	352.32	1.345	413.45
−10	Cif	(UoL, unpublished work)	1-Hexyl-3-methylimidazolium	$[N(SO_2CF_3)_2]$	1:1	130	$P\bar{1}$	1.0555	2	394.10	447.42	1.408	394.13

−1	IGUQAD	[59]	1-Hexyl-3-methylimidazolium	$[N(SO_2CF_3)_2]$	1:1	150	$P\bar{1}$	0.9788	2	401.50	447.42	1.518	401.51
−8	QIQKIK	[64]	1,2-Dimethylhydrazinium	$[N_3]$	1:1	173	*Cc*	1.1164	4	556.09	103.12	1.227	556.11
−6	TFMSAD	[65]	Oxonium monohydrate	$[CF_3SO_3]$	1:1	225	$P2_1/c$	0.689979	4	525.24	186.11	1.792	525.29
−6	TFMSAD01	[65]	Oxonium monohydrate	$[CF_3SO_3]$	1:1	85	$P2_1/c$	0.66641	4	530.15	186.11	1.855	530.18
−5	VIGMAA	[66]	1-Butyl-3-methylimidazolium	$[SO_3(OMe)]$	1:1	173	$P2_1/c$	1.313	4	443.89	250.31	1.266	443.89
~5	Cif	(UoL, unpublished work)	1-Butyl-3-methylimidazolium	$[SO_3(OMe)]$	1:1	230	$P2_1/c$	1.26447	4	448.19	250.31	1.315	448.22
−4	LAZROU	[62]	1-Hexylpyridinium	$[N(SO_2CF_3)_2]$	1:1	125	$P\bar{1}$	1.966	4	401.07	444.4	1.50	401.00
−3	IGUPUW	[59]	1-Butyl-3-methylimidazolium	$[N(SO_2CF_3)_2]$	1:1	120	$P2_1/n$	3.5064	8	412.64	419.37	1.589	412.67
−2	Cif	(UoL, unpublished work)	1-Butyl-3-methylimidazolium	$[N(SO_2CF_3)_2]$	1:1	130	$P2_1/n$	3.47894	8	413.45	419.37	1.601	413.45
−1	MESBIW	[67]	1-Ethyl-3-methylimidazolium	$[NbF_6]$	1:1	200	$P2_12_12_1$	1.123949	4	461.98	318.06	1.880	462.02
−2.5	ACASUL	[68]	1,1-Dihydroxyethylium	$[SO_3(OH)]$	1:1	231	$P2_1/c$	0.599626	4	545.43	158.13	1.752	545.48
−2	DOVQIO	[69]	Triethylsulfonium	$[I_3]$	1:1	193	*Pnma*	2.719	8	439.97	499.96	2.441	439.91
−1	DEHSOY10	[61]	Pyridinium	$[HF_2]$	1:1	163	$P2_1/m$	0.2945	2	548.07	119.14	1.34	547.70
−1	LAZRIO	[62]	1-Ethyl-3-methylimidazolium	$[BF_4]$	1:1	173	$P2_1/n$	0.92968	4	485.37	198.0	1.41	484.97
No m.pt.	LAZRIO01	[67]	1-Ethyl-3-methylimidazolium	$[BF_4]$	1:1	100	$P2_1/c$	0.9068	4	488.55	197.98	1.450	488.56
<0	SURZUA	[70]	Tris(dimethylamino) sulfonium	*cyclo*-$[P_6N_6F_{13}]$	1:1	173	$P2_12_12_1$	2.4840	4	378.78	681.17	1.821	378.77
0	GEPTAW	[57]	Trimethylammonium	$[F(HF)_2]$	1:1	123	*Pnma*	0.6497	4	533.78	119.13	1.218	533.81
~0	Cif	(UoL, unpublished work)	2-Hydroxyethyl(trimethyl) ammonium (cholinium)	$[N(SO_2CF_3)_2]$	1:1	130	$Pna2_1$	1.55018	4	425.58	384.32	1.65	425.81
0–2	LIBMEQ	[71]	*tert*-Butyl carbenium	$[Al_2Br_7]$	1:1	153	*C2/c*	5.0100	12	417.69	670.44	2.667	417.72
2	MESBOC	[67]	1-Ethyl-3-methylimidazolium	$[TaF_6]$	1:1	200	$P2_12_12_1$	1.124878	4	461.88	406.10	2.398	461.90
2	MAZXOB01	[62]	1-Butyl-3-methylimidazolium	$[PF_6]$	1:1	180	$P\bar{1}$	0.6102	2	452.28	284.2	1.55	452.54
11	MAZXOB	[72]	1-Butyl-3-methylimidazolium	$[PF_6]$	1:1	173	$P\bar{1}$	0.60496	2	453.29	284.19	1.560	453.30
5	KISNIK	[73]	1-Butyl-3-methylimidazolium	$[Sn^{II}Cl_4]$	2:1	203	$P2_1/a$	2.3119	4	1011.86	538.94	1.548	1011.81
7	LAZRUA	[62]	1-Butyl-3-methylimidazolium	$[CF_3SO_3]$	1:1	200	$P2_1/n$	2.8832	8	433.46	288.3	1.33	433.62

(*Continued*)

Table 11.1 (*Continued*)

m.pt./°C	CSD Refcode	References	Cation $[cat]^{q+}$	Anion $[A]^{p-}$	Salt Type (p:q)	Data Temp/K	Space Group	$V_{cell}/10^3$ Å^3	Z	$U_L(V)$/ kJ mol^{-1}	Formula Weight	ρ(calc)/ g cm^{-3}	$U_L(D)$/ kJ mol^{-1}
9	Cif	(UoL, unpublished work)	1-Hexyl-1-methylpyrrolidinium	$[N(SO_2CF_3)_2]$	1:1	200	$Cmc2_1$	2.11603	4	393.87	450.47	1.386	391.96
ca. 9	IQETII	[74]	1-Butyl-1-methylpyrrolidinium	$[Br_{20}]$	2:1	200	$P\bar{1}$	1.1693	2	1007.31	941.36	2.674	1007.40
ca. 9	IQETII01	[55]	1-Butyl-1-methylpyrrolidinium	$[Br_{20}]$	2:1	200	$P\bar{1}$	1.1693	2	1007.31	941.36	2.674	1007.40
ca. 10	RAQGOH	[55]	Tributylmethylammonium	$[Br_{20}]$	2:1	200	$C2/c$	5.5113	8	944.17	999.48	2.409	944.20
10	AQEYIE	[75]	1-Ethyl-3-methylimidazolium	$[SbF_6]$	1:1	200	$P2_1/c$	1.1388	4	460.41	346.92	2.024	460.47
10	OWICON	[76]	1-Ethyl-3-methylimidazolium	$[Fe^{III}Cl_4]$	1:1	283–303	$P2_1/c$	1.3171	4	443.54	308.82	1.557	443.53
10	OWICON01	[76]	1-Ethyl-3-methylimidazolium	$[Fe^{III}Cl_4]$	1:1	283–303	$P2_1$	0.6263	2	449.27	308.82	1.638	449.32
10 (dec)	No refcode	[56]	1,3-Dimethylimidazolium	$[F(HF)_2]$	1:1	173	$P2_1/n$	0.3739	2	514.11	156.15	1.387	514.11
12	Cif	(UoL, unpublished work)	1-Methyl-1-propylpyrrolidinium	$[N(SO_2CF_3)_2]$	1:1	200	$P\bar{1}$	0.87413	2	412.93	408.39	1.552	412.98
12	Cif	(UoL, unpublished work)	1-Methyl-1-propylpyrrolidinium	$[N(SO_2CF_3)_2]$	1:1	200	$Pca2_1$	3.44445	8	414.48	408.39	1.575	414.50
12	Cif	(UoL, unpublished work)	1-Methyl-1-propylpiperidinium	$[N(SO_2CF_3)_2]$	1:1	200	$P\bar{1}$	2.7606	6	407.69	422.41	1.524	407.67
12	MALCIN	[77]	1-Allyl-3-methylimidazolium	$[Al\{OCH(CF_3)_2\}_4]$	1:1	100	$P2_1/c$	3.0523	4	360.53	818.31	1.781	360.56
14–16	VIXVUU	[78]	Tetrakis(ethanenitrile)lithium	$[P_2Se_8]$	2:1	100	$Pbam$	3.6659	4	842.26	1035.93	1.877	842.30
18	No refcode	[79]	Tetrakis(ethanenitrile)silver(I)	$[Ag\{N(SO_2CF_3)_2\}_3]$	2:1	100	$P\bar{3}c1$	2.4343	2	750.12	1492.46	2.036	750.13
18–23	No refcode	[80]	Hydroxylammonium	$[N(NO_2)_2]$	Complex	223	$Pcab$	2.474	8		346.21	1.859	
19	KETMAY	[81]	1-Butyl-3-methylimidazolium	I	1:1	93	$P2_12_12_1$	1.07189	4	467.69	266.12	1.649	467.70
22	PADDEE	[82]	1,3-Dimethylimidazolium	$[N(SO_2CF_3)_2]$	1:1	173	$P\bar{1}$	1.417	4	435.36	377.31	1.769	435.40
23	KISLUU	[73]	1-Butyl-3-methylimidazolium	$[Cu^{II}Cl_4]$	2:1	113	Cc	2.2438	4	1023.78	483.80	1.432	1023.79
23–24	MARQAZ	[83]	1-Butyl-2,3-dimethylimidazolium	$[PO_2(OMe)(SeMe)]$	1:1	233	$P\bar{1}$	0.7989	2	422.35	341.25	1.419	422.40

25	ODIHOZ	[84]	2-Hydroxyethyl(trimethyl) ammonium (cholinium)	Acesulfamate	1:1	100	$P2_1/n$	1.26086	4	448.51	266.32	1.403	448.54
26	UQIKEL	[85]	1-Dodecyl-3-methylimidazolium	$[Cu^ICl_2][Cu^ICl_3]$	3:(1,2)	100	$P\bar{1}$	2.7911	2		1058.61	1.260	
27	UQIKIP	[85]	Tetrakis(1-dodecylimidazole) copper(I)	$[PF_6]$	1:1	173	$P\bar{1}$	3.2782	2	302.77	1154.09	1.169	302.77
27	UQIKUB	[85]	1-Dodecyl-3-methylimidazolium	$[Cu^I_4Cl_8]$	4:1	193	$P\bar{1}$	4.0930	2		1899.15	1.541	
27	CASKIQ	[86–88]	1-(Trimethylsilyl)pyridinium	Br	1:1	283–303	$P2_1/c$	1.08854	4	465.82	232.20	1.417	465.85
27	KEHKAL	[89]	(1-Butylimidazole)(1-ethylimidazole)silver(I)	$[N(SO_2CF_3)_2]$	1:1	100	$P2_1/c$	2.2434	4	388.28	608.34	1.801	388.28
28	KASXAF	[90]	1-(2-Hydroxyethyl) pyrrolidinium	$[C_6H_5CO_2]$	1:1	123	$C2/c$	2.5259	8	448.32	237.29	1.248	448.35
28	POYLOF	[91]	1-Ethyl-3-methylimidazolium	$[C_4F_9SO_3]$	1:1	173	$P2_1/c$	3.167	8	423.30	410.27	1.72	423.26
28–29	WOLKOY	[92]	1-Ethyl-3-trifluoromethylpyridinium	$[N(SO_2CF_3)_2]$	1:1	193	$P\bar{1}$	1.6758	4	417.33	456.31	1.809	417.37
28–30	VIBNUQ	[93]	2-Bromo-1,3-dimethoxymidazolium	$[N(SO_2CF_3)_2]$	1:1	173	$P\bar{1}$	1.7098	4	415.23	488.20	1.896	415.22
29	WUBHAC	[94]	Dimethylammonium	$[CO_2(NMe_2)]$	1:1	123	$P2_1/c$	0.75443	4	512.88	134.18	1.181	512.86
29	GETQIF	[95]	$[(Me_3NH)_2Cl]$	$[HCl_2]$	1:1	123	$P2_1$	0.6316	2	448.30	228.04	1.197	448.12
30	SAMYOW	[96]	1,3-Diethylimidazolium	$[CH_3CO_2]$	1:1	173	$P2_1/n$	1.0203	4	473.72	184.24	1.199	473.70
30	EYIZIW	[54]	1-Cyanomethyl-2-ethylpyridinium	$[N(SO_2CF_3)_2]$	1:1	100	$P\bar{1}$	1.6436	4	419.36	427.37	1.727	419.37
30	UQOCOT	[97]	2-Hydroxyethyl(trimethyl) ammonium (cholinium)	$[N(SO_2CF_3)_2]$	1:1	213	$Pna2_1$	1.576	4	423.81	384.34	1.62	423.84
30	KEHKIT	[89]	(1-Butylimidazole)(1-methylimidazole)silver(I)	$[N(SO_2CF_3)_2]$	1:1	298	$P2_1/c$	2.2767	4	386.88	594.31	1.734	386.91

TABLE 11.2 Ionic Liquids: 30°C < m.pt. < 100°C

m.pt./°C	CSD Refcode	References	Cation $[cat]^{q+}$	Anion $[A]^{p-}$	Salt Type $(p:q)$	Data Temp/K	Space Group	V_{cell}/ 10^3 Å^3	Z	$U_L(V)$/ kJ mol^{-1}	Formula Weight	ρ(calc)/ g cm^{-3}	$U_L(D)$/ kJ mol^{-1}
31	KEXBIZ	[98]	1-Allyl-3-ethylimidazolium	I	1:1	140	$P2_12_12_1$	1.0371	4	471.71	264.10	1.691	471.70
31	MALCOT	[77]	1-Ethyl-3-methylimidazolium	$[Al\{OCH(CF_3)_2\}_4]$	1:1	113	$P\bar{1}$	2.9823	4	362.52	806.30	1.796	362.55
31	MALDEK	[77]	1-Butyl-1-methylmorpholinium	$[Al\{OCH(CF_3)_2\}_4]$	1:1	113	$Pna2_1$	3.2738	4	354.60	853.39	1.731	354.60
31	ODOLOJ	[99]	1-Decyl-3-methylimidazolium	$[PF_6]$	1:1	173	$P2_1/c$	1.8445	4	407.46	368.35	1.326	407.44
32	KEXBOF	[98]	1-Allyl-3-propylimidazolium	I	1:1	140	$P\bar{1}$	1.1716	4	457.05	278.13	1.577	457.09
32	LIQNAB	[100]	1-(2-Di-*iso*-propylaminoethyl)-3-methylimidazolium	$[PF_6]$	1:1	86	$Pbca$	3.3252	8	418.15	355.31	1.419	418.14
32	TOJRAM	[101]	1-(Hydroxycarbonylmethyl) pyridinium	$[N(SO_2CF_3)_2]$	1:1	100	$P1$	0.77449	2	425.66	418.31	1.794	425.69
33	HOJHIX	[102]	Bis(*N*,*N*′-dimethylethyleneurea-*O*) trimethyltin(IV)	$[N(SO_2F)_2]$	1:1	143	$P2_1/c$	2.2892	4	386.37	572.24	1.660	386.36
33–34	DANSER	[103]	1-Ethylquinolinium	$[I_3]$	1:1	293–303	$P2_1/n$	1.5300	4	426.98	538.92	2.340	427.02
33–35	WOLLAL	[92]	1-Ethyl-4-cyanopyridinium	$[N(SO_2CF_3)_2]$	1:1	193	Cc	1.6506	4	418.91	413.34	1.663	418.91
34	SIWSOH	[104]	1-Butyl-3-methylimidazolium	$[Al\{OCH(CF_3)_2\}_4]$	1:1	173	$P2_1/c$	3.2293	4	355.75	834.35	1.716	355.76
34	HOXJUA	[105]	1-butylpyridinium	$[AlCl_4]$	1:1	170	$P2_1/c$	1.4609	4	432.00	304.99	1.387	432.05
34–36	LUSGIP	[106]	1-Pentyl-1‴-(5-bromopentyl)-1′, 1″-biferrocenium	$[I_3]$	1:1	293–303	$P\bar{1}$	1.6097	2	356.01	969.91	2.001	356.02
34.5	TFMSUL	[107]	Oxonium	$[CF_3SO_3]$	1:1	293	$P2_1/c$	0.57089	4	552.71	168.09	1.956	552.76
34.5	TFMSUL01	[107]	Oxonium	$[CF_3SO_3]$	1:1	83	$P2_1/c$	0.54179	4	560.61	168.09	2.061	560.66
35	VOJGAC	[108]	1,1-Dihydroxyethylium	$[CF_3SO_3]$	1:1	143	$P2_1/m$	0.3944	2	506.85	210.13	1.77	506.92
35	GEQYUX	[109]	1-Ethyl-3-methylimidazolium	$[C(NO_2)(NO)CN]$	1:1	293–303	$P2_1/c$	1.1093	4	463.55	225.21	1.349	463.61
35	No cif	[110]	1-Methyl-3-(trimethylsilyl-methyl) imidazolium	$[BF_4]$	1:1	100	$Pbca$	2.53329	8	448.00	256.13	1.343	448.00
36	OVUDOZ	[111]	1,3-Didodecylimidazolium	$[I_5]$	1:1	150	$P2_1/n$	7.2515	8	346.21	1040.21	1.906	346.24
36	MALDIO	[77]	1-Butylpyridinium	$[Al\{OCH(CF_3)_2\}_4]$	1:1	113	$P2_1/c$	6.089	8	360.75	831.34	1.814	360.77
36	JIRBUI	[112]	1-Methyl-3-propylimidazolium	Br	1:1	110	$Pca2_1$	0.8908	4	490.84	205.10	1.529	490.83
36–38	WOLLOZ	[92]	1-Ethyl-2-cyanopyridinium	$[N(SO_2CF_3)_2]$	1:1	298	$P2_12_12_1$	3.318	8	418.38	413.34	1.655	418.41
37	NOMXET	[52]	1-Butyl-1-methylpyrrolidinium	$[N(SO_2Me)_2]$	1:1	123	$Pbca$	3.155363	8	423.70	314.46	1.324	423.72
37	ZIZGEU	[113]	1-Ethyl-3-methylimidazolium/sodium	$[AlCl_4]$	(1,1):2	153	$P\bar{1}$	0.9847	2		471.74	1.591	
37	UDEWAC	[114]	1-Butyl-2,3-dimethylimidazolium	$[IBr_2]$	1:1	299	$P2_1/c$	1.48225	4	430.42	439.97	1.972	430.46

37	DOVQIO	[115]	Trimethylsulfonium	$[I_3]$	1:1	296	Cc	1.073	4	467.56	457.88	2.834	467.56
38	CORRUW	[116]	1,1-Dimethylhydrazinium	$[N_3]$	1:1	173	$P2_1/n$	0.56874	4	553.28	103.14	1.205	553.36
38	KUCPED	[117]	1-Ethyl-3-methylimidazolium	$[NO_3]$	1:1	293	$P2_1/n$	0.8995	4	489.59	173.2	1.279	489.61
38	XEQLOV	[118]	1-Butyl-3-methylimidazolium	$[La(NCS)_7(H_2O)]$	4:1	100	$P\bar{1}$	2.771	2		1120.44	1.343	
38	AYIBUF	[119]	1-Butyl-2, 3-dimethylimidazolium	$[Fe^{III}Cl_4]/[Fe^{II}Cl_4]$	3:(1,2)	213	$C2/c$	5.7710	4		1716.12	1.975	
38	GARYUV	[120]	1-Tetradecylpyridinium	$[Co(hfac)_3]$	1:1	140	$P\bar{1}$	2.03399	2	337.09	956.58	1.562	337.11
39	TOJREQ	[101]	1-(Hydroxycarbonylmethyl)-1-methylpyrrolidinium	$[N(SO_2CF_3)_2]$	1:1	100	$P2_1/c$	1.6114	4	421.45	424.36	1.749	421.45
39	MALCUZ	[77]	1-Ethyl-2, 3-dimethylimidazolium	$[Al\{OCH(CF_3)_2\}_4]$	1:1	120	$P2_1/n$	3.0358	4	360.99	820.33	1.795	361.01
39	MAXDOG	[121]	1-Ethyl-3-methylimidazolium	$[PO_2(OMe)(SMe)]$	1:1	100	$P\bar{1}$	0.60638	2	453.01	252.27	1.382	453.06
39.5/41	HATREB	[122]	1-Methyl-3-propylimidazolium	$[HgBr_3]$	1:1	170	$P2_1/c$	1.3472	4	440.99	565.49	2.788	441.00
40	TOJQUF	[101]	1-Hydroxycarbonylmethyl-3-methylimidazolium	$[N(SO_2CF_3)_2]$	1:1	100	$P\bar{1}$	1.5765	4	423.78	421.32	1.775	423.79
40	EZILII	[123]	1-Butyl-2, 3-dimethylimidazolium	$[BF_4]$	1:1	213	$P2_1/c$	2.41918	8	453.32	240.05	1.307	452.35
No m.pt.	EZILII01	(C. Pulham, J. Pringle, A. Parkin, S. Parsons, and D. Messenger, Cambridge Structural Database, private communication, 2005.)	1-Butyl-2,3-dimethylimidazolium	$[BF_4]$	1:1	150	$P2_1/c$	2.45601	8	451.56	240.05	1.298	451.54
40	EZILOO	[123]	1-Butyl-2, 3-dimethylimidazolium	$[PF_6]$	1:1	213	Cc	1.3104	4	444.11	298.21	1.512	444.17
No m.pt.	EZILOO01	(C. Pulham, J. Pringle, A. Parkin, S. Parsons, and D. Messenger, Cambridge Structural Database, private communication, 2005.)	1-Butyl-2, 3-dimethylimidazolium	$[PF_6]$	1:1	150	Cc	1.30248	4	444.80	298.21	1.521	444.84
40	SAXGOP	[124]	1-Butyl-2, 3-dimethylimidazolium	[SCN]	1:1	173	$P2_12_12_1$	2.4096	8	453.78	211.33	1.165	453.79

(Continued)

TABLE 11.2 (*Continued*)

m.pt./°C	CSD Refcode	References	Cation [cat]$^{q+}$	Anion [A]$^{p-}$	Salt Type (p:q)	Data Temp/K	Space Group	V_{cell}/ 10^3 Å^3	Z	U_L(V)/ kJ mol^{-1}	Formula Weight	ρ(calc)/ g cm^{-3}	U_L(D)/ kJ mol^{-1}
40–41	HIMZUZ	[125]	1-Hexadecylpyridinium	$[Co(CO)_4]$	1:1	283–303	$P2_1/c$	2.722	4	370.52	475.49	1.160	370.51
40–46	VIBNIE	[93]	1,3-Diethoxyimidazolium	$[BF_4]$	1:1	173	$Pbca$	2.2871	8	459.92	244.00	1.417	459.92
40–42	ESOFUN	[126]	*cyclo*-Hexylphosphonium	$[Ge^{II}Cl_3]$	1:1	150	$C2/c$	2.32489	8	457.98	296.08	1.679	457.10
40–55	ULOJOU	[127]	1,1-Dimethylpyrrolidinium	$[N(SO_2Me)_2]$	1:1	123	$P2_1/n$	1.2680	4	447.87	272.38	1.427	447.90
41	LUHPUA	[128]	1-Butyl-1-methylpiperidinium	$[PO_2F_2]$	1:1	100	$P2_1/c$	2.60799	8	444.67	257.26	1.310	444.65
41	OVUFIV	[111]	1,3-Didodecylimidazolium	$[SbF_6]$	1:1	150	$P2/c$	3.1382	4	358.16	641.46	1.358	358.20
41	DOQGOG	[129]	1-Methyl-3-(2-phenylethyl) imidazolium	$[N(SO_2CF_3)_2]$	1:1	100	$P2_1/n$	1.9142	4	403.73	467.41	1.622	403.75
41	SAXGIJ	[124]	1-Butyl-2, 3-dimethylimidazolium	$[N_3]$	1:1	173	$P2_1/n$	1.0951	4	465.10	195.28	1.184	465.07
41–43	ZELRUD	[130]	2-Ethylimidazolium	*d*-$[HO_2CCH(OH)CH(OH)CO_2]$	1:1	283–303	$P2_1$	1.1197	4	462.43	246.22	1.541	462.48
42	IBOLUG	[131]	Methylhydrazinium	$[NO_3]$	1:1	200	$P2_1/c$	0.47015	4	582.73	109.08	1.648	582.75
42	IHEGIM01	[132]	1,1,3,3-Tetramethylguanidinium	$[N(SO_2CF_3)_2]$	1:1	173	$P2_1/c$	1.597	4	422.40	396.34	1.678	422.39
42–43	IHEGIM	[133]	1,1,3,3-Tetramethylguanidinium	$[N(SO_2CF_3)_2]$	1:1	120	$P2_1/n$	1.5609	4	424.84	396.34	1.687	424.31
42	KEHJUE	[89]	Bis(1-butylimidazole)silver(I)	$[N(SO_2CF_3)_2]$	1:1	100	$P\bar{1}$	2.5053	4	378.00	636.39	1.425	378.00
43	UJOSER	[134]	1,3-Dimethylimidazolium	$[SO_3(OMe)]$	1:1	173	$P2_12_12_1$	0.9704	4	479.95	208.24	2.166	479.94
43	RECJUE	[135]	1,3,5-Triethyl-tetrahydro-1,3, 5-triazinium	$[I_3]$	1:1	283–303	$P\bar{1}$	0.8447	2	416.48	550.99	1.177	416.49
43	SAXGUV	[124]	1-Butyl-2,3-dimethylimidazolium	$[HC{\equiv}CO_2]$	1:1	173	$P2_1/c$	1.2549	4	449.06	222.28	1.061	449.12
44	OVUDUF	[111]	1,3-Didodecylimidazolium	$[N(CN)_2]$	1:1	150	$P\bar{1}$	1.47664	2	363.37	471.76	1.048	363.38
44	OVUFAN	[111]	1,3-Didodecylimidazolium	$[C(CN)_3]$	1:1	150	$P\bar{1}$	1.57064	2	358.08	495.78	1.859	358.07
44	EYIYIV	[54]	1-(Cyanomethyl)pyridinium	$[N(SO_2CF_3)_2]$	1:1	100	$P\bar{1}$	0.71355	2	434.57	399.32	1.773	434.62
44.5	EZILUU	[123]	1-Butyl-2,3-dimethylimidazolium	$[SbF_6]$	1:1	283–303	$P2_1/n$	1.45727	4	432.27	388.91	1.075	432.32
44.5	MOVDIL	[136]	Amidomethyltris(tetradecyl) phosphonium	Br	1:1	173	$P2_1/c$	4.70196	4	326.09	761.06	1.461	326.10
45	IBOLOA	[131]	1,1-Dimethylhydrazinium	$[NO_3]$	1:1	200	$Pna2_1$	0.60296	4	544.61	123.11	1.356	544.64
45	FOYSIW	[137]	1-Butyl-3-methylimidazolium	$[Co^{II}Br_4]$	2:1	173	$C2/c$	2.4216	4	993.60	657.01	1.802	993.63
45	IZUZAE	[138]	*rac*-1-{1-(Ethoxycarbonyl) ethyl}-3-methylimidazolium	$[N(SO_2CF_3)_2]$	1:1	223	$P2_1/c$	1.87191	4	405.97	463.38	1.644	405.97

45	UGOQAI	[139]	Tris(dimethylamino) oxosulfonium	$[HF_2]$	1:1	173	$P2_1/n$	1.06093	4	468.93	219.30	1.373	468.96
45–46	SOVKET	[140]	1-Ethylpyridinium	$[I_3]$	1:1	283–303	$P2_1/c$	1.24174	4	450.27	488.88	2.615	450.29
45/65	GAFFAV	[141]	Guanidinium	$[N(CN)_2]$	1:1	200	$P2_1/c$	1.2275	8	542.01	126.14	1.365	542.02
52/62	GAFFAV01	[141]	Guanidinium	$[N(CN)_2]$	1:1	200	$Pna2_1$	1.2267	8	542.11	126.14	1.366	542.13
45	POYLIZ	[91]	1-Ethyl-3-methylimidazolium	$[HCF_2CF_2SO_3]$	1:1	173	$P2_1/n$	1.2179	4	452.52	292.26	1.59	452.25
46	KUKSIT	[142]	1-Ethyl-1-methylpyrrolidinium	$[Br_3]$	1:1	123	$Pnma$	1.1615	4	458.08	353.94	2.024	458.09
46	OVUDIT	[111]	1,3-Didodecylimidazolium	$[I_3]$	1:1	150	$C2/c$	3.2237	4	355.89	786.41	1.620	355.89
46	TOJQOZ	[101]	1-(Ethoxycarbonylmethyl)-1-methylmorpholinium	$[N(SO_2CF_3)_2]$	1:1	100	$P2_1/c$	1.8142	4	409.14	468.41	1.715	409.16
46	RUBYOD	[143]	1-Butyl-2,3-dimethylimidazolium	$[N(C_6F_5)_2]$	1:1	193	$P\bar{1}$	1.0522	2	394.41	501.38	1.583	394.45
46	GEVGIZ	[144]	Tetrakis(1-butylimidazole) copper(II)	$[N(SO_2CF_3)_2]$	1:2	100	$P2_1/n$	9.6564	8	1117.70	1120.64	1.542	1117.78
46–48	BAQJOU	[145]	3-Chloro-3-cyclohexyltricyclo[2.1.0.0^{2,5}]pentaphosphan-3-ium	$[GaCl_4]$	1:1	153	$Cmc2_1$	5.1627	12	414.59	484.96	1.872	414.59
48	OVUFER	[111]	1,3-Didodecylimidazolium	$[B(CN)_4]$	1:1	150	$P\bar{1}$	1.68357	2	352.26	520.60	1.027	352.28
48	EYIZOC	[54]	1-(Cyanomethyl)-4-ethylpyridinium	$[N(SO_2CF_3)_2]$	1:1	100	$P2_1/c$	1.6265	4	420.46	427.37	1.745	420.47
48	NOGLUR	[146]	Hydrazinium	Tetrazolate	1:1	200	$Ccca$	1.95874	16	576.27	102.12	1.385	576.28
49	KUKSEP	[142]	1-Ethyl-3-methylimidazolium	$[Br_3]$	1:1	123	$P\bar{1}$	0.53928	2	466.93	350.88	2.161	466.96
49	YOBWAO	[147]	1-Butylpyridinium	[*nido*-$C_2B_9H_{12}$]	1:1	173	$P2_12_12_1$	1.6351	4	419.91	269.62	1.095	419.90
49	KASXIN	[90]	1-(2-Hydroxyethyl)pyrrolidinium	$[(2\text{-}HO)C_6H_4CO_2]$	1:1	123	$P2_1/c$	1.2689	4	447.78	253.29	1.326	447.82
50	No cif	(UoL, unpublished work)	2-Chloroethyl(trimethyl) ammonium	$[N(SO_2CF_3)_2]$	1:1	130	$P2_12_12_1$	1.53068	4	426.94	402.77	1.75	427.09
50	QEKHAP	[148]	1-Butyl-3-methylimidazolium	$[Au^{III}Cl_4]$	1:1	193	$P\bar{1}$	0.7384	2	430.82	477.05	2.146	430.86
50	SIKXIU	[149]	1-(2-Methoxyethyl)-3-methylimidazolium	I	1:1	140	$Pca2_1$	4.1142	#	472.73	268.09	1.731	472.73
50	JOBVEC	[150]	1-Amino-3-methyl-1,2,3-triazolium	$[N_3]$	1:1	173	$P2/c$	0.9619	6	535.65	141.16	1.462	535.66
50	DOQGIA	[129]	1-Methyl-3-(3-phenylpropyl) imidazolium	$[N(SO_2CF_3)_2]$	1:1	283–303	$P2_1/c$	2.05866	4	396.54	481.44	1.553	396.54
50	IMOYIT	[151]	1-Butyl-3-ferrocenylimidazolium	$[N(SO_2CF_3)_2]$	1:1	100	$P2_1/c$	2.3619	4	383.44	589.36	1.657	383.43

(Continued)

TABLE 11.2 (*Continued*)

m.pt./°C	CSD Refcode	References	Cation [cat]$^{q+}$	Anion [A]$^{p-}$	Salt Type (p:q)	Data Temp/K	Space Group	V_{cell}/ 10^3 Å^3	Z	U_L(V)/ kJ mol^{-1}	Formula Weight	ρ(calc)/ g cm^{-3}	U_L(D)/ kJ mol^{-1}
50	LIVWUJ	[152]	1-Methyl-1-propylpyrrolidinium	Cl	1:1	123	*Pbcn*	1.92399	8	481.05	163.68	1.130	481.05
ca. 50	JOSXIY	[153]	1-Ethyl-2-methyl-3-benzylimidazolium	$[N(SO_2CF_3)_2]$	1:1	283–303	$P2_1/n$	2.061	4	396.43	481.4	1.552	396.48
50–52	RULZOO	[154]	Triphenylarsine{chloro(phenyl)}stibonium	$[AlCl_4]$	1:1	193	*C2/c*	5.51735	8	369.33	709.30	1.708	369.36
50–56	NALGUC	[155]	*N*-Propylurotropinium	$[I_7]$	1:1	283–303	$P\bar{1}$	1.1727	2	384.09	1071.61	3.035	384.11
51	EGULOH	[156]	1-Ethyl-3-methylimidazolium	$[HF_2]$	1:1	283–303	$P2_1/m$	0.39509	2	506.62	150.18	1.262	506.60
51	FELCOP	[157]	1,3-Bis(1-methyl-1-pyrrolidinio)propane	Br	1:2	193	*Pbca*	10.108	#	1434.00	372.19	1.467	1433.93
51	EYIYUH	[54]	1-(Cyanomethyl)-2-methylpyridinium	$[N(SO_2CF_3)_2]$	1:1	100	$P2_1/n$	1.548	4	425.73	413.34	1.773	425.71
52	YAXXOL	[158]	1-Ethyl-3-methylimidazolium	$[C(C_6F_5)(SO_2CF_3)_2]$	1:1	223	$P2_1/n$	2.1395	4	392.81	556.38	1.727	392.81
52	BAVHUD	[159]	1-(Chloromethyl)-1,1-dimethylhydrazinium	$[ClO_4]$	1:1	100	$P\bar{1}$	0.40973	2	501.76	209.03	1.694	501.76
52	GEVFEU	[144]	Hexakis(1-butylimidazole)copper(II)	$[N(SO_2CF_3)_2]$	1:2	100	$P\bar{1}$	2.23673	2	1137.08	1116.54	1.658	1137.14
52.5	ACETFS	[160]	1,1-Dihydroxyethylium	$[SO_3F]$	1:1	216	$P2_1/c$	0.59385	4	546.85	160.12	1.791	546.88
52–54.5	RIHQUU	[161]	1,3-Dimethyl-1,3-diaza-2-arsenanium	$[GaCl_4]$	1:1	283–303	$Pca2_1$	1.35618	4	440.24	386.61	1.894	440.29
52/59	ESOWUF	[162]	1-Methylimidazolium	$[N(SO_2CF_3)_2]$	1:1	173	$P2_1/c$	2.6119	8	444.50	363.26	1.848	444.55
53	AQEYEA	[75]	1-Ethyl-3-methylimidazolium	$[AsF_6]$	1:1	298	$P2_1/c$	1.12304	4	462.07	300.09	1.775	462.10
53–56	NAFVEV	[163]	Bis(*N*,*N*,*N′*,*N″*,*N″*-pentamethyldiethylenetriamine)potassium	$[AlMe_4]$	1:1	283–303	$P2_12_12_1$	3.25954	4	354.97	472.82	0.964	355.03
53.5–56	FAZGUJ	[164]	*N*,*N*-Dimethyl-*N*-((methyltellurenyl)(phenyl)methylidene)ammonium	$[CF_3SO_3]$	1:1	193	$P2_1/n$	1.52343	4	427.45	424.90	1.853	427.49
54–56	ISUWIC	[165]	5-Methyl-1,3,5-dithiazinan-5-ium	$[GaCl_4]$	1:1	193	$P2_1$	0.61938	2	450.55	347.77	1.865	450.59
54–56	HIWZUJ	[166]	$[Li_2(H_2NNHMe)_3]$	$[BH_4]$	1:2	193	*C2/c*	2.3750	8	1566.11	170.91	0.956	1566.17
54–57	XISCUX	[167]	Tetrakis(decyl)phosphonium	Br	1:1	173	*C2/c*	8.68271	8	332.08	675.98	1.034	332.08
55	KUCPIH	[117]	1-Ethyl-3-methylimidazolium	$[NO_2]$	1:1	283–303	$P2_1$	0.4128	2	500.77	157.2	1.265	500.82

55	QONRIV	[168]	1*H*-Tetrazolium	$[ClO_4]$	1:1	200	$P2_1/n$	0.56034	4	555.51	170.52	2.021	555.52
56	KISMAB	[73]	1-Butyl-3-methylimidazolium	$[Ni^{II}Cl_4]$	2:1	113	Cc	2.3133	4	1011.62	478.95	1.375	1011.61
56	MIHSAY	[169]	1-(*cyclo*-Propylmethyl)-4-amino-1,2,4-triazolium	$[NO_3]$	1:1	100	$P2_1/n$	0.9253	4	485.97	201.20	1.444	485.96
56	FOZRAO	[170]	Tetraethylammonium	$[Al\{OCH(CF_3)_2\}_4]$	1:1	153	$P2_1/c$	3.2186	4	356.03	825.35	1.703	356.03
56	FOZRAO01	[170]	Tetraethylammonium	$[Al\{OCH(CF_3)_2\}_4]$	1:1	100	$P2_1/c$	3.2149	4	356.12	825.35	1.705	356.13
56	OVURON	[171]	(1,2-Dimethoxyethane)sodium	$[B\{OCH(CF_3)_2\}_4]$	Complex	100	$P\bar{1}$	1.39481	2	368.34	792.07	1.886	368.36
56–62	UFAKUH	[172]	Methyluronium	$[N(SO_2Me)_2]$	1:1	173	$P2_12_12_1$	1.013	4	474.60	247.29	1.622	474.67
57	PADDII	[82]	1,2,3-Triethylimidazolium	$[N(SO_2CF_3)_2]$	1:1	173	$P2_1/n$	1.803	4	409.77	433.42	1.597	409.81
57	HOPJIG	[173]	1-Hexadecylpyridinium	Acesulfamate	1:1	173	$P\bar{1}$	2.6242	4	373.79	466.67	1.181	373.79
57	YESSUL	[174]	(Hydroxycarbonylmethyl) trimethylammonium (protonated betaine)	$[N(SO_2CF_3)_2]$	1:1	100	$Pbca$	6.1296	16	426.82	398.33	1.727	426.86
57	MOBYAS	[175]	1-Methyl-3-[(trimethoxysilyl) methyl]imidazolium	I	1:1	193	$P2_1/n$	1.4058	4	436.24	344.23	1.626	436.22
57	UGOPOV	[139]	Tris(dimethylamino) oxosulfonium	$[SiMe_3F_2]$	1:1	173	$P2_1$	0.7702	2	426.26	291.49	1.257	426.28
57–58	AMIFEH	[176]	Trimethylphosphine(diphenyl) phosphonium	$[CF_3SO_3]$	1:1	193	$P2_1/n$	1.8939	4	404.80	410.13	1.439	404.86
58	KISMIJ	[73]	1-Butyl-3-methylimidazolium	$[Fe^{II}Cl_4]$	2:1	283–303	Cc	2.3402	4	1007.04	476.10	1.351	1007.00
58	QEKGUI	[148]	1-Ethyl-3-methylimidazolium	$[Au^{III}Cl_4]$	1:1	193	$P2_1/c$	1.2717	4	447.53	449.93	2.350	447.55
58	XAHYOW	[177]	Bis(2-azidoethyl)ammonium	$[N(CN)(NO_2)]$	1:1	100	$P\bar{1}$	0.5623	2	461.91	242.20	1.430	461.89
58	YOVVOV	[178]	5-Amino-1,3-dimethyltetrazol-3-ium	$[N(NO_2)_2]$	1:1	150	$P2_1/c$	0.92872	4	485.50	220.17	1.575	485.55
58/57	HATRIF	[122]	1-Butyl-3-methylimidazolium	$[HgBr_3]$	1:1	170	Cc	1.4211	4	435.04	579.52	2.709	435.07
58	YOYMEE	[179]	1,3,5-Trimethyl-tetrahydr*o*-1,3,5-triazinium	$[I_3]$	1:1	283–303	$Pnma$	1.3934	4	437.22	508.90	2.426	437.24
58	NEGPAS	[180]	1-Ethyl-3-methylimidazolium	$[La\{C(CN)_2NO\}_6]$	3:1	123	P(3bar)c1	2.29106	2	432.63	1036.77	1.503	
58–59	VUDPAM	[8]	1,3-Diamino-2-methylimidazolium	$[N(SO_2CF_3)_2]$	1:1	173	$P2_1/n$	1.45261	4	376.61	393.29	1.798	432.62
58–60	VIVVEC	[181]	1-Benzyl-3,5-bis(3-trifluoromethylphenyl) pyridinium	$[CF_3SO_3]$	1:1	130	$P2_1/n$	2.54377	4	465.44	607.48	1.586	376.61
58–60	HAYBUE	[182]	1-Ethyl-3-methylimidazolium	$[PF_6]$	1:1	283–303	$P2_1/c$	1.092	4	472.89	256.1	1.558	465.48
65	HAYBUE02	[99]	1-Ethyl-3-methylimidazolium	$[PF_6]$	1:1	173	$P2_1/c$	1.0272	4	414.64	256.14	1.656	472.89
59	IBEFIE	[183]	1-Ethyl-3-methylimidazolium	$[1\text{-}MeCB_{11}H_{11}]$	1:1	273	$P2_1/c$	1.7196	4	493.08	268.21	1.036	414.66

(*Continued*)

TABLE 11.2 (Continued)

m.pt./°C	CSD Refcode	References	Cation [cat]$^{q+}$	Anion [A]$^{p-}$	Salt Type (p:q)	Data Temp/K	Space Group	V_{cell}/ 10^3 Å^3	Z	U_L(V)/ kJ mol^{-1}	Formula Weight	ρ(calc)/ g cm^{-3}	U_L(D)/ kJ mol^{-1}
59	EBULAO	[184]	*N*-Amino-1-azoniacycloheptane	$[N_3]$	1:1	200	$P2_1/c$	0.8755	4	459.19	157.22	1.1927	493.09
59	QONYAU	[185]	1-(Pent-2-ynyl)-3-methylimidazolium	$[N(CN)_2]$	1:1	283–303	$P\bar{1}$	0.5753	2		215.26	1.243	459.24
59	SAMKIC	[186]	Tetrabutylammonium	$[Sm\{C(CN)_2NO\}_6]$	3:1	170	$R32$	5.9256	3		1442.08	1.212	
59–60	XISZAB	[187]	1-Butylisoquinolinium	$[GaCl_4]$	1:1	113	$P2_1/c$	1.6702	4	417.68	397.81	1.582	417.69
59–61	DAZFAJ	[188]	Piperidinium	$[B(OMe)_4]$	1:1	283–303	$Pnma$	1.31863	4	443.40	221.18	1.114	443.41
59–62	HOFTIG	[189]	1,4-Bis(1-butylimidazolium-3-ylmethyl)benzene	$[CF_3SO_3]$	1:2	100	$P\bar{1}$	1.45336	2	1256.35	650.66	1.487	1256.41
59.5–61	MIGDOV	[106]	1,1‴-Bis(6-bromohexyl)-1′,1″-biferrocenium	$[I_3]$	1:1	150	$P\bar{1}$	1.70768	2	351.09	1076.86	2.094	351.09
60	KISMUV	[73]	1-Butyl-3-methylimidazolium	$[ZnCl_4]$	2:1	283–303	Cc	2.3211	4	1010.28	485.63	1.390	1010.42
60	YIHTAL	[190]	1-Allyl-3-methylimidazolium	Br	1:1	100	$P2_1$	0.4301	2	495.37	203.08	1.568	495.39
60	HIWNOQ	[191]	1-Dodecyl-3-methylimidazolium	$[PF_6]$	1:1	123	$P2_1/a$	2.0007	4	399.34	396.40	1.32	399.66
60	ZASQAN	[192]	1,3-Dimethylimidazolium	$[CH(NO_2)_2]$	1:1	100	$P\bar{1}$	0.8896	4	491.01	202.18	1.510	491.07
60	TASZIX	[193]	(3-Methylimidazole)(trimethylamine)dihydroboronium	$[N(SO_2CF_3)_2]$	1:1	173	$P\bar{1}$	0.922	2	407.49	434.20	1.564	407.51
60	PAZPEL	[194]	Dimethyl(propyl)phenylammonium	$[I_3]$	1:1	283–303	$P2_1/n$	1.6404	4	419.57	544.98	2.207	419.60
60	FOXMEK	[195]	Pyridinium	$[HCl_2]$	1:1	153	$P\bar{1}$	0.3621	2	518.49	152.04	1.394	518.47
60	YEXNEU01	[196]	1-Methylimidazolium	$[NO_3]$	1:1	173	$P2_1/c$	0.6309	4	538.00	145.13	1.528	538.03
60	VURCAN	[197]	2-Methylimidazolium	$[HO_2C(CH_2)_6CO_2]$	1:1	120	$P\bar{1}$	0.67405	2	440.91	256.30	1.263	440.95
60–65 est	VIGLUT	[66]	1-Butyl-3-methylimidazolium	$[CH_3SO_3]$	1:1	173	$P\bar{1}$	0.60067	2	454.12	234.31	1.296	454.18
60–70	QAMBAJ	[198]	1,3-Di(benzyloxy)imidazolium	Br	1:1	173	$Pca2_1$	1.7084	4	415.32	361.24	1.404	415.30
60.5	GUTVIB	[199]	1,1,1-Trimethylhydrazinium	$[N(CN)_2]$	1:1	173	$Pbca$	1.5941	8	505.45	141.19	1.177	505.52
60–61.5	MORZAV	[200]	3,3-Diphenyltricyclo[2.1.0.0^{2,5}]pentaphosphan-3-ium	$[GaCl_4]$	1:1	153	$P\bar{1}$	2.97744	6	400.12	520.57	1.742	400.14
60–62	LUVQEZ	[201]	1-{(3-(Dimethylammonio)propyl}-2-hydroxypyridinium	$[CF_3SO_3]$	1:2	123	$P\bar{1}$	0.99501	2	1376.28	480.40	1.603	1376.22
61	AYIBIT	[119]	1-Allyl-2,3-dimethylimidazolium	Br	1:1	283–303	$P2_1/n$	1.0376	4	471.65	217.11	1.390	471.69
61	FOZRIW	[170]	Tetramethylammonium	$[Al\{OCH(CF_3)_2\}_4]$	1:1	100	$P2_1/c$	2.7822	4	368.58	769.25	1.837	368.62
61	ICOJIV	[202]	1-Methyl-3-(2-phenylethyl)imidazolium	$[BF_4]$	1:1	100	$P2_1$	0.64860	2	445.26	274.07	1.403	445.26

61	OGULEI	[203]	1-Butyl-2-methyl-3-benzylimidazolium	I	1:1	100	$P2_12_12_1$	1.550	4	425.59	356.24	1.52	425.14
61	OGULIM	[203]	1-Pentyl-2-methyl-3-benzylimidazolium	I	1:1	100	*Pbca*	6.793	#	415.94	370.26	1.44	415.37
62	KISMOP	[73]	1-Butyl-3-methylimidazolium	$[Co^{II}Cl_4]$	2:1	298	*Cc*	2.3215	4	1010.21	479.18	1.371	1010.26
62	DOQGEW	[129]	1-(2,2-Diphenylethyl)-3-methylimidazolium	$[N(SO_2CF_3)_2]$	1:1	283–303	$P2_1/n$	2.4513	4	380.00	543.5	1.47	379.84
62	ATUBAT	[132]	1,1,3,3-Tetramethylguanidinium	$[C_2H_5CO_2]$	1:1	173	$Pna2_1$	4.3679	#	465.44	189.26	1.151	465.44
62	BAVJAL	[159]	1-(Chloromethyl)-1,1-dimethylhydrazinium	$[SO_4]$	2:1	110	C_2/c	1.3518	4	1245.07	315.22	1.549	1245.18
62	VECKEU	[204]	Morpholinium	$[CF_3C(O)CHC(O)CF_3]$	1:1	86	$P2_1/n$	2.421240	8	453.22	295.19	1.620	453.27
62	NEJPOJ	[205]	Bis(1-methyl-1*H*-imidazole-3-yl)dihydroboronium	$[BH_2(CN)_2]$	1:1	296	$P\bar{1}$	0.68306	2	439.42	241.91	1.139	435.87
62–64	RAFBIL	[206]	1-Methyl-2-phenyl-3-propylimidazolinium	$[PF_6]$	1:1	173	*C2/c*	3.0434	8	427.57	348.27	1.520	427.58
62–64	FESCOV	[207]	Tetrabutylammonium	$[SiMePhF_3]$	1:1	283–303	$P2_1/c$	2.60415	4	374.48	419.69	1.070	374.46
62.5/70.5	No refcode	[208]	Hydrazinium	$[NO_3]$	1:1	120	$P2_1/n$	0.36585	4	624.49	95.07	1.726	624.51
63	No refcode	[209]	Potassium	$[N(SO_2F)_2]$	1:1	113	$P2_1/n$	1.18655	8	547.00	219.23	2.454	546.99
63	No refcode	[209]	Potassium	$[N(SO_2F)_2]$	1:1	298	$P2_1/n$	1.2232	8	542.52	219.23	2.381	542.55
63	KISMEF	[73]	1-Butyl-3-methylimidazolium	$[Mn^{II}Cl_4]$	2:1	283–303	*Cc*	2.3618	4	1003.41	475.19	1.336	1003.35
63	KENFOZ	[210]	1-(2-Methoxyethyl)-2,3-dimethylimidazolium	$[PF_6]$	1:1	173	$P\bar{1}$	0.61252	2	451.84	300.19	1.628	451.89
63	SIKXOA	[149]	1-{(2-Methoxyethyl)oxymethyl}-3-methylimidazolium	I	1:1	100	$P2_1/n$	1.13965	4	460.33	298.12	1.738	460.38
63	MIHRAX	[169]	1-Propyl-4-amino-1,2,4-triazolium	Br	1:1	100	$P\bar{1}$	0.8323	4	499.70	207.08	1.653	499.76
63	NEGPEW	[180]	1-Ethyl-2,3-dimethylimidazolium	$[Pr\{C(CN)_2NO\}_6]$	3:1	123	*Pa*(3bar)	10.62580	8		1080.85	1.351	
63	ZEJFUR	[211]	1-*iso*-Propyl-3-methyl-4-diphenylphosphorylimidazolium	$[SO_3(OH)]$	1:1	123	$P2_1/c$	2.01708	4	398.56	422.42	1.391	398.56
63–65	WOLLEP	[92]	1-Methyl-3-cyanopyridinium	$[N(SO_2CF_3)_2]$	1:1	193	$P2_1/n$	1.4357	4	433.91	399.32	1.847	433.90

(*Continued*)

TABLE 11.2 (*Continued*)

m.pt./°C	CSD Refcode	References	Cation $[cat]^{q+}$	Anion $[A]^{p-}$	Salt Type (*p*:*q*)	Data Temp/K	Space Group	V_{cell}/ 10^3 Å^3	*Z*	U_L(V)/ kJ mol^{-1}	Formula Weight	*ρ*(calc)/ g cm^{-3}	U_L(D)/ kJ mol^{-1}
63–65	RIJQAC	[212]	{Tris(dimethylamino) phosphaneselenide}iodine(I)	$[I_3]$	1:1	103	$Pna2_1$	1.8108	4	409.33	749.76	2.750	409.34
63–67	MIHQUQ	[169]	1-Ethyl-4-amino-1,2, 4-triazolium	Br	1:1	283–303	$P2_1/n$	0.7324	4	516.94	193.05	1.751	516.98
64	IBEFOK	[183]	1-Ethyl-3-methylimidazolium	$[1\text{-EtCB}_{11}H_{11}]$	1:1	198	$P2_1/c$	1.7855	4	410.77	282.24	1.050	410.79
64	YUGKAN	[213]	1,3-Didodecylimidazolium	$[ClO_4]$	1:1	200	$P2_1/n$	3.0163	4	361.54	505.16	1.112	361.53
64	RECSEY	[214]	1-Ethyl-3-methylimidazolium	$[Au(CN)_2]$	1:1	283–303	$P2_1/c$	1.0579	4	469.28	360.17	2.261	469.28
64	ZEKYOD	[215]	Tris(dimethylamino)sulfonium	$[P_3N_3F_5NPF_2NPF_2NPF_5]$	2:1	173	$P\bar{1}$	1.7701	2	854.20	864.47	1.622	854.26
64–65	QQQCWA	[216]	Tridodecylammonium	$[SO_3(OH)]$	1:1	283–303		16.9746	16	333.82	618.05	0.971	334.12
64–66	SADMOA	[217]	1-(-2-Hydroxy-2-methyl-2-phenylethyl)-3-*tert*-butylimidazolium	Cl	1:1	150	$P\bar{1}$	1.64829	4	419.06	294.81	1.188	419.08
64–66	AQUQAE	[218]	Tetrabutylphosphonium	$[SnPh_2Cl_3]$	1:1	150	$Pna2_1$	3.08695	4	359.56	638.66	1.374	359.57
64–66	VAQTEN	[219]	2-Amino-4-methylpyrimidinium	$[3\text{-FC}_6H_4CO_2]$	1:1	203	$P2_1/c$	1.1539	4	458.85	249.25	1.435	458.89
64–67	AJUWAD01	[220]	1,6-Bis(*P*,*P*,*P′*,*P′*-tetraphenylphosphinophosphonium)hexane	$[GaCl_4]$	1:2	193	*I*2/a	5.76006	4	1074.72	1247.88	1.439	1074.75
65	KEXBEV	[98]	1-Allyl-3-methylimidazolium	I	1:1	100	$P2_1$	0.48329	2	480.45	250.08	1.718	480.43
65	FIQKUL	[221]	Ethyl(methyl) diphenylammonium	$[I_3]$	1:1	283–303	*C*2/*c*	3.6712	8	407.95	593.04	2.146	407.97
65	SUSYEL	[222]	1,3-Dimethylimidazolium/lithium	$[N(SO_2CF_3)_2]$	1:1	110	$P\bar{1}$	1.11151	2	389.14	664.38	1.985	389.15
65	846081	[79]	Bis(1-ethylimidazole)silver(I)	$[N(SO_2CF_3)_2]$	1:1	100	$P2_1/n$	2.03494	4	397.68	580.28	1.894	397.69
65	FASTAV	[223]	Tetramethylformamidinium	$[Sn^{II}Cl_3]$	1:1	153	$P2_12_12_1$	1.12543	4	461.82	326.24	1.925	461.81
ca. 65	Cif	(UoL, unpublished work)	Tetrabutylammonium	$[PF_3(C_2F_5)_3]$	1:1	130	*Pbca*	5.86139	8	364.03	687.48	1.558	364.04
65–66	WOLKUE	[92]	1-Methyl-4-cyanopyridinium	$[N(SO_2CF_3)_2]$	1:1	193	$P2_1/n$	1.5048	4	428.78	399.32	1.763	428.82
65–66	VUKSEA	[224]	4-Amino-1-ethyl-3,5-dimethyl-1,2,4-triazolium	$[N(SO_2CF_3)_2]$	1:1	173	$P\bar{1}$	0.8171	2	419.96	421.35	1.71	419.82
65–66	IGIQOE	[225]	Diphenyliodonium	$[N(SO_2CF_3)_2]$	1:1	283–303	*Pbca*	3.8566	8	403.00	561.25	1.93	402.85

65–66	ISEKEX	[226]	1-Heptyl-4-(1*H*,1*H*,2*H*,2*H*-perfluorooctyl)-1,2,4-triazolium	$[N(SO_2CF_3)_2]$	1:1	233	*Cc*	6.3486	8	357.20	794.52	1.663	357.23
65–66	LUHVOZ	[227]	2-Benzylpyridinium	$[Sb^{III}{}_2Cl_{10}]$	4:2	283–303	$P\bar{1}$	2.67032	2		1278.98	1.591	
65–69	TAFZUV	[228]	Tetrabutylammonium	$[Au(CN)_2]$	1:1	283–303	*A*2/*n*	2.331	4	384.67	491.48	1.40	384.65
65	IWAHIY	[229]	Bis(ethanenitrile)copper(I)	$[N(SO_2CF_3)_2]$	1:1	100	$P\bar{1}$	0.7211	2		425.80	1.961	
	IWAHIY01	[229]	Bis(ethanenitrile)copper(I)	$[N(SO_2CF_3)_2]$	1:1	200	*C*2/*c*	1.47379	4		425.80	1.919	
66	IWAHAQ	[229]	Tetrakis(ethanenitrile)copper(I)	$[N(SO_2CF_3)_2]$	1:1	100	$P2_1/c$	7.9149	16		507.91	1.705	
	IWAHAQ01	[229]	Tetrakis(ethanenitrile)copper(I)	$[N(SO_2CF_3)_2]$	1:1	283–303	$P2_1/m$	1.0642	2		507.91	1.585	
66	MIHRUR	[169]	1-*iso*-Propyl-4-amino-1,2,4-triazolium	$[NO_3]$	1:1	100	$P\bar{1}$	0.4383	2	492.92	189.19	1.434	492.98
66	TAJCUD	[230]	1-Butyl-3-methylimidazolium	Cl	1:1	173	$P2_1/c$	0.9666	4	480.45	174.67	1.200	480.44
66	TAJCUD01	[230]	1-Butyl-3-methylimidazolium	Cl	1:1	173	$Pna2_1$	0.9611	4	481.16	174.67	1.207	481.17
66	TAJCUD02	[231]	1-Butyl-3-methylimidazolium	Cl	1:1	283–303	$P2_1/n$	0.99082	4	477.35	174.67	1.171	477.38
No m.pt.	TAJCUD03	[232]	1-Butyl-3-methylimidazolium	Cl	1:1	150	$P2_1/c$	0.9606	4	481.23	174.67	1.208	481.27
No m.pt.	TAJCUD04	(C. Pulham, J. Pringle, A. Parkin, S. Parsons, and D. Messenger, Cambridge Structural Database, private communication, 2005.)	1-Butyl-3-methylimidazolium	Cl	1:1	150	$Pna2_1$	0.94413	4	483.41	174.67	1.229	483.45
66	QOQBAZ	[233]	1-Heptyl-1-methylpyrrolidinium	$[PF_6]$	1:1	283–303	$P2_1$	1.6236	4	420.65	329.31	1.347	420.65
66	LODWOR	[234]	1-(But-2-ynyl)-3-methylimidazolium	$[N_3]$	1:1	173	$P\bar{1}$	0.4594	2	486.87	177.22	1.281	486.87
66	UWINIY	[235]	Hydroxylammonium	1-(2-Nitratoethyl)-5-nitrimino-tetrazolate	1:1	173	*C*2/*c*	1.8655	8	484.95	252.17	1.796	484.98
66	UQILAI	[85]	Bis(1-dodecylimidazole)copper(I)	$[Cu^IBr_2]$	1:1	193	$P\bar{1}$	1.7284	2	350.10	759.69	1.460	350.13
66	HOJHET	[102]	Bis(dimethylsulfoxide-*O*)trimethyltin(IV)	$[N(SO_2F)_2]$	1:1	183	$P\bar{1}$	0.9313	2	406.47	500.22	1.784	406.50
66	NEVREN	[236]	1-Ethyl-3-methylimidazolium	$[Cu\{(N(SO_2CF_3)_2\}_2]$	1:1	100	*C*2/*c*	2.4682	4	379.36	735.01	1.978	379.38

(*Continued*)

TABLE 11.2 (*Continued*)

m.pt./°C	CSD Refcode	References	Cation [cat]$^{q+}$	Anion [A]$^{p-}$	Salt Type (p:q)	Data Temp/K	Space Group	V_{cell}/ 10^3 Å^3	Z	U_L(V)/ kJ mol^{-1}	Formula Weight	ρ(calc)/ g cm^{-3}	U_L(D)/ kJ mol^{-1}
66–68	PIGHET	[237]	1-(2-Hydroxyethyl)pyridinium	Cl	1:1	93	*Pbca*	1.5263	8	511.32	159.61	1.389	511.32
66–68	YATLEM	[238]	1-(Ferrocen-1-ylmethyl)-3-methylimidazolium	$[PF_6]$	1:1	100	*Pbca*	3.20825	8	421.92	426.13	1.764	421.92
67	IBEFUQ	[183]	1-Methyl-3-octylimidazolium	$[CB_{11}H_6Cl_6]$	1:1	198	$P2_12_12_1$	2.6986	4	371.29	544.99	1.341	371.28
67	YUGJUG	[213]	1,3-Didodecylimidazolium	$[BF_4]$	1:1	200	$P\bar{1}$	3.0442	4	360.75	492.52	1.075	360.80
67	UFALAO	[172]	1,1-Dimethyluronium	$[N(SO_2Me)_2]$	1:1	178	$Pna2_1$	2.2252	8	463.19	261.32	1.560	463.21
67	WODLAD	[239]	1-Butylpyrrolidinium	$[CF_3C(O)CHC(O)C_6H_5]$	1:1	90	$P2_1/n$	1.801	4	409.89	343.38	1.266	409.87
67	PAMTAA	[240]	1-Hexyl-3-methylimidazolium	$[Co^{II}Br_3(C_9H_7N)]$	1:1	173	$P2_1/c$	2.26333	4	387.44	595.09	1.746	387.43
67	ITUSAS	[241]	1-Cyanomethyl-3-methylimidazolium	$[N(CN)_2]$	1:1	173	$P2_1/c$	0.9400	4	483.96	188.20	1.330	484.00
67	TAPBAP	[242]	1,2-Dimethylhydrazin-1,2-diium	$[SO_4]$	2:2	110	$P2_1/n$	0.582064	4		158.18	1.805	
67	EBUDOV01	[112]	1,3-Dipropylimidazolium	Br	1:1	110	$P2_1/c$	1.1729	4	456.92	233.16	1.320	456.91
No m.pt.	EBUDOV	[243]	1,3-Dipropylimidazolium	Br	1:1	110	$P2_1/c$	1.1729	4	456.92	233.16	1.320	456.91
67–68	XUTPOR	[244]	1,3-Dimethylimidazolium	$[PF_6]$	1:1	283–303	*Pbca*	1.89472	8	482.98	242.10	1.697	482.97
67–68	XUTPOR01	[244]	1,3-Dimethylimidazolium	$[PF_6]$	1:1	173	*Pbca*	1.89389	8	483.03	242.10	1.698	483.04
67–73	MUJWIX	[245]	2-Diethylamino-1,3-dimethylimidazolinium	$[SiMe_3F_2]$	1:1	173	$P2_1/n$	0.8250	2	418.95	281.47	1.133	418.96
68	SIKXEQ	[149]	1-Ethoxymethyl-3-methylimidazolium	I	1:1	140	$P2_1/n$	2.09046	8	470.75	268.09	1.704	470.80
68	QIQKEG	[64]	Methylhydrazinium	$[N_3]$	1:1	200	*Pnma*	0.47177	8	706.52	44.55	1.2545(2)	706.56
68	EBUKOB	[184]	Bis(phenylhydrazine)hydrogen	$[N_3]$	1:1	183	$P2_1/c$	1.3346	4	442.04	259.33	1.291	442.09
68	SAMKEY	[186]	Tetrabutylammonium	$[Ce\{C(CN)_2NO\}_6]$	3:1	123	$R32$	5.8859	3		1431.85	1.212	
68	HOGGOA	[246]	1-Butyl-3-methylimiazolium	$[Eu\{N(SO_2CF_3)_2\}_4]$	1:1	153	$P\bar{1}$	2.311	2	327.37	1411.78	2.029	327.39
68–69	CAFZIS	[247]	9-Aza-9-methyl-1-thioniabicyclo[3.3.1]nonane	$[I_3]$	1:1	140	$P2_12_12_1$	1.4243	4	434.79	539.00	2.51	434.65
68–69	LAZCAQ	[248]	Tris(2-methoxyethyl)ammonium	Cl	1:1	283–303	$P2_1/c$	1.29753	4	445.24	227.73	1.166	445.28
68–70	GUNRUD	[249]	Ethyltriphenylphosphonium	2,4-Dinitro-imidazolate	1:1	173	$P2_12_12_1$	2.17036	4	391.43	448.41	1.372	391.43
68–71	HOFTEC	[189]	1,4-Bis(1-methylimidazolium-3-ylmethyl)benzene	$[BF_4]$	1:2	100	$P2_1/c$	0.97258	2	1383.99	441.98	1.509	1383.98
68–72	MUTGUD	[250]	Tetrakis(ethanenitrile)silver(I)	$[Ag_3\{N(NO_2)_2\}_4]$	Catenate	193	$I4_1/a$	2.7226	4	370.50	1019.82	2.488	370.51

68.5–69.5	MIGDUB	[106]	1,1‴-Bis(5-bromopentyl)-1′,1″-biferrocenium	$[I_3]$	1:1	283–303	$P\bar{1}$	0.8180	1	354.65	1048.81	2.129	354.66
69	RECKEP	[135]	1,3,5-Tri-*iso*-propyl-tetrahydro-1,3,5-triazinium	$[I_5]$	1:1	283–303	$P2_1/c$	2.3500	4	383.91	846.88	2.394	383.94
	FOYGUC	[251]	*N,N*‴-Bis(2-ethoxyethyl)-*N,N,N*‴,*N*‴-tetramethylbutane-1,4-diammonium	$[CF_3SO_3]$	1:2	153	$P2_1/n$	1.353	2	1277.85	588.62	1.445	1277.92
	MALDAG	[77]	1-Methyl-1-propylpiperidinium	$[Al\{OCH(CF_3)_2\}_4]$	1:1	113	$P2_12_12_1$	3.2191	4	356.01	837.39	1.728	356.04
69	CEDHUQ	[252]	1-Propyl-1,2,4-triazolium	5-Nitroimino-tetrazolate	1:1	89	$P\bar{1}$	0.5927	2	455.68	241.24	1.352	455.72
69/67	HATQUQ	[122]	1-Methyl-3-propylimidazolium	$[HgCl_3]$	1:1	283–303	Cc	1.2987	4	445.13	432.14	2.210	445.14
69	SEDBUZ01	[253]	1-(3-Cyanopropyl)-3-methylimidazolium	I	1:1	140	$P\bar{1}$	0.5454	2	465.57	277.11	1.687	465.56
No m.pt.	SEDBUZ	[254]	1-(3-Cyanopropyl)-3-methylimidazolium	I	1:1	148	$P\bar{1}$	0.5464	2	465.35	277.11	1.680	465.06
69	ODIHUF	[84]	2-Hydroxyethyl(trimethyl)ammonium (cholinium)	Saccharinate	1:1	100	*Pbca*	2.7543	8	438.52	286.35	1.381	438.54
69–71	VUDPIU	[8]	1-{2-(Diethylammonio)ethyl}-3-methylimidazolium	$[N(SO_2CF_3)_2]$	1:2	233	$P\bar{1}$	1.43265	2	1260.62	743.60	1.724	1260.69
70	FIYPUZ	[255]	1-Butyl-3-methylimidazolium	$[Cu^{II}{}_3Cl_8]$	2:1	193	$P\bar{1}$	1.44157	2	927.37	752.66	1.734	927.42
70	SIKYAN	[149]	1-{2-(2-Methoxyethyl)oxyethyl}-3-methylimidazolium	I	1:1	140	$P2_1/n$	1.2374	4	450.68	312.15	1.676	450.73
70	QONROB	[168]	1*H*-Tetrazolium	$[N(NO_2)_2]$	1:1	200	$C2/c$	0.645	4	534.82	177.10	1.824	534.86
70	FUTGUX	[256]	Bis(4-methoxyphenalene-1,9-diolato)boron	$[B\{3,5\text{-}(CF_3)_2C_6H_3\}_4]$	1:1	100	$C2/c$	5.431	4	315.66	1324.46	1.620	315.68
70	YESTAS	[174]	$[(NMe_3CH_2CO_2)_2H]$	$[N(SO_2Me)_2]$	1:1	283–303	$P2_1/c$	2.6781	2		1312.13	1.627	
70	PAZPIL	[194]	Dimethyl(*iso*-propyl)phenylammonium	$[I_3]$	1:1	283–303	$P2_1/n$	1.63848	4	419.69	544.98	2.214	419.93
.70	MAPPEA	[257]	1-Hexyl-3-methylimidazolium	$[4\text{-}CH_3C_6H_4SO_3]$	1:1	173	$P2_1/c$	1.8447	4	407.45	338.46	1.219	407.49
70	724092	[258]	1,3-Dibutylbenzimidazolium	$[Fe^{III}Cl_4]$	1:1	283–303	$P2_1/c$	2.1022	4	394.51	429.02	1.356	394.56
70	ZUWCAV	[259]	Triphenylphosphonium	$[N(SO_2F)_2]$	1:1	143	$P2_1/n$	1.9443	4	402.17	443.43	1.515	402.20
70	TAJDAK01	[260]	1-Butyl-3-methylimidazolium	Br	1:1	283–303	$Pna2_1$	1.0258	4	473.06	219.13	1.419	473.09
79	TAJDAK03	[112]	1-Butyl-3-methylimidazolium	Br	1:1	110	$Pna2_1$	0.9915	4	477.27	219.13	1.468	477.29
No m.pt.	TAJDAK	[230]	1-Butyl-3-methylimidazolium	Br	1:1	173	$Pna2_1$	1.0066	4	475.39	219.13	1.446	475.41

(Continued)

TABLE 11.2 (*Continued*)

m.pt./°C	CSD Refcode	References	Cation $[cat]^{q+}$	Anion $[A]^{p-}$	Salt Type (p:q)	Data Temp/K	Space Group	V_{cell}/ 10^3 Å^3	Z	$U_L(V)$/ kJ mol^{-1}	Formula Weight	ρ(calc)/ g cm^{-3}	$U_L(D)$/ kJ mol^{-1}
No m.pt.	TAJDAK02	[243]	1-Butyl-3-methylimidazolium	Br	1:1	110	$Pna2_1$	0.9915	4	477.27	219.13	1.468	477.29
70 (dec)	KIFHOW	[261]	Aminoguanidinium	$[N(NO_2)(SF_5)]$	1:1	283–303	$P\bar{1}$	0.44495	2	490.97	262.19	1.957	490.99
70.5	No refcode	[262]	Hydrazinium	$[N_3]$	1:1			0.35417	4	630.15	75.07	1.407	
71	ZIBHUN01	[142]	1-Ethyl-3-methylimidazolium	Br	1:1	123	$P2_1/c$	0.80257	4	504.53	191.08	1.581	504.52
77	ZIBHUN02	[112]	1-Ethyl-3-methylimidazolium	Br	1:1	110	$P2_1/c$	0.8089	4	503.48	191.08	1.569	503.50
81	ZIBHUN	[263]	1-Ethyl-3-methylimidazolium	Br	1:1	283–303	$P2_1/c$	0.83316	4	499.57	191.1	1.523	499.54
71	JIMWAE	[264]	1,4-Bis(1-methylimidazolium-3-yl)butane	$[CF_3SO_3]$	1:2	100	$P\bar{1}$	1.06179	2	1354.63	518.46	1.622	1354.73
71	AWEHIT	[265]	1,10-Bis(1-methylimidazolium-3-yl)decane	$[PF_6]$	1:2	173	$P2_1/c$	1.28456	2	1293.78	594.40	1.537	1293.86
71	VECKIY	[204]	Dibutylammonium	$[CF_3C(O)CHC(O)$ (2-thienyl)]	1:1	83	$P2_1/c$	1.82576	4	408.50	351.43	1.278	408.47
71	BAZFAK	[266]	1-Aza-8-azoniabicyclo[6.6.3] heptadeca-4,11-diyne	$[(CF_3CO_2)_2H]$	1:1	200	$P2_1/c$	2.12312	4	393.55	458.40	1.434	393.56
71	OGUKUX	[203]	1-Propyl-2-methyl-3-benzylimidazolium	I	1:1	153	$P\bar{1}$	0.704	2	436.06	342.21	1.61	435.78
71–73	DEVDIR	[267]	Diethylammonium	$[N(SO_2Me)_2]$	1:1	143	$P\bar{1}$	0.5994	2	454.36	246.34	1.365	454.39
72	VUDPEQ	[8]	1,3-Dihydroxy-2-methylimidazolium	$[N(SO_2CF_3)_2]$	1:1	233	$P2_1/c$	1.42528	4	434.71	395.27	1.842	434.73
72	YOBVOB	[147]	1-Hexylpyridinium	[*nido*-$C_2B_9H_{12}$]	1:1	173	Cc	1.8927	4	404.86	296.66	1.041	404.87
72	XEWNET	[268]	1,5-Diamino-4-methyltetrazolium	$[C(NO_2)(CN)_2]$	1:1	90	$P\bar{1}$	0.50326	2	475.40	225.20	1.486	475.41
72	YOJQAQ	[269]	1,5-Diaminotetrazolium	$[N(NO_2)_2]$	1:1	200	$P2_1/c$	1.55335	8	508.94	207.14	1.771	508.92
72	MOYQUM	[270]	Trimethylselenonium	$[N_3]$	1:1	200	$P2_12_12_1$	1.28804	8	535.04	166.08	1.713	535.07
72	UCEFUF	[271]	1-Ethyl-3-methylimidazolium	[OCN]	1:1	123	$P2_1/m$	0.39972	2	505.05	153.19	1.273	505.10
72	GUFFAP	[272]	Aminoguanidinium	$[ClO_4]$	1:1	200	$P2_1/c$	0.6573	4	532.11	174.56	1.764	532.14
72	FASSOU	[223]	Tetramethylformamidinium	$[Ge^{II}Cl_3]$	1:1	123	$P2_1/n$	1.09885	4	464.69	280.17	1.693	464.67
72	BAZFEO	[266]	1-Aza-8-azoniabicyclo[6.6.4] octadeca-4,11-diyne	$[(CF_3CO_2)_2H]$	1:1	200	$P2_1/c$	2.20726	4	389.82	472.42	1.422	389.86
72	POCSIJ	[273]	Tetraethylammonium	$[BF_4]$	1:1	283–303	Cc	1.17822	4	456.39	217.06	1.224	456.44
72	POCSIJ01	[273]	Tetraethylammonium	$[BF_4]$	1:1	100	Cc	1.12427	4	461.94	217.06	1.282	461.93
72	XEFBUH	[274]	Tetramethylammonium	$[N(SO_2F)_2]$	1:1	283–303	$P2_1/m$	0.54861	2	464.86	254.28	1.539	464.86
72–73	PURKOD	[275]	1-Octadecyl-3-methylimidazolium	$[CF_3C(O)CHC(O)CF_3]$	1:1	100	$P\bar{1}$	2.8697	4	365.86	542.64	1.256	365.88

73	Cif	(UoL, unpublished work)	1-Butyl-3-methylimidazolium	$[CH_3SO_3]$	1:1	120	$P\bar{1}$	0.59728	2	454.78	234.41	1.303	454.76
73	ISEDUF	[276]	1-Ethyl-3-methylimidazolium	$[Ag(CN)_2]$	1:1	283–303	$Pbca$	2.1789	8	465.72	271.07	1.653	465.76
73	WOLLIT	[92]	1-Ethyl-3-cyanopyridinium	$[N(SO_2CF_3)_2]$	1:1	283–303	$P2_1/n$	1.64	4	419.59	413.34	1.674	419.60
73	ONOTOB	[277]	Tetrabutylphosphonium	$[B(CN)_4]$	1:1	283–303	$Pnna$	2.5665	4	375.80	374.31	0.969	375.84
73	VURKUP	[278]	1-Butyl-3-methyl-4, 5-dibromoimidazolium	$[CF_3SO_3]$	1:1	173	$P2_1/c$	1.51231	4	428.24	446.09	1.959	428.24
73	BUZSUK	[279]	Tris(2-hydroxyethyl)ammonium	$[(4\text{-}ClC_6H_4)SCH_2CO_2]$	1:1	153	$P2_1$	0.80863	2	421.07	351.85	1.445	421.08
73	VALWIQ	[280]	*N*-Methyl-*N*-(2-nitroxyethyl) ammonium	$[NO_3]$	1:1	173	$Pca2_1$	0.78908	4	506.87	183.13	1.542	506.87
73–75	MOYREX	[270]	Trimethylselenonium	[SeCN]	1:1	200	$P2_1/c$	0.75372	4	513.01	229.04	2.0185	513.04
73–79	QUSLUL	[281]	1-Hydroxypyridinium	$[N(SO_2Me)_2]$	1:1	178	$P\bar{1}$	0.5719	2	459.89	268.31	1.558	459.90
74	WEQXAR	[282]	1-Tetradecylpyridinium	$[Pd^{II}{}_2Cl_6]$	2:1	283–303	$P\bar{1}$	3.815	3	736.65	1527.1	1.994	736.67
74	UWUKON	[283]	1-Butyl-1-methylazepanium	$[CF_3CO_2]$	1:1	283–303	$Pbna$	2.98863	8	429.54	283.06	1.259	429.63
74	KUSZON	[284]	4-Cyanopyridinium	$[CF_3CO_2]$	1:1	283–303	$P2_1/n$	0.9381	4	484.22	218.13	1.544	484.20
74	LAFNOX	[285]	Butyltriphenylphosphonium	$[Bi_2I_8(Me_2CO)_2]$	2:1	283–303	$P\bar{1}$	1.59596	1	669.89	2188.08	2.277	669.97
74	YOKBOP	[286]	1-(1-Methyl-2-butyn-1-ylidene) pyrrolidinium	$[CF_3SO_3]$	1:1	283–303	$P2_1/c$	1.30841	4	444.28	285.30	1.448	444.28
74	GEVGEV	[144]	Tetrakis(1-ethylimidazole) copper(II)	$[N(SO_2CF_3)_2]$	1:2	100	$P\bar{1}$	1.0412	1	1155.69	1008.43	1.608	1155.68
74–75	ZIZJAT	[287]	*N*-(1-*O*-Methyl-2,3,4-tri-*O*-acetyl-α-D-glucopyranose-6-yl) pyridinium	$[NO_3]$	1:1	283–303	$P2_1$	1.0973	2	390.37	444.4	1.316	388.31
74–76	KEDTIX	[288]	Tributyl(ethyl)phosphonium	$[4\text{-}CH_3C_6H_4SO_3]$	1:1	283–303	$P2_1/n$	2.42589	4	380.96	402.55	1.102	380.96
75	IVASUT	[289]	1-(3-Cyanopropyl)-3-methylimidazolium	$[PF_6]$	1:1	140	$P\bar{1}$	0.5971	2	454.81	295.18	1.642	454.85
75	SIKXAM	[149]	1-Methoxymethyl-3-methylimidazolium	I	1:1	140	$P2_1/n$	0.94013	4	483.95	254.07	1.795	483.97
75	KUNYUO	[290]	1-Ethyl-3-methylimidazolium	$[V^{V}(O)F_4]$	1:1	100	$P2_1/c$	1.00744	4	475.29	254.10	1.675	475.28
75	NINGUN	[291]	Hexadecyldimethylsulfonium	Br	1:1	283–303	$P\bar{1}$	1.055	2	394.15	367.47	1.157	394.18
75	GUFFIX	[272]	Azidoformamidinium	$[ClO_4]$	1:1	200	$Pbca$	1.34958	8	528.38	185.54	1.826	528.38
75	BAZFOY	[266]	1-Aza-8-azoniabicyclo[6.6.5] nonadeca-4,11-diyne	$[(CF_3CO_2)_2H]$	1:1	200	$P\bar{1}$	1.15529	2	385.49	486.45	1.398	385.48

(Continued)

TABLE 11.2 (*Continued*)

m.pt./°C	CSD Refcode	References	Cation $[cat]^{q+}$	Anion $[A]^{p-}$	Salt Type (p:q)	Data Temp/K	Space Group	V_{cell}/ 10^3 Å^3	Z	$U_L(V)$/ kJ mol^{-1}	Formula Weight	ρ(calc)/ g cm^{-3}	$U_L(D)$/ kJ mol^{-1}
75–76	VIBNAW	[93]	1,3-Dimethoxy-2-methylimidazolium	$[PF_3(C_2F_5)_3]$	1:1	233	$P2_1/n$	2.06194	4	396.39	588.20	1.895	396.42
75–77	DEVBOV	[267]	Methylethylammonium	$[N(SO_2Me)_2]$	1:1	143	$P2_1/c$	1.0447	4	470.82	232.32	1.477	470.83
75–80	XOMQOF	[292]	$[Tl_3F_2Al\{OCH(CF_3)_2\}_3]$	$[Al\{OCH(CF_3)_2\}_4]$	1:1	160	$P\bar{1}$	2.2277	2	330.12	1874.32	2.794	330.12
75–82	EGEGAZ	[293]	1-Phenyl-4-triphenyl-phosphanylidene-4*H*-pyrazol-1-ium	$[BF_4]$	1:1	100	Cc	2.35322	4	383.77	492.25	1.389	383.77
76	RAFBAD	[206]	3-Methyl-2-phenyl-3,4,5,6-tetrahydropyrimidinium	$[CF_3SO_3]$	1:1	173	$Pbca$	2.87339	8	433.83	324.32	1.499	433.82
76	777587	[294]	1-Ethyl-3-methylimidazolium	$[(Ta^{V}Cl_5)_2O]$	2:1	223	$P2_1/c$	1.3973	2	938.93	954.74	2.269	938.94
76	MIHRIF	[169]	1-Hexyl-4-amino-1,2,4-triazolium	Br	1:1	283–303	$P2_1/c$	1.148	4	459.46	249.16	1.442	459.51
76	NIBKIU	[295]	2-Hydroxyethyl(trimethyl) ammonium (cholinium)	$[Cu^{II}Cl_4]$/Cl	3:(1,2)	100	$Pnma$	2.5674	4		553.33	1.431	
76–77	NADXIB	[296]	Decyl(tri-*tert*-butyl)phosphonium	$[BF_4]$	1:1	283–303	$Pccn$	5.1933	8	374.74	430.38	1.101	374.77
76–77	MOGZAJ	[297]	2-Hydroxyethyl(trimethyl) ammonium (cholinium)	$[N(SO_2Me)_2]$	1:1	143	$P2_12_12_1$	1.2576	4	448.81	276.37	1.460	448.86
77	ERABAA02	[99]	1-Tetradecyl-3-methyimidazolium	$[PF_6]$	1:1	173	$P2_1/c$	2.21562	4	389.46	424.45	1.272	389.44
77	ERABAA	[298]	1-Tetradecyl-3-methyimidazolium	$[PF_6]$	1:1	300	$P2_1/a$	2.33471	4	384.52	424.45	1.208	384.56
77	ERABAA01	[298]	1-Tetradecyl-3-methyimidazolium	$[PF_6]$	1:1	175	$P2_1/a$	2.20484	4	389.92	424.45	1.279	389.97
77	BAVJIT	[159]	1-(Chloromethyl)-1,1-dimethylhydrazinium	$[C_6H_2(NO_2)_3O]$	1:1	110	C_2/c	2.7291	8	439.55	337.69	1.644	439.59
77	REMMUT02	[299]	1-Ethyl-1,4-diazabicyclo[2.2.2] octanium	$[N(SO_2CF_3)_2]$	1:1	100	$P2_12_12_1$	1.67143	4	417.60	421.38	1.675	417.65
77	REMMUT	[299]	1-Ethyl-1,4-diazabicyclo[2.2.2] octanium	$[N(SO_2CF_3)_2]$	1:1	223	$P2_12_12_1$	1.71272	4	415.06	421.38	1.634	415.06
77	REMMUT01	[299]	1-Ethyl-1,4-diazabicyclo[2.2.2] octanium	$[N(SO_2CF_3)_2]$	1:1	283–303	$P2_12_12_1$	1.74443	4	413.16	421.38	1.604	413.15
77–79	IGIQIY	[225]	(2,2,2-Trifluoroethyl) phenyliodonium	$[N(SO_2CF_3)_2]$	1:1	158	$P2_1/n$	1.7809	4	411.03	567.19	2.12	411.27

78	TIRQIU	[300]	1-Vinyl-3-methylimidazolium	I	1:1	173	$P2_12_12_1$	1.706	8	496.47	236.05	1.838	496.49
78	AYICAM	[119]	1-Butyl-2, 3-dimethylimidazolium	$[SO_3(OH)]$	1:1	283–303	*Pbca*	2.4818	8	450.35	249.31	1.334	450.33
78	SIKXUG	[149]	1-{2-(2-Methoxyethyl)oxyethyl}-3-methylimidazolium	Br	1:1	100	$P2_1/c$	1.1955	4	454.69	265.16	1.473	454.69
78	MIKYEL	[301]	1-Butyl-3-methylimidazolium	$[4\text{-}MeC_6H_4SO_3]$	1:1	120	$P\bar{1}$	1.5885	4	422.97	310.41	1.298	422.98
78	MAKTAV	[302]	Dimethylbis(3-methylbenzyl) ammonium	$[N(SO_2CF_3)_2]$	1:1	153	$P2_1/n$	2.3776	4	382.82	534.53	1.493	382.82
78	RAPZIT	[303]	1-Methyl-1-propylpyrrolidinium	$[B(CN)_4]$	1:1	140	$P2_1/c$	1.4079	4	436.07	243.12	1.147	436.08
78	LOKFEW	[304]	$[(\text{Pyridine-1-oxide})_2H]$	$[ClO_4]$	1:1	283–303	$C2/c$	1.25791	4	448.78	290.66	1.535	448.82
78	YAVSET	[305]	Tetraethylammonium	$[(2\text{-}OH)C_6H_4COS]$	1:1	283–303	$Pna2_1$	1.6088	4	421.62	283.43	1.170	421.62
78	XAXROF01	[306]	Tetrabutylammonium	$[Y(BH_4)_4]$	1:1	100	$P2_1/c$	2.56444	4	375.87	390.75	1.012	375.88
78(est)	JIRCAP01	[307]	1-*iso*-Propyl-3-methylimidazolium	Br	1:1	90	$P3_2$	2.1163	9	483.89	205.10	1.448	483.88
110.5	JIRCAP	[112]	1-*iso*-Propyl-3-methylimidazolium	Br	1:1	110	$P2_12_12_1$	0.911	4	487.96	205.10	1.495	487.94
78–80	CUWYEZ	[308]	1-Butyl-4-methylpyridinium	$[PCl_6]$	1:1	283–303	$P2_1/c$	1.683	4	416.88	393.91	1.555	416.92
78–80	YAVSIX	[305]	Tetraethylammonium	$[(2\text{-}OH)C_6H_4CS_2]$	1:1	283–303	$Pna2_1$	1.6891	4	416.50	299.50	1.178	416.54
78.5–80	NALGOW	[155]	*N*-Propylurotropinium	$[I_5]$	1:1	283–303	$P2_1/n$	3.979	8	399.90	817.80	2.731	399.94
79	QICMUL	[309]	1,3-Bis-(3-cyanopropyl) imidazolium	$[PF_6]$	1:1	140	$P\bar{1}$	0.75048	2	429.06	348.24	1.541	429.07
79	LUQFIN	[310]	1,2,3-Trimethylimidazolium	$[CO_2(OMe)]$	1:1	173	$P2_1/n$	0.95113	4	482.48	186.21	1.300	482.46
79	AYIBOZ03	[311]	1-Butyl-2, 3-dimethylimidazolium	Cl	1:1	233	$R(\bar{3})$	4.9698	18	464.08	188.70	1.135	464.11
79	AYIBOZ02	[311]	1-Butyl-2, 3-dimethylimidazolium	Cl	1:1	233	$P2_1/n$	1.05778	4	469.30	188.70	1.185	469.33
100	AYIBOZ	[119]	1-Butyl-2, 3-dimethylimidazolium	Cl	1:1	283–303	$P2_1/n$	1.0558	4	469.52	188.70	1.187	469.53
No m.pt.	AYIBOZ01	(C. Pulham, J. Pringle, A. Parkin, S. Parsons, and D. Messenger, Cambridge Structural Database, private communication, 2005.)	1-Butyl-2, 3-dimethylimidazolium	Cl	1:1	150	$P2_1/n$	1.03613	4	471.82	188.70	1.210	471.88

(*Continued*)

TABLE 11.2 (*Continued*)

m.pt./°C	CSD Refcode	References	Cation $[cat]^{q+}$	Anion $[A]^{p-}$	Salt Type $(p:q)$	Data Temp/K	Space Group	$V_{cell}/10^3$ Å^3	Z	$U_L(V)$/kJ mol^{-1}	Formula Weight	ρ(calc)/g cm^{-3}	$U_L(D)$/kJ mol^{-1}
79	RETMOS	[312]	Ferrocenium	$[I_{29}]$	3:1	283–303	$P2_1/m$	3.933	2		4238.19	3.578	
79	HARJER	[196]	1,2-Dimethylimidazolium	$[NO_3]$	1:1	173	$P2_1/n$	0.75621	4	512.56	159.15	1.398	512.59
79	SURZOU	[70]	Tris(dimethylamino)sulfonium	*cyclo*-$[P_4N_4F_9]$	1:1	173	$P2_1/c$	1.9654	4	401.10	515.21	1.741	401.11
79–80	SEQKOO	[313]	3(*R*),4(*R*),5(*R*),6(*S*)-3,4,5-Trihydroxy-*cis*-1-thioniabicyclo[4.3.0]nonane	$[ClO_4]$	1:1	173	$P2_1$	1.7224	6	459.43	290.72	1.682	459.47
79–80	ISUXOJ	[165]	5-Methyl-1,3,5-dithiazinan-5-ium	$[BF_4]$	1:1	213	$P2_1/n$	0.92509	4	486.00	223.06	1.602	486.05
79–80	BAQKOV	[145]	3-Chloro-3-pentafluorophenyl tricyclo[2.1.0.0^{2,5}]penta phosphan-3-ium	$[GaCl_4]$	1:1	153	$P2_1$	0.8476	2	416.14	568.88	2.229	416.14
79–81	XOMDAE	[314]	Trimethylammonium	$[N(SO_2CF_3)_2]$	1:1	133	$P2_1/c$	1.2968	4	445.30	340.27	1.743	445.33
79–82	UYUGEB	[315]	1-Methyl-4-(benzylseleno-methyl)pyridinium	$[CF_3SO_3]$	1:1	130	$P2_1/c$	1.70584	4	415.47	426.32	1.660	415.49
79.5–82	OKEBIP	[316]	*N*,*N*-Dimethyl-*N*-((methylselenyl)(phenyl) methylidene) ammonium	$[CF_3SO_3]$	1:1	193	$P2_1/n$	1.47626	4	430.86	376.26	1.693	430.88
80	IVASON	[289]	1-(3-Cyanopropyl)-| 3-methylimidazolium	Cl	1:1	140	$P2_1/c$	0.9781	4	478.96	185.66	1.261	479.00
No m.pt.	IVASON01	[254]	1-(3-Cyanopropyl)-3-methylimidazolium	Cl	1:1	148	$P2_1/c$	0.97747	4	479.04	185.66	1.262	479.10
80	BACRAZ	[317]	1-Ethyl-3-methylimidazolium	$[1\text{-BuSnB}_{11}H_{11}]$	1:1	170	$P2_1/c$	2.0846	4	395.32	416.97	1.329	395.37
80	YOBVIV	[147]	1-Octylpyridinium	$[\textit{nido}\text{-}C_2B_9H_{12}]$	1:1	173	$P\bar{1}$	1.0403	2	395.51	325.72	1.04	395.54
80	QONVIZ	[318]	*N*,*N*′,*N*″-Triaminoguanidinium	$[N(NO_2)_2]$	1:1	100	*Pbca*	1.72357	8	495.13	211.17	1.628	495.19
80	SIGGUL	[319]	*N*,*N*′-Diaminoguanidinium	$[C(NO_2)_3]$	1:1	200	$P2_1/n$	0.9371	4	484.36	240.16	1.702	484.36
80	No refcode	[80]	Hydrazinium	$[N(NO_2)_2]$	1:1	232	$P2_1/n$	0.4998	4	573.06	139.09	1.848	573.05
80	HAWYIN	[320]	Tris(2-hydroxyethyl)ammonium	$[NO_3]$	1:1	283–303	$P2_1/c$	1.0112	4	474.82	212.20	1.393	474.77
80	MACMUA	[321]	*N*-Undecylbenzothiazolium	I	1:1	173	$P\bar{1}$	0.97005	2	402.39	417.39	1.429	402.41
80	YIQXAY	[322]	Dimethylammonium	$[NO_3]$	1:1	173	$Pna2_1$	0.55616	4	556.64	108.10	1.291	556.66

80	IHIKOA	[323]	1,3-Bis(dodecyl) benzimidazolium	$[Cd_2Cl_6]$	2:1	283–303	$P\bar{1}$	1.63992	1	662.24	1236.83	1.252	662.19
80–82	HILLET	[324]	*N*-Ethylurotropinium	$[I_8]$	2:1	283–303	$P\bar{1}$	1.6956	2	869.12	1353.74	2.651	869.09
80–82	GEWSUM	[325]	3-Hydroxy-1-*tert*-butyl-1,2-dihydropyrrolium	$[C_6H_2(NO_2)_3O]$	1:1	283–303	$P\bar{1}$	1.6782	4	417.18	368.3	1.458	417.22
80–82	FARWIF	[326]	(Pyridin-2-yl)(pyridinio-2-yl) disulfide	$[I_3]$	1:1	283–303	$P2_1$	3.3393	8	417.71	602.02	2.395	417.73
80–82	FAVSIF	[220]	{Diphenylphosphinomethyl (diphenyl)phosphino} diphenylphosphonium	$[CF_3SO_3]$	1:1	193	$P2_1$	1.73258	2	349.90	718.61	1.378	349.94
80–82	BOHSEW	[327]	Tetramethylstibonium	$[(C_6H_5CO_2)_2H]$	1:1	283–303	$Pna2_1$	2.01453	4	398.66	429.17	1.402	398.70
80–83	GOHHOA	[328]	Bis(2-pyridone)hydrogen	$[N(SO_2Me)_2]$	1:1	143	$P2_1/n$	1.59039	4	422.84	363.41	1.518	422.88
80–83	RIRWOF	[329]	1-{2-(Hydroxycarbonyl)ethyl} morpholinium	Cl	1:1	283–303	$P2_1/n$	0.97542	4	479.31	185.64	1.332	479.31
80–90	DEVFUF	[267]	Pyrrolidinium	$[N(SO_2Me)_2]$	1:1	143	$P\bar{1}$	0.5453	2	465.59	244.33	1.488	465.61
81	DUVZAV10	[263]	1-Ethyl-3-methylimidazolium	I	1:1	283–303	$P2_1/c$	0.9117	4	487.86	238.07	1.734	487.85
No m.pt.	DUVZAV	[330]	1-Ethyl-3-methylimidazolium	I	1:1		$P2_1/c$	0.91171	4	487.86	238.07	1.734	487.85
81	AXOSUC	[331]	1-Butyl-1-methylpyrrolidinium	$[Fe^{III}Cl_4]$	1:1	100	$P6_3mc$	0.76302	2	427.26	339.91	1.479	427.25
81	MEHWIG	[332]	Tetrabutylammonium	2,4-Dinitro-imidazolate	1:1	173	$P\bar{1}$	2.262	4	387.49	399.54	1.173	387.49
81	EXIGEY	[235]	Ammonium	1-(2-Nitratoethyl)-5-nitrimino-tetrazolate	1:1	173	$P2_12_12_1$	0.92683	4	485.76	236.17	1.693	485.81
81	VEPFOL02	(G.J. Reiss, Cambridge Structural Database, private communication, 2010.)	1-Ethyl-3-methylimidazolium	Cl	1:1	160	$P2_12_12_1$	3.17271	16	506.11	146.62	1.228	506.15
81	VEPFOL03	(G.J. Reiss, Cambridge Structural Database, private communication, 2010.)	1-Ethyl-3-methylimidazolium	Cl	1:1	95	$P2_12_12_1$	3.14739	16	507.18	146.62	1.238	507.24
87	VEPFOL	[333]	1-Ethyl-3-methylimidazolium	Cl	1:1	283–303	$P2_12_12_1$	3.2400	16	503.30	146.62	1.20	503.07

(*Continued*)

TABLE 11.2 (*Continued*)

m.pt./°C	CSD Refcode	References	Cation $[cat]^{q+}$	Anion $[A]^{p-}$	Salt Type (p:q)	Data Temp/K	Space Group	V_{cell}/ 10^3 Å^3	Z	$U_L(V)$/ kJ mol^{-1}	Formula Weight	ρ(calc)/ g cm^{-3}	$U_L(D)$/ kJ mol^{-1}
90	VEPFOL01	(S.Parsons, D. Sanders, A. Mount, A. Parsons and R. Johnstone, Cambridge Structural Database, private communication, 2005.)	1-Ethyl-3-methylimidazolium	Cl	1:1	190	$P2_12_12_1$	3.19165	16	505.31	146.62	1.221	505.38
81	NEQBIU	[334]	Tetramethylammonium	$[Al(i\text{-}Bu)_2F_2]$	1:1	133	*Pnma*	1.6398	4	419.60	253.35	1.026	419.60
81	HOGHAN	[246]	1-Methyl-3-propylimidazolium	$[Eu\{N(SO_2CF_3)_2\}_4]$	1:1	120	$P\bar{1}$	2.226	2	330.18	1397.75	2.085	330.18
81–84	METQOS	[335]	Tris(2-hydroxyethyl)ammonium	$[2\text{-}HC(O)C_6H_4CO_2]$	1:1	296	$P2_1/c$	1.49693	4	429.35	299.32	1.328	429.35
82	RUVYIQ	[336]	Pyridinium	$[CF_3CO_2]$	1:1	130	$P3_2$	1.8538	9	501.04	193.13	1.557	501.07
82	RECLAM	[135]	1,3,5-Tri-*tert*-butyl-tetrahydro-1,3,5-triazinium	$[I_7]$	1:1	283–303	$P2_1/c$	3.0323	4	361.09	1142.77	2.503	361.10
82	MACNAH	[321]	*N*-Decylbenzothiazolium	I	1:1	178	$P\bar{1}$	0.90299	2	409.60	403.37	1.484	409.65
82–84	MAJRIY	[337]	Ethylammonium	$[N(SO_2Me)_2]$	1:1	173	$P2_1/n$	1.9282	8	480.77	218.29	1.504	480.80
82–85	HONWEN	[338]	3-Hydroxypyridinium	$[C_6H_5CO_2]$	1:1	100	*Pc*	0.5142	2	472.74	217.22	1.403	472.77
82–85	HIRXEL	[339]	Bis(η^2-dimethyldisulfide) tetrachloroniobium(V)	$[NbCl_6]$	1:1	283–303	$P2_1/n$	2.2128	4	389.58	728.75	2.187	389.58
82.5	WEQXEV	[282]	1-Hexadecylpyridinium	$[Pd^{II}_2Cl_6]$	2:1	283–303	$P\bar{1}$	4.087	3	715.87	1579.2	1.925	715.93
83	ODOLAV	[99]	1-(2-Butyl)-3-methylimidazolium	$[PF_6]$	1:1	153	$P2_1/m$	0.62229	2	450.01	284.19	1.517	450.05
83–84	VIBMOJ	[93]	1,3-Dimethoxyimidazolium	$[PF_6]$	1:1	233	$P2_1/c$	1.08858	4	465.82	274.11	1.673	465.87
83–84	VIBMOJ01	[93]	1,3-Dimethoxyimidazolium	$[PF_6]$	1:1	173	$P2_1/n$	1.0407	4	471.28	274.11	1.750	471.34
83–84	DOJLUK	[340]	1-Benzyloxy-3-methylimidazolium	$[PF_6]$	1:1	233	$P2_1/c$	1.39946	4	436.74	334.20	1.586	436.74
83–84	COMBIP	[341]	1-Phenyl-1,2,3,4,5,6-hexamethylbenzenium	$[AlCl_4]$	1:1	163	*Pbca*	4.10996	8	396.72	408.18	1.32	396.78
83–85	CUWYOJ	[308]	1-Butyl-3-methylpyridinium	$[NbCl_6]$	1:1	283–303	$P2_1/c$	3.5059	8	412.66	455.85	1.727	412.66
83–85	MOYSEY	[342]	Caesium	$[Al\{OCH(CF_3)_2\}_4]$	1:1	180	$P2_1$	1.2365	2	379.18	828.05	2.224	379.20
83–110	QQQGRM	[343]	2,6-Dichloropyridinium	$[Sb^{III}Cl_4]$	1:1	283–303	$P2_1/m$	1.43155	4	434.23	590.38	2.739	434.24
84	SIGGIZ	[319]	*N,N′,N″*-Triaminoguanidinium	$[C(NO_2)_3]$	1:1	200	$P\bar{1}$	0.5017	2	475.78	255.18	1.689	475.79

84	SESCAV	[344]	Ammonium	[*trans*-$CH_3CH{=}CHCO_2$]	1:1	283–303	$P2_1/c$	0.59574	4	546.38	103.12	1.150	546.44
84	VIYRIE	[345]	Ethyltriphenylphosphonium	[Se_4Br_{14}]	2:1	140	$P2_1/n$	2.78814	2	709.03	2017.3	2.403	709.08
84–86	ISUWEY	[165]	5-Methyl-1,3,5-dithiazinan-5-ium	[$AlCl_4$]	1:1	193	$P2_1$	0.62029	2	450.38	305.05	1.633	450.38
85	No refcode	[346]	*N*-Butyl-*N,N*-dimethylglycine ethyl ester	I	1:1	283–303	$P2_1/c$	1.36413	4	439.59	315.19	1.535	439.63
85	IVATAA	[289]	1-(3-Cyanopropyl)-2,3-dimethylimidazolium	[PF_6]	1:1	140	$P\bar{1}$	0.6507	2	444.90	309.20	1.578	444.91
85	QONRAN	[169]	5-Aminotetrazolium	[$N(NO_2)_2$]	1:1	200	*Pc*	0.3437	2	525.77	192.12	1.856	525.76
85	MEHWOM	[332]	Tetrabutylammonium	4,5-Dinitro-imidazolate	1:1	173	$P2_1/c$	2.237	4	388.55	399.54	1.186	388.54
85	JIMVUX	[264]	1-Methyl-3-(tetrahydro-2*H*-pyran-2-ylmethyl)imidazolium	[CF_3SO_3]	1:1	100	$P2_1/n$	1.445	4	433.20	330.33	1.518	433.19
85	PYRLDI	[347]	Bis(pyridine)iodonium	[I_7]	1:1	283–303	$P2_1/c$	1.16445	2	384.75	1173.40	3.346	384.75
85	VURCIV	[197]	4-Methylimidazolium	[$HO_2C(CH_2)_6CO_2$]	1:1	120	$P2_1/n$	1.3481	4	440.91	256.30	1.263	440.95
85	VATYOG	[348]	Tetrabutylammonium/3-{(2*S*)-2-methylbutyl}thiazolium	[CF_3SO_3]	(1,1):2	100	$P2_1$	3.46219	4	349.97	696.87	1.337	349.99
85/92–94	KENTUT	[349]	Benzethonium	[NO_3]	1:1	173	$P\bar{1}$	1.3265	2	372.81	474.63	1.188	372.80
85–86	FIMQAU	[350]	1,5-Diamino-4-methyltetrazolium	[$N(NO_2)_2$]	1:1	200	$P2_12_12_1$	0.85481	4	496.20	221.14	1.718	496.19
85–87	NADXEX	[296]	Decyl(tricyclohexyl)phosphonium	[BF_4]	1:1	283–303	$P2_1/a$	3.019	4	361.47	508.49	1.119	361.50
85–87	GAMMOU	[351]	Caesium	[$N(NO_2)_2$]	Catenate	233	$P2_1/b$	1.03899	4	567.06	238.92	3.055	567.09
	No cif	[352]	Caesium	[$N(NO_2)_2$]	1:1		$P2_1$/a	1.0185	8	570.14	238.92	3.115	570.10
85–87	MORZEZ	[200]	3,3,5,5,7,7-Hexaphenyltricyclo[2.2.1.0^{2,6}]heptaphosphane-3,5,7-triium	[Ga_2Cl_7]	1:3	153	$P\bar{1}$	3.28486	2		1842.16	1.863	
85–105	MOSYEA	[353]	1,1-Dimethylpyrrolidinium	[SCN]	1:1	123	$P2_1/c$	0.8651	4	494.63	158.26	1.215	494.64
86	FUYGOW	[251]	*N,N′*-Bis(2-ethoxyethyl)-*N,N,N′,N′*-tetramethylethane-1,2-diammonium	[$N(SO_2CF_3)_2$]	1:2	153	$P\bar{1}$	3.37276	4	1213.26	822.72	1.620	1213.25
86	LEMMIA	[354]	1-Ethyl-3-methylimidazolium	[$Fe^{II}Cl_4$]	2:1	300	$I4_1/a$	3.92903	8	1078.27	420.01	1.419	1078.00
86	YANQEK	[355]	1-Methyl-4-amino-1,2,4-triazolium	[ClO_4]	1:1	84	$P3_1/c$	1.1615	6	509.34	198.58	1.703	509.33

(*Continued*)

TABLE 11.2 (Continued)

m.pt./°C	CSD Refcode	References	Cation [cat]$^{q+}$	Anion [A]$^{p-}$	Salt Type (p:q)	Data Temp/K	Space Group	V_{cell}/ 10^3 Å^3	Z	$U_L(V)$/ kJ mol^{-1}	Formula Weight	ρ(calc)/ g cm^{-3}	$U_L(D)$/ kJ mol^{-1}
86	HAMCAF	[356]	Diaminouronium	$[NO_3]$	1:1	173	$P2_1/c$	0.57063	4	552.78	153.12	1.782	552.77
86	YALNUW	[357]	1-(2-Hydroxyethyl)-3-methylimidazolium	Cl	1:1	100	$P2_1$	0.40382	2	503.69	162.62	1.337	503.67
86	BATMIU	[357a]	1,7-Bis(1-methylimidazolium-3-yl)heptane	$[BF_4]$	1:2	173	$Pca2_1$	2.04220	4	1367.59	436.02	1.418	1367.60
86	OGUKOR	[203]	1-Ethyl-2-methyl-3-benzylimidazolium	I	1:1	153	$P2_1/c$	2.731	8	439.47	328.20	1.59(7)	439.04
86–88	HEWKAW	[358]	1-Amino-3-methyl-1,2,3-triazolium	$[NO_3]$	1:1	100	$Pbca$	1.3254	8	530.95	161.14	1.615	530.96
86–88	HUHBOB	[359]	Trimethyltelluronium	[SCN]	1:1	200	$Pna2_1$	0.75634	4	512.54	230.79	2.027	512.57
86–88	XEGTUZ	[360]	^{t}BuNHP(m-N^{t}Bu)$_2$P$^+${(NH^{t}Bu)(NCO$_2$^{i}Pr)}NH(CO$_2$^{i}Pr)	$[C_6H_5CO_2]$	1:1	283–303	$P\bar{1}$	3.97352	4	338.92	672.79	1.125	338.96
86–92	CETVUS	[361]	3-Chloro-3-(di-*iso*propyl-amino)-3-phosphonio-bicyclo[3.1.0]hexane	$[AlCl_4]$	1:1	283–303	$P2_12_12_1$	2.02127	4	398.34	403.53	1.326	398.35
86/103/152	FAXVEF	[362]	1-Butylpyridinium	Cl	1:1	283–303	$P2_12_12_1$	0.98177	4	478.47	171.67	1.161	478.47
87	LOPQAI	[363]	Bis(1,3-dimethylthiourea) iodinium	$[I_3]$	1:1	203	$P2_1/c$	3.6754	8	407.83	715.95	2.588	407.86
87	LAFLAF	[364]	1-(Trimethylsilyl)-3,4-dimethylpyridinium	Br	1:1	283–303	$P2_12_12_1$	1.2595	4	448.64	260.25	1.372	448.62
87	AWEHEP	[265]	1,8-Bis(1-methylimidazolium-3-yl)-3,6-dioxaoctane	$[PF_6]$	1:2	173	$P2_1/c$	2.3280	4	1324.78	570.30	1.627	1324.78
87	KEHKEP	[89]	Bis(1-methylimidazole)silver(I)	$[N(SO_2CF_3)_2]$	1:1	100	$P2_1/c$	1.85387	4	406.95	552.26	1.979	406.98
87–88	NEMFUG	[365]	1,3-Diethyl-3,4,5,6-tetrahydropyrimidinium	$[PF_6]$	1:1	283–303	$P2_1/c$	1.2672	4	447.94	286.21	1.500	447.94
87–89	AKOXUT	[366]	Bis(pyridine)bromonium	$[CF_3SO_3]$	1:1	283–303	$P\bar{1}$	0.7478	2	429.44	387.18	1.720	429.49
88	GODNES	[367]	2,4,6-Trimethylpyridinium	$[(2\text{-}NO_2)C_6H_4CO_2]$	1:1	150	$Pbca$	2.8136	8	436.16	288.30	1.361	436.16
88	YESTIA	[174]	$[Dy_2(NMe_3CH_2CO_2)_8(H_2O)_4]$	$[N(SO_2CF_3)_2]$	1:6	100	$P\bar{1}$	2.76731	1		3007.20	1.804	
88	TUQYIN	[368]	Tris(dimethylamino)-methyliminiumphosphorane	$[CH_3CO_2]$	1:1	173	$P\bar{1}$	0.70177	2	436.41	252.30	1.195	436.53

88	ENIHIS	[369]	2-(Hydroxoiodinium) nitrobenzene	$[CF_3CO_2]$	1:1	173	*C*2/*c*	2.17513	8	465.93	379.02	2.315	465.96
88–90	VICMIE	[370]	1-Methylmorpholinium	$[Cu^{II}Cl_4]$	2:1	158	$P2_1/c$	1.75446	4	1126.55	409.67	1.551	1126.62
88–90	APURAE	[371]	1-Methyl-3-propylimidazolium	$[Ph_2PC_6H_4(4\text{-}SO_3)]$	1:1	283–303	$P2_1/c$	2.441	4	380.38	466.52	1.269	380.37
88–90	IFOFUF	[372]	*N*-Methylbenzothiadiazolium	$[CF_3SO_3]$	1:1	283–303	$P2_1/c$	1.2066	4	453.61	300.28	1.653	453.63
88–92	XISDIM	[167]	Tetradodecylammonium	Br	1:1	173	*C*2/*c*	9.6003	8	324.56	771.23	1.067	324.57
88.5	OVUDEP	[111]	1,3-Didodecylimidazolium	I	1:1	150	$P\bar{1}$	2.976	4	362.70	532.61	1.189	362.74
89	MAPPAW	[257]	1-*cyclo*-Pentyl-3-methylimidazolium	$[4\text{-}CH_3C_6H_4SO_3]$	1:1	173	$P2_1/c$	1.61396	4	421.28	322.42	1.327	421.31
89	FASSOI	[223]	Tetramethylformamidinium	$[Ga_2I_6]$	2:1	123	$P2_1$	1.37037	2	946.20	1103.21	2.674	946.30
89	883551	[144]	Tetrakis(1-methylimidazole) copper(II)	$[N(SO_2CF_3)_2]$	1:2	100	$P\bar{1}$	0.86511	1	1206.04	952.32	1.828	1206.08
89	DOCBED	[373]	1-Methyl-3-{(2*S*)-2-methylbutyl} imidazolium	Bis{(2*S*)-2-oxy-3-methylbutanoato-*O*,*O*′}borate	1:1	100	$P2_12_12_1$	2.2562	4	387.74	396.28	1.167	387.78
89–90	BEZWOS	[374]	2-Hydroxyhomotropylium	$[SbCl_6]$	1:1	208	*A*2/*m*	1.45591	4	432.38	455.62	2.079	432.41
89–91	CUWYID	[308]	1-Butyl-4-methylpyridinium	$[NbCl_6]$	1:1	283–303	$P2_1/c$	1.7613	4	412.17	455.85	1.719	412.18
89–91	HECSEN	[375]	1,1,2,3,4,5,6-Heptamethylbenzenium	$[BF_4]$	1:1	245	$P4_12_12$	1.41931	4	435.18	264.11	1.236	435.19
89–91	CEXGAN	[376]	Trimethylsulfonium	$[ICF_2CF_2OCF_2CF_2SO_3]$	1:1	283–303	$P2_1/c$	1.60279	4	422.02	500.17	2.073	422.05
90	VIFPEG	[377]	1-(3-Cyanopropyl)-3-methylimidazolium	Br	1:1	120	$P2_1/c$	0.9993	4	476.29	230.11	1.530	476.35
90	IKARER	[378]	1-Butyl-3-methylimidazolium	$[B\{C_6H_4(4\text{-}C_6F_{13})\}_4]$	1:1	150	$P4_1$	6.5141	4	303.20	1730.64	1.765	303.23
90	QONRER	[168]	5-Amino-2-methyltetrazolium	$[N(NO_2)_2]$	1:1	150	$P\bar{1}$	0.41277	2	500.78	206.15	1.659	500.83
90	YOJPUJ	[269]	5-Amino-1-methyltetrazolium	$[N(NO_2)_2]$	1:1	200	$P2_1/m$	0.41605	2	499.73	206.15	1.646	499.79
90	FIQLAS	[221]	Ethyl(methyl) diphenylammonium	$[I_5]$	1:1	283–303	$P\bar{1}$	1.1030	2	389.87	856.81	2.545	388.60
90	MOVDEH	[136]	Hydroxycarbonylmethyltris (tetradecyl)phosphonium	Br	1:1	173	$P2_1/c$	4.62014	4	327.40	762.04	1.096	327.44
90	JUYMEV	[379]	([2.2.2]Cryptand)potassium	$[I_{12}]$	2:1	283–303	$P\bar{1}$	1.6958	1	652.90	2354.03	2.306	653.04
90	QAMTOP	[380]	2-Hydroxyethyl(trimethyl) ammonium (cholinium)	Bis(salicylato) borate	1:1	123	*Pc*	1.87138	2	343.65	760.26	1.349	343.66
90	846080	[79]	(Ethanenitrile)silver(I)	$[N(SO_2CF_3)_2]$	Polymer	100	$P2_1/c$	1.19496	4	454.74	429.07	2.385	454.76

(*Continued*)

TABLE 11.2 (*Continued*)

m.pt./°C	CSD Refcode	References	Cation $[cat]^{q+}$	Anion $[A]^{p-}$	Salt Type (p:q)	Data Temp/K	Space Group	V_{cell}/ 10^3 Å^3	Z	$U_L(V)$/ kJ mol^{-1}	Formula Weight	ρ(calc)/ g cm^{-3}	$U_L(D)$/ kJ mol^{-1}
90	BEFFOH	[381]	Tetrabutylammonium	$[InCl_4]$	1:1	283–303	*Pnna*	2.4826	4	378.81	499.10	1.335	378.82
90	CUPQIN	[382]	Tetrabutylammonium	$[AlMe_2F_2]$	1:1	150	$P2_1/n$	2.22821	4	388.92	337.51	1.006	388.93
90	NAPVAB	[383]	*iso*-Propylphosphonium	$[Ge^{II}Cl_3]$	1:1	143	$P2_1/c$	0.96086	4	481.19	256.02	1.770	481.23
90–91.5	GALSAN	[384]	1-Phenylthiolanium	$[ClO_4]$	1:1	283–303	$P2_1/n$	2.41365	8	453.58	264.72	1.457	453.61
90–92	HOZMIS	[385]	1,4,7-Trimethyl-1-azonia-4, 7-diazacyclononane	$[CF_3SO_3]$	1:1	283–303	$P2_12_12_1$	1.5739	4	423.95	321.36	1.356	423.95
90–92	GOMLUQ	[386]	Tetrabutylammonium	2,3,6-Tricyano-4-fluoro-5-trifluoromethyl-phenolate	1:1	100	$P12_1/c1$	2.70357	4	371.12	496.59	1.22	371.13
90–92	UKEREI	[387]	3-(3-Butynyl)-2-(1-hexynyl) benzothiazolium	$[CF_3SO_3]$	1:1	150	$Pca2_1$	1.95235	4	401.76	417.45	1.420	401.76
90–95	QUSLOF	[281]	2,6-Dimethylpyridinium	$[N(SO_2Me)_2]$	1:1	143	$P2_1/c$	1.2958	4	445.39	280.36	1.437	445.39
90–96	XOBFUQ	[388]	1-(Prop-2-ynyl)-1-methylpyrrolidinium	Cl	1:1	173	$P2_1/n$	0.9157	4	487.30	159.65	1.158	487.31
90.5	MACNEL	[321]	*N*-Dodecylbenzothiazolium	I	1:1	178	$P\bar{1}$	1.00365	2	399.02	431.42	1.428	399.07
91	KEKVEC01	[389]	1-Ethyl-1-methylpyrrolidinium	$[N(SO_2CF_3)_2]$	1:1	213	$P\bar{1}$	0.8231	2	419.19	394.36	1.591	419.20
91	KEKVEC	[389]	1-Ethyl-1-methylpyrrolidinium	$[N(SO_2CF_3)_2]$	1:1	153	$P2_1/n$	3.1885	8	422.58	394.36	1.643	422.60
91	LEVWEP	[390]	1-Ethyl-1-methylpiperidinium	$[N(SO_2CF_3)_2]$	1:1	123	$P\bar{1}$	3.394	8	416.02	408.38	1.599	416.07
91	LEVWEP01	[390]	1-Ethyl-1-methylpiperidinium	$[N(SO_2CF_3)_2]$	1:1	223	$C2/c$	6.99708	12	384.61	408.38	1.163	384.63
91	TUSMIE	[391]	Aminoguanidinium	$[N(CN)(NO_2)]$	1:1	283–303	$P\bar{1}$	0.3533	2	521.91	161.15	1.515	521.95
91	GOBFUY	[392]	(Benzo-18-crown-6)potassium	$[I_5]/[I_7]$	2:(1,1)	283–303	$P2_1/c$	5.9280	4		2225.70	2.494	
91–92	FESCUB	[207]	Tris(dimethylamino)sulfonium	$[SiPh_2(1\text{-}C_{10}H_7)F_2]$	1:1	283–303	*Cc*	2.77911	4	368.68	511.75	1.223	368.69
91–92	SAXCAV01	[393]	1-Chloromethyl-1,2,3,4,5,6-hexamethylbenzenium	$[AlCl_4]$	1:1	*ca.* 200	$P2_1/c$	1.7978	4	410.07	380.55	1.41	410.37
No m.pt.	SAXCAV	[394]	1-Chloromethyl-1,2,3,4,5,6-hexamethylbenzenium	$[AlCl_4]$	1:1	283–303	$P2_1/n$	1.8289	4	408.32	380.57	1.38	408.18
91–92.5	PABMUA	[395]	1-Methylimidazolium	$[PO_2(OPh)_2]$	1:1	283–303	$P2_1$	0.79408	2	422.99	332.29	1.390	423.03
91–93	GUNSAK	[249]	Ethyltriphenylphosphonium	4,5-Dinitro-imidazolate	1:1	173	$P2_1/n$	2.1194	4	393.72	448.41	1.405	393.71
91–94	KIFLEQ01	[261]	Aminoguanidinium	$[N(NO_2)_2]$	1:1	283–303	*Pc*	0.3645	2	517.58	181.14	1.650	517.57
91–94	KIFLEQ	[261]	Aminoguanidinium	$[N(NO_2)_2]$	1:1	283–303	$P\bar{1}$	0.3671	2	516.60	181.14	1.639	516.65

92	MIHREB	[169]	1-*iso*-Propyl-4-amino-1,2,4-triazolium	Br	1:1	283–303	$P\bar{1}$	0.4206	2	498.30	207.08	1.635	498.31
92	CAMMAG	[351]	Ammonium	$[N(NO_2)_2]$	1:1	223	$P2_1/c$	0.4500	4	589.77	124.07	1.831	589.77
92	PILKEA	[396]	Tris(dimethylamino)sulfonium	$[SiMe_3F_2]$	1:1	283–303	$P2_1/m$	0.80448	2	421.61	275.48	2.270	421.60
92	FASTID	[223]	Tetramethylformamidinium	$[In_2Cl_6]$	2:1	123	$Pna2_1$	2.38342	4	999.83	644.70	1.797	999.95
92	BATLUF	[357a]	1,8-Bis(1-methylimidazolium-3-yl)octane	$[BF_4]$	1:2	173	$P2_1/c$	1.06834	2	1352.60	450.04	1.399	1352.63
92	HOGGUG	[246]	1-Butyl-1-methylpyrrolidinium	$[Eu\{N(SO_2CF_3)_2\}_5]$	2:1	120	$P\bar{1}$	3.1872	2	670.31	1837.23	1.914	670.29
92	WIRRIY	[397]	1-(2-Hydroxyethyl)morpholinium	$[\{2\text{-}(2,6\text{-}Cl_2C_6H_3NH)C_6H_4\}CH_2CO_2]$	1:1	283–303	$P2_1/a$	2.04587	4	397.15	427.31	1.387	397.15
92	CAVZUW	[398]	Tetraphenylphosphonium	$[As(N_3)_4]$	1:1	173	$C2$	1.31037	2	373.92	582.41	1.476	373.92
92/94	GEDLEH	[399]	1-Ethyl-3-methyl-4-nitroimidazolium	$[CF_3SO_3]$	1:1	173	$P2_1/c$	1.2079	4	453.48	305.24	1.678	453.47
92–93	YIJVOC	[400]	Benzyldimethylammonium	$[\{2,4,6\text{-}(NO_2)_3\}C_6H_2O]$ Picrate	1:1	153	$P\bar{1}$	1.62318	4	420.68	348.32	1.491	420.71
92–93	PEVDID	[401]	1-Ethyl-3-methylimidazolium	$[Ni^{II}Cl_4]$	2:1	283–303	$I4_1a$	3.87107	8	1084.51	422.84	1.451	1084.54
92–94	GUWBEF	[402]	2-Amino-5-butyl-4-methyl-1,3-thiazol-3-ium	$[NO_3]$	1:1	283–303	$P\bar{1}$	0.5790	2	458.43	233.29	1.338	458.44
92–94	SABJUC	[403]	1,4-Bis(1,2-dimethylimidazolium-3-ylmethyl)benzene	$[CF_3SO_3]$	1:2	100	$P\bar{1}$	0.62814	1	1300.69	594.55	1.572	1300.78
92–94	VIPNEN	[404]	3-(3-Ethenyl-1,2,4-oxadiazol-5-yl)-1,2,5,6-tetrahydropyridinium	$[CF_3CO_2]$	1:1	283–303	$P\bar{1}$	0.66698	2	442.10	291.23	1.45	442.11
92–94	HIWYAO	[166]	(Benzo-18-crown-6)bis(pyridine) sodium	$[BH_4]$	1:1	183	$P2_1/c$	1.30483	2	374.29	460.27	1.172	374.35
92–98	VICMEA	[370]	1-Methylmorpholinium	$[Cu^{II}Br_4]$	2:1	158	$P2_1/c$	1.86815	4	1099.52	587.47	2.089	1099.62
93	MOQBEA	[405]	1,3-Dibenzylimidazolium	$[PF_6]$	1:1	150	$P2_1/c$	3.51935	8	412.26	394.30	1.488	412.26
93	IGEDUT	[406]	(2-Pyridylthio)(2-pyridiniothio) methane	$[I_3]$	1:1	283–303	$P2_1/n$	1.77748	4	411.23	616.04	2.302	411.25
93	SALZEM	[407]	1-Amino-3-methyl-1,2,3-triazolium	5-Nitrotetrazolate	1:1	93	$P\bar{1}$	0.8918	4	490.69	213.16	1.588	490.75
93	QAMSUU	[408]	4-Amino-1,2,4-triazolium	4-(Carboxylatomethyl)-5-nitroiminotetrazolate	2:1	296	$P2_1/n$	1.48748	4	1200.39	356.31	1.591	1200.43

(*Continued*)

TABLE 11.2 (*Continued*)

m.pt./°C	CSD Refcode	References	Cation $[cat]^{q+}$	Anion $[A]^{p-}$	Salt Type (p:q)	Data Temp/K	Space Group	V_{cell}/ 10^3 Å^3	Z	$U_L(V)$/ kJ mol^{-1}	Formula Weight	ρ(calc)/ g cm^{-3}	$U_L(D)$/ kJ mol^{-1}
93(dec)	No refcode	[261]	Ammonium	$[N(NO_2)(SF_5)]$	1:1	223	$P2_1/m$	0.31138	2	539.89	205.12	2.188	539.93
93–94	MOJDIY	[409]	*N*-Methyl-1, 3-propylenediammonium	$[TlBr_5]$	2:2	173	$P2_12_12_1$	1.36855	4		693.99	3.368	
93–94	UNOKOX	[410]	[(2-Trimethylsilylethyl) dimethylamine)$_2$H]	$[B\{3,5\text{-}(CF_3)_2C_6H_3\}_4]$	1:1	130	*Pbcn*	5.47605	4	315.08	1154.88	1.401	315.10
94	MIHROL	[169]	1-Heptyl-4-amino-1,2, 4-triazolium	Br	1:1	100	$P2_1/c$	1.2704	4	447.65	263.18	1.376	447.67
94	FEMGEK	[411]	2-Amino-4, 5-dimethyltetrazolium	$[NO_3]$	1:1	85	$P2_12_12_1$	0.7673	4	510.58	176.15	1.525	510.62
94	XUJDUB	[412]	Tris(2-hydroxyethyl)ammonium	$[SeO_2(OH)]$	1:1	283–303	$P2_1$	0.5383	2	467.15	278.17	1.716	467.16
94	PILFEW	[413]	5-Hydroxy-1-methyl-3,4-dihydro-2*H*-pyrrolium	Cl	1:1	200	$P2_12_12_1$	0.69246	4	524.74	135.59	1.301	524.80
94/92	HATRAX	[122]	1-Butyl-3-methylimidazolium	$[HgCl_3]$	1:1	170	*Cc*	1.3385	4	441.72	446.20	2.214	441.72
94–95	VUKSAW	[224]	4-Amino-1-ethyl-3,5-dimethyl-1,2,4-triazolium	$[PF_6]$	1:1	100	$P2_1/n$	2.37584	8	455.43	286.17	1.60	455.44
94–95	HEBZCA01	[375]	1,1,2,3,4,5,6-Heptamethylbenzenium	$[AlCl_4]$	1:1	183	$P2_1/n$	1.72462	4	414.34	346.11	1.333	414.35
No m.pt.	HEBZCA	[414]	1,1,2,3,4,5,6-Heptamethylbenzenium	$[AlCl_4]$	1:1	283–303	$P2_1/n$	1.7897	4	410.53	346.11	1.285	410.58
No m.pt.	HEBZCA02	[415]	1,1,2,3,4,5,6-Heptamethylbenzenium	$[AlCl_4]$	1:1	123	$P2_1/n$	1.70444	4	415.56	346.11	1.349	415.59
94–95	FAVTAY	[220]	1,2-Bis(*P,P,P′,P′*-tetraphenylphosphino-phosphonium)ethane	$[GaCl_4]$	1:2	193	$P2_1/n$	2.68596	2	1091.41	1191.77	1.474	1091.51
94–95.5	VEZLUH10	[416]	1,6-Diarsa-2,5,7,10-tetrathiatricyclo[5.3.0^{1,5}.0^{6,10}] decane	$[GaCl_4]$	1:2	283–303	$P2_1/c$	1.03453	2	1363.24	757.28	2.431	1363.27
94.5–95.5	LUSGEL	[106]	1-(5-Bromopentyl)-1′,1″-biferrocenium	$[I_3]$	1:1	150	*Pnma*	2.6710	4	372.20	899.78	2.237	372.20
95	MUPHEL	[417]	1-(3-Ammoniopropyl)-3-methylimidazolium	$[NO_3]$	1:2	283–303	$Pna2_1$	1.1288	4	1586.58	251.21	1.478	1586.57

95	MEBPAL	[39]	1-Aminoimidazolium	Cl	1:1	283–303	$P2_1/n$	0.55823	4	556.08	119.56	1.423	556.15
95	RAPZOZ	[303]	Triethylpropylammonium	$[B(CN)_4]$	1:1	140	$P\bar{4}2(1)m$	0.79460	2	422.92	259.17	1.083	422.91
95	ZOYBOE	[418]	*N*-Butylurotropinium	$[I_3]$	1:1	283–303	$P2_1/m$	0.8482	2	416.05	578.02	2.263	416.06
95	JIMWEI	[264]	Tetrabutyl ammonium/1-methyl-3-(tetrahydrofuran-3-ylmethyl)imidazolium	$[CF_3SO_3]$	(1,1):2	100	$P\bar{1}$	1.79276	2		707.83	1.311	
95	HABHUN	[419]	3-Methoxyazetidinium	Cl	1:1	122	$P\bar{1}$	0.31382	2	538.76	123.58	1.308	538.80
95	HISRIK	[420]	*rac*-(1-Phenylethyl)ammonium	$[ClCH_2CO_2]$	1:1	283–303	$P2_1/c$	2.3200	8	458.23	215.68	1.235	458.25
95	UQUHEU	[421]	Bis(tetramethylene sulfone) lithium	$[BF_2(ox)]$	1:1?	110	$P2_1/n$	1.57148	4	424.12	384.12	1.623	424.10
95/104	GEDLAD	[399]	1,3-Dimethyl-4-nitroimidazolium	$[SO_3(OMe)]$	1:1	173	$P2_1/n$	1.0328	4	472.22	253.24	1.629	472.27
95–96	AQUQEI	[218]	Tetrabutylphosphonium	$[Sn(2\text{-}MeC_6H_4)_2Cl_3]$	1:1	130	$P2_1/n$	3.3679	4	352.24	666.71	1.315	352.27
95–96	VAQTOX	[219]	2-Amino-4-methylpyrimidinium	$[3\text{-}ClC_6H_4CO_2]$	1:1	203	$P2_1$	1.2005	4	454.20	265.7	1.47	454.21
95–96	KAHWOF	[422]	Tributylmethyammonium	$[Cu^I_5Br_7]$	2:1	170	$P\bar{1}$	2.06654	2	802.24	1277.83	2.054	802.35
95–96	WAWMOX	[423]	Benzyl(ethyl)ammonium	$[NO_3]$	1:1	120	$P2_1/n$	0.99337	4	477.03	198.22	1.325	477.01
95–97	DEVGIU	[267]	Piperidinium	$[N(SO_2Me)_2]$	1:1	143	$C2/c$	2.3907	8	454.70	258.35	1.44	454.75
95–97	TUNPEY	[424]	1,4-Bis(1-methylimidazolium-3-yl)butane	$[BF_4]$	1:2	283–303	$P2_1/c$	0.8962	2	1412.14	393.94	1.460	1412.22
99	TUNPEY01	[357a]	1,4-Bis(1-methylimidazolium-3-yl)butane	$[BF_4]$	1:2	173	$P2_1/c$	0.86865	2	1423.09	393.94	1.506	1423.10
95–97	HEDDAV	[425]	1,3-Dimethyl-1,3-diaza-2-phospholidinium	$[GaCl_4]$	1:1	283–303	$P2_1$	0.63637	2	447.44	328.64	1.715	447.45
95–97	REBJIS	[426]	1-Butylpyridinium	Br	1:1	283–303	$P2_12_12_1$	1.04755	4	470.47	216.12	1.370	470.47
95–98	HOPJEC	[173]	3-Hydroxy-1-(octyloxymethyl) pyridinium	Saccharinate	1:1	173	$P\bar{1}$	1.0437	2	395.19	420.51	1.338	395.20
95–98	HEBHIF	[305]	Tetraethylammonium	$[(2\text{-}OH)C_6H_4CO_2]$	1:1	283–303	$P2_1/m$	1.5464	4	425.84	267.36	1.148	425.82
96	EFOCIM	[427]	*N,N′,N″*-Triaminoguanidinium	5-Nitrotetrazolate	1:1	100	$P2_12_12_1$	0.90942	4	488.18	219.20	1.601	488.20
96	FOZROC	[170]	Tetramethylammonium	$[Al\{OC(CH_3)(CF_3)_2\}_4]$	1:1	100	$I\bar{4}$	1.5548	2	358.94	825.35	1.763	358.96
96	BEXCEM10	[428]	Piperazinediium	$[C_9H_{19}CO_2]$	1:2	283–303	$P\bar{1}$	0.67282	1	1279.51	430.67	1.063	1279.57
96	BAYCOT	[429]	Piperazinediium	$[C_{11}H_{23}CO_2]$	1:2	283–303	$P\bar{1}$	0.76932	1	1239.57	486.77	1.051	1239.70
96	BEXDAJ01	[430]	Piperazinediium	$[C_{13}H_{27}CO_2]$	1:2	283–303	$P\bar{1}$	0.86149	1	1207.21	542.88	1.046	1207.13
96	QQQDXG	[431]	Tetrapentylphosphonium	I	1:1	283–303	*Ccca*	2.511360	4	377.78	442.44	1.17	377.78
96–98	JOXZOL	[432]	2-Acetyl-9-azoniabicyclo[4.2.1] nonan-3-one	$[CF_3CO_2]$	1:1	283–303	$P\bar{1}$	1.3371	4	441.83	295.26	1.467	441.87

(*Continued*)

TABLE 11.2 (*Continued*)

m.pt./°C	CSD Refcode	References	Cation $[cat]^{q+}$	Anion $[A]^{p-}$	Salt Type (p:q)	Data Temp/K	Space Group	V_{cell}/ 10^3 Å^3	Z	U_L(V)/ kJ mol^{-1}	Formula Weight	ρ(calc)/ g cm^{-3}	U_L(D)/ kJ mol^{-1}
96–98	VUZXEU	[433]	3,5-Bis(1-butylimidazolium-3-ylmethyl)toluene	$[PF_6]$	1:2	100	$P2_1/c$	2.93366	4	1253.61	656.47	1.486	1253.58
96–98	LUWMIA	[434]	Trichloro(*cyclo*-hexyl)phosphonium	$[GaCl_4]$	1:1	123	$P2_1/c$	3.13761	8	424.30	432.02	1.829	424.31
96–98	YEDRIM	[435]	Methylammonium	$[4\text{-MeOC}_6H_4CH(CH_2NO_2)SO_3]$	1:1	283–303	$Pbca$	2.80089	8	436.66	292.31	1.386	436.65
96–100	No refcode	[436]	Hydroxylammonium	F	1:1	283–303	$Pbca$	0.8597	16	725.51	53.04	1.639	725.52
96–102	DENBON	[437]	Tetraphenylarsonium	$[C\{CF_3C(O)\}(CN)_2]$	1:1	283–303	$P2_1/c$	2.58695	4	375.08	544.40	1.398	375.11
97	LIYZEY	[438]	Methyldiphenylsulfonium	$[CF_3SO_3]$	1:1	180	$P2_1/c$	1.594	4	422.60	350.56	1.46	422.56
97	YOBVER	[147]	1-Ethyl-3-methylimidazolium	$[nido\text{-}C_2B_9H_{12}]$	1:1	125	$P\bar{1}$	0.739	2	430.73	244.57	1.10	430.84
97	KASWEI	[90]	1-(2-Hydroxyethyl)pyrrolidinium	$[\{2,5\text{-}(HO)_2\}C_6H_3CO_2]$	1:1	123	$P2_1/c$	1.27391	4	447.33	269.29	1.404	447.35
97	BAGKAV10	[439]	Piperazinediium	$[C_6H_{13}CO_2]$	1:2	283–303	$P\bar{1}$	0.52737	1	1356.83	346.51	1.091	1356.84
97	BEXDIR01	[430]	Piperazinediium	$[C_{15}H_{31}CO_2]$	1:2	283–303	$P\bar{1}$	0.95684	1	1178.27	598.99	1.04	1178.42
97	EDOTEX	[440]	1-Butyl-2,3-dimethylimidazolium	Br	1:1	120	$P2_1/n$	1.0867	4	466.02	233.16	1.425	466.03
97	IGUJID	[441]	1-(2-Hydroxyethyl)piperidinium	$[Ge(ox)_3]$	2:1	173	$P\bar{1}$	1.21762	2	991.41	597.12	1.629	991.54
97–97.5	HANFUZ	[442]	Hydrazinium	$[N(CN)_2]$	1:1	173	$Cmcm$	0.4415	4	592.87	99.11	1.491	592.88
97–98	SADMEQ	[217]	1-(2-Hydroxy-2-methylpropyl)-3-*iso*-propylimidazolium	I	1:1	283–303	$P2_1/c$	1.3564	4	440.22	310.17	1.519	440.25
97–98	SASNEG	[443]	1,5-Diaminotetrazolium	$[ClO_4]$	1:1	193	$P2_1/n$	0.7005	4	523.12	200.56	1.902	523.16
97–98	HYZMAC	[444]	Hydrazinium	$[CH_3CO_2]$	1:1	283–303	Cc	0.44839	4	590.35	92.10	1.364	590.34
97–99	XAZRIA	[445]	2-Amino-3-methylpyridinium	$[CH_3CO_2]$	1:1	283–303	$P\bar{1}$	0.43828	2	492.92	168.20	1.275	492.99
97–99	LADTAM	[446]	Nitrosonium	$[Al\{OC(Ph)(CF_3)_2\}_4]$	1:1	198	Cc	3.9304	4	339.78	1029.51	1.740	339.80
97–99	UKERUY	[387]	1-(3-Butynyl)-2-(phenylethynyl)pyrimidinium	$[CF_3SO_3]$	1:1	150	$P\bar{1}$	0.87842	2	412.43	382.35	1.446	412.48
98	ONOTUH	[277]	Ethyltriphenylphosphonium	$[B(CN)_4]$	1:1	283–303	$Pnma$	2.34429	4	384.14	406.22	1.151	384.15
98	HUVCAC	[447]	1-Methyl-4-(3,3,3-trifluoropropyl)-1,2,4-triazolium	I	1:1	203	$Pccn$	2.1748	8	465.95	307.06	1.876	465.99

98	YANPUZ	[355]	1,4-Dimethyl-3-azido-1,2,4-triazolium	$[NO_3]$	1:1	84	$P2_1/c$	0.8459	4	497.57	201.17	1.580	497.62
98	FUTHIM	[256]	Bis(3,4-dimethoxyphenalene-1,9-diolato)boron	$[B\{3,5\text{-}(CF_3)_2C_6H_3\}_4]$	1:1	223	$P\bar{1}$	3.0045	2	308.64	1384.51	1.530	308.63
98	NAGNOY	[448]	Tetrapropylammonium	$[I_3]$	1:1	283–303	$P2_1/c$	2.0049	4	399.14	567.07	1.879	399.17
98	HILLOD	[324]	H_2([2.2.2]cryptand)	$[I_8]$	2:2	283–303	$Pbcn$	3.5693	4		1393.75	2.593	
98	TANZOZ	[242]	1,2-Dimethylhydrazine-1,2-diium	$[NO_3]$	1:2	110	$P\bar{1}$	0.19539	1	1745.75	186.14	1.582	1745.84
98	EDOTIB	[440]	1-Butyl-2,3-dimethylimidazolium	I	1:1	120	$P2_1/c$	2.3205	8	458.20	280.15	1.604	458.24
98	BOJNOE	[449]	Tetramethylammonium	$[N\{P(O)(C_2F_5)_2\}_2]$	1:1	100	$P2_1/n$	4.6438	8	385.03	658.18	1.883	385.06
98–99	XISDAE	[167]	Tetraoctadecylphosphonium	I	1:1	173	$P\bar{1}$	3.81718	2	292.93	1171.85	1.02	292.97
98–100	LADBID	[450]	2-Aminopyridinium	$[HO_2C(CH_2)_8CO_2]$	2:1	120	$P2_1/n$	1.05116	2	1050.18	390.48	1.234	1050.33
98–100	XIDGEX	[451]	1,5-Dimethoxyl-3-phenyl-5a,6,7,8,9,9a-hexahydro-1*H*-1,5-benzodiazepin-5-ium	$[ClO_4]$	1:1	283–303	$Pca2_1$	3.72261	8	406.54	386.82	1.380	406.53
98–101	ZAMXIU01	[130]	1-Methylimidazolium	[HOC(O)CH(OH)CH(OH)CO_2]	1:1	283–303	$P2_1$	0.5243	2	470.36	232.19	1.471	470.40
98–101	ZAMXIU	[452]	1-Methylimidazolium	[HOC(O)CH(OH)CH(OH)CO_2]	1:1	283–303	$P2_1$	0.5244	2	470.34	232.19	1.478	470.98
98–101	WUSQUW	[453]	Tetraphenylphosphonium	$[B(C_6F_5)(N_3)_3]$	1:1	200	$P2_1/c$	2.93661	4	363.86	643.33	1.455	363.86
98.5–102	UYOCAN	[454]	2-Chloro-1-methyl-4-(phenylamino)pyridinium	$[CF_3SO_3]$	1:1	100	$P2_1/c$	1.53673	4	426.51	368.76	1.594	426.54
99	KISNEG	[73]	1-Butyl-3-methylimidazolium	$[Pt^{II}Cl_4]$	2:1	113	$P2_1/n$	1.0855	2	1037.08	615.34	1.882	1036.99
99	HEDPUB	[455]	1-Ethyl-3-methylimidazolium	$[V^{IV}(O)Cl_4]$	2:1	288–303	$P2_1/c$	1.0170	2	1063.79	431.1	1.41	1064.49
99	DOYHEF	[456]	2,2-Dimethyltriazanium	$[NO_3]$	1:1	90	$P\bar{1}$	0.31215	2	539.53	138.14	1.470	539.57
99	EHAXIV	[457]	1,3-Dibutyl-2,4,5-trimethylimidazolium	$[PF_6]$	1:1	173	$C2/c$	3.604	8	409.83	368.35	1.358	409.87
99	SALZOW	[407]	1-Amino-1,2,3-triazolium	5-Nitrotetrazolate	1:1	93	$P2_1/c$	1.5676	8	507.71	199.13	1.688	507.77
99	TABRIX	[458]	1-Benzylpyridinium	Br	1:1	188	$P2_1/c$	2.19427	8	464.87	250.13	1.514	464.87
99	TABRIX01	[458]	1-Benzylpyridinium	Br	1:1	208	$P2_1/c$	2.20192	8	464.47	250.13	1.509	464.47
99	TABRIX02	[458]	1-Benzylpyridinium	Br	1:1	218	$P2_1/c$	2.20912	8	464.07	250.13	1.504	464.07
99–100	PABNAH	[395]	Imidazolium	$[PO_2(OPh)_2]$	1:1	283–303	$P\bar{1}$	0.78239	2	424.57	318.27	1.351	424.59

(*Continued*)

TABLE 11.2 (*Continued*)

m.pt./°C	CSD Refcode	References	Cation $[cat]^{q+}$	Anion $[A]^{p-}$	Salt Type (*p*:*q*)	Data Temp/K	Space Group	V_{cell}/ 10^3 Å^3	*Z*	$U_L(V)$/ kJ mol^{-1}	Formula Weight	ρ(calc)/ g cm^{-3}	$U_L(D)$/ kJ mol^{-1}
99–100	VAPGUO	[459]	Tetrapropylammonium	$[Cu^IBr_2]$	1:1	283–303	$P2_1/n$	0.87304	2	413.06	409.71	1.559	413.11
99–100	DOCRUI	[460]	Tetrapropylammonium	$[Cu^ICl_2]$	1:1	283–303	$P2_1/n$	0.84948	2	415.90	320.81	1.254	415.89
99–102	VIBNEA	[93]	1,3-Diethoxyimidazolium	$[PF_6]$	1:1	233	*Pbca*	2.62261	8	444.04	302.16	1.531	444.09
99–103	QAGJEN	[461]	2-Amino-2-thiazolinium	*N*-Methylpyrrole-2-carboxylate	1:1	288–303	*Pbca*	2.2711	8	460.75	227.28	1.329	460.74
99.5	FUYGEM	[251]	*N*,*N‴*-Bis(2-ethoxyethyl)-*N*,*N*,*N‴*,*N‴*-tetramethylbutane-1,4-diammonium	$[PF_6]$	1:2	153	$P2_1/n$	2.5941	4	1290.79	580.41	1.495	1292.66

TABLE 11.3 Salts with m.pt. 100°C to *ca.* 110°C

m.pt./°C	CSD Refcode	References	Cation [cat]^q	Anion [A]$^{p-}$	Salt Type (p:q)	Data Temp/K	Space Group	V_{cell}/ 10^3 Å^3	Z	$U_L(V)$/ kJ mol^{-1}	Formula Weight	ρ(calc)/ g cm^{-3}	$U_L(D)$/ kJ mol^{-1}
100	AYIBOZ	[119]	1-Butyl-2,3-dimethylimidazolium	Cl	1:1	283–303	$P2_1/n$	1.0558	4	469.52	188.70	1.187	469.53
100	DOXJUW	[462]	Aminoguanidinium	5-Azidotetrazolate	1:1	200	$P\bar{1}$	0.80715	4	503.77	185.19	1.524	503.80
100	FEGDAX	[463]	Bis(dimethylamino) trifluoromethylsulfonium	$[CF_3S]$	1:1	203	$P2_1/c$	1.2363	4	450.78	290.30	1.560	450.83
100–102	PEVDEZ	[401]	1-Ethyl-3-methylimidazolium	$[Co^{II}Cl_4]$	2:1	283–303	$I4_1a$	3.8904	8	1082.41	423.08	1.445	1082.56
100–104	CACGIY	[464]	[Bis(1-ethyl-3-methylimidazol-2-ylidene)silver(I)]	$[B(CN)_4]$	1:1	233	$P2_1/c$	2.0221	4	398.30	443.07	1.455	398.29
101	WEQWUK	[282]	1-Tetradecylpyridinium	$[Pd^{II}I_4]$	2:1	283–303	$P\bar{1}$	2.294	2	768.68	1166.9	1.689	768.65
101	IBEGIF	[183]	1-Butyl-2,3-dimethylimidazolium	$[CB_{11}H_6Cl_6]$	1:1	198	$Pbca$	9.7747	16	380.28	502.93	1.367	380.30
101	FAKROZ	[465]	1-(3-Cyanopropyl)pyridinium	Cl	1:1	140	$P2_1/c$	0.9491	4	482.75	182.65	1.278	482.74
101	KUSZUT	[284]	4-Methylpyridinium	$[CF_3CO_2]$	1:1	283–303	$P2_1/n$	0.9485	4	482.83	207.15	1.451	482.88
101	BESDOS01	[466]	Tetraphenylphosphonium	$[S(NSO_2CF_3)_2(OEt)]$	1:1	283–303	$P2_1/c$	3.22079	4	355.97	710.67	1.466	356.01
101	WIRREU	[397]	1-(2-Hydroxyethyl)piperidinium	$[\{2\text{-}(2,6\text{-}Cl_2C_6H_3NH)C_6H_4\}CH_2CO_2]$	1:1	283–303	$P2_1/a$	2.10371	4	394.44	425.34	1.343	394.46
101–102	YAVRIW	[467]	2,6-Dimethylpiperidinium	$[(2\text{-}OH)C_6H_4COS]$	1:1	283–303	$P2_1/n$	1.5045	4	428.80	266.38	1.181	428.87
101–102	BAQKAH	[145]	3-Chloro-3-ethyltricyclo[2.1.0.0^{2,5}] pentaphosphan-3-ium	$[GaCl_4]$	1:1	153	$P2_1/n$	1.3941	4	437.19	430.88	2.053	437.19
101.5–103	MEQQEE	[468]	Imidazolium	$[HO_2C(CH_2)_4CO_2]$	1:1	283–303	$Pna2_1$	2.195	8	464.83	184.15	1.296	464.81
102	ZIVXEH	[469]	Dimethyldiphenylammonium	$[I_3]$	1:1	283–303	$P2_1/n$	1.7404	4	413.40	579.00	2.210	413.43
102	ROPMEO	[470]	Tetrakis(decyl)ammonium	$[B(C_6H_5)_4]$	1:1	190	$P\bar{1}$	2.9477	2	309.95	898.29	1.012	309.95
102	AQOQED	[471]	1-Hydroxycarbonylmethyl-3-decylimidazolium	Br	1:1	283–303	$P2_1/c$	1.72599	4	414.26	347.30	1.337	414.31
102	VURDAO	[197]	1,2-Dimethylimidazolium	$[HO_2C(CH_2)_3CO_2]$	1:1	120	$Pna2_1$	1.1049	4	464.02	228.25	1.372	464.03
102	TUZCEX	[472]	Butyltriphenylphosphonium	$[Hg^{II}_2Cl_6]$	2:1	283–303	$P2_1/n$	5.20768	4	729.50	1801.34	2.298	729.60
102	VATVET	[473]	2-Aminopyridinium	$[O_2CCH_2CO_2]$	2:1	283–303	$Fdd2$	2.82583	8	1224.24	292.30	1.374	1224.26
102	VATVET01	[473]	2-Aminopyridinium	$[O_2CCH_2CO_2]$	2:1	150	$Fdd2$	2.77794	8	1232.26	292.30	1.398	1232.38
102	No refcode	[209]	Potassium	$[N(SO_2F)_2]$	1:1	113	$Pcab$	2.39575	16	545.60	219.23	2.431	545.60
102	No refcode	[209]	Potassium	$[N(SO_2F)_2]$	1:1	298	$Pcab$	2.46909	16	541.18	219.23	2.359	541.20

(*Continued*)

TABLE 11.3 (Continued)

m.pt./°C	CSD Refcode	References	Cation [cat]^{q}	Anion [A]$^{p-}$	Salt Type (p:q)	Data Temp/K	Space Group	V_{cell}/ 10^3 Å^3	Z	U_L(V)/ kJ mol^{-1}	Formula Weight	ρ(calc)/ g cm^{-3}	U_L(D)/ kJ mol^{-1}
102–103	SADMUG	[217]	1-(2-Hydroxy-2-methylpropyl)-3-methylimidazolium	I	1:1	150	$Pna2_1$	1.1389	4	460.40	282.12	1.645	460.40
102–103	PIZGUA	[474]	3-*O*-Nitrato-3-ethylazetidinium	$[NO_3]$	1:1	283–303	$P2_1/c$	1.94885	8	479.43	209.16	1.426	479.48
102–103	BEPCAA	[475]	Tetrapropylammonium	$[SiPhF_4]$	1:1	283–303	$P2_1/n$	2.09053	4	395.05	367.54	1.168	395.08
102–104	TAHMAR	[476]	1-Butyl-3-methylimidazolium	$[B\{3,5\text{-}(CF_3)_2C_6H_3\}_4]$	1:1	200	$P2_1/n$	8.6253	4	285.39	2004.89	1.544	285.40
102–104	YAVRIW01	[467]	2,6-Dimethylpiperidinium	$[(2\text{-}OH)C_6H_4COS]$	1:1	283–303	$P2_12_12_1$	2.9244	8	431.90	266.38	1.215	431.96
102–104	SEPJOM	[477]	$[Ti(h^5\text{-}C_5H_5)Cl.18\text{-}Crown\text{-}6]$	$[AlMe_2Cl_2]$	1:1	283–303	$P2_1/c$	2.65591	4	372.71	540.69	1.352	372.71
102–105	ZIJNUB	[478]	Aquo(pyridine-*N*-oxide) trimethyltin(IV)	$[N(SO_2Me)_2]$	1:1	178	$P\bar{1}$	0.85916	2	414.72	449.11	1.736	414.73
102–106	No cif	[352]	Rubidium	$[N(NO_2)_2]$	1:1		$Pca2_1$	0.9479	8	581.44	191.49	2.682	581.37
103	777586	[294]	1-Butyl-4-methylpyridinium	$[Ta^{V}Cl_6]$	1:1	223	$P2_1/c$	1.7696	4	411.68	543.89	2.041	411.67
103	ITUSEW	[241]	1-Cyanomethyl-3-methylimidazolium	$[NO_3]$	1:1	173	$P\bar{1}$	0.8206	4	501.58	184.06	1.491	501.70
103	UWINOE	[235]	Diaminoguanidinium	1-(2-Nitratoethyl)-5-nitriminotetrazolate	1:1	173	$P\bar{1}$	1.2259	4	451.76	308.26	1.670	451.76
103	PEMREE	[479]	Tetrabutylammonium	$[(C_6H_5CO_2)_2H]$	1:1	283–303	$Pccn$	3.03392	4	361.04	485.7	1.063	361.03
103	RECSAU	[214]	1-Methylimidazolium	$[Au(CN)_2]$	1:1	283–303	$P\bar{1}$	0.4347	2	493.99	332.12	2.538	479.29
103	GOKKOH	[480]	1-Ethoxymethyl-3-methylimidazolium	$[Pd^{II}Cl_4]$	2:1	100	$P2_1/n$	1.08793	2	1036.17	530.59	1.620	1036.29
103	EPISIG	[481]	1-Ethyl-3-methylimidazolium	$[U^{VI}O_2(NCS)_5]$	3:1	100	Cc	3.4968	4		893.98	1.698	
103	CALZUK	[482]	2-*tert*-Butyl-3,3,5-trimethyl-1,2-diaza-3-sila-5-cyclopentenium	$[AlCl_4]$	1:1	283–303	$P2_1/n$	1.83655	4	407.90	353.15	1.277	407.90
103–104	DOZSOA	[483]	1-Ethoxyhomotropylium	$[SbCl_6]$	1:1	206	$P2_1/c$	1.6773	4	417.23	483.69	1.915	417.23
103–105	GANXEZ	[484]	Bis(2,6-dimethylpyridine)iodonium	$[ICl_2]$	1:1	120	$P\bar{1}$	0.4485	1	410.28	539.00	1.996	410.32
103–105	NUDJOM	[485]	*rac*-1-((1-(2-Naphthyl)prop-2-en-1-yl)amino)pyridinium	I	1:1	128	$P2_1/c$	1.63372	4	420.00	388.24	1.578	419.98
103–105	MUJTAN	[486]	1-{2-((4-Chlorobenzylidene)amino) ethyl}-3-methylimidazolium	$[PF_6]$	1:1	173	$P\bar{1}$	0.8174	2	419.93	393.70	1.600	419.97
103–105	AMIFIL	[176]	Chlorodiphenylphosphine (diphenyl)phosphonium	$[GaCl_4]$	1:1	283–303	$Pbca$	5.49286	8	369.72	617.35	1.493	369.74
104	SEFHER	[487]	Tetraethylammonium	$[N(SO_2CF_3)_2]$	1:1	100	$P\bar{1}$	3.5300	8	411.95	410.40	1.544	411.94

104	RECKUF	[135]	1,3,5-Tri-*tert*-butyl-tetrahydro-1,3,5-triazinium	$[I_5]$	1:1	283–303	*C*2/*m*	2.6593	4	372.60	888.96	2.220	372.60
104	BARRAN	[488]	(±)-*cis*-3-Acetoxy-1-methylthiacycohexane	$[ClO_4]$	1:1	283–303	$P2_1/n$	1.2404	4	450.40	274.73	1.471	450.41
104	QADWOJ	[489]	5-Azonia-2-oxa-spiro[4.4]nonane	$[BF_4]$	1:1	173	$P2_1/n$	0.94894	4	482.77	215.00	1.505	482.80
ca. 104	SESXAR	[490]	1-Ethyl-3-methylimidazolium	$[Sn^{II}Cl_3]$	1:1	120	$P2_12_12_1$	1.1121	4	463.25	336.21	2.008	463.26
104–105	EMUREJ	[491]	1-(Hydroxycarbonylmethyl)-1-methylmorpholinium	$[(2\text{-}OH)C_6H_4CO_2]$	1:1	140	*Pbca*	5.5919	16	436.85	285.30	1.413	436.91
104–105	EPOTHP	[492]	(−)-(1-Phenylethyl)ammonium	[POS(OEt)Et]	1:1	283–303	$P2_1$	0.78934	2	423.63	275.35	1.158	423.60
104–105	JUHPEH	[416]	1,6-Diarsa-2,5,7,10-tetramethyl-2,5,7,10-tetraazatricyclo[5.3.0^{1,5}.0^{6,10}]decane	$[GaCl_4]$	1:2	283–303	$P2_1/n$	1.21435	2	1311.33	745.18	2.038	1311.37
104–106	LEWROU	[493]	2-Aminopyrimidinium	$[(2\text{-}OH)C_6H_4CO_2]$	1:1	283–303	*Pbca*	2.2034	8	464.37	233.22	1.406	464.39
104–106	FIMWII	[494]	Tetraethylammonium	3,5-Dinitro-1,2,4-triazolate	1:1	173	$P2_12_12_1$	1.4085	4	436.02	288.31	1.36	436.07
104–106	FIMWII01	[332]	Tetraethylammonium	3,5-Dinitro-1,2,4-triazolate	1:1	173	$P2_12_12_1$	1.4085	4	436.02	288.31	1.36	436.07
104–106	FERTUS	[495]	Tetraphenylphosphonium	$[Sb(N_3)_6]$	1:1	143	*C*2/*c*	3.01411	4	361.61	713.27	1.572	361.63
104–107	AGEYIV	[496]	Ethyldi-*iso*-propylammonium	$[\{C_6H_5)_3C\}P(BH_3)(O)OH]$	1:1	213	$P2_1/n$	2.53827	4	376.80	451.37	1.181	376.81
104–107	QAGJOX	[461]	2-Aminobenzothiazolium	1-Methylpyrrole-2-carboxylate	1:1	283–303	$P2_1/c$	1.3079	4	444.33	275.32	1.398	444.33
104.5	NUTSUR	[497]	Hexamethylenediammonium	$[I_5]$	1:2	123	*C*2/*m*	1.6157	2	1225.45	1471.38	3.024	1225.44
105	DOWSIS	[498]	1-Cyanomethyl-3-methylimidazolium	$[CdCl_4]$	2:1	283–303	*Pbca*	3.8960	8	1081.81	498.51	1.700	1081.91
105	XOMDOS	[314]	Tetrapropylammonium	$[N(SO_2CF_3)_2]$	1:1	173	$P2_1/n$	2.213	4	389.57	466.51	1.400	389.58
105	PAMSUT	[240]	1-Butyl-3-methylimidazolium	$[Co^{II}Br_3(C_9H_7N)]$	1:1	173	$P\bar{1}$	1.02251	2	397.19	567.04	1.836	396.90
105	ZOYBIY	[418]	*N*-Propylurotropinium	$[I_3]$	1:1	283–303	$P2_1/c$	4.913	12	419.74	563.99	2.287	419.74
105	ZZZAPV	[499]	Hydroxylammonium	$[CF_3SO_3]$	1:1	283–303	$P6_322$	0.75777	4	512.28	183.11	1.605	512.30
105	RAGYII	[500]	1-(3-Hydroxycarbonylpropyl)-3-methylimidazolium	Cl	1:1	140	$P2_1/n$	0.9793	4	478.81	204.65	1.388	478.83
105	BOTTIO01	[197]	2-Methylimidazolium	$[HO_2C(CH_2)_3CO_2]$	1:1	120	$P\bar{1}$	0.53837	2	467.14	214.22	1.321	467.11
105	VURCUH	[197]	1,2-Dimethylimidazolium	$[HO_2C(CH_2)_2CO_2]$	1:1	120	*Pccn*	2.1238	8	468.82	214.22	1.340	468.85
105	VISKAJ	[501]	Methyltriphenylphosphonium	$[I_3]$	1:1	283–303	*Pbca*	4.22671	8	394.00	658.04	2.068	394.00

(*Continued*)

TABLE 11.3 (Continued)

m.pt./°C	CSD Refcode	References	Cation [cat]^{q}	Anion [A]$^{p-}$	Salt Type (p:q)	Data Temp/K	Space Group	V_{cell}/ 10^3 Å^3	Z	U_L(V)/ kJ mol^{-1}	Formula Weight	ρ(calc)/ g cm^{-3}	U_L(D)/ kJ mol^{-1}
105–106	VUKROJ	[224]	4-Amino-1,3,5-trimethyl-1,2,4-triazolium	$[N(SO_2CF_3)_2]$	1:1	173	$P2_1/n$	1.53434	4	426.68	407.32	1.78	427.71
105–106	VICMOK	[370]	1-Methylmorpholinium	$[Mn^{II}Cl_4]$	2:1	158	$P2_1/c$	1.76185	4	1124.73	401.06	1.512	1124.78
105–108	VICMAW	[370]	1-Methylmorpholinium	$[Co^{II}Cl_4]$	2:1	158	$P2_1/c$	1.74441	4	1129.06	405.06	1.542	1129.01
105–108	RELKAW	[502]	1-Azidosulfonylimidazolium	$[SO_3(OH)]$	1:1	173	$P2_1/c$	0.96199	4	481.05	271.26	1.873	481.07
105–118	QQQGSS	[343]	2-Fluoropyridinium	$[Sb^{III}_2Br_9]$	3:1	283–303	$C2/c$	3.09019	8		1199.96	2.702	
105/107	GEDLOR	[399]	1,3-Dimethyl-2-nitroimidazolium	$[CF_3SO_3]$	1:1	173	$P2_1/c$	1.0854	4	466.17	291.21	1.782	466.18
105–110	VUKTEB	[224]	4-Dimethylamino-1-methyl-1,2,4-triazolium	$[SO_3(OH)]$	1:1	173	$P\bar{1}$	0.48305	2	480.51	224.25	1.54	480.38
105–110	HIYCEX	[503]	1,2,4-Triazolium	$[N(SO_2Me)_2]$	1:1	143	$Pna2_1$	0.9880	4	477.70	242.28	1.629	477.74
106	KUKSOZ	[142]	1-Ethyl-1-methylpyrrolidinium	Br	1:1	123	$P2_12_12_1$	0.88905	4	491.09	194.12	1.450	491.08
106	MAKTEZ	[302]	Dimethylbis(4-methoxybenzyl) ammonium	$[N(SO_2CF_3)_2]$	1:1	153	$P2_12_12_1$	4.9345	8	379.40	566.53	1.525	379.40
106	BACQUS	[317]	1-Ethyl-3-methylimidazolium	$[1\text{-}EtSnB_{11}H_{11}]$	1:1	170	$P2_1/c$	1.8197	4	408.83	388.92	1.420	408.87
106	BARRER	[488]	(−)-(1*S*,3*S*)-*trans*-3-acetoxy-1-methylthiacycohexane	$[ClO_4]$	1:1	283–303	$P2_1$	0.6330	2	448.05	274.73	1.441	448.04
106	VACMUJ	[504]	3,3-Dinitroazetidinium	$[(2\text{-}OH)C_6H_4CO_2]$	1:1	283–303	$P2_1/n$	1.2553	4	449.02	285.22	1.509	449.03
106	VURCOB	[197]	4-Methylimidazolium	$[HO_2C(CH_2)_3CO_2]$	1:1	120	$P2_1$	1.03420	4	472.05	214.22	1.376	472.09
106	WOFJAC	[505]	([2.2.2]Cryptand)potassium	$[SiPh_3F_2]$	1:1	173	$P\bar{1}$	1.8317	2	345.38	712.99	1.293	345.41
106–107	NEMGUH	[365]	1-(Piperidinium-1-ylidene)-3-(piperidin-1-yl)-2-azapropene	$[PF_6]$	1:1	173	$Pca2_1$	3.20405	8	422.07	353.30	1.465	422.09
106–107	VUKSUQ	[224]	4-Amino-1,3,5-triethyl-1,2,4-triazolium	$[N(SO_2CF_3)_2]$	1:1	100	$Pbca$	3.68376	8	407.60	449.41	1.62	407.58
106–108	VOCZAP	[506]	2-Amino-4-phenyl-6-methylpyrimidinium	$[C_6H_5CO_2]$	1:1	100	$P\bar{1}$	0.74482	2	429.88	307.35	1.37	429.86
106–108	LIHBEK	[507]	{1-(Hydroxymethyl)propyl} ammonium	$[(4\text{-}NO_2)C_6H_4CO_2]$	1:1	283–303	$P\bar{1}$	0.63647	2	447.42	256.26	1.337	447.43
106–108	HAMVUO	[508]	1,4-Bis(1-methylimidazolium-3-yl)butane	2-Hydroxy-5-sulfonatobenzoate	2:2	283–303	$Pna2_1$	1.7970	4		408.43	1.510	
106–109	HUHCIW	[359]	Trimethyltelluronium	[SeCN]	1:1	200	$Pna2_1$	0.78682	4	507.19	277.68	2.344	507.20

106–109	LUWLUL	[434]	1,3-Dichloro-1,2,3,4-tetra-*cyclo* hexyltetraphosphetane-1,3-diium	$[Ga_2Cl_7]$	1:2	123	*Pbca*	10.07770	8	1107.07	1302.63	1.717	1107.08
107	JIZZIB	[509]	4-Methoxy-2,6-dimethyl-1-hydroxypyridinium	$[2,6\text{-}Cl_2\text{-}4\text{-}O_2NC_6H_2O]$	1:1	283–303	$P2_1/n$	1.57763	4	423.70	361.19	1.521	423.74
107	GOKKIB	[480]	1-(2-Methoxyethyl)-3-methylimidazolium	$[Pd^{II}Cl_4]$	2:1	100	$P2_1/c$	1.04676	2	1051.90	530.59	1.683	1051.84
107	NATPOO	[510]	1-Methyl-1-propylpyrrolidinium	$[Yb\{N(SO_2CF_3)_2\}_4]$	2:1	120	$P2_1/n$	5.4244	4	1089.06	1550.10	1.898	1089.08
107	SURLOG	[511]	Tris(dimethylamino)sulfonium	$[Li(C_5H_5)_2]$	1:1	173	$P\bar{1}$	0.9178	2	407.95	301.41	1.091	408.00
107–108	HRFVIY	[512]	3,6-Dioxaoctane-1,8-diammonium	$[CCl_3CO_2]$	1:2	283–303	$C2/c$	2.00226	4	1374.22	474.96	1.576	1374.34
107–108	VIVKIU	[513]	*cyclo*-Hexylammonium	$Cl,[AlCl_4]$	2:(1,1)	283–303	$P2_1/c$	2.14895	4		404.47	1.251	
107–108	FITZIR	[514]	Methyltriphenylphosphonium	$[Cu^I_2Br_4]$	2:1	283–303	$C2/c$	3.93394	4	818.52	1001.35	1.691	818.62
107.5–108	PIVFOP	[515]	Di-*cyclo*-hexyl(hydroxo) selenonium	$[4\text{-}BrC_6H_4SO_3]$	1:1	283–303	$P2_1/a$	2.02273	4	398.27	498.34	1.636	398.26
108	EFOCEI	[427]	*N,N′*-Diaminoguanidinium	5-Nitrotetrazolate	1:1	200	$P2_1$	0.8504	4	496.87	204.18	1.595	496.91
108	OREQAE	[516]	1-Ethylbenzimidazolium	$[BF_4]$	1:1	283–303	*Pnma*	1.0486	4	470.36	234.01	1.482	470.36
108	XAKLEB	[517]	1,3-Bis((*R*)-(+)-1-phenylethyl) imidazolium	$[PF_6]$	1:1	100	$P2_12_12_1$	1.91640	4	403.61	422.35	1.464	403.64
108	ASODIV	[518]	*cyclo*-Hexyltrivinylphosphonium	I	1:1	143	$P2_1/n$	1.43534	4	433.94	322.15	1.491	433.97
108	FOZQUK	[170]	Tetrabutylammonium	$[Al\{OC(CH_3)(CF_3)_2\}_4]$	1:1	100	$P2_1/c$	4.3848	4	331.33	993.67	1.505	331.33
108	WOFHOO	[505]	([2.2.2]Cryptand)potassium	$[SiPh_2F_3]$	1:1	173	$P2_1/a$	6.6128	8	353.77	654.88	1.316	353.82
108	CUHXEJ	[519]	Tetrabutylphosphonium	$[PtBr_3(C_2H_4)]$	1:1	173	$P2_1/n$	2.47267	4	379.20	722.28	1.940	379.20
108	PEVZIZ	[520]	5-Ethoxy-l-ethyl-3,4-dihydro-2-oxo-2*H*-pyrrolium	$[SbCl_6]$	1:1	283–303	$P2_1/c$	1.74394	4	413.19	490.68	1.869	413.21
108–109	YOGMEM	[521]	Tetrabutylammonium	$[C_5H_5]$ (Cyclopentadienide)	1:1	100	$P2_1/n$	2.0669	4	396.15	307.57	0.99	396.33
108–109	JORPIP	[522]	Di-*tert*-butyl(diphenylmethylene) phosphonium	$[AlCl_4]$	1:1	283–303	$P2_1/n$	2.56786	4	375.75	480.22	1.247	376.12
108–110	HOTTIU	[523]	1,3-Di-*iso*-propyl-4,5-dimethylimidazolium	*E*-2-Cyano-1-phenylethenolate	1:1	223	$P2_1/c$	1.8606	4	406.58	325.45	1.162	406.61
108–110	FAVTEC	[220]	1,2-Bis(*P,P*-dimethyl-*P′,P′*-diphenylphosphinophosphonium) ethane	$[GaCl_4]$	1:2	193	*P*1	1.01132	1	1163.40	945.51	1.549	1162.83

(*Continued*)

TABLE 11.3 (*Continued*)

m.pt./°C	CSD Refcode	References	Cation [cat]^{q}	Anion [A]$^{p-}$	Salt Type (p:q)	Data Temp/K	Space Group	V_{cell}/ 10^3 Å^3	Z	$U_L(V)$/ kJ mol^{-1}	Formula Weight	ρ(calc)/ g cm^{-3}	$U_L(D)$/ kJ mol^{-1}
108–110	RUNTOJ	[524]	[(*N*,*N*,*N*′,*N*″,*N*″-Pentamethyl diethylenetriamine)AlMe$_2$]	[AlMe$_2$Cl$_2$]	1:1	283–303	$P\bar{1}$	1.05906	2	393.78	358.31	1.124	393.83
108–110	DELFIK	[525]	2,6-Bis(2-pyridinio)pyridine	[(CF$_3$CO$_2$)$_2$H]	1:2	283–303	$P2_1/n$	2.34434	4	1322.55	575.34	1.63	1322.56
108–110	FIFBIF	[526]	Bis(4-methoxyphenyl)iodonium	[CS$_2$(NEt$_2$)]	?	283–303	$P2/c$	4.28267	8	392.73	489.44	1.518	392.73
108–110	ZOKQIZ	[527]	1-Benzylimidazolium	[PhP(O)$_2$(OH)]	1:1	283–303	$P\bar{1}$	0.78547	2	424.14	316.30	1.337	424.14
108–115	WOLKIS	[92]	1-Methyl-4-cyanopyridinium	[SO$_3$(OMe)]	1:1	193	$Pna2_1$	1.0372	4	471.70	230.25	1.474	471.68
109	RUBYUJ	[143]	1,3-Di-*iso*-propylimidazolium	[N(C$_6$F$_5$)$_2$]	1:1	283–303	$C2/c$	2.1271	4	393.37	501.38	1.566	393.40
109	LIYZAU	[438]	Dimethylphenylsulfonium	[CF$_3$SO$_3$]	1:1	180	$P2_1/c$	1.2299	4	451.38	288.3	1.557	451.40
109	KHDFRM	[528]	Potassium	[(HCO$_2$)$_2$H]	1:1	283–303	$Pbca$	0.98105	8	576.00	130.08	1.762	576.08
109	VIXWAB	[78]	Tetra(pyridine)lithium	[P$_2$Se$_8$]	2:1	200	$P\bar{1}$	1.25540	2	979.55	670.15	1.773	979.63
109	SAZMIP	[529]	Trimethyltelluronium	[AlMe$_2$Cl$_2$]	1:1	283–303	$P2_1/n$	1.23175	4	451.21	300.66	1.621	451.21
109	IGUJEZ	[441]	1-(2-Hydroxyethyl)piperidinium	[Si(ox)$_3$]	2:1	173	$P\bar{1}$	1.22754	2	988.25	552.56	1.495	988.31
109–110	VUKSOK	[224]	4-Amino-3,5-diethyl-1-methyl-1,2,4-triazolium	[N(SO$_2$CF$_3$)$_2$]	1:1	100	$P2_1/c$	5.28732	12	412.10	435.38	1.64	412.06
109–110	OCOZIQ	[530]	1-(2-Diphenylphosphinoethyl)-3-mesitylimidazolium	[BF$_4$]	1:1	180	$P\bar{1}$	1.2605	2	377.43	486.28	1.281	377.43
109–111	ISUXID	[165]	5-Methyl-1,3,5-dithiazinan-5-ium	[BCl$_4$],Cl	2:(1,1)	283–303	$P\bar{1}$	1.01082	2		460.59	1.513	
109–112	JANGUA	[531]	Benzylammonium	Cl,[AlCl$_4$]	2:(1,1)	283–303	$P\bar{1}$	2.38287	2		974.14	1.455	
109–112	VIXVOO	[78]	1-Butyl-3-methylimidazolium	[P$_2$Se$_8$]	2:1	100	$P2_1/n$	1.43541	4	1216.87	486.03	2.249	1216.92
110	GIPGOB	[532]	Tetramethylammonium	[F(HF)$_2$]	1:1	173	$Pbca$	1.43864	8	519.43	133.16	1.23	519.50
110	AXOTAJ	[331]	Methyltributylammonium	[Fe^{III}Cl$_4$]	1:1	100	$Pca2_1$	3.9642	8	400.26	398.03	1.334	400.29
110	HUHCOC	[359]	Trimethyltelluronium	[N$_3$]	1:1	200	$Pnma$	2.11614	12	522.15	214.72	2.022	522.18
110	DEFXIW	[533]	Bis(2,6-dimethylpyridine)iodonium	[IBr$_2$]	1:1	120	$P\bar{1}$	0.4626	1	407.14	627.92	2.254	407.16
110	QICMIZ	[309]	1,3-Di(cyanomethyl)imidazolium	[BF$_4$]	1:1	283–303	$P2_1/c$	0.9691	2	402.49	467.95	1.604	402.52
110	YANQIO	[355]	1,4-Dimethyl-2-*H*-1,2,4-triazolium	[I$_3$]	1:1	84	$P2_1/n$	1.0922	4	465.42	479.84	2.918	465.43
110	PABCIE	[534]	1-(2-Hydroxyethyl)pyridinium	Br	1:1	118	$Pna2_1$	0.83712	4	498.94	204.07	1.619	498.95
110	XAKLAX	[517]	1,3-Bis((*R*)-(+)-1-phenylethyl) imidazolium	[BF$_4$]	1:1	150	$P2_12_12_1$	1.80185	4	409.84	364.19	1.343	409.89
110	WONVIF	[535]	1-Ethyl-2-(1-oxy-3-oxo-4,4,5,5-tetramethylimidazolin-2-yl)-3-methylimidazolium	[N(SO$_2$CF$_3$)$_2$]	1:1	123	$P\bar{1}$	1.17427	2	383.97	546.48	1.546	384.01

110	BACJAR	[536]	4-Methylpyridinium	$[Cu^{II}Br_4]$	2:1	283–303	$P\bar{1}$	0.9359	2	1098.69	571.44	2.028	1098.79
110	CEKMOV	[537]	1,3-Bis(hydroxycarbonylmethyl) imidazolium	$[ClO_4]$	1:1	140	$P2_1/c$	1.0843	4	466.29	284.61	1.744	466.35
110	ZUGQEX01	[538]	Piperazinediium	$[C_5H_{11}CO_2]$	1:2	283–303	$P\bar{1}$	0.43589	2	1421.82	318.45	1.213	1421.82
110	JIFKIS	[539]	Methylammonium	$[NO_3]$	1:1	283–303	*Pcmn*	0.4395	4	593.61	94.07	1.422	593.67
110	DUSCOK	[540]	2-(2,4-Dinitrobenzyl)pyridinium	$[2\text{-HO-3,5-}NO_2C_6H_2CO_2]$	1:1	200	$P2_1/c$	1.98934	4	399.90	487.34	1.627	399.91
110–111	MISMAC	[541]	2-Amino-5-bromopyridinium	$[HC{\equiv}CCO_2]$	1:1	283–303	*I2/a*	1.8753	8	484.28	243.06	1.722	484.32
110–112	UCAMUG	[542]	Tri-*iso*-propyl-[*N*-(2, 6-dichlorophenyl) thiocarbamoyl]phosphonium	$[CH_3SO_3]$	1:1	110	$P2_1/c$	2.13926	4	392.82	460.42	1.430	392.87
110–112	JOXJOV	[543]	*β*,*β*-Diethoxyethenediazonium	$[SbCl_6]$	1:1	173	$Pca2_1$	3.25793	8	420.30	477.64	1.948	420.34
110/113	GEDLIH	[399]	1,2,3-Trimethyl-4-nitroimidazolium	$[N(SO_2CF_3)_2]$	1:1	173	$P\bar{1}$	1.5917	4	422.75	436.32	1.821	422.78
110–114	ATUVAM	[544]	3-(4-Methoxyphenyl)-1,2-dimethyl-4,5-dihydroimidazolium	I	1:1	183	$P2_1/n$	1.3190	4	443.37	332.18	1.673	443.41
110.5–113	SOVKIZ	[140]	1,2,4-Trimethylpyridinium	$[I_3]$	1:1	283–303	*Pbcm*	1.411	4	435.83	502.9	2.37	435.96
111	ZOYBEU	[418]	*N*-Ethylurotropinium	$[I_3]$	1:1	283–303	*Pnma*	1.4214	4	435.01	549.96	2.425	428.69
111	LABZUL	[545]	1-Butyl-2-methylbenzimidazolium	$[BF_4]$	1:1	283–303	$Pna2_1$	1.4425	4	433.39	276.09	1.271	433.39
112	HOEDSO	[546]	Oxonium	$[O_3SCH_2CH_2SO_3]$	2:1	283–303	$P2_1/c$	0.4147	2	1496.86	226.23	1.81	1496.40
112	UCOPAE	[547]	6,6′-Spirobis(1-methyl-1,5,6,7-tetrahydropyrrolo[1,2-a] imidazol-4-ium	$[N(SO_2CF_3)_2]$	1:2	283–303	*C2/c*	3.01542	4	1245.51	790.59	1.741	1245.46
112	XASJIC	[548]	4-Fluorophenyl (pentafluorophenyl)iodonium	$[BF_4]$	1:1	283–303	*C2/c*	2.9825	8	429.76	475.86	2.120	429.80

points and according to their classification, respectively, as room-temperature ionic liquids (Table 11.1), ionic liquids (Table 11.2) or salts with reported melting points in the range 101 to ~110°C (Table 11.3). Where found, a CSD refcode or a CCDC deposition number is included, along with a literature citation. 67 salts with X-ray structures melt below 30°C and are considered to be room-temperature ionic liquids. (While room-temperature ionic liquids have been described [32] as salts with melting points below 'about 298 K', we allow the definition of room temperature to extend to the upper limit used by CSD, namely, 303 K.) There is a much bigger group (>500) of additional salts that also qualify for the more conventional definition of ionic liquids. Finally, we have identified (in a less complete survey) a further 149 salts that melt up to *ca.* 10°C above that which would qualify them for inclusion as ionic liquids. Other salts of interest are discussed in the text. Late additions to the compilation are included in a short addendum.

The chapter is subdivided as follows: Section 11.2 gives a brief consideration of the crystal state, of melting and of interionic interactions; Section 11.3 gives an overview of the component anions and cations found in low-melting salts. A more detailed consideration of structures of classes of ionic liquids is found in Sections 11.4–11.8 and of the common ionic components in Sections 11.9–11.19. Finally, Section 11.21 features a short summary of the overall conclusions.

11.2 CRYSTALLINITY AND MELTING PHENOMENA

Fascination with ionic liquids in its turn stimulates curiosity concerning the identification and relative influence of those interionic forces that, in other systems, give rise to crystalline materials with high-melting points but, which for the salts under consideration, result in low-melting crystal or glass-like behaviour [33]. It is thus necessary, albeit briefly, to review the crystalline state and the phenomenon of melting.

11.2.1 Melting Phenomena

Melting is both a kinetic [34] and thermodynamic phenomenon. In its simplest form, melting involves a transformation from a (single) crystalline phase into an isotropic liquid phase at the melting temperature. As melting is observed during the process of heating of a sample from below the melting temperature to one above, the rate of heating, solid–solid phase changes and 'pre-melting' phenomena are all factors that may influence the melting process. Cooling of the melt may lead to undercooling or supercooling (with the solidification temperature defining the point at which a (possibly different) crystalline phase is formed). The role of organised assemblages of ions or molecules ('microheterogeneities') that are present prior to crystal formation ('pre-freezing') is an area of current research interest. In the limit, glass formation may be observed, a general feature of ionic liquid behaviour. Melting and solidification temperatures are often widely separated. Crystallisation may occur on heating a glassy material, sometimes resulting in different polymorphs. Crystallisation processes may be slow, evident over months, even years in some cases.

For the purposes of this chapter, two distinct circumstances need to be recognised when salts melt: the first is displayed by salts in which both anion and cation are compositionally unchanged on transferring between the crystalline and liquid states (i.e. in which melting is a so-called congruent process) [35, 36]; the second is displayed by salts in which either the anion or cation (or both) suffers compositional change on crystallisation or fusion (incongruent melting). Only in the former case is it possible to assert, a priori, that the crystalline form transforms into an ionic liquid on melting (using the particular definition we have chosen in this review). Strictly speaking, in circumstances in which composition change on melting is possible, the nature of the liquid formed from the crystalline solid would need to be explored to establish whether compositional changes have resulted in the formation of nonionic components (the liquid, therefore, not being composed entirely of ions). For example, a crystalline salt, [BH][A], formed from a base, B, and a protic acid, HA, may, depending on the nature of the combination, partially dissociate in the liquid state to give a mixture containing species other than the ions $[BH]^+$, $[A]^-$, such as B, HA or even $[HA_2]^-$. This clearly is a matter of degree, ranging from extensive dissociation (with significant vapour pressures of B and HA) to negligible. The extensive series of materials included in this review that result from protonation of a base (Section 11.4) fall into this category. We have included a number of salts containing anions of the type $[HA_2]^-$, in which A = F, Cl or RCO_2 (R = CF_3 or C_6H_5), although (other than $[H_2F_3]^-$ and $[H_3F_4]^-$) we have excluded $[H_nA_{(n+1)}]^-$, where $n > 1$, on the grounds that it is more likely that significant dissociation occurs in the liquid of salts of the latter and may be better characterised as solutions of $[cat][H_{(n-1)}A_n]$ in HA and are not therefore comprised solely of ions. Unless clearly otherwise from the context, $[cat]^{n+}$ denotes a generic cation, $[A]^{m-}$ a generic anion, X^- a generic halide ion and $[M]^{n+}$ a metal.

In addition, a series of salts has been characterised by Tebbe, Blaschette, Jones and others, in which halide ions, particularly iodide, form complexes of a range of compositions and structures with the corresponding molecular element (see Section 11.9.2), in equilibria of the type $I^- + nI_2 = [I_{(2n+1)}]^-$, where $n = 1$, 2 or 3 [37]. Related dianions are also known. Unfortunately, the stoichiometric composition of salts, such as $[cat][I_7]$, is no guide to the composition of the liquid formed on melting. To what extent is the melt better described as $[cat]^+ + [I_5]^- + I_2$, $[cat]^+ + [I_3]^- + 2I_2$ or $[cat]^+ + I^- + 3I_2$ or, indeed, a mixture of $[I_n]^-$ ($n = 1, 3, 5$ or 7) + I_2? In this case, the ambiguity is even found in the solid state, where the characterisation of a species such as $[(I_n).I_2]^-$ or

$[I_{(n+2)}]^-$ turns on the interpretation of critical I···I separations. Interestingly, some of the very lowest-melting materials included in this compilation are polybromides, $[Br_n]^-$. On the other hand, dissociation on melting of [cat][EX_n] into [cat][EX_{n-1}] and [cat]X, while a non-congruent melting, would not disqualify the material as an ionic liquid (though it would certainly complicate the discussion of the origins of the melting temperature). This situation is a common feature of the haloaluminate and related ionic liquids found in systems obtained on mixing [cat]X and Al_2X_6 in various proportions. Only in relatively few cases has this aspect of ionic liquid behaviour been fully investigated crystallographically.

The thermodynamics of melting [35, 36] is governed by enthalpic and entropic differences between the solid (crystalline) state and liquid state at equilibrium at the same temperature, subject to the overall free energy change being zero. In general, the Gibbs free energy, ΔG, for a process is given by Equation 11.1:

$$\Delta G = \Delta H - T\Delta S \tag{11.1}$$

where ΔH and ΔS are the corresponding enthalpy and entropy changes at the absolute temperature, T. When T is the fusion temperature, T_{fus}, $\Delta G = 0$, and hence T_{fus} is given by Equation 11.2:

$$T_{fus} = \frac{\Delta H_{fus}}{\Delta S_{fus}} \tag{11.2}$$

Differences in melting temperature between compounds may arise, therefore, from differences in the energies of interactions in the crystalline state and in the liquid state, as well as differences in entropic changes between liquid and solid. The complications associated with seeking an understanding of the relationship between structure and melting point of congruently melting ionic liquids are somewhat less than the situation for materials that undergo some compositional change between the crystalline and liquid states at an equilibrium melting temperature. In classical (high-melting) inorganic salts, the major contribution to the energetics of bonding in the crystal is associated with the aggregated net effects of coulombic attractions and repulsions between the component ions, determined by ion charge, ion separations and the lattice arrangement. It is to be expected that electrostatic interactions will be maintained on melting (assuming anions and cations remain unchanged), but the aggregated net effects will change as a consequence of changes in interion separations and in the number of neighbouring anions and cations. This can only be quantified (and the energy change on melting estimated) from a full knowledge of the relevant parameters in the liquid and crystalline state.

Likewise, there will be an entropic change associated with the melting process resulting from the enhanced translational and rotational freedom in the liquid state. Even in classical salts, a model of bonding based solely on electrostatics is found to be deficient. More realistic models will include contributions from covalent bonds, hydrogen bonds, π-stacking interactions and other 'secondary' [38] bonding, as well as van der Waals or London dispersion forces. These will also be seen in both solid and liquid ionic liquids, to a more significant degree when considering the nature of the combinations of anions and cations with which this chapter is concerned. While many studies discuss interionic interactions in terms of these various forces, very few have attempted to quantify their separate contribution even in the crystalline state [39]. The contributions of van der Waals, electrostatic and hydrogen bonding to the total computed lattice energies of seven *N*-aminoazolium and *N,N′*-diaminoazolium chlorides underline the dominance of electrostatic interactions and reveal van der Waals and hydrogen-bond contributions (in these particular protic materials) that are roughly similar. For example, according to Laus et al. [39], 1-aminoimidazolium chloride has a total lattice energy of 201.7 kJ mol^{-1}, made up of contributions from electrostatic, hydrogen bonding and van der Waals interactions of 150.1, 27.1 and 24.5 kJ mol^{-1}, respectively.

Major efforts, not reviewed here, use computational methods and simulations to relate liquid structure and physical property with the aim of property prediction. In her review on approaches to fully ab initio calculations on ionic liquids, Izgorodina [40] provides an excellent summary of the methods used, their strengths and weaknesses and the nature of the different forces to be estimated. In addition, there have been a number of attempts to develop correlations and simulations that might enable melting point prediction [41–46], though with mixed success. This is hardly surprising when considering the complexity of the challenge and the uncertainties in m.pt. data on which such correlations are based. From a purely technological point of view, however, any procedure that provides the basis for limiting the number of ionic liquids to be synthesised and characterised to meet a particular need will be welcome, even if precision in the predictions is modest and the possibility of excluding some material with useful characteristics may be significant.

A further comment needs to be made concerning the dynamics of processes in the solid, indeed the crystalline, state, as these can materially affect the precise nature of the (bulk) solid in equilibrium with liquid at the melting temperature. (For important insights into the processes that occur during melting and solidification of 1,3-dialkylimidazolium halides, see the work of Nishikawa and co-workers [47–50].) Solid–solid phase changes may occur on cooling the first crystalline material formed on taking the liquid below the fusion temperature. Such transitions may be reversed on heating. It is possible, therefore, that the structure of a crystalline material may (unless explicitly checked by thermophysical, spectroscopic or diffraction methods) be different from the phase in equilibrium with the liquid. The two solids may represent different polymorphic phases with different crystal structures. Polymorphism is discussed further in Section 11.2.3. In addition, some changes are associated with conformational disorder, widely represented and investigated in the study of 'rotator' phases of quaternary ammonium salts and other materials. Disorder may also be associated with the availability of a range of conformational preferences in alkyl side chains (such as in the *N*-butyl group in 1-butyl-3-methylimidazolium salts, Section 11.18), conformational isomerism (as exhibited by anions, such as bis(trifluoromethanesulfonyl)amide, Section 11.13) and as a consequence of weak interionic interactions (such as in salts containing the hexafluorophosphate ion). An inspection of the materials collected in Tables 11.1, 11.2 and 11.3 suggests that these phenomena are widespread and the consequence in any particular compound (and the energetics associated with it) is highly circumstance specific.

11.2.2 Crystallography

Instrumental and computational advances have resulted in single-crystal X-ray crystallography becoming a routine characterisation technique to the extent that many hundreds of thousands structures are now accessible and thousands are added annually. While this is so, the importance of experimental and analytical rigour for both data quality and reliable interpretation remains. A cautionary tale regarding the interpretation of poor-quality crystallographic data (using $[N_{4444}][B_{4444}]$ as an example) is given by Stilinović and Kaitner [51].

Except where stated, the crystal data assembled in Tables 11.1, 11.2 and 11.3 have been obtained from the primary research literature [52–548], usually checked for consistency with corresponding entries in the CSD. Most structures are supported (sometimes in electronic supplementary information) with a Crystallographic Information File (cif) [549] containing standard crystallographic data for both archival purposes and for sharing, the availability of which permits the consistency and integrity of the structure determination to be validated using checkCIF software (e.g. see http://checkcif.iucr.org/).

However, some data that are clearly of interest are, for a variety of reasons, only available in hard copy form in the original research article. Not all reports, particularly of earlier studies, provide data in electronic, particularly cif, format. It should be noted that we have not examined the listed structures in the detail necessary to test the validity of the crystallographic analysis. Indeed, it is possible that a significant number of these may be of sufficient interest to warrant a more detailed reinvestigation. Neither have we excluded published structures because of poor data quality.

The ability to carry out crystallography at low temperatures has also now become routine, though the growth and manipulation of crystals of materials that are liquid at room temperature have required the development of *in situ* crystallisation techniques, such as the Optical Heating and Crystallisation Device developed by Boese and colleagues [550–553]. In a notable study, Bond describes the characterisation of *in situ* crystallised co-crystals, melting between –21 and +27°C, formed between pyrazine and a series of 10 carboxylic acids [554]. The pathfinding studies of Mootz and co-workers on salt solvates [57, 61, 95, 108, 195, 368, 532, 546] should also be mentioned in this context. A description of the challenges involved in growing suitable crystals (and the persistence needed for success) can be found in the 1968 report of Jönsson and Olovsson [68] who describe the crystal structure of 1,1-dihydroxyethylium hydrogen sulfate, with a m.pt. of −2.5°C. An even earlier study [555] reports the crystal structure of $HNO_3 \cdot 3H_2O$, melting at −18.5°C.

Data mining software allows the wealth of structural data now available to be used to explore effects on a range of systems much wider than possible hitherto, putting imprecise concepts, such as weak hydrogen bonding [556, 557], on a more statistically rigorous footing, a matter of some importance in considerations of interionic interactions in crystalline ionic liquids [558]. For a more general review of hydrogen bonding in the solid state, see Ref. [559]. For an interesting perspective on C—H···C(π) interactions in cyclopentadienide salts (unfortunately reported without m.pt.), see Ref. [560].

The availability of the body of information represented by the crystal structures collected together provides an opportunity to assess the role of hydrogen bonding in interionic and supramolecular organisation in a crystalline lattice. Etter's approach to graph-set assignment of particular hydrogen-bonding motifs and their characterisation and designation as chains, rings and dimers [561, 562] has been used in the analysis of some ionic liquids, highlighting the similarities and differences, for instance, between $[(ClCH_2)Me_2NNH_2]_2[SO_4]$ and $[(ClCH_2)Me_2NNH_2][ClO_4]$ [159] and between 1,5-diaminotetrazolium nitrate and perchlorate [443]. Klapötke et al. subjected to graph-set analysis [427] the complex three-dimensional hydrogen-bonding networks that characterise the interactions between anion and cation in a series of 5-nitrotetrazolate salts, one of which (with *N,N′,N″*-triaminoguanidinium (m.pt. 96°C)) may be classified as an ionic liquid.

11.2.3 Polymorphism

Investigations into the behaviour of ionic solids are important in the search for new battery and fuel cell electrolytes. Defects, plastic phases, disorder and polymorphism in crystalline solids (and their association with ionic conductivity) and associated solid–solid equilibria are thus of considerable interest. The basis of our understanding of these phenomena and their relationship to the characteristics of the component ions includes studies of materials that are ionic liquids.

While a crystalline solid is characterised by a three-dimensional lattice, in 'plastic' phases [563], a limited amount of translational motion can be exhibited. Such phenomena have been observed in ionic salts [11, 12], including ionic liquids.

Polymorphism is a general phenomenon [564] in which a solid material may exist in one or more crystalline forms. This may be distinguished from crystalline structural disorder, which may be evident in both cation and anion. Different polymorphic modifications may arise from differences in crystal packing or from the ability of a component molecule or group to adopt more than one conformation in the crystal. Polymorphs differ in stability and may convert from one form to a more stable form at a particular temperature. Polymorphs may have different melting temperatures and solubilities that can have great significance in the manufacture and efficacy of active pharmaceutical ingredients (APIs) and can be distinguished by crystallographic and spectroscopic analysis. See, for example, a recent study [565] on the three forms of the ionic liquid dibenzoate salt of ethambutol, used in the treatment of tuberculosis. Bearing in mind the nature of the component ions that make up ionic liquids, it is not surprising, therefore, that polymorphism is a phenomenon readily and widely observed, particularly from the use of thermophysical and thermochemical techniques, such as DSC. A review of such studies is beyond the scope of this chapter. However, such work does provide an insight into both the phenomenological complexity of the phase behaviour of crystalline ionic liquids and the experimental challenges associated with achieving conditions under which a polymorph can be grown into a single crystal of good enough quality for a publishable structure to be obtained. The recent use by Mudring et al. [186] of *in situ* synchrotron X-ray powder diffraction to study $[N_{4\,4\,4\,4}][Pr\{C(CN)_2(NO)\}_6]$, a so-called 'kinetic' polymorph formed by fast cooling from the melt, illustrates this point. Such complexities may go some way to explaining why, among the several hundred crystal structures brought together in this survey, relatively few involve the characterisation of more than a single crystalline form.

Among the earliest ionic liquids for which single-crystal characterisation of polymorphs has been reported are 1,3-dialkylimidazolium chlorides and bromides [230, 566]. Monoclinic and orthorhombic forms of $[C_4mim]Cl$ are associated with differences in butyl conformation, with a computational study [567] suggesting that the free energy difference associated with a single *trans–gauche* conformational change at the NCH_2–CH_2Et bond of the butyl group (Fig. 11.2) is very small, though this change is sufficient to result in different patterns of hydrogen bonding between the cation and anion in the two crystalline forms. So far, only one form of [bmim]I has been characterised, ascribed [81] to the impact of anion size on butyl group flexibility. The two reported forms of $[C_4C_1C_1im]Cl$ [311] also differ only in the conformation of the butyl chain. A combination of single-crystal studies and powder XRD show that 1-methyl-1-propargylpyrrolidinium chloride (melting at 90–96°C) exists in at

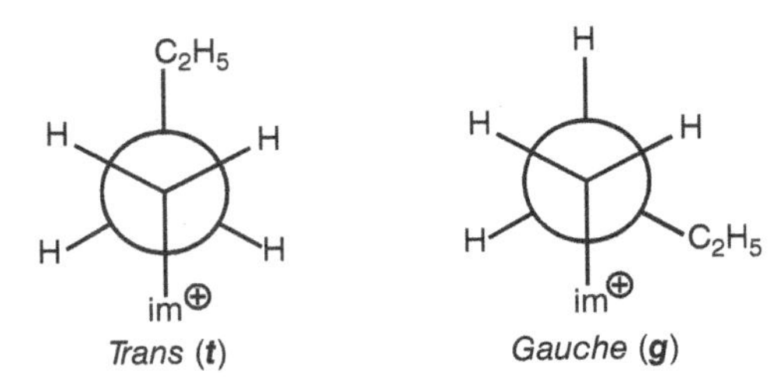

Figure 11.2 *Trans* and *gauche* conformers of the *N*-butylimidazolium group as represented by a Newman projection.

least two crystal forms [388]. Four different forms of [$^{i}C_3$mim]Br have been structurally characterised, three trigonal [307] and one orthorhombic [112], with the latter displaying a torsion angle (representing the disposition of the NC–H moiety with respect to the imidazolium ring plane) of 170.7° and the former angles of 0.3, 16.2 and 48.1°.

1,3-Dialkylimidazolium halides have also been used to study the dynamics of processes occurring during melting and crystallisation, including one very recent study [49] of [C_4mim]Br. Calorimetry showed that a 3 mg sample of [C_4mim]Br required 150 min to complete crystallisation, with conformational changes of the butyl group continuing for *ca.* 330 min after crystallisation. Such slow dynamics are likely to be the norm in such systems, particularly those exhibiting high liquid viscosity. Factors such as rates of heating and cooling and resting times are, therefore, likely to be critical in determining whether (and which) single modifications can be observed and possibly structurally characterised.

A combined study using Raman spectroscopy and powder XRD differentiated two polymorphs of [C_1mim][PF_6] [568], whereas only one form has (so far) been characterised in a single-crystal X-ray study [244]. The second form was detected using very slow scanning rate DSC. The existence of two modifications is of interest as neither anion nor cation displays conformational isomerism. The two forms arise, therefore, from crystal-packing polymorphism. What is particularly surprising is that the melting points of the two forms differ by *ca.* 50°C. By a combined use of ^{1}H NMR spectroscopy, Raman spectroscopy and calorimetry, three modifications of [C_4mim][PF_6] have also been detected, each with a different butyl group conformation [569]. The rotational dynamics of the [PF_6]$^-$ in the three polymorphs have been studied by ^{31}P NMR spectroscopy [570]. Access to new phases by the application of high pressures has also been reported for [C_4mim][PF_6] [571, 572].

The two polymorphs [76] of [C_2mim][$Fe^{III}Cl_4$] not only show two different conformations for the ethyl group relative to the imidazolium ring but also different arrangements of anion and cation. The low temperature form displays only weak interionic hydrogen bonding, while in the higher-temperature modification, cation and anion are involved in an extended hydrogen-bonding network.

The anion dinitramide, [$N(NO_2)_2$]$^-$, is found to be sufficiently flexible to adopt different conformations in the solid state such that monoclinic and triclinic forms of aminoguanidinium dinitramide, [$NH_2C(=NH_2)NHNH_2$][$N(NO_2)_2$], have been characterised as single crystals [261]. While both forms are extensively hydrogen bonded, in one modification, both oxygens of one nitro group in [$N(NO_2)_2$]$^-$ are involved in hydrogen bonding, whereas, in the other, one oxygen from each nitro group is involved.

Two polymorphs of 2,6-dimethylpiperidinium 2-hydroxythiobenzoate [467] differ in the nature of hydrogen bonding, with one (monoclinic) having an intramolecular O–H···S hydrogen bond and the other (orthorhombic; poorer quality data) an intramolecular O–H···O hydrogen bond. Two polymorphs are also known for 2-aminopyrimidinium salicylate (m.pt. 89–91 and 104–106°C) though the crystal structure of only the lower-melting form has been described [493]. The product from nicotinamide and R-mandelic acid can be isolated in two modifications, m.pt. 81 and 85°C, though these are co-crystals and not salts [573]. Two polymorphs each have been structurally characterised [574] for $[PMePh_3][PF_6]$ (having different phosphonium P···P distances reflecting differences of cation–cation interactions), $[P(CH_2Ph)Ph_3][CF_3SO_3]$ and $[1,4\text{-}(Ph_3PCH_2)_2C_6H_4][N(SO_2CF_3)_2]_2$ (in one form, the cation adopts a *cis* conformation and the anion both *cis* and *trans*, whereas, in the other, both cation and anion adopt *trans* conformations). Four crystal-packing polymorphs are now known for $[PMePh_3][I_3]$ [575].

As discussed in more detail in Section 11.13, the conformational flexibility and lability of the anion $[N(SO_2CF_3)_2]^-$ (with significant delocalisation of charge) are associated with both frustration of crystallisation and low-melting points. The use of thermochemical and spectroscopic techniques has enabled the detection of polymorphs of salts of bis(trifluoromethanesulfonyl)amide [576]. However, relatively few have been subject to crystallographic characterisation. These include those of $[C_1C_2pyr][NTf_2]$ [389, 390], $[Rmim][NTf_2]$ (R = Et, Bu or Hex) [59] (UoL, unpublished work) and the low temperature phases of $[dabcoH][NTf_2]$ (refcode UXIXAB) (three modifications, m.pt. of the high-temperature modification 147°C) and of $[dabcoH_2][NTf_2]_2$ (refcodes UXIXIJ and UXIXIJ01) (two modifications, m.pt. of high-temperature form 284°C) [577].

11.2.4 Crystal Lattice Energy

The lattice potential energy, U_L, of an assembly of ions, $[cat]_p[A]_q$, in a crystalline lattice represents, depending on the convention used, the energy associated with bringing the constituent ions from infinite separation to their respective equilibrium positions in the lattice or their infinite separation from their lattice positions. In general, U_L represents the sum of all interactions, electrostatic (E_A), and shorter-range hydrogen bonding, π–π and van der Waals interactions (aggregated in E_S), given by Equation 11.3:

$$U_L = E_A + E_S \tag{11.3}$$

For inorganic salts governed primarily by electrostatic interactions, when $E_A \gg E_S$, Madelung showed that, for one ion, the sum of attractive and repulsive forces on this ion is given by Equation 11.4:

$$E_A = \frac{Mpq}{4\pi \varepsilon d_{min}} \tag{11.4}$$

where p and q are the anion and cation charges, respectively, d_{min} is the distance to the nearest counter-ion and M is the Madelung constant [578, 579] (*n.b.* the ratio p:q defines the cation:anion ratio for the salts listed in the tables). For univalent inorganic salts, [cat][A], M is high (the value for sodium chloride is 1.748). For organic ions, typical of those found in ionic liquids, the contribution of E_S becomes more significant. Izgorodina, MacFarlane and colleagues [580] have used a numerical method to calculate Madelung constants for organic salts, [cat][A] ($p=q=1$), using crystallographic data, showing them to be significantly lower, falling in the range 1.19–1.64. Interestingly, they show that highly symmetric arrangements in the unit cell lead to high values. For example, for [$N_{1\,1\,1\,1}$][BF_4] (m.pt. 443°C), M is 1.69, whereas for the pharmaceutical, mepenzolate saccharinate, whose unit cell is made up of ion pairs, M is 1.19. The latter value belies its relatively high m.pt. (188°C) attributed [580] to strong π–π stacking between anion and cation. There is no doubt that, as argued persuasively by Izgorodina [40], reliable quantification of the contributions of such interactions requires the use of advanced and specialised methods. Nevertheless, the volume-based thermodynamics (VBT) approach discussed in Section 11.2.5 is a simple and useful method, as long as its shortcomings are understood (*n.b.* Ref. [580] also reports the crystal structures of two salts of interest (but of too high a melting temperature for inclusion in Tables 11.1, 11.2 and 11.3), namely, [C_1C_2pyr][OTs] (CSD refcode [30]: MUNKEM m.pt. 120°C) and [C_1C_4pyr][OTs] (refcode: MINKIQ m.pt. 115°C)).

Particularly in technical applications where predictability of behaviour is crucial, it is insufficient for a salt to be liquid at low temperatures simply as a consequence of frustration of crystallisation. Such a circumstance may, in fact, be an outcome of slow kinetics in a system not at thermodynamic equilibrium. Unexpected (and technically very inconvenient!) crystallisation may (and does) occur spontaneously in such systems. In others, glass formation, a common characteristic of a range of low-melting ionic liquids (though, crucially, not all), may be observed. Understanding what defines the boundary between glass formation and crystallisation is a key area of research and not simply on ionic liquids. In principle, crystallisation of a supercooled liquid at a particular temperature may still be possible, and several of the crystal structures summarised in this review relate to salts that have crystallised from the melt only after standing for periods of months or even years.

11.2.5 Volume-Based Thermodynamics (VBT)

The issue of melting phenomena has already been discussed. One (but just one) critical aspect of the thermodynamics of melting and crystallisation is a quantitative understanding of the various forces acting in the crystal, coulombic, hydrogen bonding, π stacking and van der Waals and their relative contributions to the enthalpy and entropy in the crystalline and liquid states. In this section, the focus is four terms: lattice potential energy (U_L), lattice free energy (ΔG_L), lattice enthalpy (ΔH_L) and lattice entropy (ΔS_L), their relationship with one another and estimates of values for ionic liquids that might be used (along with related (much more challenging) studies of the liquid, close to melting or crystallisation temperatures) to aid understanding of their fusion behaviour.

The VBT formalism, developed by Jenkins, Glasser and co-workers [581–586] from the original work of Bartlett [587] and Kapustinskii [588], will now be applied to the data collected in this chapter. The VBT method allows a mass of structural data for a range of salts, $[cat]_p[A]_q$, to be correlated very straightforwardly using an empirical relationship, Equation 11.5, relating the lattice potential energy with molecular volume, V_m (in units of nm^3):

$$U_L = 2I\left[\left(a / V_m^{1/3}\right) + \beta\right] \tag{11.5}$$

where α and β are the fitted coefficients depending on the stoichiometry, $p:q$, of the salt. The use of this approach here has been limited to salts for which the product $p \times q \leq 2$. V_m can be obtained from X-ray crystallographic cell volumes: $V_m = V_{cell}/Z$, where Z = number of molecules in the unit cell. I is an ionic strength factor of the chemical formula unit given by Equation 11.6:

$$I = \tfrac{1}{2}\Sigma n_i z_i^2 \tag{11.6}$$

where n_i = number of ions of type i in the formula unit having integer charge z_i. The dependency of U_L on the cube root of V_m in Equation 11.5 minimises the effect of any errors in the value of V_m used.

For [cat][A] (monovalent cation, monovalent anion: 1:1 salt), $I=1$; $\alpha=117.3\,kJ\,mol^{-1}\,nm$ and $\beta=51.9\,kJ\,mol^{-1}$. For $[cat][A]_2$ (bivalent cation, univalent anion; 1:2 salt), $I=3$; $\alpha=133.5\,kJ\,mol^{-1}\,nm$; $\beta=60.9\,kJ\,mol^{-1}$. For $[cat]_2[A]$ (univalent cation, divalent anion; 2:1 salt) $I=3$; $\alpha=165.3\,kJ\,mol^{-1}\,nm$; $\beta=-29.8\,kJ\,mol^{-1}$.

U_L can also be derived from density, ρ ($g\,cm^{-3}$), and formula or molar weight, M_m ($g\,mol^{-1}$) [582], since Equation 11.7:

$$V_m = \frac{M_m}{\rho N_A} = 1.66 \times 10^{-3}\,\frac{M_m}{\rho} \tag{11.7}$$

where $N_A = 6.02245 \times 10^{23}$ molecules mol^{-1}.

Thus, Equation 11.8 is derived from Equation 11.5:

$$U_L = \gamma\left(\frac{\rho}{M_m}\right)^{1/3} + \delta \tag{11.8}$$

For [cat][A] (1:1) salts, $\gamma=1981.2\,kJ\,mol^{-1}\,cm$ and $\delta=103.8\,kJ\,mol^{-1}$; for $[cat][A]_2$ (1:2) salts, $\gamma=6764.3\,kJ\,mol^{-1}\,cm$ and $\delta=365.4\,kJ\,mol^{-1}$; for $[cat]_2[A]$ (2:1) salts, $\gamma=8375.6\,kJ\,mol^{-1}\,cm$ and $\delta=-178.8\,kJ\,mol^{-1}$.

The VBT approach assumes that the dominant interionic interaction is coulombic. Despite this, the correlation in most cases for salts with $p:q$ of 1:1, 2:1 and 1:2 is sufficiently good for there to be value in pursuing this with ionic liquids, as has already been described by Glasser [586]. Lattice potential energies for such salts are included in the tables. Values of U_L calculated using both cell volumes, $U_L(V)$, and crystal densities, $U_L(D)$, are provided in Tables 11.1, 11.2 and 11.3. Some inconsistencies between these values point to the possible need to re-examine the original crystal data in some cases.

This section gives a brief introduction to the calculation of U_L for various values of n and x and the relationship between U_L and ΔH_L and the corresponding estimation of the absolute entropy S^o.

The lattice enthalpy, at temperature, T, for $[cat]_p[A]_q$, can be derived from the lattice potential energy using Equation 11.9 [582]:

$$\Delta H_L = U_L + F\left([cat]^{q+}, [A]^{p-}, T\right) \tag{11.9}$$

where $F = [p\{(n_M/2) - 2)\} + q\{(n_X/2) - 2\}]RT$
where n_M and n_X depend on the nature of the cation and anion, being three for monatomic ions, five for linear polyatomic ions, and six for non-linear polyatomic ions.

Most of the ionic liquids considered in this chapter are 1:1, 2:1 and 1:2 salts with non-linear polyatomic ions.
For 1:1 salts, [cat][A],

$$F = \left[1\left\{\left(\frac{6}{2}\right) - 2\right\} + 1\left\{\left(\frac{6}{2}\right) - 2\right\}\right]\mathrm{RT} = 2\mathrm{RT}\ \text{or}\ 4.96\ \mathrm{kJ\ mol^{-1}}\ \text{at}\ 298\ \mathrm{K}.$$

For 1:2 salts, $[cat][A]_2$,

$$F = \left[1\left\{\left(\frac{6}{2}\right) - 2\right\} + 2\left\{\left(\frac{6}{2}\right) - 2\right\}\right]\mathrm{RT} = 3\mathrm{RT}\ \text{or}\ 7.43\ \mathrm{kJ\ mol^{-1}}.$$

For 2:1 salts, $[cat]_2[A]$,

$$F = \left[2\left\{\left(\frac{6}{2}\right) - 2\right\} + 1\left\{\left(\frac{6}{2}\right) - 2\right\}\right]\mathrm{RT} = 3\mathrm{RT}\ \text{or}\ 7.43\ \mathrm{kJ\ mol^{-1}}.$$

Some 1:1 salts contain linear polyatomic ions (e.g. $[N_3]^-$, $[OCN]^-$, $[SCN]^-$, $[SeCN]^-$, $[Ag(CN)_2]^-$), for which n_X is 5. For 1:1 salts, [cat][A],

$$F = \left[1\left\{\left(\frac{6}{2}\right) - 2\right\} + 1\left\{\left(\frac{5}{2}\right) - 2\right\}\right]\mathrm{RT} = 1.5\mathrm{RT}\ \text{or}\ 3.72\ \mathrm{kJ\ mol^{-1}}.$$

For halide salts, $n_X = 3$. For 1:1 salts, [cat][A], of non-linear polyatomic cations,

$$F = \left[1\left\{\left(\frac{6}{2}\right) - 2\right\} + 1\left\{\left(\frac{3}{2}\right) - 2\right\}\right]\mathrm{RT} = 0.5\mathrm{RT}\ \text{or}\ 1.24\ \mathrm{kJ\ mol^{-1}}.$$

For 2:1 salts, $[cat]_2[A]$,

$$F = \left[2\left\{\left(\frac{6}{2}\right) - 2\right\} + 1\left\{\left(\frac{3}{2}\right) - 2\right\}\right]\mathrm{RT} = 1.5\mathrm{RT}\ \text{or}\ 3.72\ \mathrm{kJ\ mol^{-1}}.$$

For 1:2 salts, [cat][A]$_2$,

$$F=\left[1\left\{\left(\frac{n_M}{2}\right)-2\right\}+2\left\{\left(\frac{n_X}{2}/\right)-2\right\}\right]RT=2RT\left(4.96\,kJ\,mol^{-1}\right)$$ for non-linear polyatomic cations ($n_M = 6$) and linear polyatomic anions ($n_x = 5$) at 298 K, and 0 RT for non-linear polyatomic cations ($n_M = 6$) and monatomic anions ($n_x = 3$) at 298 K, respectively.

The VBT method [581] can also provide an estimate of the standard absolute lattice entropy, Equations 11.10 and 11.11:

$$S^\circ_{298}/\,J\,K^{-1}mol^{-1} = 1360\left(V_m\,/\,nm^3\,formula\,unit^{-1}\right)+15 \tag{11.10}$$

or

$$2.258\left[M/\left(F/g\,cm^{-3}\right)\right]+15 \tag{11.11}$$

Glasser has refined his analysis for ionic solids, producing the amended correlation for ionic liquids [586], Equation 11.12, used in this paper:

$$S^\circ_{298}/\,J\,K^{-1}mol^{-1} = 1246.5\left(V_m\,/\,nm^3\,formula\,unit^{-1}\right)+29.5 \tag{11.12}$$

Values of ΔH_L and S°_{298} may thus be derived from the data included in Tables 11.1, 11.2 and 11.3. Since U_L, ΔH_L and ΔS_L correspond to the process: $[cat^{q+}]_p[A^{p-}]_q\ (c) \rightarrow p[cat^{q+}](g) + q[A^{p-}](g)$ then ΔS_L, important for estimating lattice free energies, ΔG_L, can only be obtained by correcting the value of the absolute entropy for the entropies of the component ions in the gas phase. Values of the latter require the standard entropies of formation for the various species involved to be known. These are available for very few ions, and their calculation for complex ions included in this chapter is a non-trivial exercise. For this reason, it is not possible to derive lattice free energies directly and solely from crystallographic data.

The VBT approach, therefore, allows values of U_L, ΔH_L and $S^\circ 298$ for a wide range of salts to be obtained simply from X-ray crystallographic data. Clearly, the assumptions made give rise to errors, but for systems for which experimental or more precisely calculated data (for a series of ionic liquids, see Ref. [99]) are available, the discrepancy can be relatively small, particularly for similar materials and appear insufficiently large to negate the usefulness of the VBT approach. Indeed, many of the lattice potential energy comparisons made in this chapter involve pairs of very similar salts.

A further benefit of the many systems to which VBT may be applied is the access it provides to estimates of ion volumes, since $V = V_+ + V_-$. Where comparisons are made with ion volumes obtained using other methods, agreement is also reasonable.

The various points that should be made about the use of VBT methods can be illustrated by reference to the calculation of relevant thermodynamic properties of [C$_2$mim][ClO$_4$], an ionic liquid that melts at about 25°C [589].

Studies employing VBT in their analysis include those of [C$_n$mim][PF$_6$] (C_n = a series of alkyl groups) [99], salts of the weakly coordinating ion [Al{OCH

$(CF_3)_2\}_4]^-$ [77], *N*-aminoazolium and *N,N'*-diaminoazolium chlorides [39], spirobipyrrolidinium and related oxo-analogues [489], as well as of a series of energetic salts, including those containing *N,N*-dimethylhydrazinium [199], tetrazolium [168], 5-amino-2-methyl- and 1,3-dimethyltetrazolium [178], 1,5-diaminotetrazolium [443], dinitramide [269], nitrocyanomethanides [268, 391], triazolates and tetrazolates [252, 407].

11.2.6 Hirshfeld Surface Analysis

It is evident from the data set represented by the crystallographic studies assembled in this chapter that an analysis of crystal, molecular and interionic interactions, with a view to identifying, understanding and possibly quantifying such interactions (and doing so rigorously and objectively), is a formidable undertaking (and beyond the scope of this chapter). Such problems are well recognised [590].

The analysis of global intermolecular interactions from crystallographic data has involved the use of the so-called Hirshfeld surface (based on Hirshfeld population analysis) that maps electrostatic potential between molecules in a crystal lattice [591–595]. An example (and associated fingerprint plot) for $[C_2mim][NMes_2]$ is shown in (Fig. 11.3) taken from Ref. [52]. For each point on the surface, 'half of the electron density is due to the spherically averaged non-interacting atoms comprising the molecule, and the other half is due to those comprising the rest of the crystal' [591]. Inside the surface, the electron density of the molecule dominates those in the crystal, and distances from points on this surface to atoms inside and outside the surface may be obtained.

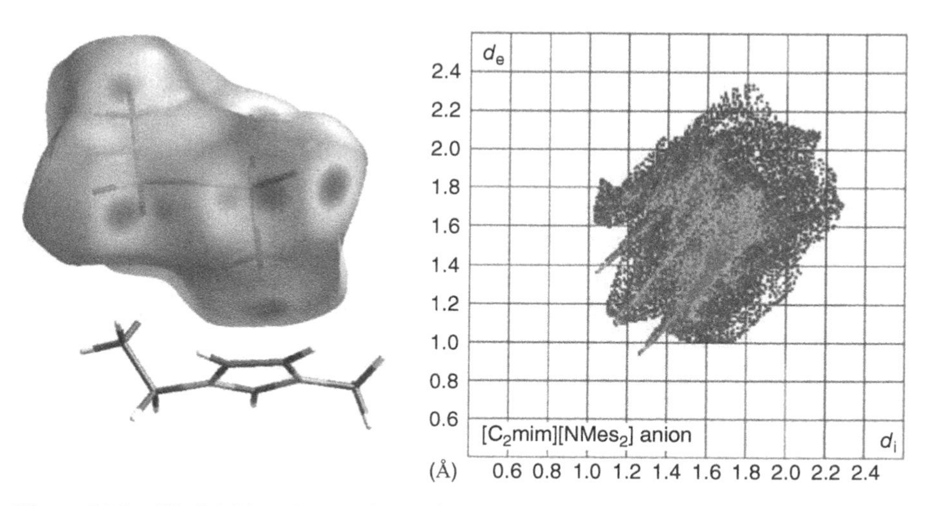

Figure 11.3 Hirshfeld surface and associated fingerprint plot of the anion for $[C_2mim][NMes_2]$. Source: Dean, P.M., Pringle, J.M., Forsyth, C.M., Scott, J.L. and MacFarlane, D.R. [52] Reproduced with permission of the Royal Society of Chemistry.

It is possible to use such distances to explore a range of close intermolecular contacts in the crystal. d_{norm} is a normalised contact distance that relates the distance between two atoms across the Hirshfeld surface to the sum of the van der Waals radii of these atoms. These may be represented as a 3D picture or as a 2D 'fingerprint' plot [596, 597] (plots of the distance from the surface to the nearest atom inside against the distance to the nearest atom outside the surface) using CrystalExplorer software [598].

There is a small number of reports of the application of the Hirshfeld approach to interionic interactions in the crystal structures of ionic liquids [52, 111, 124, 203, 574, 599]. However, some [600] have cautioned (because of weaknesses in charge partitioning) that its application should not necessarily lead to the setting aside of judgements based on the use of chemical intuition.

Dean et al. report [203] an analysis of the crystal structures of a series of *N*-alkyl-3-benzylimidazolium iodides, in which the alkyl group, R, is methyl, ethyl, propyl, butyl, 1-methylpropyl and pentyl. The methyl and 1-methylpropyl salts melt at 154 and 135°C, respectively, significantly higher than the others in the series: 86 (R = Et), 71 (Pr), 61 (Bu) and 61°C (Pent). No classical hydrogen bonding is seen. Hirshfeld surface analysis is used to 'decompose' the fraction of the surface associated with a particular type of interionic interaction in the four non-disordered salts. The analysis shows the clear dominance of C—H···H interactions. The contribution of C—H···I contacts (relative to the total Hirshfeld surface area) went in the order: methyl (9.9%), propyl (8.2%), butyl (9.6%) and 1-methylpropyl (8.6%) and of π–π contacts 6.5, 2.5, 0.1 and 0.0% in the same order of alkyl groups. Relative percentages of the contribution of C—H···H contacts increase along the same series, though no trend was apparent for C—H···π interactions. The analysis reveals a large number of contacts greater than 4 Å, consistent with less-than-optimum packing. Mean values of d_{norm} over the entire surface are used as a quantified surrogate for packing efficiency, as less efficient packing may indicate greater 'free' volume, which has been associated with lower-melting point.

A similar analysis [124] was undertaken for a series of 1-butyl-2,3-dimethylimidazolium azide, thiocyanate, propynoate, $[CuCl(CN)]^-$, mixed azide/chloride (7:3) and cyanide/chloride (1:2) salts, with C—H···H contacts between cations and anions representing 63–74% of the contributions relative to the entire surface. Similarly, C—H···N cation–anion interactions make up between 6.4% (for $[C_4C_1C_1im][CuCl(CN)]$) and 20.9% (for $[C_4C_1C_1im][N_3]$).

The relative contributions of previously identified interionic C—H···O and C—H···N hydrogen bonds and non-classical C—H···F hydrogen bonds for $[C_1C_1pyr]^+$, $[C_4C_1pyr]^+$ and $[C_2mim]^+$ salts with $[NTf_2]^-$ and $[NMes_2]^-$ have been quantified in a further detailed study by Dean et al. [52]. Two additional points of interest arise: first, the ready identification of cation–anion interactions involving the two crystallographically distinct ions found in $[C_2mim][NTf_2]$; second, the occurrence of C—H···O anion–anion interactions in the three dimesylamide salts and C—F···F anion–anion interactions in the corresponding bistriflamide salts.

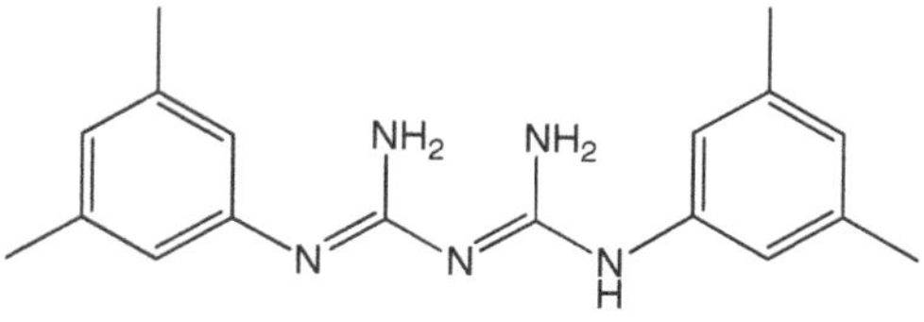

Figure 11.4 1,5-Dimexylbiguanide [599].

A series of 11 crystal structures of the phosphonium salts (m.pt. 160–172 to 240–264°C) $[PMePh_3][PF_6]$ (two polymorphs), $[PPh_4][N(CN)_2]$, $[P(CH_2Ph)Ph_3]$ [A] ([A] = [SCN], $[PF_6]$, $[CF_3SO_3]$ (two polymorphs)), $[1,4\text{-}(Ph_3PCH_2)_2C_6H_4)]$ $[A]_2$ ([A] = $[C(CN)_3]$, and $[N(SO_2CF_3)_2]$ (two polymorphs)) have also been subjected to Hirshfeld analysis [574]. Non-classical C—H···A bonding dominates cation–anion interactions; the cations associate through phenyl···phenyl interactions. The contribution of different contacts between pairs of polymorphs has also been revealed and its relevance to packing efficiency also noted.

Twenty-five of fifty-one salts (T_m 123–234°C, where observed) formed from 1,5-dimexylbiguanide (Fig. 11.4) and a series of carboxylic and sulfonic acids form glasses, some of which began to crystallise on standing for greater than 3 months. Crystal structures of six salts (three of which are non-solvated: nitrate (refcode QUPSUQ), methanoate (QUPSIE), adipate (QUPSOK)) have been solved and efforts made to relate evidence from fingerprint plots and calculated packing indices with the tendency towards glass formation [599].

It is hoped that the availability of a large number of cif files associated with the majority of structures collected in this review may allow further testing, development and refinement of these and other methods.

11.3 COMPONENT ANIONS AND CATIONS

Some basic characteristics that appear to be associated with low-melting salt behaviour and the so-called 'frustration' of crystallisation are well known [4, 601–605], including structural asymmetry, conformational flexibility, diffuseness of and shielding of ionic charge in anion and cation and differences in size and shape of constituent ions. These have been discussed in detail in original material, which is often concerned with a single compound or a relatively small group of similar compounds. From the large collection of relevant single-crystal structural data assembled here, the full range of chemical and structural types of anions, cations and their combinations that contribute to low-melting behaviour can be assembled, organised and is available to be examined for patterns and trends.

Tables 11.1, 11.2 and 11.3 [52–548] contain, respectively, 78 records for 67 salts melting below 30°C, 600 records of 553 salts melting between 30 and 100°C and data for a further 149 salts melting between 100 and 112°C. Some late additions are included in the Addendum.

Because of the conventional definition of an ionic liquid, for convenience (if not rigour or consistency), melting temperatures and ranges are given in degree Celsius. Other temperatures are quoted in Kelvin. The consolidated list of structures is presented in order of increasing reported m.pt. Where more than one structure for the same compound is reported (indicated by a common six-letter CSD refcode with a numerical suffix) and where different m.pt. are given, the structures are listed sequentially and entered at the lowest m.pt. These tables also include, for each entry, the ion charge ratio, $p:q$, literature reference (though the crystallographic data of interest may be relegated to electronic supplementary information), CSD refcode or CCDC deposition number (where available), temperature at which diffraction data were collected (and where this is reported to be at ambient temperatures, we adopt the CSD convention of recording this as 283–303 K), space group and data needed to estimate lattice potential energy using the VBT approach (Section 11.2.5), namely, cell volume, Z, calculated density and formula weight. Where discrepancies are evident between data reported in a paper and in the corresponding CSD entry, the latter is generally, though not always, relied upon.

Tables 11.4 and 11.5 list the constituent anions, $[A]^-$, and cations, $[cat]^+$ (with those presented in italics found only in salts listed in Table 11.3, that is, with melting temperatures between 100 and *ca.* 110°C). Anions and cations are grouped together in a pragmatic rational manner, classified according to charge and whether they are aromatic, alicyclic or acyclic. Further classification is somewhat arbitrary, but *N*-protonated salts are listed first, then salts containing *N*-alkyl and *N*-bound substituted alkyl groups, ring-functionalised ions, followed by other materials. The anions and cations are numbered for ease of reference. Figure 11.5 displays the basic structures for the most important series of anions and cations. Where necessary, the structures of other ions or salts are included where first mentioned in the text.

Simple inspection of Tables 11.4 and 11.5 instantly reveals the wide diversity of anions and cations to be found in crystalline low-melting salts. Closer inspection of the data in Tables 11.1, 11.2 and 11.3 reveals the diffuse territory, indeed the continuum, which exists between the extremes of salt and non-salt structures in the crystalline state.

Two basic approaches are taken in our initial analysis of the assembled material: first, the complete set or larger subsets of the data can be examined for general trends, particularly in comparison with materials that are not ionic liquids, and second, comparisons can be made between much smaller sets (even pairs) of structures, in which small, but controlled, changes have been made in composition of the anion or cation and the effect on crystal structure, particularly relating to interionic interactions, can be delineated. Space, however, dictates that only a cursory overview is possible.

Attention is given separately to salts of particular types of anions or cations, leading to some inevitable, but necessary, duplication. Those salts, such as inorganic, metal-containing, energetic, protic or chiral materials, in which the focus may be on either anion or cation (or both), are dealt with separately.

TABLE 11.4 Ionic Liquid Anions

Anion number X		p Anionic Charge
	Halides	
A1	F	1
A2	Cl	1
A3	Br	1
A4	I	1
	Protic	
A5	$[HF_2]$	1
A6	$[F(HF)_2]$	1
A7	$[F(HF)_3]$	1
A8	$[HCl_2]$	1
A9	*$[(HCO_2)_2H]$*	1
A10	$[(CF_3CO_2)_2H]$	1
A11	$[(C_6H_5CO_2)_2H]$	1
	Polyhalides and interhalides	
A12	$[Br_3]$	1
A13	$[Br_7]$	1
A14	$[Br_8]$	2
A15	$[Br_{20}]$	2
A16	*$[ICl_2]$*	1
A17	$[IBr_2]$	1
A18	$[Cl_2I_{14}]$	2
A19	$[I_3]$	1
A20	$[I_5]$	1
A21	$[I_7]$	1
A22	$[I_8]$	2
A23	$[I_{12}]$	2
A24	$[I_{29}]$	3
	Pseudohalides	
A25	$[N_3]$	1
A26	[OCN]	1
A27	[SCN]	1
A28	[SeCN]	1
A29	$[N(NO_2)_2]$	1
A30	$[N(NO_2)(SF_5)]$	1
A31	$[N(CN)(NO_2)]$	1
A32	$[N(CN)_2]$	1
A33	$[N(C_6F_5)_2]$	1
A34	$[CH(NO_2)_2]$	1
A35	$[C(NO_2)_3]$	1
A36	$[C(NO_2)(CN)_2]$	1
A37	$[C\{CF_3C(O)\}(CN)_2]$	1
A38	$[C(NO_2)(NO)CN]$	1
A39	$[C(CN)_3]$	1
A40	$[C(C_6F_5)(SO_2CF_3)_2]$	1

(*Continued*)

TABLE 11.4 (*Continued*)

Anion number X		*p* Anionic Charge
	Oxoanions and related	
A41	$[NO_2]$	1
A42	$[N(SO_2F)_2]$	1
A30	$[N(NO_2)(SF_5)]$	1
A43	$[NO_3]$	1
A44	$[PO_2F_2]$	1
A45	$[P_2Se_8]$	2
A46	*$[PhP(O)_2(OH)]$*	1
A47	$[SO_4]$	2
A48	$[SO_3(OH)]$	1
A49	$[SO_3F]$	1
A50	$[SeO_2(OH)]$	1
A51	$[ClO_4]$	1
	Inorganic ester anions	
A52	$[CO_2(OMe)]$	1
A53	$[CO_2(NMe_2)]$	1
A54	*$[CS_2(NEt_2)]$*	1
A55	$[PO_2(OPh)_2]$	1
A56	$[PO_2(OMe)(SMe)]$	1
A57	$[PO_2(OMe)(SeMe)]$	1
A58	*$[POS(OEt)Et]$*	1
A59	$[SO_3(OMe)]$	1
A60	*$[S(NSO_2CF_3)_2(OEt)]$*	1
$[RCO_2]^-$	**Carboxylates**	
A61	$[CH_3CO_2]$	1
A62	$[ClCH_2CO_2]$	1
A63	$[CF_3CO_2]$	1
A64	*$[CCl_3CO_2]$*	1
A65	$[C_2H_5CO_2]$	1
A66	$[HC{\equiv}CCO_2]$	1
A67	[*trans*-$CH_3CH{=}CHCO_2$]	1
A68	*$[C_5H_{11}CO_2]$*	1
A69	$[C_6H_{13}CO_2]$	1
A70	$[C_9H_{19}CO_2]$	1
A71	$[C_{11}H_{23}CO_2]$	1
A72	$[C_{13}H_{27}CO_2]$	1
A73	$[C_{15}H_{31}CO_2]$	1
A74	$[(4\text{-}ClC_6H_4)SCH_2CO_2]$	1
A75	$[\{2\text{-}(2,6\text{-}Cl_2C_6H_3NH)C_6H_4\}CH_2CO_2]$	1
A76	$[C_6H_5CO_2]$	1
A77	$[(2\text{-}HO)C_6H_4CO_2]$	1
A78	$[(2\text{-}HO)C_6H_4COS]$	1
A79	$[(2\text{-}HO)C_6H_4CS_2]$	1
A80	$[2\text{-}HC(O)C_6H_4CO_2]$	1
A81	$[(2\text{-}O_2N)C_6H_4CO_2]$	1
A82	*$[(4\text{-}O_2N)C_6H_4CO_2]$*	1

TABLE 11.4 (*Continued*)

Anion number X		*p* Anionic Charge
A83	[3-$FC_6H_4CO_2$]	1
A84	[3-$ClC_6H_4CO_2$]	1
A85	[{2,5-$(HO)_2$}$C_6H_3CO_2$]	1
A86	*[2-HO-3,5-$(O_2N)_2C_6H_2CO_2$]*	1
A87	1-Methylpyrrole-2-carboxylate	1
A88	*2-Hydroxy-5-sulfonatobenzoate*	2
	Dicarboxylates	
A89	*[$HO_2C(CH_2)_2CO_2$]*	1
A90	*[$HO_2C(CH_2)_3CO_2$]*	1
A91	*[$HO_2C(CH_2)_4CO_2$]*	1
A92	[$HO_2C(CH_2)_6CO_2$]	1
A93	*d*-[$HO_2CCH(OH)CH(OH)CO_2$]	1
A94	*[$O_2CCH_2CO_2$]*	2
A95	[$O_2C(CH_2)_8CO_2$]	2
$[RSO_3]^-$	**Alkyl- and arylsulfonates**	
A96	[$MeSO_3$]	1
A97	[CF_3SO_3]	1
A98	[$HCF_2CF_2SO_3$]	1
A99	[$C_4F_9SO_3$]	1
A100	[$ICF_2CF_2OCF_2CF_2SO_3$]	1
A101	[4-$CH_3C_6H_4SO_3$]	1
A102	*[4-$BrC_6H_4SO_3$]*	1
A103	[$Ph_2PC_6H_4(4\text{-}SO_3)$]	1
A104	[4-$MeOC_6H_4CH(CH_2NO_2)SO_3$]	1
A88	*2-Hydroxy-5-sulfonatobenzoate*	2
A105	*[$O_3SCH_2CH_2SO_3$]*	2
	Sulfonyl and phosphonyl amides	
A106	[$N(SO_2Me)_2$]	1
A42	[$N(SO_2F)_2$]	1
A107	[$N(SO_2CF_3)_2$]	1
A108	Acesulfamate	1
A109	Saccharinate	1
A110	[$N\{P(O)(C_2F_5)_2\}_2$]	1
A40	[$C(C_6F_5)(SO_2CF_3)_2$]	1
	$[EX_4]^-$	
A111	[BH_4]	1
A112	[$BH_2(CN)_2$]	1
A113	[BF_4]	1
A114	*[BCl_4]*	1
A115	[$BF_2(ox)$]	1
A116	[$B(OMe)_4$]	1
A117	[$B\{OCH(CF_3)_2\}_4$]	1
A118	Bis{(2*S*)-2-oxy-3-methylbutanoato-*O,O'*}borate	1

(*Continued*)

TABLE 11.4 (*Continued*)

Anion number X		*p* Anionic Charge
A119	Bis(salicylato)borate	1
A120	$[B(CN)_4]$	1
A121	$[B(C_6F_5)(N_3)_3]$	1
A122	*$[B(C_6H_5)_4]$*	1
A123	$[B\{C_6H_4(4\text{-}C_6F_{13})\}_4]$	1
A124	$[B\{3,5\text{-}(CF_3)_2C_6H_3\}_4]$	1
A114	*$[BCl_4]$*	1
A125	$[AlCl_4]$	1
A126	$[GaCl_4]$	1
A127	$[InCl_4]$	1
A128	$[Sb^{III}Cl_4]$	1
A129	$[As(N_3)_4]$	1
A117	$[B\{OCH(CF_3)_2\}_4]$	1
A130	$[Al\{OCH(CF_3)_2\}_4]$	1
A131	$[Al\{OC(CH_3)(CF_3)_2\}_4]$	1
A132	$[Al\{OC(Ph)(CF_3)_2\}_4]$	1
A133	$[AlMe_2F_2]$	1
A134	*$[AlMe_2Cl_2]$*	1
A135	$[Al(i\text{-}Bu)_2F_2]$	1
A136	$[AlMe_4]$	1
A137	$[Me_3Al(\mu^3\text{-}CH_2)(AlMe_2)_2(\mu^2\text{-}Me)]$	1
	$[EX_5]^{n-}$	
A138	*$[SiPhF_4]$*	1
A139	$[SiMePhF_3]$	1
A140	*$[SiPh_2F_3]$*	1
A141	$[SiMe_3F_2]$	1
A142	*$[SiPh_3F_2]$*	1
A143	$[SiPh_2(1\text{-}C_{10}H_7)F_2]$	1
A144	$[SnPh_2Cl_3]$	1
A145	$[Sn(2\text{-}MeC_6H_4)_2Cl_3]$	1
A146	$[TlBr_5]$	2
	$[EX_6]^{n-}/[MX_6]^{n-}$	
A147	*$[Si(ox)_3]$*	2
A148	$[Ge(ox)_3]$	2
A149	$[PF_6]$	1
A150	$[PF_3(C_2F_5)_3]$	1
A151	$[PCl_6]$	1
A152	$[AsF_6]$	1
A153	$[SbF_6]$	1
A154	$[SbCl_6]$	1
A155	*$[Sb(N_3)_6]$*	1
A156	$[NbF_6]$	1
A157	$[NbCl_6]$	1
A158	$[TaF_6]$	1
A159	*$[TaCl_6]$*	1

TABLE 11.4 (*Continued*)

Anion number X		*p* Anionic Charge
	$[MX_2]^-$	
A160	$[Cu^ICl_2]$	1
A161	$[Cu^IBr_2]$	1
A162	$[Ag(CN)_2]$	1
A163	$[Au(CN)_2]$	1
	$[MX_3]^-$	
A164	$[Ge^{II}Cl_3]$	1
A165	$[Sn^{II}Cl_3]$	1
A166	$[HgCl_3]$	1
A167	$[HgBr_3]$	1
A168	$[Cu^ICl_3]$	1
	$[MX_4]^{n-}$	
A169	$[Sn^{II}Cl_4]$	2
A170	$[ZnCl_4]$	2
A171	*$[CdCl_4]$*	2
A172	$[Mn^{II}Cl_4]$	2
A173	$[Fe^{II}Cl_4]$	2
A174	$[Co^{II}Cl_4]$	2
A175	$[Co^{II}Br_4]$	2
A176	$[Ni^{II}Cl_4]$	2
A177	$[Ni^{II}Br_4]$	2
A178	$[Cu^{II}Cl_4]$	2
A179	$[Cu^{II}Br_4]$	2
A180	*$[Pd^{II}Cl_4]$*	2
A181	*$[Pd^{II}I_4]$*	2
A182	$[Pt^{II}Cl_4]$	2
A183	$[Fe^{III}Cl_4]$	1
A184	$[Au^{III}Cl_4]$	1
	Other halometallates and oxohalometallates	
A185	$[Al_2Br_7]$	1
A186	$[Ga_2Cl_7]$	1
A187	$[Ga_2I_6]$	2
A188	$[In_2Cl_6]$	2
A189	$[Cd_2Cl_6]$	2
A190	*$[Sb^{III}{}_2Br_9]$*	3
A191	$[Sb^{III}{}_2Cl_{10}]$	4
A192	$[Se_4Br_{14}]$	2
A193	*$[Hg^{II}{}_2Cl_6]$*	2
A194	*$[Cu^I{}_2Br_4]$*	2
A195	$[Cu^{II}{}_3Cl_8]$	2
A196	$[Cu^I{}_4Cl_8]$	4
A197	$[Cu^I{}_5Br_7]$	2
A198	$[Pd^{II}{}_2Cl_6]$	2

(*Continued*)

TABLE 11.4 (*Continued*)

Anion number X		*p* Anionic Charge
A199	$[V^{V}(O)F_4]$	1
A200	$[V^{IV}(O)Cl_4]$	2
A201	$[(Ta^{V}Cl_5)_2O]$	2
	Fluorophosphazenates	
A202	*cyclo*-$[P_4N_4F_9]$	1
A203	*cyclo*-$[P_6N_6F_{13}]$	1
A204	$[P_3N_3F_5NPF_2NPF_2NPF_5]$	2
	Boranes	
A205	[*nido*-$C_2B_9H_{12}$]	1
A206	$[CB_{11}H_6Cl_6]$	1
A207	$[1\text{-}MeCB_{11}H_{11}]$	1
A208	$[1\text{-}EtCB_{11}H_{11}]$	1
A209	*$[1\text{-}EtSnB_{11}H_{11}]$*	1
A210	$[1\text{-}BuSnB_{11}H_{11}]$	1
	Coordination complexes	
A211	*$[Li(C_5H_5)_2]$*	1
A212	$[Cu\{N(SO_2CF_3)_2\}_2]$	1
A213	$[Ag\{N(SO_2CF_3)_2\}_3]$	2
A214	$[Ag_3\{N(NO_2)_2\}_4]$	1
A215	$[Co(CO)_4]$	1
A216	$[Co^{II}Br_3(C_9H_7N)]$	1
A217	$[Co^{II}(hfac)_3]$	1
A218	*$[PtBr_3(C_2H_4)]$*	1
A219	$[La(NCS)_7(H_2O)]$	4
A220	$[Sm\{C(CN)_2NO\}_6]$	3
A221	$[Ce\{C(CN)_2NO\}_6]$	3
A222	$[La\{C(CN)_2NO\}_6]$	3
A223	$[Pr\{C(CN)_2NO\}_6]$	3
A224	*$[Yb\{N(SO_2CF_3)_2\}_4]$*	2
A225	$[Eu\{N(SO_2CF_3)_2\}_4]$	1
A226	$[Eu\{N(SO_2CF_3)_2\}_5]$	2
A227	$[Y(BH_4)_4]$	1
A228	*$[U^{VI}O_2(NCS)_5]$*	3
A229	$[Bi_2I_8(Me_2CO)_2]$	2
	Energetic	
A230	$[\{2,4,6\text{-}(NO_2)_3\}C_6H_2O]$ (Picrate)	1
A231	2,4-Dinitroimidazolate	1
A232	4,5-Dinitroimidazolate	1
A233	*3,5-Dinitro-1,2,4-triazolate*	1
A234	Tetrazolate	1
A235	5-Nitrotetrazolate	1
A236	*5-Azidotetrazolate*	1
A237	5-Nitroiminotetrazolate	1
A238	1-(2-Nitratoethyl)-5-nitriminotetrazolate	1

TABLE 11.4 (*Continued*)

Anion number X		*p* Anionic Charge
A239	4-(Carboxylatomethyl)-5-nitroiminotetrazolate	2
	Diketonate	
A240	$[CF_3C(O)CHC(O)CF_3]$	1
A241	$[CF_3C(O)CHC(O)C_6H_5]$	1
A242	$[CF_3C(O)CHC(O)$(2-Thienyl)]	1
	Other organic	
A243	*$[CF_3S]$*	1
A244	2,3,6-Tricyano-4-fluoro-5-trifluoromethylphenolate	1
A245	*2,6-dichloro-4-nitrophenolate*	1
A246	*E-2-Cyano-1-phenylethenolate*	1
A247	*$[\{(C_6H_5)_3C\}P(BH_3)(O)OH]$*	1
A248	*$[C_5H_5]$ (Cyclopentadienide)*	1

The data collected are also used to estimate lattice potential energies using VBT, though space does not permit a detailed discussion. A preliminary analysis of the overall data, to test whether any trend may be evident, for instance, relating melting point with lattice potential energy, is however provided in Section 11.20.

Sections 11.4–11.19 will reveal the perhaps surprisingly few instances in which series of crystal structures of related salts are available that might allow links between structural changes in the component ions and the crystal structure of the salt to be established, thereby providing possible additional insights into ionic liquid behaviour. Indeed, most comparisons involve only pairs of salts in which the effect of a modest structural change may be delineated, such as from the introduction of a methylene group into, or a phenyl onto, an *N*-alkyl chain, or the change from a propyl to a 1-methylethyl group, or from a chloride to another halide anion.

11.3.1 Nature of Component Ions

The assembled data (using the objective, if arbitrary, criterion (m.pt.$<$100°C) to determine inclusion) provide an opportunity to test whether there is any obviously unique, but hitherto unrecognised, feature that may characterise the component ions that predispose ionic materials to have low-melting points. It is hoped, at least, that the data collected may point to under-recognised, even new, cation–anion combinations that may display ionic liquid behaviour. These are highlighted in Section 11.21. There have been few properly objective assessments of what might characterise ionic liquid composition space based on the conventional definition. Here is attempted to be gathered, probably for the first time, a complete list of ionic liquids whose structure has been investigated by single-crystal X-ray crystallography. While the compiled list excludes those

TABLE 11.5 Ionic Liquid Cations

Cation Number		q Cation Charge
	Aromatic: Imidazolium and Benzimidazolium	
[HRim]⁺	***Protic imidazolium***	
C1	Imidazolium	1
C2	2-Methylimidazolium	1
C3	4-Methylimidazolium	1
C4	2-Ethylimidazolium	1
C5	1-Aminoimidazolium	1
C6	*1-Azidosulfonylimidazolium*	1
C7	1-Methylimidazolium	1
C8	1,2-Dimethylimidazolium	1
C9	*1-Benzylimidazolium*	1
[Rmim]⁺	***N-Methyl-N′-alkylimidazolium***	
C10	1,3-Dimethylimidazolium	1
C11	1-Ethyl-3-methylimidazolium	1
C12	1-Methyl-3-propylimidazolium	1
C13	1-*iso*-Propyl-3-methylimidazolium	1
C14	1-Butyl-3-methylimidazolium	1
C15	1-(2-Butyl)-3-methylimidazolium	1
C16	1-Methyl-3-{(2*S*)-2-methylbutyl} imidazolium	1
C17	1-*cyclo*-Pentyl-3-methylimidazolium	1
C18	1-Hexyl-3-methylimidazolium	1
C19	1-Methyl-3-octylimidazolium	1
C20	1-Decyl-3-methylimidazolium	1
C21	1-Dodecyl-3-methylimidazolium	1
C22	1-Tetradecyl-3-methyimidazolium	1
C23	1-Octadecyl-3-methylimidazolium	1
[R₂im]⁺	***Symmetric 1,3-dialkylimidazolium***	
C10	1,3-Dimethylimidazolium	1
C24	1,2,3-Trimethylimidazolium	1
C25	1,3-Diethylimidazolium	1
C26	1,2,3-Triethylimidazolium	1
C27	1,3-Dipropylimidazolium	1
C28	*1,3-Di-*iso*-propylimidazolium*	1
C29	*1,3-Di-*iso*-propyl-4,5-dimethylimidazolium*	1
C30	1,3-Di-*tert*-butylimidazolium	1
C31	1,3-Dibutyl-2,4,5-trimethylimidazolium	1
C32	1,3-Didodecylimidazolium	1
C33	1,3-Dibenzylimidazolium	1
C34	*1,3-Bis((R)-(+)-(α)-methylbenzyl) imidazolium*	1
C35	1-(2-Methoxyethyl)-3-methylimidazolium	1

TABLE 11.5 (*Continued*)

Cation Number		*q* Cation Charge
C36	*1,3-Di(cyanomethyl)imidazolium*	1
C37	1,3-Bis-(3-cyanopropyl)imidazolium	1
C38	*1,3-Bis(hydroxycarbonylmethyl) imidazolium*	1
[RR′R″im]$^+$	***1,2,3-Trialkylimidazolium***	
C24	1,2,3-Trimethylimidazolium	1
C39	1-Ethyl-2,3-dimethylimidazolium	1
C40	1-Butyl-2,3-dimethylimidazolium	1
C41	1-Methyl-2-phenyl-3-propylimidazolinium	1
C42	1-Ethyl-2-methyl-3-benzylimidazolium	1
C26	1,2,3-Triethylimidazolium	1
C43	1-Propyl-2-methyl-3-benzylimidazolium	1
C44	1-Butyl-2-methyl-3-benzylimidazolium	1
C45	1-Pentyl-2-methyl-3-benzylimidazolium	1
C46	1-(3-Cyanopropyl)-2,3-dimethylimidazolium	1
C47	1-(2-Methoxyethyl)-2,3-dimethylimidazolium	1
	N-Alkenyl- or N-alkynylimidazolium	
C48	1-Vinyl-3-methylimidazolium	1
C49	1-Allyl-3-methylimidazolium	1
C50	1-Allyl-2,3-dimethylimidazolium	1
C51	1-Allyl-3-ethylimidazolium	1
C52	1-Allyl-3-propylimidazolium	1
C53	1-(Pent-2-ynyl)-3-methylimidazolium	1
C54	1-(But-2-ynyl)-3-methylimidazolium	1
	N-Side-chain-functionalised N-alkylimidazolium	
C55	1-Methyl-3-(2-phenylethyl) imidazolium	1
C56	1-Methyl-3-(3-phenylpropyl) imidazolium	1
C57	1-(2,2-Diphenylethyl)-3-methylimidazolium	1
C58	1-Methyl-3-trimethylsilylmethylimidazolium	1
C59	1-Methyl-3-[(trimethoxysilyl) methyl]imidazolium	1

(*Continued*)

TABLE 11.5 (*Continued*)

Cation Number		*q* Cation Charge
C60	1-Cyanomethyl-3-methylimidazolium	1
C61	1-(3-Cyanopropyl)-3-methylimidazolium	1
C62	1-Methoxymethyl-3-methylimidazolium	1
C63	1-Ethoxymethyl-3-methylimidazolium	1
C64	1-(2-Hydroxyethyl)-3-methylimidazolium	1
C65	1-(2-Methoxyethyl)-3-methylimidazolium	1
C66	1-{(2-Methoxyethyl)oxymethyl}-3-methylimidazolium	1
C67	1-{2-(2-Methoxyethyl)oxyethyl}-3-methylimidazolium	1
C68	*1-(2-Hydroxy-2-methylpropyl)-3-methylimidazolium*	1
C69	1-(2-Hydroxy-2-methylpropyl)-3-*iso*-propylimidazolium	1
C70	1-(3-Ammoniopropyl)-3-methylimidazolium	2
C71	1-{2-(Diethylammonio)ethyl}-3-methylimidazolium	2
C72	1-(2-Di-*iso*-propylaminoethyl)-3-methylimidazolium	1
C73	1-(Hydroxycarbonylmethyl)-3-methylimidazolium	1
C74	*1-Hydroxycarbonylmethyl-3-decylimidazolium*	1
C75	*1-(3-Hydroxycarbonylpropyl)-3-methylimidazolium*	1
C76	*rac*-1-{1-(Ethoxycarbonyl)ethyl}-3-methylimidazolium	1
C77	1-Methyl-3-(tetrahydro-2*H*-pyran-2-ylmethyl)imidazolium	1
C78	1-Methyl-3-(tetrahydrofuran-3-ylmethyl)imidazolium	1
C79	*[1-{2-((4-Chlorobenzylidene)amino)ethyl}-3-methylimidazolium*	1
C80	1-Butyl-3-ferrocenylimidazolium	1
C81	1-(-2-Hydroxy-2-methyl-2-phenylethyl)-3-*tert*-butylimidazolium	1
C82	*1-(2-Diphenylphosphinoethyl)-3-mesitylimidazolium*	1

TABLE 11.5 (*Continued*)

Cation Number		q Cation Charge
	Non-alkyl ring-functionalised imidazolium	
C83	2-Diethylamino-1, 3-dimethylimidazolinium	1
C84	*1,3-Dimethyl-2-nitromidazolium*	1
C85	1,3-Dimethyl-4-nitromidazolium	1
C86	1-Ethyl-3-methyl-4-nitroimidazolium	1
C87	1-Butyl-3-methyl-4, 5-dibromoimidazolium	1
C88	*1,2,3-Trimethyl-4-nitroimidazolium*	1
C89	1-*iso*-Propyl-3-methyl-4-diphenylphosphorylimidazolium	1
C90	*1-Ethyl-2-(1-oxy-3-oxo-4,4,5, 5-tetramethylimidazolin-2-yl)-3-methylimidazolium*	1
C91	*6,6′-Spirobis(1-methyl-1,5,6, 7-tetrahydropyrrolo[1,2-a] imidazol-4-ium*	1
	Benzimidazolium	
C92	*1-Ethylbenzimidazolium*	1
C93	*1-Butyl-2-methylbenzimidazolium*	1
C94	1,3-Dibutylbenzimidazolium	1
C95	1,3-Bis(dodecyl)benzimidazolium	1
	N-Hydroxy, N-alkoxy and N-aminoimidazolium	
C96	1,3-Dimethoxyimidazolium	1
C97	1,3-Dihydroxy-2-methylimidazolium	1
C98	1,3-Dimethoxy-2-methylimidazolium	1
C99	2-Bromo-1,3-dimethoxymidazolium	1
C100	1,3-Diethoxyimidazolium	1
C101	1-Benzyloxy-3-methylimidazolium	1
C102	1,3-Di(benzyloxy)imidazolium	1
C5	1-Aminoimidazolium	1
C103	1,3-Diamino-2-methylimidazolium	1
	Aromatic: triazolium and tetrazolium	
	1,2,3- and 1,2,4-Triazolium	
C104	1-Amino-1,2,3-triazolium	1
C105	1-Amino-3-methyl-1,2,3-triazolium	1
C106	*1,2,4-Triazolium*	1
C107	4-Amino-1,2,4-triazolium	1
C108	1-Propyl-1,2,4-triazolium	1

(*Continued*)

TABLE 11.5 (*Continued*)

Cation Number		q Cation Charge
C109	*1,4-Dimethyl-2-*H*-1,2,4-triazolium*	1
C110	1,4-Dimethyl-3-azido-1,2, 4-triazolium	1
C111	1-Methyl-4-amino-1,2,4-triazolium	1
C112	1-Methyl-4-(3,3, 3-trifluoropropyl)-1,2,4-triazolium	1
C113	*4-Dimethylamino-1-methyl-1,2, 4-triazolium*	1
C114	1-Ethyl-4-amino-1,2,4-triazolium	1
C115	1-Propyl-4-amino-1,2,4-triazolium	1
C116	1-*iso*-Propyl-4-amino-1,2,4-triazolium	1
C117	1-(*cyclo*-Propylmethyl)-4-amino-1,2,4-triazolium	1
C118	1-Hexyl-4-amino-1,2,4-triazolium	1
C119	1-Heptyl-4-amino-1,2,4-triazolium	1
C120	1-Heptyl-4-(1*H*,1*H*,2*H*,2*H*-perfluorooctyl)-1,2,4-triazolium	1
C121	*4-Amino-1,3,5-trimethyl-1,2, 4-triazolium*	1
C122	4-Amino-1-ethyl-3,5-dimethyl-1,2, 4-triazolium	1
C123	*4-Amino-3,5-diethyl-1-methyl-1,2, 4-triazolium*	1
C124	*4-Amino-1,3,5-triethyl-1,2, 4-triazolium*	1
	Tetrazolium	
C125	1*H*-Tetrazolium	1
C126	5-Aminotetrazolium	1
C127	5-Amino-1-methyltetrazolium	1
C128	5-Amino-2-methyltetrazolium	1
C129	5-Amino-1,3-dimethyltetrazolium	1
C130	2-Amino-4,5-dimethyltetrazolium	1
C131	1,5-Diaminotetrazolium	1
C132	1,5-Diamino-4-methyltetrazolium	1
	Aromatic: thiazolium and benzothiazolium	
C133	3-{(2*S*)-2-Methylbutyl}thiazolium	1
C134	2-Amino-5-butyl-4-methyl-1, 3-thiazol-3-ium	1
C135	2-Amino-2-thiazolinium	1
C136	*2-Aminobenzothiazolium*	1
C137	*N*-Methylbenzothiadiazolium	1
C138	*N*-Decylbenzothiazolium	1

TABLE 11.5 (*Continued*)

Cation Number		q Cation Charge
C139	*N*-Undecylbenzothiazolium	1
C140	*N*-Dodecylbenzothiazolium	1
C141	3-(3-Butynyl)-2-(1-hexynyl) benzothiazolium	1
	Aromatic: pyridinium, quinolinium and isoquinolinium	
[Hpy]⁺	***Protic pyridinium***	
C142	Pyridinium	1
C143	*4-Methylpyridinium*	1
C144	2-Benzylpyridinium	1
C145	*2-(2,4-Dinitrobenzyl)pyridinium*	1
C146	*2-Fluoropyridinium*	1
C147	4-Cyanopyridinium	1
C148	3-Hydroxypyridinium	1
C149	2-Aminopyridinium	1
C150	2-Amino-3-methylpyridinium	1
C151	*2-Amino-5-bromopyridinium*	1
C152	2,6-Dimethylpyridinium	1
C153	2,6-Dichloropyridinium	1
C154	2,4,6-Trimethylpyridinium	1
C155	(Pyridin-2-yl)(pyridinio-2-yl) disulfide	1
C156	(2-Pyridylthio)(2-pyridiniothio) methane	1
[Rpy]⁺	***N-alkylpyridinium***	
C157	1-Ethylpyridinium	1
C158	1-Butylpyridinium	1
C159	1-Hexylpyridinium	1
C160	1-Octylpyridinium	1
C161	1-Tetradecylpyridinium	1
C162	1-Hexadecylpyridinium	1
C163	1-(Trimethylsilyl)pyridinium	1
C164	1-(Trimethylsilyl)-3, 4-dimethylpyridinium	1
C165	1-Methyl-3-cyanopyridinium	1
C166	1-Methyl-4-cyanopyridinium	1
C167	*1,2,4-Trimethylpyridinium*	1
C168	1-(Cyanomethyl)pyridinium	1
C169	1-(Cyanomethyl)-2-methylpyridinium	1
C170	1-(Cyanomethyl)-2-ethylpyridinium	1
C171	1-(Cyanomethyl)-4-ethylpyridinium	1
C172	1-(Hydroxycarbonymethyl)pyridinium	1

(*Continued*)

TABLE 11.5 (*Continued*)

Cation Number		q Cation Charge
C173	1-(3-Cyanopropyl)pyridinium	1
C174	1-(3-Cyanopropyl)-3-methylpyridinium	1
C175	1-(3-Cyanomethyl)-2,5-dimethylpyridinium	1
C176	1-Ethyl-3-trifluoromethylpyridinium	1
C177	1-Ethyl-2-cyanopyridinium	1
C178	1-Ethyl-3-cyanopyridinium	1
C179	1-Ethyl-4-cyanopyridinium	1
C180	1-(2-Hydroxyethyl)pyridinium	1
C181	1-Butyl-3-methylpyridinium	1
C182	1-Butyl-4-methylpyridinium	1
C183	1-Benzylpyridinium	1
C184	1-Benzyl-3,5-bis(3-trifluoromethylphenyl)pyridinium	1
C185	3-Hydroxy-1-(octyloxymethyl)pyridinium	1
C186	*N*-(1-*O*-methyl-2,3,4-tri-*O*-acetyl-α-D-glucopyranose-6-yl)pyridinium	1
C187	rac-*1-((1-(2-Naphthyl)prop-2-en-1-yl)amino)pyridinium*	1
C188	1-Methyl-4-(benzylselenomethyl)pyridinium	1
C189	2-Chloro-1-methyl-4-(phenylamino)pyridinium	1
C190	1-{(3-(Dimethylammonio)propyl}-2-hydroxypyridinium	2
	Quinolinium and isoquinolinium	
C191	1-Ethylquinolinium	1
C192	1-Butylisoquinolinium	1
[HOpy]$^{+}$	***N-Hydroxylated pyridinium***	
C193	1-Hydroxypyridinium	1
C194	[(Pyridine-1-oxide)$_2$H]	1
C195	*4-Methoxy-2,6-dimethyl-1-hydroxypyridinium*	1
	Aromatic: pyrimidinium	
C196	*2-Aminopyrimidinium*	1
C197	2-Amino-4-methylpyrimidinium	1
C198	*2-Amino-4-phenyl-6-methylpyrimidinium*	1
C199	1-(3-Butynyl)-2-(phenylethynyl)pyrimidinium	1

TABLE 11.5 (*Continued*)

Cation Number		*q* Cation Charge
	Alicyclic: azetidinium	
C200	3-Methoxyazetidinium	1
C201	*3,3-Dinitroazetidinium*	1
C202	*3-O-Nitrato-3-ethylazetidinium*	1
	Alicyclic: pyrrolidinium	
[HRpyr]⁺	***Protic pyrrolidinium***	
C203	Pyrrolidinium	1
C204	1-Butylpyrrolidinium	1
C205	1-(2-Hydroxyethyl)pyrrolidinium	1
[RR′pyr]⁺	***N,N-Dialkylpyrrolidinium***	
C206	1,1-Dimethylpyrrolidinium	1
C207	1-Ethyl-1-methylpyrrolidinium	1
C208	1-Methyl-1-propylpyrrolidinium	1
C209	1-Butyl-1-methylpyrrolidinium	1
C210	1-Hexyl-1-methylpyrrolidinium	1
C211	1-Heptyl-1-methylpyrrolidinium	1
C212	1-(Hydroxycarbonylmethyl)-1-methylpyrrolidinium	1
C213	1-(Prop-2-ynyl)-1-methylpyrrolidinium	1
C214	1-Butyl-1-cyanomethylpyrrolidinium	1
C215	*5-Azonia-2-oxa-spiro[4.4]nonane*	1
	Alicyclic: piperidinium	
[HRpip]⁺	***Protic piperidinium***	
C216	Piperidinium	1
C217	*2,6-Dimethylpiperidinium*	1
C218	1-(2-Hydroxyethyl)piperidinium	1
[RR′pip]⁺	***N,N-Dialkylpiperidinium***	
C219	1-Ethyl-1-methylpiperidinium	1
C220	1-Methyl-1-propylpiperidinium	1
C221	1-Butyl-1-methylpiperidinium	1
C222	*1-(Piperidinium-1-ylidene)-3-(piperidin-1-yl)-2-azapropene*	1
	Alicyclic: morpholinium	
[RR′mor]⁺	***Morpholinium***	
C223	Morpholinium	1
C224	1-Methylmorpholinium	1
C225	1-(2-Hydroxyethyl)morpholinium	1
C226	1-{2-(Hydroxycarbonyl)ethyl} morpholinium	1

(*Continued*)

TABLE 11.5 (*Continued*)

Cation Number		*q* Cation Charge
C227	1-Butyl-1-methylmorpholinium	1
C228	*1-(Hydroxycarbonylmethyl)-1-methylmorpholinium*	1
C229	1-(Ethoxycarbonylmethyl)-1-methylmorpholinium	1
	Alicyclic: urotropinium	
C230	*N*-Ethylurotropinium	1
C231	*N*-Propylurotropinium	1
C232	*N*-Butylurotropinium	1
	Alicyclic: azepanium	
[RR′azp]$^+$	***N,N-Dialkylazepanium***	
C233	1-Butyl-1-methylazepanium	1
	Alicyclic: other heterocycles	
	Cyclic amidinium	
C234	1,3,5-Trimethyl-tetrahydro-1,3,5-triazinium	1
C235	1,3,5-Triethyl-tetrahydro-1,3,5-triazinium	1
C236	1,3,5-Tri-*iso*-propyl-tetrahydro-1,3,5-triazinium	1
C237	1,3,5-Tri-*tert*-butyl-tetrahydro-1,3,5-triazinium	1
C238	3-Methyl-2-phenyl-3,4,5,6-tetrahydropyrimidinium	1
C239	1,3-Diethyl-3,4,5,6-tetrahydropyrimidinium	1
	Other azaheterocycles	
C240	1,5-Dimethoxyl-3-phenyl-5a,6,7,8,9,9a-hexahydro-1*H*-1,5-benzodiazepin-5-ium	1
C241	*N*-Amino-1-azoniacycloheptane	1
C242	1-Aza-8-azoniabicyclo[6.6.3]heptadeca-4,11-diyne	1
C243	1-Aza-8-azoniabicyclo[6.6.4]octadeca-4,11-diyne	1
C244	1-Aza-8-azoniabicyclo[6.6.5]nonadeca-4,11-diyne	1
C245	1-Ethyl-1,4-diazabicyclo[2.2.2]octanium	1
C246	1,4,7-Trimethyl-1-azonia-4,7-diazacyclononane	1

TABLE 11.5 (*Continued*)

Cation Number		q Cation Charge
C247	3-(3-Ethenyl-1,2,4-oxadiazol-5-yl)-1,2,5,6-tetrahydropyridinium	1
C248	*3-(4-Methoxyphenyl)-1,2-dimethyl-4,5-dihydroimidazolium*	1
C249	5-Ethoxy-l-ethyl-3,4-dihydro-2-oxo-2*H*-pyrrolium	1
	Other heterocycles	
C250	1,3-Dimethyl-1,3-diaza-2-phospholidinium	1
C251	1,3-Dimethyl-1,3-diaza-2-arsenanium	1
C252	5-Methyl-1,3,5-dithiazinan-5-ium	1
C253	9-Aza-9-methyl-1-thioniabicyclo[3.3.1]nonane	1
C254	*2-*tert*-Butyl-3,3,5-trimethyl-1,2-diaza-3-sila-5-cyclopentenium*	1
C255	t-BuNHP(μ-Nt-Bu)$_2$P$^+${(NHt-Bu)(NCO$_2$i-Pr)}NH(CO$_2$i-Pr)	1
	Acyclic: hydrazinium, guanidinium and uronium	
	Hydrazinium and hydroxylammonium	
C256	Hydrazinium	1
C257	Methylhydrazinium	1
C258	1,1-Dimethylhydrazinium	1
C259	1,2-Dimethylhydrazinium [$MeNH_2NHMe]^+$	1
C260	1,2-Dimethylhydrazine-1,2-diium [$MeNH_2NH_2Me]^{2+}$	2
C261	1,1,1-Trimethylhydrazinium	1
C262	1-(Chloromethyl)-1,1-dimethylhydrazinium	1
C263	Bis(phenylhydrazine)hydrogen	1
C264	2,2-Dimethyltriazanium	1
C265	Hydroxylammonium	1
	Guanidinium	
C266	Guanidinium	1
C267	Aminoguanidinium	1
C268	*N,N′*-Diaminoguanidinium	1
C269	*N,N′,N″*-Triaminoguanidinium	1
C270	1,1,3,3-Tetramethylguanidinium	1

(*Continued*)

TABLE 11.5 (*Continued*)

Cation Number		q Cation Charge
	Amidinium, alkylidinium and diazonium	
C271	Azidoformamidinium	1
C272	Tetramethylformamidinium	1
C273	*N,N*-Dimethyl-*N*-((methyltellurenyl)(phenyl)methylidene)ammonium	1
C274	*N,N*-Dimethyl-*N*-((methylselenyl)(phenyl)methylidene)ammonium	1
C275	1-(1-Methyl-2-butyn-1-ylidene)pyrrolidinium	1
C276	β,β-*Diethoxyethenediazonium*	1
C277	1-Phenyl-4-triphenylphosphanylidene-4*H*-pyrazol-1-ium	1
	Uronium	
C278	Diaminouronium	1
C279	Methyluronium	1
C280	1,1-Dimethyluronium	1
	Acyclic: $[ER_4]^+$, E = N, P, As, Sb	
$[NH_nR_{4-n}]^+$	***Protic ammonium***	
C281	Ammonium	1
C282	Methylammonium	1
C283	Ethylammonium	1
C284	cyclo-*Hexylammonium*	1
C285	*Benzylammonium*	1
C286	(1-Phenylethyl)ammonium	1
C287	*{1-(Hydroxymethyl)propyl}ammonium*	1
C288	Dimethylammonium	1
C289	Methylethylammonium	1
C290	Diethylammonium	1
C291	Benzyl(ethyl)ammonium	1
C292	Dibutylammonium	1
C293	Bis(2-azidoethyl)ammonium	1
C294	*N*-Methyl-*N*-(2-nitroxyethyl)ammonium	1
C295	Trimethylammonium	1
C296	Triethylammonium	1
C297	Benzyldimethylammonium	1
C298	*Ethyldi*-iso-*propylammonium*	1
C299	Tris(2-hydroxyethyl)ammonium	1
C300	Tris(2-methoxyethyl)ammonium	1
C301	Tridodecylammonium	1

TABLE 11.5 (*Continued*)

Cation Number		q Cation Charge
C302	*N,N,N′*-Triethylethylenediammonium	1
[NR$_4$]$^+$	***Tetraalkylammonium***	
C303	Tetramethylammonium	1
C304	2-Hydroxyethyl(trimethyl)ammonium (cholinium)	1
C305	2-Chloroethyl(trimethyl)ammonium	1
C306	(Hydroxycarbonylmethyl)trimethylammonium (protonated betaine)	1
C307	*N*-butyl-*N,N*-dimethylglycine ethyl ester	1
C308	Dimethyl(*iso*-propyl)phenylammonium	1
C309	Dimethyl(propyl)phenylammonium	1
C310	*Dimethyldiphenylammonium*	1
C311	Dimethylbis(3-methylbenzyl)ammonium	1
C312	*Dimethylbis(4-methoxybenzyl)ammonium*	1
C313	Benzethonium	1
C314	Ethyl(methyl)diphenylammonium	1
C315	Tributylmethylammonium	1
C316	Tetraethylammonium	1
C317	Triethylpropylammonium	1
C318	Tetrapropylammonium	1
C319	Tetrabutylammonium	1
C320	*Tetrakis(decyl)ammonium*	1
C321	Tetradodecylammonium	1
[PH$_n$R$_{4-n}$]$^+$	***Protic phosphonium***	
C322	*iso*-Propylphosphonium	1
C323	*cyclo*-Hexylphosphonium	1
C324	Triphenylphosphonium	1
[PR$_4$]$^+$	***Tetraorganophosphonium***	
C325	Ethyltributylphosphonium	1
C326	Tetrabutylphosphonium	1
C327	Decyl(tri-*tert*-butyl)phosphonium	1
C328	*Di*-tert-*butyl(diphenylmethylene)phosphonium*	1
C329	Tetrapentylphosphonium	1
C330	*cyclo*-Hexyltrivinylphosphonium	1
C331	Decyl(tricyclohexyl)phosphonium	1

(*Continued*)

TABLE 11.5 (*Continued*)

Cation Number		q Cation Charge
C332	Tetrakis(decyl)phosphonium	1
C333	Amidomethyltris(tetradecyl) phosphonium	1
C334	Hydroxycarbonylmethyltris (tetradecyl)phosphonium	1
C335	Tetraoctadecylphosphonium	1
C336	Methyltriphenylphosphonium	1
C337	Ethyltriphenylphosphonium	1
C338	Butyltriphenylphosphonium	1
C339	Benzyltriphenylphosphonium	1
C340	Tetraphenylphosphonium	1
$[PX_nR_{4-n}]^+$	***Haloorganophosphonium***	
C341	Trichloro(*cyclo*-hexyl)phosphonium	1
C342	3-Chloro-3-(di-*iso*-propylamino)-3-phosphoniobicyclo[3.1.0]hexane	1
C343	Chlorotriphenylphosphonium	1
C344	Bromotriphenylphosphonium	1
C345	*Tri-*iso*-propyl-[N-(2,6-dichlorophenyl) thiocarbamoyl]phosphonium*	1
$[ER_4]+$	***Arsonium and stibonium***	
C346	Tetraphenylarsonium	1
C347	Tetramethylstibonium	1
	P-P and As-Sb bonded cations	
C348	Trimethylphosphine(diphenyl) phosphonium	1
C349	*Chlorodiphenylphosphine(diphenyl) phosphonium*	1
C350	Triphenylarsine{chloro(phenyl)} stibonium	1
C351	{Diphenylphosphinomethyl(diphen yl)phosphino} diphenylphosphonium	1
C352	3,3-Diphenyltricyclo[2.1.0.0^{2,5}] pentaphosphan-3-ium	1
C353	*3-Chloro-3-ethyltricyclo[2.1.0.0^{2,5}] pentaphosphan-3-ium*	1
C354	3-Chloro-3-*cyclo*-hexyltricyclo[2.1.0.0^{2,5}] pentaphosphan-3-ium	1
C355	3-Chloro-3-pentafluorophenyltricyclo [2.1.0.0^{2,5}]pentaphosphan-3-ium	1

TABLE 11.5 (*Continued*)

Cation Number		*q* Cation Charge
$[ER_3]^+$	**Sulfonium, selenonium and telluronium**	
C356	Trimethylsulfonium	1
C357	Triethylsulfonium	1
C358	*Dimethylphenylsulfonium*	1
C359	1-Phenylthiolanium	1
C360	Methyldiphenylsulfonium	1
C361	Hexadecyldimethylsulfonium	1
C362	(±)-cis-*3-Acetoxy-1-methylthiacycohexane*	1
C363	(−)-(*1*S,*3*S)-trans-*3-Acetoxy-1-methylthiacycohexane*	1
C364	3(*R*),4(*R*),5(*R*),6(*S*)-3,4,5-Trihydroxy-*cis*-1-thioniabicyclo[4.3.0]nonane	1
C365	Trimethylselenonium	1
C366	Trimethyltelluronium	1
C367	*Bis(dimethylamino) trifluoromethylsulfonium*	1
C368	Tris(dimethylamino)sulfonium	1
C369	Tris(dimethylamino)oxosulfonium	1
C370	*Di*-cyclo-*hexyl(hydroxo)selenonium*	1
$[XR_2]^+$	**Bromonium and iodonium**	
C371	Bis(pyridine)bromonium	1
C372	Bis(pyridine)iodonium	1
C373	*Bis(2,6-dimethylpyridine)iodonium*	1
C374	Bis(1,3-dimethylthiourea)iodonium	1
C375	Diphenyliodonium	1
C376	*Bis(4-methoxyphenyl)iodonium*	1
C377	(2,2,2-Trifluoroethyl) phenyliodonium	1
C378	*4-Fluorophenyl(pentafluorophenyl) iodonium*	1
C379	2-(Hydroxoiodinium)nitrobenzene	1
C380	{Tris(dimethylamino) phosphaneselenide}iodine(I)	1
	Carbon centred	
C381	*tert*-Butyl carbenium	1
C382	1,1,2,3,4,5,6-Heptamethylbenzenium	1
C383	1-Chloromethyl-1,2,3,4,5,6-hexamethylbenzenium	1
C384	1-Phenyl-1,2,3,4,5,6-hexamethylbenzenium	1

(*Continued*)

TABLE 11.5 (*Continued*)

Cation Number		*q* Cation Charge
C385	2-Hydroxyhomotropylium	1
C386	*1-Ethoxyhomotropylium*	1
	Protic	
C387	*Oxonium*	1
C388	Oxonium monohydrate	1
C389	Oxonium dihydrate	1
C390	1,1-Dihydroxyethylium	1
C391	H_2([2.2.2]Cryptand)	2
C392	Bis(2-pyridone)hydrogen	1
C393	5-Hydroxy-1-methyl-3, 4-dihydro-2*H*-pyrrolium	1
C394	3-Hydroxy-1-*tert*-butyl-1, 2-dihydropyrrolium	1
C395	2-Acetyl-9-azoniabicyclo [4.2.1]nonan-3-one	1
C396	Bis(phenylhydrazine)hydrogen	1
C263	[(2-Trimethylsilylethyl) dimethylamine)$_2$H]	1
C397	Tris(dimethylamino) methyliminiumphosphorane	1
C398	[(Me_3NH)$_2$Cl]	1
C399	[($NMe_3CH_2CO_2$)$_2$H]	1
	Inorganic, coordination and organometallic compounds	
C400	Lithium	1
C401	Tetrakis(ethanenitrile)lithium	1
C402	Bis(tetramethylene sulfone)lithium	1
C403	*Tetra(pyridine)lithium*	1
C404	[Li_2(H_2NNHMe)$_3$]	2
C405	Sodium	1
C406	(1,2-Dimethoxyethane)sodium	1
C407	(Benzo-18-crown-6)bis(pyridine) sodium	1
C408	Potassium	1
C409	Bis(*N*,*N*,*N*′,*N*″,*N*″-pentamethyldiethylenetriamine) potassium	1
C410	(Benzo-18-crown-6)potassium	1
C411	([2.2.2]Cryptand)potassium	1
C412	*Rubidium*	1
C413	Caesium	1
C414	Bis(ethanenitrile)copper(I)	1

TABLE 11.5 (*Continued*)

Cation Number		q Cation Charge
C415	Bis(1-dodecylimidazole)copper(I)	1
C416	Tetrakis(ethanenitrile)copper(I)	1
C417	Tetrakis(1-dodecylimidazole) copper(I)	1
C418	Tetrakis(1-methylimidazole) copper(II)	2
C419	Tetrakis(1-ethylimidazole) copper(II)	2
C420	Tetrakis(1-butylimidazole)copper(II)	2
C421	Hexakis(1-butylimidazole) copper(II)	2
C422	(Ethanenitrile)silver(I)	1
C423	Tetrakis(ethanenitrile)silver(I)	1
C424	(1-Butylimidazole)(1-methylimidazole)silver(I)	1
C425	(1-Butylimidazole)(1-ethylimidazole)silver(I)	1
C426	Bis(1-methylimidazole)silver(I)	1
C427	Bis(1-ethylimidazole)silver(I)	1
C428	Bis(1-butylimidazole)silver(I)	1
C429	*Bis(1-ethyl-3-methylimidazol-2-ylidene)silver(I)*	1
C430	Bis(4-methoxyphenalene-1,9-diolato)boron	1
C431	Bis(3,4-dimethoxyphenalene-1,9-diolato)boron	1
C432	(3-Methylimidazole) (trimethylamine) dihydroboronium	1
C433	Bis(1-methyl-1H-imidazole-3-yl) dihydroboronium	1
C434	*[(N,N,N′,N″,N″-Pentamethyldiethylenetriamine) $AlMe_2$]*	1
C435	*Aquo(pyridine-N-oxide) trimethyltin(IV)*	1
C436	Bis(*N,N′*-dimethylethyleneurea-*O*) trimethyltin(IV)	1
C437	Bis(dimethylsulfoxide-*O*) trimethyltin(IV)	1
C438	$[Tl_3F_2Al\{OCH(CF_3)_2\}_3]$	1
C439	*$[Ti(h^5\text{-}C_5H_5)Cl.18\text{-}crown\text{-}6]$*	1
C440	Bis(h^2-dimethyldisulfide) tetrachloroniobium(V)	1

(*Continued*)

TABLE 11.5 (*Continued*)

Cation Number		q Cation Charge
C441	$[Dy_2(NMe_3CH_2CO_2)_8(H_2O)_4]$	6
C442	Nitrosonium	1
C443	Ferrocenium	1
C444	1-(Ferrocen-1-ylmethyl)-3-methylimidazolium	1
C445	1-Butyl-3-ferrocenylimidazolium	1
C80	1-(5-Bromopentyl)-1′,1″-biferrocenium	1
C446	1-Pentyl-1‴-(5-bromopentyl)-1′,1″-biferrocenium	1
C447	1,1‴-Bis(5-bromopentyl)-1′,1″-biferrocenium	1
C448	1,1‴-Bis(6-bromohexyl)-1′,1″-biferrocenium	1
	Mulitiply charged	
	Dicationic imidazolium	
C70	1-(3-Ammoniopropyl)-3-methylimidazolium	2
C71	1-{2-(Diethylammonio)ethyl}-3-methylimidazolium	2
C449	1,4-Bis(1-methylimidazolium-3-yl) butane	2
C450	1,7-Bis(1-methylimidazolium-3-yl) heptane	2
C451	1,8-Bis(1-methylimidazolium-3-yl) octane	2
C452	1,10-Bis(1-methylimidazolium-3-yl) decane	2
C453	1,4-Bis(1-methylimidazolium-3-ylmethyl)benzene	2
C454	1,4-Bis(1-butylimidazolium-3-ylmethyl)benzene	2
C455	3,5-Bis(1-butylimidazolium-3-ylmethyl)toluene	2
C456	1,4-Bis(1,2-dimethylimidazolium-3-ylmethyl)benzene	2
C457	1,8-Bis(1-methylimidazolium-3-yl)-3,6-dioxaoctane	2
C458	6,6′-Spirobis(1-methyl-1,5,6,7-tetrahydropyrrolo[1,2-a]imidazol-4-ium	2
	Dicationic azonium	
C190	1-{(3-(Dimethylammonio)propyl}-2-hydroxypyridinium	2

TABLE 11.5 (*Continued*)

Cation Number		*q* Cation Charge
C260	*N,N′*-Dimethylhydrazinium $[MeNH_2NH_2Me]^{2+}$	2
C459	Piperazinediium	2
C460	*Hexamethylenediammonium*	2
C461	*3,6-Dioxaoctane-1,8-diammonium*	2
C462	*N*-Methyl-1,3-propylenediammonium	2
C463	*N,N′*-Bis(2-ethoxyethyl)-*N,N,N′,N′*-tetramethylethane-1,2-diammonium	2
C464	*N,N′′′*-Bis(2-ethoxyethyl)-*N,N,N′′′,N′′′*-tetramethylbutane-1,4-diammonium	2
C465	1,3-Bis(1-methyl-1-pyrrolidinio) propane	2
C466	*2,6-Bis(2-pyridinio)pyridine*	2
	Di- and tricationic phosphonium	
C467	*1,2-Bis(P,P-dimethyl-P′,P′-diphenylphosphinophosphonium) ethane*	2
C468	1,2-Bis(*P,P,P′,P′*-tetraphenylphosphinophosphonium) ethane	2
C469	1,6-Bis(*P,P,P′,P′*-tetraphenylphosphinophosphonium) hexane	2
C470	*1,3-Dichloro-1,2,3,4-tetracyclohexyltetraphosphetane-1, 3-diium*	2
C471	3,3,5,5,7,7-Hexaphenyltricyclo $[2.2.1.0^{2,6}]$heptaphosphane-3,5,7-triium	3
	Dicationic S^+-As, N^+-As compounds	
C472	1,6-Diarsa-2,5,7,10-tetrathiatricyclo $[5.3.0^{1,5}.0^{6,10}]$decane	2
C473	*1,6-Diarsa-2,5,7,10-tetramethyl-2,5,7,10-tetraazatricyclo*$[5.3.0^{1,5}.0^{6,10}]$*decane*	2

materials that solidify to form glasses (and which do not crystallise on subsequent heating), the wider information-gathering net has led to the identification of a large number of anions and cations, reviewed in the following text, that hitherto have been largely ignored as potential components of ionic liquids. This review, therefore, serves to draw wider attention to these anions and cations as the possible basis of new ionic liquids.

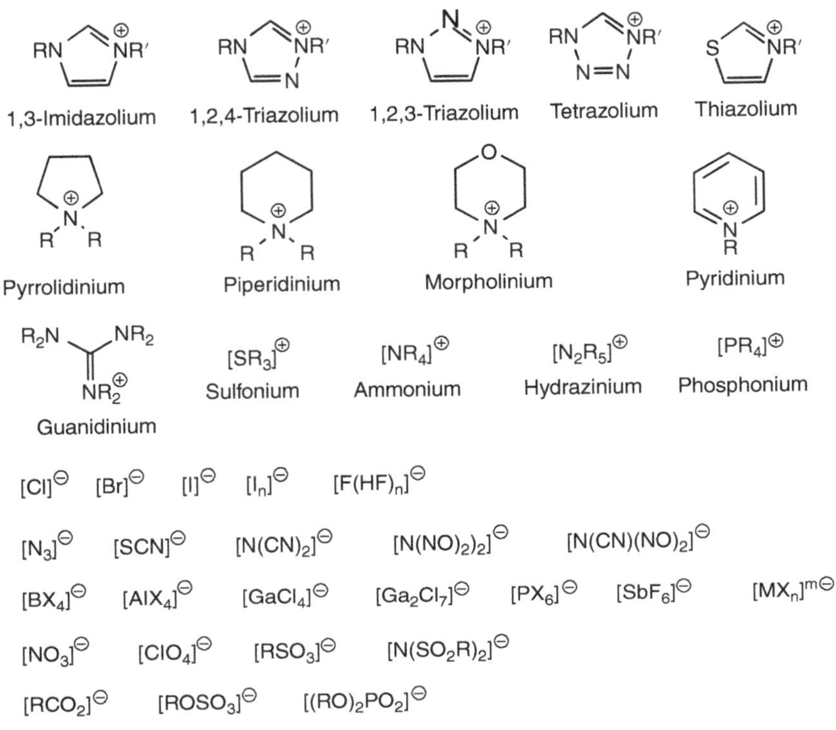

Figure 11.5 Basic structures of the most common ionic liquid anions and cations.

Our survey simply reinforces and illustrates Seddon's conclusion, from his widely quoted [606] estimate of 10^{18} possible ionic liquids, [cat][A], that a relatively small part of ionic liquid composition space has so far been investigated. Tables 11.4 and 11.5 list, respectively, the 248 anions and 473 cations found in the salts listed in Tables 11.1, 11.2 and 11.3. In the more restricted list of room-temperature ionic liquids, that is, those materials with melting temperatures of 30°C or less, 47 cations and 43 anions are found. 70 (3.5%) of the possible 2021 [A][cat] combinations have been crystallographically characterised. Salts with m.pt.<100°C contain 397 separate cations and 205 separate anions. The 623 salts for which crystal structures are reported thus represent less than 0.8% of the possible 81,385 combinations. A further 76 cations and 43 anions are found only in salts with melting temperatures in the range 100–*ca.* 110°C (and for which crystal structures are available), representing a surprisingly large additional area of relatively unexplored potential. Overall, less than 1% of the composition space has been investigated so far. Bearing in mind the large number of salts with imidazolium cations and a limited number of highly represented anions, the survey also highlights how narrow a view of ionic liquid composition (and thereby structure, dynamics and physical and chemical property) has so far been gained. And this number does not encompass the additional variation available (in principle) via salts (a further 36,000 or so) containing the cations and anions from salts with reported melting temperatures only just outside the conventional (and pointless) limit.

The simplest initial assessment of the combinations of anions and cations that might provide clues as to the delineation of those structural factors (as evidenced in their crystal and molecular structure revealed from single-crystal X-ray crystallographic studies) associated with ionic liquid behaviour arises from an examination of the nature of the individual anions and cations themselves. Factors include charge, charge distribution, composition, size, shape, symmetry, geometry and chemical and conformational lability. The crystal structures are listed of ionic liquids containing ions, the central element of which come from every group of the periodic table, other than Groups 2 and 18 (though, in the latter case, note Ref. [26] mentioned in the Introduction). The crystal structures of a number of methylmagnesium salts have been solved, though most are solvated or decompose before melting. Others, such as [(15-crown-5)(methyl)magnesium][cyclopentadienide] (m.pt. 125°C; refcode MIWHOQ) [607] and [(1,4,8,11-tetramethyl-1,4,8,11-tetraazacyclotetradecane)(methyl)magnesium][cyclopentadienide] (m.pt. 204°C; refcode FIZREM) [608], which both show the absence of close contacts between magnesium and the anion, are too high melting for inclusion in the tables.

11.3.2 Anions

A total of 248 discrete anions are collected in Table 11.4. Of these, 43 anions are components of room-temperature ionic liquids, and an additional 162 of ionic liquids and a further 43 are, as yet, only found in salts with melting temperatures in the range 100–110°C. What can be concluded from this preliminary examination of the variety of anions shown (though more detailed evaluation of the large body of information collected in this review clearly being needed) is the (unsurprising) absence of any single characteristic, apart from charge, that might be strongly associated with ionic liquid formation. Izgorodina et al. [609] has sought to relate charge delocalisation for a series of anions (using proton affinity as a surrogate) with melting point for a group of anions.

While singly charged anions predominate, dinegative anions are also common. Anions of a range of sizes, formula weights, coordination number, shapes and geometries are present. Those containing highly electronegative elements capable of engaging in strong hydrogen bonding appear to be no more or less prevalent than those in which weaker hydrogen bonding is likely. Absences may simply arise from the absence of effort to study salts containing particular anions by single-crystal X-ray crystallography.

The commonest anions (i.e. those found in at least five ionic liquids) are (in order of the number of **room-temperature ionic liquids** + **ionic liquids** in which they are found) listed as follows.

Twenty-five of the seventy salts melting below 30°C (and for which a crystal structure determination is available) contain one of just three of the foregoing

The Number of Ionic Liquids Containing the Most Common Anions.

	m.pt.: <30°C	30–100°C	>100°C
1. Bis(trifluoromethanesulfonyl)amide	20	**49**	*9*
2. Trifluoromethanesulfonate	3	**27**	*3*
3. Hexafluorophosphate	2	**26**	*3*
4. Iodide	1	**26**	*4*
5. Bromide	1	**26**	*3*
6. Nitrate	1	**22**	*3*
7. Triiodide	1	**20**	*6*
8. Tetrafluoroborate	1	**15**	*7*
9. Chloride	0	**16**	*6*
10. Bis(methanesulfonyl)amide	1	**14**	*2*
11. Dinitramide	1	**12**	*1*
12. Tetrakis(1,1,1,3,3,3-hexafluoro-*iso*-propoxy)aluminate	1	**10**	*0*
13. Tetrachlorogallate	0	**11**	*4*
14. Perchlorate	0	**11**	*3*
15. Azide	1	**9**	*1*
16. Tetrachloroaluminate	0	**8**	*4*
17. Dicyanamide	0	**6**	*0*
18. Trifluoroethanoate	0	**6**	*1*
19. Pentaiodide	0	**6**	*1*
20. $[F(HF)_2]$	4	**1**	*1*
21. Tetracyanoborate	0	**5**	*1*

ions, with one, $[N(SO_2CF_3)_2]^-$, being found in 20 room-temperature ionic liquids. The data also underline the considerable compositional diversity evident in materials that, while qualifying as ionic liquids, are under-recognised as such.

Simple (in the compositional sense) variants of anions that feature in low-melting salts, such as $[AsF_6]^-$, $[SbF_6]^-$ (and other halometallates), $[PF_3(C_2F_5)_3]^-$ and $[N(SO_2F)_2]^-$, are also found in Table 11.4. In addition (excluding solvates as a general class), salts melting at 30°C or below, listed in Table 11.1, include anions capable of displaying quite extensive further structural variation, such as sulfate and phosphate esters and 1,3-propanedionates (the latter providing conformational flexibility similar to that found in the widely studied bis(sulfonyl)amides but yet relatively understudied).

11.3.3 Cations

A total of 473 discrete cations are collected in Table 11.5. Forty-seven are found in room-temperature ionic liquids, an additional 350 in ionic liquids. A further 76 are found in salts melting between 100 and 110°C.

The commonest cations (i.e. those found in five or more ionic liquids) are (in order of the sum **room-temperature ionic liquids + ionic liquids**) as follows.

The Number of Ionic Liquids Containing the Most Common Cations.

	m.pt.: <30°C	30–100°C	>100°C
1. $[C_2mim]^+$	11	**31**	4
2. $[C_4mim]^+$	7	**19**	3
3. $[C_4C_1C_1im]^+$	1	**13**	2
4. $[N_{4444}]^+$	0	**12**	4
5. $[C_{12}C_{12}im]^+$	0	**9**	0
6. $[Cholinium]^+$	3	**4**	0
7. $[NH_2NH_3]^+$	0	**6**	0
8. $[C_1C_4pyr]^+$	3	**3**	0
9. $[S(NMe_2)_3]^+$	1	**4**	0
10. $[C_3mim]^+$	0	**5**	0
11. $[N_{1111}]^+$	0	**5**	1
12. $[N_{2222}]^+$	0	**5**	2

Six of the listed cations are found in 26 of the 70 salts (having been characterised by single-crystal X-ray diffraction) in Table 11.1. Imidazolium salts make up the greatest subgroup of both room-temperature ionic liquids and ionic liquids. Apart from cholinium, the only tetraalkylammonium salt with a m.pt. < 30°C contains the $[N_{1444}]^+$ ion. Salts of protic cations comprise a perhaps surprising 13 of the 70 room-temperature ionic liquids. A diverse range of imidazolium and pyridinium compounds have been studied. Interestingly, wholly inorganic (and structurally characterised) salts of the alkali metals have m.pt. as low as 63°C (form II of $K[N(SO_2F)_2]$ [209]).

There are many other subdivisions into which it is possible to place the salts brought together in Tables 11.1, 11.2 and 11.3. The first class is salt type: protic (Section 11.4, Table 11.6), inorganic (Section 11.5.1, Table 11.7), coordination complex and $[MX_n]$ anions (Sections 11.5.2–11.5.6, Tables 11.8, 11.9, 11.10 and 11.11), energetic (Section 11.6, Table 11.12), chiral (Section 11.7) and multiply charged (Section 11.8, Table 11.13). The second class is ion type: such as halide, polyhalide and pseudohalide (Section 11.9, Table 11.14), oxoanions and their esters (Section 11.10, Table 11.15), EX_4 and related anions (Section 11.11, Table 11.16), EX_6 and related anions (Section 11.12, Table 11.17), $[N(SO_2R)_2]$ (Section 11.13, Table 11.18), sulfonates (Section 11.14, Table 11.19), carboxylates (Section 11.15, Table 11.20), ER_4 and related cations (Section 11.16, Table 11.21), pyridinium cations (Section 11.17, Table 11.22), imidazolium cations (Section 11.18, Table 11.23) and salts of other anions and cations that do not easily fall into the foregoing categories (Section 11.19, Table 11.24).

11.4 PROTIC IONIC LIQUIDS

Protic ionic liquids are technically attractive because they represent a range of materials that may be readily prepared simply by mixing a suitable proton acceptor and a Brønsted acid. Because of the reversibility of protonation, the

TABLE 11.6 Protic Ionic Liquids

m.pt./°C			$p:q$	U_L/ kJ mol^{-1}
	Imidazolium			
99–100	Imidazolium	$[PO_2(OPh)_2]$	1:1	425
101.5–103	Imidazolium	$[HO_2C(CH_2)_4CO_2]$	1:1	465
105	2-Methylimidazolium	$[HO_2C(CH_2)_3CO_2]$	1:1	467
60	2-Methylimidazolium	$[HO_2C(CH_2)_6CO_2]$	1:1	441
106	4-methylimidazolium	$[HO_2C(CH_2)_3CO_2]$	1:1	472
85	4-Methylimidazolium	$[HO_2C(CH_2)_6CO_2]$	1:1	441
41–43	2-Ethylimidazolium	[HOC(O)CH(OH) CH(OH)CO_2]	1:1	462
105–108	3-Azidosulfonylimidazolium	$[SO_3(OH)]$	1:1	481
60	1-Methylimidazolium	$[NO_3]$	1:1	538
91–92.5	1-Methylimidazolium	$[PO_2(OPh)_2]$	1:1	423
98–101	1-Methylimidazolium	[HOC(O)CH(OH) CH(OH)CO_2]	1:1	470
52/59	1-Methylimidazolium	$[N(SO_2CF_3)_2]$	1:1	445
103	1-Methylimidazolium	$[Au(CN)_2]$	1:1	487
79	1,2-Dimethylimidazolium	$[NO_3]$	1:1	513
105	1,2-Dimethylimidazolium	$[HO_2C(CH_2)_2CO_2]$	1:1	469
102	1,2-Dimethylimidazolium	$[HO_2C(CH_2)_3CO_2]$	1:1	464
108–110	1-Benzylimidazolium	$[PhP(O)_2(OH)]$	1:1	424
95	1-(3-Ammoniopropyl)-3-methylimidazolium	$[NO_3]$	1:2	1587
	***N*-Hydroxy- and *N*-aminoimidazolium**			
72	1,3-Dihydroxy-2-methylimidazolium	$[N(SO_2CF_3)_2]$	1:1	435
95	1-Aminoimidazolium	Cl	1:1	556
58–59	1,3-Diamino-2-methylimidazolium	$[N(SO_2CF_3)_2]$	1:1	433
	Triazolium and tetrazolium			
99	1-Amino-1,2,3-triazolium	5-Nitrotetrazolate	1:1	508
105–110	1,2,4-Triazolium	$[N(SO_2Me)_2]$	1:1	478
93	4-Amino-1,2,4-triazolium	4-(Carboxylato-methyl)-5-nitro-iminotetrazolate	2:1	1200
69	1-Propyl-1,2,4-triazolium	5-Nitroimino-tetrazolate	1:1	456
110	1,4-Dimethyl-2-*H*-1,2,4-triazolium	$[I_3]$	1:1	465
70	1*H*-Tetrazolium	$[N(NO_2)_2]$	1:1	535
55	1*H*-Tetrazolium	$[ClO_4]$	1:1	556
85	5-Aminotetrazolium	$[N(NO_2)_2]$	1:1	526
90	5-Amino-1-methyltetrazolium	$[N(NO_2)_2]$	1:1	500
90	5-Amino-2-methyltetrazolium	$[N(NO_2)_2]$	1:1	501

TABLE 11.6 (*Continued*)

m.pt./°C			$p:q$	U_L/ kJ mol^{-1}
72	1,5-Diaminotetrazolium	$[N(NO_2)_2]$	1:1	509
97–98	1,5-Diaminotetrazolium	$[ClO_4]$	1:1	523
	Thiazolium and benzothiazolium			
99–103	2-Amino-2-thiazolinium	*N*-Methylpyrrole-2-carboxylate	1:1	461
104–107	2-Aminobenzothiazolium	*N*-Methylpyrrole-2-carboxylate	1:1	444
	Pyridinium			
−1	Pyridinium	$[HF_2]$	1:1	548
−17	Pyridinium	$[F(HF)_2]$	1:1	528
60	Pyridinium	$[HCl_2]$	1:1	518
82	Pyridinium	$[CF_3CO_2]$	1:1	501
101	4-Methylpyridinium	$[CF_3CO_2]$	1:1	483
110	4-Methylpyridinium	$[Cu^{II}Br_4]$	2:1	1099
65–66	2-Benzylpyridinium	$[Sb^{III}{}_2Cl_{10}]$	4:2	
110	2-(2,4-Dinitrobenzyl) pyridinium	$[2\text{-HO-3,5-}NO_2C_6H_2CO_2]$	1:1	400
105–118	2-Fluoropyridinium	$[Sb^{III}{}_2Br_9]$	3:1	
74	4-Cyanopyridinium	$[CF_3CO_2]$	1:1	484
82–85	3-Hydroxypyridinium	$[C_6H_5CO_2]$	1:1	473
102	2-Aminopyridinium	$[O_2CCH_2CO_2]$	2:1	1232/ 1234
98–100	2-Aminopyridinium	$[O_2C(CH_2)_8CO_2]$	2:1	1050
97–99	2-Amino-3-methylpyridinium	$[CH_3CO_2]$	1:1	493
110–111	2-Amino-5-bromopyridinium	$[HC{\equiv}CCO_2]$	1:1	484
90–95	2,6-Dimethylpyridinium	$[N(SO_2Me)_2]$	1:1	445
83–110	2,6-Dichloropyridinium	$[Sb^{III}Cl_4]$	1:1	434
88	2,4,6-Trimethylpyridinium	$[(2\text{-}NO_2)C_6H_4CO_2]$	1:1	436
80–82	(Pyridin-2-yl)(pyridinio-2-yl) disulfide	$[I_3]$	1:1	418
93	(2-Pyridylthio) (2-pyridiniothio)methane	$[I_3]$	1:1	411
108–110	2,6-Bis(2-pyridinio)pyridine	$[(CF_3CO_2)_2H]$	1:2	1323
	***N*-Hydroxylated pyridinium**			
73–79	1-Hydroxypyridinium	$[N(SO_2Me)_2]$	1:1	460
78	[(Pyridine-1-oxide)$_2$H]	$[ClO_4]$	1:1	449
107	4-Methoxy-2,6-dimethyl-1-hydroxypyridinium	2,6-Dichloro-4-nitrophenolate	1:1	424
	Pyrimidinium			
104–106	2-Aminopyrimidinium	$[(2\text{-}OH)C_6H_4CO_2]$	1:1	464

(*Continued*)

TABLE 11.6 (*Continued*)

m.pt./°C			p:q	U_L/ kJ mol^{-1}
64–66	2-Amino-4-methylpyrimidinium	$[3\text{-FC}_6H_4CO_2]$	1:1	459
95–96	2-Amino-4-methylpyrimidinium	$[3\text{-ClC}_6H_4CO_2]$	1:1	454
106–108	2-Amino-4-phenyl-6-methylpyrimidinium	$[C_6H_5CO_2]$	1:1	430
	Azetidinium			
95	3-Methoxyazetidinium	Cl	1:1	539
106	3,3-Dinitroazetidinium	$[(2\text{-OH})C_6H_4CO_2]$	1:1	449
102–103	3-*O*-Nitrato-3-ethylazetidinium	$[NO_3]$	1:1	479
	Pyrrolidinium			
80–90	Pyrrolidinium	$[N(SO_2Me)_2]$	1:1	466
67	1-Butylpyrrolidinium	4,4,4-Trifluoro-1-phenylbutane-1,3-dionate	1:1	410
28	1-(2-Hydroxyethyl)pyrrolidinium	$[C_6H_5CO_2]$	1:1	423
49	1-(2-Hydroxyethyl)pyrrolidinium	$[(2\text{-HO})C_6H_4CO_2]$	1:1	448
97	1-(2-Hydroxyethyl)pyrrolidinium	$[\{2,5\text{-(HO)}_2\}C_6H_3CO_2]$	1:1	447
	Piperidinium			
59–61	Piperidinium	$[B(OMe)_4]$	1:1	443
95–97	Piperidinium	$[N(SO_2Me)_2]$	1:1	455
101–102	2,6-Dimethylpiperidinium	$[(2\text{-OH})C_6H_4COS]$	1:1	429
101	1-(2-Hydroxyethyl)piperidinium	$[\{2\text{-}(2,6\text{-}Cl_2C_6H_3NH)C_6H_4\}CH_2CO_2]$	1:1	394
109	1-(2-Hydroxyethyl)piperidinium	$[Si(ox)_3]$	2:1	988
97	1-(2-Hydroxyethyl)piperidinium	$[Ge(ox)_3]$	2:1	991
	Morpholinium			
62	Morpholinium	$[CF_3C(O)CHC(O)CF_3]$	1:1	453
105–106	1-Methylmorpholinium	$[Mn^{II}Cl_4]$	2:1	1125
105–108	1-Methylmorpholinium	$[Co^{II}Cl_4]$	2:1	1129
88–90	1-Methylmorpholinium	$[Cu^{II}Cl_4]$	2:1	1127
92–98	1-Methylmorpholinium	$[Cu^{II}Br_4]$	2:1	1100
92	1-(2-Hydroxyethyl)morpholinium	$[\{2\text{-}(2,6\text{-}Cl_2C_6H_3NH)C_6H_4\}CH_2CO_2]$	1:1	397
80–83	1-{2-(Hydroxycarbonyl)ethyl)morpholinium	Cl	1:1	479

TABLE 11.6 (*Continued*)

m.pt./°C			p:q	U_L/ kJ mol^{-1}
	Other azaheterocycles			
59	*N*-Amino-1-azoniacycloheptane	$[N_3]$	1:1	493
71	1-Aza-8-azoniabicyclo[6.6.3] heptadeca-4,11-diyne	$[(CF_3CO_2)_2H]$	1:1	394
72	1-Aza-8-azoniabicyclo[6.6.4] octadeca-4,11-diyne	$[(CF_3CO_2)_2H]$	1:1	390
75	1-Aza-8-azoniabicyclo[6.6.5] nonadeca-4,11-diyne	$[(CF_3CO_2)_2H]$	1:1	385
90–92	1,4,7-Trimethyl-1-azonia-4,7-diazacyclononane	$[CF_3SO_3]$	1:1	424
	Hydrazinium and hydroxylammonium			
70.5	Hydrazinium	$[N_3]$	1:1	315
80	Hydrazinium	$[N(NO_2)_2]$	1:1	573
97–97.5	Hydrazinium	$[N(CN)_2]$	1:1	593
62.5/70.5	Hydrazinium	$[NO_3]$	1:1	625
97–98	Hydrazinium	$[CH_3CO_2]$	1:1	590
48	Hydrazinium	[Tetrazolate]	1:1	576
68	Methylhydrazinium	$[N_3]$	1:1	707
42	Methylhydrazinium	$[NO_3]$	1:1	583
38	1,1-Dimethylhydrazinium	$[N_3]$	1:1	553
45	1,1-Dimethylhydrazinium	$[NO_3]$	1:1	545
−8	1,2-Dimethylhydrazinium	$[N_3]$	1:1	556
98	1,2-Dimethylhydrazin-1,2-diium	$[NO_3]$	1:2	1746
67	1,2-Dimethylhydrazin-1,2-diium	$[SO_4]$	2:2	
60.5	1,1,1-Trimethylhydrazinium	$[N(CN)_2]$	1:1	505
62	1-(Chloromethyl)-1,1-dimethylhydrazinium	$[SO_4]$	2:1	1245
52	1-(Chloromethyl)-1,1-dimethylhydrazinium	$[ClO_4]$	1:1	502
77	1-(Chloromethyl)-1,1-dimethylhydrazinium	$[C_6H_2\{2,4,6\text{-}(NO_2)_3\}O]$ Picrate	1:1	440
68	Bis(phenylhydrazine) hydrogen	$[N_3]$	1:1	442
59	*N*-amino-1-azoniacycloheptane	$[N_3]$	1:1	493
99	2,2-Dimethyltriazanium	$[NO_3]$	1:1	540
96–100	Hydroxylammonium	F	1:1	726
18–23	Hydroxylammonium	$[N(NO_2)_2]$	Complex	
105	Hydroxylammonium	$[CF_3SO_3]$	1:1	512
66	Hydroxylammonium	1-(2-Nitratoethyl)-5-nitriminotetrazolate	1:1	485

(*Continued*)

TABLE 11.6 (*Continued*)

m.pt./°C			$p:q$	U_L/ kJ mol^{-1}
	Guanidinium			
45/65	Guanidinium	$[N(CN)_2]$	1:1	542/542
70 (dec)	Aminoguanidinium	$[N(NO_2)(SF_5)]$	1:1	491
91–94	Aminoguanidinium	$[N(NO_2)_2]$	1:1	517/518
91	Aminoguanidinium	$[N(CN)(NO_2)]$	1:1	522
72	Aminoguanidinium	$[ClO_4]$	1:1	532
100	Aminoguanidinium	5-Azidotetrazolate	1:1	504
80	*N,N′*-Diaminoguanidinium	$[C(NO_2)_3]$	1:1	484
108	*N,N′*-Diaminoguanidinium	5-Nitrotetrazolate	1:1	497
103	*N,N′*-Diaminoguanidinium	1-(2-Nitratoethyl)-5-nitriminotetrazolate	1:1	452
80	*N,N′,N″*-Triaminoguanidinium	$[N(NO_2)_2]$	1:1	495
84	*N,N′,N″*-Triaminoguanidinium	$[C(NO_2)_3]$	1:1	476
96	*N,N′,N″*-Triaminoguanidinium	5-nitrotetrazolate	1:1	488
62	1,1,3,3-Tetramethylguanidinium	$[C_2H_5CO_2]$	1:1	465
42	1,1,3,3-Tetramethylguanidinium	$[N(SO_2CF_3)_2]$	1:1	422/425
	Uronium			
86	Diaminouronium	$[NO_3]$	1:1	553
56–62	Methyluronium	$[N(SO_2Me)_2]$	1:1	475
67	1,1-Dimethyluronium	$[N(SO_2Me)_2]$	1:1	463
	Ammonium			
92	Ammonium	$[N(NO_2)_2]$	1:1	590
93(dec)	Ammonium	$[N(NO_2)(SF_5)]$	1:1	540
84	Ammonium	[*trans*-$CH_3CH{=}CHCO_2$]	1:1	546
81	Ammonium	1-(2-nitratoethyl)-5-nitriminotetrazolate	1:1	486
110	Methylammonium	$[NO_3]$	1:1	594
96–98	Methylammonium	$[4\text{-MeOC}_6H_4CH(CH_2NO_2)SO_3]$	1:1	437
82–84	Ethylammonium	$[N(SO_2Me)_2]$	1:1	481
107–108	*cyclo*-Hexylammonium	$Cl/[AlCl_4]$	2:(1,1)	
109–112	Benzylammonium	$Cl/[AlCl_4]$	2:(1,1)	
95	*rac*-(1-Phenylethyl)ammonium	$[ClCH_2CO_2]$	1:1	458
106–108	{1-(Hydroxymethyl)propyl}ammonium	$[(4\text{-NO}_2)C_6H_4CO_2]$	1:1	447
80	Dimethylammonium	$[NO_3]$	1:1	557
29	Dimethylammonium	$[CO_2(NMe_2)]$	1:1	513
75–77	Methylethylammonium	$[N(SO_2Me)_2]$	1:1	471

TABLE 11.6 (*Continued*)

m.pt./°C			$p:q$	U_L/ kJ mol^{-1}
71–73	Diethylammonium	$[N(SO_2Me)_2]$	1:1	454
95–96	Benzyl(ethyl)ammonium	$[NO_3]$	1:1	477
71	Dibutylammonium	$[CF_3C(O)CHC(O)$ (2-Thienyl)]	1:1	408
58	Bis(2-azidoethyl) ammonium	$[N(CN)(NO_2)]$	1:1	462
73	*N*-Methyl-*N*-(2-nitroxyethyl) ammonium	$[NO_3]$	1:1	507
0	Trimethylammonium	$[F(HF)_2]$	1:1	534
79–81	Trimethylammonium	$[N(SO_2CF_3)_2]$	1:1	445
−27	Triethylammonium	$[F(HF)_2]$	1:1	479
92–93	Benzyldimethylammonium	$[C_6H_2\{2,4,6\text{-}(NO_2)_3\}O]$ Picrate	1:1	421
104–107	Ethyldi-*iso*-propylammonium	$[\{C_6H_5)_3C\}P(BH_3)(O)OH]$	1:1	377
68–69	Tris(2-methoxyethyl) ammonium	Cl	1:1	445
80	Tris(2-hydroxyethyl) ammonium	$[NO_3]$	1:1	475
94	Tris(2-hydroxyethyl) ammonium	$[SeO_2(OH)]$	1:1	467
73	Tris(2-hydroxyethyl) ammonium	$[(4\text{-}ClC_6H_4)SCH_2CO_2]$	1:1	421
81–84	Tris(2-hydroxyethyl) ammonium	$[2\text{-}HC(O)C_6H_4CO_2]$	1:1	429
64–65	Tridodecylammonium	$[SO_3(OH)]$	1:1	334
−11	*N,N,N'*-Triethylethylene-diammonium	$[CF_3C(O)CHC(O)CF_3]$	1:1	413
93–94	(2-Trimethylsilylethyl) dimethylamine)$_2$H	$[B\{(3,5\text{-}(CF_3)_2)C_6H_3\}_4]$	1:1	315
29	$[(Me_3NH)_2Cl]$	$[HCl_2]$	1:1	448
70	$[\{NMe_3CH_2CO_2\}_2H]$	$[N(SO_2Me)_2]$	1:1	
104.5	Hexamethylenediammonium	$[I_5]$	1:2	1225
107–108	3,6-Dioxaoctane-1, 8-diammonium	$[CCl_3CO_2]$	1:2	1374
93–94	*N*-Methyl-1, 3-propylenediammonium	$[TlBr_5]$	2:2	
	Phosphonium			
90	*iso*-Propylphosphonium	$[Ge^{II}Cl_3]$	1:1	481
40–42	*cyclo*-Hexylphosphonium	$[Ge^{II}Cl_3]$	1:1	458
70	Triphenylphosphonium	$[N(SO_2F)_2]$	1:1	402
	Oxonium			
34.5	Oxonium	$[CF_3SO_3]$	1:1	553/561
112	Oxonium	$[O_3SCH_2CH_2SO_3]$	2:1	1497

(*Continued*)

TABLE 11.6 (*Continued*)

m.pt./°C			$p:q$	U_L/ kJ mol^{-1}
−6	Oxonium monohydrate	$[CF_3SO_3]$	1:1	525/530
−18.5	Oxonium dihydrate	$[NO_3]$	1:1	580
−2.5	1,1-Dihydroxyethylium	$[SO_3(OH)]$	1:1	545
52.5	1,1-Dihydroxyethylium	$[SO_3F]$	1:1	547
35	1,1-Dihydroxyethylium	$[CF_3SO_3]$	1:1	507
98	([2.2.2]Cryptand)H_2	$[I_8]$	2:2	
	Others			
75	Azidoformamidinium	$[ClO_4]$	1:1	528
88	2-(Hydroxoiodinium) nitrobenzene	$[CF_3CO_2]$	1:1	466
80–83	Bis(2-pyridone)hydrogen	$[N(SO_2Me)_2]$	1:1	423
94	5-Hydroxy-1-methyl-3, 4-dihydro-2*H*-pyrrolium	Cl	1:1	525
110	Piperazinediium	$[C_5H_{11}CO_2]$	1:2	1422
97	Piperazinediium	$[C_6H_{13}CO_2]$	1:2	1357
96	Piperazinediium	$[C_9H_{19}CO_2]$	1:2	1280
96	Piperazinediium	$[C_{11}H_{23}CO_2]$	1:2	1240
96	Piperazinediium	$[C_{13}H_{27}CO_2]$	1:2	1207
97	Piperazinediium	$[C_{15}H_{31}CO_2]$	1:2	1178
		Protic anions		
51	1-Ethyl-3-methylimidazolium	$[HF_2]$	1:1	507
−1	Pyridinium	$[HF_2]$	1:1	548
45	Tris(dimethylamino) oxosulfonium	$[HF_2]$	1:1	469
10(dec)	1,3-Dimethylimidazolium	$[F(HF)_2]$	1:1	514
−17	Pyridinium	$[F(HF)_2]$	1:1	528
0	Trimethylammonium	$[F(HF)_2]$	1:1	534
−27	Triethylammonium	$[F(HF)_2]$	1:1	479
110	Tetramethylammonium	$[F(HF)_2]$	1:1	519
Ref DSC	1,3-Dimethylimidazolium	$[F(HF)_3]$	1:1	495
60	Pyridinium	$[HCl_2]$	1:1	518
29	$[(Me_3NH)_2Cl]$	$[HCl_2]$	1:1	448
109	Potassium	$[(HCO_2)_2H]$	1:1	576
71	1-Aza-8-azoniabicyclo[6.6.3] heptadeca-4,11-diyne	$[(CF_3CO_2)_2H]$	1:1	394
72	1-Aza-8-azoniabicyclo[6.6.4] octadeca-4,11-diyne	$[(CF_3CO_2)_2H]$	1:1	390
75	1-Aza-8-azoniabicyclo[6.6.5] octadeca-4,11-diyne	$[(CF_3CO_2)_2H]$	1:1	385
108–110	2,6-Bis(2-pyridinio)pyridine	$[(CF_3CO_2)_2H]$	1:1	1323
103	Tetrabutylammonium	$[(PhCO_2)_2H]$	1:1	361
80–82	Tetramethylstibonium	$[(PhCO_2)_2H]$	1:1	399
108–110	1-Benzylimidazolium	$[PhP(O)_2(OH)]$	1:1	424
105–108	3-Azidosulfonylimidazolium	$[SO_3(OH)]$	1:1	481

TABLE 11.6 (*Continued*)

m.pt./°C			*p*:*q*	U_L/ kJ mol^{-1}
78	1-Butyl-2,3-dimethylimidazolium	$[SO_3(OH)]$	1:1	450
63	1-*iso*-Propyl-3-methyl-4-diphenyl-phosphorylimidazolium	$[SO_3(OH)]$	1:1	399
105–110	4-Dimethylamino-1-methyl-1,2,4-triazolium	$[SO_3(OH)]$	1:1	480
64–65	Tridodecylammonium	$[SO_3(OH)]$	1:1	334
−2.5	1,1-Dihydroxyethylium	$[SO_3(OH)]$	1:1	545
94	Tris(2-hydroxyethyl) ammonium	$[SeO_2(OH)]$	1:1	467
105	1,2-Dimethylimidazolium	$[HO_2C(CH_2)_2CO_2]$	1:1	469
105	2-Methylimidazolium	$[HO_2C(CH_2)_3CO_2]$	1:1	467
106	4-Methylimidazolium	$[HO_2C(CH_2)_3CO_2]$	1:1	472
102	1,2-Dimethylimidazolium	$[HO_2C(CH_2)_3CO_2]$	1:1	464
101.5–103	Imidazolium	$[HO_2C(CH_2)_4CO_2]$	1:1	465
60	2-Methylimidazolium	$[HO_2C(CH_2)_6CO_2]$	1:1	441
85	4-Methylimidazolium	$[HO_2C(CH_2)_6CO_2]$	1:1	441
41-43	2-Ethylimidazolium	$[HOC(O)CH(OH)CH(OH)CO_2]$	1:1	462
98–101	1-Methylimidazolium	$[HOC(O)CH(OH)CH(OH)CO_2]$	1:1	470/471
104–107	Ethyldi-*iso*-propylammonium	$[\{(C_6H_5)_3C\}P(BH_3)(O)OH]$	1:1	377

component acid and base may, in principle, be separated, purified and the salt subsequently reformed, a route to recovery and recycle not generally available for fully alkylated ionic liquids. Indeed, the ionic liquid, 1-methylimidazolium chloride, a vital component of the BASIL process, is attractive for this very reason.

Table 11.6 brings together protic salts for which a crystal structure is known. Further discussion may be found in later sections that focus on specific cation types. Crystalline inclusion compounds, co-crystals containing neutral components or solvates (see, however, Sections 11.4.1 and 11.4.4) and salts that decompose before, or on, melting are generally excluded. Since, to degrees dependent on the components and conditions, dissociation into constituent acid and base is possible on melting (or even sublimation), it may be necessary to exclude some salts, though this would require the composition of the liquid formed on melting to be known, which is not usually the case. The convention has been proposed that dissociation that occurs in the melt to an extent greater than 1% would rule a material out as an ionic liquid [610]. There has also been much discussion of the ΔpK_a (the difference between the pK_a of the acid [HA] and protonated base $[HB]^+$) needed to bring about complete protonation of a base by a protic acid. Values between 10 and 4 have been suggested.

TABLE 11.7 Wholly Inorganic Ionic Liquids

m.pt./°C			p:q	U_L/ kJ mol^{-1}
99	1-Amino-1,2,3-triazolium	5-Nitrotetrazolate	1:1	508
70	1*H*-Tetrazolium	$[N(NO_2)_2]$	1:1	535
55	1*H*-Tetrazolium	$[ClO_4]$	1:1	556
85	5-Aminotetrazolium	$[N(NO_2)_2]$	1:1	526
72	1,5-Diaminotetrazolium	$[N(NO_2)_2]$	1:1	509
97–98	1,5-Diaminotetrazolium	$[ClO_4]$	1:1	523
70.5	Hydrazinium	$[N_3]$	1:1	630
80	Hydrazinium	$[N(NO_2)_2]$	1:1	573
97–97.5	Hydrazinium	$[N(CN)_2]$	1:1	593
62.5/70.5	Hydrazinium	$[NO_3]$	1:1	625
48	Hydrazinium	Tetrazolate	1:1	576
96–100	Hydroxylammonium	F	1:1	726
18–23	Hydroxylammonium	$[N(NO_2)_2]$	Complex	
45/65	Guanidinium	$[N(CN)_2]$	1:1	542/542
91–94	Aminoguanidinium	$[N(NO_2)_2]$	1:1	517/518
91	Aminoguanidinium	$[N(CN)(NO_2)]$	1:1	522
70 (dec)	Aminoguanidinium	$[N(NO_2)(SF_5)]$	1:1	491
72	Aminoguanidinium	$[ClO_4]$	1:1	532
100	Aminoguanidinium	5-Azidotetrazolate	1:1	504
80	*N,N′*-Diaminoguanidinium	$[C(NO_2)_3]$	1:1	484
108	*N,N′*-Diaminoguanidinium	5-Nitrotetrazolate	1:1	497
80	*N,N′,N″*-Triaminoguanidinium	$[N(NO_2)_2]$	1:1	495
84	*N,N′,N″*-Triaminoguanidinium	$[C(NO_2)_3]$	1:1	476
96	*N,N′,N″*-Triaminoguanidinium	5-Nitrotetrazolate	1:1	488
75	Azidoformamidinium	$[ClO_4]$	1:1	528
86	Diaminouronium	$[NO_3]$	1:1	553
92	Ammonium	$[N(NO_2)_2]$	1:1	590
93(dec)	Ammonium	$[N(NO_2)(SF_5)]$	1:1	540
−18.5	Oxonium dihydrate	$[NO_3]$	1:1	580
63	Potassium	$[N(SO_2F)_2]$	1:1	543/547
102	Potassium	$[N(SO_2F)_2]$	1:1	541/546
102–106	Rubidium	$[N(NO_2)_2]$	1:1	581
85–87	Caesium	$[N(NO_2)_2]$	Catenate	

Arguments based solely on pK_a differences may be thought questionable as such parameters have usually been estimated in aqueous solution. Examples of protic salts have been retained in the tabulation as, more often than not, the nature of the liquid obtained on melting remains an open question in the absence of experimental data, particularly as more is to be gained by a consideration of a body of materials that includes them than one that does not.

In principle, when a Brønsted acid, HA, and Brønsted base, B, are mixed and attempts made to obtain a solid product from the mixture, three basic outcomes can be expected: (i) recrystallisation (i.e. no reaction); (ii) molecular co-crystallisation

TABLE 11.8 Group 1 Metal-Containing Ionic Liquids

m.pt./°C			$p:q$	U_L/ kJ mol^{-1}
107	Tris(dimethylamino)sulfonium	$[Li(C_5H_5)_2]$	1:1	408
65	Lithium/1,3-dimethylimidazolium	$[N(SO_2CF_3)_2]$	(1,1):2	
14–16	Tetrakis(ethanenitrile)lithium	$[P_2Se_8]$	2:1	842
95	Bis(tetramethylene sulfone) lithium	$[BF_2(ox)]$	1:1(?)	424
109	Tetra(pyridine)lithium	$[P_2Se_8]$	2:1	980
54–56	$[Li_2(H_2NNHMe)_3]$	$[BH_4]$	1:2	1566
37	Sodium/1-ethyl-3-methylimidazolium	$[AlCl_4]$	(1,1):2	
56	(1,2-Dimethoxyethane)sodium	$[B\{OCH(CF_3)_2\}_4]$	1:1	368
92–94	(Benzo-18-crown-6)bis(pyridine) sodium	$[BH_4]$	1:1	374
63	Potassium	$[N(SO_2F)_2]$	1:1	543/547
102	Potassium	$[N(SO_2F)_2]$	1:1	541/546
109	Potassium	$[(HCO_2)_2H]$	1:1	576
53–56	Bis(*N*,*N*,*N*′,*N*″,*N*″-pentamethyldiethylenetriamine) potassium	$[AlMe_4]$	1:1	355
91	(Benzo-18-crown-6)potassium	$[I_5]/[I_7]$	2:(1,1)	
90	([2.2.2]Cryptand)potassium	$[I_{12}]$	2:1	653
108	([2.2.2]Cryptand)potassium	$[SiPh_2F_3]$	1:1	354
106	([2.2.2]Cryptand)potassium	$[SiPh_3F_2]$	1:1	345
102–106	Rubidium	$[N(NO_2)_2]$	1:1	581
85–87	Caesium	$[N(NO_2)_2]$	Catenate	
83–85	Caesium	$[Al\{OCH(CF_3)_2\}_4]$	1:1	379

$[(HA)_x(B)_y$, where x and y are integers] and (iii) salt formation, [BH][A], with proton transfer. Solid solutions may also result, as described very recently for mixtures of (*R*)- and (*S*)-ethanolammonium 3-chloromandelate (Fig. 11.6) [611], some of which melt at less than 100°C and can be considered as ionic liquids (though the material for which a single-crystal structure was reported, pure (*R*)-ethanolammonium 3-chloromandelate (CSD refcode GEPNEW) melts at 122°C). Studies of associated phenomena are of both technical importance in the design and control of the characteristics of active pharmaceutical intermediates, as well as of fundamental interest in the wider topic of crystal engineering and supramolecular synthesis.

There are many reports of crystal structures of products formed from the reaction of a heterocyclic base and a carboxylic acid for which crystallography is critical, both to determine whether or not (or, in some systems, the degree to which [554, 612–614]) proton transfer has occurred, as well as to characterise the nature of intermolecular or interionic bonding. Characterisation of hydrogen-bonding networks in such systems, and in ionic liquids more generally, involves

TABLE 11.9 Coordination Complex Ionic Liquids

m.pt./°C			p:q	U_L/ kJ mol^{-1}
	Transition metal cation			
65	Bis(ethanenitrile)copper(I)	$[N(SO_2CF_3)_2]$	1:1	
66	Tetrakis(ethanenitrile)copper(I)	$[N(SO_2CF_3)_2]$	1:1	
27	Tetrakis(1-dodecylimidazole) copper(I)	$[PF_6]$	1:1	303
89	Tetrakis(1-methylimidazole) copper(II)	$[N(SO_2CF_3)_2]$	1:2	1206
74	Tetrakis(1-ethylimidazole)copper(II)	$[N(SO_2CF_3)_2]$	1:2	1156
46	Tetrakis(1-butylimidazole)copper(II)	$[N(SO_2CF_3)_2]$	1:2	1118
52	Hexakis(1-butylimidazole)copper(II)	$[N(SO_2CF_3)_2]$	1:2	1137
90	(Ethanenitrile)silver(I)	$[N(SO_2CF_3)_2]$	1:1	455
30	(1-Butylimidazole) (1-methylimidazole)silver(I)	$[N(SO_2CF_3)_2]$	1:1	387
27	(1-Butylimidazole) (1-ethylimidazole)silver(I)	$[N(SO_2CF_3)_2]$	1:1	388
87	Bis(1-methylimidazole)silver(I)	$[N(SO_2CF_3)_2]$	1:1	407
65	Bis(1-ethylimidazole)silver(I)	$[N(SO_2CF_3)_2]$	1:1	398
42	Bis(1-butylimidazole)silver(I)	$[N(SO_2CF_3)_2]$	1:1	378
100–104	[Bis(1-ethyl-3-methylimidazol-2-ylidene)silver(I)]	$[B(CN)_4]$	1:1	373
102–104	$[Ti(\eta^5\text{-}C_5H_5)Cl(18\text{-Crown-6})]$	$[AlMe_2Cl_2]$	1:1	398
88	$[Dy_2(NMe_3CH_2CO_2)_8(H_2O)_4]$	$[N(SO_2CF_3)_2]$	1:6	
	Transition metal cation	**Transition metal anion**		
66	Bis(1-dodecylimidazole)copper(I)	$[Cu^IBr_2]$	1:1	350
18	Tetrakis(ethanenitrile)silver(I)	$[Ag\{N(SO_2CF_3)_2\}_3]$	2:1	750
68–72	Tetrakis(ethanenitrile)silver(I)	$[Ag_3\{N(NO_2)_2\}_4]$	Catenate	
82–85	Bis(η^2-dimethyldisulfide) tetrachloroniobium(V)	$[NbCl_6]$	1:1	390
		Transition metal anion		
40–41	1-Hexadecylpyridinium	$[Co(CO)_4]$	1:1	371
67	1-Hexyl-3-methylimidazolium	$[Co^{II}Br_3(C_9H_7N)]$	1:1	387
105	1-Butyl-3-methylimidazolium	$[Co^{II}Br_3(C_9H_7N)]$	1:1	397
38	1-Tetradecylpyridinium	$[Co^{II}(hfac)_3]$	1:1	337
108	Tetrabutylphosphonium	$[PtBr_3(C_2H_4)]$	1:1	379
	Main group metal cation			
108–110	[(*N*,*N*,*N′*,*N″*,*N″*-Pentamethyldiethylenetriamine) $AlMe_2$]	$[AlMe_2Cl_2]$	1:1	394
102–105	Aquo(pyridine-*N*-oxide) trimethyltin(IV)	$[N(SO_2Me)_2]$	1:1	415
33	Bis(*N*,*N′*dimethylethyleneurea-*O*) trimethyltin(IV)	$[N(SO_2F)_2]$	1:1	386
66	Bis(dimethylsulfoxide-*O*) trimethyltin(IV)	$[N(SO_2F)_2]$	1:1	406
75–80	$[Tl_3F_2Al\{OCH(CF_3)_2\}_3]$	$[Al\{OCH(CF_3)_2\}_4]$	1:1	330

TABLE 11.10 Lanthanide and Actinide-Containing Ionic Liquids

m.pt./°C			p:q	U_L/kJ mol^{-1}
38	1-Butyl-3-methylimidazolium	$[La(NCS)_7(H_2O)]$	4:1	
59	Tetrabutylammonium	$[Sm\{C(CN)_2NO\}_6]$	3:1	
68	Tetrabutylammonium	$[Ce\{C(CN)_2NO\}_6]$	3:1	
58	1-Ethyl-3-methylimidazolium	$[La\{C(CN)_2NO\}_6]$	3:1	
63	1-Ethyl-2,3-dimethylimidazolium	$[Pr\{C(CN)_2NO\}_6]$	3:1	
107	1-Methyl-1-propylpyrrolidinium	$[Yb\{N(SO_2CF_3)_2\}_4]$	2:1	1089
68	1-Butyl-3-methylimidazolium	$[Eu\{N(SO_2CF_3)_2\}_4]$	1:1	327
81	1-Methyl-3-propylimidazolium	$[Eu\{N(SO_2CF_3)_2\}_4]$	1:1	330
92	1-Butyl-1-methylpyrrolidinium	$[Eu\{N(SO_2CF_3)_2\}_5]$	2:1	670
78	Tetrabutylammonium	$[Y(BH_4)_4]$	1:1	376
103	1-Ethyl-3-methylimidazolium	$[U^{VI}O_2(NCS)_5]$	3:1	
88	$[Dy_2(NMe_3CH_2CO_2)_8(H_2O)_4]$	$[N(SO_2CF_3)_2]$	1:6	

topological analysis, such as that developed by Etter [561] and others [562]. A full discussion lies beyond the scope of this chapter. While most salts are high melting, systems in which salt formation and a melting point below 100°C are reported are included in Table 11.2. These include salts from 2-aminopyridine and alkylimidazoles with carboxylic and dicarboxylic acids [450, 473]. Other related materials are included in Table 11.3 for those that melt in the range 100–110°C.

Protic ionic liquids are listed in Table 11.6 generally in order of cation number. In some instances, where more than one structural feature is present in an ion, the salt may appear twice. For instance, the $[CF_3SO_3]^-$ salt of the azaheterocycle, *N*-amino-1-azoniacycloheptane, is also listed with hydrazinium compounds. The m.pt. provides a link to further information in Tables 11.1, 11.2 and 11.3, including crystallographic data, its source and, where available, the CSD ref-code or CCDC deposition number. Table 11.6 (as with all the tables dealing with selected categories of ionic liquids) includes values of the lattice potential energies, $U_L(V)$ and $U_L(D)$, derived from crystal data for salts with p:q of 1:1, 1:2 and 2:1, listed in Tables 11.1, 11.2 and 11.3 and calculated according to the methods described in Section 11.2.4. The data are used in the discussion to three significant figures only, in light of the assumptions made in their derivation.

11.4.1 Salts From Carboxylic Acids

Non-protic carboxylate salts are dealt with in Section 11.14. Low-melting salts of the lower carboxylic acids are well known ($[NH_4][CH_3CO_2]$, for instance, melts at 114°C), but because of a tendency towards glass formation, relatively few have been characterised crystallographically. $[NH_4]$[*trans*-$CH_3CH{=}CHCO_2$], with a m.pt. of 84°C (U_L 546 kJ mol^{-1}), is the lowest-melting ammonium carboxylate, structurally characterised by single-crystal X-ray

TABLE 11.11 Metalloanionic Ionic Liquids

m.pt./°C			$p:q$	U_L/kJ mol^{-1}
		Anion containing: $[MX_2]^-$		
99–100	Tetrapropylammonium	$[Cu^{I}Cl_2]$	1:1	416
99–100	Tetrapropylammonium	$[Cu^{I}Br_2]$	1:1	413
73	1-Ethyl-3-methylimidazolium	$[Ag(CN)_2]$	1:1	466
64	1-Ethyl-3-methylimidazolium	$[Au(CN)_2]$	1:1	469
65–69	Tetrabutylammonium	$[Au(CN)_2]$	1:1	385
103	1-Methylimidazolium	$[Au(CN)_2]$	1:1	
		Anion containing: $[MX_3]^-$		
40–42	*cyclo*-Hexylphosphonium	$[Ge^{II}Cl_3]$	1:1	458
90	*iso*-Propylphosphonium	$[Ge^{II}Cl_3]$	1:1	481
72	Tetramethylformamidinium	$[Ge^{II}Cl_3]$	1:1	465
65	Tetramethylformamidinium	$[Sn^{II}Cl_3]$	1:1	462
ca 104	1-Ethyl-3-methylimidazolium	$[Sn^{II}Cl_3]$	1:1	463
69/67	1-Methyl-3-propylimidazolium	$[HgCl_3]$	1:1	445
94/92	1-Butyl-3-methylimidazolium	$[HgCl_3]$	1:1	442
39.5/41	1-Methyl-3-propylimidazolium	$[HgBr_3]$	1:1	441
58/57	1-Butyl-3-methylimidazolium	$[HgBr_3]$	1:1	435
		Anion containing: $[M^{II}X_4]^{2-}$		
5	1-Butyl-3-methylimidazolium	$[Sn^{II}Cl_4]$	2:1	1012
60	1-Butyl-3-methylimidazolium	$[ZnCl_4]$	2:1	1010
105	1-Cyanomethyl-3-methylimidazolium	$[CdCl_4]$	2:1	1082
63	1-Butyl-3-methylimidazolium	$[Mn^{II}Cl_4]$	2:1	1003
105–106	1-Methylmorpholinium	$[Mn^{II}Cl_4]$	2:1	1125
58	1-Butyl-3-methylimidazolium	$[Fe^{II}Cl_4]$	2:1	1007
86	1-Ethyl-3-methylimidazolium	$[Fe^{II}Cl_4]$	2:1	1078
62	1-Butyl-3-methylimidazolium	$[Co^{II}Cl_4]$	2:1	1010
100–102	1-Ethyl-3-methylimidazolium	$[Co^{II}Cl_4]$	2:1	1082
105–108	1-Methylmorpholinium	$[Co^{II}Cl_4]$	2:1	1129
45	1-Butyl-3-methylimidazolium	$[Co^{II}Br_4]$	2:1	994
56	1-Butyl-3-methylimidazolium	$[Ni^{II}Cl_4]$	2:1	1012
92–93	1-Ethyl-3-methylimidazolium	$[Ni^{II}Cl_4]$	2:1	1085
88–90	1-Methylmorpholinium	$[Cu^{II}Cl_4]$	2:1	1127
92–98	1-Methylmorpholinium	$[Cu^{II}Br_4]$	2:1	1100
110	4-Methylpyridinium	$[Cu^{II}Br_4]$	2:1	1099
103	1-Ethoxymethyl-3-methylimidazolium	$[Pd^{II}Cl_4]$	2:1	1036
107	1-(2-Methoxyethyl)-3-methylimidazolium	$[Pd^{II}Cl_4]$	2:1	1052
101	1-Tetradecylpyridinium	$[Pd^{II}I_4]$	2:1	769
99	1-Butyl-3-methylimidazolium	$[Pt^{II}Cl_4]$	2:1	1037

TABLE 11.11 (*Continued*)

m.pt./°C			p:q	U_L/kJ mol^{-1}
		Anion containing: $[EX_4]^-$or$[M^{III}X_4]^-$		
83–110	2,6-Dichloropyridinium	$[Sb^{III}Cl_4]$	1:1	434
92	Tetraphenylphosphonium	$[As(N_3)_4]$	1:1	374
10	1-Ethyl-3-methylimidazolium	$[Fe^{III}Cl_4]$	1:1	444/449
70	1,3-Dibutylbenzimidazolium	$[Fe^{III}Cl_4]$	1:1	395
81	1-Butyl-1-methylpyrrolidinium	$[Fe^{III}Cl_4]$	1:1	427
110	Tributylmethylammonium	$[Fe^{III}Cl_4]$	1:1	400
38	1-Butyl-2,3-dimethylimidazolium	$[Fe^{III}Cl_4]/[Fe^{II}Cl_4]$	3:(1,2)	
58	1-Ethyl-3-methylimidazolium	$[Au^{III}Cl_4]$	1:1	448
50	1-Butyl-3-methylimidazolium	$[Au^{III}Cl_4]$	1:1	431
		Anion containing: $[EX_5]^-$		
102–103	Tetrapropylammonium	$[SiPhF_4]$	1:1	395
62–64	Tetrabutylammonium	$[SiMePhF_3]$	1:1	374
108	([2.2.2]Cryptand)potassium	$[SiPh_2F_3]$	1:1	354
67–73	2-Diethylamino-1,3-dimethylimidazolinium	$[SiMe_3F_2]$	1:1	419
92	Tris(dimethylamino)sulfonium	$[SiMe_3F_2]$	1:1	422
57	Tris(dimethylamino)oxosulfonium	$[SiMe_3F_2]$	1:1	426
106	([2.2.2]Cryptand)potassium	$[SiPh_3F_2]$	1:1	345
91–92	Tris(dimethylamino)sulfonium	$[SiPh_2(1\text{-}C_{10}H_7)F_2]$	1:1	369
64–66	Tetrabutylphosphonium	$[SnPh_2Cl_3]$	1:1	360
95–96	Tetrabutylphosphonium	$[Sn(2\text{-}MeC_6H_4)_2Cl_3]$	1:1	352
93–94	*N*-Methyl-1,3-propylenediammonium	$[TlBr_5]$	2:2	
		Anion containing: $[MX_6]^-$		
109	1-(2-Hydroxyethyl)piperidinium	$[Si(ox)_3]$	2:1	988
97	1-(2-Hydroxyethyl)piperidinium	$[Ge(ox)_3]$	2:1	991
41	1,3-Didodecylimidazolium	$[SbF_6]$	1:1	358
44.5	1-Butyl-2,3-dimethylimidazolium	$[SbF_6]$	1:1	432
108	5-Ethoxy-l-ethyl-3,4-dihydro-2-oxo-2*H*-pyrrolium	$[SbCl_6]$	1:1	413
110–112	*β*,*β*-Diethoxyethenediazonium	$[SbCl_6]$	1:1	420
89–90	2-Hydroxyhomotropylium	$[SbCl_6]$	1:1	432
103–104	1-Ethoxyhomotropylium	$[SbCl_6]$	1:1	417
104–106	Tetraphenylphosphonium	$[Sb(N_3)_6]$	1:1	362
–1	1-Ethyl-3-methylimidazolium	$[NbF_6]$	1:1	462
83–85	1-Butyl-3-methylpyridinium	$[NbCl_6]$	1:1	413
89–91	1-Butyl-4-methylpyridinium	$[NbCl_6]$	1:1	412
82–85	Bis(η^2-dimethyldisulfide) tetrachloroniobium(V)	$[NbCl_6]$	1:1	390
2	1-Ethyl-3-methylimidazolium	$[TaF_6]$	1:1	462
103	1-Butyl-4-methylpyridinium	$[TaCl_6]$	1:1	412

(*Continued*)

TABLE 11.11 (*Continued*)

m.pt./°C			$p:q$	U_L/kJ mol^{-1}
		Anion containing: miscellaneous		
105–118	2-Fluoropyridinium	$[Sb^{III}_2Br_9]$	3:1	
65–66	2-Benzylpyridinium	$[Sb^{III}_2Cl_{10}]$	4:2	
84	Ethyltriphenylphosphonium	$[Se_4Br_{14}]$	2:1	709
80	1,3-Bis(dodecyl)benzimidazolium	$[Cd_2Cl_6]$	2:1	662
102	Butyltriphenylphosphonium	$[Hg^{II}_2Cl_6]$	2:1	730
74	1-Tetradecylpyridinium	$[Pd^{II}_2Cl_6]$	2:1	737
82.5	1-Hexadecylpyridinium	$[Pd^{II}_2Cl_6]$	2:1	716
75	1-Ethyl-3-methylimidazolium	$[V^{V}(O)F_4]$	1:1	475
99	1-Ethyl-3-methylimidazolium	$[V^{IV}(O)Cl_4]$	1:1	1064
76	1-Ethyl-3-methylimidazolium	$[(Ta^{V}Cl_5)_2O]$	2:1	939
74	Butyltriphenylphosphonium	$[Bi_2I_8(Me_2CO)_2]$	2:1	670
		Anion -containing: $[Cu_nX_m]^{y-}$ (X = Cl, Br)		
99–100	Tetrapropylammonium	$[Cu^{I}Cl_2]$	1:1	416
26	1-Dodecyl-3-methylimidazolium	$[Cu^{I}Cl_2]$/$[Cu^{I}Cl_3]$	3:(1,2)	
66	Bis(1-dodecylimidazole)copper(I)	$[Cu^{I}Br_2]$	1:1	350
99–100	Tetrapropylammonium	$[Cu^{I}Br_2]$	1:1	413
23	1-Butyl-3-methylimidazolium	$[Cu^{II}Cl_4]$	2:1	1024
88–90	1-Methylmorpholinium	$[Cu^{II}Cl_4]$	2:1	1127
76	2-Hydroxyethyl(trimethyl) ammonium (cholinium)	$[Cu^{II}Cl_4]$/Cl	3:(1,2)	
92–98	1-Methylmorpholinium	$[Cu^{II}Br_4]$	2:1	1100
110	4-Methylpyridinium	$[Cu^{II}Br_4]$	2:1	1099
107–108	Methyltriphenylphosphonium	$[Cu^{I}_2Br_4]$	2:1	819
70	1-Butyl-3-methylimidazolium	$[Cu^{II}_3Cl_8]$	2:1	927
27	1-Dodecyl-3-methylimidazolium	$[Cu^{I}_4Cl_8]$	4:1	
95–96	Tributylmethylammonium	$[Cu^{I}_5Br_7]$	2:1	802

diffraction [344]. Other carboxylates for which single-crystal studies have been undertaken, and having an established ionic formulation, include ethanoate (with hydrazinium [444], m.pt. at 97–98°C, U_L 590 kJ mol^{-1}; tris(dimethylamino) iminiumphosphorane, m.pt. 88°C, U_L 436 kJ mol^{-1} [368] and 2-amino-3-methylpyridinium, m.pt. 97–99°C, U_L 493 kJ mol^{-1}) [445], monochloroethanoate (with 1-phenylethylammonium, m.pt. 95°C, U_L 458 kJ mol^{-1}) [420], propanoate (with 1,1,3,3-tetramethylguanidinium, m.pt. 62°C, U_L 465 kJ mol^{-1}) [132], benzoate (with 3-hydroxypyridinium, m.pt. 82–85°C, U_L 473 kJ mol^{-1} [338]) and trifluoroethanoate (with 4-cyanopyridinium, m.pt. 74°C, U_L 484 kJ mol^{-1} [284]; pyridinium, m.pt. 82°C, U_L 501 kJ mol^{-1} [336] and from the *O*-protonation by trifluoroethanoic acid of 2-(oxoiodo)nitrobenzene, m.pt. 88°C, U_L 466 kJ mol^{-1}) [369]. An X-ray crystallographic investigation of complexes formed from 2,4,6-trimethylpyridine and a series of benzoic acids [367] reveals

TABLE 11.12 Energetic Salts

m.pt./°C			$p:q$	U_L/ kJ mol^{-1}
	1,2,3- and 1,2,4-Triazolium			
99	1-Amino-1,2,3-triazolium	5-Nitrotetrazolate	1:1	508
50	1-Amino-3-methyl-1,2,3-triazolium	[N_3]	1:1	536
86–88	1-Amino-3-methyl-1,2,3-triazolium	[NO_3]	1:1	531
93	1-Amino-3-methyl-1,2,3-triazolium	5-Nitrotetrazolate	1:1	491
105–110	1,2,4-Triazolium	[$N(SO_2Me)_2$]	1:1	478
93	4-Amino-1,2,4-triazolium	4-(Carboxylatomethyl)-5-nitroiminotetrazolate	2:1	1200
69	1-Propyl-1,2,4-triazolium	5-Nitroiminotetrazolate	1:1	456
110	1,4-Dimethyl-2-*H*-1,2,4-triazolium	[I_3]	1:1	465
98	1,4-Dimethyl-3-azido-1,2,4-triazolium	[NO_3]	1:1	498
86	1-Methyl-4-amino-1,2,4-triazolium	[ClO_4]	1:1	509
105–110	4-Dimethylamino-1-methyl-1,2,4-triazolium	[$SO_3(OH)$]	1:1	480
63-67	1-Ethyl-4-amino-1,2,4-triazolium	Br	1:1	517
63	1-Propyl-4-amino-1,2,4-triazolium	Br	1:1	500
92	1-*iso*-Propyl-4-amino-1,2,4-triazolium	Br	1:1	498
66	1-*iso*-Propyl-4-amino-1,2,4-triazolium	[NO_3]	1:1	493
56	1-(*cyclo*-Propylmethyl)-4-amino-1,2,4-triazolium	[NO_3]	1:1	486
76	1-Hexyl-4-amino-1,2,4-triazolium	Br	1:1	459
94	1-Heptyl-4-amino-1,2,4-triazolium	Br	1:1	448
105–106	4-Amino-1,3,5-trimethyl-1,2,4-triazolium	[$N(SO_2CF_3)_2$]	1:1	427
65–66	4-Amino-1-ethyl-3,5-dimethyl-1,2,4-triazolium	[$N(SO_2CF_3)_2$]	1:1	420
94–95	4-Amino-1-ethyl-3,5-dimethyl-1,2,4-triazolium	[PF_6]	1:1	455
109–110	4-Amino-3,5-diethyl-1-methyl-1,2,4-triazolium	[$N(SO_2CF_3)_2$]	1:1	412
106–107	4-Amino-1,3,5-triethyl-1,2,4-triazolium	[$N(SO_2CF_3)_2$]	1:1	408
	Tetrazolium			
70	1*H*-Tetrazolium	[$N(NO_2)_2$]	1:1	535
55	1*H*-Tetrazolium	[ClO_4]	1:1	556
85	5-Aminotetrazolium	[$N(NO_2)_2$]	1:1	526
90	5-Amino-1-methyltetrazolium	[$N(NO_2)_2$]	1:1	500
90	5-Amino-2-methyltetrazolium	[$N(NO_2)_2$]	1:1	501
58	5-Amino-1,3-dimethyltetrazolium	[$N(NO_2)_2$]	1:1	486
94	2-Amino-4,5-dimethyltetrazolium	[NO_3]	1:1	511
72	1,5-Diaminotetrazolium	[$N(NO_2)_2$]	1:1	509
97–98	1,5-Diaminotetrazolium	[ClO_4]	1:1	523
85–86	1,5-Diamino-4-methyltetrazolium	[$N(NO_2)_2$]	1:1	496
72	1,5-Diamino-4-methyltetrazolium	[$C(NO_2)(CN)_2$]	1:1	475

(Continued)

TABLE 11.12 (*Continued*)

m.pt./°C			$p:q$	U_L/ kJ mol^{-1}
	Ammonium			
92	Ammonium	$[N(NO_2)_2]$	1:1	590
81	Ammonium	1-(2-Nitratoethyl)-5-nitriminotetrazolate	1:1	486
110	Methylammonium	$[NO_3]$	1:1	594
58	Bis(2-azidoethyl)ammonium	$[N(CN)(NO_2)]$	1:1	462
73	*N*-Methyl-*N*-(2-nitroxyethyl) ammonium	$[NO_3]$	1:1	507
80	Tris(2-hydroxyethyl)ammonium	$[NO_3]$	1:1	475
104–106	Tetraethylammonium	3,5-Dinitro-1,2,4-triazolate	1:1	436
81	Tetrabutylammonium	2,4-Dinitroimidazolate	1:1	387
85	Tetrabutylammonium	4,5-Dinitroimidazolate	1:1	389
	Phosphonium			
68–70	Ethyltriphenylphosphonium	2,4-Dinitroimidazolate	1:1	391
91–93	Ethyltriphenylphosphonium	4,5-Dinitroimidazolate	1:1	394
	Others			
106	3,3-Dinitroazetidinium	$[(2\text{-OH})C_6H_4CO_2]$	1:1	449
102–103	3-*O*-Nitrato-3-ethylazetidinium	$[NO_3]$	1:1	479
59	*N*-Amino-1-azoniacycloheptane	$[N_3]$	1:1	493
99	2,2-Dimethyltriazanium	$[NO_3]$	1:1	540
75	Azidoformamidinium	$[ClO_4]$	1:1	528
86	Diaminouronium	$[NO_3]$	1:1	553
92–93	Benzyldimethylammonium	$[C_6H_2\{2,4,6\text{-}(NO_2)_3\}O]$	1:1	421
80–82	3-Hydroxy-1-*tert*-butyl-1,2-dihydropyrrolium	$[C_6H_2\{2,4,6\text{-}(NO_2)_3\}O]$	1:1	417
102–106	Rubidium	$[N(NO_2)_2]$	1:1	581
85–87	Caesium	$[N(NO_2)_2]$	1:1	567
		Picrate		
92–93	Benzyldimethylammonium	$[C_6H_2\{2,4,6\text{-}(NO_2)_3\}O]$	1:1	421
80–82	3-Hydroxy-1-*tert*-butyl-1,2-dihydropyrrolium	$[C_6H_2\{2,4,6\text{-}(NO_2)_3\}O]$	1:1	417
		Tetrazolate		
81	Tetrabutylammonium	2,4-Dinitroimidazolate	1:1	387
68–70	Ethyltriphenylphosphonium	2,4-Dinitroimidazolate	1:1	391
85	Tetrabutylammonium	4,5-Dinitroimidazolate	1:1	389
91–93	Ethyltriphenylphosphonium	4,5-Dinitroimidazolate	1:1	394
104–106	Tetraethylammonium	3,5-Dinitro-1,2,4-triazolate	1:1	436
48	Hydrazinium	Tetrazolate	1:1	576
99	1-Amino-1,2,3-triazolium	5-Nitrotetrazolate	1:1	508
93	1-Amino-3-methyl-1,2,3-triazolium	5-Nitrotetrazolate	1:1	491
108	*N,N'*-Diaminoguanidinium	5-Nitrotetrazolate	1:1	497

TABLE 11.12 (*Continued*)

m.pt./°C			p:q	U_L/ kJ mol^{-1}
96	*N,N′,N″*-Triaminoguanidinium	5-Nitrotetrazolate	1:1	488
100	Aminoguanidinium	5-Azidotetrazolate	1:1	504
69	1-Propyl-1,2,4-triazolium	5-Nitroiminotetrazolate	1:1	456
81	Ammonium	1-(2-Nitratoethyl)-5-nitriminotetrazolate	1:1	486
66	Hydroxylammonium	1-(2-Nitratoethyl)-5-nitriminotetrazolate	1:1	485
103	*N,N′*-Diaminoguanidinium	1-(2-Nitratoethyl)-5-nitriminotetrazolate	1:1	452
93	4-Amino-1,2,4-triazolium	4-(Carboxylatomethyl)-5-nitroiminotetrazolate	2:1	1200

that the product formed with benzoic acid itself (m.pt. 66°C; refcode GODNAO) is a hydrogen-bonded molecular complex, whereas that from 2-nitrobenzoic acid (m.pt. 88°C) may be categorised as a salt. Methanoic, benzoic and trifluoroethanoic acids give rise to crystalline salts of the anions $[(RCO_2)_2H]^-$ (R = H, Ph, or CF_3), with $[(PhCO_2)_2H]^-$ being found with tetramethylstibonium (m.pt. 80–82°C, U_L 399 kJ mol^{-1}) [327] and tetrabutylammonium (m.pt. 103°C, U_L 361 kJ mol^{-1}) [479]. Potassium hydrogen dimethanoate, $K[(HCO_2)_2H]$, melts at 109°C, with U_L = 576 kJ mol^{-1} [528]. Diprotonated piperazinium is found in a series of crystalline salts with $[RCO_2]^-$, R = C_6H_{13}, C_9H_{19}, $C_{10}H_{21}$, $C_{13}H_{27}$ or $C_{15}H_{21}$, all with m.pt. in the range 96–98°C [428–430, 439]. The effect of ring size in a series of 'inside'-protonated bicyclic 1,8-diazabicyclo[6.6.*n*]alka-4,11-diynes (n = 3–5, as their $[(CF_3CO_2)_2H]$ salts; Fig. 11.7) is discussed in Section 11.19.5 [266]. Salts of the formally analogous $[X(HX)_n]^-$ salts, where X = F or Cl and n = 1 or 2, are dealt with separately in Section 11.4.4. Finally, the bis(4-amino-1,2,4-triazolium) salt of the dianion 4-(carboxylatomethyl)-5-nitroiminotetrazolate (m.pt. 93°C) [408] should be noted.

11.4.2 Salts From Non-Carboxylic Weak Acids

Several protic ionic liquids formed from nitrogen bases and non-carboxylic weak acids have been crystallographically characterised. These include a subset of the large group of energetic materials described further in Section 11.6. Protic materials containing tetrazolate (with hydrazinium, m.pt. 48°C, U_L 576 kJ mol^{-1}[146]), 5-nitrotetrazolate (with *N,N′,N″*-triaminoguanidinium, m.pt. 96°C, U_L 488 kJ mol^{-1} [427] and 1-amino-1,2,3-triazolium, m.pt. 99°C, U_L 508 kJ mol^{-1} [407]), 5-nitroiminotetrazolate (with 1-propyl-1,2,4-tetrazolium, m.pt. 69°C, U_L 457 kJ mol^{-1} [252]) and 1-(2-nitratoethyl)-5-nitriminotetrazolate (with hydroxylammonium, m.pt. 66°C, U_L 485 kJ mol^{-1}[235] and ammonium, m.pt. 81°C, U_L 486 kJ mol^{-1} [235]) are all ionic liquids. Aminoguanidinium 5-azidotetrazolate melts at 100°C (U_L 504 kJ mol^{-1}) [462].

TABLE 11.13 Salts of Multiply Charged Ions

m.pt./°C			U_L/ kJ mol^{-1}
	Dinegative anions $p{:}q = 2{:}1$		
Liq	Benzyltriphenylphosphonium	$[Br_8]$	752
ca. 9	1-Butyl-1-methylpyrrolidinium	$[Br_{20}]$	1007
ca. 10	Tributylmethylammonium	$[Br_{20}]$	944
Liq	Chlorotriphenylphosphonium	$[Cl_2I_{14}]$	688
80–82	*N*-Ethylurotropinium	$[I_8]$	869
90	([2.2.2]Cryptand)potassium	$[I_{12}]$	653
14–16	Tetrakis(ethanenitrile)lithium	$[P_2Se_8]$	842
109	Tetra(pyridine)lithium	$[P_2Se_8]$	980
110.5	1-Butyl-3-methylimidazolium	$[P_2Se_8]$	1217
62	1-(Chloromethyl)-1, 1-dimethylhydrazinium	$[SO_4]$	1245
102	2-Aminopyridinium	$[O_2CCH_2CO_2]$	1224/1232
98–100	2-Aminopyridinium	$[HO_2C(CH_2)_8CO_2]$	1050
112	Oxonium	$[O_3SCH_2CH_2SO_3]$	1497
109	1-(2-Hydroxyethyl)piperidinium	$[Si(ox)_3]$	988
97	1-(2-Hydroxyethyl)piperidinium	$[Ge(ox)_3]$	991
5	1-Butyl-3-methylimidazolium	$[Sn^{II}Cl_4]$	1012
60	1-Butyl-3-methylimidazolium	$[ZnCl_4]$	1010
105	1-Cyanomethyl- 3-methylimidazolium	$[CdCl_4]$	1082
63	1-Butyl-3-methylimidazolium	$[Mn^{II}Cl_4]$	1003
105.5	*N*-Methylmorpholinium	$[Mn^{II}Cl_4]$	1125
86	1-Ethyl-3-methylimidazolium	$[Fe^{II}Cl_4]$	1078
58	1-Butyl-3-methylimidazolium	$[Fe^{II}Cl_4]$	1007
101	1-Ethyl-3-methylimidazolium	$[Co^{II}Cl_4]$	1082
62	1-Butyl-3-methylimidazolium	$[Co^{II}Cl_4]$	1010
106.5	*N*-Methylmorpholinium	$[Co^{II}Cl_4]$	1129
45	1-Butyl-3-methylimidazolium	$[Co^{II}Br_4]$	994
92–93	1-Ethyl-3-methylimidazolium	$[Ni^{II}Cl_4]$	1085
56	1-Butyl-3-methylimidazolium	$[Ni^{II}Cl_4]$	1012
23	1-Butyl-3-methylimidazolium	$[Cu^{II}Cl_4]$	1024
88–90	*N*-Methylmorpholinium	$[Cu^{II}Cl_4]$	1127
92–98	*N*-Methylmorpholinium	$[Cu^{II}Br_4]$	1100
110	4-Methylpyridinium	$[Cu^{II}Br_4]$	1099
103	1-Ethoxymethyl- 3-methylimidazolium	$[Pd^{II}Cl_4]$	1036
107	1-(2-Methoxyethyl)- 3-methylimidazolium	$[Pd^{II}Cl_4]$	1052
101	1-Tetradecylpyridinium	$[Pd^{II}I_4]$	769
99	1-Butyl-3-methylimidazolium	$[Pt^{II}Cl_4]$	1037
89	Tetramethylformamidinium	$[Ga_2I_6]$	946
92	Tetramethylformamidinium	$[In_2Cl_6]$	
84	Ethyltriphenylphosphonium	$[Se_4Br_{14}]$	709
102	Butyltriphenylphosphonium	$[Hg^{II}{}_2Cl_6]$	730
107.5	Methyltriphenylphosphonium	$[Cu^{I}{}_2Br_4]$	819

TABLE 11.13 (*Continued*)

m.pt./°C			U_L/ kJ mol^{-1}
70	1-Butyl-3-methylimidazolium	$[Cu^{II}_3Cl_8]$	927
95–96	Tributylmethyammonium	$[Cu^{I}_5Br_7]$	802
74	1-Tetradecylpyridinium	$[Pd^{II}_2Cl_6]$	737
82.5	1-Hexadecylpyridinium	$[Pd^{II}_2Cl_6]$	716
99	1-Ethyl-3-methylimidazolium	$[V^{IV}(O)Cl_4]$	1064
76	1-Ethyl-3-methylimidazolium	$[(Ta^{V}Cl_5)_2O]$	939
64	Tris(dimethylamino)sulfonium	$[P_3N_3F_5NPF_2NPF_2NPF_5]$	854
18	Tetrakis(ethanenitrile)silver(I)	$[Ag\{(CF_3SO_2)_2N\}_3]$	750
107	1-Methyl-1-propylpyrrolidinium	$[Yb\{N(SO_2CF_3)_2\}_4]$	1089
92	Butylmethylpyrrolidinium	$[Eu\{N(SO_2CF_3)_2\}_5]$	670
74	Butyltriphenylphosphonium	$[Bi_2I_8(Me_2CO)_2]$	670
93	4-Amino-1,2,4-triazolium	4-(Carboxylatomethyl)-5-nitroiminotetrazolate	1200
	Dipositive cations $p:q=1:2$		
95	1-(3-Ammoniopropyl)-3-methylimidazolium	$[NO_3]$	1587
69–71	1-{2-(Diethylammonio)ethyl}-3-methylimidazolium	$[N(SO_2CF_3)_2]$	1261
60–62	1-{(3-(Dimethylammonio)propyl}-2-hydroxypyridinium	$[CF_3SO_3]$	1376
98	1,2-Dimethylhydrazine-1,2-diium	$[NO_3]$	1746
54–56	$[Li_2(H_2NNHMe)_3]$	$[BH_4]$	1566
89	Tetrakis(1-methylimidazole) copper(II)	$[N(SO_2CF_3)_2]$	1206
74	Tetrakis(1-ethylimidazole) copper(II)	$[N(SO_2CF_3)_2]$	1156
46	Tetrakis(1-butylimidazole) copper(II)	$[N(SO_2CF_3)_2]$	1118
52	Hexakis(1-butylimidazole) copper(II)	$[N(SO_2CF_3)_2]$	1137
95–97	1,4-Bis(1-methylimidazolium-3-yl)butane	$[BF_4]$	1412
99	1,4-Bis(1-methylimidazolium-3-yl)butane	$[BF_4]$	1423
71	1,4-Bis(1-methylimidazolium-3-yl)butane	$[CF_3SO_3]$	1355
86	1,7-Bis(1-methylimidazolium-3-yl)heptane	$[BF_4]$	1368
92	1,8-Bis(1-methylimidazolium-3-yl)octane	$[BF_4]$	1353
71	1,10-Bis(1-methylimidazolium-3-yl)decane	$[PF_6]$	1294

(*Continued*)

TABLE 11.13 (*Continued*)

m.pt./°C			U_L/ kJ mol^{-1}
68–71	1,4-Bis(1-methylimidazolium-3-ylmethyl)benzene	$[BF_4]$	1384
59–62	1,4-Bis(1-butylimidazolium-3-ylmethyl)benzene	$[CF_3SO_3]$	1256
96–98	3,5-Bis(1-butylimidazolium-3-ylmethyl)toluene	$[PF_6]$	1254
92–94	1,4-Bis(1,2-dimethylimidazolium-3-ylmethyl)benzene	$[CF_3SO_3]$	1301
87	1,8-Bis(1-methylimidazolium-3-yl)-3,6-dioxaoctane	$[PF_6]$	1325
112	6,6′-Spirobi(1-methyl-1,5,6,7-tetrahydropyrrolo[1,2-a]imidazol-4-ium	$[N(SO_2CF_3)_2]$	1245
110	Piperazinium	$[C_5H_{11}CO_2]$	1422
97	Piperazinium	$[C_6H_{13}CO_2]$	1357
96	Piperazinium	$[C_9H_{19}CO_2]$	1280
96	Piperazinium	$[C_{11}H_{23}CO_2]$	1240
96	Piperazinium	$[C_{13}H_{27}CO_2]$	1207
97	Piperazinium	$[C_{15}H_{31}CO_2]$	1178
104.5	Hexamethylenediammonium	$[I_5]$	1225
107.5	3,6-Dioxaoctane-1,8-diammonium	$[CCl_3CO_2]$	1374
86	*N,N*′-Bis(2-ethoxyethyl)-*N,N,N*′,*N*′-tetramethylethane-1,2-diammonium	$[N(SO_2CF_3)_2]$	1213
69	*N,N*′′′-Bis(2-ethoxyethyl)-*N,N,N*′′′,*N*′′′-tetramethylbutane-1,4-diammonium	$[CF_3SO_3]$	1278
99.5	*N,N*′′′-Bis(2-ethoxyethyl)-*N,N,N*′′′,*N*′′′-tetramethylbutane-1,4-diammonium	$[PF_6]$	1292
51	1,3-Bis(1-methyl-1-pyrrolidinio)propane	Br	1434
109	1,2-Bis(*P,P*-dimethyl-*P*′,*P*′-diphenylphosphinophosphonium)ethane	$[GaCl_4]$	1163
94–95	1,2-Bis(*P,P,P*′,*P*′-tetraphenylphosphinophosphonium)ethane	$[GaCl_4]$	1091
64–67	1,6-Bis(*P,P,P*′,*P*′-tetraphenylphosphinophosphonium)hexane	$[GaCl_4]$	1075
107.5	1,3-Dichloro-1,2,3,4-tetracyclohexyltetraphosphetane-1,3-diium	$[Ga_2Cl_7]$	1107
94–95.5	1,6-Diarsa-2,5,7,10-tetrathiatricyclo[5.3.0^{1,5}.0^{6,10}]decane	$[GaCl_4]$	1363

TABLE 11.13 (*Continued*)

m.pt./°C			U_L/ kJ mol^{-1}
104.5	1,6-Diarsa-2,5,7,10-tetramethyl-2,5,7,10-tetraazatricyclo[5.3.0^{1,5}.0^{6,10}]decane	$[GaCl_4]$	1311
	Salts with $p \times q > 2$		$p{:}q$
98	H_2([2.2.2]Cryptand)	$[I_8]$	2:2
67	*N,N'*-Dimethylhydrazinium	$[SO_4]$	2:2
107	1,4-Bis(1-methylimidazolium-3-yl)butane	2-Hydroxy-5-sulfonatobenzoate	2:2
93–94	*N*-Methyl-1,3-propylenediammonium	$[TlBr_5]$	2:2
79	Ferrocenium	$[I_{29}]$	3:1
105–118	2-Fluoropyridinium	$[Sb^{III}{}_2Br_9]$	3:1
59	Tetrabutylammonium	$[Sm\{C(CN)_2NO\}_6]$	3:1
68	Tetrabutylammonium	$[Ce\{C(CN)_2NO\}_6]$	3:1
58	1-Ethyl-3-methylimidazolium	$[La\{C(CN)_2NO\}_6]$	3:1
63	1-Ethyl-2,3-dimethylimidazolium	$[Pr\{C(CN)_2NO\}_6]$	3:1
103	1-Ethyl-3-methylimidazolium	$[U^{VI}O_2(NCS)_5]$	3:1
85–87	3,3,5,5,7,7-Hexaphenyltricyclo[2.2.1.0^{2,6}]heptaphosphane-3,5,7-triium	$[Ga_2Cl_7]$	1:3
65–66	2-Benzylpyridinium	$[Sb^{III}{}_2Cl_{10}]$	4:1
27	1-Dodecyl-3-methylimidazolium	$[Cu^{I}{}_4Cl_8]$	4:1
38	1-Butyl-3-methylimidazolium	$[La(NCS)_7(H_2O)]$	4:1
88	$[Dy_2(NMe_3CH_2CO_2)_8(H_2O)_4]$	$[(CF_3SO_2)_2N]$	1:6
	Mixed salts/co-crystals		
65	1,3-Dimethylimidazolium/lithium	$[N(SO_2CF_3)_2]$	(1,1):2
37	1-Ethyl-3-methylimidazolium/sodium	$[AlCl_4]$	(1,1):2
85	Tetrabutylammonium/3-[(2*S*)-2-methylbutyl]thiazolium	$[CF_3SO_3]$	(1,1):2
95	Tetrabutylammonium/1-methyl-3-(tetrahydrofuran-3-ylmethyl)imidazolium	$[CF_3SO_3]$	(1,1):2
91	(Benzo-18-crown-6)potassium	$[I_5]/[I_7]$	2:(1,1)
107.5	*cyclo*-Hexylammonium	$Cl/[AlCl_4]$	2:(1,1)
110.5	Benzylammonium	$Cl/[AlCl_4]$	2:(1,1)
110	5-Methyl-1,3,5-dithiazinan-5-ium	$Cl/[BCl_4]$	2:(1,1)
26	1-Dodecyl-3-methylimidazolium	$[Cu^{I}Cl_2]/[Cu^{I}Cl_3]$	2:(1,1)
38	1-Butyl-2,3-dimethylimidazolium	$[Fe^{III}Cl_4]/[Fe^{II}Cl_4]$	1:(1,2)
76	2-Hydroxyethyl(trimethyl)ammonium (cholinium)	$Cl/[Cu^{II}Cl_4]$	3:(1,2)

TABLE 11.14 Halides, Poly- and Interhalides and Pseudohalides

m.pt./°C			$p:q$	U_L/ kJ mol^{-1}
		Fluoride		
96–100	Hydroxylammonium	F	1:1	726
		Chloride		
95	1-Aminoimidazolium	Cl	1:1	556
81	1-Ethyl-3-methylimidazolium	Cl	1:1	503–507
66	1-Butyl-3-methylimidazolium	Cl	1:1	477–483
79/100	1-Butyl-2,3-dimethylimidazolium	Cl	1:1	464–472
80	1-(3-Cyanopropyl)-3-methylimidazolium	Cl	1:1	479/479
86	1-(2-Hydroxyethyl)-3-methylimidazolium	Cl	1:1	504
105	1-(3-Hydroxycarbonylpropyl)-3-methylimidazolium	Cl	1:1	479
64–66	1-(-2-Hydroxy-2-methyl-2-phenylethyl)-3-*tert*-butylimidazolium	Cl	1:1	419
86/103/152	1-Butylpyridinium	Cl	1:1	478
101	1-(3-Cyanopropyl)pyridinium	Cl	1:1	483
66–68	1-(2-Hydroxyethyl)pyridinium	Cl	1:1	511
95	3-Methoxyazetidinium	Cl	1:1	539
50	1-Methyl-1-propylpyrrolidinium	Cl	1:1	481
90–96	1-(Prop-2-ynyl)-1-methylpyrrolidinium	Cl	1:1	487
68–69	Tris(2-methoxyethyl)ammonium	Cl	1:1	445
94	5-Hydroxy-1-methyl-3,4-dihydro-2*H*-pyrrolium	Cl	1:1	525
		Bromide		
71	1-Ethyl-3-methylimidazolium	Br	1:1	505
36	1-Methyl-3-propylimidazolium	Br	1:1	491
78(est)/110.5	1-*iso*-Propyl-3-methylimidazolium	Br	1:1	484/488
70/79	1-Butyl-3-methylimidazolium	Br	1:1	473–477
67	1,3-Dipropylimidazolium	Br	1:1	457
97	1-Butyl-2,3-dimethylimidazolium	Br	1:1	466
60	1-Allyl-3-methylimidazolium	Br	1:1	495
61	1-Allyl-2,3-dimethylimidazolium	Br	1:1	472
90	1-(3-Cyanopropyl)-3-methylimidazolium	Br	1:1	476
78	1-{2-(2-Methoxyethyl)oxyethyl}-3-methylimidazolium	Br	1:1	455
102	1-Hydroxycarbonylmethyl-3-decylimidazolium	Br	1:1	414
60–70	1,3-Di(benzyloxy)imidazolium	Br	1:1	415

TABLE 11.14 (*Continued*)

m.pt./°C			p:q	U_L/ kJ mol^{-1}
63–67	1-Ethyl-4-amino-1,2,4-triazolium	Br	1:1	517
63	1-Propyl-4-amino-1,2,4-triazolium	Br	1:1	500
92	1-Isopropyl-4-amino-1,2, 4-triazolium	Br	1:1	498
76	1-Hexyl-4-amino-1,2,4-triazolium	Br	1:1	459
94	1-Heptyl-4-amino-1,2, 4-triazolium	Br	1:1	448
95–97	1-Butylpyridinium	Br	1:1	470
27	1-(Trimethylsilyl)pyridinium	Br	1:1	466
87	1-(Trimethylsilyl)-3, 4-dimethylpyridinium	Br	1:1	449
110	1-(2-Hydroxyethyl)pyridinium	Br	1:1	499
99	1-Benzylpyridinium	Br	1:1	464–465
106	1-Ethyl-1-methylpyrrolidinium	Br	1:1	491
88–92	Tetradodecylammonium	Br	1:1	325
54–57	Tetrakis(decyl)phosphonium	Br	1:1	332
44.5	Amidomethyltris(tetradecyl) phosphonium	Br	1:1	326
90	Hydroxycarbonylmethyltris (tetradecyl)phosphonium	Br	1:1	327
75	Hexadecyldimethylsulfonium	Br	1:1	394
51	1,3-Bis(1-methyl- 1-pyrrolidinio)propane	Br	1:2	1434
		Iodide		
81	1-Ethyl-3-methylimidazolium	I	1:1	488
19	1-Butyl-3-methylimidazolium	I	1:1	468
88.5	1,3-Didodecylimidazolium	I	1:1	363
98	1-Butyl-2,3-dimethylimidazolium	I	1:1	458
86/89	1-Ethyl-2-methyl- 3-benzylimidazolium	I	1:1	439
71/75	1-Propyl-2-methyl- 3-benzylimidazolium	I	1:1	436
61/64	1-Butyl-2-methyl- 3-benzylimidazolium	I	1:1	425
61/65	1-Pentyl-2-methyl- 3-benzylimidazolium	I	1:1	416
78	1-Vinyl-3-methylimidazolium	I	1:1	496
65	1-Allyl-3-methylimidazolium	I	1:1	480
31	1-Allyl-3-ethylimidazolium	I	1:1	472
32	1-Allyl-3-propylimidazolium	I	1:1	457
57	1-Methyl-3-[(trimethoxysilyl) methyl]imidazolium	I	1:1	436
69	1-(3-Cyanopropyl)- 3-methylimidazolium	I	1:1	465/466

(*Continued*)

TABLE 11.14 (*Continued*)

m.pt./°C			$p:q$	U_L/ kJ mol^{-1}
75	1-Methoxymethyl-3-methylimidazolium	I	1:1	484
68	1-Ethoxymethyl-3-methylimidazolium	I	1:1	471
50	1-(2-Methoxyethyl)-3-methylimidazolium	I	1:1	473
63	1-{(2-Methoxyethyl)oxymethyl}-3-methylimidazolium	I	1:1	460
70	1-{2-(2-Methoxyethyl) oxyethyl}-3-methylimidazolium	I	1:1	451
102–103	1-(2-Hydroxy-2-methylpropyl)-3-methylimidazolium	I	1:1	460
97–98	1-(2-Hydroxy-2-methylpropyl)-3-*iso*-propylimidazolium	I	1:1	440
98	1-Methyl-4-(3,3,3-trifluoropropyl)-1,2,4-triazolium	I	1:1	466
82	*N*-Decylbenzothiazolium	I	1:1	410
80	*N*-Undecylbenzothiazolium	I	1:1	402
90.5	*N*-Dodecylbenzothiazolium	I	1:1	399
103–105	*rac*-1-((1-(2-Naphthyl)prop-2-en-1-yl)amino)pyridinium	I	1:1	420
110–114	3-(4-Methoxyphenyl)-1,2-dimethyl-4,5-dihydroimidazolium	I	1:1	443
85	*N*-Butyl-*N,N*-dimethylglycine ethyl ester	I	1:1	440
96	Tetrapentylphosphonium	I	1:1	378
108	*cyclo*-Hexyltrivinylphosphonium	I	1:1	434
98–99	Tetraoctadecylphosphonium	I	1.1	293
		Polybromides		
49	1-Ethyl-3-methylimidazolium	$[Br_3]$	1:1	467
46	1-Ethyl-1-methylpyrrolidinium	$[Br_3]$	1:1	458
ca. 9	1-Butyl-1-methylpyrrolidinium	$[Br_{20}]$	2:1	1007
ca. 10	Tributylmethylammonium	$[Br_{20}]$	2:1	944
Liq	Benzyltriphenylphosphonium	$[Br_8]$	2:1	752
Liq	Bromotriphenylphosphonium	$[Br_7]$	1:1	378
		Polyiodides		
46	1,3-Didodecylimidazolium	$[I_3]$	1:1	356
110	1,4-Dimethyl-2-*H*-1,2,4-triazolium	$[I_3]$	1:1	465
80–82	(Pyridin-2-yl)(pyridinio-2-yl) disulfide	$[I_3]$	1:1	418

TABLE 11.14 (*Continued*)

m.pt./°C			$p:q$	U_L/ kJ mol^{-1}
93	(2-Pyridylthio)(2-pyridiniothio) methane	$[I_3]$	1:1	411
45–46	1-Ethylpyridinium	$[I_3]$	1:1	450
110.5–113	1,2,4-Trimethylpyridinium	$[I_3]$	1:1	436
33–34	1-Ethylquinolinium	$[I_3]$	1:1	427
111	*N*-Ethylurotropinium	$[I_3]$	1:1	432
105	*N*-Propylurotropinium	$[I_3]$	1:1	420
95	*N*-Butylurotropinium	$[I_3]$	1:1	416
58	1,3,5-Trimethyl-tetrahydro-1,3, 5-triazinium	$[I_3]$	1:1	437
43	1,3,5-Triethyl-tetrahydro-1,3, 5-triazinium	$[I_3]$	1:1	416
68–69	9-Aza-9-methyl- 1-thioniabicyclo[3.3.1]nonane	$[I_3]$	1:1	435
70	Dimethyl(isopropyl) phenylammonium	$[I_3]$	1:1	420
60	Dimethyl(propyl) phenylammonium	$[I_3]$	1:1	420
102	Dimethyldiphenylammonium	$[I_3]$	1:1	413
65	Ethyl(methyl) diphenylammonium	$[I_3]$	1:1	408
98	Tetrapropylammonium	$[I_3]$	1:1	399
105	Methyltriphenylphosphonium	$[I_3]$	1:1	394
37	Trimethylsulfonium	$[I_3]$	1:1	468
−2	Triethylsulfonium	$[I_3]$	1:1	440
87	Bis(1,3-dimethylthiourea) iodonium	$[I_3]$	1:1	408
63–65	{Tris(dimethylamino) phosphaneselenide}iodine(I)	$[I_3]$	1:1	409
94.5–95.5	1-(5-Bromopentyl)-1′, 1″-biferrocenium	$[I_3]$	1:1	372
34–36	1-Pentyl-1‴-(5-bromopentyl)-1′, 1″-biferrocenium	$[I_3]$	1:1	356
68.5–69.5	1,1‴-Bis(5-bromopentyl)-1′, 1″-biferrocenium	$[I_3]$	1:1	355
59.5–61	1,1‴-Bis(6-bromohexyl)-1′, 1″-biferrocenium	$[I_3]$	1:1	351
36	1,3-Didodecylimidazolium	$[I_5]$	1:1	346
78.5–80	*N*-Propylurotropinium	$[I_5]$	1:1	400
69	1,3,5-Tri-*iso*-propyl- tetrahydro-1,3,5-triazinium	$[I_5]$	1:1	384
104	1,3,5-Tri-*tert*-butyl- tetrahydro-1,3,5-triazenium	$[I_5]$	1:1	373
90	Ethyl(methyl) diphenylammonium	$[I_5]$	1:1	389

(*Continued*)

TABLE 11.14 (*Continued*)

m.pt./°C			$p:q$	U_L/ kJ mol^{-1}
104.5	Hexamethylenediammonium	$[I_5]$	1:2	1225
50–56	*N*-Propylurotropinium	$[I_7]$	1:1	384
82	1,3,5-Tri-^{t}butyltetrahydro-1,3,5-triazinium	$[I_7]$	1:1	361
85	Bis(pyridine)iodonium	$[I_7]$	1:1	385
80–82	*N*-Ethylurotropinium	$[I_8]$	2:1	869
98	H_2([2.2.2]Cryptand)	$[I_8]$	2:2	
91	(Benzo-18-crown-6)potassium	$[I_5][I_7]$	1:(1,1)	
90	([2.2.2]Cryptand)potassium	$[I_{12}]$	2:1	653
79	Ferrocenium	$[I_{29}]$	3:1	
		Interhalides		
103–105	Bis(2,6-dimethylpyridine) iodonium	$[ICl_2]$	1:1	410
37	1-Butyl-2,3-dimethylimidazolium	$[IBr_2]$	1:1	430
110	Bis(2,6-dimethylpyridine) iodonium	$[IBr_2]$	1:1	407
Liq	Chlorotriphenylphosphonium	$[Cl_2I_{14}]$	2:1	688
	Pseudohalides (I)			
41	1-Butyl-2,3-dimethylimidazolium	$[N_3]$	1:1	465
66	1-(But-2-ynyl)-3-methylimidazolium	$[N_3]$	1:1	487
50	1-Amino-3-methyl-1,2,3-triazolium	$[N_3]$	1:1	536
59	*N*-Amino-1-azoniacycloheptane	$[N_3]$	1:1	493
70.5	Hydrazinium	$[N_3]$	1:1	630
68	Methylhydrazinium	$[N_3]$	1:1	707
38	1,1-Dimethylhydrazinium	$[N_3]$	1:1	553
−8	1,2-Dimethylhydrazinium	$[N_3]$	1:1	556
68	Bis(phenylhydrazine)hydrogen	$[N_3]$	1:1	442
72	Trimethylselenonium	$[N_3]$	1:1	535
110	Trimethyltelluronium	$[N_3]$	1:1	522
72	1-Ethyl-3-methylimidazolium	[OCN]	1:1	505
40	1-Butyl-2,3-dimethylimidazolium	[SCN]	1:1	454
85–105	1,1-Dimethylpyrrolidinium	[SCN]	1:1	495
86–88	Trimethyltelluronium	[SCN]	1:1	513
73–75	Trimethylselenonium	[SeCN]	1:1	513
106–109	Trimethyltelluronium	[SeCN]	1:1	507
	Pseudohalides (II)	**Dinitramide**		
70	1*H*-Tetrazolium	$[N(NO_2)_2]$	1:1	535
85	5-Aminotetrazolium	$[N(NO_2)_2]$	1:1	526
90	5-Amino-1-methyltetrazolium	$[N(NO_2)_2]$	1:1	500/501
90	5-Amino-2-methyltetrazolium	$[N(NO_2)_2]$	1:1	501
58	5-Amino-1,3-dimethyltetrazol-3-ium	$[N(NO_2)_2]$	1:1	486

TABLE 11.14 (*Continued*)

m.pt./°C			p:q	U_L/ kJ mol^{-1}
72	1,5-Diaminotetrazolium	$[N(NO_2)_2]$	1:1	509
85–86	1,5-Diamino-4-methyltetrazolium	$[N(NO_2)_2]$	1:1	496
80	Hydrazinium	$[N(NO_2)_2]$	1:1	573
18–23	Hydroxylammonium	$[N(NO_2)_2]$	Complex	
91–94	Aminoguanidinium	$[N(NO_2)_2]$	1:1	517/518
80	*N,N′,N″*-Triaminoguanidinium	$[N(NO_2)_2]$	1:1	495
92	Ammonium	$[N(NO_2)_2]$	1:1	590
102–106	Rubidium	$[N(NO_2)_2]$	1:1	581
85–87	Caesium	$[N(NO_2)_2]$	Catenate	
		Nitrocyanamide		
91	Aminoguanidinium	$[N(CN)(NO_2)]$	1:1	522
58	Bis(2-azidoethyl)ammonium	$[N(CN)(NO_2)]$	1:1	462
		Dicyanamide		
44	1,3-Didodecylimidazolium	$[N(CN)_2]$	1:1	363
59	1-(Pent-2-ynyl)-3-methylimidazolium	$[N(CN)_2]$	1:1	459
67	1-Cyanomethyl-3-methylimidazolium	$[N(CN)_2]$	1:1	484
97–97.5	Hydrazinium	$[N(CN)_2]$	1:1	593
60.5	1,1,1-Trimethylhydrazinium	$[N(CN)_2]$	1:1	505
45/65	Guanidinium	$[N(CN)_2]$	1:1	542
52/62	Guanidinium	$[N(CN)_2]$	1:1	542
		Others		
109	1,3-Di-*iso*-propylimidazolium	$[N(C_6F_5)_2]$	1:1	393
46	1-Butyl-2,3-dimethylimidazolium	$[N(C_6F_5)_2]$	1:1	394
60	1,3-Dimethylimidazolium	$[CH(NO_2)_2]$	1:1	491
80	*N,N′*-Diaminoguanidinium	$[C(NO_2)_3]$	1:1	484
84	*N,N′,N″*-Triaminoguanidinium	$[C(NO_2)_3]$	1:1	476
72	1,5-Diamino-4-methyltetrazolium	$[C(NO_2)(CN)_2]$	1:1	475
35	1-Ethyl-3-methylimidazolium	[C(NO2)(NO)CN]	1:1	464
44	1,3-Didodecylimidazolium	$[C(CN)_3]$	1:1	358
96–102	Tetraphenylarsonium	$[CF_3C(O)C(CN)_2]$	1:1	375

Energetic salts of the pseudohalides [615], $[N_3]^-$, $[N(NO_2)_2]^-$, $[C(NO_2)_3]^-$, $[N(CN)_2]^-$ and $[N(NO_2)CN)]^-$ are also found in the protic salt subclass, with $[NH_2MeNHMe][N_3]$ being one of the lowest-melting ionic liquids known (m.pt. −8°C, U_L 556 kJ mol^{-1}) for which a single-crystal X-ray structure has been obtained [64]. Interestingly, $[NH_3NH_2][N_3]$ (a structure without cif file [262]) is reported [616] to melt at 70.5°C, with a value of U_L 630 kJ mol^{-1}, consistent with a higher melting temperature. Other protic inorganic pseudohalide salts are

TABLE 11.15 Oxoanions and Their Esters

m.pt./°C			$p:q$	U_L/ kJ mol^{-1}
		Nitrite		
55	1-Ethyl-3-methylimidazolium	$[NO_2]$	1:1	501
		Nitrate		
60	1-Methylimidazolium	$[NO_3]$	1:1	538
79	1,2-Dimethylimidazolium	$[NO_3]$	1:1	513
38	1-Ethyl-3-methylimidazolium	$[NO_3]$	1:1	490
103	1-Cyanomethyl-3-methylimidazolium	$[NO_3]$	1:1	502
95	1-(3-Ammoniopropyl)-3-methylimidazolium	$[NO_3]$	1:2	1587
86–88	1-Amino-3-methyl-1,2,3-triazolium	$[NO_3]$	1:1	531
98	1,4-Dimethyl-3-azido-1,2,4-triazolium	$[NO_3]$	1:1	498
66	1-iso-Propyl-4-amino-1,2,4-triazolium	$[NO_3]$	1:1	493
56	1-(Cyclopropylmethyl)-4-amino-1,2,4-triazolium	$[NO_3]$	1:1	486
94	2-Amino-4,5-dimethyltetrazolium	$[NO_3]$	1:1	511
92–94	2-Amino-5-butyl-4-methyl-1,3-thiazol-3-ium	$[NO_3]$	1:1	458
74–75	*N*-(1-*O*-Methyl-2,3,4-tri-*O*-acetyl-*α*-D-glucopyranose-6-yl) pyridinium	$[NO_3]$	1:1	389
102–103	3-*O*-Nitrato-3-ethylazetidinium	$[NO_3]$	1:1	479
62.5/70.5	Hydrazinium	$[NO_3]$	1:1	625
42	Methylhydrazinium	$[NO_3]$	1:1	583
45	1,1-Dimethylhydrazinium	$[NO_3]$	1:1	545
98	1,2-Dimethylhydrazine-1,2-diium $[MeNH_2NH_2Me]^{2+}$	$[NO_3]$	1:2	1746
99	2,2-Dimethyltriazanium	$[NO_3]$	1:1	540
86	Diaminouronium	$[NO_3]$	1:1	553
110	Methylammonium	$[NO_3]$	1:1	594
80	Dimethylammonium	$[NO_3]$	1:1	557
95–96	Benzyl(ethyl)ammonium	$[NO_3]$	1:1	477
73	*N*-Methyl-*N*-(2-nitroxyethyl) ammonium	$[NO_3]$	1:1	507
80	Tris(2-hydroxyethyl)ammonium	$[NO_3]$	1:1	475
85/92–94	Benzethonium	$[NO_3]$	1:1	373
−18.5	Oxonium dihydrate	$[NO_3]$	1:1	580
		Perchlorate		
64	1,3-Didodecylimidazolium	$[ClO_4]$	1:1	362
110	1,3-Bis(hydroxycarbonylmethyl) imidazolium	$[ClO_4]$	1:1	466

TABLE 11.15 (*Continued*)

m.pt./°C			p:q	U_L/ kJ mol^{-1}
86	1-Methyl-4-amino-1,2, 4-triazolium	$[ClO_4]$	1:1	509
55	1*H*-Tetrazolium	$[ClO_4]$	1:1	556
97–98	1,5-Diaminotetrazolium	$[ClO_4]$	1:1	523
78	[(Pyridine-1-oxide)$_2$H]	$[ClO_4]$	1:1	449
98–100	1,5-Dimethoxyl-3-phenyl- 5a,6,7,8,9,9a-hexahydro-1*H*-1, 5-benzodiazepin-5-ium	$[ClO_4]$	1:1	407
52	1-(Chloromethyl)-1,1- dimethylhydrazinium	$[ClO_4]$	1:1	502
72	Aminoguanidinium	$[ClO_4]$	1:1	532
75	Azidoformamidinium	$[ClO_4]$	1:1	528
90–91.5	1-Phenylthiolanium	$[ClO_4]$	1:1	454
104	(±)-*cis*-3-Acetoxy-1- methylthiacycohexane	$[ClO_4]$	1:1	450
106	(−)-(1*S*,3*S*)-*trans*-3-Acetoxy- 1-methylthiacycohexane	$[ClO_4]$	1:1	448
79-80	3(*R*),4(*R*),5(*R*),6(*S*)-3,4, 5-Trihydroxy-*cis*-1- thioniabicyclo[4.3.0]nonane	$[ClO_4]$	1:1	459
		Sulfate		
67	1,2-Dimethylhydrazine-1, 2-diium $[MeNH_2NH_2Me]^{2+}$	$[SO_4]$	2:2	
62	1-(Chloromethyl)-1, 1-dimethylhydrazinium	$[SO_4]$	2:1	1245
105–108	1-Azidosulfonylimidazolium	$[SO_3(OH)]$	1:1	481
78	1-Butyl-2,3-dimethylimidazolium	$[SO_3(OH)]$	1:1	450
63	1-iso-Propyl-3-methyl-4- diphenylphosphorylimidazolium	$[SO_3(OH)]$	1:1	399
105–110	4-Dimethylamino-1-methyl-1,2, 4-triazolium	$[SO_3(OH)]$	1:1	480
64–65	Tridodecylammonium	$[SO_3(OH)]$	1:1	334
−2.5	1,1-Dihydroxyethylium	$[SO_3(OH)]$	1:1	545
52.5	1,1-Dihydroxyethylium	$[SO_3F]$	1:1	547
94	Tris(2-hydroxyethyl)ammonium	$[SeO_2(OH)]$	1:1	467
43	1,3-Dimethylimidazolium	$[SO_3(OMe)]$	1:1	480
−5	1-Butyl-3-methylimidazolium	$[SO_3(OMe)]$	1:1	444
95/104	1,3-Dimethyl-4-nitromidazolium	$[SO_3(OMe)]$	1:1	472
108–115	1-Methyl-4-cyanopyridinium	$[SO_3(OMe)]$	1:1	472
		Phosphate		
41	1-Butyl-1-methylpiperidinium	$[PO_2F_2]$	1:1	445
108–110	1-Benzylimidazolium	$[PhP(O)_2(OH)]$	1:1	424
99–100	Imidazolium	$[PO_2(OPh)_2]$	1:1	425
91–92.5	1-Methylimidazolium	$[PO_2(OPh)_2]$	1:1	423

(*Continued*)

TABLE 11.15 (*Continued*)

m.pt./°C			$p:q$	U_L/ kJ mol^{-1}
39	1-Ethyl-3-methylimidazolium	$[PO_2(OMe)(SMe)]$	1:1	453
23–24	1-Butyl-2,3-dimethylimidazolium	$[PO_2(OMe)(SeMe)]$	1:1	422
104–105	(1-Phenylethyl)ammonium	[POS(OEt)Et]	1:1	424
		Carbonate		
79	1,2,3-Trimethylimidazolium	$[CO_2(OMe)]$	1:1	482
108–110	Bis(4-methoxyphenyl)iodonium	$[CS_2(NEt_2)]$	Complex	

listed: $[NH_3OH][N(NO_2)_2]$ [80] (m.pt. 18–23°C [617]; see the following text); $[NH_2NH_3][N(NO_2)_2]$ [80] (m.pt. *ca.* 80°C, U_L 573 kJ mol^{-1} [617]); $[NH_2NH_3][N(CN)_2]$ (m.pt. 97–97.5°C, U_L 593 kJ mol^{-1}) [442], [tetrazolium]$[N(NO_2)_2]$ (m.pt. 70°C, U_L 535 kJ mol^{-1}) [168] and [5-aminotetrazolium]$[N(NO_2)_2]$ (m.pt. 85°C, U_L 526 kJ mol^{-1}) [168]. All of the salts [*N*-aminoguanidinium]$[N(NO_2)CN]$ [391] and $[N(NO_2)_2]$ (two polymorphs) [261], [*N,N′*-diaminoguanidinium] and [*N,N′,N″*-triaminoguanidinium]$[C(NO_2)_3]$ [319] and [*N,N′,N″*-triaminoguanidinium]$[N(NO_2)_2]$ [318] melt in the range 80–94°C. Attention is also drawn to the compound $[NH_4][N(NO_2)(SF_5)]$ [261]. While it decomposes on melting at 70°C, and strictly speaking falls outside the conventional definition of ionic liquid, it is included to draw attention to the possible wider use of this anion. The crystals of both hydrazinium dinitramide and hydroxylammonium dinitramide are reported [80] to contain hydrogen-bond-linked $[NH_3NH_2]^+$ or $[NH_3OH]^+$ with $[N(NO_2)_2]^-$, whereas the hydroxylammonium salt contains, in addition, neutral NH_2OH and zwitterionic $[NH_3O]$. However, in none of these cases has the extent of reformation of the component acid and base in the liquid been studied.

Protic ionic liquids obtained by *N*-protonation with 1,3-diketones include some very low-melting salts and are clearly worthy of further study. For example, *N,N,N′*-triethylethylenediammonium hexafluoropentane-2,4-dionate [63] melts at −11°C (U_L 413 kJ mol^{-1}). Morpholinium hexafluoroacetylacetonate [204], 1-butylpyrrolidinium 4,4,4-trifluoro-1-phenylbutane-1,3-dionate [239] and dibutylammonium 4,4,4-trifluoro-1-(2-thienoyl)-1,3-butanedionate [204] all melt between 62 and 71°C, with U_L in the range 409–453 kJ mol^{-1}.

11.4.3 Salts From Strong Acids

O-Protonation of some weak oxygen bases gives rise to some low-melting salts. Studies of the phase behaviour of mixtures of carboxylic acids and strong protic acids reveal the formation of 1,1-dihydroxyethylium ($[CH_3C(OH)_2]^+$) salts of $[SO_3(OH)]^-$ (m.pt. −2.5°C, U_L 545 kJ mol^{-1}) [68], $[SO_3F]^-$ (m.pt. 52.5°C, U_L 547 kJ mol^{-1}) [160] and $[CF_3SO_3]^-$ (m.pt. 35°C, U_L 507 kJ mol^{-1}) [108]. Pyridine *N*-oxide forms an *N*-hydroxy compound, [1-HOpy]$[NMes_2]$

TABLE 11.16 $[EX_4]^-$ and Related Anions

m.pt./°C			p:q	U_L/ kJ mol^{-1}
		E = Al, Ga, or In X = Cl, Br, or alkyl		
37	1-Ethyl-3-methylimidazolium/sodium	$[AlCl_4]$	(1,1):1	
34	1-Butylpyridinium	$[AlCl_4]$	1:1	432
84-86	5-Methyl-1,3,5-dithiazinan-5-ium	$[AlCl_4]$	1:1	450
103	2-*tert*-Butyl-3,3,5-trimethyl-1, 2-diaza-3-sila-5-cyclopentenium	$[AlCl_4]$	1:1	408
107–108	*cyclo*-Hexylammonium	$[AlCl_4]$/Cl	1:(1,1)	
109–112	Benzylammonium	$[AlCl_4]$/Cl	1:(1,1)	
108–109	Di-*tert*-butyl(diphenylmethylene) phosphonium	$[AlCl_4]$	1:1	376
86–92	3-Chloro-3-(di-*iso*-propylamino)-3-phosphonio-bicyclo[3.1.0]hexane	$[AlCl_4]$	1:1	398
50-52	Triphenylarsine{chloro(phenyl)} stibonium	$[AlCl_4]$	1:1	369
94–95	1,1,2,3,4,5,6-Heptamethylbenzenium	$[AlCl_4]$	1:1	414
91–92	1-Chloromethyl-1,2,3,4,5, 6-hexamethylbenzenium	$[AlCl_4]$	1:1	408/410
83–84	1-Phenyl-1,2,3,4,5,6-hexamethylbenzenium	$[AlCl_4]$	1:1	397
59–60	1-Butylisoquinolinium	$[GaCl_4]$	1:1	418
95–97	1,3-Dimethyl-1,3-diaza-2-phospholidinium	$[GaCl_4]$	1:1	447
52–54.5	1,3-Dimethyl-1,3-diaza-2-arsenanium	$[GaCl_4]$	1:1	440
54–56	5-Methyl-1,3,5-dithiazinan-5-ium	$[GaCl_4]$	1:1	451
96–98	Trichloro(*cyclo*-hexyl)phosphonium	$[GaCl_4]$	1:1	424
103–105	Chlorodiphenylphosphine(diphenyl) phosphonium	$[GaCl_4]$	1:1	370
60–61.5	3,3-Diphenyltricyclo[2.1.0.0^{2,5}] pentaphosphan-3-ium	$[GaCl_4]$	1:1	400
101–102	3-Chloro-3-ethyltricyclo[2.1.0.0^{2,5}] pentaphosphan-3-ium	$[GaCl_4]$	1:1	437
46–48	3-Chloro-3-cyclohexyltricyclo[2.1.0.0^{2,5}] pentaphosphan-3-ium	$[GaCl_4]$	1:1	415
79–80	3-Chloro-3-pentafluorophenyltricyclo [2.1.0.0^{2,5}]pentaphosphan-3-ium	$[GaCl_4]$	1:1	416
108–110	1,2-Bis(*P*,*P*-dimethyl-*P*′,*P*′-diphenylphosphinophosphonium) ethane	$[GaCl_4]$	1:2	1163
94–95	1,2-Bis(*P*,*P*,*P*′,*P*′-tetraphenylphosphinophos phonium)ethane	$[GaCl_4]$	1:2	1091

(*Continued*)

TABLE 11.16 (*Continued*)

m.pt./°C			p:q	U_L/ kJ mol^{-1}
64–67	1,6-Bis(P,P,P',P'-tetraphenylphosphinophosphonium)hexane	$[GaCl_4]$	1:2	1075
94–95.5	1,6-Diarsa-2,5,7,10-tetrathiatricyclo[5.3.0^{1,5}.0^{6,10}]decane	$[GaCl_4]$	1:2	1363
104–105	1,6-Diarsa-2,5,7,10-tetramethyl-2,5,7,10-tetraazatricyclo[5.3.0^{1,5}.0^{6,10}]decane	$[GaCl_4]$	1:2	1311
90	Tetrabutylammonium	$[InCl_4]$	1:1	379
liq	1,3-Di-*tert*-butylimidazolium	$[Me_3Al(\mu^3\text{-}CH_2)(AlMe_2)_2(\mu^2\text{-}Me)]$	1:1	370
81	Tetramethylammonium	$[Al(^iBu)_2F_2]$	1:1	420
90	Tetrabutylammonium	$[AlMe_2F_2]$	1:1	389
109	Trimethyltelluronium	$[AlMe_2Cl_2]$	1:1	451
108–110	[(N,N,N',N'',N''-Pentamethyldiethylenetriamine)$AlMe_2$]	$[AlMe_2Cl_2]$	1:1	394
102–104	$[Ti(\eta^5\text{-}C_5H_5)Cl(18\text{-Crown-}6)]$	$[AlMe_2Cl_2]$	1:1	373
53–56	Bis(N,N,N',N'',N''-pentamethyldiethylenetriamine)potassium	$[AlMe_4]$	1:1	355
		$[E_2X_7]^-$; $[E_2X_6]^{2-}$		
0-2	*tert*-Butyl carbenium	$[Al_2Br_7]$	1:1	418
106–109	1,3-Dichloro-1,2,3,4-tetra-*cyclo*-hexyltetraphosphetane-1,3-diium	$[Ga_2Cl_7]$	1:2	1107
85–87	3,3,5,5,7,7-Hexaphenyltricyclo[2.2.1.0^{2,6}]heptaphosphane-3,5,7-triium	$[Ga_2Cl_7]$	1:3	
89	Tetramethylformamidinium	$[Ga_2I_6]$	2:1	946
92	Tetramethylformamidinium	$[In_2Cl_6]$	2:1	
		E = B or Al X = OR		
56	(1,2-Dimethoxyethane)sodium	$[B\{OCH(CF_3)_2\}_4]$	1:1	368
31	1-Ethyl-3-methylimidazolium	$[Al\{OCH(CF_3)_2\}_4]$	1:1	363
34	1-Butyl-3-methylimidazolium	$[Al\{OCH(CF_3)_2\}_4]$	1:1	356
39	1-Ethyl-2,3-dimethylimidazolium	$[Al\{OCH(CF_3)_2\}_4]$	1:1	361
12	1-Allyl-3-methylimidazolium	$[Al\{OCH(CF_3)_2\}_4]$	1:1	361
36	1-Butylpyridinium	$[Al\{OCH(CF_3)_2\}_4]$	1:1	361
69	1-Methyl-1-propylpiperidinium	$[Al\{OCH(CF_3)_2\}_4]$	1:1	356
31	1-Butyl-1-methylmorpholinium	$[Al\{OCH(CF_3)_2\}_4]$	1:1	355
61	Tetramethylammonium	$[Al\{OCH(CF_3)_2\}_4]$	1:1	369
56	Tetraethylammonium	$[Al\{OCH(CF_3)_2\}_4]$	1:1	356/356
83–85	Caesium	$[Al\{OCH(CF_3)_2\}_4]$	1:1	379
75–80	$[Tl_3F_2Al\{OCH(CF_3)_2\}_3]$	$[Al\{OCH(CF_3)_2\}_4]$	1:1	330
96	Tetramethylammonium	$[Al\{OC(CH_3)(CF_3)_2\}_4]$	1:1	359

TABLE 11.16 (*Continued*)

m.pt./°C			$p:q$	U_L/ kJ mol^{-1}
108	Tetrabutylammonium	$[Al\{OC(CH_3)(CF_3)_2\}_4]$	1:1	331
97–99	Nitrosonium	$[Al\{OC(Ph)(CF_3)_2\}_4]$	1:1	340
		E = B X = F, Cl, CN, R, H, or OR		
54–56	$[Li_2(H_2NNHMe)_3]$	$[BH_4]$	1:2	1566
92–94	(Benzo-18-crown-6)bis(pyridine) sodium	$[BH_4]$	1:1	374
62	Bis(1-methyl-1*H*-imidazole-3-yl) dihydroboronium	$[BH_2(CN)_2]$	1:1	438
–1	1-Ethyl-3-methylimidazolium	$[BF_4]$	1:1	485
67	1,3-Didodecylimidazolium	$[BF_4]$	1:1	361
110	1,3-Bis((*R*)-(+)-1-phenylethyl) imidazolium	$[BF_4]$	1:1	410
110	1,3-Di(cyanomethyl)imidazolium	$[BF_4]$	1:1	403
40	1-Butyl-2,3-dimethylimidazolium	$[BF_4]$	1:1	452/453
61	1-Methyl-3-(2-phenylethyl)imidazolium	$[BF_4]$	1:1	445
35	1-Methyl-3-trimethylsilylmethylimidazolium	$[BF_4]$	1:1	448
109–110	1-(2-Diphenylphosphinoethyl)-3-mesitylimidazolium	$[BF_4]$	1:1	377
108	1-Ethylbenzimidazolium	$[BF_4]$	1:1	470
111	1-Butyl-2-methylbenzimidazolium	$[BF_4]$	1:1	433
40–46	1,3-Diethoxyimidazolium	$[BF_4]$	1:1	460
104	5-Azonia-2-oxa-spiro[4.4]nonane	$[BF_4]$	1:1	483
79–80	5-Methyl-1,3,5-dithiazinan-5-ium	$[BF_4]$	1:1	486
75–82	1-Phenyl-4-triphenylphosphanylidene-4*H*-pyrazol-1-ium	$[BF_4]$	1:1	384
72	Tetraethylammonium	$[BF_4]$	1:1	462
76–77	Decyl(tri-*tert*-butyl)phosphonium	$[BF_4]$	1:1	375
85–87	Decyl(tri-*cyclo*-hexyl)phosphonium	$[BF_4]$	1:1	361
112	4-Fluorophenyl(pentafluorophenyl) iodonium	$[BF_4]$	1:1	430
89–91	1,1,2,3,4,5,6-Heptamethylbenzenium	$[BF_4]$	1:1	435
95–97	1,4-Bis(1-methylimidazolium-3-yl) butane	$[BF_4]$	1:2	1412
86	1,7-Bis(1-methylimidazolium-3-yl) heptane	$[BF_4]$	1:2	1368
92	1,8-Bis(1-methylimidazolium-3-yl) octane	$[BF_4]$	1:2	1353
68–71	1,4-Bis(1-methylimidazolium-3-ylmethyl)benzene	$[BF_4]$	1:2	1384
109–111	5-Methyl-1,3,5-dithiazinan-5-ium	$[BCl_4]/Cl$	1:(1,1)	
95	Bis(tetramethylene sulfone)lithium	$[BF_2(ox)]$	1:1?	424
59–61	Piperidinium	$[B(OMe)_4]$	1:1	443

(*Continued*)

TABLE 11.16 (*Continued*)

m.pt./°C			p:q	U_L/ kJ mol^{-1}
56	(1,2-Dimethoxyethane)sodium	$[B\{OCH(CF_3)_2\}_4]$	1:1?	368
90	2-Hydroxyethyl(trimethyl)ammonium (cholinium)	Bis(salicylato) borate	1:1	344
48	1,3-Didodecylimidazolium	$[B(CN)_4]$	1:1	352
78	1-Methyl-1-propylpyrrolidinium	$[B(CN)_4]$	1:1	436
95	Triethylpropylammonium	$[B(CN)_4]$	1:1	423
73	Tetrabutylphosphonium	$[B(CN)_4]$	1:1	376
98	Ethyltriphenylphosphonium	$[B(CN)_4]$	1:1	384
100–104	Bis(1-ethyl-3-methylimidazol-2-ylidene)silver(I)	$[B(CN)_4]$	1:1	398
98–101	Tetraphenylphosphonium	$[B(C_6F_5)(N_3)_3]$	1:1	364
102	Tetrakis(decyl)ammonium	$[B(C_6H_5)_4]$	1:1	310
90	1-Butyl-3-methylimidazolium	$[B\{C_6H_4(4\text{-}C_6F_{13})\}_4]$	1:1	303
102–104	1-Butyl-3-methylimidazolium	$[B\{3,5\text{-}(CF_3)_2C_6H_3\}_4]$	1:1	285
93–94	[(2-Trimethylsilylethyl) dimethylamine)$_2$H]	$[B\{3,5\text{-}(CF_3)_2C_6H_3\}_4]$	1:1	315
70	Bis(4-methoxyphenalene-1, 9-diolato)boron	$[B\{3,5\text{-}(CF_3)_2C_6H_3\}_4]$	1:1	316
98	Bis(3,4-dimethoxyphenalene-1, 9-diolato)boron	$[B\{3,5\text{-}(CF_3)_2C_6H_3\}_4]$	1:1	309
		Others		
92	Tetraphenylphosphonium	$[As(N_3)_4]$	1:1	374
83–110	2,6-Dichloropyridinium	$[Sb^{III}Cl_4]$	1:1	434

(m.pt. 73–79°C, U_L 460 kJ mol^{-1}) [281] and a bridged material $[(pyO)_2H][ClO_4]$ (m.pt. 78°C, U_L 449 kJ mol^{-1}) [304]. 2-Pyridone is protonated by $HNMes_2$ [328] and *N*-methylpyrrolidone [413] by HCl to give the corresponding *O*-protonated salts, m.pt. 80–83°C (U_L 423 kJ mol^{-1}) and 94°C (U_L 525 kJ mol^{-1}), respectively.

For strong protic acids, HA, where A = is inorganic (Cl^-, $[NO_3]^-$, $[ClO_4]^-$, $[SO_3(OH)]^-$, $[SO_3F]^-$, $[N(SO_2F)_2]^-$, ½$[SO_4]^{2-}$) or organic ($[CF_3SO_3]^-$, $[N(SO_2CF_3)_2]^-$, $[N(SO_2Me)_2]^-$ or $[PO_2(OPh)_2]^-$), the tendency of the salt to dissociate into its component acid and base is low. In the former group are ionic liquids that are wholly inorganic. For example, tetrazolium perchlorate (m.pt. 55°C; U_L 556 kJ mol^{-1}) [168], hydrazinium nitrate [208] (m.pt. reportedly [618] 62.5°C (Form II) or 70.5°C (Form I); U_L 624 kJ mol^{-1}), aminoguanidinium perchlorate (m.pt. 72°C; U_L 532 kJ mol^{-1}) [272], diaminouronium nitrate (m.pt. 86°C; U_L 553 kJ mol^{-1}) [356] and 1,5-diaminotetrazolium perchlorate (m.pt. 97–98°C; U_L 523 kJ mol^{-1}) [443] have all been crystallographically characterised, largely in programmes seeking to develop novel energetic materials. In addition, *N*-methyl- and *N,N*-dimethyl-hydrazinium nitrates (m.pt. 42°C; U_L 583 kJ mol^{-1} and m.pt. 45°C; U_L 545 kJ mol^{-1}, respectively) [131], a related triazanium salt,

TABLE 11.17 $[EX_6]^-$ and Related Anions

m.pt./°C			$p:q$	U_L/kJ mol^{-1}
		E = P X = F, C_2F_5 or Cl		
67–68	1,3-Dimethylimidazolium	$[PF_6]$	1:1	483/483
58–60	1-Ethyl-3-methylimidazolium	$[PF_6]$	1:1	465/473
2	1-Butyl-3-methylimidazolium	$[PF_6]$	1:1	452/452
83	1-(2-Butyl)-3-methylimidazolium	$[PF_6]$	1:1	450
31	1-Decyl-3-methylimidazolium	$[PF_6]$	1:1	407
60	1-Dodecyl-3-methylimidazolium	$[PF_6]$	1:1	400
77	1-Tetradecyl-3-methyimidazolium	$[PF_6]$	1:1	385–390
99	1,3-Dibutyl-2,4, 5-trimethylimidazolium	$[PF_6]$	1:1	410
93	1,3-Dibenzylimidazolium	$[PF_6]$	1:1	412
108	1,3-Bis((*R*)-(+)-1-phenylethyl) imidazolium	$[PF_6]$	1:1	404
79	1,3-Bis-(3-cyanopropyl)imidazolium	$[PF_6]$	1:1	429
40	1-Butyl-2,3-dimethylimidazolium	$[PF_6]$	1:1	444/445
62–64	1-Methyl-2-phenyl- 3-propylimidazolinium	$[PF_6]$	1:1	428
85	1-(3-Cyanopropyl)-2, 3-dimethylimidazolium	$[PF_6]$	1:1	445
63	1-(2-Methoxyethyl)-2, 3-dimethylimidazolium	$[PF_6]$	1:1	452
75	1-(3-Cyanopropyl)- 3-methylimidazolium	$[PF_6]$	1:1	455
32	1-(2-Di-*iso*-propylaminoethyl)- 3-methylimidazolium	$[PF_6]$	1:1	418
103–105	[1-{2-((4-Chlorobenzylidene)amino) ethyl}-3-methylimidazolium	$[PF_6]$	1:1	420
83–84	1,3-Dimethoxyimidazolium	$[PF_6]$	1:1	466/471
99–102	1,3-Diethoxyimidazolium	$[PF_6]$	1:1	444
83–84	1-Benzyloxy-3-methylimidazolium	$[PF_6]$	1:1	437
94–95	4-Amino-1-ethyl-3,5-dimethyl-1,2, 4-triazolium	$[PF_6]$	1:1	455
66	1-Heptyl-1-methylpyrrolidinium	$[PF_6]$	1:1	421
106–107	1-(Piperidinium-1-ylidene)- 3-(piperidin-1-yl)-2-azapropene	$[PF_6]$	1:1	422
87–88	1,3-Diethyl-3,4,5,6- tetrahydropyrimidinium	$[PF_6]$	1:1	448
27	Tetrakis(1-dodecylimidazole) copper(I)	$[PF_6]$	1:1	303
66–68	1-(Ferrocen-1-ylmethyl)- 3-methylimidazolium	$[PF_6]$	1:1	422
71	1,10-Bis(1-methylimidazolium- 3-yl)decane	$[PF_6]$	1:2	1294

(*Continued*)

TABLE 11.17 (*Continued*)

m.pt./°C			p:q	U_L/kJ mol^{-1}
96–98	3,5-Bis(3-butylimidazolium-1-ylmethyl)toluene	$[PF_6]$	1:2	1254
87	1,8-Bis(1-methylimidazolium-3-yl)-3,6-dioxaoctane	$[PF_6]$	1:2	1325
99.5	*N,N'''*-Bis(2-ethoxyethyl)-*N,N,N''',N'''*-tetramethylbutane-1,4-diammonium	$[PF_6]$	1:2	1292
75–76	1,3-Dimethoxy-2-methylimidazolium	$[PF_3(C_2F_5)_3]$	1:1	396
2	1-Butyl-1-methylpyrrolidinium	$[PF_3(C_2F_5)_3]$	1:1	
ca. 65	Tetrabutylammonium	$[PF_3(C_2F_5)_3]$	1:1	
78–80	1-Butyl-4-methylpyridinium	$[PCl_6]$	1:1	417
		E = As or Sb; X = F or Cl		
53	1-Ethyl-3-methylimidazolium	$[AsF_6]$	1:1	462
10	1-Ethyl-3-methylimidazolium	$[SbF_6]$	1:1	460
41	1,3-Didodecylimidazolium	$[SbF_6]$	1:1	358
44.5	1-Butyl-2,3-dimethylimidazolium	$[SbF_6]$	1:1	432
108	5-Ethoxy-l-ethyl-3,4-dihydro-2-oxo-2*H*-pyrrolium	$[SbCl_6]$	1:1	413
110–112	β,β-Diethoxyethenediazonium	$[SbCl_6]$	1:1	420
89–90	2-Hydroxyhomotropylium	$[SbCl_6]$	1:1	432
103–104	1-Ethoxyhomotropylium	$[SbCl_6]$	1:1	417
104–106	Tetraphenylphosphonium	$[Sb(N_3)_6]$	1:1	362
		E = Nb or Ta; X = F or Cl		
–1	1-Ethyl-3-methylimidazolium	$[NbF_6]$	1:1	462
2	1-Ethyl-3-methylimidazolium	$[TaF_6]$	1:1	462
83–85	1-Butyl-3-methylpyridinium	$[NbCl_6]$	1:1	413
89–91	1-Butyl-4-methylpyridinium	$[NbCl_6]$	1:1	412
82–85	Bis(h^2-dimethyldisulfide)tetrachloroniobium(V)	$[NbCl_6]$	1:1	390
103	1-Butyl-4-methylpyridinium	$[TaCl_6]$	1:1	412

2,2-dimethyltriazanium nitrate (m.pt. 99°C; U_L 540 kJ mol^{-1}) [456] and the 1:2 salt *N,N'*-dimethylhydrazin-1,2-diium dinitrate (m.pt. 98°C) [242] have all been studied. The single-crystal structure of $[MeNH_2NMeH_2][SO_4]$ [242], one of the few ionic liquids (m.pt. 67°C) containing both a doubly charged cation and a doubly charged anion, has also been reported. $[AlCl_4]^-$ has been characterised in mixed salts with chloride, $[cat]_2[AlCl_4]Cl$, where $[cat]^+ = [CyNH_3]^+$ [513] and $[PhCH_2NH_3]^+$ [531] (m.pt. 107–108 and 109–112°C, respectively). A few low-melting salts of the hydronium (or oxonium) ion, $[H_3O]^+$, and its hydrates are reported. $[H_3O][CF_3SO_3]$ melts congruently at 34.5°C and its crystal structure has been determined [107] at 293 and 83 K (giving two values of U_L, 553 and

TABLE 11.18 Sulfonyl and Phosphonyl Amides

m.pt./°C			p:q	U_L/kJ mol^{-1}
Liq	1-Ethyl-3-methylimidazolium	$[N(SO_2Me)_2]$	1:1	446
105–110	1,2,4-Triazolium	$[N(SO_2Me)_2]$	1:1	478
90–95	2,6-Dimethylpyridinium	$[N(SO_2Me)_2]$	1:1	445
73–79	1-Hydroxypyridinium	$[N(SO_2Me)_2]$	1:1	460
80–90	Pyrrolidinium	$[N(SO_2Me)_2]$	1:1	466
40–55	1,1-Dimethylpyrrolidinium	$[N(SO_2Me)_2]$	1:1	448
37	1-Butyl-1-methylpyrrolidinium	$[N(SO_2Me)_2]$	1:1	424
95–97	Piperidinium	$[N(SO_2Me)_2]$	1:1	455
56–62	Methyluronium	$[N(SO_2Me)_2]$	1:1	475
67	1,1-Dimethyluronium	$[N(SO_2Me)_2]$	1:1	463
82–84	Ethylammonium	$[N(SO_2Me)_2]$	1:1	481
75–77	Methylethylammonium	$[N(SO_2Me)_2]$	1:1	471
71–73	Diethylammonium	$[N(SO_2Me)_2]$	1:1	454
76–77	2-Hydroxyethyl(trimethyl) ammonium (cholinium)	$[N(SO_2Me)_2]$	1:1	449
80–83	Bis(2-pyridone)hydrogen	$[N(SO_2Me)_2]$	1:1	423
70	$[(NMe_3CH_2CO_2)_2H]$	$[N(SO_2Me)_2]$	1:1(*S*)	
102–105	Aquo(pyridine-N-oxide) trimethyltin(IV)	$[N(SO_2Me)_2]$	1:1	415
72	Tetramethylammonium	$[N(SO_2F)_2]$	1:1	465
70	Triphenylphosphonium	$[N(SO_2F)_2]$	1:1	402
63	Potassium	$[N(SO_2F)_2]$	1:1	541–547
33	Bis(*N,N'*dimethylethyleneurea-*O*) trimethyltin(IV)	$[N(SO_2F)_2]$	1:1	386
66	Bis(dimethylsulfoxide-*O*) trimethyltin(IV)	$[N(SO_2F)_2]$	1:1	406
	Imidazolium			
52/9	1-Methylimidazolium	$[N(SO_2CF_3)_2]$	1:1	445
22	1,3-Dimethylimidazolium	$[N(SO_2CF_3)_2]$	1:1	435
65	1,3-Dimethylimidazolium/lithium	$[N(SO_2CF_3)_2]$	1:1	389
−26	1-Ethyl-3-methylimidazolium	$[N(SO_2CF_3)_2]$	1:1	420–426
−3	1-Butyl-3-methylimidazolium	$[N(SO_2CF_3)_2]$	1:1	413/413
−10	1-Hexyl-3-methylimidazolium	$[N(SO_2CF_3)_2]$	1:1	394/402
57	1,2,3-Triethylimidazolium	$[N(SO_2CF_3)_2]$	1:1	410
ca. 50	1-Ethyl-2-methyl-3-benzylimidazolium	$[N(SO_2CF_3)_2]$	1:1	396
41	1-Methyl-3-(2-phenylethyl) imidazolium	$[N(SO_2CF_3)_2]$	1:1	404
50	1-(3-Phenylpropyl)-3-methylimidazolium	$[N(SO_2CF_3)_2]$	1:1	397
62	1-(2,2-Diphenylethyl)-3-methylimidazolium	$[N(SO_2CF_3)_2]$	1:1	380
69–71	1-{2-(Diethylammonio)ethyl}-3-methylimidazolium	$[N(SO_2CF_3)_2]$	1:2	1261

(*Continued*)

TABLE 11.18 (*Continued*)

m.pt./°C			$p:q$	U_L/kJ mol^{-1}
40	1-(Hydroxycarbonylmethyl)-3-methylimidazolium	$[N(SO_2CF_3)_2]$	1:1	424
45	*rac*-1-{1-(Ethoxycarbonyl)ethyl}-3-methylimidazolium	$[N(SO_2CF_3)_2]$	1:1	406
50	1-Butyl-3-ferrocenylimidazolium	$[N(SO_2CF_3)_2]$	1:1	383
110/113	1,2,3-Trimethyl-4-nitroimidazolium	$[N(SO_2CF_3)_2]$	1:1	423
110	1-Ethyl-2-(1-oxy-3-oxo-4,4,5,5-tetramethylimidazolin-2-yl)-3-methylimidazolium	$[N(SO_2CF_3)_2]$	1:1	384
112	6,6′-Spirobis(1-methyl-1,5,6,7-tetrahydropyrrolo[1,2-a]imidazol-4-ium	$[N(SO_2CF_3)_2]$	1:2	1245
72	1,3-Dihydroxy-2-methylimidazolium	$[N(SO_2CF_3)_2]$	1:1	435
28–30	2-Bromo-1,3-dimethoxymidazolium	$[N(SO_2CF_3)_2]$	1:1	415
58–59	1,3-Diamino-2-methylimidazolium	$[N(SO_2CF_3)_2]$	1:1	433
	Triazolium			
65–66	1-Heptyl-4-(1*H*,1*H*,2*H*,2*H*-perfluorooctyl)-1,2,4-triazolium	$[N(SO_2CF_3)_2]$	1:1	357
105–106	4-Amino-1,3,5-trimethyl-1,2,4-triazolium	$[N(SO_2CF_3)_2]$	1:1	427
65–66	4-Amino-1-ethyl-3,5-dimethyl-1,2,4-triazolium	$[N(SO_2CF_3)_2]$	1:1	420
109–110	4-Amino-3,5-diethyl-1-methyl-1,2,4-triazolium	$[N(SO_2CF_3)_2]$	1:1	412
106–107	4-Amino-1,3,5-triethyl-1,2,4-triazolium	$[N(SO_2CF_3)_2]$	1:1	408
	Pyridinium			
−4	1-Hexylpyridinium	$[N(SO_2CF_3)_2]$	1:1	401
63–65	1-Methyl-3-cyanopyridinium	$[N(SO_2CF_3)_2]$	1:1	434
65–66	1-Methyl-4-cyanopyridinium	$[N(SO_2CF_3)_2]$	1:1	429
44	1-(Cyanomethyl)pyridinium	$[N(SO_2CF_3)_2]$	1:1	435
51	1-(Cyanomethyl)-2-methylpyridinium	$[N(SO_2CF_3)_2]$	1:1	426
30	1-(Cyanomethyl)-2-ethylpyridinium	$[N(SO_2CF_3)_2]$	1:1	419
48	1-(Cyanomethyl)-4-ethylpyridinium	$[N(SO_2CF_3)_2]$	1:1	420
32	1-(Hydroxycarbonymethyl) pyridinium	$[N(SO_2CF_3)_2]$	1:1	426
Liq	1-(3-Cyanopropyl)pyridinium	$[N(SO_2CF_3)_2]$	1:1	416
Liq	1-(3-Cyanopropyl)-3-methylpyridinium	$[N(SO_2CF_3)_2]$	1:1	412
Liq	1-(3-Cyanomethyl)-2,5-dimethylpyridinium	$[N(SO_2CF_3)_2]$	1:1	419
28–29	1-Ethyl-3-trifluoromethylpyridinium	$[N(SO_2CF_3)_2]$	1:1	417
36–38	1-Ethyl-2-cyanopyridinium	$[N(SO_2CF_3)_2]$	1:1	418
73	1-Ethyl-3-cyanopyridinium	$[N(SO_2CF_3)_2]$	1:1	420
33–35	1-Ethyl-4-cyanopyridinium	$[N(SO_2CF_3)_2]$	1:1	419

TABLE 11.18 (*Continued*)

m.pt./°C			p:q	U_L/kJ mol^{-1}
	Pyrrolidinium, piperidinium and morpholinium			
91	1-Ethyl-1-methylpyrrolidinium	$[N(SO_2CF_3)_2]$	1:1	419/423
12	1-Methyl-1-propylpyrrolidinium	$[N(SO_2CF_3)_2]$	1:1	413/414
−11	1-Butyl-1-methylpyrrolidinium	$[N(SO_2CF_3)_2]$	1:1	409
9	1-Hexyl-1-methylpyrrolidinium	$[N(SO_2CF_3)_2]$	1:1	393
39	1-(Hydroxycarbonylmethyl)-1-methylpyrrolidinium	$[N(SO_2CF_3)_2]$	1:1	421
91	1-Ethyl-1-methylpiperidinium	$[N(SO_2CF_3)_2]$	1:1	385/416
12	1-Methyl-1-propylpiperidinium	$[N(SO_2CF_3)_2]$	1:1	408
46	1-(Ethoxycarbonylmethyl)-1-methylmorpholinium	$[N(SO_2CF_3)_2]$	1:1	409
77	1-Ethyl-1,4-diazabicyclo[2.2.2]octanium	$[N(SO_2CF_3)_2]$	1:1	413–418
42	1,1,3,3-Tetramethylguanidinium	$[N(SO_2CF_3)_2]$	1:1	422/425
	Ammonium			
79–81	Trimethylammonium	$[N(SO_2CF_3)_2]$	1:1	445
~0	2-Hydroxyethyl(trimethyl)ammonium (cholinium)	$[N(SO_2CF_3)_2]$	1:1	426
50	2-Chloroethyl(trimethyl)ammonium	$[N(SO_2CF_3)_2]$	1:1	427
57	(Hydroxycarbonylmethyl)trimethylammonium (protonated betaine)	$[N(SO_2CF_3)_2]$	1:1	427
78	Dimethylbis(3-methylbenzyl)ammonium	$[N(SO_2CF_3)_2]$	1:1	383
106	Dimethylbis(4-methoxybenzyl)ammonium	$[N(SO_2CF_3)_2]$	1:1	379
104	Tetraethylammonium	$[N(SO_2CF_3)_2]$	1:1	412
105	Tetrapropylammonium	$[N(SO_2CF_3)_2]$	1:1	390
86	*N,N′*-Bis(2-ethoxyethyl)-*N,N,N′,N′*-tetramethylethane-1,2-diammonium	$[N(SO_2CF_3)_2]$	1:2	1213
65–66	Diphenyliodonium	$[N(SO_2CF_3)_2]$	1:1	403
77–79	(2,2,2-Trifluoroethyl)phenyliodonium	$[N(SO_2CF_3)_2]$	1:1	411
65	Bis(ethanenitrile)copper(I)	$[N(SO_2CF_3)_2]$	1:1	
66	Tetrakis(ethanenitrile)copper(I)	$[N(SO_2CF_3)_2]$	1:1	
89	Tetrakis(1-methylimidazole)copper(II)	$[N(SO_2CF_3)_2]$	1:2	1206
74	Tetrakis(1-ethylimidazole)copper(II)	$[N(SO_2CF_3)_2]$	1:2	1156
46	Tetrakis(1-butylimidazole)copper(II)	$[N(SO_2CF_3)_2]$	1:2	1118
52	Hexakis(1-butylimidazole)copper(II)	$[N(SO_2CF_3)_2]$	1:2	1137
90	(Ethanenitrile)silver(I)	$[N(SO_2CF_3)_2]$	Polymer	455

(*Continued*)

TABLE 11.18 (*Continued*)

m.pt./°C			p:q	U_L/kJ mol^{-1}
30	(1-Butylimidazole)(1-methylimidazole)silver(I)	$[N(SO_2CF_3)_2]$	1:1	387
27	(1-Butylimidazole)(1-ethylimidazole)silver(I)	$[N(SO_2CF_3)_2]$	1:1	388
87	Bis(1-methylimidazole)silver(I)	$[N(SO_2CF_3)_2]$	1:1	
65	Bis(1-ethylimidazole)silver(I)	$[N(SO_2CF_3)_2]$	1:1	398
42	Bis(1-butylimidazole)silver(I)	$[N(SO_2CF_3)_2]$	1:1	378
60	(3-Methylimidazole)(trimethylamine) dihydroboronium	$[N(SO_2CF_3)_2]$	1:1	407
88	$[Dy_2(NMe_3CH_2CO_2)_8(H_2O)_4]$	$[N(SO_2CF_3)_2]$	1:6	
57	1-Hexadecylpyridinium	Acesulfamate	1:1	374
25	2-Hydroxyethyl(trimethyl) ammonium (cholinium)	Acesulfamate	1:1	449
95–98	3-Hydroxy-1-(octyloxymethyl) pyridinium	Saccharinate	1:1	395
69	2-Hydroxyethyl(trimethyl) ammonium (cholinium)	Saccharinate	1:1	439
98	Tetramethylammonium	$[N\{P(O)(C_2F_5)_2\}_2]$	1:1	385
52	1-Ethyl-3-methylimidazolium	$[C(C_6F_5)(SO_2CF_3)_2]$	1:1	393

561 kJ mol^{-1}, from the higher- and lower-temperature data, respectively). A high-melting oxonium salt of a disulfonic acid, $[H_3O]_2[O_3SCH_2CH_2SO_3]$, m.pt. 112°C [546], is also worth mentioning. Single crystals of an oxonium monohydrate, $[H_5O_2][CF_3SO_3]$ (m.pt. −6°C; U_L 525 kJ mol^{-1} from data collected at 225 K and U_L 530 kJ mol^{-1} from data collected at 85 K) [65], and an oxonium dihydate, $[H_7O_3][NO_3]$ (m.pt. −18.5°C; U_L 580 kJ mol^{-1}) [60], have also been crystallographically characterised using X-ray diffraction. While the lattice energies of the diphenylphosphate salts of imidazolium (m.pt. 99–100°C; U_L 425 kJ mol^{-1}) [395] and 1-methylimidazolium (m.pt. 91–92.5°C; U_L 423 kJ mol^{-1}) [395] are similar, they differ in their solid-state structure, the former existing as hydrogen-bonded chains and the latter as hydrogen-bonded ion pairs.

11.4.4 Salts of $[X(HX)_n]^-$ (X = Halide)

$[NMe_3H][H_2F_3]$ (m.pt. 0°C, U_L 534 kJ mol^{-1}) [57] may be compared with the corresponding triethylammonium salt, $[NEt_3H][H_2F_3]$ (m.pt. −27°C, U_L 479 kJ mol^{-1}) [57]. Both crystallise in the orthorhombic system. In both, the N—H proton is involved in strong hydrogen bonding to the central fluoride of the $[H_2F_3]^-$ ion, resulting in a discrete ion-pairing arrangement. Some evidence of weak C—H···F hydrogen bonding is seen, though the authors note the difficulty of discriminating between this and 'mere packing contacts'. The lattice

TABLE 11.19 Alkyl- and Arylsulfonates

m.pt./°C			$p:q$	U_L/ kJ mol^{-1}
60–65 est	1-Butyl-3-methylimidazolium	[CH_3SO_3]	1:1	454
110–112	Tri-*iso*-propyl-[*N*-(2,6-Dichlorophenyl) thiocarbamoyl]phosphonium	[CH_3SO_3]	1:1	393
–26	1-Ethyl-3-methylimidazolium	[CF_3SO_3]	1:1	459
7	1-Butyl-3-methylimidazolium	[CF_3SO_3]	1:1	434
85	1-Methyl-3-(tetrahydro-2*H*-pyran-2-ylmethyl)] imidazolium	[CF_3SO_3]	1:1	433
95	Tetrabutyl ammonium/ 1-methyl-3-(tetrahydrofuran-3-ylmethyl)imidazolium	[CF_3SO_3]	(1,1):1	
73	1-Butyl-3-methyl-4, 5-dibromoimidazolium	[CF_3SO_3]	1:1	428
92/94	1-Ethyl-3-methyl-4-nitroimidazolium	[CF_3SO_3]	1:1	453
105/107	1,3-Dimethyl-2-nitromidazolium	[CF_3SO_3]	1:1	466
85	Tetrabutylammonium/3-[(2*S*)-2-methylbutyl]thiazolium	[CF_3SO_3]	(1,1):1	
88–90	*N*-Methylbenzothiadiazolium	[CF_3SO_3]	1:1	454
90–92	3-(3-Butynyl)-2-(1-hexynyl) benzothiazolium	[CF_3SO_3]	1:1	402
58–60	1-Benzyl-3,5-bis(3-trifluoromethylphenyl) pyridinium	[CF_3SO_3]	1:1	377
79–82	1-Methyl-4-(benzylselenomethyl) pyridinium	[CF_3SO_3]	1:1	415
98.5–102	2-Chloro-1-methyl-4-(phenylamino)pyridinium	[CF_3SO_3]	1:1	427
60–62	1-{(3-(Dimethylammonio) propyl}-2-hydroxypyridinium	[CF_3SO_3]	1:2	1376
97–99	1-(3-Butynyl)-2-(phenylethynyl) pyrimidinium	[CF_3SO_3]	1:1	412
76	3-Methyl-2-phenyl-3,4,5,6-tetrahydropyrimidinium	[CF_3SO_3]	1:1	434
90–92	1,4,7-Trimethyl-1-azonia-4,7-diazacyclononane	[CF_3SO_3]	1:1	424
105	Hydroxylammonium	[CF_3SO_3]	1:1	512
53.5–56	*N,N*-Dimethyl-*N*-((methyltellurenyl)(phenyl) methylidene)ammonium	[CF_3SO_3]	1:1	427

(*Continued*)

TABLE 11.19 (*Continued*)

m.pt./°C			$p:q$	U_L/ kJ mol^{-1}
74	1-(1-Methyl-2-butyn-1-ylidene) pyrrolidinium	$[CF_3SO_3]$	1:1	444
57–58	Trimethylphosphine(diphenyl) phosphonium	$[CF_3SO_3]$	1:1	405
80–82	(Diphenylphosphinomethyl (diphenylphosphino) diphenylphosphonium	$[CF_3SO_3]$	1:1	350
109	Dimethylphenylsulfonium	$[CF_3SO_3]$	1:1	451
97	Methyldiphenylsulfonium	$[CF_3SO_3]$	1:1	423
87–89	Bis(pyridine)bromonium	$[CF_3SO_3]$	1:1	429
34.5	Oxonium	$[CF_3SO_3]$	1:1	553/561
−6	Oxonium monohydrate	$[CF_3SO_3]$	1:1	525/530
35	1,1-Dihydroxyethylium	$[CF_3SO_3]$	1:1	507
71	1,4-Bis(1-methylimidazolium-3-yl)butane	$[CF_3SO_3]$	1:2	1355
59–62	1,4-Bis(1-butylimidazolium-3-ylmethyl)benzene	$[CF_3SO_3]$	1:2	1256
92–94	1,4-Bis(1,2-dimethylimidazolium-3-ylmethyl)benzene	$[CF_3SO_3]$	1:2	1301
69	*N,N'''*-Bis(2-ethoxyethyl)-*N,N,N''',N'''*-tetramethylbutane-1,4-diammonium	$[CF_3SO_3]$	1:2	1278
45	1-Ethyl-3-methylimidazolium	$[HCF_2CF_2SO_3]$	1:1	452
28	1-Ethyl-3-methylimidazolium	$[C_4F_9SO_3]$	1:1	423
89–91	Trimethylsulfonium	$[ICF_2CF_2OCF_2CF_2SO_3]$	1:1	422
89	1-Cyclopentyl-3-methylimidazolium	$[4\text{-}CH_3C_6H_4SO_3]$	1:1	421
70	1-Hexyl-3-methylimidazolium	$[4\text{-}CH_3C_6H_4SO_3]$	1:1	407
74–76	Tributyl(ethyl)phosphonium	$[4\text{-}CH_3C_6H_4SO_3]$	1:1	381
107.5–108	Dicyclohexyl(hydroxo) selenonium	$[4\text{-}BrC_6H_4SO_3]$	1:1	398
88–90	1-Methyl-3-propylimidazolium	$[Ph_2PC_6H_4(4\text{-}SO_3)]$	1:1	380
96–98	Methylammonium	$[4\text{-}MeOC_6H_4CH(CH_2NO_2)SO_3]$	1:1	437
106–108	1,4-Bis(1-methylimidazolium-3-yl)butane	22-Hydroxy-5-sulfonatobenzoate	2:2	
112	Oxonium	$[O_3SCH_2CH_2SO_3]$	2:1	1497

potential energy of the ethyl salt is significantly less than that for the methyl salt (three methylene residues leading to a reduction of 55 kJ mol^{-1} in U_L) that may be linked (in this instance) with a significant reduction in m.pt. The loss of hydrogen bonding in $[NMe_4][H_2F_3]$ (m.pt. 110°C, U_L 519 kJ mol^{-1} [532];

TABLE 11.20 Carboxylates

m.pt./°C			p:q	U_L/ kJ mol^{-1}
30	1,3-Diethylimidazolium	$[CH_3CO_2]$	1:1	474
97–98	Hydrazinium	$[CH_3CO_2]$	1:1	590
97–99	2-Amino-3-methylpyridinium	$[CH_3CO_2]$	1:1	493
88	Tris(dimethylamino) methyliminiumphosphorane	$[CH_3CO_2]$	1:1	436
62	1,1,3,3-Tetramethylguanidinium	$[C_2H_5CO_2]$	1:1	465
95	*rac*-(1-Phenylethyl)ammonium	$[ClCH_2CO_2]$	1:1	458
82	Pyridinium	$[CF_3CO_2]$	1:1	501
101	4-Methylpyridinium	$[CF_3CO_2]$	1:1	483
74	4-Cyanopyridinium	$[CF_3CO_2]$	1:1	484
74	1-Butyl-1-methylazepanium	$[CF_3CO_2]$	1:1	430
92–94	3-(3-Ethenyl-1,2,4-oxadiazol-5-yl)-1,2,5,6-tetrahydropyridinium	$[CF_3CO_2]$	1:1	442
88	2-(Hydroxoiodinium)nitrobenzene	$[CF_3CO_2]$	1:1	466
96–98	2-Acetyl-9-azoniabicyclo [4.2.1]nonan-3-one	$[CF_3CO_2]$	1:1	442
107–108	3,6-Dioxaoctane-1, 8-diammonium	$[CCl_3CO_2]$	1:2	1374
43	1-Butyl-2, 3-dimethylimidazolium	$[HC{\equiv}CCO_2]$	1:1	449
110–111	2-Amino- 5-bromopyridinium	$[HC{\equiv}CCO_2]$	1:1	484
84	Ammonium	[*trans*-$CH_3CH{=}CHCO_2$]	1:1	546
110	Piperazinediium	$[C_5H_{11}CO_2]$	1:2	1422
97	Piperazinediium	$[C_6H_{13}CO_2]$	1:2	1357
96	Piperazinediium	$[C_9H_{19}CO_2]$	1:2	1280
96	Piperazinediium	$[C_{11}H_{23}CO_2]$	1:2	1240
96	Piperazinediium	$[C_{13}H_{27}CO_2]$	1:2	1207
97	Piperazinediium	$[C_{15}H_{31}CO_2]$	1:2	1178
73	Tris(2-hydroxyethyl)ammonium	$[(4\text{-}ClC_6H_4)SCH_2CO_2]$	1:1	421
101	1-(2-Hydroxyethyl)piperidinium	$[\{2\text{-}(2,6\text{-}Cl_2C_6H_3NH)C_6H_4\}CH_2CO_2]$	1:1	394
92	1-(2-Hydroxyethyl)morpholinium	$[\{2\text{-}(2,6\text{-}Cl_2C_6H_3NH)C_6H_4\}CH_2CO_2]$	1:1	397
82–85	3-Hydroxypyridinium	$[C_6H_5CO_2]$	1:1	473
106–108	2-Amino-4-phenyl-6-methylpyrimidinium	$[C_6H_5CO_2]$	1:1	430
28	1-(2-Hydroxyethyl)pyrrolidinium	$[C_6H_5CO_2]$	1:1	448
86–88	$^{t}BuNHP(\mu\text{-}N^{t}Bu)_2P^+\{(NH^{t}Bu)(NCO_2{}^{i}Pr)\}NH(CO_2{}^{i}Pr)]$	$[C_6H_5CO_2]$	1:1	339
104–106	2-Aminopyrimidinium	$[(2\text{-}HO)C_6H_4CO_2]$	1:1	464
106	3,3-Dinitroazetidinium	$[(2\text{-}HO)C_6H_4CO_2]$	1:1	449
49	1-(2-Hydroxyethyl)pyrrolidinium	$[(2\text{-}HO)C_6H_4CO_2]$	1:1	448

(*Continued*)

TABLE 11.20 (*Continued*)

m.pt./°C			*p*:*q*	U_L/ kJ mol^{-1}
95–98	Tetraethylammonium	$[(2\text{-HO})C_6H_4CO_2]$	1:1	426
104–105	1-(Hydroxycarbonylmethyl)-1-methylmorpholinium	$[(2\text{-HO})C_6H_4CO_2]$	1:1	437
88	2,4,6-Trimethylpyridinium	$[(2\text{-}O_2N)C_6H_4CO_2]$	1:1	436
106–108	{1-(Hydroxymethyl)propyl} ammonium	$[(4\text{-}O_2N)C_6H_4CO_2]$	1:1	447
64–66	2-Amino-4-methylpyrimidinium	$[3\text{-}FC_6H_4CO_2]$	1:1	459
95–96	2-Amino-4-methylpyrimidinium	$[3\text{-}ClC_6H_4CO_2]$	1:1	454
81–84	Tris(2-hydroxyethyl)ammonium	$[2\text{-}HC(O)C_6H_4CO_2]$	1:1	429
97	1-(2-Hydroxyethyl)pyrrolidinium	$[\{2,5\text{-}(HO)_2\}C_6H_3CO_2]$	1:1	447
110	2-(2,4-Dinitrobenzyl)pyridinium	$[2\text{-HO-}3,5\text{-}(O_2N)_2C_6H_2CO_2]$	1:1	400
99–103	2-Amino-2-thiazolinium	*N*-Methylpyrrole-2-carboxylate	1:1	461
104–107	2-Aminobenzothiazolium	*N*-Methylpyrrole-2-carboxylate	1:1	444
106–108	1,4-Bis(1-methylimidazolium-3-yl)butane	2-Hydroxy-5-Sulfonatobenzoate	2:2	
95–98	Tetraethylammonium	$[(2\text{-HO})C_6H_4CO_2]$	1:1	426
78	Tetraethylammonium	$[(2\text{-HO})C_6H_4COS]$	1:1	422
78–80	Tetraethylammonium	$[(2\text{-HO})C_6H_4CS_2]$	1:1	417
101–102	2,6-Dimethylpiperidinium	$[(2\text{-HO})C_6H_4COS]$	1:1	429
102–104	2,6-Dimethylpiperidinium	$[(2\text{-HO})C_6H_4COS]$	1:1	432
105	1,2-Dimethylimidazolium	$[HO_2C(CH_2)_2CO_2]$	1:1	469
102	1,2-Dimethylimidazolium	$[HO_2C(CH_2)_3CO_2]$	1:1	464
105	2-Methylimidazolium	$[HO_2C(CH_2)_3CO_2]$	1:1	467
106	4-Methylimidazolium	$[HO_2C(CH_2)_3CO_2]$	1:1	472
101.5–103	Imidazolium	$[HO_2C(CH_2)_4CO_2]$	1:1	465
60	2-Methylimidazolium	$[HO_2C(CH_2)_6CO_2]$	1:1	441
85	4-Methylimidazolium	$[HO_2C(CH_2)_6CO_2]$	1:1	441
41–43	2-Ethylimidazolium	*d*-$[HO_2CCH(OH)CH(OH)CO_2]$	1:1	462
98–101	1-Methylimidazolium	*d*-$[HO_2CCH(OH)CH(OH)CO_2]$	1:1	470/471
102	2-Aminopyridinium	$[O_2CCH_2CO_2]$	2:1	1224/1232
98–100	2-Aminopyridinium	$[O_2C(CH_2)_8CO_2]$	2:1	1050
71	1-Aza-8-azoniabicyclo[6.6.3]heptadeca-4,11-diyne	$[(CF_3CO_2)_2H]$	1:1	394
72	1-Aza-8-azoniabicyclo[6.6.4]octadeca-4,11-diyne	$[(CF_3CO_2)_2H]$	1:1	390
75	1-Aza-8-azoniabicyclo[6.6.5]nonadeca-4,11-diyne	$[(CF_3CO_2)_2H]$	1:1	385
103	Tetrabutylammonium	$[(PhCO_2)_2H]$	1:1	361
80–82	Tetramethylstibonium	$[(PhCO_2)_2H]$	1:1	399
109	Potassium	$[(HCO_2)_2H]$	1:1	576
108–110	2,6-Bis(2-pyridinio)pyridine	$[(CF_3CO_2)_2H]$	1:2	1323

TABLE 11.21 $[ER_4]^+$ and $[ER_3]^+$ Ionic Liquids

m.pt./°C			p:q	U_L/ kJ mol^{-1}
	Tetraalkylammonium			
110	Tetramethylammonium	$[F(HF)_2]$	1:1	519
72	Tetramethylammonium	$[N(SO_2F)_2]$	1:1	465
98	Tetramethylammonium	$[N\{P(O)(C_2F_5)_2\}_2]$	1:1	385
61	Tetramethylammonium	$[Al\{OCH(CF_3)_2\}_4]$	1:1	369
96	Tetramethylammonium	$[Al\{OC(CH_3)(CF_3)_2\}_4]$	1:1	359
81	Tetramethylammonium	$[Al(i\text{-}Bu)_2F_2]$	1:1	420
76–77	2-Hydroxyethyl(trimethyl) ammonium (cholinium)	$[N(SO_2Me)_2]$	1:1	449
~0/30	2-Hydroxyethyl(trimethyl) ammonium (cholinium)	$[N(SO_2CF_3)_2]$	1:1	424
25	2-Hydroxyethyl(trimethyl) ammonium (cholinium)	Acesulfamate	1:1	449
69	2-Hydroxyethyl(trimethyl) ammonium (cholinium)	Saccharinate	1:1	439
90	2-Hydroxyethyl(trimethyl) ammonium (cholinium)	Bis(salicylato)borate	1:1	344
76	2-Hydroxyethyl(trimethyl) ammonium (cholinium)	$[Cu^{II}Cl_4]$/Cl	3:(1,2)	
50	2-Chloroethyl(trimethyl) ammonium	$[N(SO_2CF_3)_2]$	1:1	
57	(Hydroxycarbonylmethyl) trimethylammonium (protonated betaine)	$[N(SO_2CF_3)_2]$	1:1	427
70	$[(NMe_3CH_2CO_2)_2H]$	$[N(SO_2CF_3)_2]$	1:1	
85	*N*-Butyl-*N,N*-dimethylglycine ethyl ester	I	1:1	440
70	Dimethyl(*iso*-propyl) phenylammonium	$[I_3]$	1:1	420
60	Dimethyl(propyl) phenylammonium	$[I_3]$	1:1	420
102	Dimethyldiphenylammonium	$[I_3]$	1:1	413
78	Dimethylbis(3-methylbenzyl) ammonium	$[N(SO_2CF_3)_2]$	1:1	383
106	Dimethylbis(4-methoxybenzyl)ammonium	$[N(SO_2CF_3)_2]$	1:1	379
85/92–94	Benzethonium	$[NO_3]$	1:1	373
65	Ethyl(methyl) diphenylammonium	$[I_3]$	1:1	408
90	Ethyl(methyl) diphenylammonium	$[I_5]$	1:1	389
ca. 10	Tributylmethylammonium	$[Br_{20}]$	2:1	944
110	Tributylmethylammonium	$[Fe^{III}Cl_4]$	1:1	400
95–96	Tributylmethylammonium	$[Cu^I_5Br_7]$	2:1	802
95–98	Tetraethylammonium	$[(2\text{-}OH)C_6H_4CO_2]$	1:1	426
78	Tetraethylammonium	$[(2\text{-}OH)C_6H_4COS]$	1:1	422

(*Continued*)

TABLE 11.21 (*Continued*)

m.pt./°C			$p:q$	U_L/ kJ mol^{-1}
78–80	Tetraethylammonium	$[(2\text{-OH})C_6H_4CS_2]$	1:1	417
104	Tetraethylammonium	$[N(SO_2CF_3)_2]$	1:1	412
72	Tetraethylammonium	$[BF_4]$	1:1	456/462
56	Tetraethylammonium	$[Al\{OCH(CF_3)_2\}_4]$	1:1	356/356
104–106	Tetraethylammonium	3,5-Dinitro-1,2, 4-triazolate	1:1	436
95	Triethylpropylammonium	$[B(CN)_4]$	1:1	423
98	Tetrapropylammonium	$[I_3]$	1:1	399
105	Tetrapropylammonium	$[N(SO_2CF_3)_2]$	1:1	390
102–103	Tetrapropylammonium	$[SiPhF_4]$	1:1	395
99–100	Tetrapropylammonium	$[Cu^IBr_2]$	1:1	413
99–100	Tetrapropylammonium	$[Cu^ICl_2]$	1:1	416
103	Tetrabutylammonium	$[(C_6H_5CO_2)_2H]$	1:1	361
90	Tetrabutylammonium	$[InCl_4]$	1:1	379
108	Tetrabutylammonium	$[Al\{OC(CH_3)(CF_3)_2\}_4]$	1:1	331
90	Tetrabutylammonium	$[AlMe_2F_2]$	1:1	389
62–64	Tetrabutylammonium	$[SiMePhF_3]$	1:1	374
ca. 65	Tetrabutylammonium	$[PF_3(C_2F_5)_3]$	1:1	364
65–69	Tetrabutylammonium	$[Au(CN)_2]$	1:1	385
59	Tetrabutylammonium	$[Sm\{C(CN)_2NO\}_6]$	3:1	
68	Tetrabutylammonium	$[Ce\{C(CN)_2NO\}_6]$	3:1	
78	Tetrabutylammonium	$[Y(BH_4)_4]$	1:1	376
81	Tetrabutylammonium	2,4-Dinitroimidazolate	1:1	387
85	Tetrabutylammonium	4,5-Dinitroimidazolate	1:1	389
90–92	Tetrabutylammonium	2,3,6-Tricyano-4-fluoro-5-trifluoromethylphenolate	1:1	371
108–109	Tetrabutylammonium	$[C_5H_5]$ (Cyclopentadienide)	1:1	396
85	Tetrabutylammonium/3-[(2*S*)-2-methylbutyl]thiazolium	$[CF_3SO_3]$	(1,1):1	
95	Tetrabutyl ammonium/ 1-methyl-3-(tetrahydrofuran-3-ylmethyl)imidazolium	$[CF_3SO_3]$	(1,1):1	
102	Tetrakis(decyl)ammonium	$[B(C_6H_5)_4]$	1:1	310
88–92	Tetradodecylammonium	Br	1:1	325
53.5–56	*N*,*N*-Dimethyl-*N*-((methyltellurenyl)(phenyl) methylidene)ammonium	$[CF_3SO_3]$	1:1	427
69	*N*,*N*′′′′-Bis(2-ethoxyethyl)-*N*,*N*,*N*′′′′,*N*′′′′-tetramethylbutane-1, 4-diammonium	$[CF_3SO_3]$	1:2	1278
79.5–82	*N*,*N*-Dimethyl-*N*-((methylselenyl)(phenyl) methylidene)ammonium	$[CF_3SO_3]$	1:1	431
86	*N*,*N*′-Bis(2-ethoxyethyl)-*N*,*N*,*N*′,*N*′-tetramethylethane-1,2-diammonium	$[N(SO_2CF_3)_2]$	1:2	1213

TABLE 11.21 (*Continued*)

m.pt./°C			$p:q$	U_L/ kJ mol^{-1}
99.5	N,N'''-Bis(2-ethoxyethyl)-N,N,N''',N'''-tetramethylbutane-1,4-diammonium	$[PF_6]$	1:2	1292
	Pyrrolidinium			
85–105	1,1-Dimethylpyrrolidinium	[SCN]	1:1	495
40–55	1,1-Dimethylpyrrolidinium	$[N(SO_2Me)_2]$	1:1	448
106	1-Ethyl-1-methylpyrrolidinium	Br	1:1	491
46	1-Ethyl-1-methylpyrrolidinium	$[Br_3]$	1:1	458
91	1-Ethyl-1-methylpyrrolidinium	$[N(SO_2CF_3)_2]$	1:1	419/423
50	1-Methyl-1-propylpyrrolidinium	Cl	1:1	481
12	1-Methyl-1-propylpyrrolidinium	$[N(SO_2CF_3)_2]$	1:1	413/414
78	1-Methyl-1-propylpyrrolidinium	$[B(CN)_4]$	1:1	436
107	1-Methyl-1-propylpyrrolidinium	$[Yb\{N(SO_2CF_3)_2\}_4]$	2:1	1089
ca. 9	1-Butyl-1-methylpyrrolidinium	$[Br_{20}]$	2:1	1007
37	1-Butyl-1-methylpyrrolidinium	$[N(SO_2Me)_2]$	1:1	424
−11	1-Butyl-1-methylpyrrolidinium	$[N(SO_2CF_3)_2]$	1:1	409
81	1-Butyl-1-methylpyrrolidinium	$[Fe^{III}Cl_4]$	1:1	427
92	1-Butyl-1-methylpyrrolidinium	$[Eu\{N(SO_2CF_3)_2\}_5]$	2:1	670
9	1-Hexyl-1-methylpyrrolidinium	$[N(SO_2CF_3)_2]$	1:1	393
66	1-Heptyl-1-methylpyrrolidinium	$[PF_6]$	1:1	421
39	1-(Hydroxycarbonylmethyl)-1-methylpyrrolidinium	$[N(SO_2CF_3)_2]$	1:1	421
90–96	1-(Prop-2-ynyl)-1-methylpyrrolidinium	Cl	1:1	487
Liq	1-Butyl-1-cyanomethylpyrrolidinium	$[N(SO_2CF_3)_2]$	1:1	402
74	1-(1-Methyl-2-butyn-1-ylidene)pyrrolidinium	$[CF_3SO_3]$	1:1	444
51	1,3-Bis(1-methyl-1-pyrrolidinio)propane	Br	1:2	1434
	Piperidinium			
91	1-Ethyl-1-methylpiperidinium	$[N(SO_2CF_3)_2]$	1:1	385/416
12	1-Methyl-1-propylpiperidinium	$[N(SO_2CF_3)_2]$	1:1	408
69	1-Methyl-1-propylpiperidinium	$[Al\{OCH(CF_3)_2\}_4]$	1:1	356
41	1-Butyl-1-methylpiperidinium	$[PO_2F_2]$	1:1	445
106–107	1-(Piperidinium-1-ylidene)-3-(piperidin-1-yl)-2-azapropene	$[PF_6]$	1:1	422
	Morpholinium			
31	1-Butyl-1-methylmorpholinium	$[Al\{OCH(CF_3)_2\}_4]$	1:1	355
104–105	1-(Hydroxycarbonylmethyl)-1-methylmorpholinium	$[(2\text{-}OH)C_6H_4CO_2]$	1:1	437

(*Continued*)

TABLE 11.21 (*Continued*)

m.pt./°C			$p:q$	U_L/ kJ mol^{-1}
46	1-(Ethoxycarbonylmethyl)-1-methylmorpholinium	$[N(SO_2CF_3)_2]$	1:1	409
	Tetra-alkyl- and aryl phosphonium			
74–76	Ethyltributylphosphonium	$[4\text{-}CH_3C_6H_4SO_3]$	1:1	381
73	Tetrabutylphosphonium	$[B(CN)_4]$	1:1	376
64–66	Tetrabutylphosphonium	$[SnPh_2Cl_3]$	1:1	360
95–96	Tetrabutylphosphonium	$[Sn(2\text{-}MeC_6H_4)_2Cl_3]$	1:1	352
108	Tetrabutylphosphonium	$[PtBr_3(C_2H_4)]$	1:1	379
76–77	Decyl(tri-*tert*-butyl) phosphonium	$[BF_4]$	1:1	375
108–109	Di-*tert*-butyl (diphenylmethylene) phosphonium	$[AlCl_4]$	1:1	376
96	Tetrapentylphosphonium	I	1:1	378
108	Cyclohexyltrivinylphosphonium	I	1:1	434
85–87	Decyl(tricyclohexyl)phosphonium	$[BF_4]$	1:1	361
54–57	Tetrakis(decyl)phosphonium	Br	1:1	332
44.5	Amidomethyltris(tetradecyl) phosphonium	Br	1:1	326
90	Hydroxycarbonylmethyltris (tetradecyl)phosphonium	Br	1:1	327
98–99	Tetraoctadecylphosphonium	I	1:1	293
105	Methyltriphenylphosphonium	$[I_3]$	1:1	394
107–108	Methyltriphenylphosphonium	$[Cu^{I}_2Br_4]$	2:1	819
98	Ethyltriphenylphosphonium	$[B(CN)_4]$	1:1	384
84	Ethyltriphenylphosphonium	$[Se_4Br_{14}]$	2:1	709
68–70	Ethyltriphenylphosphonium	2,4-Dinitroimidazolate	1:1	391
91–93	Ethyltriphenylphosphonium	4,5-Dinitroimidazolate	1:1	394
102	Butyltriphenylphosphonium	$[Hg^{II}_2Cl_6]$	2:1	730
74	Butyltriphenylphosphonium	$[Bi_2I_8(Me_2CO)_2]$	2:1	670
Liq	Benzyltriphenylphosphonium	$[Br_8]$	2:1	752
101	Tetraphenylphosphonium	$[S(NSO_2CF_3)_2(OEt)]$	1:1	356
98–101	Tetraphenylphosphonium	$[B(C_6F_5)(N_3)_3]$	1:1	364
92	Tetraphenylphosphonium	$[As(N_3)_4]$	1:1	374
104–106	Tetraphenylphosphonium	$[Sb(N_3)_6]$	1:1	362
96–98	Trichloro(cyclohexyl) phosphonium	$[GaCl_4]$	1:1	424
Liq	Chlorotriphenylphosphonium	$[Cl_2I_{14}]$	2:1	688
Liq	Bromotriphenylphosphonium	$[Br_7]$	1:1	378
110–112	Tri-*iso*-propyl-[N-(2,6-dichlorophenyl) thiocarbamoyl]phosphonium	$[CH_3SO_3]$	1:1	393
	Arsonium and stibonium			
96–102	Tetraphenylarsonium	$[C\{CF_3C(O)\}(CN)_2]$	1:1	375
80–82	Tetramethylstibonium	$[(C_6H_5CO_2)_2H]$	1:1	399

TABLE 11.21 (***Continued***)

m.pt./°C			$p:q$	U_L/ kJ mol^{-1}
	Trialkylsulfonium, selenium and tellurium			
37	Trimethylsulfonium	$[I_3]$	1:1	468
89–91	Trimethylsulfonium	$[ICF_2CF_2OCF_2CF_2SO_3]$	1:1	422
−2	Triethylsulfonium	$[I_3]$	1:1	440
109	Dimethylphenylsulfonium	$[CF_3SO_3]$	1:1	451
97	Methyldiphenylsulfonium	$[CF_3SO_3]$	1:1	423
75	Hexadecyldimethylsulfonium	Br	1:1	394
72	Trimethylselenonium	$[N_3]$	1:1	535
73–75	Trimethylselenonium	[SeCN]	1:1	513
110	Trimethyltelluronium	$[N_3]$	1:1	522
86–88	Trimethyltelluronium	[SCN]	1:1	513
106–109	Trimethyltelluronium	[SeCN]	1:1	507
109	Trimethyltelluronium	$[AlMe_2Cl_2]$	1:1	451
107.5–108	Di-*cyclo*-hexyl(hydroxo) selenonium	$[4\text{-}BrC_6H_4SO_3]$	1:1	398
	Dimethylaminosulfonium and sulfoxonium			
100	Bis(dimethylamino) trifluoromethylsulfonium	$[CF_3S]$	1:1	451
92	Tris(dimethylamino)sulfonium	$[SiMe_3F_2]$	1:1	422
91–92	Tris(dimethylamino)sulfonium	$[SiPh_2(1\text{-}C_{10}H_7)F_2]$	1:1	369
79	Tris(dimethylamino)sulfonium	*cyclo*-$[P_4N_4F_9]$	1:1	401
<0	Tris(dimethylamino)sulfonium	*cyclo*-$[P_6N_6F_{13}]$	1:1	379
64	Tris(dimethylamino)sulfonium	$[P_3N_3F_5NPF_2NPF_2NPF_5]$	2:1	854
107	Tris(dimethylamino)sulfonium	$[Li(C_5H_5)_2]$	1:1	408
45	Tris(dimethylamino) oxosulfonium	$[HF_2]$	1:1	469
57	Tris(dimethylamino) oxosulfonium	$[SiMe_3F_2]$	1:1	426

the crystal structure is of a low-temperature form stable below 85°C; a further solid–solid transition is seen at 94°C), associated with the introduction of the fourth methyl group into $[NMe_3H][H_2F_3]$, results in a marked increase in melting temperature but a reduction in lattice potential energy. A 2:3 adduct of trimethylamine and hydrogen chloride has been characterised crystallographically [95] as $[(NMe_3H)_2Cl][HCl_2]$, with two cations linked to a single chloride ion by N—H···Cl hydrogen bonds. The Cl···Cl separation in $[HCl_2]^-$ is 3.136 Å. Further studies by Mootz et al. of the phase diagrams of (pyridinium halide + hydrogen halide) have established the congruence of melting and the presence of discrete ions in the melt. A series of complexes of composition py.nHF has been characterised, with the structures of crystals, grown by a zone-melting technique [619] reported for $n = 1$, 3 and 4 [61]. Interestingly, when $n = 1$, the solid compound, melting at −31°C, is better described as a co-crystal rather than as

TABLE 11.22 Pyridinium and Related Salts

m.pt./°C			$p:q$	U_L/kJ mol^{-1}
	***N*-Alkylpyridinium**			
45–46	1-Ethylpyridinium	$[I_3]$	1:1	450
−4	1-Hexylpyridinium	$[N(SO_2CF_3)_2]$	1:1	401
86/103/152	1-Butylpyridinium	Cl	1:1	478
95–97	1-Butylpyridinium	Br	1:1	470
34	1-Butylpyridinium	$[AlCl_4]$	1:1	432
36	1-Butylpyridinium	$[Al\{OCH(CF_3)_2\}_4]$	1:1	361
49	1-Butylpyridinium	[*nido*-$C_2B_9H_{12}$]	1:1	420
72	1-Hexylpyridinium	[*nido*-$C_2B_9H_{12}$]	1:1	405
80	1-Octylpyridinium	[*nido*-$C_2B_9H_{12}$]	1:1	396
74	1-Tetradecylpyridinium	$[Pd^{II}_2Cl_6]$	2:1	737
101	1-Tetradecylpyridinium	$[Pd^{II}I_4]$	2:1	769
38	1-Tetradecylpyridinium	$[Co(hfac)_3]$	1:1	337
57	1-Hexadecylpyridinium	Acesulfamate	1:1	374
82.5	1-Hexadecylpyridinium	$[Pd^{II}_2Cl_6]$	2:1	716
40–41	1-Hexadecylpyridinium	$[Co(CO)_4]$	1:1	371
99	1-Benzylpyridinium	Br	1:1	464–465
	***N*-Alkyl quinolinium and isoquinolinium**			
33–34	1-Ethylquinolinium	$[I_3]$	1:1	427
59–60	1-Butylisoquinolinium	$[GaCl_4]$	1:1	418
	Ring or alkyl group functionalised pyridinium			
27	1-(Trimethylsilyl)pyridinium	Br	1:1	466
87	1-(Trimethylsilyl)-3, 4-dimethylpyridinium	Br	1:1	449
63–65	1-Methyl-3-cyanopyridinium	$[N(SO_2CF_3)_2]$	1:1	434
108–115	1-Methyl-4-cyanopyridinium	$[SO_3(OMe)]$	1:1	472
65–66	1-Methyl-4-cyanopyridinium	$[N(SO_2CF_3)_2]$	1:1	429
110.5–113	1,2,4-Trimethylpyridinium	$[I_3]$	1:1	436
44	1-(Cyanomethyl)pyridinium	$[N(SO_2CF_3)_2]$	1:1	435
51	1-(Cyanomethyl)-2-methylpyridinium	$[N(SO_2CF_3)_2]$	1:1	426
30	1-(Cyanomethyl)-2-ethylpyridinium	$[N(SO_2CF_3)_2]$	1:1	419
48	1-(Cyanomethyl)-4-ethylpyridinium	$[N(SO_2CF_3)_2]$	1:1	420
32	1-(Hydroxycarbonymethyl) pyridinium	$[N(SO_2CF_3)_2]$	1:1	426
101	1-(3-Cyanopropyl)pyridinium	Cl	1:1	483
Liq	1-(3-Cyanopropyl)pyridinium	$[N(SO_2CF_3)_2]$	1:1	416
Liq	1-(3-Cyanopropyl)-3-methylpyridinium	$[N(SO_2CF_3)_2]$	1:1	412
Liq	1-(Cyanomethyl)-2, 5-dimethylpyridinium	$[N(SO_2CF_3)_2]$	1:1	419
28–29	1-Ethyl-3-trifluoromethylpyridinium	$[N(SO_2CF_3)_2]$	1:1	417

TABLE 11.22 ***(Continued)***

m.pt./°C			p:q	U_L/kJ mol^{-1}
36–38	1-Ethyl-2-cyanopyridinium	$[N(SO_2CF_3)_2]$	1:1	418
73	1-Ethyl-3-cyanopyridinium	$[N(SO_2CF_3)_2]$	1:1	420
33–35	1-Ethyl-4-cyanopyridinium	$[N(SO_2CF_3)_2]$	1:1	419
66–68	1-(2-Hydroxyethyl)pyridinium	Cl	1:1	511
110	1-(2-Hydroxyethyl)pyridinium	Br	1:1	499
83–85	1-Butyl-3-methylpyridinium	$[NbCl_6]$	1:1	413
78–80	1-Butyl-4-methylpyridinium	$[PCl_6]$	1:1	417
89–91	1-Butyl-4-methylpyridinium	$[NbCl_6]$	1:1	412
103	1-Butyl-4-methylpyridinium	$[TaCl_6]$	1:1	412
58–60	1-Benzyl-3,5-bis (3-trifluoromethylphenyl) pyridinium	$[CF_3SO_3]$	1:1	377
95–98	3-Hydroxy-1-(octyloxymethyl) pyridinium	Saccharinate	1:1	395
74–75	*N*-(1-*O*-Methyl-2,3,4-tri-*O*-acetyl-*α*-D-glucopyranose-6-yl) pyridinium	$[NO_3]$	1:1	389
103–105	*rac*-1-((1-(2-Naphthyl)prop-2-en-1-yl)amino)pyridinium	I	1:1	420
79–82	1-Methyl-4-(benzylselenomethyl) pyridinium	$[CF_3SO_3]$	1:1	415
98.5–102	2-Chloro-1-methyl-4-(phenylamino)pyridinium	$[CF_3SO_3]$	1:1	427
60–62	1-{(3-(Dimethylammonio) propyl}-2-hydroxypyridinium	$[CF_3SO_3]$	1:2	1376

a salt. Related investigations on py.*n*HCl have been reported, with $[pyH][Cl(HCl)_5]$ melting at −73°C [195]. Structural comparisons are possible between $[Hpy][HCl_2]$ (m.pt. 60°C; U_L 518 kJ mol^{-1}) [195] and $[Hpy][HF_2]$ (m.pt. −1°C; U_L 548 kJ mol^{-1}) [61]. In crystals of [emim][H(HF)] (m.pt. 51°C; U_L 507 kJ mol^{-1}), grown from a melt at room temperature [156], in addition to strong cation–anion hydrogen bonding, pillars of cations are linked by hydrogen bonds involving the ring C-4 proton and the ring π-electrons of adjacent cations. Very recently, low-melting $[C_1mim][F(HF)_3]$ (U_L 495 kJ mol^{-1}) and $[C_1mim][F(HF)_2]$ (U_L 514 kJ mol^{-1}) have been structurally characterised [56]. The ionic plastic crystals, *N*,*N*-dimethylpyrrolidinium fluorohydrogenate, $[C_1C_1pyr][F(HF)_2]$ (m.pt. 52°C) and *N*-ethyl-*N*-methylpyrrolidinium fluorohydrogenate $[C_2C_1pyr][F(HF)_2]$ (m.pt. 30°C), have been characterised by powder XRD [620].

The structure of $[S(O)(NMe_2)_3][F(HF)]$ (m.pt. 45°C; U_L 469 kJ mol^{-1}) [139] is known. The crystal structure of a salt of composition $[Hmim][Br(HBr)_2]$ (refcode HACBOE), formed from [Hmim]Br and HBr, has also been reported [621], though without a m.pt. The related salts of $[A(HA)_n]^-$, when $A = RCO_2$ and $n = 1$, have been discussed in Section 11.4.1.

TABLE 11.23 Salts of Imidazolium Cations

m.pt./°C			$p:q$	U_L/ kJ mol^{-1}
	Protic imidazolium cations	**Anion**		
99–100	Imidazolium	$[PO_2(OPh)_2]$	1:1	425
101.5–103	Imidazolium	$[HO_2C(CH_2)_4CO_2]$	1:1	465
105	2-Methylimidazolium	$[HO_2C(CH_2)_3CO_2]$	1:1	467
60	2-Methylimidazolium	$[HO_2C(CH_2)_6CO_2]$	1:1	441
106	4-Methylimidazolium	$[HO_2C(CH_2)_3CO_2]$	1:1	472
85	4-Methylimidazolium	$[HO_2C(CH_2)_6CO_2]$	1:1	441
41–43	2-Ethylimidazolium	*d*-$[HO_2CCH(OH)CH(OH)CO_2]$	1:1	462
95	1-Aminoimidazolium	Cl	1:1	556
105–108	1-Azidosulfonylimidazolium	$[SO_3(OH)]$	1:1	481
60	1-Methylimidazolium	$[NO_3]$	1:1	538
91–92.5	1-Methylimidazolium	$[PO_2(OPh)_2]$	1:1	423
98–101	1-Methylimidazolium	*d*-$[HO_2CCH(OH)CH(OH)CO_2]$	1:1	470/471
52/59	1-Methylimidazolium	$[N(SO_2CF_3)_2]$	1:1	445
103	1-Methylimidazolium	$[Au(CN)_2]$	1:1	487
79	1,2-Dimethylimidazolium	$[NO_3]$	1:1	513
105	1,2-Dimethylimidazolium	$[HO_2C(CH_2)_2CO_2]$	1:1	469
102	1,2-Dimethylimidazolium	$[HO_2C(CH_2)_3CO_2]$	1:1	464
108–110	1-Benzylimidazolium	$[PhP(O)_2(OH)]$	1:1	424
	Dipositively charged imidazolium cations			
95	1-(3-Ammoniopropyl)-3-methylimidazolium	$[NO_3]$	1:2	1587
69–71	1-{2-(Diethylammonio)ethyl}-3-methylimidazolium	$[N(SO_2CF_3)_2]$	1:2	1261
106–108	1,4-Bis(1-methylimidazolium-3-yl)butane	2-Hydroxy-5-sulfonatobenzoate	2:2	
71	1,4-Bis(1-methylimidazolium-3-yl)butane	$[CF_3SO_3]$	1:2	1355
95–99	1,4-Bis(1-methylimidazolium-3-yl)butane	$[BF_4]$	1:2	1412/1423
86	1,7-Bis(1-methylimidazolium-3-yl)heptane	$[BF_4]$	1:2	1368
92	1,8-Bis(1-methylimidazolium-3-yl)octane	$[BF_4]$	1:2	1353
71	1,10-Bis(1-methylimidazolium-3-yl)decane	$[PF_6]$	1:2	1294
68–71	1,4-Bis(1-methylimidazolium-3-ylmethyl)benzene	$[BF_4]$	1:2	1384
59–62	1,4-Bis(1-butylimidazolium-3-ylmethyl)benzene	$[CF_3SO_3]$	1:2	1256
96–98	3,5-Bis(1-butylimidazolium-3-ylmethyl)toluene	$[PF_6]$	1:2	1254
92–94	1,4-Bis(1,2-dimethylimidazolium-3-ylmethyl)benzene	$[CF_3SO_3]$	1:2	1301

TABLE 11.23 (*Continued*)

m.pt./°C			$p:q$	U_L/ kJ mol^{-1}
87	1,8-Bis(1-methylimidazolium-3-yl)-3,6-dioxaoctane	$[PF_6]$	1:2	1325
112	6,6′-Spirobis(1-methyl-1,5,6,7-tetrahydropyrrolo [1,2-a]imidazol-4-ium	$[N(SO_2CF_3)_2]$	1:2	1245
	$[C_4mim]^+$			
66	1-Butyl-3-methylimidazolium	Cl	1:1	477–483
70–79	1-Butyl-3-methylimidazolium	Br	1:1	473–477
19	1-Butyl-3-methylimidazolium	I	1:1	468
109–112	1-Butyl-3-methylimidazolium	$[P_2Se_8]$	2:1	1217
−5/−5	1-Butyl-3-methylimidazolium	$[SO_3(OMe)]$	1:1	444
60–65 est	1-Butyl-3-methylimidazolium	$[CH_3SO_3]$	1:1	454
73	1-Butyl-3-methylimidazolium	$[CH_3SO_3]$	1:1	
7	1-Butyl-3-methylimidazolium	$[CF_3SO_3]$	1:1	434
78	1-Butyl-3-methylimidazolium	$[4\text{-}MeC_6H_4SO_3]$	1:1	423
−3/−2	1-Butyl-3-methylimidazolium	$[N(SO_2CF_3)_2]$	1:1	413
90	1-Butyl-3-methylimidazolium	$[B\{C_6H_4(4\text{-}C_6F_{13})\}_4]$	1:1	303
102–104	1-Butyl-3-methylimidazolium	$[B\{(3,5\text{-}CF_3)_2C_6H_3\}_4]$	1:1	285
34	1-Butyl-3-methylimidazolium	$[Al\{OCH(CF_3)_2\}_4]$	1:1	356
2	1-Butyl-3-methylimidazolium	$[PF_6]$	1:1	452/453
94/92	1-Butyl-3-methylimidazolium	$[HgCl_3]$	1:1	442
58/57	1-Butyl-3-methylimidazolium	$[HgBr_3]$	1:1	435
5	1-Butyl-3-methylimidazolium	$[Sn^{II}Cl_4]$	2:1	1012
60	1-Butyl-3-methylimidazolium	$[ZnCl_4]$	2:1	1010
63	1-Butyl-3-methylimidazolium	$[Mn^{II}Cl_4]$	2:1	1003
58	1-Butyl-3-methylimidazolium	$[Fe^{II}Cl_4]$	2:1	1007
62	1-Butyl-3-methylimidazolium	$[Co^{II}Cl_4]$	2:1	1010
45	1-Butyl-3-methylimidazolium	$[Co^{II}Br_4]$	2:1	994
56	1-Butyl-3-methylimidazolium	$[Ni^{II}Br_4]$	2:1	1012
23	1-Butyl-3-methylimidazolium	$[Cu^{II}Cl_4]$	2:1	1024
99	1-Butyl-3-methylimidazolium	$[Pt^{II}Cl_4]$	2:1	1037
50	1-Butyl-3-methylimidazolium	$[Au^{III}Cl_4]$	1:1	431
70	1-Butyl-3-methylimidazolium	$[Cu^{II}_3Cl_8]$	2:1	927
105	1-Butyl-3-methylimidazolium	$[Co^{II}Br_3(C_9H_7N)]$	1:1	397
38	1-Butyl-3-methylimidazolium	$[La(NCS)_7(H_2O)]$	4:1	
68	1-Butyl-3-methylimidazolium	$[Eu\{N(SO_2CF_3)_2\}_4]$	1:1	327
	$[C_4C_1C_1im]^+$			
79/100	1-Butyl-2, 3-dimethylimidazolium	Cl	1:1	464–472
97	1-Butyl-2, 3-dimethylimidazolium	Br	1:1	466
98	1-Butyl-2, 3-dimethylimidazolium	I	1:1	458
37	1-Butyl-2, 3-dimethylimidazolium	$[IBr_2]$	1:1	430
41	1-Butyl-2, 3-dimethylimidazolium	$[N_3]$	1:1	465

(*Continued*)

TABLE 11.23 (*Continued*)

m.pt./°C			$p:q$	U_L/ kJ mol^{-1}
40	1-Butyl-2,3-dimethylimidazolium	[SCN]	1:1	454
46	1-Butyl-2,3-dimethylimidazolium	$[N(C_6F_5)_2]$	1:1	394
78	1-Butyl-2,3-dimethylimidazolium	$[SO_3(OH)]$	1:1	450
23–24	1-Butyl-2,3-dimethylimidazolium	$[PO_2(OMe)(SMe)]$	1:1	422
43	1-Butyl-2,3-dimethylimidazolium	$[HC^{\circ}CCO_2]$	1:1	449
40	1-Butyl-2,3-dimethylimidazolium	$[BF_4]$	1:1	452/453
40	1-Butyl-2,3-dimethylimidazolium	$[PF_6]$	1:1	444/445
44.5	1-Butyl-2,3-dimethylimidazolium	$[SbF_6]$	1:1	432
101	1-Butyl-2,3-dimethylimidazolium	$[CB_{11}H_6Cl_6]$	1:1	380
38	1-Butyl-2,3-dimethylimidazolium	$[Fe^{III}Cl_4]/[Fe^{II}Cl_4]$	3:(1,2)	
	$[C_2mim]^+$			
81–90	1-Ethyl-3-methylimidazolium	Cl	1:1	503–507
71–81	1-Ethyl-3-methylimidazolium	Br	1:1	500–505
81	1-Ethyl-3-methylimidazolium	I	1:1	488/488
51	1-Ethyl-3-methylimidazolium	$[HF_2]$	1:1	507
49	1-Ethyl-3-methylimidazolium	$[Br_3]$	1:1	467
72	1-Ethyl-3-methylimidazolium	[OCN]	1:1	505
35	1-Ethyl-3-methylimidazolium	[C(NO2)(NO)CN]	1:1	464
52	1-Ethyl-3-methylimidazolium	$[C(C_6F_5)(SO_2CF_3)_2]$	1:1	393
55	1-Ethyl-3-methylimidazolium	$[NO_2]$	1:1	501
38	1-Ethyl-3-methylimidazolium	$[NO_3]$	1:1	490
39	1-Ethyl-3-methylimidazolium	$[PO_2(OMe)(SMe)]$	1:1	453
−26	1-Ethyl-3-methylimidazolium	$[CF_3SO_3]$	1:1	459
45	1-Ethyl-3-methylimidazolium	$[HCF_2CF_2SO_3]$	1:1	490
28	1-Ethyl-3-methylimidazolium	$[C_4F_9SO_3]$	1:1	423
Liq	1-Ethyl-3-methylimidazolium	$[N(SO_2Me)_2]$	1:1	446
−26/−2	1-Ethyl-3-methylimidazolium	$[N(SO_2CF_3)_2]$	1:1	426/426
−1	1-Ethyl-3-methylimidazolium	$[BF_4]$	1:1	485/489
37	1-Ethyl-3-methylimidazolium/ sodium	$[AlCl_4]$	(1,1):2	
31	1-Ethyl-3-methylimidazolium	$[Al\{OCH(CF_3)_2\}_4]$	1:1	363
58–65	1-Ethyl-3-methylimidazolium	$[PF_6]$	1:1	465/473
53	1-Ethyl-3-methylimidazolium	$[AsF_6]$	1:1	462
10	1-Ethyl-3-methylimidazolium	$[SbF_6]$	1:1	460
−1	1-Ethyl-3-methylimidazolium	$[NbF_6]$	1:1	462
2	1-Ethyl-3-methylimidazolium	$[TaF_6]$	1:1	462
73	1-Ethyl-3-methylimidazolium	$[Ag(CN)_2]$	1:1	466
64	1-Ethyl-3-methylimidazolium	$[Au(CN)_2]$	1:1	469
ca. 104	1-Ethyl-3-methylimidazolium	$[Sn^{II}Cl_3]$	1:1	463
86	1-Ethyl-3-methylimidazolium	$[Fe^{II}Cl_4]$	2:1	1078
100–102	1-Ethyl-3-methylimidazolium	$[Co^{II}Cl_4]$	2:1	1082
92–93	1-Ethyl-3-methylimidazolium	$[Ni^{II}Cl_4]$	2:1	1085
10	1-Ethyl-3-methylimidazolium	$[Fe^{III}Cl_4]$	1:1	444/449
58	1-Ethyl-3-methylimidazolium	$[Au^{III}Cl_4]$	1:1	448
75	1-Ethyl-3-methylimidazolium	$[V^{V}(O)F_4]$	1:1	475
99	1-Ethyl-3-methylimidazolium	$[V^{IV}(O)Cl_4]$	2:1	1064
76	1-Ethyl-3-methylimidazolium	$[(Ta^{V}Cl_5)_2O]$	2:1	939
97	1-Ethyl-3-methylimidazolium	[*nido*-$C_2B_9H_{12}$]	1:1	431

TABLE 11.23 (*Continued*)

m.pt./°C			$p:q$	U_L/ kJ mol^{-1}
59	1-Ethyl-3-methylimidazolium	$[1\text{-MeCB}_{11}H_{11}]$	1:1	415
64	1-Ethyl-3-methylimidazolium	$[1\text{-EtCB}_{11}H_{11}]$	1:1	411
80	1-Ethyl-3-methylimidazolium	$[1\text{-BuSnB}_{11}H_{11}]$	1:1	395
106	1-Ethyl-3-methylimidazolium	$[1\text{-EtSnB}_{11}H_{11}]$	1:1	409
66	1-Ethyl-3-methylimidazolium	$[Cu\{(CF_3SO_2)_2N_2\}_2]$	1:1	379
58	1-Ethyl-3-methylimidazolium	$[La\{C(CN)_2NO\}_6]$	3:1	
103	1-Ethyl-3-methylimidazolium	$[U^{VI}O_2(NCS)_5]$	3:1	
	Other [Rmim]$^+$			
36	1-Methyl-3-propylimidazolium	Br	1:1	491
88–90	1-Methyl-3-propylimidazolium	$[Ph_2PC_6H_4(4\text{-}SO_3)]$	1:1	380
69/67	1-Methyl-3-propylimidazolium	$[HgCl_3]$	1:1	445
39.5/41	1-Methyl-3-propylimidazolium	$[HgBr_3]$	1:1	441
81	1-Methyl-3-propylimidazolium	$[Eu\{N(SO_2CF_3)_2\}_4]$	1:1	330
78(est)/ 110.5	1-*iso*-Propyl-3-methylimidazolium	Br	1:1	484/488
83	1-(2-Butyl)-3-methylimidazolium	$[PF_6]$	1:1	450
91	1-Methyl-3-[(2*S*)-2-methylbutyl] imidazolium	[Bis{(2*S*)-2-oxy-3-methylbutanoato-*O*,*O*′}borate]	1:1	388
89	1-*cyclo*-Pentyl-3-methylimidazolium	$[4\text{-MeC}_6H_4SO_3]$	1:1	421
70	1-Hexyl-3-methylimidazolium	$[4\text{-MeC}_6H_4SO_3]$	1:1	407
	1-Hexyl-3-methylimidazolium	$[N(SO_2CF_3)_2]$	1:1	402
67	1-Hexyl-3-methylimidazolium	$[Co^{II}Br_3(C_9H_7N)]$	1:1	387
67	1-Methyl-3-octylimidazolium	$[CB_{11}H_6Cl_6]$	1:1	371
31	1-Decyl-3-methylimidazolium	$[PF_6]$	1:1	407
60	1-Dodecyl-3-methylimidazolium	$[PF_6]$	1:1	400
26	1-Dodecyl-3-methylimidazolium	$[Cu^ICl_2][Cu^ICl_3]$	2:(1,1)	
27	1-Dodecyl-3-methylimidazolium	$[Cu^I_4Cl_8]$	4:1	
77	1-Tetradecyl-3-methyimidazolium	$[PF_6]$	1:1	385/390
72–73	1-Octadecyl-3-methylimidazolium	$[CF_3C(O)CHC(O)CF_3]$	1:1	366
	[RR′im]$^+$ and [RR′(2-R″)im]$^+$ (with phenyl-substituted R or R″)			
62–64	1-Methyl-2-phenyl-3-propylimidazolinium	$[PF_6]$	1:1	428
ca. 50	1-Ethyl-2-methyl-3-benzylimidazolium	$[N(SO_2CF_3)_2]$	1:1	396
86/89	1-Ethyl-2-methyl-3-benzylimidazolium	I	1:1	439
71/75	1-Propyl-2-methyl-3-benzylimidazolium	I	1:1	436
61/64	1-Butyl-2-methyl-3-benzylimidazolium	I	1:1	425
61/65	1-Pentyl-2-methyl-3-benzylimidazolium	I	1:1	416

(*Continued*)

TABLE 11.23 (*Continued*)

m.pt./°C			$p:q$	U_L/ kJ mol^{-1}
41	1-Methyl-3-(2-phenylethyl) imidazolium	$[N(SO_2CF_3)_2]$	1:1	404
61	1-Methyl-3-(2-phenylethyl) imidazolium	$[BF_4]$	1:1	445
50	1-Methyl-3-(3-phenylpropyl) imidazolium	$[N(SO_2CF_3)_2]$	1:1	397
62	1-(2,2-Diphenylethyl)-3-methylimidazolium	$[N(SO_2CF_3)_2]$	1:1	380
	[RR′im]$^+$ and [RR′(2-Me)im]$^+$ (R or R′ alkenyl or alkynyl)			
78	1-Vinyl-3-methylimidazolium	I	1:1	496
60	1-Allyl-3-methylimidazolium	Br	1:1	495
65	1-Allyl-3-methylimidazolium	I	1:1	480
12	1-Allyl-3-methylimidazolium	$[Al\{OCH(CF_3)_2\}_4]$	1:1	361
61	1-Allyl-2,3-dimethylimidazolium	Br	1:1	472
31	1-Allyl-3-ethylimidazolium	I	1:1	472
32	1-Allyl-3-propylimidazolium	I	1:1	457
59	1-(Pent-2-ynyl)-3-methylimidazolium	$[N(CN)_2]$	1:1	459
66	1-(But-2-ynyl)-3-methylimidazolium	$[N_3]$	1:1	487
	[RR′(2-R″)im]$^+$ (excluding phenyl-substituted R and R″)			
79	1,2,3-Trimethylimidazolium	$[CO_2(OMe)]$	1:1	482
39	1-Ethyl-2,3-dimethylimidazolium	$[Al\{OCH(CF_3)_2\}_4]$	1:1	361
63	1-Ethyl-2,3-dimethylimidazolium	$[Pr\{C(CN)_2NO\}_6]$	3:1	
62–64	1-Methyl-2-phenyl-3-propylimidazolinium	$[PF_6]$	1:1	428
ca. 50	1-Ethyl-2-methyl-3-benzylimidazolium	$[N(SO_2CF_3)_2]$	1:1	396
86/89	1-Ethyl-2-methyl-3-benzylimidazolium	I	1:1	439
57	1,2,3-Triethylimidazolium	$[N(SO_2CF_3)_2]$	1:1	410
85	1-(3-Cyanopropyl)-2,3-dimethylimidazolium	$[PF_6]$	1:1	445
63	1-(2-Methoxyethyl)-2,3-dimethylimidazolium	$[PF_6]$	1:1	452
	[RRim]$^+$, [RR(2-R″)im]$^+$ and related cations			
10 (dec)	1,3-Dimethylimidazolium	$[F(HF)_2]$	1:1	514
Ref DSC	1,3-Dimethylimidazolium	$[F(HF)_3]$	1:1	495
60	1,3-Dimethylimidazolium	$[CH(NO_2)_2]$	1:1	491
43	1,3-Dimethylimidazolium	$[SO_3(OMe)]$	1:1	480
22	1,3-Dimethylimidazolium	$[N(SO_2CF_3)_2]$	1:1	435
67–68	1,3-Dimethylimidazolium	$[PF_6]$	1:1	483/483
65	1,3-Dimethylimidazolium/lithium	$[N(SO_2CF_3)_2]$	(1,1):2	389
79	1,2,3-Trimethylimidazolium	$[CO_2(OMe)]$	1:1	482

TABLE 11.23 (*Continued*)

m.pt./°C			$p:q$	U_L/ kJ mol^{-1}
30	1,3-Diethylimidazolium	$[CH_3CO_2]$	1:1	474
67	1,3-Dipropylimidazolium	Br	1:1	457
109	1,3-Di-*iso*-propylimidazolium	$[N(C_6F_5)_2]$	1:1	393
108–110	1,3-Di-*iso*-propyl-4, 5-dimethylimidazolium	*E*-2-Cyano- 1-phenylethenolate	1:1	407
Liq	1,3-Di-*tert*-butylimidazolium	$[Me_3Al(m^3\text{-}CH_2)(AlMe_2)_2(m^2\text{-}CH_3)]$	1:1	370
99	1,3-Dibutyl-2,4, 5-trimethylimidazolium	$[PF_6]$	1:1	410
93	1,3-Dibenzylimidazolium	$[PF_6]$	1:1	412
88.5	1,3-Didodecylimidazolium	I	1:1	363
46	1,3-Didodecylimidazolium	$[I_3]$	1:1	356
36	1,3-Didodecylimidazolium	$[I_5]$	1:1	346
44	1,3-Didodecylimidazolium	$[N(CN)_2]$	1:1	363
44	1,3-Didodecylimidazolium	$[C(CN)_3]$	1:1	358
64	1,3-Didodecylimidazolium	$[ClO_4]$	1:1	362
67	1,3-Didodecylimidazolium	$[BF_4]$	1:1	361
48	1,3-Didodecylimidazolium	$[B(CN)_4]$	1:1	352
41	1,3-Didodecylimidazolium	$[SbF_6]$	1:1	358
110	1,3-Bis((*R*)-(+)-1-phenylethyl) imidazolium	$[BF_4]$	1:1	410
108	1,3-Bis((*R*)-(+)-1-phenylethyl) imidazolium	$[PF_6]$	1:1	404
110	1,3-Di(cyanomethyl)imidazolium	$[BF_4]$	1:1	403
79	1,3-Bis-(3-cyanopropyl) imidazolium	$[PF_6]$	1:1	429
110	1,3-Bis(hydroxycarbonylmethyl) imidazolium	$[ClO_4]$	1:1	466
	[RR′im]$^+$ cations with functionalised R or R′			
35	1-Methyl-3- trimethylsilylmethylimidazolium	$[BF_4]$	1:1	448
57	1-Methyl-3-[(trimethoxysilyl) methyl]imidazolium	I	1:1	436
67	1-Cyanomethyl- 3-methylimidazolium	$[N(CN)_2]$	1:1	484
103	1-Cyanomethyl- 3-methylimidazolium	$[NO_3]$	1:1	502
105	1-Cyanomethyl- 3-methylimidazolium	$[CdCl_4]$	2:1	1082
80	1-(3-Cyanopropyl)- 3-methylimidazolium	Cl	1:1	479/479
90	1-(3-Cyanopropyl)- 3-methylimidazolium	Br	1:1	476
69	1-(3-Cyanopropyl)- 3-methylimidazolium	I	1:1	465/466

(*Continued*)

TABLE 11.23 (*Continued*)

m.pt./°C			$p:q$	U_L/ kJ mol^{-1}
75	1-(3-Cyanopropyl)-3-methylimidazolium	$[PF_6]$	1:1	455
75	1-Methoxymethyl-3-methylimidazolium	I	1:1	484
68	1-Ethoxymethyl-3-methylimidazolium	I	1:1	471
103	1-Ethoxymethyl-3-methylimidazolium	$[Pd^{II}Cl_4]$	2:1	1036
86	1-(2-Hydroxyethyl)-3-methylimidazolium	Cl	1:1	504
50	1-(2-Methoxyethyl)-3-methylimidazolium	I	1:1	473
107	1-(2-Methoxyethyl)-3-methylimidazolium	$[Pd^{II}Cl_4]$	2:1	1052
63	1-{(2-Methoxyethyl)oxymethyl}-3-methylimidazolium	I	1:1	460
78	1-{2-(2-Methoxyethyl)oxyethyl}-3-methylimidazolium	Br	1:1	455
70	1-{2-(2-Methoxyethyl)oxyethyl}-3-methylimidazolium	I	1:1	451
102–103	1-(2-Hydroxy-2-methylpropyl)-3-methylimidazolium	I	1:1	460
97–98	1-(2-Hydroxy-2-methylpropyl)-3-*iso*-propylimidazolium	I	1:1	440
95	1-(3-Ammoniopropyl)-3-methylimidazolium	$[NO_3]$	1:2	1587
32	1-(2-Di-*iso*-propylaminoethyl)-3-methylimidazolium	$[PF_6]$	1:1	418
40	1-(Hydroxycarbonylmethyl)-3-methylimidazolium	$[N(SO_2CF_3)_2]$	1:1	424
102	1-Hydroxycarbonylmethyl-3-decylimidazolium	Br	1:1	414
105	1-(3-Hydroxycarbonylpropyl)-3-methylimidazolium	Cl	1:1	479
45	*rac*-1-{1-(Ethoxycarbonyl)ethyl}-3-methylimidazolium	$[N(SO_2CF_3)_2]$	1:1	406
85	1-Methyl-3-(tetrahydro-2*H*-pyran-2-ylmethyl)]imidazolium	$[CF_3SO_3]$	1:1	433
95	Tetrabutyl ammonium/1-methyl-3-(tetrahydrofuran-3-ylmethyl) imidazolium	$[CF_3SO_3]$	Cocryst	
103–105	[1-{2-((4-Chlorobenzylidene)amino)ethyl}-3-methylimidazolium	$[PF_6]$	1:1	420
66–68	1-(Ferrocen-1-ylmethyl)-3-methylimidazolium	$[PF_6]$	1:1	422
50	1-Butyl-3-ferrocenylimidazolium	$[N(SO_2CF_3)_2]$	1:1	383
64–66	1-(-2-Hydroxy-2-methyl-2-phenylethyl)-3-*tert*-butylimidazolium	Cl	1:1	419

TABLE 11.23 (*Continued*)

m.pt./°C			p:q	U_L/ kJ mol^{-1}
109–110	1-(2-Diphenylphosphinoethyl)-3-mesitylimidazolium	$[BF_4]$	1:1	377
	$[RR'(X)im]^+$, X a non-alkyl group			
67–73	2-Diethylamino-1,3-dimethylimidazolinium	$[SiMe_3F_2]$	1:1	419
105/107	1,3-Dimethyl-2-nitromidazolium	$[CF_3SO_3]$	1:1	466
95/104	1,3-Dimethyl-4-nitromidazolium	$[SO_3(OMe)]$	1:1	472
92/94	1-Ethyl-3-methyl-4-nitroimidazolium	$[CF_3SO_3]$	1:1	453
73	1-Butyl-3-methyl-4,5-dibromoimidazolium	$[CF_3SO_3]$	1:1	428
110/113	1,2,3-Trimethyl-4-nitroimidazolium	$[N(SO_2CF_3)_2]$	1:1	423
63	1-*iso*-Propyl-3-methyl-4-diphenyl-phosphorylimidazolium	$[SO_3(OH)]$	1:1	399
110	1-Ethyl-2-(1-oxy-3-oxo-4,4,5,5-tetramethylimidazolin-2-yl)-3-methylimidazolium	$[N(SO_2CF_3)_2]$	1:1	384
112	6,6′-Spirobis(1-methyl-1,5,6,7-tetrahydropyrrolo[1,2-a]imidazol-4-ium	$[N(SO_2CF_3)_2]$	1:2	1245
	***N*-Hydroxy, *N*-alkoxyl and *N*-aminoimidazolium cations**			
83–84	1,3-Dimethoxyimidazolium	$[PF_6]$	1:1	466
72	1,3-Dihydroxy-2-methylimidazolium	$[N(SO_2CF_3)_2]$	1:1	435
75–76	1,3-Dimethoxy-2-methylimidazolium	$[PF_3(C_2F_5)_3]$	1:1	396
28–30	2-Bromo-1,3-dimethoxymidazolium	$[N(SO_2CF_3)_2]$	1:1	415
40–46	1,3-Diethoxyimidazolium	$[BF_4]$	1:1	460
99–102	1,3-Diethoxyimidazolium	$[PF_6]$	1:1	444
83–84	1-Benzyloxy-3-methylimidazolium	$[PF_6]$	1:1	437
60–70	1,3-Di(benzyloxy)imidazolium	Br	1:1	415
58–59	1,3-Diamino-2-methylimidazolium	$[N(SO_2CF_3)_2]$	1:1	433

11.4.5 Protic Salts of Other Acids and Bases

In addition to the classical *N*-protonated alkylammonium, alkylimidazolium salts and those of pyridinium, the following materials, containing less-common cations, are noted: $[P^iPrH_3][GeCl_3]$ (m.pt. 90°C, U_L 458 kJ mol^{-1}) [383], $[PCyH_3][GeCl_3]$ (m.pt. 40–42°C, U_L 481 kJ mol^{-1}) [126], $[PPh_3H][N(SO_2F)_2]$ (m.pt. 70°C, U_L 402 kJ mol^{-1}) [259] and $[Hpip][B(OMe)_4]$ (m.pt. 59–61°C, U_L 443 kJ mol^{-1}) [188].

It is somewhat ironic that the single-crystal X-ray structure of neither of the salts, oft quoted as the original ionic liquids, namely, ethanolammonium

TABLE 11.24 Additional Ionic Liquids

m.pt./°C			p:q	U_L/ kJ mol^{-1}
	Carbon-centred ions			
0–2	*tert*-Butyl carbenium	[Al_2Br_7]	1:1	418
89–91	1,1,2,3,4,5,6-Heptamethylbenzenium	[BF_4]	1:1	435
94–95	1,1,2,3,4,5,6-Heptamethylbenzenium	[$AlCl_4$]	1:1	414
No m.pt.	1,1,2,3,4,5,6-Heptamethylbenzenium	[$AlCl_4$]	1:1	411
No m.pt.	1,1,2,3,4,5,6-Heptamethylbenzenium	[$AlCl_4$]	1:1	416
91–92	1-Chloromethyl-1,2,3,4,5, 6-hexamethylbenzenium	[$AlCl_4$]	1:1	410
No m.pt.	1-Chloromethyl-1,2,3,4,5, 6-hexamethylbenzenium	[$AlCl_4$]	1:1	408
83–84	1-Phenyl-1,2,3,4,5, 6-hexamethylbenzenium	[$AlCl_4$]	1:1	397
89–90	2-Hydroxyhomotropylium	[$SbCl_6$]	1:1	432
103–104	1-Ethoxyhomotropylium	[$SbCl_6$]	1:1	417
	Boronium salts			
70	Bis(4-methoxyphenalene-1, 9-diolato)boron	[$B\{3,5\text{-}(CF_3)_2C_6H_3\}_4$]	1:1	316
98	Bis(3,4-dimethoxyphenalene-1, 9-diolato)boron	[$B\{3,5\text{-}(CF_3)_2C_6H_3\}_4$]	1:1	309
60	(3-Methylimidazole) (trimethylamine) dihydroboronium	[$N(SO_2CF_3)_2$]	1:1	407
62	Bis(1-methyl-1*H*-imidazole-3-yl) dihydroboronium	[$BH_2(CN)_2$]	1:1	438
	Carborane and stannaborane salts			
97	1-Ethyl-3-methylimidazolium	[nido-$C_2B_9H_{12}$]	1:1	431
80	1-Octylpyridinium	[nido-$C_2B_9H_{12}$]	1:1	396
67	1-Methyl-3-octylimidazolium	[$CB_{11}H_6Cl_6$]	1:1	371
101	1-Butyl-2,3-dimethylimidazolium	[$CB_{11}H_6Cl_6$]	1:1	380
59	1-Ethyl-3-methylimidazolium	[1-$MeCB_{11}H_{11}$]	1:1	415
64	1-Ethyl-3-methylimidazolium	[1-$EtCB_{11}H_{11}$]	1:1	411
106	1-Ethyl-3-methylimidazolium	[1-$EtSnB_{11}H_{11}$]	1:1	409
80	1-Ethyl-3-methylimidazolium	[1-$BuSnB_{11}H_{11}$]	1:1	395
	Bromonium and iodonium salts			
87–89	Bis(pyridine)bromonium	[CF_3SO_3]	1:1	429
85	Bis(pyridine)iodonium	[I_7]	1:1	385
103–105	Bis(2,6-dimethylpyridine)iodonium	[ICl_2]	1:1	410
110	Bis(2,6-dimethylpyridine)iodonium	[IBr_2]	1:1	407
87	Bis(1,3-dimethylthiourea)iodinium	[I_3]	1:1	408
65–66	Diphenyliodonium	[$N(SO_2CF_3)_2$]	1:1	403
108–110	Bis(4-methoxyphenyl)iodonium	[$CS_2(NEt_2)$]	Salt?	393
77–79	(2,2,2-Trifluoroethyl)phenyliodonium	[$N(SO_2CF_3)_2$]	1:1	411
112	4-Fluorophenyl(pentafluorophenyl) iodonium	[BF_4]	1:1	430
88	2-(Hydroxoiodinium)nitrobenzene	[CF_3CO_2]	1:1	466
63–65	{Tris(dimethylamino) phosphaneselenide}iodine(I)	[I_3]	1:1	409

TABLE 11.24 (*Continued*)

m.pt./°C			$p:q$	U_L/ kJ mol^{-1}
	P–P and As–Sb bonded cations			
57–58	Trimethylphosphine(diphenyl) phosphonium	$[CF_3SO_3]$	1:1	405
103–105	Chlorodiphenylphosphine(diphenyl) phosphonium	$[GaCl_4]$	1:1	370
50–52	Triphenylarsine{chloro(phenyl)} stibonium	$[AlCl_4]$	1:1	369
80–82	{Diphenylphosphinomethyl(diphenyl) phosphino}diphenylphosphonium	$[CF_3SO_3]$	1:1	350
60–61.5	3,3-Diphenyltricyclo[2.1.0.0^{2,5}] pentaphosphan-3-ium	$[GaCl_4]$	1:1	400
101–102	3-Chloro-3-ethyltricyclo[2.1.0.0^{2,5}] pentaphosphan-3-ium	$[GaCl_4]$	1:1	437
46–48	3-Chloro-3-cyclohexyltricyclo[2.1.0.0^{2,5}] pentaphosphan-3-ium	$[GaCl_4]$	1:1	415
79–80	3-Chloro-3-pentafluorophenyltricyclo [2.1.0.0^{2,5}]pentaphosphan-3-ium	$[GaCl_4]$	1:1	416
	Di- and tricationic phosphonium			
108–110	1,2-Bis(*P,P*-dimethyl-*P′,P′*-diphenylphosphinophosphonium) ethane	$[GaCl_4]$	1:2	1163
94–95	1,2-Bis(*P,P,P′,P′*-tetraphenylphosphinophosphonium) ethane	$[GaCl_4]$	1:2	1091
64–67	1,6-Bis(*P,P,P′,P′*-tetraphenylphosphinophosphonium) hexane	$[GaCl_4]$	1:2	1075
106–109	1,3-Dichloro-1,2,3,4-tetracyclohexyltetraphosphetane-1, 3-diium	$[GaCl_4]$	1:2	1107
85–87	3,3,5,5,7,7-Hexaphenyltricyclo [2.2.1.0^{2,6}]heptaphosphane-3,5,7-triium	$[Ga_2Cl_7]$	1:3	
	Other aza-heterocycles			
108	1-Ethylbenzimidazolium	$[BF_4]$	1:1	470
111	1-Butyl-2-methylbenzimidazolium	$[BF_4]$	1:1	433
70	1,3-Dibutylbenzimidazolium	$[Fe^{III}Cl_4]$	1:1	395
80	1,3-Bis(dodecyl)benzimidazolium	$[Cd_2Cl_6]$	2:1	662
104–106	2-Aminopyrimidinium	$[(2\text{-}OH)C_6H_4CO_2]$	1:1	464
64–66	2-Amino-4-methylpyrimidinium	$[3\text{-}FC_6H_4CO_2]$	1:1	459
95–96	2-Amino-4-methylpyrimidinium	$[3\text{-}ClC_6H_4CO_2]$	1:1	454
106–108	2-Amino-4-phenyl-6-methylpyrimidinium	$[C_6H_5CO_2]$	1:1	430
97–99	1-(3-Butynyl)-2-(phenylethynyl) pyrimidinium	$[CF_3SO_3]$	1:1	412
95	3-Methoxyazetidinium	Cl	1:1	539

(*Continued*)

TABLE 11.24 (*Continued*)

m.pt./°C			$p:q$	U_L/ kJ mol^{-1}
102–103	3-*O*-Nitrato-3-ethylazetidinium	$[NO_3]$	1:1	479
106	3,3-Dinitroazetidinium	$[(2\text{-}OH)C_6H_4CO_2]$	1:1	449
104	5-Azonia-2-oxa-spiro[4.4]nonane	$[BF_4]$	1:1	483
74	1-Butyl-1-methylazepanium	$[CF_3CO_2]$	1:1	430
58	1,3,5-Trimethyl-tetrahydro-1, 3,5-triazinium	$[I_3]$	1:1	437
43	1,3,5-Triethyl-tetrahydro-1, 3,5-triazinium	$[I_3]$	1:1	416
69	1,3,5-Tri-*iso*-propyl-tetrahydro-1, 3,5-triazinium	$[I_5]$	1:1	384
82	1,3,5-Tri-*tert*-butyl-tetrahydro-1, 3,5-triazinium	$[I_7]$	1:1	361
104	1,3,5-Tri-*tert*-butyl-tetrahydro-1, 3,5-triazinium	$[I_5]$	1:1	373
76	3-Methyl-2-phenyl-3,4,5, 6-tetrahydropyrimidinium	$[CF_3SO_3]$	1:1	434
87–88	1,3-Diethyl-3,4,5,6-tetrahydropyrimidinium	$[PF_6]$	1:1	448
98–100	1,5-Dimethoxyl-3-phenyl-5a,6,7,8,9,9a-hexahydro-1*H*-1, 5-benzodiazepin-5-ium	$[ClO_4]$	1:1	407
59	*N*-Amino-1-azoniacycloheptane	$[N_3]$	1:1	493
71	1-Aza-8-azoniabicyclo[6.6.3] heptadeca-4,11-diyne	$[(CF_3CO_2)_2H]$	1:1	394
72	1-Aza-8-azoniabicyclo[6.6.4] octadeca-4,11-diyne	$[(CF_3CO_2)_2H]$	1:1	390
75	1-Aza-8-azoniabicyclo[6.6.5] nonadeca-4,11-diyne	$[(CF_3CO_2)_2H]$	1:1	385
77	1-Ethyl-1,4-diazabicyclo[2.2.2] octanium	$[N(SO_2CF_3)_2]$	1:1	413–418
90–92	1,4,7-Trimethyl-1-azonia-4, 7-diazacyclononane	$[CF_3SO_3]$	1:1	424
92–94	3-(3-Ethenyl-1,2,4-oxadiazol-5-yl)-1,2,5,6-tetrahydropyridinium	$[CF_3CO_2]$	1:1	442
110–114	3-(4-Methoxyphenyl)-1,2-dimethyl-4,5-dihydroimidazolium	I	1:1	443
108	5-Ethoxy-l-ethyl-3,4-dihydro-2-oxo-2*H*-pyrrolium	$[SbCl_6]$	1:1	413
94	5-Hydroxy-1-methyl-3,4-dihydro-2*H*-pyrrolium	Cl	1:1	525
80–82	3-Hydroxy-1-*tert*-butyl-1,2-dihydropyrrolium	$[\{2,4,6\text{-}(NO_2)_3\}C_6H_2O]$ Picrate	1:1	417
96–98	2-Acetyl-9-azoniabicyclo[4.2.1] nonan-3-one	$[CF_3CO_2]$	1:1	442
112	6,6′-Spirobis(1-methyl-1,5,6, 7-tetrahydropyrrolo[1,2-a] imidazol-4-ium	$[N(SO_2CF_3)_2]$	1:2	1245
96	Piperazinediium	$[C_9H_{19}CO_2]$	1:2	1280
96	Piperazinediium	$[C_{11}H_{23}CO_2]$	1:2	1240

TABLE 11.24 (*Continued*)

m.pt./°C			p:q	U_L/ kJ mol^{-1}
96	Piperazinediium	$[C_{13}H_{27}CO_2]$	1:2	1207
97	Piperazinediium	$[C_6H_{13}CO_2]$	1:2	1357
97	Piperazinediium	$[C_{15}H_{31}CO_2]$	1:2	1178
110	Piperazinediium	$[C_5H_{11}CO_2]$	1:2	1422
	Thiazolium and benzothiazolium			
85	3-{(2*S*)-2-Methylbutyl}thiazolium/ tetrabutylammonium	$[CF_3SO_3]$	(1,1):1	350
92–94	2-Amino-5-butyl-4-methyl-1, 3-thiazol-3-ium	$[NO_3]$	1:1	458
99–103	2-Amino-2-thiazolinium	*N*-Methylpyrrole-2-carboxylate	1:1	461
104–107	2-Aminobenzothiazolium	1-Methylpyrrole-2-carboxylate	1:1	444
88–90	*N*-Methylbenzothiadiazolium	$[CF_3SO_3]$	1:1	454
82	*N*-Decylbenzothiazolium	I	1:1	410
80	*N*-Undecylbenzothiazolium	I	1:1	402
90.5	*N*-Dodecylbenzothiazolium	I	1:1	399
90–92	3-(3-Butynyl)-2-(1-hexynyl) benzothiazolium	$[CF_3SO_3]$	1:1	402
	Silicon, sulfur, phosphorus and arsenic heterocycles			
95–97	1,3-Dimethyl-1,3-diaza-2-phospholidinium	$[GaCl_4]$	1:1	447
52–54.5	1,3-Dimethyl-1,3-diaza-2-arsenanium	$[GaCl_4]$	1:1	440
54–56	5-Methyl-1,3,5-dithiazinan-5-ium	$[GaCl_4]$	1:1	451
68–69	9-Aza-9-methyl-1-thioniabicyclo [3.3.1]nonane	$[I_3]$	1:1	435
103	2-*tert*-Butyl-3,3,5-trimethyl-1,2-diaza-3-sila-5-cyclopentenium	$[AlCl_4]$	1:1	408
86–88	t-BuNHP(μ-Nt-Bu)$_2$P$^+${(NHt-Bu) (NCO$_2$i-Pr)}NH(CO$_2$i-Pr)	$[C_6H_5CO_2]$	1:1	339
90–91.5	1-Phenylthiolanium	$[ClO_4]$	1:1	454
104	(±)-*cis*-3-Acetoxy-1-methylthiacycohexane	$[ClO_4]$	1:1	450
106	(−)-(1*S*,3*S*)-*trans*-3-Acetoxy-1-methylthiacycohexane	$[ClO_4]$	1:1	448
79–80	3(*R*),4(*R*),5(*R*),6(*S*)-3,4,5-Trihydroxy-*cis*-1-thioniabicyclo[4.3.0]nonane	$[ClO_4]$	1:1	459
94–95.5	1,6-Diarsa-2,5,7,10-tetrathiatricyclo[5.3.0^{1,5}.0^{6,10}]decane	$[GaCl_4]$	1:2	1363
104–105	1,6-Diarsa-2,5,7,10-tetramethyl-2,5,7,10-tetraazatricyclo [5.3.0^{1,5}.0^{6,10}]decane	$[GaCl_4]$	1:2	1311
	Diketonate			
72–73	1-Octadecyl-3-methylimidazolium	$[CF_3C(O)CHC(O)CF_3]$	1:1	366
62	Morpholinium	$[CF_3C(O)CHC(O)CF_3]$	1:1	453

(*Continued*)

TABLE 11.24 (*Continued*)

m.pt./°C			p:q	U_L/ kJ mol^{-1}
−11	*N,N,N′*-Triethylethylenediammonium	$[CF_3C(O)CHC(O)CF_3]$	1:1	413
67	1-Butylpyrrolidinium	$[CF_3C(O)CHC(O)C_6H_5]$	1:1	410
71	Dibutylammonium	$[CF_3C(O)CHC(O)$ (2-thienyl)]	1:1	408
	Unclassified salts			
100	Bis(dimethylamino) trifluoromethylsulfonium	$[CF_3S]$	1:1	451
90–92	Tetrabutylammonium	2,3,6-Tricyano-4-fluoro-5-trifluoromethylphenolate	1:1	371
107	4-Methoxy-2,6-dimethyl-1-hydroxypyridinium	2,6-Dichloro-4-nitrophenolate	1:1	424
108–110	1,3-Di-*iso*-propyl-4,5-dimethylimidazolium	*E*-2-Cyano-1-phenylethenolate	1:1	407
104–107	Ethyldi-*iso*-propylammonium	$[\{C_6H_5)_3C\}P(BH_3)(O)OH]$	1:1	377
108–109	Tetrabutylammonium	$[C_5H_5]$ (Cyclopentadienide)	1:1	396

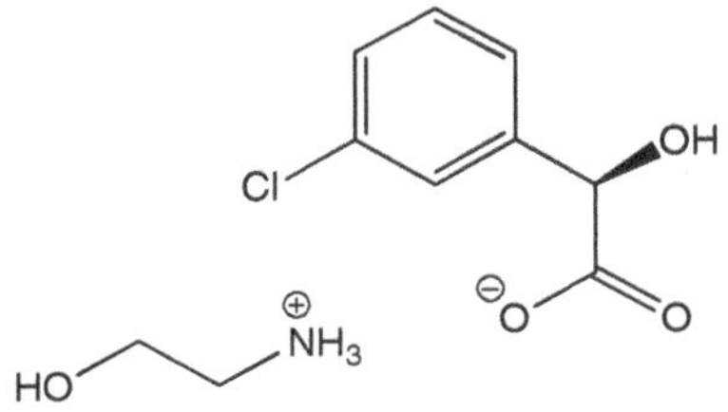

Figure 11.6 (*R*)-ethanolammonium 3-chloromandelate [611].

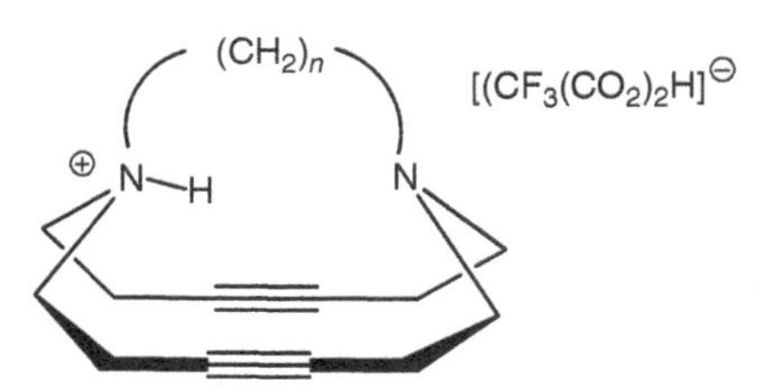

Figure 11.7 1-Aza-8-azoniabicyclo[6.6.*n*]alka-4,11-diyne hydrogenbis(trifluoroethanoate) (n=3, 4 or 5) [266].

nitrate (m.pt. 52°C) and ethylammonium nitrate (m.pt. 14°C) first described, respectively, in 1888 and 1914, has been reported (*n.b.* see the Addendum). Those of $[NMe_2H_2][NO_3]$ [322] (m.pt. 80°C [622], U_L 557 kJ mol^{-1}) and $[NMeH_3][NO_3]$ (m.pt. 110°C, U_L 594 kJ mol^{-1}) [539] are known. Other values for the melting point of the latter have been reported variously as 105°C and 108–108.5°C. The effect on melting point of the introduction of the

$HOCH_2CH_2$— moiety manifests itself in a number of materials included in this review, including the protic $[NH(CH_2CH_2OH)_3][NO_3]$ that melts at 80°C (U_L 475 kJ mol^{-1}) [320].

11.5 METAL-CONTAINING IONIC LIQUIDS

Ionic liquids based on metal-containing anions and cations, including low-melting coordination complexes, are an area of active investigation. Such materials include the classic haloaluminate salts and related materials. Discussion of the latter is concentrated in Section 11.11.1 for salts of the anion $[EX_4]^-$, in which E may be metal, metalloid or non-metal. Likewise, mention will also be made of metal-containing $[EX_6]^-$ salts in Section 11.5.6.5.

Ten structurally characterised metal-containing salts (discussed further in the following text) melt at or below 30°C and fall into the category of room-temperature ionic liquids. These are $[C_2mim][NbF_6]$ (m.pt. −1°C, U_L 462 kJ mol^{-1}), $[C_2mim][TaF_6]$ (m.pt. 2°C, U_L 462 kJ mol^{-1}), $[C_4mim]_2[Sn^{II}Cl_4]$ (m.pt. 5°C), $[C_2mim][FeCl_4]$ (m.pt. 10°C, U_L 444 and 449 kJ mol^{-1}; for two polymorphs), $[C_4mim]_2[Cu^{II}Cl_4]$ (m.pt. 23°C), $[C_{12}mim]_3[CuCl_2][CuCl_3]$ (m.pt. 26°C), $[C_{12}mim]_4[Cu_4Cl_8]$ (m.pt. 27°C), $[Cu^I(C_{12}im)_4][PF_6]$ (m.pt. 27°C), $[Ag(C_4im)(C_2im)][N(SO_2CF_3)_2]$ (m.pt. 27°C, U_L 388 kJ mol^{-1}) and $[Ag(C_4im)(C_2im)][N(SO_2CF_3)_2]$ (m.pt. 30°C, U_L 387 kJ mol^{-1}).

Note should be taken of a small number of ionic liquids than can be classified as organometallic. Those containing Al—C, Si—C, Sn—C and Ag—C bonds are dealt with in this or other sections, where appropriate. Three aza-cations [164, 315, 316], in which Se—C and Te—C bonds are to be found, but in which the Group 16 element is not the centre of charge, have been characterised (as trifluoromethanesulfonates). In addition, attention is drawn, for completeness, to structurally characterised salts, whose reported melting temperatures qualify them for inclusion in this survey, and which contain $[Co(CO)_4]^-$ [125] or cyclopentadienide, the latter either as a free anion [521] or complexed to lithium [511], titanium(III) [292] or iron(III) [106, 151, 238, 312].

11.5.1 Wholly Inorganic Ionic Liquids

Some eutectic mixtures of high-melting salts melt at temperatures that formally qualify them as ionic liquids. Indeed, in 1978, the then National Bureau of Standards in the United States published [623] a compilation of the m.pt. and composition of *ca.* 6000 inorganic eutectics. About 300 of these are reported to have a m.pt. between −138 and 100°C. Ignoring mixtures based on salt hydrates, the lowest-melting entry is a mixture of $K[HCO_2]$ (48.3 mole%)–$K[SCN]$ (34.9 mole%)–$K[NO_3]$ (16.8 mole%) that melts at 55.5°C [624], though the inclusion of methanoate raises a question regarding the definition of 'inorganic'. The compilation contains a 1937 report of $Cs[NO_3]$ (68 mole%)–$Pb(NO_3)_2$ (32 mole%), having a m.pt. of 68°C, that is the lowest-melting 'wholly' inorganic salt mixture among these early reports. More recently, low-temperature eutectics have been detected for glass-forming mixtures of $AlCl_3$–M[SCN] (M = Li, Na or K) [625] and of $AlCl_3$–Na[N(CN)$_2$] [626]. Alkali bis(fluorosulfonyl)amides, M[N(SO$_2$F)$_2$] (M = Li, Na, K or Cs, reported [627] m.pt. 130, 106, 102 and 113°C, respectively), form binary eutectics [627] with eutectic temperatures as low as 52°C for Na[N(SO$_2$F)$_2$] (47 mole%)–Cs[N(SO$_2$F)$_2$] (53 mole%) and ternary eutectics with eutectic temperatures as low as 38°C (for Li[N(SO$_2$F)$_2$] (30 mole%)–Na[N(SO$_2$F)$_2$] (40 mole%)–Cs[N(SO$_2$F)$_2$] (30 mole%) [628]).

Salts of inorganic anions and cations that do not contain C—H bonds, which melt below 110°C, and for which a crystal structure has been reported are included in Table 11.7. Surprisingly, perhaps, there are over 30 of them. These include those with the cations $[H_5O_3]^+$, $[NH_3OH]^+$, $[NH_4]^+$, $[NH_2NH_3]^+$, potassium, rubidium and caesium. If C—N-bonded systems not having C—H bonds are permitted, included also is a series of guanidinium and amino-, diamino- and triaminoguanidinium salts, as well as salts of diaminouronium and azidoformamidinium. The compilation contains five hydrazinium salts and four $[(H_2NNH)_2CNH_2]^+$ salts. The heterocycles 1*H*-tetrazolium, 1-amino-1,2,3-triazolium, 5-aminotetrazolium and 1,5-diaminotetrazolium are also represented. The commonest anion is dinitramide (nine entries, if the rubidium salt, melting at 102–106°C is permitted), with the related $[N(CN)(NO_2)]^-$, $[N(CN)_2]^-$, $[N(NO_2)(SF_5)]^-$ and $[C(NO_2)_3]^-$ being found in a further six salts between them. There are four perchlorates and three nitrates, with fluoride, azide and $[N(SO_2F)_2]^-$ found in one salt each. There are four salts of heterocyclic anions, tetrazolate, 5-azidotetraolate and 5-nitrotetrazolate, the latter found in three salts.

Excluding the oxonium dihydrate nitrate salt (m.pt. −18.5°C) [60], the lowest-melting wholly inorganic material, with a m.pt. of 18–23°C [617], is hydroxylammonium dinitramide [80] though (as discussed in Section 11.6.1) in the crystal, in addition to the two eponymous ions, neutral hydroxylamine and its zwitterionic $[^+NH_3O^-]$ isomer are seen. Interestingly, the crystal structure of the monopropellant $[NH_3OH][NO_3]$ is reported [629], and while no reliable melting point appears to be available, it is believed to melt at less than 100°C. Despite its relatively high m.pt., $[NH_3OH]F$ is important [436] as the only salt included in Tables 11.1, 11.2 and 11.3 containing an unsolvated fluoride ion. Early crystallographic studies of the formally ionic liquids, hydrazinium fluoride (m.pt. 100.5°C [630]) [631], chloride (m.pt. 92°C [632]) [633] and bromide (m.pt. 86.5–88.5°C [634]) [635], have also been reported, though no cif records are available.

In a recent report, the individual salts M[N(SO$_2$F)$_2$], M = Na, K or Cs, have been subjected to a multi-technique study, including single-crystal and powder X-ray diffraction, DSC and Raman spectroscopy, revealing the existence of three polymorphs for the sodium salt and two each for the potassium and caesium salts [209], discussed further in the next section. The caesium [351, 352] and rubidium [352] salts (no cifs) of the analogous anion, $[N(NO_2)_2]^-$, are also relatively low melting [636–638]. The higher-melting ammonium, lithium and potassium dinitramide salts have also been structurally characterised [351]. Interestingly, only in the case of the lithium salt is the anion found in the minimum energy C_2 conformation.

11.5.2 Ionic Liquids of the Group 1 Metals

The melting points of ionic liquids are often contrasted with those of alkali metal halides, particularly the archetypical sodium chloride, which has a m.pt. of 804°C. The m.pt. (°C) of MX (M = Li, Na, K or Rb and X = F, Cl, Br or I) are taken from Ubbelohde [639].

	F	Cl	Br	I
Li	870	613	547	446
Na	980	804	755	651
K	880	775	733	723
Rb	760	715	684	640

It may be surprising, therefore, to discover structurally characterised inorganic salts of the alkali metals having melting points as low as 63°C (X-ray data collected at 113 K (U_L = 547 kJ mol^{-1}) and at 298 K (U_L = 543 kJ mol^{-1})), as found for form II of K[N(SO_2F)$_2$] obtained on cooling the melt [209]. A polymorph stable at room temperature (Form I, melting at 102°C; data collected at 113 K (U_L = 546 kJ mol^{-1}) and 298 K (U_L = 541 kJ mol^{-1})) was also characterised. In both forms, the bis(fluorosulfonyl)amide ion adopts a *cisoid* conformation. Even lower melting (14–16°C) is the coordination complex, [LiL_4]$_2$[P_2Se_8] (L = CH_3CN) [78]: when L = py, the m.pt. = 109°C. In fact, low-melting salts of the other alkali metals, Na^+, Rb^+ and Cs^+, have all been characterised, as seen in Table 11.8. Structural characterisation is particularly critical in establishing the constitution of the composite species and the nature of the interionic interactions, to determine whether the degree of association between anionic and cationic components begins to challenge the notion of what may be described as a salt. The structure of [Na(dimethoxyethane)][B{OCH(CF_3)$_2$}$_4$] (m.pt. 56°C) [171] suggests that it is better considered as a coordination complex rather than a salt, whereas [Na(18-crown-6)py$_2$][BH_4] (m.pt. 92–94°C, U_L = 374 kJ mol^{-1}) [166] can be considered a salt. The latter study is of interest as, in the related [Na(BH_4)(18-crown-5)(py)$_{0.5}$] (refcode HIWXUH) [166] melting at 109–112°C, the [BH_4]$^-$ acts as a bidentate ion towards sodium and the compound is not a salt in the solid state. In passing, the crystal structure of [C_1mim][BH_4] has been described [640], refcode CANTES, but the study does not include the measurement of a melting point. Structures containing anions of the type [EX_4] are discussed in Section 11.11. The structures of potassium-polyether complexes with polyiodide anions [379, 392] and caesium salts of

$[N(NO_2)_2]^-$ [351, 352] (m.pt. 85–87°C [638]) and $[Al\{OCH(CF_3)_2\}_4]^-$ (m.pt. 83–85°C) [342] have been reported. $Rb[N(NO_2)_2]$ melts at 102–106°C [352].

It is also worth noting briefly that the reported m.pt. of $Na[N(NO_2)_2]$ and $Na[N(SO_2F)_2]$ are 101–107°C [638] for the former (though no crystal structure has been reported) and 110°C (for form III) and 118°C (for form I) [209] (with a crystal structure) for the latter.

11.5.3 Salts of Transition and Main Group Metal-Containing Complex Cations

Sections 11.5.3 and 11.5.4 cover ionic liquids containing metal coordination complexes, either as cations or as both cations and anions. Those containing anionic metal coordination complexes are also included in Table 11.9. However, they may also be referred to in Section 11.5.6 where metal halometallates and oxohalometallates are discussed.

Interest in metal complex chemistry using ionic liquids as synthetic media, and as media to carry out metal-catalysed reactions, has excited much interest, particularly as a result of the involatility of many ionic liquids that, potentially, can aid the separation of volatile products. However, the solubility of metal salts with conventional counter-anions in ionic liquids tends to be low. In addition, transition metal coordination complexes tend to give salts that are relatively high melting. This has motivated the synthesis and study of a range of metal complexes associated with ions now widely found in ionic liquids, including 1,3-dialkylimidazolium and bis(trifluoromethanesulfonyl)amide. It is beyond the scope of this chapter to review the chemistry and applications of these materials, other than to note the increasing range of complexes with melting temperatures less than 100°C now available that have been fully characterised by X-ray diffraction methods, and to highlight key features of their crystal structures. The importance should also be noted of the need to establish the nature of such complexes present in the liquid formed on melting.

Recently reported cationic complexes (listed in order of increasing melting temperatures in Table 11.9) include $[Cu^I(C_{12}im)_4][PF_6]$ (m.pt. 27°C, U_L 303 kJ mol^{-1}) [85], $[Ag(C_4im)(C_2im)][N(SO_2CF_3)_2]$ (m.pt. 27°C, U_L 388 kJ mol^{-1}) [89], $[Ag(C_4im)(mim)][N(SO_2CF_3)_2]$ (m.pt. 30°C, U_L 387 kJ mol^{-1}) [89], $[Ag(C_4im)_2][N(SO_2CF_3)_2]$ (m.pt. 42°C, U_L 378 kJ mol^{-1}) [89], $[Ag(C_2im)_2][N(SO_2CF_3)_2]$ (m.pt. 65°C, U_L 398 kJ mol^{-1}) [79], $[Cu^I(CH_3CN)_2][N(SO_2CF_3)_2]$ (m.pt. 65°C) [229], $[Cu^I(CH_3CN)_4][N(SO_2CF_3)_2]$ (m.pt. 65°C) [229], $[Ag(mim)_2][N(SO_2CF_3)_2]$ (m.pt. 87°C, U_L 407 kJ mol^{-1}) [89] and $[Ag(CH_3CN)_2][N(SO_2CF_3)_2]$ (m.pt. 90°C, U_L 455 kJ mol^{-1}) [144].

Vander Hoogerstraete et al. [144] have very recently extended this series to include 1-alkylimidazole complexes of copper(II), cobalt(II) and nickel(II), but only the 1:2 (p:q) salts with $[N(SO_2CF_3)_2]^-$ of $[Cu^{II}(C_4im)_4]^{2+}$ (m.pt. 46°C, U_L 1118 kJ mol^{-1}), $[Cu^{II}(C_4im)_6]^{2+}$ (m.pt. 52°C, U_L 1137 kJ mol^{-1}), $[Cu^{II}(C_2im)_4]^{2+}$ (m.pt. 74°C, U_L 1156 kJ mol^{-1}) and $[Cu^{II}(mim)_4]^{2+}$ (m.pt. 89°C, U_L 1206 kJ mol^{-1}) meet the requirement for inclusion in this chapter.

The palette of ligands employed is thus limited, so far, to materials such as *N*-alkylimidazoles and nitriles. The reports of two low-melting $[SnMe_3L_2][N(SO_2F)_2]$ (L = *O*-bonded *N,N'*-dimethylethyleneurea; m.pt. 33°C, $U_L = 386\,kJ\,mol^{-1}$; L = *O*-bonded dimethylsulfoxide, m.pt. 66°C, $U_L = 406\,kJ\,mol^{-1}$) complexes [102] that have been crystallographically characterised may point to other ligand–metal–anion combinations that may be worthy of investigation.

While many coordination complexes are relatively high melting and fall outside the scope of the compilation, of interest, nevertheless, is an analysis of published and new structural solutions for $[M(NCMe)_4][A]$, where M = Cu or Ag; $[A]^- = [BF_4]^-$, $[ClO_4]^-$, $[PF_6]^-$, of relevance because it was shown [641] that the underlying basic structures are virtually identical, being made up of alternating cationic and anionic sheets with cation–anion columns perpendicular to the sheets maintained by C—H···F,O interactions.

Of associated relevance are investigations that show that the imidazolium cation of the ionic liquid may be reactive towards dissolved transition metal catalysts, with the isolation and characterisation of a carbene-platinum complex [642] (CSD refcode TETVOC), m.pt. 193°C, from one such system. This provides a link to the extensive literature [643–645] on imidazolylidene carbenes and related metal imidazolylidene complexes. More recently, the crystal structures of $[Ag(1\text{-ethyl-3-methylimidazol-2-ylidene})_2][B(CN)_4]$ (m.pt. 100–104°C) [464] and the lower-melting $[Ag(1\text{-ethyl-3-methylimidazol-2-ylidene})_2][N(SO_2CF_3)_2]$ (refcode BAHZAN) [646] (m.pt. *ca.* 40°C) have been reported. There are over 3500 imidazolylidene structures in the CSD, though the great majority of entries are neutral compounds. Most salts that do not retain solvent on crystallisation are either high melting, decompose on or before melting, or have no m.pt. reported for them. One low-melting salt (not metal containing) is, however, noted: 1-phenyl-4-(triphenylphosphanylidene)-4*H*-pyrazol-1-ium tetrafluoroborate (m.pt. 75–82°C) [293], with the higher melting bis(1,3-di-*iso*-propyl-4,5-dimethylimidazol-2-lyidene)selenium bis(trifluoromethanesulfonate) (FUJTOU; m.pt. 102–105°C) [647] and its tellurium analogue (SURCOY; decomposing at 110–112°C) [648] included here for completeness.

11.5.4 Salts with both Transition Metal-Containing Complex Cations and Anions

Similarly, only modest diversity (as far as the component metals and anions are concerned) can be found among other structurally characterised salts in which both anion and cation are metal containing. These include $[Ag(CH_3CN)_4]_2[Ag\{(CF_3SO_2)_2N\}_3]$ (m.pt. 18°C) [79], $[Cu^I(C_{12}im)_2][Cu^IBr_2]$ (m.pt. 66°C) [85] and $[Ag(CH_3CN)_4][Ag_3\{N(NO_2)_2\}_4]$ (m.pt. 68–72°C) [250]. The structurally characterised dimethyldisulfide complex, $[Nb^V(\eta^2\text{-MeSSMe})_2Cl_4][Nb^VCl_6]$ [339], with an eight-coordinate dodecahedral cation, purified by sublimation at 80°C/0.01 mm Hg and melting at 82–85°C, is one of a series of η^2-RSSR salts.

These examples show the compositional and structural variability evident even within this narrow subclass of ionic liquids. Mention has already been made that the imidazolium cation is reactive, particularly towards transition metals. In addition, structures of metal-containing counter-anions (e.g. in which silver is coordinated to $[N(SO_2CF_3)_2]^-$ in a trinuclear species [250]) establish that the anion most closely associated with ionic liquid behaviour, $[N(SO_2CF_3)_2]^-$, itself cannot be considered wholly 'innocent' [649] in the presence of transition metal ions, a factor of some importance in seeking to understand reaction chemistry and the nature of reactive intermediates in ionic liquids. The nature of melts formed from these salts would, therefore, be an important aspect to explore further, to establish whether nonionic species may be generated close to the melting point. Whether long-lived ion pairs should be considered 'neutral species', or simply represent an indication of short-range order in the melt, is something of a moot point (though the importance of ion pairing in the solid state has been pointed out [650]).

11.5.5 Lanthanide- and Actinide-Containing Ionic Liquids

The first reported lanthanide (Ln) compounds that formally qualify as ionic liquids as far as their melting temperatures are concerned are $[C_4mim]^+$ salts with anions of the type $[Ln(NCS)_x(H_2O)_y]^{(3-x)}$ ($x = 6, 7$ or 8; $y = 2, 1$ or 0) prepared in $[C_4mim][SCN]$. The crystal structure of $[C_4mim]_4[La(NCS)_7(H_2O)]$ (m.pt. 38°C) reveals [118] eight-coordinate lanthanum(III) with a distorted square *anti*-prismatic coordination geometry. The coordinated water is strongly hydrogen bonded to a sulfur of an adjacent anion, an interaction believed to be associated with the tendency to crystallise as opposed to forming a glass. UV–visible spectroscopy on the solid and liquid phases supports the view that the bonding arrangement in the anion survives the melting process.

The crystal structures of three anionic lanthanide complexes of europium(III) (with $[N(SO_2CF_3)_2]^-$) have been reported [246], with formally 1:1 materials being isolated with imidazolium cations, $[C_4mim][Eu\{N(SO_2CF_3)_2\}_4]$ and $[C_3mim][Eu\{N(SO_2CF_3)_2\}_4]$ (melting at 68 and 81°C, respectively, with corresponding values of U_L of 327 and 330 kJ mol^{-1}), on the one hand, and a 2:1 salt, $[C_4C_1pyr]_2[Eu\{N(SO_2CF_3)_2\}_5]$ (m.pt. 92°C), on the other. All melt congruently. In the three structures, nine europium-bound oxygens describe a distorted tricapped trigonal prism. However, the formally monoanionic complexes achieve nine coordination by forming a dimer, $[Eu_2\{[N(SO_2CF_3)_2\}_8]^{2-}$, in which one chelating bistriflamide anion on each europium(III) acts as a monodentate bridge to a neighbour. In addition, the conformations of the five bistriflamide ligands in the $[C_4mim]^+$ salt are (*trans*oid)$_4$(*cis*oid) and in the $[C_3mim]^+$ salts are (*trans*oid)(*cis*oid)$_4$. Attempts to crystallise $[C_4mim]_2[Eu\{N(SO_2CF_3)_2\}_5]$ from a melt of this stoichiometry yielded (after much effort) only $[C_4mim][Eu\{N(SO_2CF_3)_2\}_4]$. $[C_3C_1pyr][Yb\{N(SO_2CF_3)_2\}_4]$, melting at 107°C, has been characterised structurally [510] as a single crystal. Recently, the anion $\{C(CN)_2(NO)\}^-$ has been employed to generate [180, 186] a series of complexes of formulation $[cat]_3[M\{C(CN)_2(NO)\}_6]$ (M = La, Sm, Ce or Pr), where $[cat]^+$ is, respectively, $[C_2mim]^+$, $[N_{4444}]^+$, $[C_2C_1C_1im]^+$ or $[N_{4444}]^+$. Crystal structures have been reported for $[N_{4444}]_3[Ce\{C(CN)_2(NO)\}_6]$ (m.pt. onset in first and second DSC heating cycles are 84.5 and 68°C, respectively) [186] and $[NBu_4]_3[Sm\{C(CN)_2(NO)\}_6]$ [186] (similarly, 69°C (first) and 59°C (second)). A more detailed study [186] of the praseodymium analogue, $[N_{4444}]_3[Pr\{C(CN)_2(NO)\}_6]$, shows that a thermodynamically less stable polymorph may be formed on crystallisation in the DSC. $[C_2mim]_3[La\{C(CN)_2(NO)\}_6]$ (m.pt. 58°C) [180] and $[C_2C_1C_1im]_3[Pr\{C(CN)_2(NO)\}_6]$ (m.pt. 63°C (first) and 61°C (second)) [180] both show involvement of the C(4)–H and C(5)–H protons of the imidazolium cation in (albeit rather long) hydrogen bonding. $[N_{4444}][Y(BH_4)_4]$ has also been characterised [306] and melts at 78°C (with, interestingly, the melt reportedly stable over a range of 150°C).

Remarkably, the structurally characterised [174] dysprosium-betaine complex, $[Dy_2(NMe_3CH_2CO_2)_8(H_2O)_4][N(SO_2CF_3)_2]_6$ (FW = 3007.20), is reported to melt at 88°C. There is, as yet, no actinide ionic liquid characterised using X-ray crystallography. The lowest-melting such material is $[C_2mim]_3[UO_2(NCS)_5]$ (m.pt. 103°C) [481]. A series of $[NR_4][UO_2(NCS)_5]$ (R = Me, Et or Pr) have been characterised crystallographically, though no m.pt. data are provided [651].

11.5.6 Metal-Containing Anions

Coordination numbers of 2–6 are found for mononuclear metal complex ionic liquid anions, mostly with halide ions.

11.5.6.1 Two Coordination Digonal complexes of Ag(I) and Au(I), studied crystallographically, include the $[Au(CN)_2]^-$ salts of $[N_{4444}]^+$ (m.pt. 65–69°C) [228], [1-methylimidazolium]$^+$ (m.pt. 103°C) [214], $[C_2mim]^+$ (m.pt. 64°C; U_L 469 kJ mol^{-1}) [276] and $[C_2mim][Ag(CN)_2]$ (m.pt. 73°C; U_L 466 kJ mol^{-1}) [276]. [1-methylimidazolium] $[Au(CN)_2]$ and $[C_2mim][Au(CN)_2]$ differ in the extent of Au–Au interactions, with $[AuCN)_2]^-$ forming dimers with a Au···Au contact at 3.553 Å in the former and infinite chains of the anion in the latter, with Au···Au separations greater than 4.9 Å. The silver salt, $[C_2mim][Ag(CN)_2]$, also contains 1-dimensional anionic chains, with Ag···Ag separations of 3.226 Å (shorter than the corresponding van der Waals distance of 3.44 Å) and C–H···π interactions between cations. In $[N_{4444}][AuCN)_2]$, the shortest Au···Au contact is 8.05 Å [228].

11.5.6.2 Three Coordination A series of complexes containing trigonal pyramidal $[SnCl_3]^-$ (with the cations $[C_2mim]^+$ (m.pt. *ca.* 104°C) [490] and tetramethylformamidinium (m.pt. 65°C) [223], $[GeCl_3]^-$ (with *iso*-propylphosphonium, m.pt. 40–42°C [126]), have been structurally characterised. Hydrogen bonding between anion and cation is noted, though for the $[GeCl_3]^-$ structures, this and other interionic interactions (including dimerisation of the anion) are weak. Anion association in the analogous indium compound, [tetramethylform amidinium]$_2$[In_2Cl_6] (m.pt. 92°C), involves a metal–metal bond 2.719 Å long [223]. The group of salts, [C_3mim][$HgCl_3$] (m.pt. 69°C (first heating)/67°C (second); U_L 445 kJ mol^{-1}), [C_3mim][$HgBr_3$] (m.pt. 39.5°C (first heating)/41°C (second); U_L 441 kJ mol^{-1}), [C_4mim][$HgCl_3$] (m.pt. 94°C (first heating)/92°C (second); U_L 442 kJ mol^{-1}) and [C_4mim][$HgBr_3$] (m.pt. 58°C (first heating)/57°C (second); U_L 435 kJ mol^{-1}), allow the effect on crystal structure of small compositional changes to be evaluated [122]. Three in this series are isostructural, with chains of edge-sharing trigonal bipyramidal anions. [C_3mim][$HgBr_3$] is the odd one out, containing anions made up of edge-sharing tetrahedra. A dinuclear anion, [Hg_2Cl_6]$^{2-}$, is also found as the [$PBuPh_3$]$^+$ salt (m.pt. 102°C) [472].

11.5.6.3 Four Coordination The largest group of metal-containing ionic liquids are salts of four-coordinate anions, including tetrahedral chloro- and bromometallates of transition metals of the first series and of divalent p-block metals, as well as square-planar complexes of heavier members of the transition series. The latter include salts with anions containing palladium(II): [C_{14}py]$_2$[A], where [A] = [PdI_4]$^{2-}$ (m.pt. 101°C) [282] and [Pd_2Cl_6]$^{2-}$ (m.pt. 74°C) [282], [($EtOCH_2$)mim]$_2$[$PdCl_4$] (m.pt. 103°C) [480] and [($MeOCH_2CH_2$)mim]$_2$[$PdCl_4$] (m.pt. 107°C) [480]; platinum(II): [C_4mim]$_2$[$PtCl_4$] (m.pt. 99°C) [73], an analogue of Zeise's salt, [P_{4444}][$PtBr_3(CH_2{=}CH_2)$] [519] (m.pt. 108°C); and gold(III): [C_2mim][$Au^{III}Cl_4$] [148] (m.pt. 58°C) and [C_4mim][$Au^{III}Cl_4$] [148] (m.pt. 50°C). A salt obtained from [C_6C_6im]Br and K_2[$PdCl_4$], analysing for [C_6C_6im]$_2$[$PdBr_{1.78}Cl_{2.22}$] (CCDC deposition number 898618) and containing a partially exchanged tetrachloropalladate anion, melts at the low temperature of 34°C [652]. The structures of [C_3C_3im][$AuCl_4$] (m.pt. 98°C), [$^tC_4{}^tC_4$im][$AuCl_4$] (m.pt. 132°C) and of [$^tC_4{}^tC_4$im]$_2$[$PdCl_4$] (m.pt. 204°C) were reported in the same study. The structures of a series: [C_4mim]$_2$[MCl_4], M = Mn(II) (m.pt. 63°C; U_L 1003 kJ mol^{-1}), Fe(II) (m.pt. 58°C; U_L 1007 kJ mol^{-1}), Co(II) (m.pt. 62°C; U_L 1010 kJ mol^{-1}), Ni(II) (m.pt. 56°C; U_L 1012 kJ mol^{-1}), Cu(II) (m.pt. 23°C; U_L 1024 kJ mol^{-1}), Zn (m.pt. 60°C; U_L 1010 kJ mol^{-1}) and Sn(II) (m.pt. 5°C; U_L 1012 kJ mol^{-1}), the lowest-melting structurally characterised tetrachlorometallate ionic liquid, provide, in a single study [73], a comparison of the subtle effects associated with increasing numbers of d-electrons. Three types of interactions influence packing, despite their apparent weakness, namely, C—H···Cl hydrogen bonding, C—H···π interactions involving the imidazolium ring as well as π–π stacking.

Single crystals of related materials, 2:1 [C_2mim]$^+$ salts of [$CoCl_4$]$^{2-}$ (m.pt. 100–102°C; U_L 1082 kJ mol^{-1}) [401], [$NiCl_4$]$^{2-}$ (m.pt. 92–93°C; U_L 1085 kJ mol^{-1}) [401] and [$Fe^{II}Cl_4$]$^{2-}$ (m.pt. 86°C; U_L 1078 kJ mol^{-1}) [354] and the [C_4mim]$^+$ salt of [$CoBr_4$]$^{2-}$ (m.pt. 45°C; U_L 994 kJ mol^{-1}) [137], have also been subjected to X-ray crystallographic study. Monoanionic [$Fe^{III}Cl_4$]$^-$ is found with [emim]$^+$ (m.pt. 10°C, two polymorphs; U_L = 444 and 449 kJ mol^{-1}) [76], [C_4C_4im]$^+$ (m.pt. 70°C; U_L 395 kJ mol^{-1}) [258], [C_4C_1pyr]$^+$ (m.pt. 81°C; U_L 427 kJ mol^{-1}) [331], [N_{1444}]$^+$ (m.pt. 110°C; U_L 400 kJ mol^{-1}) [331] and in the mixed valent salt, [$C_4C_1C_1$im]$_3$[$Fe^{III}Cl_4$][$Fe^{II}Cl_4$], melting at 38°C [119].

The six salts of the type [N-methylmorpholinium]$_2$[MX_4] (X = Cl or Br for M = Mn(II), Co(II) and Cu(II), two of which melt below 100°C and two between 100 and 110°C) were established [370] to be isomorphous. The [$MnBr_4$]$^{2-}$ and [$CoBr_4$]$^{2-}$ salts melt at 135–137 and 133–137°C, respectively.

Iron-containing salts are reported, containing either Fe(II), Fe(III) or in both oxidation states. The crystal structures of high- and low-temperature forms of [C_2mim][$FeCl_4$], which melts at 10°C, differ in the conformation of the N-bound ethyl group and the arrangement of anions and cations [76]. The low-temperature form displays weaker hydrogen bonding compared with the extended hydrogen bonding in the high-temperature modification. No π–π stacking was seen in either polymorph, though such interactions are evident in [C_4mim]$_2$[$FeCl_4$] (m.pt. 58°C) [73]. The mixed salt, [$C_4C_1C_1$im]$_3$[$Fe^{III}Cl_4$][$Fe^{II}Cl_4$], melting at 38°C [119], has also been described. Polymorphism is discussed in Section 11.2.3.

11.5.6.4 Five Coordination Five coordination is manifested in both the cation (see Table 11.9) and the anion of Sn(IV)-containing ionic liquids (as well as in other materials with m.pt. in the range 100–110°C). The lowest melting is [$SnMe_3$(O-N,N'-dimethyleneurea)$_2$][$N(SO_2F)_2$] (m.pt. 33°C) [102]. [PBu_4][$SnPh_2Cl_3$], melting at 64–66°C (U_L = 356 kJ mol^{-1}) [218], is one of the lowest-melting [PR_4]$^+$ salts known for which a crystal structure is available.

Anions of the type [SiR_nF_{5-n}]$^-$ (R = Me or Ph; n = 1, 2 or 3) give a variety of low-melting salts, with [$SiMe_3F_2$]$^-$ giving salts with [$S(O)(NMe_2)_3$]$^+$ (m.pt. 57°C; U_L 426 kJ mol^{-1}) [139], [C_1(2-NMe_2)C_1im]$^+$ (m.pt. 67–73°C; U_L 419 kJ mol^{-1}) [245] and [$S(NMe_2)_3$]$^+$ (m.pt. 92°C [653]; U_L 422 kJ mol^{-1}) [396]. Three salts containing [$SiPh_nF_{5-n}$]$^-$, n = 1 (with [$N_{4\,4\,4\,4}$]$^+$) [475, 654], 2 or 3 (both with [K([2.2.2]cryptand)]$^+$) [505], have also been structurally characterised, although all melt in the range 100–110°C.

A further example of the compositional and structural complexity that can arise from salt formation associated with apparently minor changes in a counter-cation is provided by the system [diammonium]Br_2 and $TlBr_3$ [409]. This gives a range of products with the stoichiometry [diammonium][$TlBr_5$]. The choice of cation can lead to salts containing (variously) Br^-, [$TlBr_4$]$^-$, [$TlBr_5$]$^{2-}$ or [$TlBr_6$]$^{3-}$. For instance, [$EtMe_2NCH_2CH_2NMe_2Et$]$^{2+}$ gives a salt containing [$TlBr_5$]$^{2-}$ (as a distorted square antiprism) (m.pt. 141°C), whereas, in [$MeNH_2CH_2CH_2CH_2NH_3$][$TlBr_5$] (the only ionic liquid in this group, melting at 95–96°C), a long inter-anionic Tl···Br interaction leads to a distorted octahedral geometry at Tl(III).

11.5.6.5 Six Coordination The six-coordinate anion, [$NbCl_6$]$^-$, already met in [Nb^V(η^2-MeSSMe)$_2Cl_4$][$NbCl_6$] in Section 11.5.4, is also found in salts with [C_4mβpy] (m.pt. 83–85°C; U_L 413 kJ mol^{-1}) and [C_4m_γpy] (m.pt. 89–91°C; U_L 412 kJ mol^{-1}), both of which have been structurally characterised [308]. [C_2mim][$NbCl_6$], which melts at 161–163°C, was prepared in the same study. Among the lowest-melting salts so far characterised are [C_2mim][NbF_6] [67] and [C_2mim]

$[TaF_6]$ [67], which melt at −1°C (U_L 462 kJ mol^{-1}) and +2°C (U_L 462 kJ mol^{-1}), respectively. For the record, the relevant entries in Table 11.3 of Ref. [73] incorrectly give m.pt. of 545 and 548 K (or 272 and 275°C), respectively, for these salts, citing Ref. [655]. However, Ref. [655] correctly gives the m.pt. as 272 and 275 K. A more detailed discussion of $[EX_6]^-$ salts is found in Section 11.12, and a more detailed discussion of the role of the cation, particularly of 1-alkylpyridinium and 1,3-dialkylimidazolium, is found in Sections 11.17 and 11.18, respectively.

The structures of *N*-(2-hydroxyethyl)piperidinium salts with the doubly negatively charged anions, $[E(ox)_3]^{2-}$, in which E = Si (m.pt. 109°C; U_L 988 kJ mol^{-1}) or Ge (m.pt. 97°C; U_L 991 kJ mol^{-1}) [441], show the limited consequences of a replacement of one Group 14 element by a heavier member. The *N*-(2-hydroxyethyl)pyrrolidinium analogue, with E = Si, melts at 121°C.

11.5.6.6 Other Complex Anions In addition, more complex coordination arrangements and nuclearities can arise in metal-containing anions, providing another dimension to the variability that may be associated with ionic liquid (sometimes eutectic) behaviour (or even to the frustration of crystallisation on cooling a melt). For instance, copper(II) can be found (in order of increasing melting temperatures) in $[C_4mim]_2[CuCl_4]$ (m.pt. 23°C) [73], $[C_4mim]_2[Cu_3Cl_8]$ (m.pt. 70°C) [255] and in the mixed salt $[Me_3NCH_2CH_2OH]_3[CuCl_4]Cl$ (m.pt. 76°C) [295], whereas copper(I) is found in $[C_{12}mim]_3[CuCl_2][CuCl_3]$ (m.pt. 26°C) [85], $[C_{12}mim]_4[Cu_4Cl_8]$ (m.pt. 27°C) [85] and $[N_{1\,4\,4\,4}]_2[Cu_5Br_7]$ [422], as well as the simpler salts $[N_{3\,3\,3\,3}][CuCl_2]$ [460] and $[N_{3\,3\,3\,3}][CuBr_2]$ [459] (both of which melt at 99–100°C).

11.6 ENERGETIC IONIC LIQUIDS

Research and development for new propellants and explosives have included an exploration of chemical variation [656–658] designed to improve the energetics, kinetics, processability and handling of these important materials. This has included investigations into series of inorganic and organic salts. Of the many examples reported, a relatively small proportion has been shown both to have melting points below 100°C, to be stable at or near their melting points, and to be without solvents of crystallisation. Fortunately, single-crystal structure determinations are used both for characterisation purposes in many of these studies, as well as to provide density, cell volume and the number of formula units per unit cell for use in performance data prediction. It should also be noted that, because of the very nature of so-called 'energetic' salts, their wider use as ionic liquids in more general applications may be limited. They are nevertheless important in that they encompass a group of additional anions and cations that can provide complementary insights into the nature of interionic structure and bonding.

The definition of what is, or is not, an energetic salt depends on a range of technical parameters that are not relevant to this chapter. Materials included in Table 11.12 have, therefore, been selected on chemical grounds rather than on any criterion related to their application. Furthermore, energetic salts fall into a number of subgroups that overlap with other categories brought together in other tables. To avoid unnecessary duplication, Section 11.6.1 should thus be read in conjunction with the hydrazinium and hydroxylammonium entries in Table 11.6, and Section 11.6.2 with the guanidinium entries in the same table. Table 11.7 lists wholly inorganic materials discussed in Section 11.5.1, which include low-melting salts that may be considered 'energetic'. 1,5-Diaminotetrazolium perchlorate [443] has been discussed in Section 11.4, along with other protic salts. Table 11.12 lists a further 70 structures for an additional 69 energetic salts, mostly containing cations and anions derived from triazoles and tetrazoles. The structures of energetic anions and cations not already shown in Figure 11.5 are brought together in Figure 11.8.

11.6.1 Hydrazinium and Hydroxylammonium Salts

An evaluation of the series $[NH_2NH_3][NO_3]$ [208] (m.pt. 62.5/70.5°C; U_L 624 kJ mol^{-1}) [618], hydrazinium tetrazolate (m.pt. 48°C; U_L 576 kJ mol^{-1}; poor structural refinement) [146], $[MeNH_2NH_2][[NO_3]$ (m.pt. 42°C; U_L 583 kJ mol^{-1}) [131], $[MeNH_2NHMe][NO_3]$ (m.pt. 45°C; U_L 545 kJ mol^{-1}) [131] and the 1:2 salt, $[Me_2NHNH_3][NO_3]_2$ (m.pt. 98°C; U_L 1746 kJ mol^{-1}) [242], as well as the azides $[MeNH_2NH_2][N_3]$ (m.pt. 68°C; U_L 707 kJ mol^{-1})[64], $[Me_2NHNHMe][N_3]$ (m.pt. 38°C; U_L 553 kJ mol^{-1}) [116] and $[MeNH_2NHMe][N_3]$ (m.pt. −8°C; U_L 556 kJ mol^{-1}) [64], provides an opportunity to examine the consequences of the replacement of a proton by a methyl group. As might be expected, hydrogen bonding between cation and anion is stronger for protonated NH^+ compared with non-protonated N, as evidenced from shorter O···N distances. The low m. pt. of the last azide salt [64] is ascribed to the orientation of methyl groups between layers formed by hydrogen-bonded hydrazinium and azide. Other series should reveal the consequences of changes in anion, though comparisons with additional salts not included in this survey would be of value. There are five hydroxylammonium salts (with F^- [436], $[CF_3SO_3]^-$ (m.pt. 105°C; U_L 512 kJ mol^{-1}) [499], $[NO_3]^-$ [629], $[N(NO_2)_2]^-$ [80] (m.pt. 18–23°C [617]) and 1-(2-nitratoethyl)-5-nitriminotetrazolate (m.pt. 66°C; U_L 485 kJ mol^{-1}) [235] that have been crystallographically characterised as ionic liquids (although no reliable m.pt. for $[NH_3OH][NO_3]$, a highly reactive monopropellant, appears to be available). The crystal structure of $[NH_3OH][N(NO_2)_2]$ is reported [80] to contain hydrogen-bond-linked cations and anions, as well as cationic $[NH_3OH]^+$ and neutral zwitterionic $[^+NH_3O^-]$ species also involved in hydrogen bonding. For this reason, calculated values of U_L are not given. $[NH_3OH]F$, with a m.pt. of 98°C and only just qualifying as an ionic liquid, nevertheless is significant as the only simple fluoride included in this chapter [436]. Hydrazinium fluoride [631], chloride [633] and bromide [635] have been briefly mentioned in Section 11.5.1.

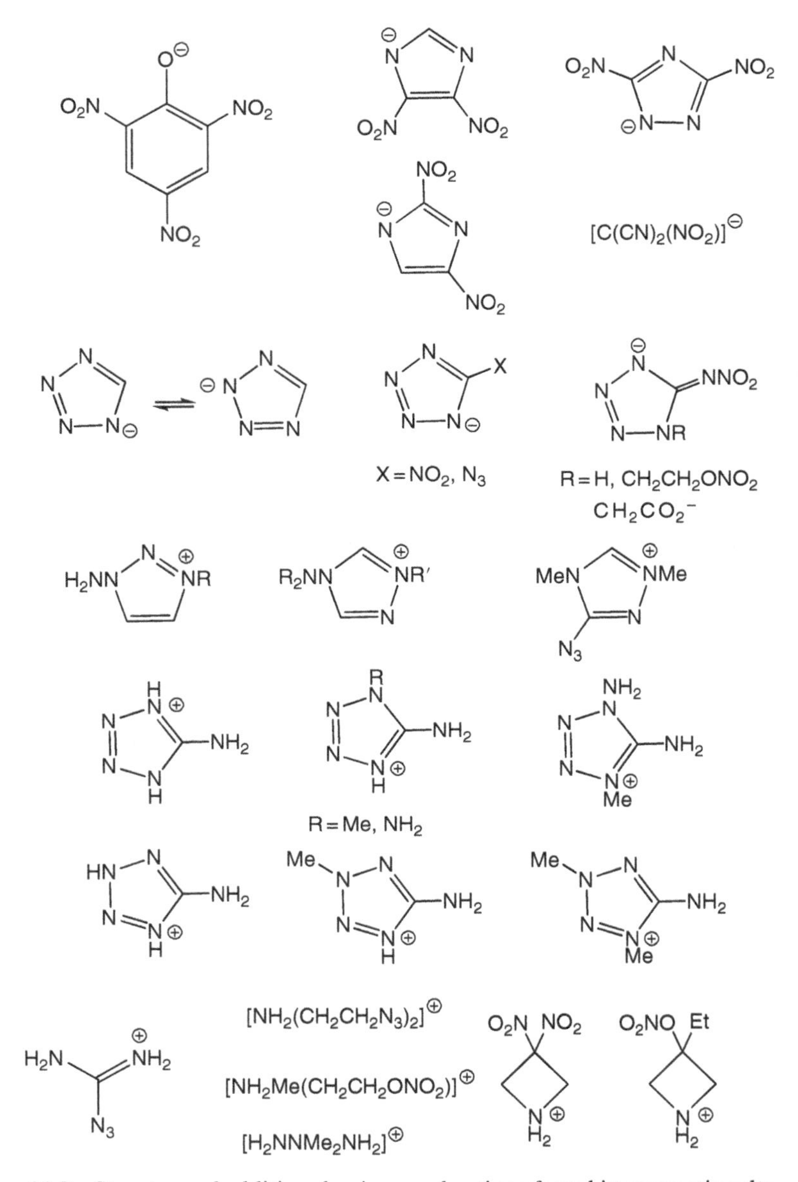

Figure 11.8 Structure of additional anions and cations found in energetic salts.

11.6.2 Guanidinium Salts

Eight crystal structures of guanidinium and mono-, di- and triaminoguaninidinium energetic salts are reported, not including the 1,1,3,3-tetramethylguanidinium ethanoate and $[N(SO_2CF_3)_2]^-$ ionic liquids. Apart from the structural consequences of changes in anion, $[A]^-$, in the series of salts $[(H_2NNH)(H_2N)C{=}NH_2][A]$, $[A]^- = [ClO_4]^-$ [272], $[N(CN)(NO_2)]^-$ [391] and $[N(NO_2)_2]^-$ [261], the only one-to-one comparisons possible are between the $[C(NO_2)_3]^-$ salts of diaminoguanidinium [319] and triaminoguanidinium [319], respectively, m.pt. 80°C; 484 U_L kJ mol^{-1} and m.pt. 84°C; U_L 476 kJ mol^{-1}, and between the triaminoguanidinium salts of $[N(NO_2)_2]^-$ (m.pt. 80°C; U_L 495 kJ mol^{-1}) [318] and $[C(NO_2)_3]^-$ [319]. In the two forms of dinitramide salt, $[(H_2NNH)(H_2N)C{=}NH_2][N(NO_2)_2]$ (m.pt. 91–94°C; monoclinic: U_L 518 kJ mol^{-1}; triclinic: U_L 517 kJ mol^{-1}) [261], crystal packing [318] accommodates ten different hydrogen bonds with all N—H and both oxygen atoms and the central nitrogen atom of the anion, building a 3D network.

11.6.3 Triazolium and Tetrazolium Salts

An extensive series of energetic salts of 1,2,3- and 1,2,4-triazolium and tetrazolium cations and imidazolate and tetrazolate anions (all with melting temperatures >50°C) has been studied crystallographically. The azolates are also considered in Section 11.6.4.

Introducing a 3-methyl group into 1-amino-1,2,3-triazolium 5-nitrotetrazolate [407] decreases the m.pt. from 99 to 93°C, while lattice potential energy, U_L, decreases from 508 to 491 (or 17 kJ mol^{-1} per $-CH_2-$ group). Replacement of the anion in 1-amino-3-methyl-1,2,3-triazolium 5-nitrotetrazolate by $[N_3]^-$ [150] or $[NO_3]^-$ [358] reduces the m.pt. to 50 and 86–88°C, respectively, while U_L increases to 536 and 531 kJ mol^{-1}, respectively.

A cation homologous series is evident for 1-alkyl-5-amino-1,2,4-triazolium, alkyl = methyl [355], ethyl [169], propyl [169], 1-methylethyl [169], *cyclo*-propylmethyl [169], hexyl [169] and heptyl [169], though a direct comparison is only possible with the bromide salts [169] when alkyl (R) is Et, Pr, Me_2CH, Hex or Hept, where the consequences of successive introduction of methylene groups can be seen.

R	m.pt./°C	U_L/kJ mol^{-1}	$U_L(Et)-U_L(R)$
Et	63–67	517	0
Pr	63	500	17
Me_2CH	92	498	19
Hex	76	460	57
Hept	94	448	69

A one-to-one comparison is possible [169] for the 1-alkyl-5-amino-1,2,4-triazolium nitrate (m.pt. 66°C; U_L 493 kJ mol^{-1}) and bromide (m.pt. 92°C; U_L 498 kJ mol^{-1}) salts, when the alkyl is Me_2CH.

Dinitramide salts, and those of related pseudohalides, are also discussed in Section 11.9.3. A further five $[N(NO_2)_2]^-$ salts with various tetrazolium cations provide an opportunity to assess the impact of variations in the latter on interionic interactions. These are (i) [5-amino-1-methyl-1*H*-tetrazol-4-ium][$N(NO_2)_2$] [269], which has a layer structure with all dinitramide oxygens involved in strong hydrogen bonding (with one short C–H···O contact of 2.33Å). The dinitramide is completely planar, an extremely rare, possibly unique, arrangement; (ii) [2-methyl-5-aminotetrazolium][$N(NO_2)_2$] [168]; (iii) [5-amino-1,3-dimethyltetrazol-3-ium][$N(NO_2)_2$] [178], which has only weak interactions between cation and anion with considerable disorder in the anion, both static (resulting from the lack of a single energetically preferred arrangement of cations and anions) and dynamic (resulting from the lack of strong directional interactions between O_2 and O_3 of the dinitramide anion and the cation), not previously observed and linked with low m.pt. The planar cation and pyramidal anion interact to form three strong hydrogen bonds. Further topographical analysis is provided in the paper; (iv) [1,5-diamino-1*H*-tetrazol-4-ium][$N(NO_2)_2$] [269] in which the packing is dominated by a 2D hydrogen-bonded network and (v) [1,5-diamino-4-methyl-1*H*-tetrazolium][$N(NO_2)_2$] [350] in which anions and cations form a zigzag strand connected by strong hydrogen bonds in a 3D network. The dinitramide takes up a near-perpendicular position with respect to the cation, with short interionic oxygen–nitrogen separations of 3.02–3.18Å, despite the ring nitrogens being negatively or zero charged with the positive charge located on the carbons and the NH hydrogens.

The effects of the various introduction of amino and alkyl groups on lattice potential energies can be seen from the following.

	m.pt./°C	U_L/kJ mol^{-1}	References
[Tetrazolium][$N(NO_2)_2$]	70	535	[168]
[5-Aminotetrazolium][$N(NO_2)_2$]	85	526	[168]
[5-Amino-1-methyl-1*H*-tetrazol-4-ium][$N(NO_2)_2$]	75	500	[269]
[2-Methyl-5-aminotetrazolium][$N(NO_2)_2$]	90	501	[168]
[5-Amino-1,3-dimethyltetrazol-3-ium][$N(NO_2)_2$]	58	485	[178]
[1,5-Diamino-1*H*-tetrazol-4-ium][$N(NO_2)_2$]	72	509	[269]
[1,5-Diamino-4-methyl-1*H*-tetrazolium][$N(NO_2)_2$]	85–86	496	[350]

A series of dinitramide salts, [ER_3][$N(NO_2)_2$] (E = S, Se or Te; R = Me or Ph), has been described, though only the structures of $[SePh_3]^+$ and $[TePh_3]^+$ salts are reported [659]. Other than a description of the methyl compounds as 'oily', no information regarding any of their melting temperatures is provided. The $[EMe_3]^+$ salts reportedly decompose more quickly than their phenyl analogues.

11.6.4 Azolates and Other Energetic Anions

Azaheterocyclic anions incorporated into so-called energetic salts are shown in Figure 11.8. Most are associated with protic cations. Dinitramide salts and those of related pseudohalides are also discussed in Section 11.9.3.

Hydrazinium tetrazolate (m.pt. 48°C; U_L 576 kJ mol^{-1}) [146], a derivative of the 1*H* tautomer (Fig. 11.8), was one of the seven inorganic tetrazolate salts characterised by X-ray diffraction. It crystallises as very thin plates, giving poor data. The ammonium salt, m.pt. 43°C, was characterised as a crystalline hydrate. The consequences of replacing a proton by a methyl group can be seen from a comparison of 1-amino-1,2,3-triazolium 5-nitrotetrazolate (m.pt. 99°C; U_L 508 kJ mol^{-1}) [407] and 1-amino-3-methyl-1,2,3-triazolium 5-nitrotetrazolate (m.pt. 93°C; U_L 491 kJ mol^{-1}) [407], showing the loss of 17 kJ mol^{-1} in lattice potential energy as a consequence of introducing a —CH_2— moiety. The corresponding di- and triaminoguanidinium salts [427] have been discussed briefly in Section 11.2.2. Many of the ten salts of 5-azidotetrazolate reported by Klapötke and Stierstorfer [462] explode before melting. However, that with hydrazinium (while violently decomposing at 136°C) is less sensitive than the corresponding alkali metal salts, apparently because of the presence of a strong hydrogen-bonded network in the crystal. Aminoguanidinium azidotetrazolate melts at 100°C. A value of U_L of 504 kJ mol^{-1} can be derived from the crystallographic data [462]. The nitroimino group of the anion in 1-propyl-1,2,4-triazolium 5-nitroiminotetrazolate (m.pt. 69°C; U_L 356 kJ mol^{-1}) [252] is twisted about 20° from the plane of the anion ring. A strong intramolecular hydrogen bond is evident between the tetrazolate ring N—H and an oxygen of the NO_2 group. In addition to a series of weak interionic C—H···O hydrogen bonds, another strong hydrogen bond is seen between the cationic N—H and the imino-N. 1-(2-Nitratoethyl)-5-nitriminotetrazolate salts with ammonium (m.pt. 81°C; U_L 486 kJ mol^{-1}), hydroxylammonium (m.pt. 66°C; U_L 485 kJ mol^{-1}) and *N,N′*-diaminoguanidinium (m.pt. 103°C; U_L 452 kJ mol^{-1}) [235] and a single 4-(carboxylatomethyl)-5-nitroiminotetrazolate salt (with 4-amino-1,2,4-triazolium (m.pt. 93°C; U_L 1200 kJ mol^{-1}) [408] and of interest as a 2:1 salt) have all been structurally characterised (along with a series of related, but nonionic liquid, salts).

The sole triazolate salt characterised by X-ray crystallography is tetraethylammonium 3,5-dinitro-1,2,4-triazolate (m.pt. 104–106°C (taken from the experimental section of [494], though a different value is found in Table 11.1); U_L 436 kJ mol^{-1}) [332, 494], reported in a major study in which 28 azolate salts were prepared. Most are high melting, though [C_4mim][3,5-dinitro-1,2,4-triazolate], which melts at 33°C, was not subject to an X-ray study. A quartet of compounds, not strictly speaking energetic compounds, $[N_{4444}]^+$ and $[PEtPh_3]^+$ salts of 2,4- and 4,5-dinitroimidazolate, provide an additional set for comparison. The two $[N_{4444}]^+$ salts [332] melt at 81 and 85°C, respectively, though each may be distinguished by its packing arrangement in the crystal. In the 2,4-salt, layers of cation and anion sheets at van der Waals distances are seen, whereas the

4,5-salt forms a close-packed arrangement with each ion surrounded by five counter-ions. The $[PEtPh_3]^+$ salts [249] melt at 68–70 and 91–93°C, respectively, both displaying weak interactions between cation C—H and ring N and O atoms of the anion. It should be pointed out that there is a discrepancy between the original paper [249] and the CCDC record (refcode GUNRUD), in that the former reports the compound as a hydrate, whereas the latter does not include H_2O in the formulation.

	[2,4-Dinitroimidazolate]$^-$		[4,5-Dinitroimidazolate]$^-$		References
	m.pt./°C	U_L/kJ mol^{-1}	m.pt./°C	U_L/kJ mol^{-1}	
$[N_{4444}]^+$	81	387	85	389	[332]
$[PEtPh_3]^+$	68–70	391	91–93	394	[249]

Included here for completeness are two salts containing the picrate ion, $[2,4,6\text{-}(NO_2)_3C_6H_2O]^-$. The structures of both confirm proton transfer, one containing $[NHMe_2(CH_2Ph)]^+$ (m.pt. 92–93°C; U_L 420 kJ mol^{-1}) [400], the other 3-hydroxy-1-*tert*-butyl-1,2-dihydropyrrolium (m.pt. 80–82°C; U_L 417 kJ mol^{-1}) [325] and interionic hydrogen bonding. In the latter, the OH group is hydrogen bonded to the phenolate oxygen of the anion. An unusual deformation in the $—CMe_3$ group is also observed.

11.6.5 Other Salts

Note should be taken of studies of single compounds whose analogues would expand both the range of ionic liquids and the potential for new insights arising from a study of their structure and properties. These include [1-aminoimidazolium]Cl (m.pt. 95°C; U_L 556 kJ mol^{-1}) [39] and [*N*-amino-1-azoniacycloheptane] $[N_3]$ (m.pt. 59°C; U_L 493 kJ mol^{-1}) [184]. It has been suggested [456] that elongation of one of the two N—N distances in the cation of [2,2-dimethyltriazanium] $[NO_3]$ (m.pt. 99°C; U_L 540 kJ mol^{-1}) is associated with the strength and pattern of the relevant hydrogen bonds. The structure of the $[C(NO_2)_2(CN)]^-$ salt (which decomposes on heating) is reported in the same study. Two salts of the azetidinium cation, [3,3-dinitroazetidinium][salicylate] [504] and [3-*O*-nitrato-3-ethylazetidinium]$[NO_3]$ [474], have melting points just beyond the ionic liquid limit, at 106 and 102–103°C, respectively.

11.7 CHIRAL IONIC LIQUIDS

The synthesis and study of chiral ionic liquids have been spurred by their potential in chiral recognition, enantiomer separation and in asymmetric synthesis [660, 661]. However, there have been relatively few single-crystal X-ray studies of single enantiomers of chiral ionic liquids. These are summarised in the following text, along with those of low-melting racemates and other relevant salts.

Interestingly, while chirality may be found either in anion or cation, a chiral structure can arise even when neither anion nor cation is chiral. For instance, the (high-melting) pyridinium trifluoromethanesulfonate crystallises [662] in the chiral space group $P4_32_12$, brought about by a combination of coulombic and directional hydrogen bonding.

Single crystals of ionic liquids with chiral cations have mostly been investigated as racemates, including [1-{1-(ethoxycarbonyl)ethyl}-3-methylimidazolium][NTf_2] (m.pt. 45°C) [138], [1-(2-hydroxy-2-methyl-2-phenylethyl)-3-*tert*-butylimidazolium]Cl (m.pt. 64–66°C) [217], [(1-phenylethyl)ammonium][CH_2ClCO_2] (m.pt. 95°C) [420], [{1-(2-naphthyl)prop-2-en-1-yl]amino}pyridinium]I (m.pt. 103–105°C) [485] and [6,6′-spirobi(1-methyl-1,5,6,7-tetrahydropyrrolo[1,2-a]imidazol-4-ium] $[NTf_2]_2$ (m.pt. 112°C; Fig. 11.9) [547]. Exceptions include [1,3-di-(*R*)-(+)-methylbenzylimidazolium][A], $[A]^- = [PF_6]^-$ (m.pt. 108°C) or $[BF_4]^-$ (m.pt. 110°C) [517] (cation prepared from reaction of *R*-(+)-1-phenethylamine with glyoxal and methanal) and the cation, [1,1-dibenzyl-3,4-dihydroxypyrrolidinium]$^+$ (synthesised from L(+)-tartaric acid), and structurally characterised as the bromide salt (CCDC-710820), though melting at 205–206°C [663]. The corresponding $[NTf_2]^-$ salt melts at 50–53°C, but its crystal structure has not been investigated. The chiral salts (*R*)- and (*S*)-[ethanolammonium][3-chloromandelate] [611] (Fig. 11.6) melt individually at 122°C. The pure (*R*)-salt has been subjected to a single-crystal X-ray study. Interestingly, mixtures of the two forms, containing 50–75 mole% of the (*R*)-isomer, melt at 100°C. A racemic mixture of [1-(*cis*-2-hydroxy-*cyclo*-pentyl)-3-benzylimidazolium]Br (Fig. 11.10) (and one of a series of related materials), for which a crystal structure has been described [664] (though there is no associated CSD refcode), is reported to melt at 300°C, whereas the pure

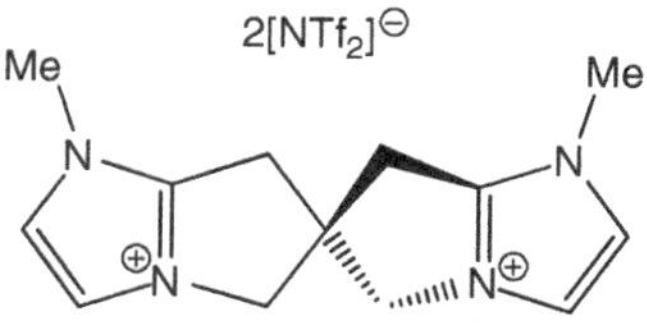

Figure 11.9 6,6′-Spiro(1-methyl-1,5,6,7-tetrahydropyrrolo[1,2-a]imidazol-4-ium bis-triflamide [547].

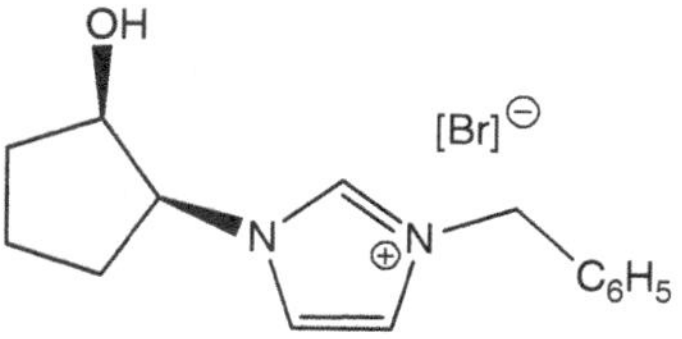

Figure 11.10 1-(*cis*-2-hydroxy-*cyclo*-pentyl)-3-benzylimidazolium bromide [664].

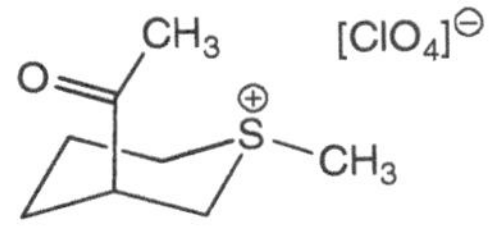

CH_3 $[ClO_4]^{\ominus}$
O
$\overset{\oplus}{S}$—CH_3

Figure 11.11 (−)-*trans*-3-Acetoxy-1-methylthiane perchlorate [488].

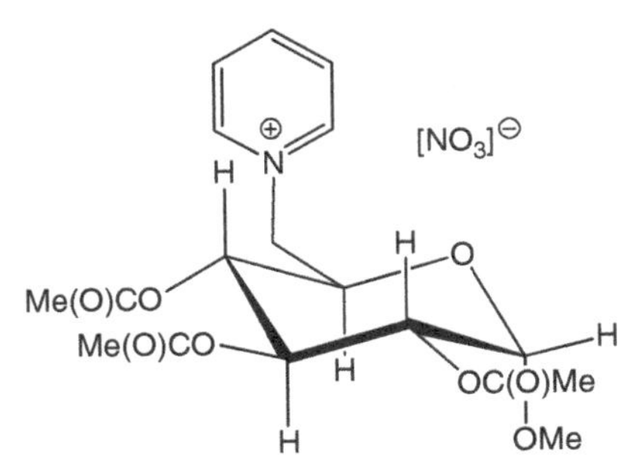

Figure 11.12 *N*-(1-*O*-Methyl-2,3,4-tri-*O*-acetyl-*α*-D-glucopyranos-6-yl)pyridinium nitrate [287].

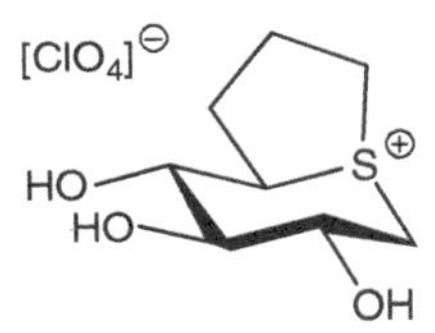

Figure 11.13 3(*R*),4(*S*),5(*R*),6(*S*)-3,4,5-trihydroxy-*cis*-1-thionabicyclo[4.3.0]nonane perchlorate [313].

(*R*,*S*)-enantiomer is reported to melt at 19°C. The chiral salts [1-{(*S*,*S*)-2-hydroxy-*cyclo*-hexyl}-3-benzylimidazolium]X, X = Cl (refcode KIHNEV) or Br (refcode KIHNIZ) [665], have been characterised, but they melt at 219 and 188°C, respectively.

The absolute configuration of [(−)-*trans*-acetoxy-1-methylthiane][ClO_4], m.pt. 106°C (Fig. 11.11), has been shown to be (1*S*,3*S*) [488]. The racemic *cis* compound (m.pt. 104°C) was also characterised. The structures of the salts [*N*-(1-*O*-methyl-2,3,4-tri-*O*-acetyl-*α*-D-glucopyranose-6-yl)pyridinium][NO_3] (m.pt. 74–75°C) [287] (Fig. 11.12), [3(*R*),4(*R*),5(*R*),6(*S*)-3,4,5-trihydroxy-*cis*-1-thioniabicyclo[4.3.0]nonane][ClO_4] (m.pt. 79–80°C) [313] (Fig. 11.13) and that of the co-crystal [N_{4444}]/[3-{(2*S*)-2-methylbutyl}thiazolium][OTf]$_2$ (m.pt. 85°C) [348] have also been reported. A structural study of [3-ethoxymethyl-1-(1*R*,2*S*,5*R*)-(−)-menthoxymethylimidazolium]Cl (CCDC-786077), with a m.pt. of 55–58°C, reveals it to be a monohydrate [666].

Structurally characterised ionic liquids with chiral anions are less common. Examples include [HC_2im][d-tartrate] (m.pt. 41–43°C; U_L 462 kJ mol^{-1}) [130], [Hmim][d-tartrate] (m.pt. 98–101°C; U_L 470 kJ mol^{-1}) [130, 452] and

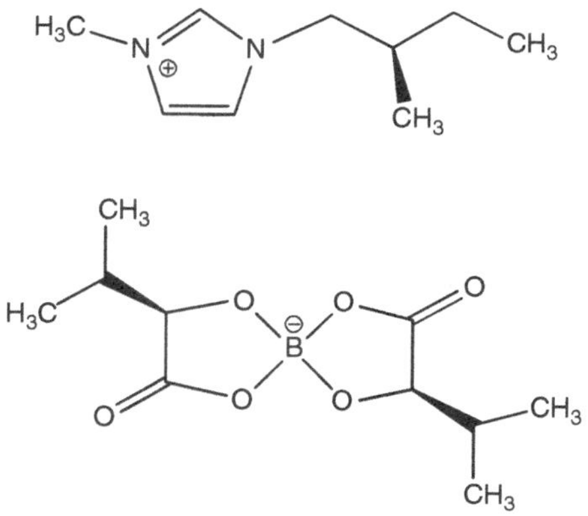

Figure 11.14 [(1-Methyl-3-{(2*S*)-2-methylbutyl}imidazolium][bis{(2*S*)-2-oxy-3-methylbutanoato-*O,O′*}borate [373].

[1-methyl-3-{(2*S*)-2-methylbutyl}imidazolium][bis{(2*S*)-2-oxy-3-methylbutanoato-*O,O′*}borate] (Fig. 11.14) (m.pt. 89°C) [373]. The absolute configurations of the anion *O*-ethyl ethylphosphonothioate was determined from the crystal structure of [(–)-(1-phenylethyl)ammonium][POS(OEt)Et] (m.pt. 104–105°C) [492].

11.8 MULTIPLY CHARGED ANIONS AND CATIONS

A further dimension of ionic liquid composition space is represented by the charges on the anion (p) and cation (q). While the great majority of salts, $[cat]_p[A]_q$, listed in Tables 11.1, 11.2 and 11.3, combine singly charged cations with singly charged anions, a surprising number of ionic liquids, including some that may be considered room-temperature ionic liquids, contain ions with non-single charges, with p taking values of 2, 3 and 4 and q of 2, 3 and 6. Such salts are collected in Table 11.13 (including those melting between 100 and *ca.* 110°C and which are noted without further comment).

Table 11.13 is dominated by salts of p:q ratio of 2:1 and 1:2, the former containing doubly charged anions, the latter doubly charged cations. Many of the salts are discussed coincidentally under other headings. While there are nine entries in Table 11.1 of the type $[cat]_2[A]$, five of these are polyhalides containing the ions $[Br_8]^{2-}$, $[Br_{20}]^{2-}$ and $[Cl_2I_{14}]^{2-}$, which may better be considered to be solvates [55, 74]. However, $[C_4mim]_2[SnCl_4]$ (m.pt. 5°C; U_L 1012 kJ mol^{-1}) and $[C_4mim]_2[CuCl_4]$ (m.pt. 23°C; U_L 1024 kJ mol^{-1}) are authentic room-temperature ionic liquids [73]. These are two of an extended series of $[C_4mim]_2[MX_4]$ for M = Cu(II), Co(II), Ni(II), Fe(II), Zn(II), Mn(II) or Pt(II). The structures of an $[I_8]^{2-}$ salt (of *N*-ethylurotropinium; m.pt. 80–82°C [324]) and an $[I_{12}]^{2-}$ (of [K([2.2.2]crypt)]; m.pt. 90°C [379]) have also been described. Other structurally characterised coordination complexes include salts of

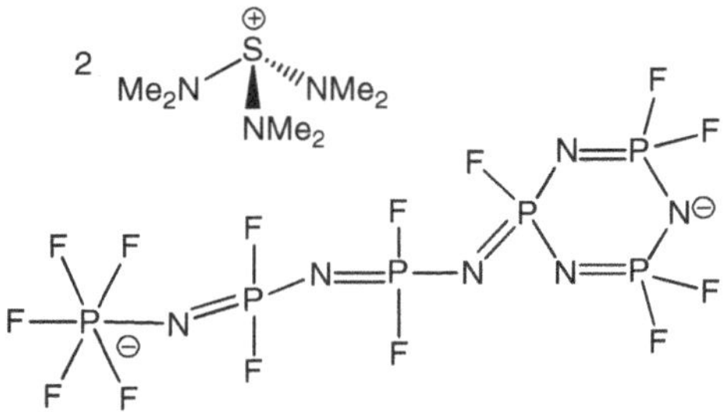

Figure 11.15 The tris(dimethylamino)sulfonium salt of the phosphazenate dianion, $[P_3N_3F_5NPF_2NPF_2NPF_5]^{2-}$ [215].

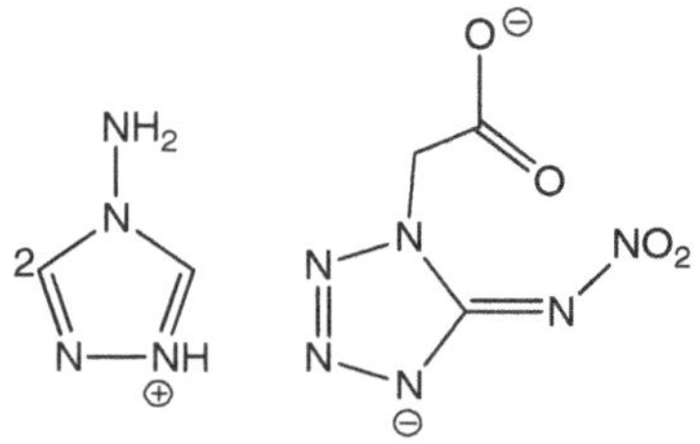

Figure 11.16 Bis(4-amino-1,2,4-triazolium) 4-(carboxylatomethyl)-5-nitroiminotetrazolate [408].

lithium and silver, $[Li(MeCN)_4]_2[P_2Se_8]$ [78] and $[Ag(MeCN)_4]_2[Ag\{N(CF_3SO_2)_2\}_3]$ [79], that melt at the surprisingly low temperatures of 14–16 and 18°C, respectively.

There is just one sulfate salt, $[cat]_2[SO_4]$, where $[cat]^+ = $ [1-(chloromethyl)-1,1-dimethylhydrazinium]$^+$, m.pt. 62°C [159] (but see the following text). While most dinegatively charged anions are metal centred (including a europium salt, $[C_4C_1pyr]_2[Eu^{III}\{N(SO_2CF_3)_2\}_5]$ (m.pt. 92°C) [246]), worthy of special note are $[S(NMe_2)_3]_2[P_3N_3F_5NPF_2NPF_2NPF_5]$ (Fig. 11.15; m.pt. 64°C [215]) (the analogous $[S(NMe_2)_3]$[*cyclo*-$P_6N_6F_{13}$] melts < 0°C), [4-amino-1,2,4-triazolium]$_2$ [4-(carboxylatomethyl)-5-nitroiminotetrazolate] (Fig. 11.16; m.pt. 93°C) [408] and [(1-(2-hydroxyethyl)piperidinium)][tris(oxolato)germanium(IV)] (m.pt. 97°C) [441].

There are, currently, no room-temperature ionic liquids of the type $[cat][A]_2$, though this probably is a consequence of the absence of appropriate structural characterisation. The lowest-melting examples are $[Cu^{II}(bim)_4][N(SO_2CF_3)_2]$ (m.pt. 46°C) [144] and [1,3-bis(1-methyl-1-pyrrolidinio)propane]Br_2 (m.pt. 51°C) [157]. Salts of dipositive cations (mostly with conventional anions) include diprotonated heterocycles, such as piperazine (predominantly a series of carboxylate salts [428–430, 439]), diprotonated alkylhydrazines as well as cations in which two monopositive moieties are linked together. For example, a series of bis(imidazolium) salts are known; the lowest reported melting

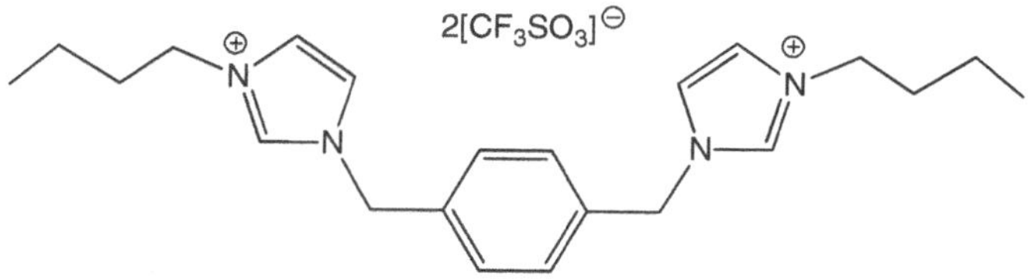

Figure 11.17 1,4-Bis(1-butylimidazolium-3-ylmethyl)benzene bis(trifluoromethanesulfonate) [189].

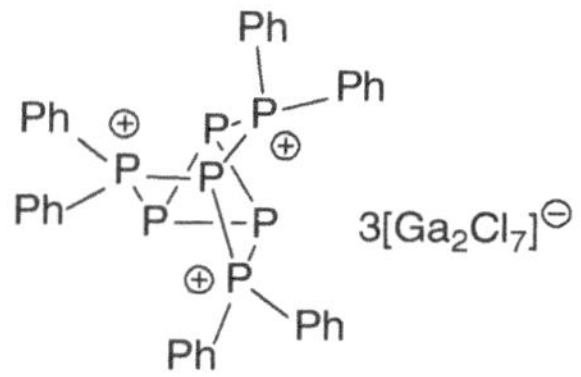

Figure 11.18 3,3,5,5,7,7-Hexaphenyltricyclo[2.2.1.0^{2,6}]heptaphosphan-3,5,7-triium tri(heptachlorodigallate) [200].

temperature (59–62°C) is for [1,4-bis(1-butylimidazolium-3-ylmethyl)benzene] [trifluoromethanesulfonate]$_2$ (Fig. 11.17) [189].

There are four 2:2 salts: [1,2-dimethylhydrazin-1,2-diium][SO_4] (m.pt. 67°C) [242], [*N*-methyl-1,3-propylenediammonium][$TlBr_5$] (m.pt. 93–94°C) [409], [H_2([2.2.2]cryptand)][I_8] (m.pt. 98°C) [324] and [1,4-bis(1-methylimidazolium-3-yl)butane][2-hydroxy-5-sulfonatobenzoate] (m.pt. 106–108°C) [508].

There is one salt of a tripositive cation: [3,3,5,5,7,7-hexaphenyltricyclo[2.2.1.0^{2,6}]heptaphosphane-3,5,7-triium][Ga_2Cl_7] (Fig. 11.18), melting at 85–87°C [200], and seven of dipositive anions. There are four lanthanide salts, [cat]$_3$[M^{III}{C(CN)$_2$NO}$_6$] (M = La, Sm, Pr or Ce) [180, 186] and [cat]$^+$ = [C_2mim]$^+$, [$C_2C_1C_1$im]$^+$ or [N_{4444}]$^+$, the lowest melting (just) being [C_2mim]$_3$[La{C(CN)$_2$NO}$_6$] [180] (m.pt. 58°C). [C_{12}mim]$_4$[$Cu^I{}_4Cl_8$] [85] melts at 27°C, [C_4mim]$_4$ [La(NCS)$_7$(H_2O)] [118] (38°C), the tentatively formulated [(2-benzyl)pyH]$_4$[$Sb^{III}{}_2Cl_{10}$] [227] (65–66°C) and [$Dy_2(NMe_3CH_2CO_2)_8(H_2O)_4$][$NTf_2$]$_6$ [174] (m.pt. 88°C). All of these examples widen the scope of what may be considered to be an ionic liquid and highlight the questions that arise when seeking to understand the forces that influence their structure and interionic interactions.

11.9 HALIDES, POLYHALIDES AND PSEUDOHALIDES

11.9.1 Halides

Halide salts of quaternary ammonium and phosphonium cations, and those derived from heterocyclic bases (both aromatic and alicyclic), have long been known (Table 11.14). Most are relatively high melting (and would not be considered ionic liquids), though the structural characteristics of quaternary ammonium halides have been studied by those with interests in plastic crystals and so-called 'rotator' phases [7]. Many decompose at or near their melting temperatures, a characteristic that would limit interest in them as reaction media [667]. Indeed, it is probably as a result of the view, widely held in the past [668], that such materials may be intrinsically unstable that the onset of the recent massive interest in ionic liquids was delayed.

However, the use of quaternary ammonium (and pyridinium) halides to modify the vapour pressure of volatile reactants, such as the halogens or the hydrogen halides, or the use of highly acidic media to investigate (and possibly isolate) reactive intermediates such as carbenium ions [25], prompted interest in the physical characteristics and phase behaviour of relevant systems. In an important series of papers, Mootz and colleagues have characterised the phase behaviour and crystal structures of a series of salts and hydrogen-bonded adducts, with melting points as low as −98°C (175 K), arising from various combinations of aliphatic and aromatic amines (as well as of ethers and carboxylic acids) with hydrogen halides and other acids. These are discussed further in this review only insofar as the materials described qualify for inclusion in this survey. Mootz's crystallographic work relies upon the development (with Boese and others [619]) of an *in situ* zone-melting technique [57, 550–553] for growing single crystals for crystallographic study that has more recently been used [58, 62] for the investigation of difficult-to-crystallise ionic liquids.

It is worth noting, in passing, that Mootz's work is predated by that of G. A. Jeffrey and co-workers who, in a series of papers [669, 670], have described an extensive series of quaternary ammonium halide hydrates with low-melting points. A number of these have been structurally characterised, but as they melt to give liquids that are not comprised solely of ions (indeed, they are aqueous solutions), they are only briefly mentioned in the following text.

Simple halide-containing ionic liquid salts are widely studied as, usually, they are straightforward to prepare. They are of both fundamental relevance in exploring the physical and structural characteristics of ionic liquids as well as of importance as intermediates in the preparation of other ionic liquids.

Interestingly, there is, currently, only one report of a crystal structure of a 'naked' fluoride ion-containing ionic liquid, namely, $[NH_3OH]F$ melting at 97–98°C (lattice potential energy, $U_L = 726\,kJ\,mol^{-1}$) [436]. The structure of $[NH_2NH_3]F$, melting at 100.5°C, $U_L = 709\,kJ\,mol^{-1}$, is described in Ref. [631]. There are many reports of crystallographic characterisation of organic fluoride (and other halide) salt hydrates and water clathrates, most notably from the 1960s with the series of papers by G. A. Jeffrey and co-workers. See, for example, $[N(^iPrCH_2CH_2)_4]F.(H_2O)_n$ (CSD refcode TAAMFH) [670] with a melting point of 31°C. More recently, the salts $[C_4mim]F \cdot H_2O$ (NAZHUS) [671] and $[C_1mim]F \cdot 0.5H_2O$ (PAJBIN) [672] have been reported (though without melting points). 'Naked' fluoride salts, such as $[NMeH_3]F$ (MEAMMF) [673] and 1,1,3,3,5,5-hexamethylpiperidinium fluoride (ZAPZUL) [674], are likewise reported without melting temperatures. Both $[(HO_2CCH_2)_2im]F$ (CEKMIP)

[537] and $[N\{C(CH_2OH)_3\}H_3]F$ (BUZGEI) [675] melt well above 110°C, with decomposition. While tetramethylphosphonium fluoride (UJIJAY) [676] is a salt with a melting point greater than 120°C, it may be sublimed at 35°C in a dynamic vacuum. Studies of the gas phase establish that neutral $[PMe_4F]$ is present, highlighting the importance of a full characterisation of phases other than the solid when attempting to characterise and understand ionic liquid behaviour. Worthy of note (though none of the materials are low melting) are investigations into the differences in the solid state for the series of materials of stoichiometry $[EPh_4N_3]$ (E = P, As, Sb or Bi), which suggest [677] that the main factor determining whether a salt $[EPh_4][N_3]$ (E = P) or a covalent pentacoordinate molecule $[EPh_4N_3]$ (E = Sb or Bi) is favoured at ambient temperature is, surprisingly, the first ionisation potential of $[EPh_4]^+$ rather than factors such as lattice or sublimation energies. The results are consistent with crystallographic analyses for E = P (CCDC deposition number 263416) [678]and Sb (CCDC deposition number 263415) [678]. The outcome suggests that E = As, which represents a borderline case, though the crystallographic study establishes salt formation [679].

The lowest-melting non-cyclic $[ER_4]X$, of any type (with a crystal structure), is a protic salt, $[N(CH_2CH_2OMe)_3H]Cl$ [248], that melts at 68–69°C (U_L 445 kJ mol^{-1}). The lowest-melting chloride, bromide and iodide salts in this collection are, respectively, $[C_3C_1pyr]Cl$ (m.pt. 50°C; U_L 481 kJ mol^{-1}) [152], an *N*-silylated pyridinium, $[1\text{-}SiMe_3py]Br$ (m.pt. 27°C; U_L 466 kJ mol^{-1}) [86–88], from a class not highly represented among ionic liquids, and $[C_4mim]I$ (with a melting point of 19°C; U_L 468 kJ mol^{-1}) [81].

11.9.1.1 Ammonium, Phosponium and Sulfonium Salts Crystallographically characterised ionic liquids, [cat][A], with halide counter-anions are listed in Table 11.14. Consistent with the view that $[ER_4]X$ (R = alkyl, E = N or P) [667] are relatively high melting, only one quaternary ammonium salt and three tetraalkylphosphonium salts with unsubstituted alkyl groups, namely, $[N_{12\,12\,12\,12}]Br$ (m.pt. 88–92°C) [167], $[P_{10\,10\,10\,10}]Br$ (m.pt. 54–57°C) [167], $[P_{5\,5\,5\,5}]I$ (96°C) [431] and $[P_{18\,18\,18\,18}]I$ (m.pt. 98–99°C) [167], are included in the tabulation. Single crystals of $[S_{2\,2\,16}]Br$, m.pt. 75°C, have also been studied crystallographically [291].

11.9.1.2 Imidazolium Salts: Effect of Halide Several 1,3-dialkylimidazolium halides have been intensely structurally investigated, not only as a consequence of polymorphism but also as subjects of studies of the dynamics of processes occurring during melting and crystallisation and associated butyl group conformer lability. These aspects have been introduced in Section 11.2.

While there are many 'one-off' studies of individual compounds, a significant number of ionic liquids in this subclass, being 1,3-dialkylimidazolium salts, provide an opportunity for a comparison of salts with the same cation and a number of halide anions and of materials that either make up homologous series or differ by the presence or absence of a particular functionality.

	m.pt./°C	U_L/kJ mol^{-1}	References
[C_2mim]Cl	81–90	503–507	[333] (G.J. Reiss, Cambridge Structural Database, private communication, 2010; S. Parsons, D. Sanders, A. Mount, A. Parsons and R. Johnstone, Cambridge Structural Database, private communication, 2005.)
[C_4mim]Cl	66	477–483	[230–232, 566] (C. Pulham, J. Pringle, A. Parkin, S. Parsons, and D. Messenger, Cambridge Structural Database, private communication, 2005.)
[C_2mim]Br	71–81	500–505	[142]
[C_4mim]Br	70–79	473–477	[112, 230, 243, 260]
[C_2mim]I	81	488	[263, 330]
[C_4mim]I	19	468	[81]

A consideration of these compounds and the very varied reports of their respective melting points requires comment (SciFinder, for instance, lists 30 individual m.pt. for [C_4mim]Br spanning the range 37–38 to 73°C, some quoted to four significant figures; 12 m.pt. appear for [C_2mim]Br). While a number of these can be dismissed as of low reliability, there is sufficient variability in the data that remain to sound a note of caution in attempting to relate structure with melting temperature, and vice versa. Interestingly, one attempt to generate correlations of ionic liquid melting points [31], discussed briefly in Section 11.2, does not include halide-containing materials.

Many of the structural studies of halide salts are complemented by spectroscopic evidence relating to the nature of hydrogen bonding involving cation and halide anion. All ring protons of imidazolium halides appear able to involve themselves in hydrogen bonding. However, the hydrogen-bonding arrangement in [C_2mim]Cl is much more complicated (and giving rise to much more disorder) than that for either [C_2mim]Br or [C_2mim]I, even though their m.pt. are similar.

Monoclinic and orthorhombic forms of [C_4mim]Cl, studied by two groups [230, 566], exemplify the subtle structural changes that may arise from changes in side-chain conformation (see Section 11.18.1). The C(2)—H···Br contact in [C_4mim]Br [243] of 2.46 Å is markedly shorter than those in the related 1,3-dialkylimidazolium salts, [C_2mim]Br (2.78 Å) [142], [C_3C_3im]Br (2.62 Å) [112, 243] and [$^iC_3{}^iC_3$im]Br (2.88 Å) [112, 243].

Changing the halide counter-ion in [($NCCH_2CH_2CH_2$)mim]Cl (m.pt. 80°C; U_L 479 kJ mol^{-1}) [254, 289], [($NCCH_2CH_2CH_2$)mim]Br (m.pt. 90°C; U_L 476 kJ mol^{-1}) [377] and [($NCCH_2CH_2CH_2$)mim]I (m.pt. 69°C; U_L 466 kJ mol^{-1}) [253] leads to a modest decline in lattice potential energy as the ionic radius of the anion increases. A smaller decline is seen when [($CH_3OCH_2CH_2OCH_2CH_2$)mim]Br (m.pt. 78°C; U_L 455 kJ mol^{-1}) and [($CH_3OCH_2CH_2OCH_2CH_2$)mim]I (m.pt. 70°C; U_L 451 kJ mol^{-1}) are compared [149].

	m.pt./°C	U_L/kJ mol^{-1}	References
$[C_4mim]Cl$	66	477–483	[230–232, 566] (C. Pulham, J. Pringle, A. Parkin, S. Parsons, and D. Messenger, Cambridge Structural Database, private communication, 2005.)
$[C_4C_1C_1im]Cl$	79/100	464–472	[119, 311] (C. Pulham, J. Pringle, A. Parkin, S. Parsons, and D. Messenger, Cambridge Structural Database, private communication, 2005.)
$[C_4C_1C_1im]Br$	97	466	[440, 680]
$[C_4C_1C_1im]I$	98	458	[440, 680]

While having similar melting points, $[C_4C_1C_1im]Br$ and $[C_4C_1C_1im]I$ differ in that π-π stacking involving the imidazolium moieties is present in crystals of the former but not the latter.

11.9.1.3 Homologous Series and Side-Chain Effects Golovanov and colleagues [112] have discussed the melting points of six 1,3-dialkylimidazolium bromides, [Rmim]Br, where R = Et (m.pt. 77°C; U_L 503 kJ mol^{-1}), Pr (m.pt. 36°C; U_L 491 kJ mol^{-1}), ^{i}Pr (m.pt. 110.5°C; U_L 488 kJ mol^{-1}) or Bu (m.pt. 79°C; U_L 477 kJ mol^{-1}), and $[R'_2im]Br$, where R′ = Pr (m.pt. 67°C; U_L 457 kJ mol^{-1}) or ^{i}Pr (m.pt. 132°C; U_L 465 kJ mol^{-1}), in terms of the number of contacts involving the anion and the number of anions close to the cation as revealed from the crystal structure. The lowest-melting salt is associated with the largest number of such contacts formed by bromide ions, with the smallest number of such contacts exhibited by the salt with the highest melting point. These observations highlight the importance of specific interactions in the crystal in determining lattice energies that are not taken account of in VBT approaches. A second polymorph of $[^iC_3mim]Br$ [307] (m.pt. 78°C; U_L 484 kJ mol^{-1}) has been characterised in which the asymmetric unit contains three independent conformers, with torsion angles (defined by (ring)C-N–CMe_2-H) of 0.3, 16.2 and 48.1° compared with 170.7° for the polymorph characterised by Golovanov and colleagues.

A comparison can be made between $[C_2mim]I$ [263, 330] (m.pt. 81°C; U_L 488 kJ mol^{-1}) and [(vinyl)mim]I [300] (m.pt. 78°C; U_L 496 kJ mol^{-1}), the unit cell of the latter containing two imidazolium units, one with a *cis*- and the other with a *trans*-geometry defined by the C(2)—N—C=C torsion angle.

$[C_3mim]Br$ (m.pt. 36°C; U_L 491 kJ mol^{-1}) [243], two polymorphs of $[^iC_3mim]Br$ (m.pt. 78/110.5°C; U_L 484/488 kJ mol^{-1}) [243, 307] and $[(CH_2{=}CHCH_2)mim]Br$ (m.pt. 60°C; U_L 495 kJ mol^{-1}) [190] reveal the effects of small changes in an *N*-bound group and the difficulty of relating these to melting point differences.

The modest compositional change along the series [98] $[(CH_2{=}CHCH_2)mim]I$ (m.pt. 65°C; U_L 480 kJ mol^{-1}), $[(CH_2{=}CHCH_2)C_2im]I$ (m.pt. 31°C; U_L 472 kJ mol^{-1}) and $[(CH_2{=}CHCH_2)C_3im]I$ (m.pt. 32°C; U_L 457 kJ mol^{-1})

nevertheless results in significant differences in crystal-packing interactions, with the first two showing π–π interactions between the allyl moiety and the imidazolium ring that are absent in the third member of the series. The latter (but not the former) shows π–π stacking between imidazolium rings.

The introduction of unsaturation into the side chain is further manifested [190] in the comparison between 1-allyl-, 1-propargyl- (117°C), 1-(2-butynyl)- (131°C) and 1-(2-pentynyl)-3-methylimidazolium bromides (66°C), though the latter was not subjected to a single-crystal X-ray diffraction study.

Changing the terminal group, Z, in $[(ZCH_2CH_2CH_2)mim]Cl$, Z = CH_3 (m.pt. 66°C; U_L 477–483 kJ mol^{-1}) [230–232] (C. Pulham, J. Pringle, A. Parkin, S. Parsons, and D. Messenger, Cambridge Structural Database, private communication, 2005.), CN (m.pt. 80°C; U_L 479 kJ mol^{-1}) [254, 289] or CO_2H (m.pt. 105°C; U_L 479 kJ mol^{-1}) [500], leads to increases in melting point with little effect on lattice potential energy.

Vygodskii and co-workers [377] studied differences in cation–anion interactions between $[C_4mim]Br$ and $[(NCCH_2CH_2CH_2)mim]Br$ and noted the absence of specific contacts involving the nitrile group, concluding that, while the number of C—H···Br contacts was the same (seven), their strength was higher for the nitrile compound. Saha and Hamaguchi [254] compare $[(NCCH_2CH_2CH_2)mim]$ Cl and $[(NCCH_2CH_2CH_2)mim]Cl \cdot H_2O$, revealing the profound role played by water in determining the side-chain conformation, stabilising a rare *trans–gauche* arrangement, though without evident direct interaction with the cation.

The main difference noted [149] between *N*-alkyl- and *N*-ether-functionalised imidazolium cations is the preference for *gauche* conformation at (ring) $N—CH_2—O—CH_2—$ or (ring)$N—CH_2—CH_2—O—$ compared with *anti* conformation for (ring)$N—CH_2—CH_2—CH_2—$. The absence of intramolecular hydrogen bonding involving the ether oxygen atoms was also noted. A comparison with analogous hexafluorophosphate salts [210] concluded that hydrogen bonding between cation and anion in the halide salts is a contributing rather than a determining factor. The lattice potential energy is reduced by 9 kJ mol^{-1} when an additional $—CH_2—$ moiety is inserted adjacent to the nitrogen to which the ether-containing group is bound, whereas this reduction is slightly larger (11–13 kJ mol^{-1}) when the substitution occurs on the terminal carbon of this group [149].

	m.pt./°C	U_L/kJ mol^{-1}
$[(CH_3OCH_2)mim]I$	75	484
$[(CH_3CH_2OCH_2)mim]I$	68	471
$[(CH_3OCH_2CH_2)mim]I$	50	473
$[(CH_3OCH_2CH_2OCH_2)mim]I$	63	460
$[(CH_3OCH_2CH_2OCH_2CH_2)mim]I$	70	451

The series of *N*-benzylimidazolium iodides [203], $[(PhCH_2)C_1C_nim]I$, $n = 2$ (m.pt. 86°C; U_L 439 kJ mol^{-1}), 3 (m.pt. 71°C; U_L 436 kJ mol^{-1}), 4 (m.pt. 61°C; U_L

426 kJ mol^{-1}) or 5 (m.pt. 61°C; U_L 416 kJ mol^{-1}), in which one, two and three $-CH_2-$ groups reduce lattice potential energy by 3, 13 and 23 kJ mol^{-1}, is discussed in more detail in Section 11.2.6 dealing with Hirshfeld plots. The less disordered structures, $[(PhCH_2)C_1C_1im]I$ (m.pt. 154°C) and $[(PhCH_2)C_1{}^sC_4im]I$ (135°C), were also studied.

Three benzothiazolium iodides may be compared [321]: $[C_{10}btz]I$ (m.pt. 82°C; U_L 410 kJ mol^{-1}), $[C_{11}btz]I$ (m.pt. 80°C; U_L 402 kJ mol^{-1}) and $[C_{12}btz]I$ (m.pt. 90.5°C; U_L 399 kJ mol^{-1}) for which U_L is reduced by 8 and then 3 kJ mol^{-1}.

11.9.2 Interhalides and Polyhalides

Addition of halogens, X_2, to halide salts [cat][X′] gives rise to a range of complexes of varying stability. A series of polybromide salts has recently been characterised [55, 74], containing the anions $[Br_8]^{2-}$, $[Br_9]^-$ and $[Br_{20}]^{2-}$, some of which are liquid at room temperature. These are included in Table 11.1. The $[Br_9]^-$ anion is one of a series of materials of general formula $[X'(X_2)_n]$, where n = an integer, other examples including $[IBr_2]^-$, $[Br_3]^-$, $[I_3]^-$, $[I_5]^-$and $[I_7]^-$, some of which have been characterised in the crystalline state and have melting points that bring them within the definition of ionic liquid. The salt $[I(py)_2][ICl_2]$ [484], with a m.pt. of 103–105°C falls just outside the specified limit. Further series of iodo compounds are also known in which m molecules of iodine complex with n iodide ions, to form anions, $[(I)_n(I_2)_m]^{n-}$, that include $[I_8]^{2-}$, $[I_{12}]^{2-}$ and $[I_{29}]^{3-}$, with the latter in the form of a ferrocenium salt [312], $[Fe(\eta^5\text{-}C_5H_5)_2]_3[I_{29}]$, with a formula weight of 4238.19, melting at 79°C. The analysis of the crystal structure suggests that the I_{29} species is built up of a highly cross-linked network of $[I_3]^-$ and $[I_5]^-$ units and bridging iodine molecules, though the composition of the melt is unknown. The polyiodide species in two complexed potassium salts of composition $[K(L)]_2[I_{12}]$, with L = benzo-18-crown-6 [392] or [2.2.2]cryptand [379], are suggested best to be formulated as salts of $[I_5]^-$ and $[I_7]^-$ in the former case and of $[I_{12}]^{2-}$ in the latter, highlighting the range of subtle bonding effects that influence the stability of particular ionic arrangements in the solid state. The structure and bonding of these interesting systems have been widely investigated and well reviewed [37], with pioneering investigations by Tebbe, Kloo and others. Bearing in mind the noted diverse structures displayed by the polyiodides, included cross-linking as well as sometimes weak interactions between ion and iodine molecules, the proper formulation of at least some of this subclass as solvates or adducts underlines the importance of a detailed appreciation of both crystal structure and the composition of the equilibrated liquid phase formed on melting. This does not prevent polyhalide salts being of intense current attention in their technical application as components of dye-sensitised fuel cells [681].

By far the most abundant group of materials in this subclass are the triiodides, with 19 entries, including an interesting triethylsulfonium triodide [69] that melts at −2°C [115], with a lattice potential energy (U_L) of 440 kJ mol^{-1} (as discussed in Section 11.2.4). The corresponding trimethylsulfonium salt [115] melts at 37°C with $U_L = 468\,kJ\,mol^{-1}$. The structure of $[SMe_3][I_3]$ has been studied in both the crystalline and liquid states. 1-Ethylpyridinium triodide [140] and 1-ethylquinolinium triodide [103] are also low melting, the latter with a m.pt. of 33–34°C. $[C_2py][I_3]$ (m.pt. 45–46°C) has been known since 1895 [682]. While there is a decrease in lattice energy when the counter-ion is changed from that in $[NEtMePh_2][I_3]$ [221] (m.pt. 65°C; $U_L = 408\,kJ\,mol^{-1}$) to that in $[NEtMePh_2][I_5]$ [221] (m.pt. 90°C; $U_L = 390\,kJ\,mol^{-1}$), the melting point increases. $[PMePh_3][I_3]$ melts at 105°C [501].

While a comparison between $[C_2mim]Br$ (m.pt. 71–81°C; U_L 500–505 $kJ\,mol^{-1}$) and $[C_2mim][Br_3]$ (m.pt. 46°C; $U_L = 458\,kJ\,mol^{-1}$) is possible [142], those between $[C_2mim]I$ and $[C_2mim][I_3]$ and between $[C_4mim]I$ and $[C_4mim]]I_3]$ are not, because of the absence of structural information for the corresponding triodides. However, we know that $[C_4mim][I_3]$ (among a number of related triodides [140]) is a liquid at room temperature. The crystal structure of $[C_4C_1C_1im][IBr_2]$ [114], melting at 37°C, has also been investigated.

Two series of cations that yield relatively low-melting (and structurally characterised) polyiodide salts are the trialkyl-tetrahydro-1,3,5-triaziniums and the *N*-alkylurotropiniums (Fig. 11.19). A comparison can be made between *N*-propylurotropinium salts of $[I_3]^-$ [418] (m.pt. 105°C; U_L 420 $kJ\,mol^{-1}$), $[I_5]^-$ (m.pt. 78.5–80°C; $U_L = 400\,kJ\,mol^{-1}$) [155] and $[I_7]^-$ (m.pt. 50–56°C; $U_L = 384\,kJ\,mol^{-1}$) [155]. While the triazinium salts require a hydrogen abstraction reaction involving iodine in their preparation, the urotropinium salts may be obtained from a straightforward alkylation. Apart from two relevant papers investigating (higher-melting) energetic salts [683, 684], urotropinium salts appear to be a relatively under-explored source of ionic liquids.

Several other polyiodide salts with cations not conventionally associated with ionic liquid behaviour are noted for completeness. These include the salt, bis(1,3-dimethylthiourea)iodinium triidodide [363], that melts at 87°C. The crystal structure shows that the closest approach of the positively charged iodine with any atom of the triiodide counter-ion is at, or greater than, the sum of the van der Waals radii for two iodine atoms, suggesting little interaction. The structure of the long-known $[I(py)_2][I_7]$ [682] has been described [347].

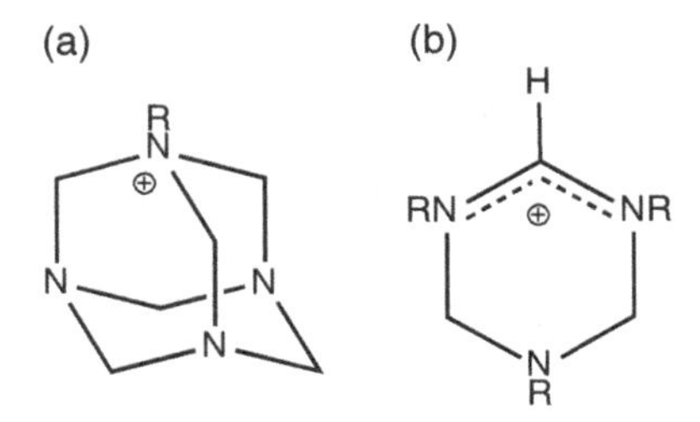

Figure 11.19 (a) *N*-Alkylurotropinium cation and (b) trialkyltetrahydro-1,3,5-triazinium cation [155, 418].

11.9.3 Pseudohalides

The term pseudohalogen was first introduced in 1925 (for a recent review, see Ref. [615]). Pseudohalides, classically, have been taken to include linear species, such as $[CN]^-$, $[OCN]^-$, $[SCN]^-$ and $[N_3]^-$, as well as $[NCSe]^-$ and $[NCTe]^-$. As far as we are aware, no cyanide salt with a reported m.pt. < 100°C has been subjected to a single-crystal structural evaluation. Surprisingly, only 18 structures containing cyanide as a counter-ion were retrieved in a search (20 September 2012) of the CCDC chemical structural database. These include $[NMe_4][CN]$ (no m.pt.; refcode HIWWOZ [685]), $[NEt_4][CN]$ (m.pt. 240°C; refcode MOCFAM [686]) and two 1,3-dialkylimidazolium salts, $[(^iC_3)_2(4,5\text{-}Me_2)im][CN]$ (refcode MENYOU) and $[^iC_3C_1(4,5\text{-}Me_2)im][CN]$ (refcode MENYUA) [687], though neither of these are a m.pt. reported. Crystal structures of ionic liquids containing the remaining ions are listed in Table 11.14 and discussed in the following text.

The pseudohalogen concept has been extended to include non-linear resonance-stabilised amide and methanide species of the type $[NXY]^-$ and $[CXYZ]^-$, in which X, Y and Z are electron-withdrawing groups, typically —CN, $—NO_2$ and —NO, but can include $—CF_3$, $—C_6F_5$, —C(O)R, $—S(O)_2R$ and $—P(O)R_2$. The sulfonylamides are sufficiently numerous to be dealt with separately (see Section 11.13). Brand et al. [615] list the 18 possible methanides, when XYZ can be made up of H, CN, NO_2 and NO. It is beyond the scope of this article to review such species (even their crystal chemistry). However, those salts that melt below 110°C and for which single-crystal structures are reported are included in Table 11.14.

As methanide, amide and azide are components of some protic and/or energetic salts, some structures have already been discussed in Sections 11.4 and 11.6. The crystal structures of a further three azide-containing ionic liquids (and one, $[TeMe_3][N_3]$ [359], with a m.pt. of 110°C ($U_L = 522\,kJ\,mol^{-1}$)) have been reported, including $[SeMe_3][N_3]$ [270] (72°C; $535\,kJ\,mol^{-1}$), [1-(but-2-ynyl)-3-methylimidazolium]$[N_3]$ (m.pt. 66°C; $U_L = 487\,kJ\,mol^{-1}$) [234] and [1-butyl-2,3-dimethylimidazolium]$[N_3]$ (m.pt. 41°C; $U_L = 465\,kJ\,mol^{-1}$) [124]. The latter may be compared with [1-butyl-2,3-dimethylimidazolium][SCN] (m.pt. 40°C; $U_L = 454\,kJ\,mol^{-1}$) [124], both discussed by Laus et al. in a key paper (considered additionally in Section 11.2.6 on Hirshfeld surface analysis). $[C_1C_1pyr]$[SCN] (m.pt. 85–105°C; $U_L = 495\,kJ\,mol^{-1}$) [353] has also been described, having four close contacts (i.e. less than the sum of the van der Waals radii) between cation hydrogens and anion nitrogen and two involving sulfur. The number of such contacts (and their strength) and packing efficiency (linked to relative size of anion and cation) are thought [353] to affect the melting point. The salts $[TeMe_3]$[SCN] (m.pt. 86–88°C; $U_L = 513\,kJ\,mol^{-1}$) [359], $[SeMe_3]$[SeCN] (m.pt. 73–75°C; $U_L = 513\,kJ\,mol^{-1}$) [270] and $[TeMe_3]$[SeCN] (m.pt. 106–109°C; $U_L = 507\,kJ\,mol^{-1}$) [359] are also discussed in Section 11.16.5. [1-ethyl-3-methylimidazolium][OCN], melting at 72°C, is the only cyanate ionic liquid with a single-crystal structure determination [271]. This reveals layers of coplanar anions and cations with weak C—H···O and C—H···N interactions. The layers stack vertically with alternate cations and anions (with the shortest anion–cation contact being between a ring carbon and the central carbon of the anion at 3.217 Å). The authors draw attention to the similarity of this arrangement to that in $[C_2mim][NO_2]$. There are only seven additional $[OCN]^-$ structures in the CCDC database, including $[PPh_4][OCN]$ (refcode MARWOT) [688]. No m.pt. is reported. $[N_{4444}][OCN]$ melts at 86–88°C [689], though no crystal structure is available. MacFarlane et al. suggest [271] that the limited use of cyanates in ionic liquids may be associated with the anion's chemical reactivity.

The largest group in this category are the 14 salts (and 16 structures) of the dinitramide anion (one, $Rb[N(NO_2)_2]$, melting at 102–106°C [690]), most of which have already been discussed with the energetic salts in Section 11.6. Six $[N(CN)_2]^-$ and two $[N(CN)(NO_2)]^-$ salts qualify for inclusion in this compilation. [*N*-aminoguanidinium][A] may be compared when $[A]^- = [N(NO_2)_2]^-$ (m.pt. 91–94°C; form 1: U_L 518 kJ mol^{-1}; form 2: U_L 517 kJ mol^{-1}) [261] and $[N(CN)(NO_2)]^-$ (m.pt. 91°C; U_L 522 kJ mol^{-1}) [391]. One-to-one comparisons may be made between $[NH_3NH_2][N(NO_2)_2]$ (m.pt. 80°C [614]; U_L 573 kJ mol^{-1}) [80] and $[NH_3NH_2][N(CN)_2]$ (m.pt. 97–97.5°C; U_L 593 kJ mol^{-1}) [442] and between the latter and $[Me_3NNH_2][N(CN)_2]$ (m.pt. 60.5°C; U_L 505 kJ mol^{-1}) [199]. Three 1,3-dialkylimidazolium $[RC_nim]$ salts of $[N(CN)_2]^-$ have been studied, where $R = CH_2C{\equiv}CCH_2CH_3$, $n = 1$ (m.pt. 59°C) [185]; $R = NCCH_2$, $n = 1$ (m.pt. 67°C) [241] and $R = C_{10}H_{21}$, $n = 10$ (m.pt. 44°C) [111]. The latter salt was part of a systematic comparison [111] of a series, $[(C_{10})_2im][A]$, $[A]^- = [N_3]^-$, $[N(CN)_2]^-$, $[C(CN)_3]^-$ (m.pt. 44°C) and $[B(CN)_4]^-$ (48°C), also discussed in Section 11.18.6. The effect of an additional *N*-amino group can be seen from a comparison of [*N,N'*-diaminoguanidinium] [319] and [*N,N',N''*-triaminoguanidinium]$[C(NO_2)_3]$ [319], m.pt. 80 and 84°C with U_L 484 and 476 kJ mol^{-1}, respectively. The only other pairwise comparison arises between the $[N(C_6F_5)_2]^-$ salts with $[C_4C_1C_1im]^+$ (m.pt. 46°C) [143] and $[({}^iC_3)_2im]^+$ (m.pt. 109°C) [143], with the same formula weight, for which U_L changes only from 394 to 393 kJ mol^{-1}. Other methanide salts include $[C_2mim][C(NO_2)(NO)(CN)]$ (m.pt. 35°C) [109], $[C_2mim][C(C_6F_5)(SO_2CF_3)_2]$ (m.pt. 52°C) [158], $[C_1mim][CH(NO_2)_2]$ (m.pt. 60°C) [192], [1,5-diamino-4-methyltetrazolium]$[C(NO_2)(CN)_2]$ (m.pt. 72°C) [268] and $[AsPh_4][C\{C(O)CF_3\}(CN)_2]$ (m.pt. 96–102°C) [437].

11.10 OXOANIONS AND THEIR ESTERS

Table 11.15 contains details of salts of nitrate, nitrite (a single entry, $[C_2mim][NO_2]$), perchlorate, sulfate and hydrogensulfate, as well as of the esters, $[CO_2(OMe)]^-$, $[SO_3(OMe)]^-$, $[[PO_2(OMe)(SMe)]^-$ and $[PO_2(OMe)(SeMe)]^-$.

The majority of salts (28 out of 45) involve cations that are protic and/or energetic and their structures have already been discussed in Sections 11.4 or 11.6.

There is only one pair of compounds in this group from which the structural effect of a single compositional factor can be ascertained, namely: [emim][NO_3] (m.pt. 38°C; U_L 490 kJ mol^{-1}) [117] and [emim][NO_2] [117] (m.pt. 55°C; U_L 501 kJ mol^{-1}). Both are early exemplars [117] of cation–anion hydrogen bonding via the C(2)—H, though a 2D network is apparent for the former and an ion-pair arrangement for the latter. Packing in the two cases is different, with cation stacking in the nitrate salt and hexagonal close packing in the nitrite salt.

The structures of [C_1mim][SO_3(OMe)] [134] (m.pt. 43°C; U_L 480 kJ mol^{-1}) and [C_4mim][SO_3(OMe)] (m.pt. −5°C [134]; U_L 444 kJ mol^{-1}) [66] may be compared, with the three additional methylenes leading to a reduction of 36 kJ mol^{-1} in VBT-estimated lattice potential energy. The former salt is characterised by hydrogen-bonded ribbon-like chains of cations and anions, with short contacts between ring hydrogens and non-methylated oxygens of the anion. There are no hydrogen bonds between ribbons. The butyl chain appears to prevent such cation–anion interactions with a surprising absence of interaction of the acidic C(2)—H with the anion.

The anions [PO_2(OMe)(SMe)]$^-$ and [PO_2(OMe)(SeMe)]$^-$ are associated, respectively, with [C_2mim]$^+$ [121] (m.pt. 39°C; U_L 453 kJ mol^{-1}) and [$C_4C_1C_1$im]$^+$ (m.pt. 23–24°C; U_L 422 kJ mol^{-1}) [5]. The lower m.pt. of the latter, compared with that of [$C_4C_1C_1$im][$PO_2(OMe)_2$] (m.pt. 54°C; Ref. 16 in [83]), is associated with the lower symmetry of the phosphoroselenate. These materials are of interest in the development of processing solvents for cellulosic materials. The structure of two protic ionic liquids, based on the diphenylphosphate ion, is discussed in Section 4.3.

While a number of dialkylphosphate ionic liquids are known (e.g. [C_1mim][$PO_2(OMe)_2$] [691, 692], [C_2mim][$PO_2(OEt)_2$] [692] and [C_4mim][$PO_2(OBu)_2$] [692]), they form glasses on cooling, and crystallisation occurs neither on cooling nor heating.

11.11 $[EX_4]^-$ OR $[E_2X_7]^-$

The ionic liquids discussed in this section are collected in Table 11.16.

11.11.1 E = B, Al, Ga or In; X = Cl

The archetypal ionic liquid is one based on haloaluminate(III) ions, particularly [$AlCl_4$]$^-$, investigated by pioneers of this topic [13] during the 1970s, taking inspiration from earlier industrial interest in the electrotechnological applications [693, 694] of melts obtained from mixtures of pyridinium halides and aluminium trichloride, as well as in the processing of cellulose [695].

Low-melting temperatures were associated with eutectic formation. A surprisingly small number of related congruently melting stoichiometric chloroaluminate compounds has been structurally characterised. These include

$[C_4py][AlCl_4]$ (m.pt. 34°C; U_L 432 kJ mol^{-1}) [105] and the co-crystallised mixed salt $Na[C_2mim]Cl_2$ [113] (melting at 37°C), the lowest-melting chloroaluminate salts collected in Table 11.16. Two other mixed salts containing $[AlCl_4]^-$ considered in Section 11.4.3 melt above 100°C.

The other $[AlCl_4]^-$ salts that formally qualify as ionic liquids arise from the isolation of three benzenium carbocations, $[C_6Me_6R][AlCl_4]$: R = Ph (m.pt. 83–84°C, U_L 397 kJ mol^{-1}) [341], R = $ClCH_2$ (m.pt. 91–92°C; U_L = 408–410 kJ mol^{-1}) [393, 394] and R = CH_3 (m.pt. 94–95°C; U_L 411–416 kJ mol^{-1}) [375, 414, 415], resulting from Friedel–Crafts chemistry and from cations containing Group 15 element-to-Group 15 element bonds, the lowest melting of which is $[Ph_3AsSbPhCl][AlCl_4]$ (m.pt. 50–52°C) [154]. A more extensive series of related compounds (including dipositively charged cations) involves the use of the heavier $[GaCl_4]^-$ as counter-anion. The single isoquinolinium ionic liquid with a crystal structure, isoquinolinium tetrachlorogallate [187], melts at 59–60°C (U_L 418 kJ mol^{-1}), with only weak interionic hydrogen bonding. The structure of $[C_2mim][AlBr_4]$ has been reported [263] [refcode ZIBJEZ], though without a melting point. However, a phase diagram of the system, $[C_2mim]Br$-$[AlBr_3]$, has been reported [696], from which it is possible to estimate that the 1:1 salt has a m.pt. of 46–48°C.

Only one pairwise comparison is possible between ionic liquids containing $[ECl_4]^-$, E = Al or Ga, involving protic salts of the heterocyclic cation, 5-methyl-1,3,5-dithiazinan-5-ium [165], with $[AlCl_4]^-$ (m.pt. 84–86°C; U_L 450 kJ mol^{-1}; Fig. 11.20) and $[GaCl_4]^-$ (m.pt. 54–56°C; U_L 451 kJ mol^{-1}). In both, the anion sits atop the cation, with the ring N—H hydrogen bonding with two halides of the anion. Related $[BF_4]^-$ (m.pt. 79–80°C; U_L 486 kJ mol^{-1}) and $[AlBr_4]^-$ salts (m.pt. 83–85°C) were also characterised [165], though the latter co-crystallises as $[LH][L][AlBr_4]$, where L is the neutral dithiazinane.

Of all the known $[NR_4][EX_4]$ salts, in which R is an alkyl group, E is a Group 13 element and X = Cl, Br or I (and for which a single-crystal structure has been reported), only $[N_{4444}][InCl_4]$ [381], with a m.pt. of 90°C (U_L 379 kJ mol^{-1}) [697], qualifies for inclusion. While tetrachloroborate ionic liquids have been studied [698], there are no crystal structures of such materials (though that of a mixed salt, [5-methyl-1,3,5-dithiazinan-5-ium]Cl$[BCl_4]$, melting at 109–110°C, is noted [165]).

Just one $[Ga_2Cl_7]^-$ salt, for which a single-crystal X-ray study is available, melts below 100°C, the 1:3 salt, 3,3,5,5,7,7-hexaphenyltricyclo[2.2.1.0^{2,6}]heptaphosphane-3,5,7-triium heptachlorodigallate (m.pt. 85–87°C; Fig. 11.18) [200]. No $[Al_2Cl_7]^-$ or $[Al_2Br_7]^-$ salt qualifies for inclusion (but see Section 11.22.2).

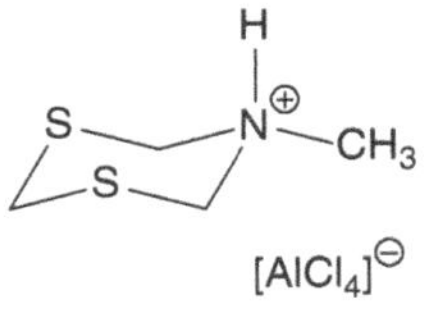

Figure 11.20 5-Methyl-1,3,5-dithiazinan-5-ium tetrachloroaluminate [165].

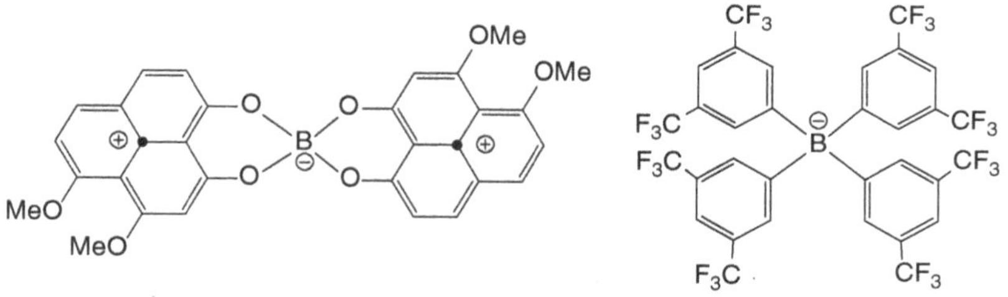

Figure 11.21 Bis(2,4-dimethoxyphenalene-1,9-diolato)boron tetrakis{3,5-bis(trifluoromethyl)phenyl}borate [256].

11.11.2 Organoaluminate and Organoborate Anions

For completeness, attention is drawn to salts of organoaluminate anions $[N_{4444}]$ $[AlMe_2F_2]$ [382] (m.pt. 90°C; U_L 389 kJ mol^{-1}) and $[N_{1111}][Al^iBu_2F_2]$ [334] (m.pt. 81°C; U_L 420 kJ mol^{-1}). The structures of three salts of $[AlMe_2Cl_2]^-$ [477, 524, 529] are known, but these melt in the range 100–110°C. The crystal of $[K(Me_5dien)]$ $[AlMe_4]$ (Me_5dien = *N,N,N′,N″,N″*-pentamethyldiethylenetriamine) contains discrete ions [163] and melts at 53–56°C (U_L 355 kJ mol^{-1}). Mention should also be made of $[(^tC_4)_2im][Me_3Al(\mu^3\text{-}CH_2)(AlMe_2)_2(\mu^2\text{-}Me)]$, reported [53] to be a liquid at room temperature.

Salts of low-melting organoborates include $[N_{10\,10\,10\,10}][BPh_4]$ [470], $[C_4mim]$ $[B\{(3,5\text{-}CF_3)_2C_6H_3\}_4]$ [476], [bis(3,4-dimethoxyphenalene-1,9-diolato)boron] $[B\{(3,5\text{-}CF_3)_2C_6H_3\}_4]$ (Fig. 11.21) and $[C_4mim][B\{C_6H_4(4\text{-}C_6F_{13})\}_4]$ [378]. Worthy of note, also, is the compound $[PPh_4][B(C_6F_5)(N_3)_3]$ [453] (m.pt. 98–101°C), one of only two (structurally characterised) tetraphenylphosphonium compounds with m.pt. < 100°C.

11.11.3 $[E(OR)_4]^-$, E = Al or B

Apart from their continued use as counter-anions in the isolation of novel structures, interest in the anions $[ECl_4]^-$, E = Al or Ga, in ionic liquid chemistry is waning, primarily because of the availability of much more hydrolytically robust anions, first $[BF_4]^-$ and $[PF_6]^-$ and later $[N(SO_2CF_3)_2]^-$. Salts of $[BF_4]^-$ are dealt with in Section 11.11.4.

In addition, an exploration of other aluminium-centred anions, in which interaction with cations can be more subtly varied, has produced a range of low-melting salts and interesting structural types. Part of this is stimulated by an interest in minimising complexation of metal (and other coordinatively unsaturated) centres by the anionic component of ionic liquids used as reaction solvents. The terms 'innocent' and weakly coordinating have been used to describe such minimally interacting species, and the topic has been reviewed [649].

Thirteen materials are tabulated containing the anions $[Al(OR)_4]^-$, R = $CH(CF_3)_2$, $C(Me)(CF_3)_2$ or $C(Ph)(CF_3)_2$, associated with tetraalkylammonium or imidazolium cations and single examples of piperidinium [77] and morpholium [77] salts, as well as, interestingly, of Cs^+ [342] (m.pt. 83–85°C) and nitrosonium [446] (m.pt. 97–99°C). The degree of 'non-innocence' of the anion can be appreciated by the isolation of the salt $[Tl_3F_2Al\{OCH(CF_3)_2\}_3][Al\{OCH(CF_3)_2\}_4]$ [292], also an ionic liquid, with a melting point of 75–80°C.

The consequences of replacement of $[AlCl_4]^-$ by $[Al\{OCH(CF_3)_2\}_4]^-$ can be gauged in the pair of salts with $[C_4py]^+$ [77, 105] (m.pt. 34 and 36°C; U_L 432 and 361 kJ mol^{-1}, respectively). It was concluded [77] that the interionic contacts for these fluoroalkoxyaluminate salts were generally longer than for salts involving other ionic liquid-forming anions with the same cation.

Comparisons are possible between $[C_n mim][Al\{OCH(CF_3)_2\}_4]$, $n=2$ (m.pt. 31°C; U_L 363 kJ mol^{-1}) [77] and $n=4$ (m.pt. 34°C; U_L 356 kJ mol^{-1}) [104], between the $[C_2mim]^+$ and $[C_2C_1C_1im]$ salts (m.pt. 39°C; U_L 361 kJ mol^{-1}) [77], and between $[N_{1111}]^+$ (m.pt. 61°C; U_L 369 kJ mol^{-1}) [170] and $[N_{2222}]^+$ (m.pt. 56°C; U_L 356 kJ mol^{-1}) [170] salts. No other simple pairwise comparison is possible. There is a melting point difference of 115°C between the closely related salts, $[C_4mim][Al\{OCH(CF_3)_2\}_4]$ (m.pt. 34°C; U_L 432 kJ mol^{-1}) and $[C_4mim][Al\{OC(CF_3)_3\}_4]$ (m.pt. 149°C) [104].

The crystal structure of piperidinium tetramethoxyborate (m.pt. 59–61°C), reported in 1982 [188], reveals alternate layer of cations and anions loosely linked by hydrogen bonds. More recently, a single $[B\{OCH(CF_3)_2\}_4]^-$ compound, that is, with [Na(dimethoxyethane)]$^+$ (m.pt. 56°C) [171], has been structurally characterised (using crystals grown by sublimation at *ca.* 315 K). However, close contacts between both O and F of the anion and Na^+ suggest that it should be formulated as a coordination complex rather than as a salt (at least in the crystal).

11.11.4 E = B, X = F

Perhaps surprisingly, only 11 singly charged and 5 doubly charged cations are associated with single-crystal structurally characterised ionic liquids with $[BF_4]^-$ as counter-ion, mostly with imidazolium cations. Some imidazolium tetrafluoroborates, such as the long-known $[C_4mim][BF_4]$, have yet to be crystallised. Also surprising, perhaps, considering the interest in pyridinium and tetrafluoroboate ionic liquids, is the absence of reports of the crystal structure of an ionic liquid of the type $[Rpy][BF_4]$, in which R is any *N*-bound —C— or —O— linked group.

Apart from a series of salts $[(mim)(CH_2)_x(mim][BF_4]_2$, $x=4$ (m.pt. 99°C; U_L 1423 kJ mol^{-1}), 7 (m.pt. 86°C; U_L 1366 kJ mol^{-1}) or 8 (m.pt. 92°C; U_L 1353 kJ mol^{-1}) [357a], the only simple pairwise comparisons involving a $[BF_4]^-$ anion are between two carbenium salts: $[C_6Me_7][EX_4]$ {E = Al, X = Cl (m.pt. 94–95°C; U_L 411–416 kJ mol^{-1}) [375, 414, 415]; E = B, X = F (m.pt. 89–91°C; U_L 435 kJ mol^{-1}) [375]} and the 5-methyl-1,3,5-dithiazinan-5-ium salts [165] described earlier in this section.

$[(C_{12})_2im][BF_4]$ differs from its perchlorate analogue (m.pt. 64°C; U_L 362 kJ mol^{-1}) [213] in the arrangement of the (all-staggered) inter-digitated alkyl groups. [1,3-diethoxyimidazolium][BF_4] [93] and [$P(C_{10})R_3$][BF_4] (R = cyclohexyl or ^{t}butyl) [296] are also noted.

11.11.5 E = B, X = CN

The structures of six $[B(CN)_4]^-$ salts (with six different cations) are described. A one-to-one comparison between [$(C_{12})_2$im][BF_4] (m.pt. 67°C, U_L 361 kJ mol^{-1}) [213] and [$(C_{12})_2$im][$B(CN)_4$] (m.pt. 48°C; U_L 352 kJ mol^{-1}) [111] is possible.

11.11.6 E = B, X = H

Interestingly, the crystal structure of the tetraborohydride compound [C_1mim] [BH_4] (refcode CANTES) [640] has been reported (though without a m.pt. that would permit its inclusion (or exclusion) from our tabulation). However, the salt [Na(18-crown-6)py_2][BH_4] melts at 92–94°C, with U_L 374 kJ mol^{-1} [166]. The structure of [$Li_2(H_2NNHMe)_3$][BH_4]$_2$ (m.pt. 54–56°C) is reported in the same study. Closely related materials [166] are not salts (but still low melting).

11.11.7 E = As, X = N_3

Of a series of [PPh_4][$E(N_3)_4$] and [PPh_4]$_2$[$E(N_3)_5$] salts, E = As, Sb or Bi, only [PPh_4][$As(N_3)_4$], melting at 92°C (dec 125°C), falls into the category of materials considered in this chapter [398]. No significant cation–anion or anion–anion interactions are seen in the crystal. [PPh_4][$B(C_6F_5)(N_3)_3$] [453] has previously been noted.

11.11.8 Other $[EX_4]^-$ and $[EX_2Y_2]^-$ Salts

[1-Methyl-3-{(2*S*)-2-methylbutyl}imidazolium][bis{(2*S*)-2-oxy-3-methylbutanoato-*O*,*O′*} borate]) (m.pt. 89°C; Fig. 11.14) [373] and [Li(tetramethylene sulfone)$_2$][BF_2(oxalate)] (m.pt. 95°C) [421] represent the only low-melting $[EX_4]^-$ anion-containing salts for which crystal structures have been reported and which contain chelated dicarboxylates. However, in the latter case, the crystal is made up of eight-membered rings in which two Li^+ are bridged by oxygen atoms from the sulfone and coordinated to oxalate and monodentate sulfone oxygens and should not be characterised as an ionic liquid.

Bis(1-methyl-1*H*-imidazole-3-yl)dihydroboronium dicyanoborohydride (m.pt. 62°C) is the sole representative of a series of eight hypergolic [(1-alkyl-1*H*-imidazole-3-yl)dihydroboronium][A] salts, alkyl = Me or allyl; $[A]^- = [NO_3]^-$, $[N(CN)_2]^-$, $[N(CN)(NO_2)]^-$ or $[BH_2(CN)_2]^-$, recently synthesised, for which a crystal structure is reported [205]. Six of the seven other materials are liquids.

11.12 $[EX_6]^-$

The ionic liquids discussed in this section are collected in Table 11.17.

11.12.1 E = P, X = F

The now classic contributions of Wilkes and Zaworotko [117] paved the way for the rapid growth of interest in ionic liquids, particularly those containing $[PF_6]^-$ and $[BF_4]^-$ (for the latter, see Section 11.11.4), that were easier to handle because of their much reduced (though still appreciable) water sensitivity. The $[PF_6]^-$ ion was of particular interest because of its tendency to engage in only weak hydrogen bonding [191, 401, 699].

A wide range of $[PF_6]^-$ ionic liquids is now available, including 27 (four with dipositive cations) with m.pt. < 100°C that have been characterised using single-crystal X-ray diffraction and a further three with m.pt. between 100 and 110°C. Imidazolium cations predominate (21 of 27).

In a key paper, Rogers, Seddon and their colleagues [99] have analysed systematically single-crystal structural data from 12 1,3-dialkylimidazolium salts with the $[PF_6]^-$ counter-anion, including eight that are formally ionic liquids. The latter set includes 1,3-dimethylimidazolium hexafluorophosphate, for which melting points of 130°C [99] and 67–68°C [700] are reported. A more detailed investigation (DSC at 20 mK s^{-1}, with Raman spectroscopy and X-ray powder diffraction) reveals two forms of this salt, a β form (m.pt. 41.2°C) and an α form (m.pt. 91.2°C). These two crystal (non-plastic) polymorphs are believed [568] to be related by a structural relaxation rather than by a change of solid phase. The very large difference in m.pt. of these two forms is ascribed [568] to entropic factors associated with anion–cation interactions and should temper optimism in the search for rules that might enable melting point prediction from structural considerations alone.

Table 11.17 includes the following (quoted m.pt. from a single study [99]): $[C_n mim][PF_6]$; $n = 1$ (m.pt. 67–68°C; U_L 483 kJ mol^{-1}) [244], 2 (m.pt. 65°C; U_L 473 kJ mol^{-1}) [99, 182], 4 (m.pt. 11°C; U_L 453 kJ mol^{-1}) [62, 72], 4 (m.pt. 83°C; U_L 450 kJ mol^{-1}) [99], 10 (m.pt. 31°C; U_L 407 kJ mol^{-1}) [99], 12 (m.pt. 60°C; U_L 399 kJ mol^{-1}) [191] or 14 (m.pt. 77°C; U_L 389–390 kJ mol^{-1}) [99, 298]. The Rogers–Seddon study [99] also includes the higher-melting $[PF_6]^-$ salts of $[C_2C_1C_1im]^+$ (m.pt. 201°C), $[C_2C_1C_1(4,5\text{-}Me_2)im]^+$ (119°C), $[{}^tC_4mim]^+$ (160°C) and $[({}^iC_3)_2im]^+$ (135°C). $[C_4mim][PF_6]$ has the shortest C—H···F separation (2.36 Å) in the series, $[C_n mim][PF_6]$, compared with 2.64, 2.58 or 2.58 Å, for $n = 1, 2$ or 10, respectively (noted from [72]). CH_2 addition to the *N*-alkyl group leads to a loss of 10 kJ mol^{-1} in U_L on going from $[C_1mim][PF_6]$ to $[C_2mim][PF_6]$ with little change in reported melting point. Replacing $CH_3CH_2CH_2CH_2$— by $(CH_3)_2CH_2CH_2$— causes little change in lattice potential energy but has a marked effect on melting point. Dibrov and Kochi [72] also noted, for $[C_4mim][PF_6]$, that two different inter-planar cation arrangements could be seen in the crystal, whose separations (15.6Å) were similar to the extent of local order seen by small-angle X-ray scattering (SAXS) in the liquid [701].

Using VBT methods (and a published value of V_- for $[PF_6]^-$), Rogers, Seddon and colleagues [99] were able to evaluate the consequence of changes in n and m for $[C_nC_m im][PF_6]$ on the ion volume of the cation and values of U_L and lattice free energy, ΔG_L, with (for shorter alkyl groups) good agreement between values of U_L obtained from X-ray crystal data and those obtained by computational methods, serving to reinforce the view that interactions in the solid are largely electrostatic. They excluded the longer-chain compounds from the lattice energy calculations because the validity of Equation 11.1 is likely to be reduced as van der Waals interactions become more important when $n \geq 10$. U_L is the largest for the smallest cation and the smallest for the largest cation. C—H···F interionic contacts are weak, at best, and appear not to compromise the validity of the assumptions surrounding Equation 11.1.

For imidazolium ionic liquids, it has long been known that replacement of the proton at the ring C(2) by an alkyl group, such as CH_3, can lead to an increase in m.pt. by as much as 50°C, even for strong hydrogen-bonding anions [702]. A comparison of $[C_4mim][PF_6]$ (m.pt. 2°C; U_L 452 kJ mol^{-1} [62]; m.pt. 11°C; U_L 453 kJ mol^{-1} [72]) and $[C_4C_1C_1im][PF_6]$ (m.pt. 40°C; U_L 444 kJ mol^{-1}; X-ray data collected at 213 K [123]; U_L 445 kJ mol^{-1}; X-ray data collected at 150 K (C. Pulham, J. Pringle, A. Parkin, S. Parsons, and D. Messenger, Cambridge Structural Database, private communication, 2005.)) reveals the consequence (loss of 8–9 kJ mol^{-1} of lattice potential energy) from the insertion of a CH_2 group into the most acidic C—H bond. The authors of the latter study (which included structural studies of $[C_4C_1C_1im][SbF_6]$ (m.pt. 44.5°C; U_L 432 kJ mol^{-1}) and $[C_4C_1C_1im][BF_4]$ (m.pt. 40°C; U_L 453 kJ mol^{-1}; X-ray data collected at 213 K [123], U_L 452 kJ mol^{-1}; X-ray data collected at 150 K (C. Pulham, J. Pringle, A. Parkin, S. Parsons, and D. Messenger, Cambridge Structural Database, private communication, 2005.)) conclude that, in this case compared, for instance, with conclusions from the study of dialkylimidazolium carboranes [183] (see Section 11.19.2), inefficient packing is not a requirement for low-melting behaviour of ionic liquids. Instead, cation–cation repulsion is observed for the two smaller anions, though this is not seen for the $[SbF_6]^-$ salt. U_L values for $[(NCCH_2CH_2CH_2)mim][PF_6]$ (m.pt. 75°C; U_L 455 kJ mol^{-1}) [289] and $[(NCCH_2CH_2CH_2)C_1C_1im][PF_6]$ (m.pt. 85°C; U_L 445 kJ mol^{-1}) [289] differ by a roughly similar amount (10 kJ mol^{-1}) as a consequence of replacing C—H by C—Me at the 2-position of the imidazolium ring. U_L for $[C_4mim][PF_6]$ and $[(NCCH_2CH_2CH_2)C_1im][PF_6]$ differ by only 3 kJ mol^{-1} and for $[C_4C_1C_1im][PF_6]$ and $[(NCCH_2CH_2CH_2)C_1C_1im][PF_6]$ by less than 1 kJ mol^{-1}, consistent with limited secondary bonding involving the CN group [289].

Data for $[(PhCH_2)_2im][PF_6]$ [405] (m.pt. 93°C; U_L 412 kJ mol^{-1}) and $[\{PhCH(CH_3)\}_2im][PF_6]$ (m.pt. 108°C; U_L 404 kJ mol^{-1}) [517] reveal the effects of insertion of CH_2 into N-bound side chains (loss of *ca.* 4 kJ mol^{-1} per CH_2 group). In the latter case, a short C—H···F separation of 2.204 Å is seen, with C—H···F distances to the α-methyl group of 2.408 Å being ascribed to weak hydrogen bonding. The conformational arrangement of the PhCH(Me) group changes with anion, bringing about more efficient anion–cation packing. The structure of $[(PhCH_2)_2im][PF_6]$ has also been compared with those of (water-containing) $[(PhCH_2)_2im]Cl$ and $[(PhCH_2)_2im]Br$ [405], the authors noting only weak hydrogen bonding and the absence of π–π stacking in the $[PF_6]^-$ salt.

Laus and his colleagues are responsible for a valuable and extensive series of studies of imidazolium derivatives that includes the crystallographic characterisation of a series of *N*-hydroxy, alkoxy and amino salts. Among these, a pairwise comparison may be made between $[(RO)_2im][PF_6]$ [93]; data collected at 233 K]: R = Me (*syn* form) (m.pt. 83–84°C; U_L 466 kJ mol^{-1}) and R = Et [[93]; data collected at 233 K] (m.pt. 99–102°C; U_L 444 kJ mol^{-1}). While U_L for the diethoxy salt is less per additional $-CH_2-$ by 11 kJ mol^{-1} compared with the dimethoxy salt, its m.pt. is *ca.* 18°C higher. The corresponding datum for the *anti* form of $[MeO)_2im][PF_6]$ [93], collected at 173 K, is U_L 471 kJ mol^{-1}.

Replacing $-CH_2-$ by $-O-$ in a butyl side chain in $[bm(2\text{-}Me)im][PF_6]$ [123] (C. Pulham, J. Pringle, A. Parkin, S. Parsons, and D. Messenger, Cambridge Structural Database, private communication, 2005.) (m.pt. 40°C; U_L 444 kJ mol^{-1}) (data collected at 213 K), U_L 445 kJ mol^{-1} (data collected at 150 K), to give $[(MeOCH_2CH_2)m(2\text{-}Me)im][PF_6]$ [210] (m.pt. 63°C; U_L 452 kJ mol^{-1}), results in an increase of 8 kJ mol^{-1} in U_L and a increase of 23°C in m.pt. Changes in the orientation and position of the oxygen-containing side chain are ascribed [210] to the minimisation of lone-pair repulsions.

11.12.2 $[PF_3(C_2F_5)_3]^-$

Despite its greater cost, the anion, $[PF_3(C_2F_5)_3]^-$, is of some interest as a component of ionic liquids because of its enhanced hydrolytic robustness compared with $[PF_6]^-$. However, we are aware of only two ionic liquids containing $[PF_3(C_2F_5)_3]^-$ that have been characterised crystallographically, namely, $[(MeO)_2(2\text{-}Me)im][PF_3(C_2F_5)_3]$ melting at 75–76°C [93] and the as yet unpublished $[N_{4\,4\,4\,4}][PF_3(C_2F_5)_3]$ (m.pt. *ca.* 65°C) (UoL, unpublished work). For completeness, the only other salt for which a crystal structure is available is $[N_{1\,1\,1\,1}][PF_3(C_2F_5)_3]$ (m.pt. *ca.* 120°C) (UoL, unpublished work). In all three cases, the octahedral *mer*-isomer of the anion, rather than the *fac* isomer, is found.

11.12.3 E = Sb, As, Nb or Ta; X = F or Cl

The availability of the structures containing the heavier Group 15 elements, Sb and As, and Nb and Ta from Group 5, allows an extensive comparison (see also Section 11.5.6.5) of $[C_2mim][SbF_6]$, $[C_2mim][AsF_6]$, $[C_2mim][NbF_6]$ and $[C_2mim][TaF_6]$, the last two displaying particularly low-melting points. $[C_4C_1C_1im][SbF_6]$ may be compared with the $[PF_6]^-$ analogue.

	m.pt./°C	U_L/kJ mol^{-1}	References
$[C_2mim][PF_6]$	65	473	[99, 182]
$[C_2mim][SbF_6]$	10	460	[75]
$[C_2mim][AsF_6]$	53	462	[75]
$[C_2mim][NbF_6]$	−1	462	[67, 655]
$[C_2mim][TaF_6]$	2	462	[67, 655]

The listed compounds, $[C_2mim][EF_6]$, fall into two different isostructural classes: E = Nb or Ta and E = P, Sb or As, [67, 75]. In the latter class, only a few weak hydrogen bonds are seen: the shortest H···F separation is 2.51 Å for $[C_2mim][SbF_6]$ (data collected at 200 K). On the other hand, the former group displays a range of short H···F separations (2.40 Å in $[C_2mim][NbF_6]$ being the shortest), and hydrogen bonds are judged to be stronger. Neither these trends nor variations in lattice potential energies reflect differences in melting point.

We also note, surprisingly, that there are no crystal structures of [*N*-alkylpyridinium]$[PF_6]$ ionic liquids so far reported. However, the one $[PCl_6]^-$ ionic liquid is a pyridinium salt. This is $[C_4m_\gamma py][PCl_6]$ (m.pt. 78–80°C; U_L 417 kJ mol^{-1}) [308], which may be compared with the related salts, $[C_4mpy][NbCl_6]$ (m.pt. 89–91°C; U_L 412 kJ mol^{-1}) [308], $[C_4m_\beta py][NbCl_6]$ (m.pt. 83–85°C; U_L 413 kJ mol^{-1}) [308] and $[C_4m_\gamma py][TaCl_6]$ (m.pt. 103°C; U_L 412 kJ mol^{-1}) [294]. The structures of $[C_4m_\gamma py][NbCl_6]$ and $[C_4m_\beta py][NbCl_6]$ differ by the presence of a single ion pair in the asymmetric unit in the former and two in the latter. The $[C_2mim]^+$ salts, $[C_2mim][PCl_6]$ and $[C_2mim][NbCl_6]$, for which crystal structures are also available, melt at 161–163 and 146–149°C, respectively [308]. Hydrogen bonding between anion and cation in this series is judged to be weak or absent.

11.12.4 E = Sb, X = N_3

Worthy of note as a 'one off' is $[PPh_4][Sb(N_3)_6]$ [495], one of the lowest-melting (crystallographically characterised) tetraphenylphosphonium salts. The compound could be melted at 104–106°C without decomposition. The reduction in shock sensitivity is ascribed to the suppression of detonation propagation arising from the presence of a large cation. The structure shows little (non-coulombic) interaction between anion and cation.

11.13 SALTS OF BIS(SULFONYL)AMIDES, $[N(SO_2R)_2]^-$

11.13.1 Comparisons Between Salts of $[N(SO_2CF_3)_2]^-$ and $[N(SO_2CH_3)_2]^-$

Table 11.18 assembles 66 bis(trifluoromethanesulfonyl)amide (bistriflamide, $[NTf_2]^-$) ionic liquids, $[cat][NTf_2]$, fully characterised in the crystalline state by X-ray diffraction methods (and a further nine with m.pt. 100–110°C). The first notable observation is that bistriflamide salts make up a significant proportion of materials with very low-melting temperatures that, combined with the anion's chemical, particularly hydrolytic, thermal and electrochemical robustness, underpins the considerable fundamental and technological interest in them. Of the 15 bis(methanesulfonyl)amide (dimesylamide; $[NMes_2]^-$) salts, [cat] $[NMes_2]$, all with m.pt. < 100°C (Table 11.18), three offer the possibility of a direct comparison with their fluorinated analogues, [cat]$[NTf_2]$, when $[cat]^+$ = 1-butyl-1-methylpyrrolidinium [52, 62], 2-hydroxyethyl(trimethyl)ammonium (cholinium) [297] (UoL, unpublished work) and $[C_2mim]^+$ [52, 58, 59]. MacFarlane's group reports the crystal structure of $[C_2mim][NMes_2]$ (a liquid at room temperature) [52], but thermophysical investigations using DSC provide neither a melting point nor evidence of a crystallisation process during either heating or cooling. In fact, crystals of both $[C_2mim][NMes_2]$ and $[C_4C_1pyr]$ $[NMes_2]$ were obtained only after more than 2 years' storage [52].

While comparisons involving higher-melting salts (such as between $[C_1C_1pyr]$ $[NTf_2]$ (m.pt. 132°C; refcode XOMDIM) [314] and $[C_1C_1pyr][NMes_2]$ [127]) fall outside the scope of this chapter, they will be considered in the following text as a complement to the three comparisons mentioned previously. Pringle et al. [127] report that $[C_1C_1pyr][NMes_2]$ melts over a wide temperature range, 40–55°C. These observations clearly show that the absence of a melting point (or an imprecise value) is not simply a consequence of a failure to measure one.

Most of the dimesylamide salts listed (and many more for which melting points are not available) are protic ionic liquids and have been discussed in Section 11.4. Other salts are known (and have been structurally characterised, see Refs. [52, 127, 281], but either are high melting (e.g. $[H_2im][NMes_2]$; refcode: QUSMAS, melting at 120°C [281]) or have no melting point reported. The $[NMes_2]^-$ salts are generally more viscous [127] than their $[NTf_2]^-$ analogues, with both displaying interactions in the crystal between anion oxygens and cation hydrogens. In the case of $[NMes_2]^-$ salts, interactions are also seen between anion C—H and the oxygens of neighbouring anions. Despite stronger electrostatic interactions in $[C_1C_1pyr][NMes_2]$ compared with $[C_1C_1pyr][NTf_2]$, as reflected in an 18 kJ mol^{-1} difference in lattice potential energies, the m.pt. of the latter is significantly higher than the former.

	m.pt./°C	U_L/kJ mol^{-1}	References
$[C_1C_1pyr][NTf_2]$	132	430	[314]
$[C_1C_1pyr][NMes_2]$	40–55	448	[127]
$[C_4C_1pyr][NTf_2]$	−11	409	[62]
$[C_4C_1pyr][NMes_2]$	37	424	[52]
$[NMe_3CH_2CH_2OH][NTf_2]$	30	424	(UoL, unpublished work)
$[NMe_3CH_2CH_2OH][NMes_2]$	76–77	449	[297]
$[C_2mim][NTf_2]$	−26/−2	426–427	[58, 59]
$[C_2mim][NMes_2]$	Liq	446	[52]

The reverse is true for $[C_4C_1pyr]^+$ and cholinium salts. U_L is greater for the dimesylamide salt in each case, being larger by 15, 25 and 20–21 kJ mol^{-1} for the pyrrolidinium, cholinium, and 1-ethyl-3-methylimidazolium salts, respectively. This simply highlights the complex relationship between melting point, solid-state structure, and lattice energetics.

The crystal structures of six salts derived from $[HNMes_2]$ and a series of secondary alkylamines have been reported [267] in which the anion adopts a *trans*oid conformation, though the authors point out that the salts display 'unpredictable' packing arrangements. Four of the salts are protic ionic liquids: $[NH_2Et_2][NMes_2]$ (m.pt. 71–73°C; U_L 454 kJ mol^{-1}), $[NH_2MeEt][NMes_2]$ (m.pt. 75–77°C; U_L 471 kJ mol^{-1}), $[H_2pyr][NMes_2]$ (m.pt. 80–90°C; U_L 466 kJ mol^{-1}) and $[H_2pip][NMes_2]$ (m.pt. 95–97°C; U_L 458 kJ mol^{-1}). The additional $-CH_2-$ in the hetero-alicyclic ring reduces the lattice potential energy by 8 kJ mol^{-1}. All the crystal data were collected at 143 K. The structure of $[NEtH_3][NMes_2]$ has also been reported [337].

11.13.2 $[N(SO_2R)_2]^-$ Salts, $R = C_2F_5$ or C_3F_7, and Related Ions

In addition to salts of $[N(SO_2R)_2]^-$, $R = CH_3$, CF_3 or F (those of the latter are discussed in Section 11.5.1), there is a small number of 'one-off' studies of materials, where $R = C_2F_5$ [274] and C_3F_7 [274]. In addition, the structure of the analogues $[C_6F_5C(SO_2CF_3)_2]^-$ [158] and $[N\{P(O)(C_2F_5)_2\}_2]^-$ [449] may be compared with those of the corresponding $[C_2mim][NTf_2]$ [58, 59] and $[N_{1\,1\,1\,1}][N(SO_2C_2F_5)_2]$ [274] salts, respectively.

Repeated cycling of $[C_3C_1pyr][A]$, $[C_4C_1pyr][A]$ and $[(MeOCH_2CH_2OMe)C_1pyr][A]$, $[A]^- = [N(SO_2F)(SO_2CF_3)]^-$ to 150°C fails to bring about crystallisation, associated with the enhanced asymmetry of the anion [703]. A crystal structure of $[PPh_4][N(SO_2F)(SO_2CF_3)]$ (refcode TUQQIG) has been reported [704], though without a m.pt.

11.13.3 Anion Conformation

Aspects of the structure that focus on the cation are considered in Sections 11.17 (pyridinium) and 11.18 (imidazolium). The focus of this section is on the anion. Many of the studies on $[N(SO_2CF_3)_2]^-$ ionic liquids underline the point that the anion is characterised by extensive charge delocalisation that reduces specific interionic interactions, along with conformational flexibility and lability, usually characterised by a C—S···S—C torsion angle between planes defined by C(1)—S(1)—N and N—S(2)—C(2), as shown in Figure 11.22. For $[NTf_2]^-$, conformers can be described as *syn*periplanar, *anti*periplanar, *syn*-clinal or *anti*clinal, according to the value and sign of the angle between the two C—S—N planes, for the two CF_3SO_2 units linked by the sp^3 nitrogen atom. The atoms that define the torsion angle are not, therefore, directly connected to one another. Additional conformational flexibility is conferred on the

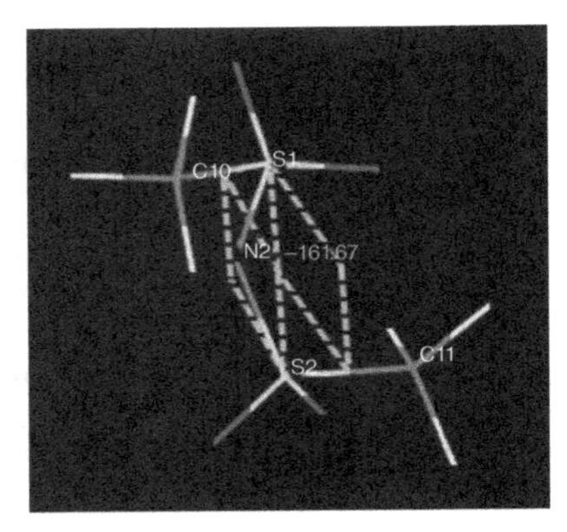

Figure 11.22 A representation, using CCDC software Mercury 3.1, of the $[N(SO_2CF_3)_2]^-$ in $[C_4C_1pyr][NTf_2]$ (refcode LAZREK) [62] showing the C–S···S–C torsion angle (−161.7°) between planes defined by C(10)–S(1)–N and N–S(2)–C(11).

anion as a result [705, 706]. While the computed potential energy surface is complex, the four conformers (two sets of mirror images) are generally categorised either as *anti* or '*trans*oid' (~ C_2 symmetry) or *gauche* or '*cis*oid' (C_1 symmetry). The *trans*oid conformer is calculated to be more stable than the *cis*oid isomer, though the energy barriers that permit an interchange between them are sufficiently small to allow the adoption of the conformer that best fits the crystal structure [705]. However, each mirror conformer does not readily interconvert with its mirror image and both are usually present in crystals in equal proportions [705]. An exception noted by Lopes et al. [705] is $[C_4C_1pyr][NTf_2]$ [62].

As noted in an analysis by Laus et al. [8] of 143 bistriflamide data sets from the CSD (version 5.29, January 2008, 436384 entries; excluding disordered structures), 83 *trans*oid and 60 *cis*oid structures (45 *syn*periplanar, 15 *syn*clinal, 12 *anti*clinal, 71 *anti*periplanar) were noted. A torsion angle of ±170±5° is preferred in 45 out of the 143 structures.

11.13.3.1 Imidazolium Salts $[1,3\text{-Dialkylimidazolium}][NTf_2]$ and $[N\text{-alkylpyridinium}][NTf_2]$ salts are among the lowest-melting ionic liquids for which single-crystal characterisation by X-ray diffraction has been reported. Because of the compositional and structural variation possible in the *N*-bound groups in the imidazolium series and ring-bound groups in the pyridinium series, opportunities to search for clues as to the origin of their physical properties (and of any departure from physical property estimates arising from the use of theoretical or computational models) arise from the assemblage of information collected in this chapter.

The significant number of studies of imidazolium bistriflamide salts permits a series of comparisons that may reveal the consequences of modest changes in salt composition.

	m.pt./°C	U_L/kJ mol^{-1}	Torsion Angle/°		References
$[C_1C_1im][NTf_2]$	22	435	32.88	31.47	[82]
$[C_2C_1im][NTf_2]$(230 K)	−26	427	26.85	27.04	[58]
(120 K)	−2	426	176.16	−167.58	[59]
$[C_4C_1im][NTf_2]$	−3	413	15.61	34.44	[59]
$[C_6C_1im][NTf_2]$	−1	402	28.60		[59]

An early study of $[C_1C_1im][NTf_2]$ reports C–S⋯S–C torsion angles that point to a *cis* anion conformation [82]. The stabilisation of the higher-energy conformer (ΔH(*trans–cis*) = 3.5–4.5 kJ mol^{-1} found from Raman spectra measured at 328 and 348 K [59]) is associated [82] with C–H⋯O bonding between anion and cation – the C(2)–H to the sulfonyl oxygen atoms of two anion units. $[C_2mim][NTf_2]$ (at 230 K) shows the same anion conformation, whereas a more detailed study [59] reveals additional polymorphs, one of which is characterised by X-ray crystallography as having the anion in the *trans* conformation. All the structures show weak O⋯π and F⋯π interactions. For longer alkyl chains in which the *cis* conformation is also evident, flexibility in the alkyl group leads to additional polymorphism as well as structural disorder in the crystal.

	m.pt./°C	U_L/kJ mol^{-1}	Torsion Angle/°		References
$[C_2mim][NTf_2]$ (230 K)	−26	427	26.85	27.04	[58]
$[(PhCH_2CH_2)mim][NTf_2]$	41	404	136.42		[129]
$[Ph_2CHCH_2)mim][NTf_2]$	62	380	177.01		[129]
$[(PhCH_2CH_2CH_2)mim][NTf_2]$	50	397	174.42	(Disorder)	[129]

Introduction of phenyl residues into the *N*-bound side chain [129] reduces U_L, successively, by 23 and 24 kJ mol^{-1} (with the m.pt. increasing substantially as U_L falls). An additional methylene residue leads to a decrease of 7 kJ mol^{-1} in lattice potential energy and to a higher melting point. The 2-phenylethyl and 3-phenylpropyl compounds are shown to have layer structures, in contrast to the diphenyl compound, structural arrangements that may be associated with density differences between the three compounds.

	m.pt./°C	U_L/kJ mol^{-1}	Torsion Angle/°		References
$[C_2mim][NTf_2]$	−26	427	26.85	27.04	[58]
$[(HO_2CCH_2)mim][NTf_2]$	40	424	171.72	177.68	[101]
$[(EtO_2CCH_2CH_2)mim][NTf_2]$	45	406	33.74		[138]

The 1-hydroxycarbonylmethyl compound, $[(HO_2CCH_2)mim][NTf_2]$ (m.pt. 40°C; U_L 424 kJ mol^{-1}) [101], forms strongly hydrogen-bonded ion pairs in the crystal, with O—H···O distances in the range 1.75–1.89 Å, along with a range of C—H···O and C—H···F contacts lying between 2.35 and 2.65 Å and 2.37 and 2.80 Å, respectively. The anion adopts a *trans* conformation, with the torsion angle being 171.72 and 177.68°, in contrast to 33.74° for the $EtO_2CCH_2CH_2$ analogue [138]. The latter compound was prepared from (*S*)-ethyl lactate, giving an enantiopure ionic liquid, the melt from which deposited crystals of the racemate on standing in air for 2 months.

The following *N*-amino and *N*-hydroxycompounds are characterised as protic ionic liquids and follow the general tendency of 2-methylimidazolium salts to crystallise readily [8]. Both salts contain $[NTf_2]^-$ anions in the *trans* conformation. Unusually, strong hydrogen bonds (O···N separation of 2.75 Å, O—H···N of 1.95 Å) are found between N—OH and the nitrogen of $[NTf_2]^-$.

	m.pt./°C	U_L/kJ mol^{-1}	Torsion Angle/°	References
$[(H_2N)_2(2\text{-Me})im][NTf_2]$	58–59	433	171.63	[8]
$[(HO)_2(2\text{-Me})im][NTf_2]$	72	435	179.42	[8]

11.13.3.2 Pyridinium Salts Seddon and colleagues have prepared an extensive series [92] of 37 pyridinium salts with $[NTf_2]^-$ (including 14 ionic liquids) to explore the consequences of substituting electron-withdrawing groups, —CN and —CF_3, into the pyridinium ring. Six were characterised by singe-crystal X-ray crystallography, with anion disorder noted in two salts and a disordered CF_3 in [1-ethyl-3-trifluoromethylpyridinium]$[NTf_2]$. The asymmetric unit was made up of either one or two ion pairs. A significant change may be imposed on the anion as a consequence of changing the position of the cyano group in the pyridine ring (with little effect on the m.pt.): in the 1-methyl-4-cyanopyridinium salt, $[NTf_2]^-$ adopts the *trans* conformation, whereas in the 3-cyano-analogue, the higher-energy *cis* conformation is seen. In both the 1-ethyl analogues, however, the anion adopts the *trans* structure (though their melting points differ by *ca.* 40°C), with the ethyl group twisted perpendicular to the plane of the pyridine ring. The polarisation of the pyridine by the electron-withdrawing CN group is manifested in changes to the short contacts involving ring protons – particularly at C(5)—H in the case of 3-cyano substitution – some to adjacent cations, though the latter interactions are not observed for the 3-CF_3 pyridinium compound.

	m.pt./°C	U_L/kJ mol^{-1}	Torsion Angle		References
[1-Methyl-3-cyanopyridinium]$[NTf_2]$	63–65	434	13.34		[92]
[1-Methyl-4-cyanopyridinium]$[NTf_2]$	65–66	429	175.80	(Disorder)	[92]
[1-Ethyl-2-cyanopyridinium]$[NTf_2]$	36–38	418	−176.36	171.66	[92]
[1-Ethyl-3-cyanopyridinium]$[NTf_2]$	73	420	175.96		[92]

(Continued)

	m.pt./°C	U_L/kJ mol^{-1}	Torsion Angle		References
[1-Ethyl-4-cyanopyridinium][NTf_2]	33–35	419	170.77	(Disorder)	[92]
[1-Ethyl-3-trifluoromethyl-pyridinium][NTf_2]	28–29	417	173.12	−160.40	[92]

In another extensive comparison, Binnemans and co-workers [54] report a series of 25 1-(cyanoalkyl)pyridinium, pyrrolidinium and piperidinium ionic liquids, seven of which have been studied crystallographically. For both [1-cyanomethylpyridinium][NTf_2] and its 2-methylpyridinium analogue, separations of *ca.* 2.6 Å are observed between a pyridine C—H and the cyano group of an adjacent cation. However, no significant change is evident in the conformation of $[NTf_2]^-$. In all cases, the $NCCH_2$— group lies in the same plane as the pyridine ring, whereas the plane of the $NCCH_2CH_2CH_2$— lies at 76 and 82.3° to that of the pyridine ring in the two cyanopropylpyridinium salts. The latter two salts are both lower melting than their cyanomethyl counterparts. The packing is discussed in terms of various hydrogen bonding and π interactions involving both the pyridine ring and the anion. Pyridine ring substitution of H by Me or Et results in both increased and decreased melting point that appear not to be reflected in VBT-estimated lattice potential energy. Introducing successive —CH_2— groups into the pyridine ring of the 1-cyanomethylpyridinium series reduces U_L by 9 (H to 2-Me), 7 (2-Me to 2,5-Me_2) and 7 (2-Me to 2-Et) kJ mol^{-1}. Introducing —CH_2— at the 3-position of the pyridine ring of [1-cyanopropylpyridinium][NTf_2] reduces U_L by 4 kJ mol^{-1}. Introducing a —CH_2CH_2— moiety into the *N*-bound cyanomethyl group reduces U_L by 19 kJ mol^{-1}, somewhat greater than the 15–16 kJ mol^{-1} when introducing a similar entity into the 2- or 4-position of the pyridine ring of the cyanomethylpyridinium salt.

	m.pt./°C	U_L/kJ mol^{-1}	Torsion Angle/°		References
[1-(Cyanomethyl)py][NTf_2]	44	435	167.49		[54]
[1-(Cyanomethyl)(2-Me)py][NTf_2]	51	426	167.01		[54]
[1-Cyanomethyl(2,5-Me_2)py][NTf_2]	Liq	419	−175.96		[54]
[1-Cyanomethyl(2-Et)py][NTf_2]	30	419	168.40	−163.99	[54]
[1-(Cyanomethyl)(4-Et)py][NTf_2]	48	420	153.34		[54]
[*N*-Carboxymethyl-pyridinium][NTf_2]	32	426	175.81	175.44	[101]
[1-(3-Cyanopropyl)py][NTf_2]	Liq	416	−178.92		[54]
[1-(3-Cyanopropyl)(3-Me)py][NTf_2]	Liq	412	−179.34		[54]

11.13.3.3 Pyrrolidinium and Piperidinium Salts Successive introduction of three methylene and one ethylene moiety into the *N*-methyl residue of $[C_1C_1pyr][NTf_2]$ (m.pt. 132°C [127]; U_L 430 kJ mol^{-1}; anion conformer: C_2; refcode XOMDIM [314]) to give $[C_nC_1pyr][NTf_2]$ ($n=2$; m.pt. 91°C; U_L 419 kJ mol^{-1}; phase III, data collected at 213 K: U_L 423 kJ mol^{-1}; low-temperature phase IV, data collected at 153 K; all C_2) [389], ($n=3$; m.pt. 12°C; U_L 413, 414 kJ mol^{-1}; data for both forms collected at 200 K; C_2) (UoL, unpublished work), ($n=4$; m.pt. −11°C; U_L 409 kJ mol^{-1}; C_2) [62] and ($n=6$; m.pt. 9°C; U_L 394 kJ mol^{-1}; C_1) (UoL, unpublished work) results in reduction in U_L by 7–11, 6–10, 4–5 and then 16 kJ mol^{-1}. Disorder is seen in these salts associated with the additional conformational arrangements possible in side chain, ring and anion. For $n=1, 2, 3$ or 4, only a *trans*oid (C_2) arrangement of the anion is seen in the crystalline forms of the salts. In $[C_6mpyr][NTf_2]$, however, the anion is found in a *cis*oid (C_1) conformation. For $[C_2C_1pyr][NTf_2]$, the lower-temperature structure is an ordered phase, whereas a solid–solid transition at 188 K gives a plastic crystalline phase, whose structure was determined at 153 K. The phase behaviour of these and related salts are of considerable interest. Detailed discussion is beyond the scope of this chapter.

A series of low-melting $[(cyanoalkyl)C_1pyr][NTf_2]$ (cyanoalkyl = $-(CH_2)_nCN$, $n=1, 2, 4$ or 5) salts have been synthesised [707] for use as ligands to cobalt(II). While none gave single crystals, the analogue $[C_4(NCCH_2)pyr][NTf_2]$ gave crystals (U_L 402 kJ mol^{-1}; refcode GUNHAZ) and was characterised by X-ray diffraction. This showed weak interactions between the anion and the more acidic cation α-protons, with the anion adopting a *trans*oid conformation. Unfortunately, no melting point was reported. In $[C_4C_1pyr][NTf_2]$ (m.pt. −11°C; U_L 409 kJ mol^{-1}) [62], the anion is also found in a *trans*oid conformation. The replacement of NCH_2—H by N—CH_2—CN results in a modest reduction in lattice potential energy of 7 kJ mol^{-1}.

Replacement of NCH_2—H by NCH_2—CO_2H, reduces the lattice potential energy for $[C_1C_1pyr][NTf_2]$ (m.pt. 132°C [127]; U_L 430 kJ mol^{-1} [314]; refcode XOMDIM) by 9 kJ mol^{-1} for $[(1\text{-carboxymethyl})C_1pyr][NTf_2]$ (m.pt. 39°C; U_L 421 kJ mol^{-1}) [101]. In both cases, the anion is *trans*oid.

Two piperidine salts with $[NTf_2]^-$ counter-anions have been characterised by X-ray crystallography: [1-methyl-1-propylpiperidinium]$[NTf_2]$ (m.pt. 12°C; U_L 408 kJ mol^{-1}) (UoL, unpublished work) and [1-ethyl-1-methyl-piperidinium]$[NTf_2]$ (m.pt. 91°C; U_L 416 kJ mol^{-1}; $P\bar{1}$ form, data collected at 123 K): U_L 385 kJ mol^{-1}; C2/c form, data collected at 223 K) [390].

11.13.3.4 Ammonium and Phosphonium Salts Surprisingly, no X-ray crystal structures of the type $[PR_4][NTf_2]$ or $[PR_4][NMes_2]$ were found in a search of the CSD. The structure of the protic salt $[(Ph_3AsO)_2H][NMes_2]$ was found (refcode ZEMFIG), though it melts at 146–148°C [708].

The structures of three simple [alkylammonium]$[NTf_2]$ salts have been reported, with trimethylammonium (m.pt. 79–81°C; U_L 445 kJ mol^{-1}) [314], tetraethylammonium (m.pt. 104°C; U_L 412 kJ mol^{-1}) [487] and tetrapropylammonium (m.pt. 105°C; U_L 390 kJ mol^{-1}) cations [314]. A crystal structure of the

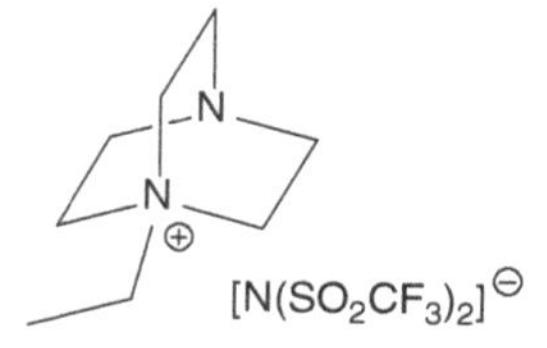

Figure 11.23 1-Ethyl-1,4-diazabicyclo[2.2.2]octanium bis(trifluoromethanesulfonyl) amide [299].

supercooled room-temperature phase II of $[N_{2\,2\,2\,2}][NTf_2]$, undertaken at 100K, shows typical disorder in the cation (Raman spectroscopy suggests that, in the low-temperature phase III, the cations become ordered). In phase II, the anion is also disordered. The four anions in the asymmetric unit are disordered but not between the *cis*oid and *trans*oid conformations. They are, instead, all *trans*oid, though two of the four anions adopt, unusually (seen also in $[C_2C_1pyr][NTf_2]$ and $[C_2C_1pip][NTf_2]$ [389]), two different *trans*oid conformations. Forsyth et al. report crystal structures of $[N_{0\,1\,1\,1}][NTf_2]$ and $[N_{3\,3\,3\,3}][NTf_2]$ [314], noting one short cation–anion separation and several others significantly longer, whereas for the corresponding iodide salt (U_L for $[N_{3\,3\,3\,3}]I = 427\,kJ\,mol^{-1}$; data from refcode HEVMEA), a different distribution of such separations is seen. These differences in size and shape result in different packing efficiencies [314] that may contribute to m.pt. differences.

Three of eight [quaternary ammonium]$[NTf_2]$ salts are reported [302] to be liquid below room temperature: $[NR_3(CH_2CH_2OEt)][NTf_2]$ (R = Et, Bu or Pent). The anion adopts the *trans* conformation in each of the crystal structures of [dimethylbis(3-methylbenzyl)ammonium]$[NTf_2]$ (m.pt. 78°C; U_L 383 kJ mol^{-1}), [dimethylbis(4-methoxybenzyl)ammonium]$[NTf_2]$ (m.pt. 106°C; U_L 379 kJ mol^{-1}) and of [dimethylbis(4-methylbenzyl)ammonium]$[NTf_2]$ (m.pt. 131°C; U_L 384 kJ mol^{-1}; refcode MAKTID).

The dabco salt, [1-ethyl-1,4-diazabicyclo[2.2.2]octanium]$[NTf_2]$ (Fig. 11.23), undergoes a solid–solid transition at 36°C and melts at 77°C. X-ray data were collected at 100, 223 and 295 K (with associated values of U_L = 418, 415 and 413 kJ mol^{-1}) [299]. The anion has the *trans*oid conformation. Related data obtained for the high-melting [1-ethyl-1,4-diazabicyclo[2.2.2]octanium]Br revealed greater distortion of the cation in the latter. This ring strain in the $[NTf_2]^-$ salt reduces as the temperature increases.

11.14 SULFONATES, $[RSO_3]^-$

The ionic liquids discussed in this section are collected in Table 11.19.

11.14.1 Trifluoromethanesulfonate Salts (R = CF_3)

The 29 trifluoromethanesulfonate ('triflate' or $[OTf]^-$) salts, melting under 100°C with a crystal structure described, contain a disparate group of cations, with a limited number of pairs from which to make structural comparisons.

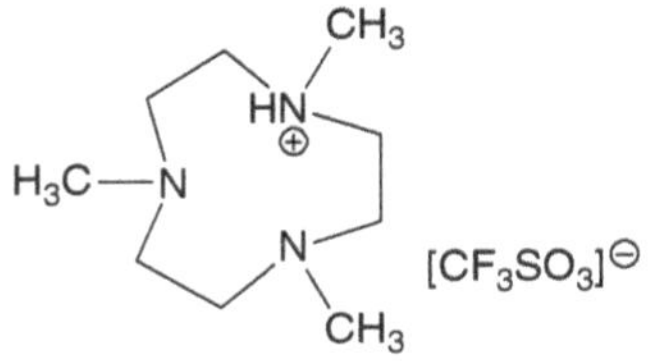

Figure 11.24 1,4,7-Trimethyl-1-azonia-4,7-diazacyclononane trifluoromethanesulfonate [385].

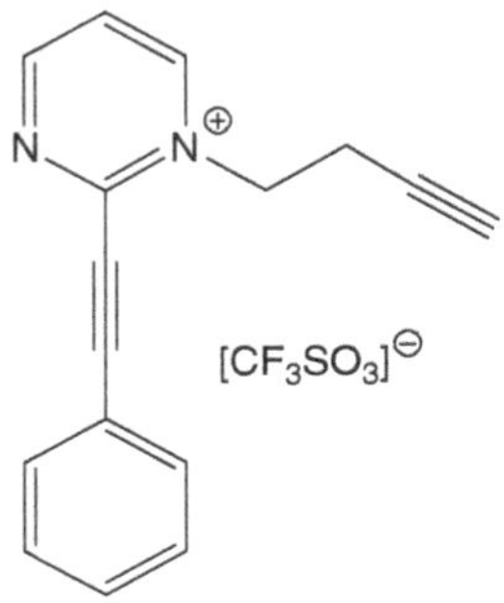

Figure 11.25 1-(3-Butynyl)-2-(phenylethynyl)pyrimidinium trifluoromethanesulfonate [387].

As triflate is widely used in organic chemistry to effect crystallisation, it is not too surprising, therefore, that many of the listed compounds are adventitious ionic liquids (and their structures 'one offs'), highlighting again the wide variety of relatively low-melting salts that may be accessible. These include [1-benzyl-3,5-bis(3-trifluoromethylphenyl)pyridinium][OTf] (m.pt. 58–60°C) reported in a natural product synthesis investigation [181], [1,4,7-trimethyl-1-azonia-4,7-diazacyclononane][OTf] (m.pt. 90–92°C; Fig. 11.24) [385] and [trimethylphosphine(diphenyl)phosphonium][OTf] (m.pt. 57–58°C) [176]. The quaternary diyne, [1-(3-butynyl)-2-(phenylethynyl)pyrimidinium][OTf] (m.pt. 97–98°C; Fig. 11.25) was crystallographically characterised in a programme studying [2+2+2] cycloadditions involving acetylene [387]. The structure of [hydroxylammonium][OTf] (melting at 105°C, with U_L 512 kJ mol^{-1}) was reported in 1977, in support of a vibrational spectroscopic investigation [499].

[$C_1C_1C_1$im][OTf] (m.pt. 129°C; refcode OFADUU) [709] and [$C_2C_1C_1$im][OTf] [709] (m.pt. 113°C; refcode: OFAFAC) have also been characterised, but neither are, formally speaking, ionic liquids. The structure of the interesting, but hydrolytically sensitive, [$(Me_3Si)_2$im][OTf] [710] (refcode: RAHZEH) has been described, but no m.pt. is available.

Among the imidazolium salts, [C_2mim][OTf] (m.pt. −26°C; U_L 459 kJ mol^{-1}) [58] may be compared with [C_4mim][OTf] (m.pt. 7°C; U_L 433 kJ mol^{-1}) [62].

The introduction of two methylenes into the ethyl chain leads to a 26 kJ mol^{-1} drop in lattice potential energy, though with a significant increase in melting point. The effect of ring bromination may be evident in the structural differences between [C_4mim][OTf] and [C_4C_1(4,5-Br_2)im][OTf] {m.pt. 73°C (on first heating only; gives a glass on cooling from the melt and shows only a glass transition on subsequent heating); U_L 428 kJ mol^{-1}} [278]. The effect on U_L is quite modest. The consequences of ring nitration can be seen from a comparison of [C_2mim][OTf] [58] and [C_2C_1(4-NO_2)im][OTf] (m.pt. 92°C; U_L 453 kJ mol^{-1}) [399] (see also Section 11.18.8).

There are relatively few pyridinium salts in this class. Several pyridinium salts with heavier p-block elements in side chains (Si, S or Se) have been crystallographically characterised and are discussed in Section 11.17. One or two melt below 100°C, though none contain the triflate counter-anion. However, in one of these reports [454], some of the starting materials have been isolated as triflate salts. [2,4-Dichloro-1-methylpyridinium][OTf], [4-chloro-2-dimethylamino-1-methylpyridinium][OTf] and [2-chloro-1-methyl-4-phenylaminopyridinium][OTf] were isolated, with m.pt. of 96–98, 57–59 and 98.5–102°C, respectively, though only the latter was subjected to single-crystal study (U_L 427 kJ mol^{-1}).

The only other pair worthy of particular note is [$SMePh_2$][OTf] (m.pt. 97°C; U_L 423 kJ mol^{-1}) [438] and [SMe_2Ph][OTf] [438], though the latter, melting at 109°C (U_L 451 kJ mol^{-1}), is outside the specified range of melting temperatures to qualify as an ionic liquid. These structures are reported [438] to display similar interactions between anion and cation, though with the S—Me interacting preferentially with the SO_3 end of the anion and S-Ph with the CF_3 end.

11.14.2 Other Fluoroalkylsulfonate Salts

The effect of changes in the anion can be seen in the trio of ionic liquids: [C_2mim][OTf] (m.pt. −26°C; U_L 459 kJ mol^{-1}) [58], [C_2mim][$C_4F_9SO_3$] (m.pt. 28°C; U_L 423 kJ mol^{-1}) [91] and [C_2mim][$HCF_2CF_2SO_3$] (m.pt. 45°C; U_L 453 kJ mol^{-1}) [91]. Chinese workers report [376] the structure of [SMe_3][$ICF_2CF_2OCF_2CF_2SO_3$] that melts at 89–91°C.

11.14.3 Methanesulfonate Salts (R = CH_3)

The only methanesulfonate ([OMes]$^-$) ionic liquid so far reported is [C_4mim][OMes] (m.pt. estimated to be 60–65°C; U_L 454 kJ mol^{-1}) [66], which can be compared with [C_4mim][OTf] (m.pt. 7°C; U_L 433 kJ mol^{-1}) [62], showing the higher lattice energy and m.pt. for the former. The structures of [C_4mim][OMes] and [C_4mim][SO_3(OMe)] (m.pt. −5°C [134]; U_L 444 kJ mol^{-1}) [66] may be compared. In both, the butyl group adopts the *tt* conformation (see Section 11.15).

11.14.4 Arylsulfonate Salts

Three 4-toluenesulfonate ionic liquids have been crystallographically characterised, one associated with the 1-cyclopentyl-3-methylimidazolium cation [257].

11.15 CARBOXYLATES

The ionic liquids discussed in this section are collected in Table 11.20. The crystal structures of protic ionic liquids derived from carboxylic acids have already been brought together in Table 11.6 and discussed in Section 11.4.1. Of the 67 salts listed in Table 11.20, only 11 are not formally the result of the reaction of a base and a carboxylic acid. The only crystal structure for a stibonium salt with a m.pt. < 100°C is that of $[SbMe_4][(PhCO_2)_2H]$, reported to melt at 80–82°C [327]. The structure of 1-(hydroxycarbonylmethyl)-1-methylmorpholinium salicylate (m.pt. 104–105°C) is noted [491].

The lowest-melting (protic or non-protic) carboxylate salt for which a crystal structure is reported is, perhaps surprisingly, a benzoate [90] with the cation 1-(2-hydroxyethyl)pyrrolidinium (m.pt. 28°C; U_L 448 kJ mol^{-1}). $[C_2C_2im]$[ethanoate], melting at 30°C (U_L 474 kJ mol^{-1}) [96], is the lowest melting of four ethanoate salts and the only non-protic example. In the latter, the C···O separation between C(2)—H and a carboxylate oxygen is 2.16 Å, indicating a hydrogen bond, the consequence of which is significantly different C—O bond lengths in the ethanoate group (1.258 and 1.245 Å). The single entries for propanoate, chloroethanoate and trichloroethanoate are associated with protic cations.

Among seven trifluoroethanoate salts, the only non-protic ionic liquid is [1-butyl-1-methylazepanium]$[CF_3CO_2]$ (m.pt. 74°C; Fig. 11.26) [283], with a value of U_L 474 kJ mol^{-1}, the lowest of the seven. Close anion–cation contacts involve the carboxylate oxygens and cationic C—H groups, α and β to the ring N. The mean O···H separation is 2.51 Å (compared with the sum of the van der Waals radii of 2.70 Å). In crystals of [1-butyl-2,3-dimethylimidazolium] $[HC{\equiv}CCO_2]$ (m.pt. 43°C; U_L 449 kJ mol^{-1}), obtained from the melt, the butyl group adopts a ***tt*** conformation and the C≡C—H hydrogen bonds to an adjacent anion [124].

Changes resulting from the replacement of O by S in a carboxylate may be revealed in a comparison of $[N_{2222}]$[ArCXY], where Ar = 2-hydroxyphenyl; X,

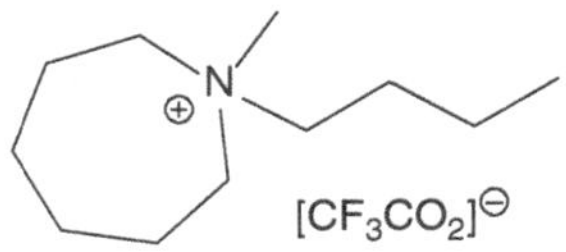

Figure 11.26 1-Butyl-1-methylazepanium trifluoroethanoate [283].

Y = O, S [305], giving $[N_{2222}][(2\text{-OH})C_6H_4CO_2]$ (m.pt. 95–98°C; U_L 426 kJ mol^{-1}), $[N_{2222}][(2\text{-OH})C_6H_4COS]$ (m.pt. 78°C; U_L 422 kJ mol^{-1}) and $[N_{2222}][(2\text{-OH})C_6H_4CS_2]$ (m.pt. 78–80°C; U_L 417 kJ mol^{-1}). Changes in anion size and electronegativity associated with the substitution may account for the trend in lattice potential energies. The cation conformation is D_{2d} in all three cases. In the anion, changes to the dihedral angle between the 2-hydroxyphenyl ring and the CXY group, as X and Y are changed from O to S, are much smaller than for the corresponding $[C_6H_5CXY]$ analogues, ascribed [305] to intramolecular hydrogen bonding in the former cases.

11.16 $[ER_4]^+$ AND $[E'R_3]^+$

The ionic liquids discussed in this section are collected in Table 11.21. $[ER_4]^+$ salts, where E = N (including the alicyclic heterocycles, pyrrolidinium, piperidinium and morpholinium), P, As or Sb, and $[E'R_3]^+$ salts, where E′ = S, Se or Te, are component cations in a large group of ionic liquids. Table 11.21 lists the 150 that have been structurally characterised, with a further 42 melting between 100 and 110°C. There are 51 acyclic quaternary ammonium ionic liquids (and 13 that melt between 100 and 110°C), 19 (7 that melt between 100 and 110°C) phosphonium and 9 (6 that melt between 100 and 110°C) ternary sulfonium, selenonium and telluronium ionic liquids. A further 8 (1 that melts between 100 and 110°C) salts of the type $[S(NMe_2)_3][A]$ and $[S(O)(NMe_2)_3][A]$ are also reviewed, as are 27 (4 that melt between 100 and 110°C) *N,N*-dialkylpyrrolidinium, piperidinium and morpholinium salts. Some aspects of the protic and solvated members of this group will be considered in the following text (though additional discussion can be found in Section 11.4). Pyridinium salts and the salts of other nitrogen-containing heterocycles are considered in Sections 11.17 and 11.19.5, respectively.

11.16.1 Protonated Alkylammonium Salts

These have already been collected in Table 11.6. The crystal structures and m.pt. for each member of the series $[NMe_nH_{4-n}][NO_3]$ (n = 0–4) have been reported, though only when n = 2 is the salt formally an ionic liquid. Ammonium nitrate, well studied because of its technological importance, may exist in one of five polymorphs (at atmospheric pressure) before melting at 169–170°C [711, 712]. Partial data are reported for $[N_{0111}][NO_3]$ [713] and $[N_{1111}][NO_3]$ [714, 715], though these are also both high melting [716]. Crystal structures of protonated primary and secondary ammonium nitrates may be compared: $[N_{0001}][NO_3]$ [539] (m.pt. 110°C; U_L 594 kJ mol^{-1}) and $[N_{0011}][NO_3]$ (m.pt. 80°C [622]; U_L 557 kJ mol^{-1}) [322], with the introduction of the second —CH_2— reducing lattice potential energy by 37 kJ mol^{-1} and m.pt. by 30°C. In $[N_{0001}][NO_3]$ (at room temperature), the cation has two equidistant neighbouring anions, with H···O separations of 1.87 and 2.19 Å, with all the nitrate oxygens taking part in hydrogen bonding. Solid–solid phase transitions occur at 355 and 245 K. In $[N_{0\,0\,1\,1}][NO_3]$ (at 173 K), one H···O separation of 1.98 Å is seen, with four other values greater than 2.25 Å.

$[N_{0\,1\,1\,1}][H_2F_3]$ (m.pt. 0°C; U_L 534 kJ mol^{-1}) [57] may be compared with $[N_{0\,1\,1\,1}][N(SO_2CF_3)_2]$ (m.pt. 79–81°C; U_L 445 kJ mol^{-1}) [314], where a large reduction in lattice potential energy is not, in this case, associated with a reduction in melting point. In the latter salt, the cation engages in weak bifurcated hydrogen bonding with two $[NTf_2]^-$ oxygens, though the relevant distances are close to the sum of the van der Waals radii, suggesting the dominant interactions are coulombic. The size and shape of $[NTf_2]^-$ (see Section 11.13.3) are believed [314] to render packing in the crystal less efficient than with spherical ions, such as iodide. The loss of hydrogen bonding associated with the introduction of the fourth methyl group into $[N_{0\,1\,1\,1}][H_2F_3]$, to give $[N_{1\,1\,1\,1}][H_2F_3]$ (m.pt. 110°C; U_L 519 kJ mol^{-1}; the crystal structure is of a low-temperature form stable below 85°C; a further solid–solid transition is seen at 94°C) [532], results in a marked increase in melting temperature but a reduction in lattice potential energy.

11.16.2 Tetraalkylammonium Salts

Tetraalkylammonium salts of the type $[NR_4][A]$, where $[A]^- = [ClO_4]^-$, $[IO_4]^-$ or $[BF_4]^-$, have been of some interest as plastic crystals [11, 12, 563], in which the approximately spherical nature of the cation gives rise to significant disorder in the crystal. Historically, as has already been noted, more general interest in $[NR_4][A]$ as liquids has been limited [1, 2] because of the observation of their relatively high-melting temperatures and low thermal stability in the liquid state and the belief that such materials were intrinsically unstable. The lowest-melting tetraalkylammonium halide for which a single-crystal X-ray structure has been reported (studied along with seven other $[NR_4]X$ and $[PR_4]X$ salts in which R is the same C_{10}–C_{18} linear alkyl chain [167]) appears to be $[N_{12\,12\,12\,12}]Br$, which melts at 88–92°C [167]. Interestingly, in the solid state of such materials, the alkyl chain can adopt higher-energy *gauche* conformations, the energy requirement for which being compensated by intramolecular van der Waals interactions between chains, worth about 8 kJ per CH_2 [167].

In this subgroup, there are five tetramethylammonium ionic liquids, $[NMe_4][A]$, whose crystal structures have been determined, in which (in order of increasing melting temperatures) $[A]^- = [Al\{OCH(CF_3)_2\}_4]^-$ (m.pt. 61°C) [170], $[N(SO_2F)_2]^-$ (m.pt. 72°C) [274], $[Al(^iBu)_2F_2]^-$ (m.pt. 81°C) [334], $[Al\{OC(CH_3)(CF_3)_2\}_4]^-$ (m.pt. 96°C) [170] and $[N\{P(O)(C_2F_5)_2\}_2]^-$ (m.pt. 98°C) [449]. By comparison, the m.pt. for $[NMe_4][N(SO_2CF_3)_2]$ is 134°C [274]. Interestingly, $[NMe_3H][N(SO_2CF_3)_2]$ [314] melts at 79–81°C and the $[NEt_4]^+$ [487] and $[NPr_4]^+$ [314] salts at 104 and 105°C, respectively, and for which the lattice potential energies, $U_L = 412$ and 390 kJ mol^{-1}, may be calculated from structural data.

There has been much interest in anions of the type $[Al(OR)_4]^-$ as weakly coordinating anions, a topic that has been reviewed [649]. The pair $[N_{1\,1\,1\,1}][Al\{OCH(CF_3)_2\}_4]$ (m.pt. 61°C; U_L 369 kJ mol^{-1}) [170] and $[N_{1\,1\,1\,1}][Al\{OC(CH_3)$

$(CF_3)_2\}_4]$ (m.pt. 96°C; U_L 359 kJ mol^{-1}) [170] provides an opportunity to compare the subtle effects of an additional CH_2-residue. The Al—O distances in these two salts are essentially the same, though the anion packing (C-centred vs. I-centred lattices) and cation environment (greater distortion of an octahedral environment for the *iso*-propoxy compound compared with that for the *tert*-butoxy compound) show distinct differences. The average ion volumes [77, 170, 583, 717] of these two anions are calculated to be 0.577 and 0.658 nm^3, respectively. Only weak H···F contacts between isolated cations and anions are evident. These salts are seen [170], therefore, as model systems in the study of weakly coordinating anions. Interestingly, $[N_{1\,1\,1\,1}][Al\{OC(CF_3)_3\}_4]$ (anion volume 0.736 nm^3 (calculated from the $[N_{2\,2\,2\,2}]$ salt) melts at 320°C [170].

The crystal structures of $[N_{1\,1\,1\,1}][N(SO_2C_nF_{2n+1})_2]$ have been reported for $n=1$–3, in a single study [274]. However, their melting temperatures are all greater than 120°C. Nevertheless, as this is one of the few sets of data that may reveal trends along this series, it is discussed in the $[NTf_2]^-$ section, Section 11.13.3.

A series of $[N_{2222}][A]$ salts may be compared, in which $[A]^- = [Al\{OCH(CF_3)_2\}_4]^-$ (m.pt. 56°C; U_L 356 kJ mol^{-1}) [170]; $[BF_4]^-$ (m.pt. 72°C [718]; data at RT: U_L 456 kJ mol^{-1}; data at 100 K: U_L 462 kJ mol^{-1} [273]; note added in proof – this is not in accord with SciFinder that gives a range of m.pt. much higher than this, typical of which is 382°C [719]); and $[N(SO_2CF_3)_2]^-$ [487] (m.pt. 104°C; U_L 412 kJ mol^{-1}).

$[N_{3\,3\,3\,3}][Cu^IBr_2]$ [459] is a minor product from a solution that first yields $[N_{3333}]_2[Cu_4Br_6]$. A salt with yet another composition is obtained when $[N_{2222}]^+$ is used as the counter-cation. $[N_{3\,3\,3\,3}][Cu^IBr_2]$ and $[N_{3\,3\,3\,3}][Cu^ICl_2]$ [460] both melt at 99–100°C with very similar values of U_L, 413 and 416 kJ mol^{-1}, respectively. The X–Cu–X angles in the two salts are also the same: 178.4° (X = Br) and 178.5° (X = Cl). Comparisons with salts obtained with other cations suggest [460] that the coordination number of copper and anion catenation both increase with decreasing cation size.

Table 11.21 also reveals a disparate set of ionic liquids containing $[N_{4\,4\,4\,4}]^+$, including $[N_{4\,4\,4\,4}]_n[A]$, $n=3$, $[A]^{3-} = [Sm\{C(CN)_2NO\}_6]^{3-}$ (m.pt. 59°C) [186] and $[Ce\{C(CN)_2NO\}_6]^{3-}$ (m.pt. 68°C) [186]; $n=1$, $[A]^- = [Si(C_6H_5)(CH_3)F_3]^-$ (m.pt. 62–64°C) [207]; $[PF_3(C_2F_5)_3]^-$ (m.pt. *ca.* 65°C) (UoL, unpublished work); $[Au(CN)_2]^-$ (m.pt. 65–69°C) [228]; $[Y(BH_4)_4]^-$ (m.pt. 78°C) [306]; 2,4-dinitroimidazolate (m.pt. 81°C) [332]; 4,5-dinitroimidazolate (85°C) [332]; $[InCl_4]^-$ (m.pt. 90°C [697]) [381]; $[AlMe_2F_2]^-$ (m.pt. 90°C) [382]; 2,3,6-tricyano-4-fluoro-5-trifluoromethylphenolate (m.pt. 90–92°C) [386]; $[Al\{OC(CH_3)(CF_3)_2\}_4]^-$ (m.pt. 108°C) [170] and $[C_5H_5]^-$ (m.pt. 108–109°C) [521].

One-to-one comparisons reveal possible consequences of replacing Sm by Ce in salts of trinegatively charged anions [186] and of moving a nitro residue from the 2- to the 4-position in an imidazolate ring [332]. While isomorphous, $[N_{4\,4\,4\,4}]_3[Sm\{C(CN)_2NO\}_6]$ and $[N_{4\,4\,4\,4}]_3[Ce\{C(CN)_2NO\}_6]$ show interesting thermochemical behaviour, with a *ca.* 50°C temperature difference in melting and crystallisation temperature. A lower-melting temperature is revealed during a second heating and cooling cycle. The authors suggest that a thermodynamically less stable polymorph may form on crystallisation in the DSC.

Importantly, they establish that the composition of solid and liquid is the same. Temperature-dependent (273–350 K) synchrotron X-ray powder diffraction studies (on a praseodymium analogue) support the interpretation that these observations [186] are associated with conformational changes in the cation butyl groups.

The structure of $[N_{4\,4\,4\,4}]$[2,4-dinitroimidazolate] (U_L = 387 kJ mol^{-1}) [332] may be compared with that of $[N_{4444}]$[4,5-dinitroimidazolate] (U_L = 389 kJ mol^{-1}) [332]; this reference also includes non-crystallographic studies of four other $[N_{4444}]^+$ salts of energetic anions. Interestingly, these two salts showed different thermochemical behaviour, with the 2,4-dinitroimidazolate salt showing simple, and the 4,5-analogue more complicated, behaviour. The 2,4-salt crystallises in the triclinic and the 4,5-salt in the monoclinic system. The former salt shows a difference in the close contacts between the cation and anion in the two ion pairs in the unit cell. The two salts are also distinguished by the degree of twist of the NO_2 group out of the plane of the imidazolate ring, being only slight (<6°) for the 2,4-salt but 32.7 and 18.8° for the 4,5-isomer, for which greater steric crowding is expected. The packing arrangements are also quite different, the former forming a layered arrangement and the latter having each ion surrounded by five counter-ions. There is, in the 4,5-salt, significant hydrogen bonding between the $-NO_2$ residues and $N-CH_2$ hydrogens on the cation.

The salt $[N_{4\,4\,4\,4}][BPh_4]$ (m.pt. 110.5–112°C) deserved special mention because of a useful paper that highlights the difficulties of analysing diffraction data and the errors that might arise [51].

The modest impact of changing one alkyl groups attached to *N* from propyl to *iso*-propyl can be seen from a comparison of $[NMe_2(Pr)Ph][I_3]$ [194] (m.pt. 60°C; U_L 420 kJ mol^{-1}) and $[NMe_2(^iPr)Ph][I_3]$ (m.pt. 70°C; U_L 420 kJ mol^{-1}) [194].

Cholinium chloride, $[NMe_3(CH_2CH_2OH)]Cl$, is readily available, very cheap and, being an essential nutrient, is generally accepted as safe [720]. This has excited interest in its use as a component of ionic liquids (though of course the nature of the anion is likely to alter the toxicity profile of the salt). Cholinium chloride itself melts with decomposition at 302°C. Some cholinium-related ionic liquids, based on cations of the type $[NMe_3(CH_2X)]^+$, X = CH_2OH [84, 97, 295, 297, 380] (UoL, unpublished work), CH_2Cl (UoL, unpublished work) or CO_2H [174], have been structurally characterised and those melting below 100°C are included in Table 11.21. These include a number of structurally related pairs that permit an examination of interionic interactions and other structural features that may provide clues to special characteristics, including $[NMe_3(CH_2CH_2OH)][A]$, $[A]^-$ = acesulfamate (Fig. 11.27; m.pt. 25°C; U_L 449 kJ mol^{-1}) [84], $[N(SO_2CF_3)_2]^-$ (m.pt. 30°C; U_L 424 kJ mol^{-1}) [97]; saccharinate (Fig. 11.27; m.pt. 69°C; U_L 439 kJ mol^{-1}) [84]; $[N(SO_2CH_3)_2]^-$ (m.pt. 76–77°C; U_L 449 kJ mol^{-1}) [297] and [bis(salicylato)borate]$^-$ (m.pt. 90°C; U_L 344 kJ mol^{-1}) [380]. In the $[NTf_2]^-$ salt, the cholinium cation adopts a folded conformation (with a N–C–C–O torsion angle of 63°) while the anion is found in the lower

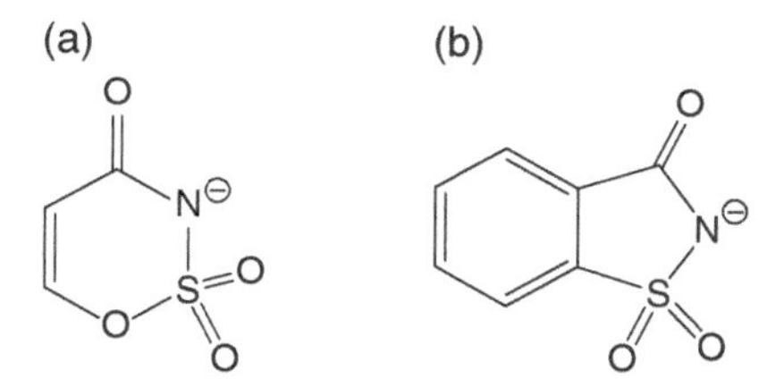

Figure 11.27 The anions (a) acesulfamate and (b) saccharinate [84].

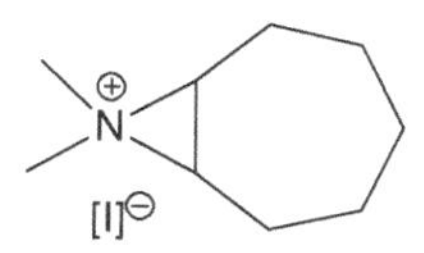

Figure 11.28 8,8-Dimethyl-8-azoniabicyclo[5.1.0]octane iodide [721].

energy *trans*oid conformation. A strong hydrogen bond (O(H)···O(S) = 2.84 Å; H···O 2.05 Å) is found between the OH proton and a sulfonyl oxygen. The arrangement is similar to that found in the non-fluorinated analogue [297], though with differing involvement of CF_3 and CH_3 in weaker interionic bonding. Attention has been drawn [297] to a so-called 'tripod' pattern, $^+N(CH_2—H\cdots)_3\cdots O$, between protons on each of the three cation methyls and an anion oxygen in the dimesylamide salt. A similar arrangement is described for $[NMe_3(CH_2CO_2H)][N(SO_2CH_3)_2]$ (though this has a melting range of 90–160°C [297]).

$[NMe_3(CH_2CO_2H)][N(SO_2CF_3)_2]$ crystallises slowly from the melt. Its crystal structure (m.pt. 57 or 69°C; U_L 427 kJ mol^{-1}) [174] reveals strong anion–cation hydrogen bonding with d(O···O) of 2.72 A, d(OH···O) 1.89 A. Interestingly, when the anion is $[PF_6]^-$, $[OTf]^-$ or $[C_6H_5CO_2]^-$, much higher-melting solids (m.pt. 159, 125, 144°C, respectively) are obtained. Attempts to crystallise $[NMe_3(CH_2CO_2H)][N(SO_2CF_3)_2]$ from solution in $[C_4mim][NTf_2]$ gives an inclusion compound instead. The latter, of composition $[\{NMe_3(CH_2CO_2H)\}_3\{NMe_3(CH_2CO_2\}][N(SO_2CF_3)_2]_3$, (m.pt. 70°C) contains the betaine zwitterion.

11.16.3 Alicyclic Quaternary Ammonium Compounds

Quaternary ammonium salts containing 3-, 4-, 5-, 6- and 7-membered saturated rings have been structurally characterised, some of which are ionic liquids.

Three fused ring *N*,*N*-dimethylaziridinium iodides, [(8,8-dimethyl-8-azoniabicyclo(5.1.0)octane]I (Fig. 11.28; refcode DMAZOI) [721], [9,9-dimethyl-9-azoniabicyclo(6.1.0)nonane]I (refcode MAZNOI) [722] and [*cis*-13,13-dimethyl-13-azoniabicyclo(10.1.0)tridecane]I [723] (MAZDCI10), have been characterised using X-ray crystallography, though all are high melting. Very recently,

$[NTf_2]^-$ and $[N(CN)_2]^-$ salts of [1-butyl-1-methylaziridinium]$^+$ and [1-ethyl-1-propylaziridinium]$^+$ have been reported [724], but so far, their crystal structures have not been investigated. [*N*,*N*-dialkylazetidinium]$^+$ salts have already been discussed in Section 11.6 on energetic salts.

More interest has been shown in ionic liquids derived from the 5- and 6-membered saturated nitrogen heterocycles, pyrrolidine and piperidine. *N*,*N*-dialkyl salts from these secondary amines display polymorphism (Section 11.2.3) as well as a propensity to frustrate attempts to crystallise them. Related materials (with some ring unsaturation, two heteroatoms or with fused ring structures) are discussed in Section 11.19.5.

A comparison of the consequences of *N*-alkylation is possible from a consideration of the differences in the crystal structures of the pair $[H_2pyr][N(SO_2CH_3)_2]$ (U_L 466 kJ mol^{-1}) [267] and $[Me_2pyr][N(SO_2CH_3)_2]$ (U_L 448 kJ mol^{-1}) [127] that melt at 80–90 and 40–55°C, respectively, for which alkylation results in a reduction in both m.pt. and U_L. The range recorded for the melting temperatures is a result of careful observation of the thermophysical properties of these compounds, a group that shows a propensity for glass and plastic crystal behaviour. Blaschette et al. [267] have studied the crystal structures of six secondary ammonium salts with dimesylamide (salts with $[NMe_2H_2]^+$ (m.pt. 126°C), $[NMeEtH_2]^+$ (m.pt. 75–77°C; U_L 471 kJ mol^{-1}), $[NEt_2H_2]^+$ (m.pt. 71–73°C; U_L 454 kJ mol^{-1}), [pyrrolidinium]$^+$ (m.pt. 80–90°C; U_L 466 kJ mol^{-1}), [piperidinium]$^+$ (m.pt. 95–97°C; U_L 455 kJ mol^{-1}) and [morpholinium]$^+$ (m.pt. 125–128°C)) and observe that, while the anion displays approximate C_2 symmetry in the six structures, a 'variety of unpredictable packing patterns' arise governed by the steric demands of moieties not involved in hydrogen bonding.

The consequences of replacing CH_3 by CF_3 should be evident in structural differences between the two pairs $[C_1C_1pyr][A]$, $[A]^- = [N(SO_2CH_3)_2]^-$ [127] and $[N(SO_2CF_3)_2]^-$ (XOMDIM) [314], m.pt. 40–55 and 132°C, respectively, and $[C_4C_1pyr][A]$, $[A]^- = [N(SO_2CH_3)_2]^-$ [52] and $[N(SO_2CF_3)_2]^-$ [62], m.pt. 37 and −11°C, respectively. These four compounds are compared in detail in [52] using Hirshfeld surface analysis, discussed further in Section 11.2.6. The major role played by C—H···O interactions was confirmed, but more subtle effects, such as C—F···π and C—H···π interactions, as well as the role of anion–anion interactions, were recognised.

While salts $[C_nC_1pyr][A]$ ($n = 1$–4, 6 or 7) are known, a series comparison is only possible when $[A]^- = [N(SO_2CF_3)_2]^-$ and $n = 2$ [389, 390], 3 (UoL, unpublished work), 4 [62] or 6 (UoL, unpublished work) that melt at 91 (data for two polymorphs: U_L 419 kJ mol^{-1} (collected at 213 K); U_L 423 kJ mol^{-1} (collected at 153 K)), 12, −11, 5 and 9°C, respectively. Because of their potential as battery electrolytes, mixtures of $Li[NTf_2]$ and $[C_nC_1pyr][NTf_2]$ have been investigated, including by X-ray crystallography [725]. The crystal structures of $Li_2[C_5C_1pyr][NTf_2]_3$ (m.pt. 76°C; refcode PAQTAE) and $Li_2[C_4C_1pyr][NTf_2]_3$ (PAQSUX) are characterised [725] by anion–anion cross-linking resulting from coordination to lithium ions.

The crystal structures of a relatively smaller number of piperidinium ionic liquids are known, with a single comparison possible between $[C_2C_1pip][N(SO_2CF_3)_2]$ [390] and $[C_3C_1pip][N(SO_2CF_3)_2]$ (UoL, unpublished work), melting at 91 and 12°C, respectively, with the corresponding values of the lattice potential energy of 416 and 385 kJ mol^{-1} for the two forms of the former and 408 kJ mol^{-1} for the latter. Complex phase behaviour and associated polymorphism of $[C_2C_1pyr][N(SO_2CF_3)_2]$ [389] (m.pt. 91°C) and $[C_2C_1pip][N(SO_2CF_3)_2]$ [390] (m.pt. 91°C) have been studied by Truelove and colleagues and are discussed in Section 11.2.3.

The effect of introducing an additional —CH_2— or —O— into the ring may be seen in the pyrrolidinium–piperidinium–morpholinium trio of salts. Single crystals of the dimesylamide salts of $[H_2pip]^+$ (m.pt. 95–97°C) and $[H_2mor]^+$ (m.pt. 125–128°C) were structurally characterised [267] along with $[H_2pyr][NMes_2]$ (m.pt. 80–90°C). These have already been discussed in terms of unpredictable packing arrangements. The consequences can be seen of changes in the nature of M and X in 1-methylmorpholinium salts of the dinegatively charged anions, $[MX_4]^{2-}$, when M = Mn(II), Co(II) or Cu(II) and X = Cl or Br, for a series of salts that melt at or above the 100°C boundary [370]. The structure of $[C_4C_1pip][PO_2F_2]$ (m.pt. 41°C) has also been described [128].

A notable study [397] of isomorphous salts of [2-(2,6-dichlorophenylamino) phenyl ethanoate]$^-$ (the *anti*-inflammatory API, diclofenac) with [*N*-(2-hydroxethyl)piperidinium]$^+$ (m.pt. 101°C; U_L 394 kJ mol^{-1}), [*N*-(2-hydroxethyl)-morpholinium]$^+$ (m.pt. 92°C; U_L 397 kJ mol^{-1}) and [*N*-(2-hydroxethyl) piperazinium]$^+$ (m.pt. 121°C; U_L 395 kJ mol^{-1}) allows a rare comparison of the structural consequences of replacing, in the heterocyclic ring, the 4-CH_2- moiety by 4-NH and 4-O. The twist angles between the two phenyl rings in the anion along the same series are 66.8, 68.3 and 66.8°. All the cations adopt the chair conformation, with the N—C—C—O torsion angles that describe the side-chain orientation being 52.4, 51.2 and 53.4°, respectively. Hydrogen bonding between carboxyl CO_2 and O—H and NH groups on the cation results in hydrogen-bonded pairs that, the authors speculate [397], survive into the melt. The additional O and N atoms in the cation are not involved in additional hydrogen bonding. There is little correlation evident between interionic coulombic potential energy and melting point, though this study shows that both ΔH_{fus} and ΔS_{fus} increase as CH_2 is replaced by O and NH, suggesting additional anion–cation hydrogen bonding in the melt for the latter two compounds. An analogous pyrrolidinium salt has also been studied [726], though no melting point is given.

11.16.4 Phosphonium Salts

Protic and solvated ionic liquids in this subgroup have been discussed in Section 11.4. The lowest-melting tetraalkylphosphonium salt, crystallographically characterised, is $[P_{10\,10\,10\,10}]Br$ (m.pt. 54–57°C; U_L 332 kJ mol^{-1}) [167]. The related $[P_{10\,10\,10}CH_2C(O)NH_2]Br$ and $[P_{10\,10\,10}CH_2CO_2]Br$ melt at 44.5 and

90°C [136], with U_L of 326 and 327 kJ mol^{-1}, respectively. $[PEtPh_3]^+$ partnered with the unusual doubly negatively charged anion, $[Se^{III}{}_4Br_{14}]^{2-}$, melts at 84°C and is found, from crystal data [345], to have a lattice potential energy of 709 kJ mol^{-1}. $[PMePh_3][I_3]$ melts at 105°C [501].

Three $[P_{4444}][A]$ may be compared, where $[A]^- = [SnPh_2Cl_3]^-$ (m.pt. 64–66°C; U_L 360 kJ mol^{-1}) [218], $[B(CN)_4]^-$ (m.pt. 73°C; U_L 376 kJ mol^{-1}) [277] or $[Sn(2\text{-}MeC_6H_4)_2Cl_3]^-$ (m.pt. 95–96°C; U_L 352 kJ mol^{-1}) [218]. Surprisingly, there are no data (within the ionic liquids assembled in this chapter) that permit comparisons between salts of the type [NRR′R″R‴][A] and [PRR′R″R‴][A].

The only tetramethylated cation among the heavier Group 15 elements that qualifies as an ionic liquid and that has been crystallographically characterised is $[SbMe_4][(PhCO_2)_2H]$ (m.pt. 80–82°C; U_L 399 kJ mol^{-1}) [327].

There are no tetraphenylammonium salts with which to compare $[PPh_4][A]$. The lowest melting of the latter (m.pt. 98–101°C; U_L 364 kJ mol^{-1}) has $[A]^- = [B(C_6F_5)(N_3)_3]^-$ [453]. The crystal structure of $[AsPh_4][CF_3C(O)C(CN)_2]$ (m.pt. 96–102°C; U_L 375 kJ mol^{-1}) [437] has been described.

Aluminium(III) chloride extracts chloride ion from the phosphonium ylide, $[^tBu_2PCl(CPh_2)]$, to give the salt, $[^tBu_2P{=}CPh_2][AlCl_4]$, which melts at 108–109°C [522].

11.16.5 Sulfonium and Other Group 16 Salts

Relatively few low-melting salts of the type $[E'R_3][A]$ (E′ = S, Se or Te) have been structurally characterised by X-ray crystallography, probably because of the nature of the precursors, R_2E', and the relative reactivity of the salts formed from them. However, three one-to-one comparisons can be made:

A comparison [115] of $[SMe_3][I_3]$ (m.pt. 37°C; U_L 468 kJ mol^{-1}) and $[SEt_3][I_3]$ (m.pt. −2°C; U_L 440 kJ mol^{-1}) in the solid state reveals that change of alkyl substituent results in only very minor changes in the local structure. In this comparison, lattice energies follow the m.pt. changes. Importantly, the studies show the retention of the composition in the melt. Similar interionic interactions are seen [438] in $[SMe_2Ph][CF_3SO_3]$ (m.pt. 109°C; U_L 451 kJ mol^{-1}) and $[SMePh_2][CF_3SO_3]$ (m.pt. 97°C; U_L 423 kJ mol^{-1}). The related acyclic compound, [1-phenylthiolanium]$[ClO_4]$ (m.pt. 90–91.5°C; U_L 454 kJ mol^{-1}) [384], is noted.

The structures of $[SeMe_3][A]$ [270] and $[TeMe_3][A]$ [359] may be compared for $[A]^-$ = $[SeCN]^-$ or $[N_3]^-$. The selenium compounds melt at 73–75°C (U_L = 513 kJ mol^{-1}) and 72°C (535 kJ mol^{-1}), respectively, and the tellurium analogues at 106–109°C (507 kJ mol^{-1}) and 110°C (522 kJ mol^{-1}).

In addition, single crystals of low-melting aminosulfonium salts of the type $[S(NMe_2)_3][A]$ and $[S(O)(NMe_2)_3][A]$ have been studied. Particularly notable examples have $[A]^-$ as a fluorophosphazenate anion [70, 215] or a pentacoordinate siliconate [139, 207, 396]. $[S(NMe_2)_3]$[*cyclo*-$P_6N_6F_{13}$] [70] is especially low melting (m.pt. < 0°C; U_L 379 kJ mol^{-1}), whereas an analogue with a smaller ring size $[S(NMe_2)_3]$[*cyclo*-$P_4N_4F_9$] [70] melts at 79°C, with U_L = 401 kJ mol^{-1}, though the main interest of the study was the characterisation of anion structure and bonding [70]. The same cation is found in the 'sandwich' salts, $[S(NMe_2)_3][Li(C_5H_5)_2]$ [511], that melts at 107°C, and $[S(NMe_2)_3]_2[C_5H_5][Na(C_5H_5)_2]$ (refcode: SURLUM; unstable at room temperature, but melting at 85°C). A 2:1 salt, $[S(NMe_2)_3]_2[P_3N_3F_5NPF_2NPF_2NPF_5]$ [215], melting at 64°C, has also been structurally characterised (Fig. 11.15).

Comparison [139, 396] of salts of the fluoride donor, $[S(NMe_2)_3][Si(CH_3)_3F_2]$ (m.pt. 92°C [653]; U_L 422 kJ mol^{-1}) and $[S(O)(NMe_2)_3][Si(CH_3)_3F_2]$ (m.pt. 57°C, slowly decomposes at room temperature; U_L 426 kJ mol^{-1}), reveals the consequence of introducing a sulfoxo moiety into the structure and showing, in both cases, C—H···F—Si contacts between anion and cation *N*-methyl. Comparisons with salts of these cations with other anions, such as $[HF_2]^-$ (m.pt. 45°C) and $[AsF_6]^-$ (m.pt. 252°C) [139], suggest that the cation symmetry (C_3 vs. *Cs*) may well be determined by anion–cation interactions. For $[S(NMe_2)_3][Si(CH_3)_3F_2]$, calculation of the charge distribution on the trigonal bipyramidal anion shows the Si to be the locus of +1.66 e charge and the fluorines 0.65 e negative [396]. $[S(NMe_2)_3][Si(C_6H_5)_2(1\text{-}C_{10}H_7)F_2]$ (m.pt. 91–92°C; U_L 369 kJ mol^{-1}) [207] and $[S(O)(NMe_2)_3][HF_2]$ [139] (m.pt. 45°C; U_L 469 kJ mol^{-1}) have both been characterised crystallographically.

11.17 PYRIDINIUM SALTS

The ionic liquids discussed in this section are collected in Table 11.22.

A range of protonated pyridines feature in salts whose structures are of interest in efforts to understand the role of hydrogen bonding in crystal engineering. Structural comparisons can be made between the protic salts $[Hpy][HCl_2]$ (m.pt. 60°C; U_L 518 kJ mol^{-1}) [195] and $[Hpy][HF_2]$ (m.pt. −1°C; U_L 548 kJ mol^{-1}) [61] and between $[H(4\text{-}X)py][CF_3CO_2]$, X = NC (m.pt. 74°C [727]) [284], H (m.pt. 82°C) [336] or CH_3 (m.pt. 101°C [727]) [284]. Values of U_L (see Refs. [585, 586] and discussion in Section 11.2.5) in the latter sequence are 484, 501 and 483 kJ mol^{-1} respectively, suggesting no obvious relationship with melting point. Pyridinium trifluoroethanoate [336] consists of three ion pairs in the solid, each with two hydrogen bonds: one between the pyridinium N—H and a carboxyl oxygen, the second between the second carboxyl oxygen and the pyridine C—H in the 2-position. The length of the N—H···O bond in this motif ($R_2^2(7)$ in Etter's notation [561, 562]) in the three compounds, 2.587 (X = NC), 2.681 (X = H) and 2.704 Å (X = Me), reflects the base strength of the pyridine. Other *N*-protonated pyridine ionic liquid structures have been listed in Table 11.6 and have been discussed with other protic materials in Section 11.4.

The lowest-melting structurally characterised *N*-alkyl pyridinium ionic liquid is $[C_6py][NTf_2]$ [62], melting at −4°C. Surprisingly, this is the only unsubstituted *N*-alkylpyridinium salt with this anion (though see the following text) for which a crystal structure has been reported. The crystal structures of $[C_4py]Cl$ [362] and $[C_4py]Br$ [426] have been described, with U_L = 478 and 470 kJ mol^{-1}, respectively. However, three melting points have been reported for each salt: 86, 103 and 162°C for the former and 95–97, 99 and 220–221°C for the latter. In both salts, the ions are linked by a network of C—H···Cl^- or C—H···Br^- interactions. The structure of [1-benzylpyridinium]Br [458] has been studied at 188, 208 and 218 K, with corresponding values of lattice potential energy of 465, 464.5 and 464 kJ mol^{-1}.

A series of salts between the weakly coordinating carboranyl anion, [*nido*-$C_2B_9H_{12}$]$^-$ and [$C_n$py]$^+$, with $n = 4$ (m.pt. 49°C; U_L 420 kJ mol^{-1}), 6 (m.pt. 72°C; U_L 405 kJ mol^{-1}), or 8 (m.pt. 80°C; U_L 396 kJ mol^{-1}) have been structurally characterised [147]. The additional —CH_2CH_2— moieties reduce lattice potential energy (by 15 and 9 kJ mol^{-1}, respectively) as m.pt. increases. The butyl and hexyl salts crystallise with a single ion pair in the asymmetric unit. For the [C_8py]$^+$ salt, cation disorder is seen, with the data being of poor quality. Particularly for the longer alkyl chains, association between alkyl groups is seen, leading to hydrocarbon-rich domains. No other simple pairwise comparison arises from a consideration of the crystal structures of these unsubstituted pyridines. However, [C_2py][I_3] (m.pt. 45–46°C; U_L 458 kJ mol^{-1}) [140] (one of an extensive series of alkylpyridinium triiodide salts, including materials liquid at room temperature [140]) and [C_2quin][I_3] (m.pt. 33–34°C; U_L 427 kJ mol^{-1}) [103] may be compared. In both salts, the [I_3]$^-$ ions are linked in chains with short contacts (compared with the sum of the van der Waals radii, 4.30 Å) between pairs of terminal iodines, 4.1088 Å for the pyridinium and 3.805 and 4.184 Å for the quinolinium salt.

Other *N*-substituted pyridinium compounds have been studied in which functionality is introduced into the *N*-bound alkyl group. However, first of all, we can observe that no comparisons are possible involving *N*-alkylpyridinium salts, [Rpy][A], unfunctionalised in the pyridine ring and in which R represents alkyl homologues and the cation is associated with an identical anion, [A]$^-$. Second, there are limited pairwise comparisons between salts with the same cation (e.g. [($HOCH_2CH_2$)py]X, where $X^- = Cl^-$ (m.pt. 66–68°C; U_L 511 kJ mol^{-1}) [237] and Br^- (m.pt. 110°C; U_L 499 kJ mol^{-1}) [534]; [($NCCH_2CH_2CH_2$)py][A], where $[A]^- = Cl^-$ (m.pt. 101°C; U_L 483 kJ mol^{-1}) [465] and [NTf_2]$^-$ (U_L 415 kJ mol^{-1}; reported to be a liquid at room temperature) [54].

[($HOCH_2CH_2$)py]Cl was studied as part of an investigation of the influence of electrostatic interactions on conformation and reactivity [237], though the effect is difficult to separate from associated hydrogen bonding. Na[(FCH_2CH_2)py][OTs]$_2$ (m.pt. 207–210°C; refcode: PIGGUI) and [(FCH_2CH_2)(4-NMe_2)py][OTs] (m.pt. 118–120°C; refcode: PIGHAP) were prepared in the same study. There is a preference for *gauche* conformation in the *N*-substituent in both of the latter higher melting salts, with the N—C—C—F torsion angle being 68.1 and 59.1°, respectively. Fluorine appears not to engage in intramolecular hydrogen bonding. The analogous N—C—C—O torsion angle in [($HOCH_2CH_2$)py]Cl is 61.2°. In [($HOCH_2CH_2$)py]Br [534], it is 63.6°. Bromide ions are located between the pyridinium rings and are also involved in hydrogen bonding to the terminal OH, with *d*(O—H···Br) = 2.386 A.

The next level of complexity considered arises from substitution in the pyridine ring. Here, we find a number of structures [92] with a common anion, $[NTf_2]^-$, which permits two series of close comparisons to be made of the consequences of introducing electron-withdrawing groups into the pyridine ring and into the *N*-alkyl side chain. $[C_1(Y)py][NTf_2]$, Y=3-CN (m.pt. 63–65°C; U_L 434 kJ mol^{-1}), may be compared with Y=4-CN (m.pt. 65–66°C; U_L 429 kJ mol^{-1}), though the data for the 4-cyano-salt are of poor quality. Crystals could not be obtained for the 2-cyano-salt. While there appears to be little difference in these parameters, the change of ring position of CN, nevertheless, has a profound effect on the conformation of the $[NTf_2]^-$ anion, adopting the lower-energy *trans*oid conformation in the 3-isomer (C—S—N—S—C dihedral angle 175.8°) and the *cis*oid conformation (dihedral angle 13.3°) in the 4-isomer [92]. The change is associated with a sheet structure for the 4-cyano-compound, with the layers of anions and cations separated by CF_3-rich regions. Cation–cation hydrogen bonds involving ring C—H and a neighbouring —CN nitrogen are also evident. The same group studied a second series of compounds [92], namely, $[e(Y)py][NTf_2]$, where Y=3-CF_3 (m.pt. 28–29°C; U_L 417 kJ mol^{-1}; rotational disorder in CF_3), 2-CN (m.pt. 36–38°C; U_L 418 kJ mol^{-1}), 3-CN (m.pt. 73°C; U_L 420 kJ mol^{-1}) and 4-CN (m.pt. 33–35°C; U_L 419 kJ mol^{-1}; anion disorder). In each of these compounds, the ethyl group is twisted perpendicular to the pyridinium plane and the anion adopts the *trans* conformation. The salt with Y=—CF_3 is distinguishable from the other materials studied by the absence of cation–cation hydrogen bonding. In the latter series, the four compounds have essentially the same value of the lattice energy, as estimated from the crystallographic data via VBT methods, whereas there is a wide range of melting point, highlighting the importance of other, non-coulombic (and possibly non-solid-state) origins of melting point differences. The crystallographic consequences of an additional N—CH_2- group can be seen in the comparison of $[C_1(3\text{-CN})py][NTf_2]$ and $[C_2(3\text{-CN})py][NTf_2]$ [92], with U_L reduced by 14 kJ mol^{-1}.

Anion effects can be noted in the series based on $[C_4(4\text{-Me})py][A]$, where $[A]^-=[PCl_6]^-$ (m.pt. 78–80°C; U_L 417 kJ mol^{-1}) [308], $[NbCl_6]^-$ (m.pt. 89–91°C; U_L 412 kJ mol^{-1}) [308] and $[TaCl_6]^-$ (m.pt. 103°C; U_L 412 kJ mol^{-1}) [294]. In the first two compounds, sheets of cations and anions pack together at van der Waals separations, with little significant hydrogen bonding between anion and cation. The butyl group adopts an all *cis* conformation with a C(8)—C(9) torsion angle of −47.6° for the $[PF_6]^-$ salt, −25.4° for the $[NbCl_6]^-$ salt and −79.4° for the $[TaCl_6]^-$ salt.

A further series [54], in which the effect of substitution of CN into the *N*-alkyl group can be seen, is $[(NCCH_2)py][NTf_2]$ (m.pt. 44°C; U_L 435 kJ mol^{-1}), $[(NCCH_2)(2\text{-Me})py][NTf_2]$ (m.pt. 51°C; U_L 426 kJ mol^{-1}), $[(NCCH_2)(2\text{-Et})py][NTf_2]$ (m.pt. 30°C; U_L 419 kJ mol^{-1}), $[(NCCH_2)(4\text{-Et})py][NTf_2]$ (m.pt. 48°C; U_L 420 kJ mol^{-1}) and $[(NCCH_2)(2,5\text{-Me}_2)py][NTf_2]$ (liquid; U_L 419 kJ mol^{-1}). The —CH_2CN groups lie within 1–5° of the plane of the pyridine ring, with cations interacting via CN···H bonds *ca.* 2.6 Å in length. The F_3C—S···S—CF_3 torsion angle for the anion in these five salts are, respectively, 167.5, 167.0, 168.4 and 164.0, 163.4 and 176.0° [54].

In both $[(NCCH_2CH_2CH_2)py][NTf_2]$ (liquid; U_L 416 kJ mol^{-1}) [54] and $[(NCCH_2CH_2CH_2)(3\text{-}Me)py][NTf_2]$ (liquid; U_L 412 kJ mol^{-1}) [54], the *N*-cyanopropyl chain lies perpendicular to the pyridine ring, with evidence of CN···π packing interactions between the cyano group and the pyridine ring. The $[NTf_2]^-$ anion adopts the *trans* conformation in both salts.

Two silylated materials can also be compared, revealing the consequences of replacing H at the 3- and 4-positions of the pyridine nucleus by Me: $[(Me_3Si)py]Br$ (m.pt. 27°C; U_L 466 kJ mol^{-1}) [86–88] and $[(Me_3Si)(3,4\text{-}Me_2)py]Br$ (m.pt. 87°C; U_L 449 kJ mol^{-1}; crystals grown by sublimation) [364]. The structural analysis establishes that the Si—Br separation is *ca.* 0.5 Å greater in the pyridinium salt than the sum of the van der Waals radii and 0.3 Å greater in the 3,4-lutidine compound. *N*-silylimidazolium compounds are noted in Section 11.18.10.

11.18 IMIDAZOLIUM SALTS

Table 11.23 contains nearly 200 structurally characterised salts based on the imidazolium cation (with a further 36 that melt between 100 and 110°C). These cover the complete range of subtypes, including 1:1, 2:1, 1:2 (with two imidazolium moieties connected by linking groups), 3:1 and 4:1 salts, protic compounds, 1,3-dialkyl- derivatives, [RR′im][A], in which R and R′ are simple or functionalised alkyl groups or less common ($-SiR_3$, -OR or -NR substituents), as well as cations containing ring functionality. Protonated imidazole and 1-alkylimidazole salts are discussed in Section 11.4, and salts of multiply charged ions in Section 11.8. While the data for $[C_nmim]X$ (where $n = 2$ or 4; X = Cl, Br or I) have already been discussed in Section 11.9.1 dealing with halide ions, Section 11.18.1 considers this again, briefly, from the perspective of the cation.

As structurally characterised imidazolium salts are a sizeable subgroup of ionic liquids, patterns or trends that may be evident could illuminate thinking about structure–property relationships, the most basic one of course, being melting point [41]. As discussed in Section 11.2, attempts to develop predictive correlations for melting point of ionic liquids have met with limited success [41–46, 567, 601]. Nevertheless, we will use the crystallographic data to estimate lattice potential energy, U_L, for both imidazolium ionic liquids as a group and 1-butyl-3-methylimidazolium salts as a subset and to test whether a correlation with melting temperature (from experiment, with all the attendant uncertainties previously highlighted) can be seen. A further characteristic of the molecular structure of alkylimidazolium salts in the solid state is the conformation of the alkyl group [680, 728–730]. The conformation of the *N*-alkyl residue may be a sensitive indicator of the local environment, with torsion angles available from many *N*-butyl salts from crystal structure data, including those for ionic liquids. These are also discussed in the following text. There are, no doubt, other patterns that may become evident from more detailed analysis of the original data for the compounds listed.

11.18.1 1-Alkyl-3-methylimidazolium Salts

Because they are easily prepared from readily available precursors, there has been a large number of 1,3-dialkyimidazolium salts reported. While the data should provide an opportunity to seek clues as to the effect of changes in *N*-bound groups, as well as the effect of ring substitution, particularly in the 2-position, trends associated with sequential introductions of methylene groups into alkyl chains are difficult to delineate because, despite there being 36 entries for 1,3-dimethylimidazolium salts and 14 for 1-methyl-3-propylimidazolium salts in the CSD, there are only four single-crystal structures for materials that qualify as ionic liquids containing the former [82, 134, 192, 244] and five containing the latter [112, 122, 246, 371]. As a result, there are few straightforward pairwise comparisons that can be made between salts of a cation of interest with a common counter-anion. Nevertheless, the availability of structural data for 37 1-ethyl-3-methylimidazolium (Table 11.23) and 27 1-butyl-3-methylimidazolium (Table 11.23) salts provides an opportunity for a series of comparisons involving salts $[C_2mim][A]$ and $[C_4mim][A]$, where $[A]^- = [CF_3SO_3]^-$, $[Au^{III}Cl_4]^-$ or $[Al\{OCH(CF_3)_2\}_4]^-$, and the 2:1 compounds $[Rmim]_2[Fe^{II}Cl_4]$ or $[Rmim]_2[Co^{II}Cl_4]$ (dealt with in Section 11.8 or 11.5.6.3).

A comparison of $[C_2mim][OTf]$ (m.pt. −26°C; U_L 459 kJ mol^{-1}) [58] and $[C_4mim][OTf]$ (m.pt. 7°C; U_L 433 kJ mol^{-1}) [62] suggests that introducing an additional $-CH_2CH_2-$ into *N*-Et reduces U_L by 26 kJ mol^{-1}. The butyl groups in the two unique cations adopt the *gauche* (***g***) conformation at N—**C**$_\alpha$—**C**$_\beta$—C—C and *trans* (***t***) at N—C—**C**$_\beta$—**C**$_\gamma$—C, abbreviated to ***gt***. This arrangement is a common feature found in crystals of butylimidazolium ionic liquids and a ***gt*** conformation is observed, unless otherwise stated. Introducing $-CH_2CH_2-$ into $[C_2mim][AuCl_4]$ (m.pt. 58°C; U_L 448 kJ mol^{-1}) [148] reduces U_L of $[C_4mim][AuCl_4]$ (m.pt. 50°C; U_L 531 kJ mol^{-1}) [148] by 17 kJ mol^{-1}. In both salts, the alkyl groups are disordered [148]. In both, the anions make up a linear arrangements of alternating corner-to-face $[AuCl_4]^-$ chains, with only weak C—H···Cl cation–anion interactions.

When $[A]^- = [Al\{OCH(CF_3)_2\}_4]^-$, the $[C_2mim]^+$ [77] and $[C_4mim]^+$ [104] salts melt at 31 and 34°C, respectively, with U_L reduced by 7 kJ mol^{-1}, though the validity of VBT for systems such as these, for which the electrostatic point charge model may be questionable and which may be significantly affected by secondary bonding interactions, suggesting that the significance of U_L (and their differences) for such compounds may be limited.

The longest sequence of 1-alkyl-3-methylimidazolium salts with a single anion involves those with $[PF_6]^-$: $[C_1mim]^+$, $[C_2mim]^+$, $[C_4mim]^+$, $[C_{10}mim]^+$, $[C_{12}mim]^+$ and $[C_{14}mim]^+$. These have been discussed in detail by Reichert et al. [99], both in relation to the nature of the interaction between cation and anion, as well as to the validity of estimating U_L and ΔG_L using VBT methods. The authors also analysed C—H···F cation–anion close contacts showing them, generally, to be very weak and more likely to arise from coulombic interactions rather than from 'true' hydrogen bonds.

The crystalline structure of $[C_1mim][NTf_2]$ is characterised by the adoption of *cis*oid conformation for the anion, the oxygens of which are involved in bifurcated C—H···O hydrogen bonds with the cation C(2)-proton. Paulechka et al. have used a combination of thermochemical IR and X-ray structural studies to investigate the crystalline modifications of $[C_2mim][NTf_2]$, $[C_4mim][NTf_2]$ and $[C_6mim][NTf_2]$ [59], distinguishing a second polymorph of $[C_2mim][NTf_2]$. One of the butyl groups in the unit cell of $[C_4mim][NTf_2]$ is disordered between ***gt*** and ***g't*** conformations, with the anion adopting the *cis*oid conformation. These two forms (crystal data collected at 120 K for one and 230 K for the other) have essentially identical U_L (426 kJ mol^{-1}), though they may be distinguished by the anion conformation, being *cis*oid [58] for one and *trans*oid [59] for the other. Yet a third form (U_L = 420 kJ mol^{-1}), with *trans*oid $[NTf_2]^-$, has recently been observed (UoL, unpublished work). Introducing a $—CH_2—$ into the NCH_3 group of $[C_1mim][NTf_2]$ (m.pt. 22°C; U_L 435 kJ mol^{-1}) [82] reduces U_L by 8–9 or 15 kJ mol^{-1}, respectively, depending on the form of $[C_2mim][NTf_2]$ with which the comparison is being made. The lattice potential energies of the forms of $[C_2mim][NTf_2]$ themselves differ by 6–7 kJ mol^{-1}. Introducing $—CH_2CH_2—$ makes further reductions in U_L of 13–14 ($[C_2mim][NTf_2]$ (form 1) to $[C_4mim][NTf_2]$ (m.pt. −3°C; U_L 413 kJ mol^{-1}) [59] (UoL, unpublished work)) and 11 kJ mol^{-1} ($[C_4mim][NTf_2]$ to $[C_6mim][NTf_2]$ (m.pt. −1°C; U_L 402 kJ mol^{-1}) [59] (UoL, unpublished work)). $[NTf_2]^-$ adopts a *cis*oid conformation in $[C_6mim][NTf_2]$.

The halide salts for the lighter members of this series are relatively high melting. $[C_nmim]X$ salts, with $n \geq 6$, are likely to fall outside the range of melting point considered for inclusion in this survey.

In the series of bromide salts of $[C_2mim]^+$ [112, 142, 263], $[C_3mim]^+$ [112] and $[C_4mim]^+$ [112, 230, 260], the introduction of $—CH_2—$ successively into a linear alkyl chain reduces U_L by 9–14 from ethyl (U_L 500–504 kJ mol^{-1}) to propyl (U_L 491 kJ mol^{-1}) and a further 14–18 kJ mol^{-1} from propyl to butyl (U_L 473–477 kJ mol^{-1}). The three structures for $[C_4mim]Br$ all show a ***gt*** conformation for the butyl group.

A difference in lattice energies roughly similar to that seen for the corresponding bromide salts is seen for the pair $[C_2mim]Cl$ (U_L 503–507 kJ mol^{-1}) and $[C_4mim]Cl$ (U_L 477–483 kJ mol^{-1}). The X-ray data for $[C_2mim]Cl$ were collected at different temperatures: room temperature (503 kJ mol^{-1}) [333], 190 K (505 kJ mol^{-1}) (S.Parsons, D. Sanders, A. Mount, A. Parsons and R. Johnstone, Cambridge Structural Database, private communication, 2005.), 160 K (506 kJ mol^{-1}) (G.J. Reiss, Cambridge Structural Database, private communication, 2010.) and 95 K (507 kJ mol^{-1}) (G.J. Reiss, Cambridge Structural Database, private communication, 2010.). In $[C_4mim]Cl$, it would appear that a delicate balance between coulombic, interionic hydrogen bonding, C—H···π and van der Waals interactions leads to the existence of two polymorphs, distinguishable by orientations in the butyl chain, ***gt*** and ***tt***. [230–232] (C. Pulham, J. Pringle, A. Parkin, S. Parsons, and D. Messenger, Cambridge Structural Database, private communication, 2005.).

Monoclinic and orthorhombic forms of [C_4mim]Cl, studied by two groups [230, 566], exemplify the subtle structural changes that may arise from changes in side-chain conformation. Interestingly, one study points to the monoclinic form as the more stable [566], the other the orthorhombic [230]. A computational study [567] suggests that the monoclinic form is the more stable, but the free energy difference between them (associated with a single *trans–gauche* conformational difference in the butyl group) is very small. Such small energy barriers between such conformations can result in the frustration of the extended packing necessary for crystal formation, a phenomenon that characterises a number of ionic liquids.

11.18.2 Isomeric *N*-Alkyl Groups

The difference in lattice potential energy between [C_3mim]Br (m.pt. 36°C; U_L 491 kJ mol^{-1}) [112] with a linear propyl group and data for (two) *iso*-propyl (1-methylethyl) salts, [$^iC_3$mim]Br (m.pt. 78°C; U_L 484 kJ mol^{-1} [307] and m.pt. 110.5°C; U_L 488 kJ mol^{-1}) [112] is 7 and 3 kJ mol^{-1}, respectively, though melting points differ quite markedly, with the more branched isomer being the higher melting. The two *iso*-propyl salts are distinguished by the conformation of the –$CHMe_2$ group with respect to the imidazolium ring [112, 307], even though the bromide ion positions with respect to the cation are nearly the same. The *N*-butyl salts also differ in melting point (again, the higher-melting salt contains the more branched alkyl group), and U_L reducing by 3 kJ mol^{-1} as –$CH_2CH_2CH_2CH_3$ (m.pt. 2 or 11°C; U_L = 453 kJ mol^{-1} [62]) is replaced by –$CH_2CH(CH_3)_2$ (m.pt. 83°C, U_L 450 kJ mol^{-1} [99]) in [C_4mim][PF_6].

11.18.3 Introduction of Phenyl Group into Side Chain

Successive introduction of phenyl groups at the terminal carbon of the *N*-ethyl group reduces U_L by 22–23 and then a further 24 kJ mol^{-1} along the series [C_2mim][NTf_2] [58, 59], [($PhCH_2CH_2$)mim][NTf_2] (m.pt. 41°C; U_L 404 kJ mol^{-1}) [129] and [(Ph_2CHCH_2)mim][NTf_2] (m.pt. 62°C; U_L 380 kJ mol^{-1}) [129]. Extending the alkyl chain by one –CH_2– unit, comparing [($PhCH_2CH_2$)mim][NTf_2] and [($PhCH_2CH_2CH_2$)mim][NTf_2] (m.pt. 50°C; U_L 397 kJ mol^{-1}) [129] reduces U_L by 7 kJ mol^{-1}.

For the [BF_4]$^-$ salts, [C_2mim][BF_4] (m.pt. –1°C; U_L 485 kJ mol^{-1}; data collected at 173 K [62]; U_L 489 kJ mol^{-1} (data collected at 100 K [67])) and [($PhCH_2CH_2$)mim][BF_4] (m.pt. 61°C; U_L 445 kJ mol^{-1}) [202], a single phenyl residue reduces U_L by 40–44 kJ mol^{-1}, a much more marked effect than that seen for the [NTf_2]$^-$ analogues.

When the anion is [PF_6]$^-$, introduction of one phenyl residue into each of the two methyl groups of [C_1mim][PF_6] (m.pt. 67–78°C (another report [99] gives the m.pt. as 130°C); U_L 483 kJ mol^{-1}) [244] to give [($PhCH_2$)$_2$im][PF_6] (m.pt. 93°C; U_L 412 kJ mol^{-1}) [405] reduces U_L by 71 kJ mol^{-1}, the latter being

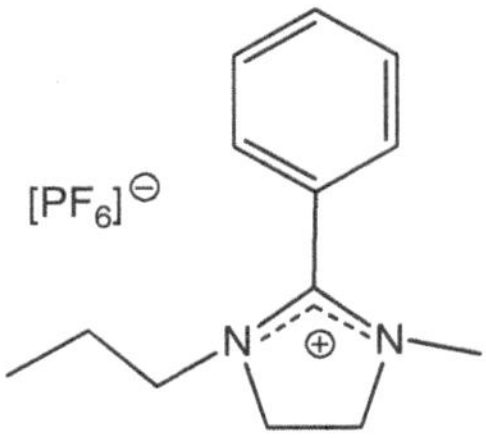

$[PF_6]^{\ominus}$

Figure 11.29 1-Methyl-2-phenyl-3-propylimidazolinium hexafluorophosphate [206].

significantly more than introducing two phenyl groups onto the terminal carbon of the ethyl group in $[C_2mim][NTf_2]$ (46–47 kJ mol^{-1}). Unfortunately, the crystal structure of $[(PhCH_2)mim][PF_6]$ has not been determined.

The structure of [1-methyl-2-phenyl-3-propylimidazolinium]$[PF_6]$ (m.pt. 62–64°C, Fig. 11.29), one of a series of nine crystal structures from 53 amidinium and related salts [206], displays no specific anion–cation interactions. The phenyl ring and the imidazolinium rings are tilted by 51° with respect to one another. While this could provide a means of assessing the impact of imidazolium ring hydrogenation, the corresponding structural data for [1-methyl-2-phenyl-3-propylimidazolium]$[PF_6]$ appear not to be available.

11.18.4 Introduction of Unsaturation into Side Chain

The structural effect of introducing –C=C– into an *N*-bound alkyl group can be seen in a comparison of $[(CH_2{=}CHCH_2)mim]Br$ (m.pt. 60°C; U_L 495 kJ mol^{-1}) [190] and $[C_3mim]Br$ (m.pt. 36°C; U_L 491 kJ mol^{-1}) [112]. The series $[(CH_2{=}CH)mim]I$ (m.pt. 78°C; U_L 496 kJ mol^{-1}) [300], $[C_2mim]I$ (m.pt. 81°C; U_L 488 kJ mol^{-1}) [263], $[(CH_2{=}CHCH_2)mim]I$ (m.pt. 65°C; U_L 480 kJ mol^{-1}), $[(CH_2{=}CHCH_2)C_2im]I$ (m.pt. 31°C; U_L 472 kJ mol^{-1}) and $[(CH_2{=}CHCH_2)C_3im]I$ (m.pt. 32°C; U_L 457 kJ mol^{-1}) [98] may also be compared. Side-chain unsaturation appears to have only a modest effect on U_L, with the impact on lattice potential energies being greater in the vinyl–ethyl pair (a difference of 8 kJ mol^{-1}) than in the allyl–propyl pair (a difference of 4 kJ mol^{-1}), though with differing effects on melting point. U_L is also reduced, respectively, by 8 and then a further 15 kJ mol^{-1} by the successive introduction of $-CH_2-$ groups into an alkyl side chain. The effect of introducing a methylene group between C and N in a vinylimidazolium salt is more substantial, with U_L reduced by 16 kJ mol^{-1} on going from $[(CH_2{=}CH)mim]I$ to $[(CH_2{=}CHCH_2)mim]I$. The lattice of $[(CH_2{=}CHCH_2)C_3im]I$ differs from that of both $[(CH_2{=}CHCH_2)mim]I$ and $[(CH_2{=}CHCH_2)C_2im]I$ with the presence of 'extensive' [98] π–π stacking between imidazolium rings in the former and its absence in the latter. The crystal structure of $[C_3mim]I$ (m.pt. 17°C) has not yet been reported. The lowest-melting allylimidazolium salt for which a crystal structure is available is $[(CH_2{=}CHCH_2)mim][Al\{OCH(CF_3)_2\}_4]$ [77], which melts at 12°C.

The Cambridge Structural Database contains at least 16 records of imidazolium salts containing the —C≡C— moiety in an *N*-bound side chain. Most are reported either without a melting point or with melting temperatures outside the range covered by this review. Among the latter, which may be of interest, are 1-methyl-3-propynyl- and 1-(but-2-ynyl)-3-methylimidazolium bromides [190] (m.pt. 117 and 131 °C, respectively; refcodes YIHTEP and YIHTIT). Two 1,3-bis(propynyl)imidazolum salts are reported, with the counter-anions $[BPh_4]^-$ (m.pt. 185°C, refcode AVUGUT) [731] and Br^- (no m.pt., refcode MIXVOF) [732]. The analogous chloride salt melts at 142°C [732]. Two proper ionic liquid alkynylimidazolium salts have been crystallographically characterised, [1-(pent-2-ynyl)-3-methylimidazolium][$N(CN)_2$] [185] (m.pt. 59°C) and [1-(but-2-ynyl)-3-methylimidazolium][N_3] [234] (m.pt. 66°C). Very recently, the structure of [1-propynyl-2,3-dimethylimidazolium]Br [733] has been reported, though without a m.pt.

11.18.5 Replacement of C(2)—H by C(2)—CH_3

The role of hydrogen bonding has been much discussed in relation to ionic liquid behaviour and interionic interactions, both in the liquid and solid state. The authors of [119] (a study that includes a series of 1-butyl-2,3-dimethylimidazolium salts) conclude that, in crystal packing in the 2-methylimidazolium derivatives, hydrogen bonding dominates over coulombic and dispersive forces. They also pointed out the difficulty of devising simple correlations between melting point, the number of C—H···X contacts and anion charge density.

Studies have extended to a consideration of weaker secondary bonding involving the relatively acidic 2-proton in imidazolium ionic liquids. A number of comparisons are therefore possible between 1,3-disubstituted imidazolium salts and their 2-alkylated analogues. There are four possible direct comparisons of [bmim][A] with [$C_4C_1C_1$im][A], when $[A]^- = Cl^-$, Br^-, I^- or $[PF_6]^-$. In addition, [(CH_2=$CHCH_2$)mim]Br (m.pt. 60°C; U_L 495 kJ mol^{-1}; data collected at 100 K) [190] may be compared with [(CH_2=$CHCH_2$)C_1C_1im]Br (m.pt. 61 °C; U_L 472 kJ mol^{-1}; data collected at 293 K) [119] and [($NCCH_2CH_2CH_2$)mim][PF_6] (m.pt. 75°C; U_L 455 kJ mol^{-1}) with [($NCCH_2CH_2CH_2$)C_1C_1im][PF_6] (m.pt. 85°C; U_L 445 kJ mol^{-1}) [289]. Interestingly, the N-$CH_2CH_2CH_2$CN groups adopt different conformations in the crystal [289], ***g*** at N—$\mathbf{C_\alpha}$—$\mathbf{C_\beta}$—C—CN and ***t*** at N—C—$\mathbf{C_\beta}$—$\mathbf{C_\gamma}$—CN (***gt***) for the former and ***tt*** for the latter.

Two forms of [C_4mim]Cl have been isolated, one from solution, the other from the melt. They differ only in the conformation of the butyl group, ***gt*** or ***tt***, resulting in different patterns of hydrogen bonding between cation and chloride [230]. This report was the first direct evidence of the role of alkyl chain conformation in the frustration of crystal packing in ionic liquids. Three other X-ray crystallographic studies of [C_4mim]Cl have been reported [231, 232] (C. Pulham, J. Pringle, A. Parkin, S. Parsons, and D. Messenger, Cambridge Structural Database, private communication, 2005.). Saha et al. [231], in a combined study of the liquid and crystalline state using X-ray crystallography and Raman spectroscopy, suggest that the aggregates of [C_4mim]Cl seen in the crystal survive in the melt. The structure of [$C_4C_1C_1$im]Cl (data collected at room temperature) reported by Kölle and Dronskowski [119] reveals a ***gt*** conformation for the alkyl chain with an ion-pair dimer arrangement being proposed from a consideration of the stacking of imidazolium rings. This study examined a number of related [$C_4C_1C_1$im][A] salts, with the authors concluding that a simple relationship between m.pt. and features such as the number of C—H···A contacts was not evident. Two forms of [$C_4C_1C_1$im]Cl, both isolated from propanone solution and differing in butyl group conformation, ***gt*** and ***tt***, have been reported, without discussion, in the electronic supplementary information associated with [311]. A fourth study (C. Pulham, J. Pringle, A. Parkin, S. Parsons, and D. Messenger, Cambridge Structural Database, private communication, 2005.) can be found in a private communication deposted in the CSD. Lattice potential energies derived by VBT methods are 477–483 and 464–472 kJ mol^{-1} for [bmim]Cl and [$C_4C_1C_1$im]Cl, respectively.

The three sets of data for [C_4mim]Br (m.pt. 70°C; U_L 473 kJ mol^{-1}; data collected at 283–303 K) [260], (m.pt. 79°C; U_L 477 kJ mol^{-1}; data collected at 110 K) [112] and (m.pt. 79°C; U_L 475 kJ mol^{-1}; data collected at 173 K) [230] may be compared with the corresponding data for [$C_4C_1C_1$im]Br (m.pt. 97°C [680]; U_L 466 kJ mol^{-1}; data collected at 120 K) [440]. The butyl group adopts the ***gt*** conformation in each case. The same data for [bmim]I (m.pt. 19°C; U_L 468 kJ mol^{-1}; data collected at 93 K; ***gt*** butyl conformation) [81] may be compared with corresponding data for [$C_4C_1C_1$im]I (m.pt. 98°C [680]; U_L 458 kJ mol^{-1}; data collected at 120 K; ***tt,tt*** butyl conformation) [440]. The cations in [$C_4C_1C_1$im]Br and [$C_4C_1C_1$im]Cl stack through π–π interactions, though these are absent in the crystal of the iodide salt.

Introduction of a methylene group into [C_4mim][PF_6] to give [$C_4C_1C_1$im][PF_6] sees an increase in m.pt. from 2 or 11 to 40°C, while the lattice potential energy changes from 452–453 kJ mol^{-1} [62, 72] to 444–445 kJ mol^{-1} [123] (C. Pulham, J. Pringle, A. Parkin, S. Parsons, and D. Messenger, Cambridge Structural Database, private communication, 2005.). The conformation of the *N*-butyl residue (***gt***) is unaffected.

Overall, U_L covers a relatively narrow range, considering the large spread of melting points. However, with the exception of the allylimidazolium pair, for which the reduction is somewhat larger, for all pairs, the reduction in lattice potential energy as a consequence of replacing C(2)—H by C(2)—Me is *ca.* 10 kJ mol^{-1}.

The series of iodide salts of 1-C_nH_{2n+1}-2-methyl-3-benzylimidazolium, where $n = 2, 3, 4$ or 5, has been studied [203], with U_L decreasing from 439 to 436 to 426 and to 416 kJ mol^{-1}. The introduction of successive —CH_2— groups leads to a 3, 10 and 10 kJ mol^{-1} reduction in lattice potential energy. In this instance, the changes appear to parallel reductions in melting point. A *sec*-butyl analogue melts at 135°C, consistent with other observations that suggest that the more highly branched the alkyl group, the higher is its melting point. In the butyl and pentyl salts, the *N*-alkyl chains adopt different conformations, *trans* (***t***) conformation at N—$\mathbf{C_\alpha}$—$\mathbf{C_\beta}$—C—C and *gauche* (***g***) at N—C—$\mathbf{C_\beta}$—$\mathbf{C_\gamma}$—C for the butyl derivative and ***gg*** and ***tt*** in the two pentyl cations in $[(bz)C_1C_5im]I$. The cation–anion interactions have been discussed in terms of Hirshfeld surface analysis summarised in Section 11.2.6.

11.18.6 Symmetrical Dialkylimidazolium Salts

A further subgroup of imidazolium salts contains the same alkyl residue on both ring nitrogen atoms of imidazolium, namely, compounds of the type $[R_2im][A]$. Table 11.23 lists the 19 salts of seven alkyl members of this series, with R = Me [82, 134, 192, 244], Et [96], Pr, [112, 243], ^{i}Pr [143, 523], Bu [457], ^{t}Bu [53] or $C_{12}H_{25}$ [111, 213], though it is perhaps surprising that no pair of compounds has been studied crystallographically with the same anion. On the other hand, the structures of the pairs $[(PhCH_2)_2im][PF_6]$ (m.pt. 93°C; U_L 412 kJ mol^{-1}) [405], $[(PhCHMe)_2im][PF_6]$ (m.pt. 108°C; U_L 404 kJ mol^{-1}) [517] and $[(PhCHMe)_2im][BF_4]$ (m.pt. 110°C; U_L 410 kJ mol^{-1}) [517] may be compared. The additional methylene group, α to the ring nitrogen, reduces U_L by 8 kJ mol^{-1}. When $[PF_6]^-$ is replaced by $[BF_4]^-$ when paired with $[(PhCHMe)_2im]^+$ U_L increases by 6 kJ mol^{-1}. The absence of π–π stacking and weak hydrogen bonding in $[(PhCH_2)_2im][PF_6]$ is ascribed to the size of the anion [405]. Dupont and colleagues [517] interpret the structures of the $[BF_4]^-$ and $[PF_6]^-$ ($[A]^-$) salts of the 1,3-bis(1-phenylethyl)imidazolium cation ($[cat]^+$) in terms of a polymeric supramolecular arrangement of the type $\{[cat]_3[A]^{2+}[cat][A]_3]^{2-}\}$.

The nine $[(C_{12})_2im][A]$ salts [111, 213] provide an opportunity of assessing the structural impact of changing $[A]^-$ among I^- (m.pt. 88.5°C), $[I_3]^-$ (m.pt. 46°C), $[I_5]^-$ (m.pt. 36°C), $[SbF_6]^-$ (m.pt. 41°C), $[BF_4]^-$ (m.pt. 67°C) [213], $[B(CN)_4]^-$ (m.pt. 48°C), $[C(CN)_3]^-$ (m.pt. 44°C), $[N(CN)_2]^-$ (m.pt. 44°C) and $[ClO_4]^-$ (m.pt. 64°C) [213]. Bearing in mind the failure of VBT in generating lattice potential energies consistent with experimentally and more rigorously computed values for imidazolium salts with alkyl chains C_nH_{2n+1}, when $n > 8$, the fact that U_L values for these nine compounds (m.pt. 36–88.5°C), with such a diversity of anions, span the range 346–363 kJ mol^{-1}, suggests that interionic interactions are influenced by factors other than coulombic. The two C_{12} side chains adopt three different molecular arrangements, giving rod, 'V' and 'U' shapes, dependent on the anion, $[A]^-$. In addition, side-chain conformation (represented by the torsion angles, N—$\mathbf{C}_\alpha$—$\mathbf{C}_\beta$—C—C and N—C—$\mathbf{C}_\beta$—$\mathbf{C}_\gamma$—C) varies, all *trans* for $[A]^- = [I_3]^-$, $[I_5]^-$, $[SbF_6]^-$, $[B(CN)_4]^-$ and $[ClO_4]^-$ and *gauche-trans* for $[C(CN)_3]^-$ and $[N(CN)_2]^-$. For I^- and $[BF_4]^-$, three out of the four dodecyl groups adopt ***tt*** and one ***gt*** conformations. These structures have also been subjected to Hirshfeld surface analysis (see Section 11.2.6).

11.18.7 Functionalised Alkyl Groups

For a variety of reasons, there is growing interest in introducing functionality into the imidazolium ring and into *N*-bound alkyl groups. Most of the materials that have been structurally characterised are 'one-offs'. For instance, $[R_2im][A]$ (R = $NCCH_2$—, HO_2CCH_2— and $NCCH_2CH_2CH_2$—) have been studied, associated with $[BF_4]^-$ (m.pt. 110°C) [309], $[ClO_4]^-$ (m.pt. 110°C) [537] and $[PF_6]^-$ (m.pt. 79°C) [309] counter-ions, respectively. However, some pairs permit a limited assessment of structural impact of these changes. For example, the consequences of the introduction of an additional $-CH_2-$ group and the replacement of $-CH_2-$ by —O— can be investigated using these data. $[(MeOCH_2CH_2OCH_2)mim]I$ (m.pt. 63°C; U_L 460 kJ mol^{-1}) can be compared with $[(MeOCH_2CH_2OCH_2CH_2)mim]I$ (m.pt. 70°C; U_L 451 kJ mol^{-1}) [149] and $[(MeOCH_2)mim]I$ (m.pt. 75°C; U_L 484 kJ mol^{-1}) with $[(MeOCH_2CH_2)mim]I$ (m.pt. 50°C; U_L 473 kJ mol^{-1}) [149]. An additional methylene group α to the ring nitrogen again reduces lattice potential energies by amounts similar to those found previously.

Converting $-O-CH_3$ to $-O-C_2H_5$ to give $[(EtOCH_2)mim]I$ (m.pt. 68°C; U_L 471 kJ mol^{-1}) [149] yields a very slightly bigger effect (though probably at the limit of significance). Shifting the ether linkage to a position one more remote from the imidazolium ring ($[(MeOCH_2CH_2)mim]I$ to $[(EtOCH_2)mim]I$) has only a marginal effect on lattice potential energies, though the melting point changes from 50 to 68°C, while a more significant reduction in U_L can be seen from the rare comparison involving the 2:1 salts $[(MeOCH_2CH_2)mim]_2[PdCl_4]$ (m.pt. 107°C; U_L 1052 kJ mol^{-1}) and $[(EtOCH_2)mim]_2[PdCl_4]$ (m.pt. 103°C; U_L 1036 kJ mol^{-1}) [480].

Replacement of I by Br in $[(MeOCH_2CH_2OCH_2CH_2)mim]I$ to give $[(MeOCH_2CH_2OCH_2CH_2)mim]Br$ [149] results in a very modest increase in U_L, from 451 to 455 kJ mol^{-1}, whereas the effect on this parameter is greater in the pair $[(NCCH_2)mim][A]$, $[A] = [N(CN)_2]$ and $[NO_3]$ [241]. Here, U_L increases from 484 to 502 kJ mol^{-1} and the melting point from 67 to 103°C.

11.18.8 Substitution in the Imidazolium Ring

The effect of introducing a ring nitro group can be seen from a comparison of two $[MeOSO_3]^-$ salts, $[C_1mim][MeOSO_3]$ (m.pt. 43°C; U_L 480 kJ mol^{-1}) [134] and $[C_1C_1(4\text{-}NO_2)im][MeOSO_3]$ (m.pt. 95/104°C; U_L 472 kJ mol^{-1}) [399], the crystal data having been collected at the same temperature. The consequences of ring nitration can also be seen from a comparison of $[C_2mim][OTf]$ [58] and $[C_2C_1(4\text{-}NO_2)][OTf]$ [399] (m.pt. 92°C; U_L 453 kJ mol^{-1}). It is worth pointing out that the crystal structures of the analogous nitro-compounds $[C_1C_1(2\text{-}NO_2)im][OTf]$ (m.pt. 112–114°C; refcode: GEDLOR), $[C_1C_1(2\text{-}Me\text{-}4\text{-}NO_2)im][NTf_2]$ (m.pt. 111–112°C) and $[C_2C_1(4\text{-}NO_2)im][OTf]$ (m.pt. 114–117°C; refcode: GEDLEN) are reported in the same study.

The effect of ring bromination may be evident in the structural differences between $[C_4mim][OTf]$ and $[C_4C_1(4,5\text{-}Br_2)im][OTf]$ [278] (m.pt. 73°C; on first heating only – gives a glass on cooling from the melt and shows only a glass transition on subsequent heating; U_L 428 kJ mol^{-1}). The impact on U_L is modest. A 4,5-diodo analogue is also known (m.pt. 125°C; refcode VURLAW) [278], which (from solution-crystallised material) also forms a glass on cooling from the melt, though on subsequent warming this crystallises to give the same polymorph.

Whittell et al. report the crystal structure of $[(^tC_4)_2(4\text{-}SiMe_3)im][OTf]$ (refcode XAGSUV) [734], though without a m.pt.

11.18.9 *N*-Amino and *N*-Hydroxyimidazolium Salts

Eight ionic liquid *N*-amino and *N*-hydroxy salts of imidazole have been prepared and characterised, permitting the following comparisons in which hydrogen-bond donor and acceptor groups are linked directly to the imidazolium ring nitrogens: $[(H_2N)_2(2\text{-}Me)im][NTf_2]$ (m.pt. 58–59°C; U_L 433 kJ mol^{-1}); data collected at 173 K) [8] with $[(HO)_2(2\text{-}Me)im][NTf_2]$ (m.pt. 72°C; U_L 435 kJ mol^{-1}; data collected at 233 K). The consequences of adding two $-CH_2-$ groups can be seen in the comparison of $[(MeO)_2im][PF_6]$ (m.pt. 83–84°C; U_L 466 kJ mol^{-1}) with $[(EtO)_2im][PF_6]$ (m.pt. 99–102°C; U_L 444 kJ mol^{-1}) [93]. Replacing $[PF_6]^-$ in the latter by $[BF_4]^-$ results in an increase in U_L by 16 kJ mol^{-1} and a reduction in m.pt. to 40–46°C [93]. The crystal structures of [2-bromo-1,3-dimethoxymidazolium]$[NTf_2]$ [93] (m.pt. 28–30°C) and [1,3-di(benzyloxy)-imidazolium]Br [198] (m.pt. 60–70°C) have also been reported.

11.18.10 *N*-Silylated Imidazolium Salt

Very recently, [1-methyl-3-trimethylsilylimidazolium]$[BF_4]$ (m.pt. 35°C) was crystallised under pressure and its crystal structure (at ambient pressure) determined [110]. The structures of $[(Me_3Si)_2im][OTf]$ (refcode RAHZEH) and $[(Me_3Si)_2im]Br$ (refcode NIDFAU [735]) are reported without m.pt., though the latter sublimes at 28–29°C under vacuum. They are of particular interest as they highlight the lability of compounds of the type $[(Me_3Si)_2im][A]$ that are in equilibrium with $[(Me_3Si)im]$ and $[Me_3SiA]$. As a result, in such cases, fluorinated anions, such as $[BF_4]^-$ and $[PF_6]^-$, cannot be used as $[A]^-$.

11.19 SALTS OF CARBON-, BORON-, HALOGEN- AND PNICTOGEN-CENTRED CATIONS AND OF SULFUR- AND ARSENIC-CONTAINING HETEROCYCLES

There are, in Table 11.24, over 60 salts (with 66 crystal structures) melting below 100°C and formally qualifying as ionic liquids but which contain cations (a total of 74, some multiply charged) not generally associated with ionic liquid behaviour. A further nineteen are listed with m.pt. in the range 101–110°C. Most are partnered with conventional anions and may have been surveyed in the relevant section of this chapter. Here, the focus is on the cation. Many are, synthetically, relatively inaccessible or have chemical characteristics that may make them unattractive in ionic liquid applications. However, the remainder may not and may provide the starting point for the discovery and study of materials hitherto unevaluated. They are discussed briefly in the following text, only outstanding or novel structural features being noted. Details can be found in either the chemical structural database (via the refcode) or from the original reference.

11.19.1 Carbon-Centred and Related Cations

Mootz and co-workers have characterised the acetatidium salt, $[CH_3C(OH)_2][CF_3SO_3]$, and the incongruently-melting $[(CH_3C(OH)_2.HOC(O)CH_3)][CF_3SO_3]$ obtained from the system $CH_3COOH–CF_3SO_3H$ [108]. The crystal structures of $[CH_3C(OH)_2][SO_3(OH)]$ [68] and $[CH_3C(OH)_2][SO_3F]$ [160] have also been reported. Corresponding hydrogen bonds in the hydrogen sulfate salt are shorter than those in the fluorosulfate salt, ascribed [160] to the weakening of oxygen hydrogen-bonding acceptor strength when OH is replaced by F (which itself does not participate in hydrogen bonding). No C–H···O hydrogen bonding is seen. The melting point of $[CH_3C(OH)_2][SO_3(OH)]$ is reported to be 2.5°C [68]. This is significantly lower than that of either ethanoic acid (16.6°C) or sulfuric acid (10.4°C) and is ascribed to the persistence of hydrogen bonding in the liquid state [68].

It is probable that a very early preparation of an ionic liquid (if not its characterisation) occurred on the isolation of the so-called 'red oil' in Friedel–Crafts reactions [736]. While the isolation and structural characterisation of carbocations and other reactive cations from strongly acidic media are an extensive and important area of study, only those materials that are stable at ambient temperatures and which have a reported m.pt.<100°C are listed in Table 11.24. These include $[C_6Me_7]^+$ [375, 414, 415], $[C_6Me_6CH_2Cl]^+$ [393, 394] and $[C_6Me_6Ph]^+$ [341], all except $[C_6Me_7][BF_4]$ [375], as the $[AlCl_4]^-$ salts.

The carbocation in [2-hydroxyhomotropylium]$[SbCl_6]$ [374] (m.pt. 89–90°C; [737] Fig. 11.30a) is found to be homoaromatic, whereas that in the related [1-ethoxyhomotropylium]$[SbCl_6]$ (m.pt. 103–104°C; [483] Fig. 11.30b) is not. Cation–anion interactions are confined to ionic and van der Waals interactions. Both of these salts are unstable in air.

Also of interest, but not included in Table 11.24, are the following. The diazonium salt $[(EtO)_2C{=}CH(N_2)][SbCl_6]$, while melting at 110–112°C, is shown to have C–O bond lengths shorter and C–C longer than expected from

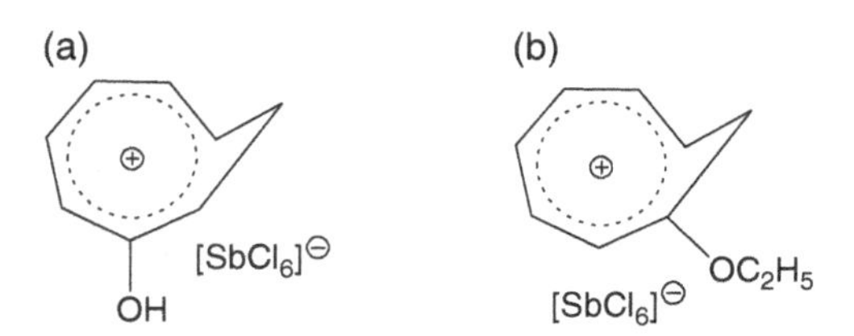

Figure 11.30 (a) 2-Hydroxyhomotropylium hexachloroantimonate [374] and (b) 1-ethoxyhomotropylium hexachloroantimonate [483].

the simple valence bond formulation, suggesting [543] oxo-stabilised carbenium as a possible bonding model. A delocalised model is also proposed for the tetramethylformamidinium cation, $[Me_2NCNMe_2]_n[A]$, where $n=1$, $[A]^- = [SnCl_3]^-$ (m.pt. 65°C; U_L 462 kJ mol^{-1}) or $[GeCl_3]^-$ (m.pt. 72°C; U_L 465 kJ mol^{-1}); $n=2$, $[A]^{2-} = [In_2Cl_6]^{2-}$ (m.pt. 92°C; U_L 1000 kJ mol^{-1}) or $[Ga_2Cl_6]^{2-}$ (m.pt. 89°C; U_L 946 kJ mol^{-1}) [223]. The anions $[SnCl_3]^-$ and $[GeCl_3]^-$ associate in the structure via, respectively, three and two weak E···Cl contacts, with an additional weak Ge···N contact. The indium and gallium anions contain In—In and Ga—Ga bonds. On the other hand, the highly explosive azidoformamidinium perchlorate, m.pt. 75°C (dec. 125°C) [272], contains an *N*-centred cation, $[H_2N{-}C(N_3){=}NH_2^+]$.

11.19.2 Boron-Centred and Cations and Carboranes

The 'boronium' ionic liquid, $[BH_2(L')(L'')][NTf_2]$ (L′ = 3-methylimidazole, L″ = trimethylamine), melting at 60°C, has been structurally characterised [193], revealing only weak cation–anion interactions and the anion in the low-energy *trans* conformation. The structure of the $[BH_2(CN)_2]^-$ salt of [bis(1-methyl-1*H*-imidazole-3-yl)dihydroboronium]$^+$ [205], with B—C as opposed to B—N bonds in the latter and melting at 62°C, has also been reported. [Bis(4-methoxyphenalene-1,9-diolato)boron]$[B\{(3,5-(CF_3)_2)C_6H_3\}_4]$ (m.pt. 70°C) and [bis(3,4-dimethoxyphenalene-1,9-diolato)boron]$[B\{(3,5-(CF_3)_2)C_6H_3\}_4]$ (m.pt. 98°C; Fig. 11.21), with high formula weights (1324.46 and 1384.51, respectively) [256], were both synthesised in a study of neutral radical molecular conductors. Details of the single-crystal structures are included, without discussion, in the supplementary information provided with the paper.

11.19.3 Halogen-Centred Cations

The crystal structure of a bromonium ionic liquid $[Br(py)_2][OTf]$, m.pt. 87–89°C [366], reveals arrays of nearly parallel cations repeated in such a way that the 1,2-bond of a pyridine in one array lies over the 3,4-bond of a pyridine in a parallel array, the average separation between the rings being less than the sum of the van der Waals distances. The salt containing the corresponding iodonium cation, $[I(py)_2][I_7]$, was first reported (with its m.pt. of 85°C) in 1895 [682], and its structure reported in 1961 [347]. The structures of the related iodonium salts, $[I(2,6\text{-}Me_2py)_2][ICl_2]$ (m.pt. 103–105°C) [484], $[I(2,6\text{-}Me_2py)_2][IBr_2]$ (m.pt. 110°C) [533] and $[I(1,3\text{-dimethylthiourea})_2][I_3]$ (m.pt. 87°C) [363], have also been described. In the latter case, there appears to be little interaction between the iodonium centre and the anion. On the other hand, the structure of $[(Me_2N)_3PSeI][I_3]$ (m.pt. 63–65°C) has been discussed [212] in terms of a three-centre, four-electron formalism based on the I(I) atom (with the description set in the context of an extensive series of related compounds). The authors conclude that the structure comprises $[(Me_2N)_3PSeI]^+$ cations, iodide anions and diiodine 'solvate' molecules in close contact with I^-. It is a moot point whether this material can properly be considered an ionic liquid. Similarly, the interaction between [bis(4-methoxyphenyl)iodonium]$^+$ and $[CS_2(NMe_2)]^-$ in $[I\{(C_6H_4(4\text{-}OMe)\}_2S_2CNMe_2]$ has been described as 'secondary' bonding [526]. What appear to be organo-iodinium ionic liquids, $[IPh_2][NTf_2]$ (m.pt. 65–66°C) and $[I(CH_2CF_3)Ph][NTf_2]$ (m.pt. 77–79°C) [225], in fact involve distorted square-planar coordination at iodine, with I···O separations of *ca.* 2.92–2.94 Å, and are best not formulated as salts (at least in the solid). The anion in the former compound adopts the *cis* conformation (torsion angle 38.2°) and that in the latter *trans* (169.1°).

11.19.4 Pnictogen-Centred Cations

Phosphonium ionic liquids have already been discussed in Sections 11.3 and 11.16.4. This section is concerned with salts of cations containing more than one Group 15 element (other than nitrogen). Salts of the type $[R_3PPR'_2]^+$ (and other complexes) may be obtained from molten reaction media comprising mixtures of $[PPh_2Cl]$ and gallium(III) chloride. Among a large number reported, several are, considering their high formula weight, relatively low melting (and unsolvated in the solid state). These monocationic phosphinophosphonium compounds (with X-ray single-crystal diffraction studies confirming their ionic formulation) include $[Me_3PPPh_2][OTf]$ (m.pt. 57–58°C) [176], $[Ph_2(Cl)PPPh_2][GaCl_4]$ (m.pt. 103–105°C) [176], $[Ph_2PCH_2PPh_2PPh_2][OTf]$ (m.pt. 80–82°C) [220] and an arsinostibonium analogue $[Ph_3AsSbPhCl][AlCl_4]$ (m.pt. 50–52°C) [154], as well as the dicationic bis(phosphinophosphonium) compounds, $[Ph_2PP(Ph)_2(CH_2)_nP(Ph)_2PPh_2][GaCl_4]_2$ ($n=2$, m.pt. 94–95°C; $n=6$, m.pt. 64–67°C) [220] and $[Ph_2PP(Me)_2CH_2CH_2P(Me)_2PPh_2][GaCl_4]_2$ (m.pt. 108–110°C) [220]. Cyclic compounds have also been prepared (also using a molten medium), including $[P_5Ph_2][GaCl_4]$ (m.pt. 60–61.5°C) and the tricationic $[P_7Ph_6][Ga_2Cl_7]_3$ (m.pt. 85–87°C) (Fig. 11.18), both structurally crystallised [200].

11.19.5 Salts of Other Aza-Heterocycles

$[(CF_3CO_2)_2H]^-$ salts of monoprotonated bicyclic 1,8-diazabicyclo[6.6.n]alka-4,11-diynes (Fig. 11.7), $n=3$ (m.pt. 71°C), $n=4$ (m.pt. 72°C) and $n=5$ (m.pt. 75°C), and a $[CF_3CO_2]^-$ salt ($n=6$, m.pt. 125°C) (as well as higher-melting diprotonated salts) [266] all involve 'inside' protonation of one of the bridgehead nitrogens and no significant hydrogen bonding to the other bridgehead nitrogen. The transannular N···N separation increases from 3.197 ($n=3$) to 3.721 ($n=4$), 4.088 ($n=5$) and 4.405 Å ($n=6$). U_L decreases modestly from 394 to 390 and to 385 kJ mol^{-1} as n increases from 3 to 5.

The crystal structure (and *in situ* variable-temperature powder diffraction using synchrotron radiation) of [1-ethyl-1,4-diazabicyclo[2.2.2]octanium] $[NTf_2]$ (Fig. 11.23; m.pt. 77°C) has been studied at 100, 223 and 295 K [299] and compared with that of the corresponding bromide salt (m.pt. 194°C). The anion adopts the *trans*oid conformation, torsion angle 165.5°. Ions pack with weak CH···O and C–H···F interactions below the sum of the relevant van der Waals radii. The preferred [299] measure of interionic distance, that is, between the quaternary nitrogen in dabco and the nitrogen of the anion, increases with temperature: 4.3882 (100 K), 4.440 (223 K) and 4.478 Å (295 K). U_L reduces correspondingly as temperature increases: 418, 415 and 413 kJ mol^{-1}

The following salts are shown in Figures 11.31, 11.32, 11.33, 11.34, 11.35, 11.36, 11.37, 11.38, 11.39 and 11.40: [3-methyl-2-phenyl-3,4,5,6-tetrahydropyrimidinium] [OTf] (another amidinium salt, [1-methyl-2-phenyl-3-propylimidazolinium] [PF_6] (m.pt. 62–64°C; Fig. 11.29) has already been discussed in Section 11.18.3) [206], [1,3-diethyl-3,4,5,6-tetrahydropyrimidinium][PF_6] (Fig. 11.32) [365], [3-hydroxy-1-*tert*-butyl-1,2-dihydropyrrolium][picrate] [325], [1,4,7-trimethyl-1-azonia-4,7-diazacyclononane][OTf] [385], [3-(3-ethenyl-1,2,4-oxadiazol-5-yl)-1,2,5,6-tetrahydropyridinium] [CF_3CO_2] [404], [5-hydroxy-1-methyl-3,4-dihydro-2*H*-pyrrolium]Cl (the hydrochloride salt of *N*-methyl pyrrolidone) [413], [2-acetyl-9-azoniabicyclo[4.2.1]nonan-3-one][CF_3CO_2] [432] and [1,5-dimethoxyl-3-phenyl-5a,6,7,8,9,9a-hexahydro-1*H*-1,5-benzodiazepin-5-ium][ClO_4]

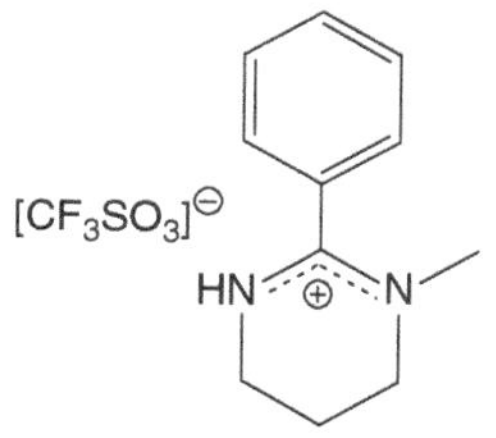

Figure 11.31 3-Methyl-2-phenyl-3,4,5,6-tetrahydropyrimidinium trifluoromethanesulfonate [206].

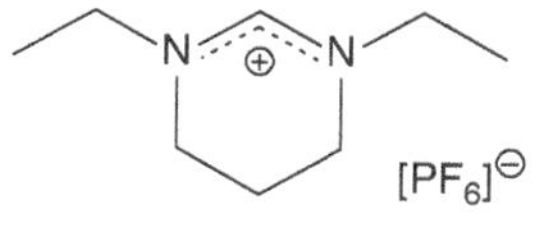

Figure 11.32 1,3-Diethyl-3,4,5,6-tetrahydropyrimidinium hexafluorophosphate [365].

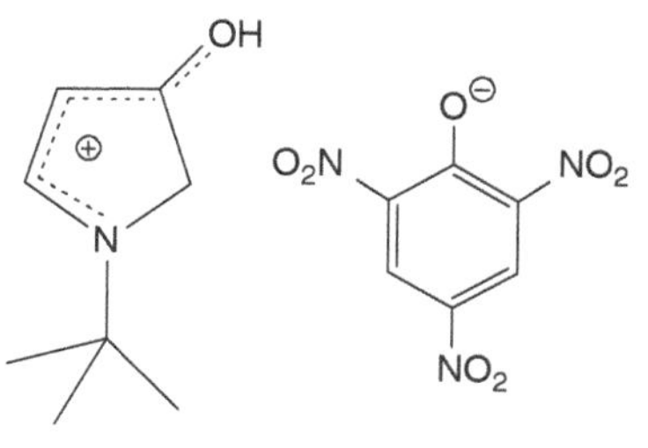

Figure 11.33 3-Hydroxy-1-*tert*-butyl-1,2-dihydropyrrolium picrate [325].

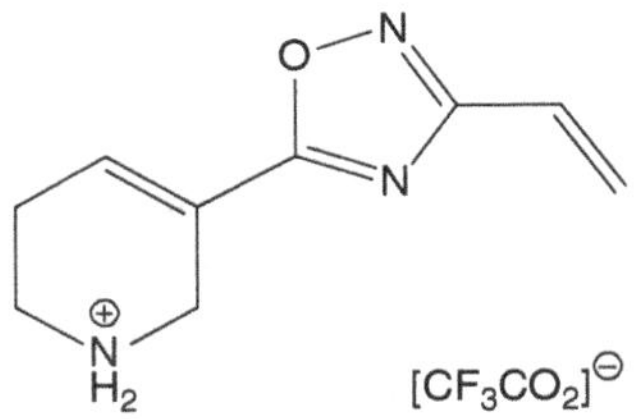

Figure 11.34 3-(3-Ethenyl-1,2,4-oxadiazol-5-yl)-1,2,5,6-tetrahydropyridinium trifluoroethanoate [404].

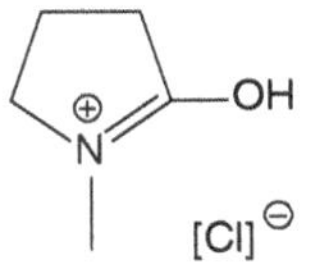

Figure 11.35 5-Hydroxy-1-methyl-3,4-dihydropyrrolium chloride [413].

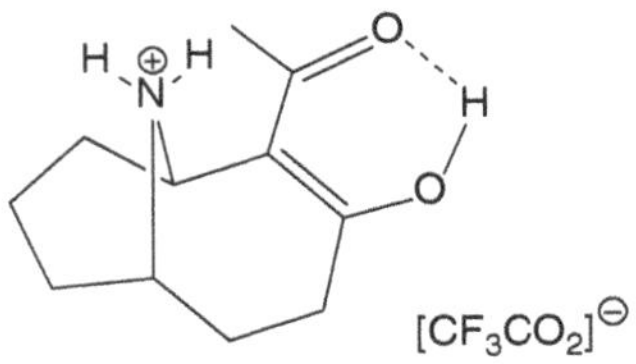

Figure 11.36 2-Acetyl-9-azoniabicyclo[4.2.1]nonan-3-one trifluoroethanoate [432].

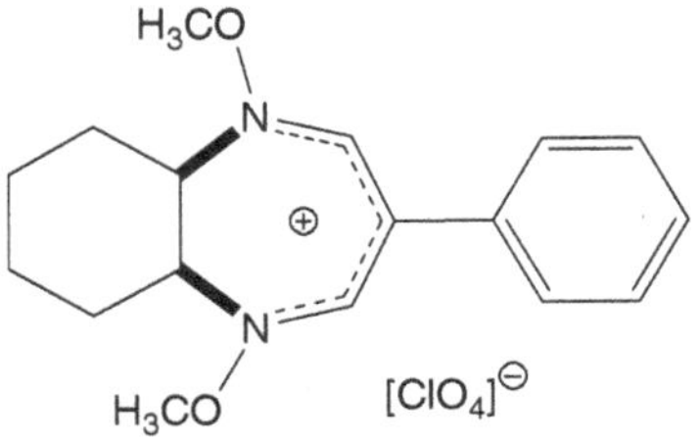

Figure 11.37 (1,5-Dimethoxyl-3-phenyl)-5a,6,7,8,9,9a-hexahydro-1*H*-1,5-benzodiazepin-5-ium perchlorate [451].

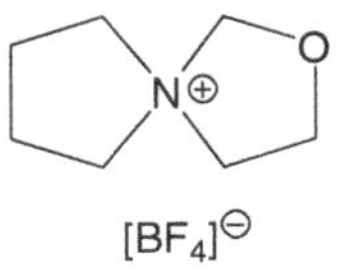

Figure 11.38 5-Azonia-2-oxa-spiro[4.4]nonane tetrafluoroborate [489].

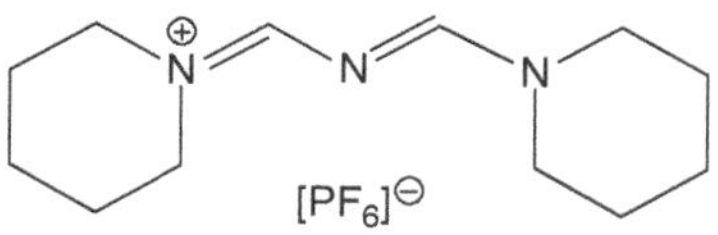

Figure 11.39 1-(Piperidinium-1-ylidene)-3-(piperidin-1-yl)-2-azapropene hexafluorophosphate [365].

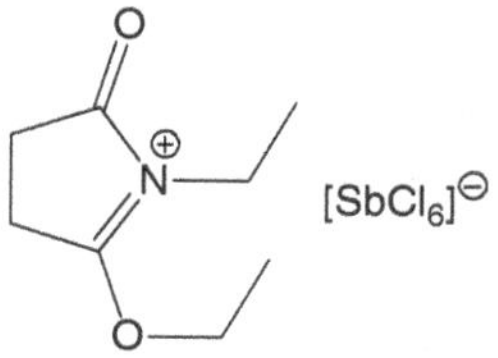

Figure 11.40 5-Ethoxy-l-ethyl-3,4-dihydro-2-oxo-2*H*-pyrrolium hexachloroantimonate [520].

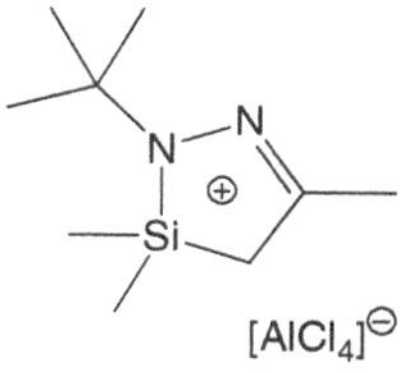

Figure 11.41 2-*Tert*-butyl-3,3,5-trimethyl-1,2-diaza-3-sila-5-cyclopentenium tetrachloroaluminate [482].

[451] all formally qualify as ionic liquids, with m.pt. in the range 76–100°C and for which a crystal structure is reported. The salts [5-azonia-2-oxa-spiro[4.4]nonane][BF_4] [489] (m.pt. 104°C), [1-(piperidinium-1-ylidene)-3-(piperidin-1-yl)-2-azapropene][PF_6] [365] (m.pt. 106–107°C) and [5-ethoxy-l-ethyl-3,4-dihydro-2-oxo-2*H*-pyrrolium][$SbCl_6$] [520] (m.pt. 108°C) highlight cations that may be of interest (with searches of the Cambridge Structural Database failing to reveal other salts of these cations melting below 100°C). The content of [489] is discussed further in Section 11.2.5, in the context of VBT.

11.19.6 Salts of Silicon-, Sulfur- and Arsenic-Containing Heterocycles

Single crystals of the radical cation salt, [2-*tert*-butyl-3,3,5-trimethyl-1,2-diaza-3-sila-5-cyclopentenium][$AlCl_4$] (m.pt. 103°C; Fig. 11.41) [482], have been studied by X-ray diffraction.

The crystal structure of [(–)-(1*S*,3*S*)-*trans*-3-acetoxy-1-methylthiacycohexane][ClO_4] (m.pt. 106°C; Fig. 11.11) establishes the absolute configuration of the chiral centre [488]. The structure of the *cis*-compound, [(±)-*cis*-3-acetoxy-1-methylthiacycohexane][ClO_4], (m.pt. 104°C), is also reported. X-ray crystallography confirms the high-energy axial OH conformation preference of the alkaloid analogue [3(*R*),4(*R*),5(*R*),6(*S*)-3,4,5-trihydroxy-*cis*-1-thioniabicyclo[4.3.0]nonane][ClO_4] (Fig. 11.13) (m.pt. 79–80°C) [313].

The structures of three of the 39 *N*-alkylbenzothiazolium salts, [Rbztz]I, have been reported, where R = decyl (m.pt. 82°C), undecyl (m.pt. 80°C) and dodecyl (m.pt. 90.5°C) [321]. Alkyl chains interdigitate; the cation centre and anion form polar domains, with short cation S···I contacts.

The salts of the cation, [5-methyl-1,3,5-dithiazinan-5-ium]$^+$ [165], with [$AlCl_4$]$^-$ (m.pt. 84–86°C; Fig. 11.20) and [$GaCl_4$]$^-$ (m.pt. 54–56°C) and a mixed salt with [BCl_4]$^-$ and Cl^- (m.pt. 109–111°C) have been structurally characterised. The mixed salt [tetrabutylammonium][3-{(2*S*)-2-methylbutyl}thiazolium] $[OTf]_2$ [348] melts at 85°C. Strong interactions between anion and cation, with a S···O contact of 3.15 Å, in [2-amino-5-butyl-4-methyl-1,3-thiazol-3-ium] [NO_3] (m.pt. 92–94°C; Fig. 11.42) [402], are reinforced by additional N–H···O hydrogen bonds. A 9-aza-9-methyl-1-thioniabicyclo[3.3.1]nonane salt, characterised structurally as the triiodide (m.pt. 68–69°C; Fig. 11.43) [247], precipitates during the aqueous oxidation of 5-(methylamino)-l-thiacyclooctane.

The structures of [1,3-dimethyl-1,3-diaza-2-arsenanium][$GaCl_4$] (m.pt. 52–54.5°C; Fig. 11.44) [161], [1,3-dimethyl-1,3-diaza-2-phospholidinium] [$GaCl_4$] (m.pt. 95–97°C; Fig. 11.45) [425], [1,3-dichloro-1,2,3,4-tetracyclohexyltetraphosphetane-1,3-diium][Ga_2Cl_7]$_2$ (m.pt. 106–109°C; Fig. 11.46) [434] and

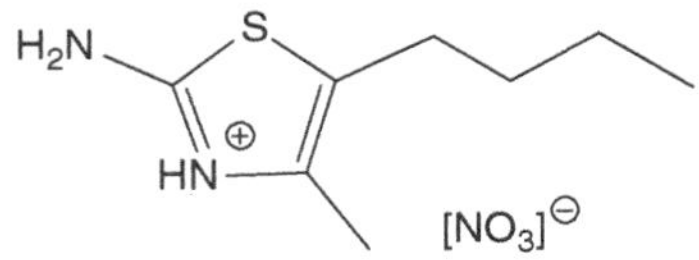

Figure 11.42 2-Amino-5-butyl-4-methyl-1,3-thiazol-3-ium nitrate [402].

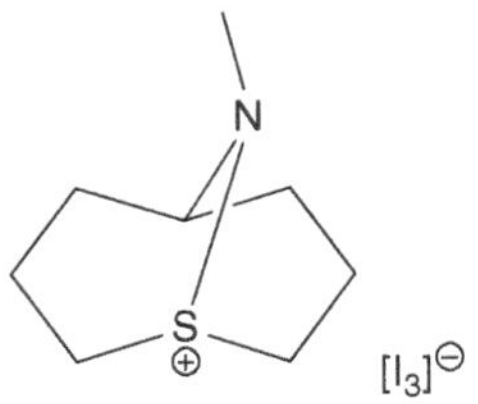

Figure 11.43 9-Aza-9-methyl-1-thionabicyclo[3.3.1]nonane triiodide [247].

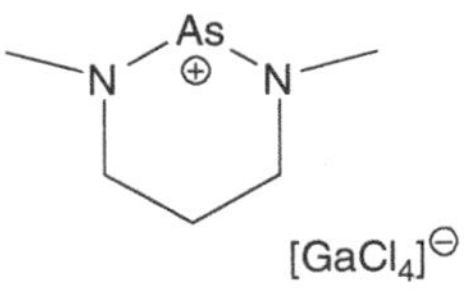

Figure 11.44 1,3-Dimethyldiaza-2-arsenanium tetrachlorogallate [161].

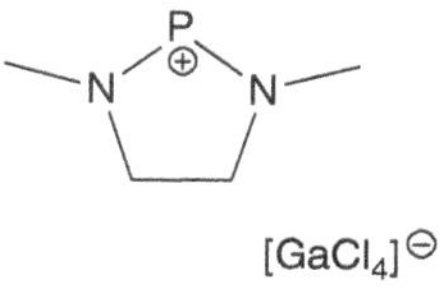

Figure 11.45 1,3-Dimethyl-1,3-diaza-2-phospholidinium tetrachlorogallate [425].

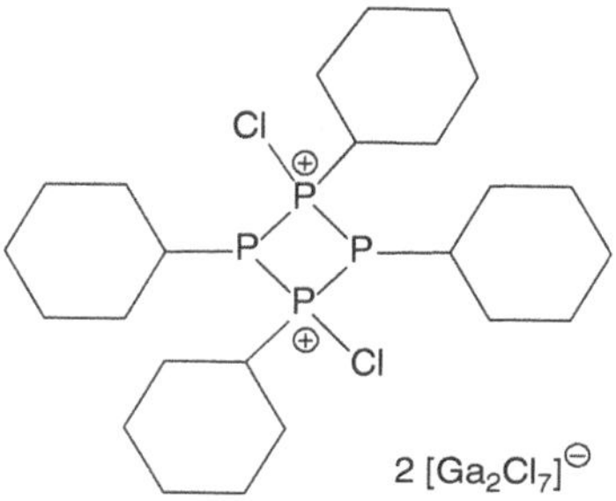

Figure 11.46 1,3-Dichloro-1,2,3,4-tetracyclohexyltetraphosphetane-1,3-diium bis(heptachloro digermanate) [434].

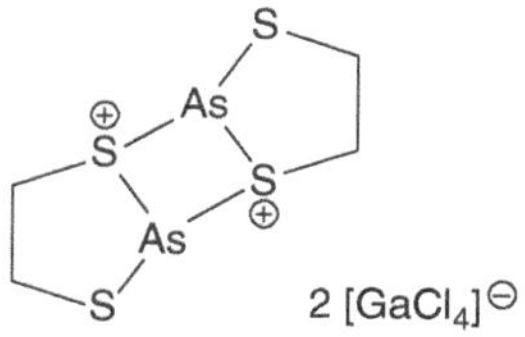

Figure 11.47 1,6-Diarsa-2,5,7,10-tetrathiatricyclo[5.3.0^{1,5}.0^{6,10}]decane di(tetrachlorogallate) [416].

[1,6-diarsa-2,5,7,10-tetrathiatricyclo[5.3.0^{1,5}.0^{6,10}]decane][$GaCl_4$]$_2$ (m.pt. 104–105°C; Fig. 11.47) [416] are noted for completeness.

11.20 TRENDS IN THE RELATIONSHIP BETWEEN U_L AND MELTING TEMPERATURE

A detailed search for trends in the relationship between lattice potential energy and melting temperatures will be reported elsewhere. Nevertheless, a preliminary global analysis of data for 1:1 salts from Tables 11.1, 11.2 and 11.3 is presented. To enable such plots to be undertaken, some adjustment of the data is necessary. For instance, melting temperatures are converted to a number (melting 'point') with a precision of 0.5°C. The midpoint of melting temperature ranges is used. Lattice potential energy data are uncorrected for any effect associated with differences in data collection temperature.

The associated plots of U_L versus m.pt. are found in Figure 11.48a–c. Trend lines drawn through the widely scattered data are included for information only. The weak overall trend is further shown from the average values for U_L/kJ mol^{-1} for compounds having melting points that fall into the following intervals: 456 (−29 to −10°C), 472 (−10–0°C), 440 (1–10°C), 420 (11–20°C), 427 (21–30°C), 431 (31–40°C), 433 (41–50°C), 441 (51–60°C), 438 (61–70°C), 441 (71–80°C), 443 (81–90°C), 439 (91–100°C) and 431 (101–110°C).

In addition, related plots are presented for a small number of subsets of selected ionic liquids. Data for salts containing Cl^-, Br^-, I^- and $[I_3]^-$ and for

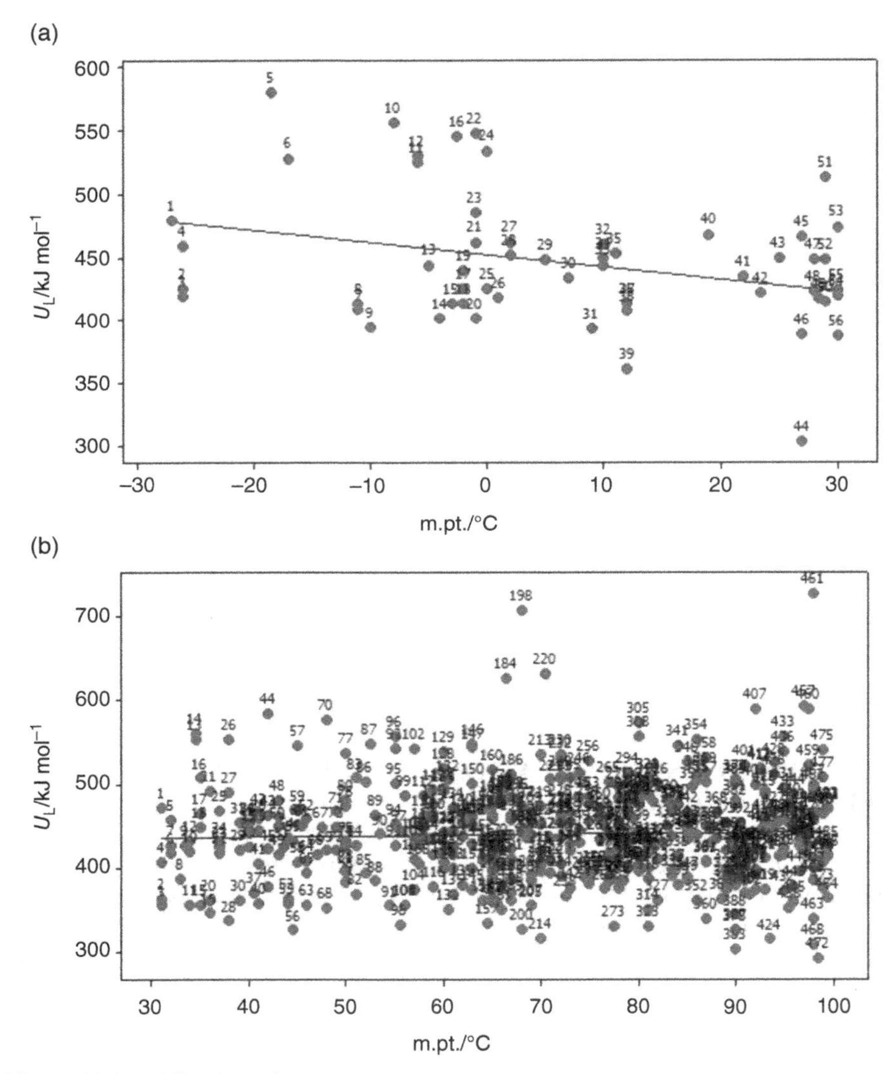

Figure 11.48 (*Continued*)

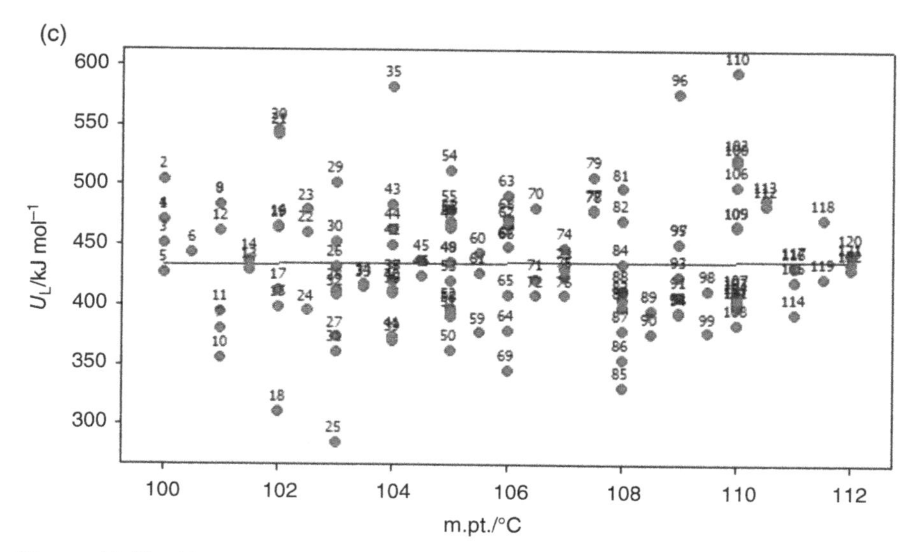

Figure 11.48 Plots of lattice potential energy (U_L/kJ mol^{-1}) versus melting point, calculated from crystallographic data for 1:1 salts, [cat][A] for (a) room-temperature ionic liquids (m.pt. < 30°C) from Table 11.1, (b) ionic liquids (m.pt. 30–99°C) from Table 11.2 and (c) salts with m.pt. 100–112°C.

materials containing ions often found in ionic liquids, such as $[NTf_2]^-$ and $[C_2mim]^+$, taken respectively from Tables 11.14, 11.18 and 11.23, have been plotted. These are shown in Figure 11.49a–g.

The effect of anion size is evident in the comparison of data for chloride, bromide and iodide salts, as is the difference between salts of $[NTf_2]^-$ and $[NMes_2]^-$. However, the plots highlight the dangers of relying on single precise values for melting temperature. Further analyses should find ways of incorporating such data uncertainty in such parameters. In addition, the availability of many data for higher-melting materials, where included, may skew the plots.

11.21 CONCLUSIONS

Despite limiting for inclusion salts that are stable on melting and at ambient temperatures, which crystallise without solvent of crystalisation and which have reported melting temperatures, a substantial body of data on the crystallography of ionic liquids has been assembled and should provide the basis for more detailed investigation. Crystallographically characterised ionic liquids (according to the conventional definition) contain 205 different anions and 397 different cations. Forty-three different anions and 47 different cations are found in salts with m.pt. < 30°C (room-temperature ionic liquids). The ionic liquids tabulated occupy a relatively small fraction of the possible composition space represented by the large number of individual anions and cations found in materials with m.pt. < 100°C.

Ionic liquids have been characterised crystallographically that contain anions and cations whose central atom can be drawn from each group of the periodic table other than from Groups 2 and 18. The structures of an increasing number of ionic liquid coordination complexes are now appearing, though the palette of ligands used is currently limited.

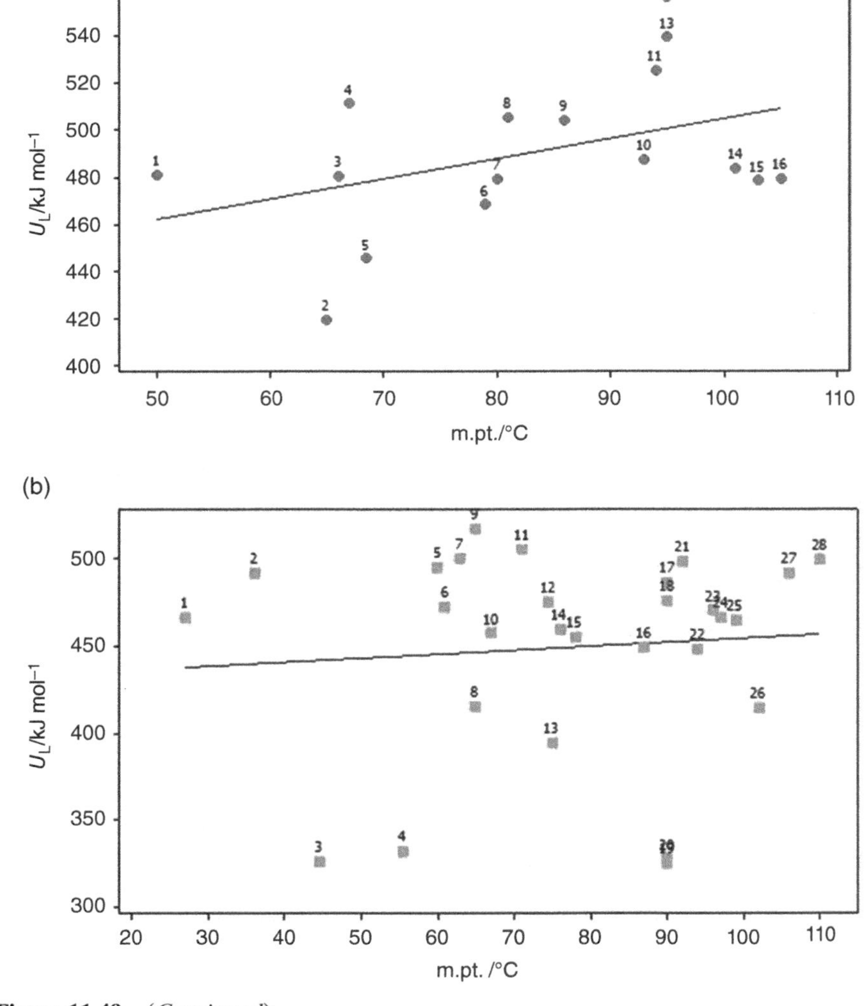

Figure 11.49 (*Continued*)

The most illuminating structural studies on ionic liquids have used combinations of techniques to investigate phase behaviour, conformational change and associated ion–ion interactions in the solid and liquid states. Generally, however, the composition of the liquid formed on melting a salt, necessary to confirm that it consists entirely of ions, has been investigated in relatively few systems.

The overall picture provided by the structural studies surveyed confirms the general picture regarding ionic liquid behaviour, namely, the importance of charge delocalisation and component ion flexibility in the frustration of

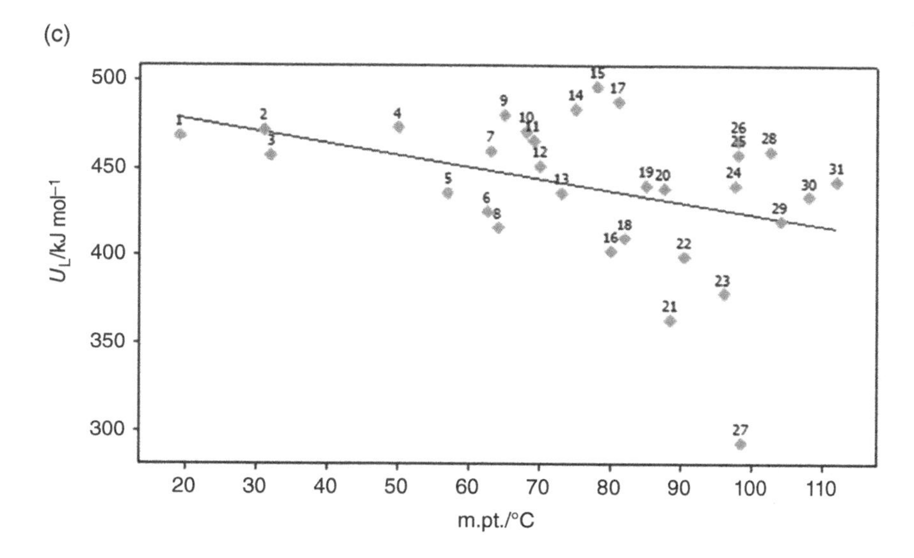

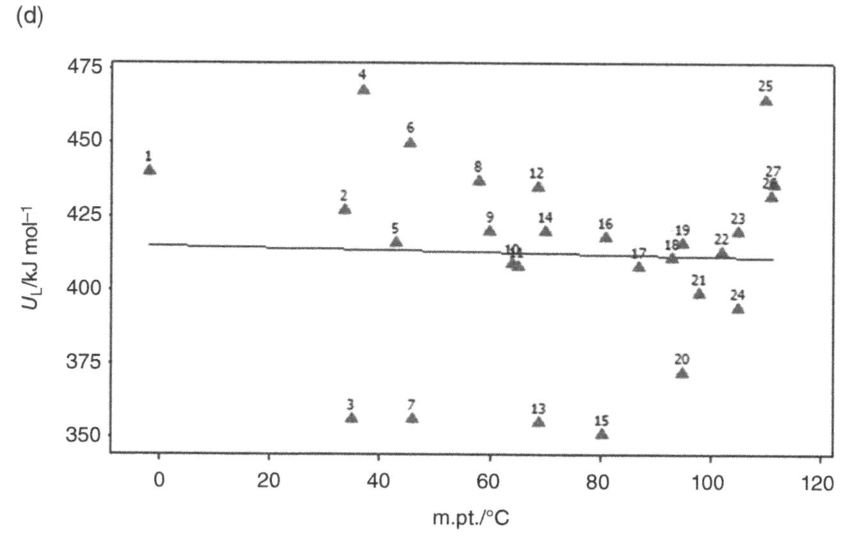

Figure 11.49 (*Continued*)

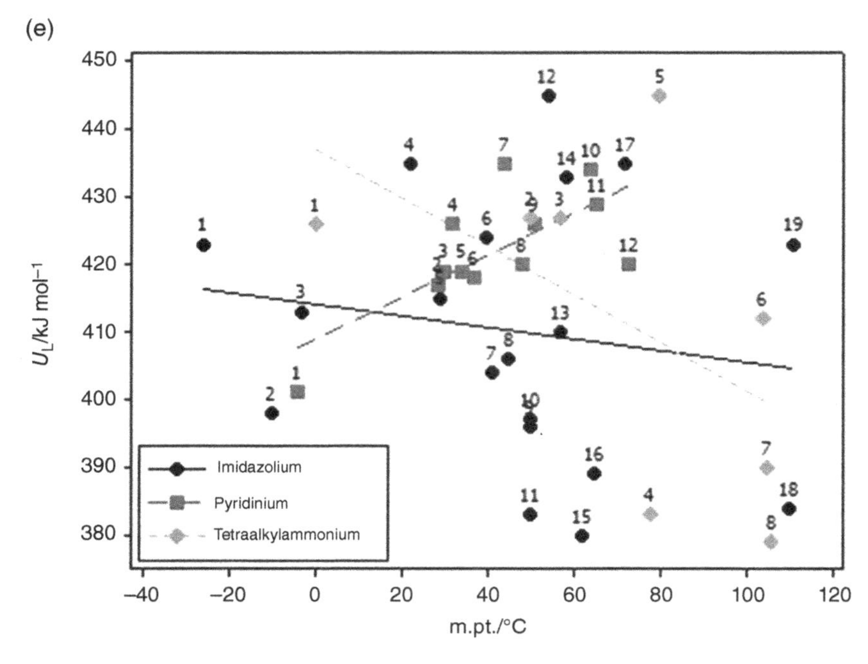

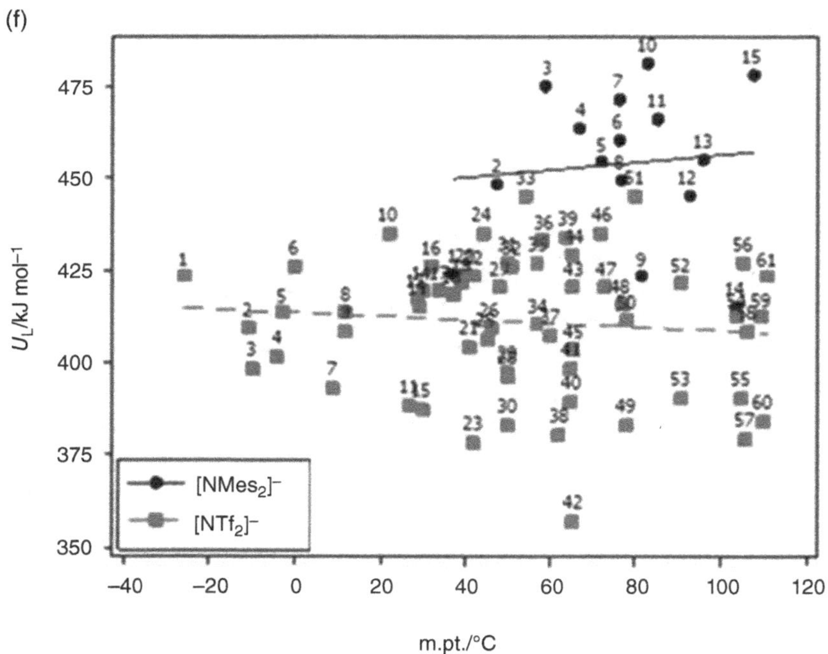

Figure 11.49 (*Continued*)

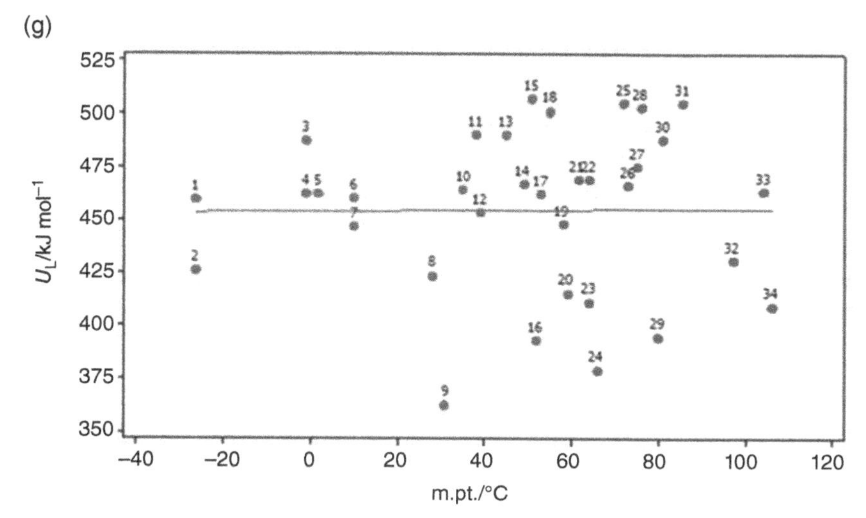

Figure 11.49 Plots of lattice potential energy (U_L/kJ mol^{-1}) versus melting point, calculated from crystallographic data for 1 : 1 halide salts, [cat]X (Table 11.14) for (a) X = Cl [U_L^{mean} 489.9 kJ mol^{-1}, U_L^{min} 419 kJ mol^{-1}, U_L^{max} 556 kJ mol^{-1}, σ 33.7 kJ mol^{-1}, n = 16]; (b) X = Br [U_L^{mean} 449.0 kJ mol^{-1}, U_L^{min} 325 kJ mol^{-1}, U_L^{max} 517 kJ mol^{-1}, σ 57.8 kJ mol^{-1}, n = 28]; (c) X = I [U_L^{mean} 439.5 kJ mol^{-1}, U_L^{min} 293 kJ mol^{-1}, U_L^{max} 496 kJ mol^{-1}, σ 41.8 kJ mol^{-1}, n = 31] and (d) X = I_3 [U_L^{mean} 412.3 kJ mol^{-1}, U_L^{min} 351 kJ mol^{-1}, U_L^{max} 468 kJ mol^{-1}, σ 31.7 kJ mol^{-1}, n = 27]. Related plots for (e) [cat][NTf_2] salts (Table 11.18), cat = imidazolium, pyridinium or ammonium; (f) [cat][NTf_2] and [cat][$NMes_2$] (Table 11.18), [NTf_2]$^-$ salts [U_L^{mean} 410.6 kJ mol^{-1}, U_L^{min} 357 kJ mol^{-1}, U_L^{max} 445 kJ mol^{-1}, σ 18.4 kJ mol^{-1}, n = 61] and [$NMes_2$]$^-$ salts [U_L^{mean} 453.8 kJ mol^{-1}, U_L^{min} 415 kJ mol^{-1}, U_L^{max} 481 kJ mol^{-1}, σ 20.4 kJ mol^{-1}, n = 15]; (g) [C_2mim][A] salts (Table 11.23) [U_L^{mean} 454.3 kJ mol^{-1}, U_L^{min} 363 kJ mol^{-1}, U_L^{max} 507 kJ mol^{-1}, σ 38.1 kJ mol^{-1}, n = 34].

crystallisation. However, too few studies combine investigations of both the crystalline and liquid states for a single system at or near the melting point. More work is needed to estimate the relative contributions to lattice energies of coulombic, dispersion and hydrogen-bonding interactions.

Particular anions (such as 1,3-diketonates and their analogues) and cations (such as alkylurotropinium) are surprisingly under-represented and provide opportunities for the synthesis, characterisation and study of new materials.

Bearing in mind the low m.pt. of some salts with extensive networks of strong hydrogen bonds, a simplistic view regarding the role of hydrogen bonding in ionic liquids should be avoided. Using crystallographic data, an extensive set of lattice potential energy values can be obtained using VBT approaches. There appears to be no discernible trend in plots of calculated lattice potential energy and melting point that may not be the result of confounding experimental or structural factors. Further analysis is needed, including the calculation of lattice free energies, though these require the determination of the absolute entropies for a range of complex ions not currently available.

11.22 ADDENDUM

The following papers have not been reviewed in the foregoing but are important enough to note. Most are very recent, but some have only come to light as a result of recent searches of the literature. They report single-crystal X-ray structures of compounds that could have qualified them for inclusion in Tables 11.1, 11.2 and 11.3. They are listed in order of reference number. Some additional publications are also included that would have merited comment in the text. In the latter case, they are grouped under a relevant keyword heading.

11.22.1 Additional Structurally Characterised Ionic Liquids

[738] Liquid $[\{C_6H_3(1,3,5\text{-}(CH_3)_3)\}H][\{(CF_3)_3CO\}_3Al\text{–}F\text{–}Al\{OC(CF_3)_3\}_3]$ (REDVON).

[739] Liquid (explodes at 75°C). $[N(CH_3)_4][As(N_3)_4]$. Crystal structures also determined for $[PPh_4][As(N_3)_4]$, $[N(PPh_3)_2][As(N_3)_4]$, $[NMe_4][Sb(N_3)_4]$, $[PPh_4][Sb(N_3)_4]$ and $[N(PPh_3)_2][Sb(N_3)_4]$.

[740] Liquid 1,3-dihexyl-2-iodo-4,5-dimethylimidazolium iodide (XEZREB). A series of related materials were all found to contain water of crystallisation.

[741] Liquid $[N_{4444}][Br_9]$ (HAZWAI) is one of a series of $[NR_4][Br_9]$ (R = Me, Et, Pr or Bu), all with m.pt. < 45°C.

[742, 743] $[H(MeOH)_2][BF_4]$, m.pt. –41°C (MEHTFB/MEHTFB10). $[H_5O_2][BF_4]$, m.pt. –34°C (no refcode). $[H_3O][BF_4]$, m.pt. +52°C (no refcode).

[744] $[EtNH_3][NO_3]$, m.pt. 13°C {SELWAJ (260 K); SELWAJ01 (150 K); SELWAJ02 (100 K)}.

[745] $[C_6mazp][NTf_2]$, m.pt. 25°C (BASVIC). $[C_4mazp]I$, m.pt. 87°C (BASVEY).

[746] $[C_2mim]$[tetracyanopyrrolide], m.pt. 30°C (PEMTUY). $[N_{1111}]$[tetracyanopyrrolide], m.pt. 119°C (PEMTOS).

[747] Coordination complexes of the type $[AgL_2][NTf_2]$, where L = $^iBuNH_2$ (m.pt. 31°C; BIQWAB), $^tBuNH_2$ (m.pt. 62°C; BIQWOP) or piperidine (m.pt. 100°C; BIQVUU). When L = 1,2-diaminoethane (m.pt. 21°C; BIQVOO), the salt is polymeric in the solid state.

[748] Bis[(benzyl)ammonium] 5-sulfonatosalicylate, m.pt. 31–33°C (ADAPUR).

[749] 1,3-Di(ethoxy)imidazolium bistriflamide, m.pt. 32–34°C (CCDC 903895) (previously reported to be a liquid at ambient temperatures). 1,3-Di(methoxy)imidazolium tetrachloroferrate(III), m.pt. 37°C (CCDC 903893). 1,3-Di(ethoxy)imidazolium tetrachloroferrate(III), m.pt. 39°C (CCDC 903894).

[750] $[C_{11}C_2im]Br$, m.pt. 38°C; KIPFIA, is the only one of the seven salts of the type $[(C_n)(C_m)im]Br$ for which $n + m = 13$ that shows no thermotropic liquid crystal behaviour. The single-crystal structure of $[C_{13}imH]Br$ (melting to an isotropic liquid at 142°C) is also reported.

[751] $[(HOCH_2CH_2)mim][PF_6]$, m.pt. *ca.* 40°C. Part of a wider study of hydrogen bonding in ionic liquids.

[752] $[(MeO_2CCH_2)mim][FeCl_4]$, m.pt. 42°C (WETNAM).

(Continued)

[753] 1,3-Diamino-1,2,3-triazolium nitrate, m.pt. 45°C (FIBGEE). 1,3-Diamino-1,2,3-triazolium nitrotetrazolate-2-oxide, m.pt. 50°C (FIBGAA).

[754] 1-Methyl-3-(3-sulfopropyl)imidazolium 4-toluenesulfonate, m.pt. *ca.* 47°C (VEGKID), is a viscous liquid that solidifies slowly over a period of months at room temperature. A weak endothermic peak in the DSC at *ca.* 47°C is assigned to a 'poorly defined melting transition'.

[755] $[H_2NNMe_2(CH_2CN)][ClO_4]$, m.pt. *ca.* 47°C (TAZMIS).

[756] $[C_4mim][B\{OCH(CF_3)_2\}_4]$, m.pt. 49°C [171] (DEZJUP). The relationship of the mean hole volume, derived from positron annihilation lifetime spectroscopy, and volumes, the scaled molecular volume and van der Waals volume, derived from X-ray diffraction data was explored for a series of $[C_4mim]X$ ionic liquids, where $X = Cl^-$, $[BF_4]^-$, $[PF_6]^-$, $[OTf]^-$, $[NTf_2]^-$, $[PF_3(C_2F_5)_3]^-$, $[N\{P(O)(C_2F_5)_2\}_2]^-$ or $[B\{OCH(CF_3)_2\}_4]^-$.

[757] $[N_{4444}]_2$[L-tartrate], m.pt. 50°C (FEHHUX (CSD record shows salt to be a heptahydrate.

[758] $[C_2mpyr][C(CN)(NO)C(O)NH_2]$, m.pt. 52°C (CCDC-926167). Carbamoylcyano-(nitroso)methanide (from the reaction of water with $[C(CN)_2(NO)]^-$) is capable of adopting either *syn* or *anti* conformations. $[C_nmpyr][C(CN)(NO)C(O)NH_2]$ {$n = 3$ (88°C) or 4 (58°C) and $[P_{122}\,i_4]^+$ (m.pt. 53°C) (CCDC-926168-926170)} have also been crystallogaphically characterised. The previously reported $[N_{1111}]^+$ salt (HAMGIN) decomposes at *ca.* 150°C [759].

[760] 1-(2-(Cyclohexyloxy)-2-oxoethyl)pyridinium tetrafluoroborate, m.pt. 65°C (RIJCOE). 1-(2-(Cyclohexyloxy)-2-oxoethyl)pyridinium trifluoromethylsulfonate, m.pt. 76°C (RIJCIY). 1-(2-(Cyclohexyloxy)-2-oxoethyl)pyridinium hexafluorophosphate, m.pt. 105°C (RIJCEU) [749].

[761] The mixed salt tetrakis(triethylammonium) bis(6-dicyanomethylidene-1,3-dinitrohexa-2,4-dien-1-ide) hexafluorosilicate, m.pt. 65–66°C (GELKAL), is isolated when malononitrile reacts with 2,4-dinitrofluorobenzene in the presence of triethylamine on contact with glass or silicon dioxide, in addition to the expected triethylammonium 6-dicyanomethylidene-1,3-dinitrohexa-2,4-dien-1-ide.

[762] $[Fe^{III}(\eta^5\text{-}C_5Me_4C_6H_{13})(\eta^5\text{-}C_5Me_4H)][NTf_2]$, m.pt. 67°C (DEZMOM). $[Fe^{III}(\eta^5\text{-}C_5Me_4C_{10}H_{21})(\eta^5\text{-}C_5Me_4H)][NO_3]$, m.pt. 80–82°C (DEZLOL). $[Fe^{III}(\eta^5\text{-}C_5Me_4C_6H_{13})(\eta^5\text{-}C_5Me_4H)][NO_3]$, m.pt. 109°C (DEZMAY). The related $[Fe^{III}(\eta^5\text{-}C_5Me_4R)(\eta^5\text{-}C_5Me_4H)][PF_6]$ are all high melting.

[763] 4-(Trimethylammonio)benzaldehyde bis(trifluoromethanesulfonyl)amide, m.pt. 67°C (VOJLUD).

[764] $[N_{4444}][Bi_2I_8(dmso)_2]$, m.pt. 76°C (GELGOV). See also $[PBuPh_3][Bi_2I_8(Me_2CO)_2]$ [285].

[765] $[1,3,5\text{-}Me_3C_6H_2Te(PPh_3)][OTf]$, m.pt. 76°C (GENROI).

[766] 1-Octylquinolinium dodecylsulfate, m.pt. 76°C (MEHJEQ) (non-mesomorphic), part of an extensive study of *N*-alkylquinolinium and *N*-alkylisoquinolinium ionic liquid crystals.

[767] $[Li(tmen)_2][(C_6F_5)_2I]$, m.pt. 77°C (dec) (DOXMUY).

(*Continued*)

(Continued)

[768]	Propyltributylphosphonium chloride, m.pt. 78°C (KAXDIY).
[769]	[C_1mim][4-nitroimidazolate], m.pt. 78°C (FEXMIG); [C_1mim] [4-nitro-1,2,3-triazolate], m.pt. 85°C (FEXLUR); [C_1mim] [4,5-dinitroimidazolate], m.pt. 96°C (FEXMAY); [C_1mim] [2-methyl-5-nitroimidazolate], m.pt. 112°C (FEXMIC).
[770]	[R_2P_5][$GaCl_4$] (R = ethyl), m.pt. 79°C (WIJCID), and (R = cyclohexyl), m.pt. 106–108°C (WIHZUK), are salts of cage polyphosphorus cations.
[771]	[C_4C_2benzimidazolium][BF_4], m.pt. 81–83°C (CEFTIS), is a crystalline solid in contrast to [C_4mim][BF_4], which has yet to be crystallised.
[772]	[Cu{$NH(CH_2CH_2OH)_2$}{$NH(CH_2CH_2OH)(CH_2CH_2O)$}][$CF_3SO_3$], m.pt. 82°C (no cif available).
[773]	(*S*,*E*)-2-[(*S*)-fluoro(phenyl)methyl]-1-[(*E*)-3-phenylallylidene] pyrrolidinium hexafluoroantimonate, m.pt. 82–83°C (HOYSUK01).
[774]	Tris-(2-hydroxyethyl)ammonium 4-chlorophenoxyethanoate, m.pt. 82–84°C (KELVEE).
[775]	Three of a series of 13 structurally characterised naphthalene-1,8- and acenaphthalene-5,6-diyl chalconium triflates (m.pt. 82–84°C, REKPUU), (m.pt. 94–96°C, REKMUR) and (m.pt. 92–94°C, REKMOL) qualify as ionic liquids according to the conventional definition.
[776]	1-Methyl-3-phenylimidazolium triflate, m.pt. 82–84°C (VETHOT). A cyclopropeniooxophosphine tetrafluoroborate (m.pt. 110–112°C, VETJAH) was also structurally characterised.
[777]	[1-Hydroxyimidazolium]$_2$[SO_4], m.pt. 86°C (IFUVUC). [1-Hydroxyimidazolium][NO_3], m.pt. 92–93°C (IFUVOW). The crystal structures of these and [1-hydroxyimidazolium]X (X = Cl or Br) have been subject to Hirshfeld surface analysis.
[778]	[(C_6H_6)RuCl{bis(methylthio)propane}][NTf_2], m.pt. 92°C (KIMJUN).
[779]	[($NCCH_2CH_2$)aim]Br, m.pt. 92°C (YIHPOW). [N_{1222}][Al{OC_6H_4(4-CN)$_4$}], m.pt. 99°C (YIHQAJ), is the lowest-melting member of a series of salts containing [$Al(OR)_4$]$^-$, in which R = —C_6H_4(4-CN).
[780]	[$P(NMe_2)_3CH_2CH(OEt)_2$]Br, m.pt. 97°C (XENDAX), a triamidophosphonium acetal.
[781]	1,4,5-Trimethyl-3-perfluorooctyl-1,2,4-triazolium tetrafluoroborate, m.pt. 98°C (EVIBUG).
[782]	[PPh_3Et]$_2$[$Cu^{II}{}_2Cl_6$], m.pt. 98°C (EPEDAF).
[783]	[$PPh_3(C_5H_{11})$]$_2$I[I_3], m.pt. 98°C (XEMPUC).
[784]	Acetylcholine hydrogen(+)-tartrate (form I), m.pt. 98–99°C (ACHTAR10); acetylcholine hydrogen(±)-tartrate, m.pt. 100–103°C (ACHTAR10); acetylcholine hydrogen(+)-tartrate (form II), m.pt. 104–108°C (ACHTAR11).
[785]	[*N*-methyl-(*N*-3-ammoniopropyl)pyrollidinium][NTf_2]$_2$, m.pt. 99°C (FEHHEH).
[786]	Dimethyl(2-oxo-2-ferrocenylethyl)sulfonium iodide, m.pt. 100°C (VELNEH).
[787]	[$(Me_3Si)_2NCN(SiMe_3)$][$B(C_6F_5)_4$], m.pt. 104°C (KEPQIH), and [$(Me_3Si)_2NNC(SiMe_3)$][$B(C_6F_5)_4$], m.pt. 108°C, catalyse the isomerisation of [$(Me_3Si)_2C{=}N{=}N$] and [$(Me_3Si)_2N{-}N{\equiv}C$].
[788]	[(1-$PyCH_2CH_2CH_2$)mim][$O(FeCl_3)_2$], m.pt. 104–105°C (GAZJEY) is a low-melting 2:2 salt.

(Continued)

[789]	(2,2-Dimethyl-3-methylidenenorbonane)triethylphosphonium perchlorate, m.pt. 107–110°C (XEMXEU).
[790]	Aminoguanidinium 2,4-dicyano-1,2,3-triazolate, m.pt. 108°C (HOXVOG).
[791]	$[Cu^{II}(L)Cl][NTf_2]$ (L = bis(2-dimethylaminoethyl)-((1-methylimidazol-2-yl) methyl)-amine), m.pt. 108°C, extending the ligand palette for transition metal coordination complex ionic liquids to include quadridentate tripodal chelates.
[792]	$[C_2mim][NiBr_3(1\text{-}mim)]$, m.pt. 108°C.
[793]	Tris(benzylguanidinium triflate, m.pt. 109°C (QEHFUG).
[794]	$[(Cy_2N)PCl][GaCl_4]$, m.pt. 110–111°C (BEDKOM).
[795]	1-Octadecyl-2,3,4,6,7,8,9,10-octahydropyrimido(1,2-a)azepin-1-ium iodide, m.pt. 111–112°C (MEJXEF).

11.22.2 Additional Structures of Interest

[796]	$[C_4mim][B(C_6F_5)_4]$, m.pt. 145°C (CSD reports this to be a CH_2Cl_2 solvate, crystallisation temperature 97°C) (PENREH); $[C_4C_1C_1im][B(C_6F_5)_4]$, m.pt. 151°C (crystallisation temperature 70°C) (PENRIL); $[Li(NCMe)_4][B(C_6F_5)_4]$ (YENMIP).
[797]	1-Carboxymethyl-3-methylimidazolium chloride, m.pt. 199°C (KERNOM).
[798]	$[C_{18}mim]_3[Al_2F_9](CH_2Cl_2)_{1.754}$ (LELXOR), formed in an attempt to grow crystals of $[C_{18}mim][AlF_4]$, contains a face-sharing anion.
[799]	$Li[C_2mim][(CF_3SO_3)]_2$ (NOFQIJ); prepared by cooling melt from 70°C; explicit value of m.pt. not given.
[800]	1,3-Dimethylimidazolium hexanitratolanthanate, m.pt. 115°C. Others low m.pt. materials described, such as 1-butyl-3-methylimidazolium hexanitratolanthanate, m.pt. 29°C, but without crystal structure.
[801]	$[C_2mim][PbCl_3]$ (SEZDEI) grown from the melt of [emim]Cl and $PbCl_2$.
[802, 803]	Two diazonium salts of the bistriflamide analogue, bis(trifluoromethanesulfonyl)methanide ($[CH(SO_2CF_3)_2]^-$), have been crystallographically characterised: $[PhN_2][CHTf_2]$ (refcode LAKGUZ) and $[4\text{-}TolN_2][CHTf_2]$ (refcode LAKGUZ10). Both are reported to decompose on melting, respectively at 74 and 76°C, respectively.
[804]	Crystal structures of 17 salts of the type [ERPh]X or $[ER_2]X_2$ (where R = 2,3-bis(di-*iso*propylamino)-*cyclo*-propenium, E = O, S, Se or Te and X^- = $[BF_4]^-$ or $[OTf]^-$) are reported but without melting point. Related aza-compounds, including $[NR_3][BF_4]_3$, all high melting, have also been structurally characterised [775, 805].
[806]	What appear to be the first crystal structures of a salt containing the cyclic perfluoroalkylenesulfonylamide anion, from 4,4,5,5,6,6-hexafluorodihydro-1,3,2-dithiazinane 1,1,3,3-tetraoxide, $[A]^-$, have been reported for [cobalticenium][A], which melt at greater than 370°C (CCDC 915718, 915719). The related $[Co(\eta^5\text{-}C_5H_5)_2][N(SO_2C_nF_{2n+1})_2]$, n = 1–4, all with m.pt. > 138°C, have been structurally characterised. Interestingly, the corresponding $[Co(\eta^5\text{-}C_5H_4Et)_2]^+$ salts melt in the range 6–14.5°C, though no X-ray studies are available. Similarly, $[Co(\eta^5\text{-}C_5H_4Et)_2][A]$ melts at 22°C [807].

(*Continued*)

(Continued)

[808]	While not an ionic liquid (being a CH_2Br_2 solvate), attention is nevertheless drawn to the recent publication of a landmark crystal structure of [2-norbornyl][Al_2Br_7], for which *in situ* annealing and cooling permitted the identification of a bridged non-classical form of the cation.
[809]	[CPh_3][Nb{Al{OCH(CF_3)$_2$}$_4$}$_6$] (DIFLOV) is believed to decompose at 98°C.
[810]	While not a study of ionic liquids, this work should be noted for its extensive study of salts of 4-hydroxyphenethylamine (tyramine). The crystal structures of 38 salts are reported (though without melting temperatures).

11.22.3 Polymorphism

[811]	Further studies of polymorphs of [C_4mim][PF_6] (MAZXOB02-MAZXOB06).
[812]	Further studies of pressure-induced crystallisation, this time of [C_4mim][PF_6].
[813]	Decompression-induced crystal polymorphism of *N,N*-diethyl-*N*-methyl-*N*-(2-methoxyethyl)ammonium tetrafluoroborate.

11.22.4 Plastic Crystals

[814]	Phase and conductivity studies of solid cholinium triflate (m.pt. 140–150°C) [815], using synchrotron XRD and positron annihilation lifetime spectroscopy, investigate the relationship between conductivity and lattice defects.

11.22.5 Chiral Salts

[816]	1,2-(*α*-*N,N*-Dimethylammoniotetramethylene)ferrocene hydrogentartrate (WATCOJ) is reported in the CSD record as having a m.pt. of 17–18°C. However, reference to the paper suggests that it is the free base, 1,2-(*α*-*N,N*-dimethylaminotetramethylene) ferrocene, that melts at this temperature. The paper does not report the m.pt. of the salt.
[817]	Chiral triazolium salts and ionic liquids.
[818]	Chiral thiouronium salts, including the crystal structure of *S*-butyl-*N,N'*-bis(dehydroabietyl)thiouronium bromide (m.pt. 128°C) (TEQXOE).

11.22.6 Protic Salts

[819, 820]	A very recent perspective on ionicity particularly in protic ionic liquids [821], discusses the difficulty, highlighted in this review, of inferring the nature of intermolecular interactions (in both solid and liquid) simply on the basis of stoichiometry. While the paper considers solvates and co-crystals, as well as ionic liquids, its conclusion that predictions of property (whether ionicity or melting point) from composition will generally be uncertain is one that should be seen as generally valid.

[822] Molecular and crystal structure of 1-amino-X-pyrazinium mesitylenesulfonates.

[823] Anion–cation interaction energies in protic ionic liquids.

[824] The physical and structural consequences of adsorption of water by an ionic liquid is well exemplified by Rogers et al., who report that the anhydrous ionic liquid procainium ethanoate [4-$H_2NC_6H_4C(O)OCH_2CH_2NEt_2H$][$O_2CMe$] ($T_g$ −25°C) forms a crystalline dihydrate (m.pt. 52°C) on exposure to moisture.

11.22.7 Energetic Salts

Synthesis and characterisation of the following salts have been reported:

[825] Dihydroxylammonium 5,5′-bistetrazole-1,1′-diolate

[826] 1,4,5-Triaminotetrazolium cation ($CN_7H_6^+$)

[827] Salts of 1- and 2-oxido-5-cyanotetrazolate

[828] 1-Amino-3-nitroguanidine (ANQ) in high-performance ionic energetic materials

11.22.8 Melting, Melting Point Estimation and Prediction

[829] Prediction of melting point by molecular-based models for ionic liquids includes a list of the melting points of greater than 700 salts.

[830] Prediction of melting points using generative topographic maps.

[831] A QSPR model for melting point prediction of ionic liquids is reported.

[832] Zhang and Maginn use crystal structure prediction to provide the starting point for a free energy-based melting point calculation.

[833] Is the liquid or the solid phase responsible for the low-melting points of ionic liquids? Alkyl-chain-length dependence of thermodynamic properties of [$C_n$mim][NTf_2].

[834] [C_2mim]Cl and [C_4mim]Cl form a eutectic, with a eutectic temperature of 42°C, a melting point depression of 46°C compared with pure [C_2mim]Cl.

ACKNOWLEDGEMENTS

I would like to thank Professor H. D. B. Jenkins, University of Warwick, for helpful comments on volume-based thermodynamics (Section 11.2.5) and the University of Liverpool for the provision of facilities.

REFERENCES

1 Ubbelohde, A.R., *The Molten State of Matter: Melting and Crystal Structure* (John Wiley & Sons, Ltd, Chichester, 1978).

2 Sundermeyer, W., Fused salts and their use as reaction media, *Angew. Chem. Int. Ed. Engl.* **4**, 222–238 (1965).

3 Lind Jr., J.E., Abdel-Rehim, H.A.A. and Rudich, S.W., Structure of organic melts[1], *J. Phys. Chem.* **70**, 3610–3619 (1966).

4 Gordon, J.E. and Subbarao, G.N., Fused organic salts. 8. Properties of molten straight-chain isomers of tetra-n-pentylammonium salts, *J. Am. Chem. Soc.* **100** (24), 7445–7454 (1978).

5 Russina, O., Triolo, A., Gontrani, L. and Caminiti, R., Mesoscopic structural heterogeneities in room-temperature ionic liquids, *J. Phys. Chem. Lett.* **3** (1), 27–33 (2012).

6 Triolo, A., Russina, O., Bleif, H.-J. and Di Cola, E., Nanoscale segregation in room temperature ionic liquids, *J. Phys. Chem. B* **111** (18), 4641–4644 (2007).

7 Ubbelohde, A.R., *The Molten State of Matter: Melting and Crystal Structure* (John Wiley & Sons, Ltd, Chichester, 1978), pp. 119–128.

8 Bentivoglio, G., Schwärzler, A., Wurst, K., Kahlenberg, V., Nauer, G., Bonn, G., Schottenberger, H. and Laus, G., Hydrogen bonding in the crystal structures of new imidazolium triflimide protonic ionic liquids, *J. Chem. Crystallogr.* **39** (9), 662–668 (2009).

9 Xu, W., Cooper, E.I. and Angell, C.A., Ionic liquids: Ion mobilities, glass temperatures, and fragilities, *J. Phys. Chem. B* **107** (25), 6170–6178 (2003).

10 Xu, W., Wang, L.M., Nieman, R.A. and Angell, C.A., Ionic liquids of chelated orthoborates as model ionic glassformers, *J. Phys. Chem. B* **107** (42), 11749–11756 (2003).

11 Pringle, J.M., Howlett, P.C., MacFarlane, D.R. and Forsyth, M., Organic ionic plastic crystals: Recent advances, *J. Mater. Chem.* **20** (11), 2056–2062 (2010).

12 MacFarlane, D.R., Meakin, P., Amini, N. and Forsyth, M., Structural studies of ambient temperature plastic crystal ion conductors, *J. Phys. Condens. Matter* **13** (36), 8257–8267 (2001).

13 Wilkes, J.S., A short history of ionic liquids – from molten salts to neoteric solvents, *Green Chem.* **4** (2), 73–80 (2002).

14 Reinsborough, V., Physical chemistry of molten organic salts, *Rev. Pure Appl. Chem.* **18**, 281–290 (1968).

15 Gordon, J.E., *Applications of Fused Salts in Organic Chemistry*, Ed. D.B. Denny (Marcel Dekker, New York, 1969).

16 Angell, C.A., Fused salts, *Annu. Rev. Phys. Chem.* **22**, 429–464 (1971).

17 Coker, T.G., Wunderlich, B. and Janz, G.J., Melting mechanisms of ionic salts: Tetra-*n*-amyl-ammonium thiocyanate, *Trans. Faraday Soc.* **65**, 3361–3368 (1969).

18 Plechkova, N.V. and Seddon, K.R., Applications of ionic liquids in the chemical industry, *Chem. Soc. Rev.* **37** (1), 123–150 (2008).

19 Su, L., Zhu, X., Wang, Z., Cheng, X., Wang, Y., Yuan, C., Chen, Z., Ma, C., Li, F., Zhou, Q. and Cui, Q., In situ observation of multiple phase transitions in low-melting ionic liquid BMIM BF4 under high pressure up to 30 GPa, *J. Phys. Chem. B* **116** (7), 2216–2222 (2012).

20 Kaminski, K., Adrjanowicz, K., Wojnarowska, Z., Dulski, M., Wrzalik, R., Paluch, M., Kaminska, E. and Kasprzycka, A., Do intermolecular interactions control crystallization abilities of glass-forming liquids? *J. Phys. Chem. B* **115** (40), 11537–11547 (2011).

21 Mudring, A.V., Solidification of ionic liquids: Theory and techniques, *Aust. J. Chem.* **63** (4), 544–564 (2010).

22 Dean, P.M., Pringle, J.M. and MacFarlane, D.R., Structural analysis of low melting organic salts: Perspectives on ionic liquids, *Phys. Chem. Chem. Phys.* **12** (32), 9144–9153 (2010).

23 Matsumoto, K. and Hagiwara, R., Structural characteristics of alkylimidazolium-based salts containing fluoroanions, *J. Fluor. Chem.* **128** (4), 317–331 (2007).

24 Dupont, J., On the solid, liquid and solution structural organization of imidazolium ionic liquids, *J. Braz. Chem. Soc.* **15** (3), 341–350 (2004).

25 Laube, T., Interactions between carbocations and anions in crystals, *Chem. Rev.* **98** (4), 1277–1312 (1998).

26 Smith, G.L., Mercier, H.P.A. and Schrobilgen, G.J., Synthesis of [F_3SNXeF][AsF_6] and structural study by multi-NMR and Raman spectroscopy, electronic structure calculations, and X-ray crystallography, *Inorg. Chem.* **46** (4), 1369–1378 (2007).

27 Mata, L., Avenoza, A., Busto, J.H., Corzana, F. and Peregrina, J.M., Quaternary chiral $\beta^{2,2}$-amino acids with pyridinium and imidazolium substituents, *Chem. Eur. J.* **18** (49), 15822–15830 (2012).

28 Binnemans, K., Ionic liquid crystals, *Chem. Rev.* **105** (11), 4148–4204 (2005).

29 Allen, F.H., The Cambridge Structural Database: A quarter of a million crystal structures and rising, *Acta Crystallogr.* **B58**, 380–388 (2002).

30 Web, http://cds.rsc.org (accessed 11 July 2015).

31 Preiss, U.P., Beichel, W., Erle, A.M.T., Paulechka, Y.U. and Krossing, I., Is universal, simple melting point prediction possible? *ChemPhysChem* **12** (16), 2959–2972 (2011).

32 Silvester, D.S. and Compton, R.G., Electrochemistry in room temperature ionic liquids: A review and some possible applications, *Z. Phys. Chem.* **220** (10–11), 1247–1274 (2006).

33 Holbrey, J.D. Rogers, R.D., Mantz, R.A., Trulove, P.C., Cocalia, V.A., Visser, A.E., Anderson, J.L., Anthony, J.L., Brennecke, J.F., Maginn, E.J., Welton, T. and Mantz, R.A., Physicochemical Properties, in *Ionic Liquids in Synthesis*, Eds. P. Wasserscheid and T. Welton, 2nd Edit., Vol. **1** (Wiley-VCH, New York, 2008), pp. 57–72.

34 Pas, S.J., Dargusch, M.S. and MacFarlane, D.R., Crystallisation kinetics of some archetypal ionic liquids: Isothermal and non-isothermal determination of the Avrami exponent, *Phys. Chem. Chem. Phys.* **13** (25), 12033–12040 (2011).

35 Atkins, P. and de Paula, J., *Atkins' Physical Chemistry*, 9th Edit. (Oxford University Press, Oxford, 2010).

36 Ubbelohde, A.R., *The Molten State of Matter: Melting and Crystal Structure* (John Wiley & Sons, Ltd, Chichester, 1978), pp. 6–53.

37 Svensson, P.H. and Kloo, L., Synthesis, structure, and bonding in polyiodide and metal iodide-iodine systems, *Chem. Rev.* **103** (5), 1649–1684 (2003).

38 Alcock, N.W., Secondary bonding to nonmetallic elements, *Adv. Inorg. Chem. Radiochem.* **15**, 1–58 (1972).

39 Laus, G., Kahlenberg, V., Többens, D.M., Jetti, R.K.R., Griesser, U.J., Schütz, J., Kristeva, E., Wurst, K. and Schottenberger, H., Lattice architecture and

hydrogen-bonding networks of *N*-aminoazolium and *N,N′*-diaminoazolium chlorides, *Cryst. Growth Des.* **6** (2), 404–410 (2006).

40 Izgorodina, E.I., Towards large-scale, fully ab initio calculations of ionic liquids, *Phys. Chem. Chem. Phys.* **13** (10), 4189–4207 (2011).

41 Katritzky, A.R., Jain, R., Lomaka, A., Petrukhin, R., Maran, U. and Karelson, M., Perspective on the relationship between melting points and chemical structure, *Cryst. Growth Des.* **1** (4), 261–265 (2001).

42 López-Martin, I., Burello, E., Davey, P.N., Seddon, K.R. and Rothenberg, G., Anion and cation effects on imidazolium salt melting points: A descriptor modelling study, *ChemPhysChem* **8** (5), 690–695 (2007).

43 Torrecilla, J.S., Rodriguez, F., Bravo, J.L., Rothenberg, G., Seddon, K.R. and López-Martin, I., Optimising an artificial neural network for predicting the melting point of ionic liquids, *Phys. Chem. Chem. Phys.* **10** (38), 5826–5831 (2008).

44 Preiss, U., Bulut, S. and Krossing, I., In silico prediction of the melting points of ionic liquids from thermodynamic considerations: A case study on 67 salts with a melting point range of 337 degrees C, *J. Phys. Chem. B* **114** (34), 11133–11140 (2010).

45 Lazzús, J.A., A group contribution method to predict the melting point of ionic liquids, *Fluid Phase Equilib.* **313**, 1–6 (2012).

46 Zhang, Y. and Maginn, E.J., A comparison of methods for melting point calculation using molecular dynamics simulations, *J. Chem. Phys.* **136** (14), 144116 (2012).

47 Nishikawa, K., Wang, S.L., Katayanagi, H., Hayashi, S., Hamaguchi, H.O., Koga, Y. and Tozaki, K.I., Melting and freezing behaviors of prototype ionic liquids, 1-butyl-3-methylimidazolium bromide and its chloride, studied by using a nano-watt differential scanning calorimeter, *J. Phys. Chem. B* **111** (18), 4894–4900 (2007).

48 Endo, T. and Nishikawa, K., Isomer populations in liquids for 1-isopropyl-3-methylimidazolium bromide and its iodide and their conformational changes accompanying the crystallizing and melting processes, *J. Phys. Chem. A* **112** (33), 7543–7550 (2008).

49 Nishikawa, K., Wang, S.L., Endo, T. and Tozaki, K., Melting and crystallization behaviors of an ionic liquid, 1-isopropyl-3-methylimidazolium bromide, studied by using nanowatt-stabilized differential scanning calorimetry, *Bull. Chem. Soc. Jpn.* **82** (7), 806–812 (2009).

50 Imanari, M., Fujii, K., Endo, T., Seki, H., Tozaki, K.-i. and Nishikawa, K., Ultraslow dynamics at crystallization of a room-temperature ionic liquid, 1-butyl-3-methylimidazolium bromide, *J. Phys. Chem. B* **116** (13), 3991–3997 (2012).

51 Stilinović, V. and Kaitner, B., Comparative refinement of correct and incorrect structural models of tetrabutylammonium tetrabutylborate – pitfalls arising from poor-quality data, *Acta Crystallogr.* **A66**, 441–445 (2010).

52 Dean, P.M., Pringle, J.M., Forsyth, C.M., Scott, J.L. and MacFarlane, D.R., Interactions in bisamide ionic liquids-insights from a Hirshfeld surface analysis of their crystalline states, *New J. Chem.* **32** (12), 2121–2126 (2008).

53 Schmitt, A.L., Schnee, G., Welter, R. and Dagorne, S., Unusual reactivity in organoaluminium and NHC chemistry: deprotonation of $AlMe_3$ by an NHC moiety involving the formation of a sterically bulky NHC-$AlMe_3$ Lewis adduct, *Chem. Commun.* **46** (14), 2480–2482 (2010).

54 Lethesh, K.C., Van Hecke, K., Van Meervelt, L., Nockemann, P., Kirchner, B., Zahn, S., Parac-Vogt, T.N., Dehaen, W. and Binnemans, K., Nitrile-functionalized pyridinium, pyrrolidinium, and piperidinium ionic liquids, *J. Phys. Chem. B* **115** (26), 8424–8438 (2011).

55 Wolff, M., Okrut, A. and Feldmann, C., $[(Ph)_3PBr][Br_7]$, $[(Bz)(Ph)_3P]_2[Br_8]$, $[(n\text{-}Bu)_3MeN]_2[Br_{20}]$, $[C_4MPyr]_2[Br_{20}]$, and $[(Ph)_3PCl]_2[Cl_2I_{14}]$: Extending the horizon of polyhalides via synthesis in ionic liquids, *Inorg. Chem.* **50** (22), 11683–11694 (2011).

56 Enomoto, T., Nakamori, Y., Matsumoto, K. and Hagiwara, R., Ion-ion interactions and conduction mechanism of highly conductive fluorohydrogenate ionic liquids, *J. Phys. Chem. C* **115** (10), 4324–4332 (2011).

57 Wiechert, D., Mootz, D., Franz, R. and Siegemund, G., Amine-poly(hydrogen fluoride) solid complexes: New studies of formation, crystal structures, and $H_{n-1}F_n^-$ ion diversity, *Chem. Eur. J.* **4** (6), 1043–1047 (1998).

58 Choudhury, A.R., Winterton, N., Steiner, A., Cooper, A.I. and Johnson, K.A., *In situ* crystallization of ionic liquids with melting points below −25°C, *CrystEngComm* **8** (10), 742–745 (2006).

59 Paulechka, Y.U., Kabo, G.J., Blokhin, A.V., Shaplov, A.S., Lozinskaya, E.I., Golovanov, D.G., Lyssenko, K.A., Korlyukov, A.A. and Vygodskii, Y.S., IR and X-ray study of polymorphism in 1-alkyl-3-methylimidazolium bis(trifluoromethanesulfonyl) imides, *J. Phys. Chem. B* **113** (28), 9538–9546 (2009).

60 Taesler, I., Delaplane, R.G. and Olovsson, I., Hydrogen-bond studies. 94. Diaquaoxonium ion in nitric-acid trihydrate, *Acta Crystallogr.* **B31**, 1489–1492 (1975).

61 Boenigk, D. and Mootz, D., Fluorides and fluoro acids. Part 18. The system pyridine hydrogen-fluoride at low-temperatures – formation and crystal-structures of solid complexes with very strong NHF and FHF hydrogen-bonding, *J. Am. Chem. Soc.* **110** (7), 2135–2139 (1988).

62 Choudhury, A.R., Winterton, N., Steiner, A., Cooper, A.I. and Johnson, K.A., In situ crystallization of low-melting ionic liquids, *J. Am. Chem. Soc.* **127** (48), 16792–16793 (2005).

63 Neculai, D., Neculai, A.M., Roesky, H.W., Magull, J. and Bunkóczi, G., Synthesis and structure of a new fluorinated beta-ketoiminato ligand and its lithium derivative, *J. Fluor. Chem.* **118** (1–2), 131–134 (2002).

64 Hammerl, A., Holl, G., Hübler, K., Kaiser, M., Klapötke, T.M. and Mayer, P., Methylated derivatives of hydrazinium azide, *Eur. J. Inorg. Chem.* (3), 755–760 (2001).

65 Delaplane, R.G., Lundgren, J.O. and Olovsson, I., Hydrogen-bond studies. 97. Crystal-structure of trifluoromethanesulfonic acid dihydrate, $H_5O_2^+CF_3SO_3^-$, at 225 and 85 K, *Acta Crystallogr.* **B31**, 2202–2207 (1975).

66 Santos, C.S., Rivera-Rubero, S., Dibrov, S. and Baldelli, S., Ions at the surface of a room-temperature ionic liquid, *J. Phys. Chem. C* **111** (21), 7682–7691 (2007).

67 Matsumoto, K., Hagiwara, R., Mazej, Z., Benkič, P. and Žemva, B., Crystal structures of frozen room temperature ionic liquids, 1-ethyl-3-methylimidazolium tetrafluoroborate ($EMImBF_4$), hexafluoroniobate ($EMImNbF_6$) and hexafluorotantalate

($EMImTaF_6$), determined by low-temperature X-ray diffraction, *Solid State Sci.* **8** (10), 1250–1257 (2006).

68 Jönsson, P.-G. and Olovsson, I., Hydrogen bond studies. XIX. The crystal structure of the 1:1 addition compound of acetic acid with sulphuric acid, $CH_3C(OH)_2^+HSO_4^-$, *Acta Crystallogr.* **B24**, 559–564 (1968).

69 Svensson, P.H. and Kloo, L., A vibrational spectroscopic, structural and quantum chemical study of the triiodide ion, *J. Chem. Soc. Dalton Trans.* (14), 2449–2455 (2000).

70 Lork, E., Böhler, D. and Mews, R., Fluorophosphazenate ions: A route to complexation of fluoride ions, *Angew. Chem. Int. Ed.* **34** (23–24), 2696–2698 (1995).

71 Scholz, F., Himmel, D., Scherer, H. and Krossing, I., Superacidic or not ...? Synthesis, characterisation, and acidity of the room-temperature ionic liquid $[C(CH_3)_3]^+[Al_2Br_7]^-$, *Chem. Eur. J.* **19** (1), 109–116 (2013).

72 Dibrov, S.M. and Kochi, J.K., Crystallographic view of fluidic structures for room-temperature ionic liquids: 1-butyl-3-methylimidazolium hexafluorophosphate, *Acta Crystallogr.* **C62**, o19–o21 (2006).

73 Zhong, C., Sasaki, T., Jimbo-Kobayashi, A., Fujiwara, E., Kobayashi, A., Tada, M. and Iwasawa, Y., Syntheses, structures, and properties of a series of metal ion-containing dialkylimidazolium ionic liquids, *Bull. Chem. Soc. Jpn.* **80** (12), 2365–2374 (2007).

74 Wolff, M., Meyer, J. and Feldmann, C., $[C_4MPyr_2][Br_{20}]$: Ionic-liquid-based synthesis of a three-dimensional polybromide network, *Angew. Chem. Int. Ed.* **50** (21), 4970–4973 (2011).

75 Matsumoto, K., Hagiwara, R., Yoshida, R., Ito, Y., Mazej, Z., Benkič, P., Žemva, B., Tamada, O., Yoshino, H. and Matsubara, S., Syntheses, structures and properties of 1-ethyl-3-methylimidazolium salts of fluorocomplex anions, *Dalton Trans.* (1), 144–149 (2004).

76 Bäcker, T., Breunig, O., Valldor, M., Merz, K., Vasylyeva, V. and Mudring, A.-V., In-situ crystal growth and properties of the magnetic ionic liquid $[C_2mim][FeCl_4]$, *Cryst. Growth Des.* **11** (6), 2564–2571 (2011).

77 Bulut, S., Klose, P., Huang, M.M., Weingärtner, H., Dyson, P.J., Laurenczy, G., Friedrich, C., Menz, J., Kümmerer, K. and Krossing, I., Synthesis of room-temperature ionic liquids with the weakly coordinating $[Al(OR^F)_4]^-$ anion ($R^F = C(H)(CF_3)_2$) and the determination of their principal physical properties, *Chem. Eur. J.* **16** (44), 13139–13154 (2010).

78 Rotter, C., Schuster, M., Kidik, M., Schön, O., Klapötke, T.M. and Karaghiosoff, K., Structural and NMR spectroscopic investigations of chair and twist conformers of the $P_2Se_8^{2-}$ anion, *Inorg. Chem.* **47** (5), 1663–1673 (2008).

79 Schaltin, S., Brooks, N.R., Stappers, L., Van Hecke, K., Van Meervelt, L., Binnemans, K. and Fransaer, J., High current density electrodeposition from silver complex ionic liquids, *Phys. Chem. Chem. Phys.* **14** (5), 1706–1715 (2012).

80 Gilardi, R. and Butcher, R.J., A new class of flexible energetic salts, part 6: The structures of the hydrazinium and hydroxylammonium salts of dinitramide, *J. Chem. Crystallogr.* **30** (9), 599–604 (2000).

81 Nakakoshi, M., Shiro, M., Fujimoto, T., Machinami, T., Seki, H., Tashiro, M. and Nishikawa, K., Crystal structure of 1-butyl-3-methylimidazolium iodide, *Chem. Lett.* **35** (12), 1400–1401 (2006).

82 Holbrey, J.D., Reichert, W.M. and Rogers, R.D., Crystal structures of imidazolium bis(trifluoromethanesulfonyl)amide 'ionic liquid' salts: The first organic salt with a *cis*-TFSI anion conformation, *Dalton Trans.* 2267–2271 (2004).

83 Froschauer, C., Wurst, K., Laus, G., Weber, H.K. and Schottenberger, H., Crystal structure of 1-butyl-2,3-dimethylimidazolium O,Se-dimethyl selenophosphate, $[C_9H_{17}N_2][C_2H_6O_3PSe]$, a new ionic liquid, *Z. Kristallogr. NCS* **227** (1), 21–23 (2012).

84 Nockemann, P., Thijs, B., Driesen, K., Janssen, C.R., Van Hecke, K., Van Meervelt, L., Kossmann, S., Kirchner, B. and Binnemans, K., Choline saccharinate and choline acesulfamate: Ionic liquids with low toxicities, *J. Phys. Chem. B* **111** (19), 5254–5263 (2007).

85 Stricker, M., Linder, T., Oelkers, B. and Sundermeyer, J., Cu(I)/(II) based catalytic ionic liquids, their metallo-laminate solid state structures and catalytic activities in oxidative methanol carbonylation, *Green Chem.* **12** (9), 1589–1598 (2010).

86 Hensen, K., Zengerly, T., Pickel, P. and Klebe, G., $[Me_3Si(py)]^+X^-$ (X = Br, I) X-ray structural analysis of the 1 : 1-adducts of bromotrimethylsilane or iodotrimethylsilane and pyridine, *Angew. Chem. Int. Ed. Engl.* **22** (9), 725–726 (1983).

87 Hensen, K., Zengerly, T., Pickel, P. and Klebe, G., $[Me_3Si(py)]^+X^-$ (X = Br, I): Röntgen-Strukturanalyse der 1 : 1-Addukte von Brom- und Iodtrimethylsilan mit Pyridin, *Angew. Chem. Int. Ed. Engl.* **95**, 739 (1983).

88 Hensen, K., Zengerly, T., Pickel, P. and Klebe, G., $[Me_3Si(py)]^+X^-$ (X = Br, I): Röntgen-Strukturanalyse der 1 : 1-Addukte von Brom- und Iodtrimethylsilan mit Pyridin, *Angew. Chem. Int. Ed. Engl.* **22**, 973–984 (1983).

89 Brooks, N.R., Schaltin, S., Van Hecke, K., Van Meervelt, L., Fransaer, J. and Binnemans, K., Heteroleptic silver-containing ionic liquids, *Dalton Trans.* **41** (23), 6902–6905 (2012).

90 Stoimenovski, J., Dean, P.M., Izgorodina, E.I. and MacFarlane, D.R., Protic pharmaceutical ionic liquids and solids: Aspects of protonics, *Faraday Discuss.* **154**, 335–352 (2012).

91 Harmer, M.A., Junk, C.P., Rostovtsev, V.V., Marshall, W.J., Grieco, L.M., Vickery, J., Miller, R. and Work, S., Catalytic reactions using superacids in new types of ionic liquids, *Green Chem.* **11** (4), 517–525 (2009).

92 Hardacre, C., Holbrey, J.D., Mullan, C.L., Nieuwenhuyzen, M., Reichert, W.M., Seddon, K.R. and Teat, S.J., Ionic liquid characteristics of 1-alkyl-*n*-cyanopyridinium and 1-alkyl-*n*-(trifluoromethyl)pyridinium salts, *New J. Chem.* **32** (11), 1953–1967 (2008).

93 Laus, G., Schwärzler, A., Schuster, P., Bentivoglio, G., Hummel, M., Wurst, K., Kahlenberg, V., Lörting, T., Schütz, J., Peringer, P., Bonn, G., Nauer, G. and Schottenberger, H., N,N′-Di(alkyloxy)imidazolium salts: New patent-free ionic liquids and NHC precatalysts, *Z. Naturforsch. B* **62** (3), 295–308 (2007).

94 Kreher, U., Raston, C.L., Strauss, C.R. and Nichols, P.J., *N,N*-dimethylammonium *N′,N′*-dimethylcarbamate, *Acta Crystallogr.* **E58**, o948–o949 (2002).

95 Deeg, A., Dahlems, T. and Mootz, D., Crystal structure of trimethylamine-hydrogen chloride (2/3), $[(Me_3NH)_2Cl][HCl_2]$, *Z. Krystallogr. NCS* **212** (3), 401–402 (1997).

96 Gurau, G., Rodríguez, H., Kelley, S.P., Janiczek, P., Kalb, R.S. and Rogers, R.D., Demonstration of chemisorption of carbon dioxide in 1,3-dialkylimidazolium acetate ionic liquids, *Angew. Chem. Int. Ed. Engl.* **50** (50), 12024–12026 (2011).

97 Nockemann, P., Binnemans, K., Thijs, B., Parac-Vogt, T.N., Merz, K., Mudring, A.-V., Menon, P.C., Rajesh, R.N., Cordoyiannis, G., Thoen, J., Leys, J. and Glorieux, C., Temperature-driven mixing-demixing behavior of binary mixtures of the ionic liquid choline bis(trifluoromethylsulfonyl)imide and water, *J. Phys. Chem. B* **113** (5), 1429–1437 (2009).

98 Fei, Z., Kuang, D., Zhao, D., Klein, C., Ang, W.H., Zakeeruddin, S.M., Grätzel, M. and Dyson, P.J., A supercooled imidazolium iodide ionic liquid as a low-viscosity electrolyte for dye-sensitized solar cells, *Inorg. Chem.* **45** (26), 10407–10409 (2006).

99 Reichert, W.M., Holbrey, J.D., Swatloski, R.P., Gutowski, K.E., Visser, A.E., Nieuwenhuyzen, M., Seddon, K.R. and Rogers, R.D., Solid-state analysis of low-melting 1,3-dialkylimidazolium hexafluorophosphate salts (ionic liquids) by combined x-ray crystallographic and computational analyses, *Cryst. Growth Des.* **7** (6), 1106–1114 (2007).

100 Ye, C., Xiao, X.C., Twamley, B., LaLonde, A.D., Norton, M.G. and Shreeve, J.M., Basic ionic liquids: Facile solvents for carbon-carbon bond formation reactions and ready access to palladium nanoparticles, *Eur. J. Org. Chem.* (30), 5095–5100 (2007).

101 Nockemann, P., Thijs, B., Parac-Vogt, T.N., Van Hecke, K., Van Meervelt, L., Tinant, B., Hartenbach, I., Schleid, T., Ngan, V.T., Nguyen, M.T. and Binnemans, K., Carboxyl-functionalized task-specific ionic liquids for solubilizing metal oxides, *Inorg. Chem.* **47** (21), 9987–9999 (2008).

102 Hiemisch, O., Henschel, D., Blaschette, A. and Jones, P.G., Polysulfonylamines. CXIII. Coordination compounds derived from trimethyltin(IV) di(fluorosulfonyl) amide: Ionic complexes with monodentate uncharged ligands, *Z. Anorg. Allg. Chem.* **625** (8), 1391–1394 (1999).

103 Kazheva, O.N., Aleksandrov, G.G., D'yachenko, O.A., Chernov'yants, M.S., Lykova, E.O., Tolpygin, I.E. and Raskita, I.M., Crystal and molecular structures of *N*-ethylquinolinium triiodide, *Russ. J. Coord. Chem.* **30**, 599–603 (2004).

104 Timofte, T., Pitula, S. and Mudring, A.-V., Ionic liquids with perfluorinated alkoxy-aluminates, *Inorg. Chem.* **46** (26), 10938–10940 (2007).

105 Mahjoor, P. and Latturner, S.E., Synthesis and structural characterization of $[bpyr]_4[V_4O_4Cl_{12}]$ and $[bpyr]_4[Bi_4Cl_{16}]$ grown in ionic liquid $[bpyr][AlCl_4]$ (bpyr = 1-butylpyridinium), *Cryst. Growth. Des.* **9** (3), 1385–1389 (2009).

106 Dong, T.Y., Chang, L.S., Lee, G.H. and Peng, S.M., Pronounced effects of zero-point energy difference on intramolecular electron transfer in asymmetric mixed-valence biferrocenium cations: Structural, EPR, and ^{57}Fe Mössbauer characteristics, *Organometallics* **21** (20), 4192–4200 (2002).

107 Spencer, J.B. and Lundgren, J.O., Hydrogen-bond studies. LXXIII. Crystal-structure of trifluoromethanesulfonic acid monohydrate, $H_3O^+CF_3SO_3^-$, at 298 and 83°K, *Acta Crystallogr.* **B29**, 1923–1928 (1973).

108 Bartmann, K. and Mootz, D., Fluorides and fluoro acids. XXVIII. On formation and structure of crystalline adducts composed of an acid, CH_3COOH, and a superacid, CF_3SO_3H as well as HF, *Z. Anorg. Allg. Chem.* **601** (10), 31–40 (1991).

109 Brand, H., Liebman, J.F., Schulz, A., Mayer, P. and Villinger, A., Nonlinear, resonance-stabilized pseudohalides: From alkali methanides to ionic liquids of methanides, *Eur. J. Inorg. Chem.* **2006** (21), 4294–4308 (2006).

110 Hawelek, L., Shirota, H., Kusz, J., Grzybowska, K., Mierzwa, M., Paluch, M., Burian, A. and Ziolo, J., High-pressure crystallization of 1-methyl-3-trimethylsilylmethylimidazolium tetrafluoroborate ionic liquid, *Chem. Phys. Lett.* **546**, 150–152 (2012).

111 Wang, X., Vogel, C.S., Heinemann, F.W., Wasserscheid, P. and Meyer, K., Solid-state structures of double-long-chain imidazolium ionic liquids: Influence of anion shape on cation geometry and crystal packing, *Cryst. Growth Des.* **11** (5), 1974–1988 (2011).

112 Golovanov, D.G., Lyssenko, K.A., Vygodskii, Y.S., Lozinskaya, E.I., Shaplov, A.S. and Antipin, M.Y., Crystal structure of 1,3-dialkyldiazolium bromides, *Russ. Chem. Bull.* **55** (11), 1989–1999 (2006).

113 Boon, J.A., Carlin, R.T., Elias, A.M. and Wilkes, J.S., Dialkylimidazolium-sodium chloroaluminate ternary salt system – phase-diagram and crystal-structure, *J. Chem. Crystallogr.* **25** (2), 57–62 (1995).

114 Gorlov, M., Pettersson, H., Hagfeldt, A. and Kloo, L., Electrolytes for dye-sensitized solar cells based on interhalogen ionic salts and liquids, *Inorg. Chem.* **46** (9), 3566–3575 (2007).

115 Bengtsson, L.A., Oskarsson, A., Stegemann, H. and Redeker, A., The structure of $(CH_3)_3SI_3$ – comparison between the structure in the solid and liquid-state, *Inorg. Chim. Acta* **215** (1–2), 33–40 (1994).

116 Klapötke, T.M., Nöth, H., Schwenk-Kircher, H., Walther, W.H. and Holl, G., Synthesis and X-ray structure of 1,1-dimethylhydrazinium azide, *Polyhedron* **18** (5), 717–719 (1999).

117 Wilkes, J.S. and Zaworotko, M.J., Air and water stable 1-ethyl-3-methylimidazolium based ionic liquids, *Chem. Commun.* (**13**), 965–967 (1992).

118 Nockemann, P., Thijs, B., Postelmans, N., Van Hecke, K., Van Meervelt, L. and Binnemans, K., Anionic rare-earth thiocyanate complexes as building blocks for low-melting metal-containing ionic liquids, *J. Am. Chem. Soc.* **128** (42), 13658–13659 (2006).

119 Kölle, P. and Dronskowski, R., Hydrogen bonding in the crystal structures of the ionic liquid compounds butyldimethylimidazolium hydrogen sulfate, chloride, and chloroferrate(II,III), *Inorg. Chem.* **43** (9), 2803–2809 (2004).

120 Zhang, P., Gong, Y., Lv, Y., Guo, Y., Wang, Y., Wang, C. and Li, H., Ionic liquids with metal chelate anions, *Chem. Commun.* **48** (17), 2334–2336 (2012).

121 Hummel, M., Froschauer, C., Laus, G., Röder, T., Kopacka, H., Hauru, L.K.J., Weber, H.K., Sixta, H. and Schottenberger, H., Dimethyl phosphorothioate and phosphoroselenoate ionic liquids as solvent media for cellulosic materials, *Green Chem.* **13** (9), 2507–2517 (2011).

122 Mallick, B., Metlen, A., Nieuwenhuyzen, M., Rogers, R.D. and Mudring, A.-V., Mercuric ionic liquids: $[C_nmim][HgX_3]$, where $n = 3, 4$ and X = Cl, Br, *Inorg. Chem.* **51** (1), 193–200 (2012).

123 Kölle, P. and Dronskowski, R., Synthesis, crystal structures and electrical conductivities of the ionic liquid compounds butyldimethylimidazolium tetrafluorobo-

rate, hexafluorophosphate and hexafluoroantimonate, *Eur. J. Inorg. Chem.* (11), 2313–2320 (2004).

124 Laus, G., Bentivoglio, G., Kahlenberg, V., Wurst, K., Nauer, G., Schottenberger, H., Tanaka, M. and Siehl, H.-U., Conformational flexibility and cation-anion interactions in 1-butyl-2,3-dimethylimidazolium salts, *Cryst. Growth Des.* **12** (4), 1838–1846 (2012).

125 Deng, F.-G., Hu, B., Sun, W., Chen, J. and Xia, C.-G., Novel pyridinium based cobalt carbonyl ionic liquids: synthesis, full characterization, crystal structure and application in catalysis, *Dalton Trans.* (38), 4262–4267 (2007).

126 Apostolico, L., Mahon, M.F., Molloy, K.C., Binions, R., Blackman, C.S., Carmalt, C.J. and Parkin, I.P., The reaction of $GeCl_4$ with primary and secondary phosphines, *Dalton Trans.* (3), 470–475 (2004).

127 Pringle, J.M., Golding, J., Baranyai, K., Forsyth, C.M., Deacon, G.B., Scott, J.L. and MacFarlane, D.R., The effect of anion fluorination in ionic liquids – physical properties of a range of bis(methanesulfonyl)amide salts, *New J. Chem.* **27** (10), 1504–1510 (2003).

128 Matsumoto, K. and Hagiwara, R., A new series of ionic liquids based on the difluorophosphate anion, *Inorg. Chem.* **48** (15), 7350–7358 (2009).

129 Stracke, M.P., Migliorini, M.V., Lissner, E., Schrekker, H.S., Back, D., Lang, E.S., Dupont, J. and Gonçalves, R.S., Electrochemical methodology for determination of imidazolium ionic liquids (solids at room temperature) properties: Influence of the temperature, *New J. Chem.* **33** (1), 82–87 (2009).

130 Fuller, J., Carlin, R.T., Simpson, L.J. and Furtak, T.E., Incorporation of imidazolium cations into an enantiomeric tartrate host lattice – designing new nonlinear-optical materials, *Chem. Mater.* **7** (5), 909–919 (1995).

131 de Bonn, O., Hammerl, A., Klapötke, T.M., Mayer, P., Piotrowski, H. and Zewen, H., Plume deposits from bipropellant rocket engines: Methylhydrazinium nitrate and *N,N*-dimethylhydrazinium nitrate, *Z. Anorg. Allg. Chem.* **627** (8), 2011–2015 (2001).

132 King, A.W.T., Asikkala, J., Mutikainen, I., Järvi, P. and Kilpeläinen, I., Distillable acid-base conjugate ionic liquids for cellulose dissolution and processing, *Angew. Chem. Int. Ed.* **50** (28), 6301–6305 (2011).

133 Berg, R.W., Riisager, A., Van Buu, O.N., Fehrmann, R., Harris, P., Tomaszowska, A.A. and Seddon, K.R., Crystal structure, vibrational spectroscopy and ab initio density functional theory calculations on the ionic liquid forming 1,1,3,3-tetramethylguanidinium bis{(trifluoromethyl)sulfonyl}amide, *J. Phys. Chem. B* **113** (26), 8878–8886 (2009).

134 Holbrey, J.D., Reichert, W.M., Swatloski, R.P., Broker, G.A., Pitner, W.R., Seddon, K.R. and Rogers, R.D., Efficient, halide free synthesis of new, low cost ionic liquids: 1,3-dialkylimidazolium salts containing methyl- and ethyl-sulfate anions, *Green Chem.* **4** (5), 407–413 (2002).

135 Stegemann, H., Oprea, A., Nagel, K. and Tebbe, K.F., Studies on polyhalides. XXVIII. Further investigations on 1,3,5-trialkyl-tetrahydro-1,3,5-triazinium polyiodides RTazI(x): Preparation and crystal structure analysis of the compounds EtTazI(3), i-PrTazI(x) with x=3 and 5 and t-BuTazI(x) with x=(1),3,5 and 7, *Z. Anorg. Allg. Chem.* **623** (1), 89–103 (1997).

136 Ma, K., Lee, K.-M., Minkova, L. and Weiss, R.G., Design criteria for ionic liquid crystalline phases of phosphonium salts with three equivalent long n-alkyl chains, *J. Org. Chem.* **74** (5), 2088–2098 (2009).

137 Kozlova, S.A., Verevkin, S.P., Heintz, A., Peppel, T. and Köckerling, M., Paramagnetic ionic liquid 1-butyl-3-methylimidazolium tetrabromidocobaltate(II): activity coefficients at infinite dilution of organic solutes and crystal structure, *J. Chem. Eng. Data* **54** (5), 1524–1528 (2009).

138 Jodry, J.J. and Mikami, K., New chiral imidazolium ionic liquids: 3D-network of hydrogen bonding, *Tetrahedron. Lett.* **45** (23), 4429–4431 (2004).

139 Wessel, J., Behrens, U., Lork, E., Borrmann, T., Stohrer, W.D. and Mews, R., Tris(dimethylamino)oxosulfoniumdifluorotrimethylsilicate,$(Me_2N)_3SO^+Me_3SiF_2^-$ (TAOS fluoride), *Inorg. Chem.* **41** (18), 4715–4721 (2002).

140 Christie, S., Dubois, R.H., Rogers, R.D., White, P.S. and Zaworotko, M.J., Air stable liquid clathrates – solid-state structure and hydrocarbon solubility of organic cation triiodide salts, *J. Incl. Phenom. Mol. Recognit. Chem.* **11** (2), 103–114 (1991).

141 Lotsch, B.V. and Schnick, W., Towards novel C—N materials: Crystal structures of two polymorphs of guanidinium dicyanamide and their thermal conversion into melamine, *New J. Chem.* **28** (9), 1129–1136 (2004).

142 Dean, P.M., Clare, B.R., Armel, V., Pringle, J.M., Forsyth, C.M., Forsyth, M. and MacFarlane, D.R., Structural characterization of novel ionic salts incorporating trihalide anions, *Aust. J. Chem.* **62** (4), 334–340 (2009).

143 Linder, T. and Sundermeyer, J., Three novel anions based on pentafluorophenyl amine combined with two new synthetic strategies for the synthesis of highly lipophilic ionic liquids, *Chem. Commun.* (20), 2914–2916 (2009).

144 Vander Hoogerstraete, T., Brooks, N.R., Norberg, B., Wouters, J., Van Hecke, K., Van Meervelt, L. and Binnemans, K., Crystal structures of low-melting ionic transition-metal complexes with *N*-alkylimidazole ligands, *CrystEngComm* **14** (15), 4902–4911 (2012).

145 Holthausen, M.H., Feldmann, K.-O., Schulz, S., Hepp, A. and Weigand, J.J., Formation of cationic $[RP_5Cl]^+$-cages via insertion of $[RPCl]^+$-cations into a P—P bond of the P_4 tetrahedron, *Inorg. Chem.* **51** (6), 3374–3387 (2012).

146 Klapötke, T.M., Stein, M. and Stierstorfer, J., Salts of 1*H*-tetrazole – synthesis, characterization and properties, *Z. Anorg. Allg. Chem.* **634** (10), 1711–1723 (2008).

147 Dymon, J., Wibby, R., Kleingardner, J., Tanski, J.M., Guzei, I.A., Holbrey, J.D. and Larsen, A.S., Designing ionic liquids with boron cluster anions: Alkylpyridinium and imidazolium [*nido*-$C_2B_9H_{11}$] and [*closo*-$CB_{11}H_{12}$] carborane salts, *Dalton Trans.* (22), 2999–3006 (2008).

148 Hasan, M., Kozhevnikov, I.V., Siddiqui, M.R.H., Steiner, A. and Winterton, N., Gold compounds as ionic liquids. Synthesis, structures, and thermal properties of *N,N′*-dialkylimidazolium tetrachloroaurate salts, *Inorg. Chem.* **38** (25), 5637–5641 (1999).

149 Fei, Z., Ang, W.H., Zhao, D., Scopelliti, R., Zvereva, E.E., Katsyuba, S.A. and Dyson, P.J., Revisiting ether-derivatized imidazolium-based ionic liquids, *J. Phys. Chem. B* **111** (34), 10095–10108 (2007).

150 Schneider, S., Hawkins, T., Rosander, M., Mills, J., Brand, A., Hudgens, L., Warmoth, G. and Vij, A., Liquid azide salts, *Inorg. Chem.* **47** (9), 3617–3624 (2008).

151 Miura, Y., Shimizu, F. and Mochida, T., Preparation, properties, and crystal structures of organometallic ionic liquids comprising 1-ferrocenyl-3-alkylimidazolium-based salts of bis(trifluoromethanesulfonyl)amide and hexafluorophosphate, *Inorg. Chem.* **49** (21), 10032–10040 (2010).

152 Dean, P.M., Pringle, J.M. and MacFarlane, D.R., 1-Methyl-1-propylpyrrolidinium chloride, *Acta Crystallogr.* **E64**, o637 (2008).

153 Golding, J.J., Macfarlane, D.R., Spiccia, L., Forsyth, M., Skelton, B.W. and White, A.H., Weak intermolecular interactions in sulfonamide salts: Structure of 1-ethyl-2-methyl-3-benzylimidazolium bis[(trifluoromethyl)sulfonyl]amide, *Chem. Commun.* (15), 1593–1594 (1998).

154 Conrad, E., Burford, N., McDonald, R. and Ferguson, M.J., Coordination of arsine ligands as a general synthetic approach to rare examples of arsenic-antimony and arsenic-bismuth bonds, *J. Am. Chem. Soc.* **131** (14), 5066–5067 (2009).

155 Tebbe, K.F. and Nagel, K., Studies on polyhalides. XXVI. On *N*-propylurotropinium polyiodides UrPrI(x) with x = 5 and 7: Crystal structures of a pentaiodide and a heptaiodide, *Z. Anorg. Allg. Chem.* **622** (8), 1323–1328 (1996).

156 Matsumoto, K., Tsuda, T., Hagiwara, R., Ito, Y. and Tamada, O., Structural characteristics of 1-ethyl-3-methylimidazolium bifluoride: HF-deficient form of a highly conductive room temperature molten salt, *Solid State Sci.* **4** (1), 23–26 (2002).

157 Anderson, J.L., Ding, R.F., Ellern, A. and Armstrong, D.W., Structure and properties of high stability geminal dicationic ionic liquids, *J. Am. Chem. Soc.* **127** (2), 593–604 (2005).

158 Hasegawa, A., Ishikawa, T., Ishihara, K. and Yamamoto, H., Facile synthesis of aryl- and alkyl-bis(trifluoromethylsulfonyl)methanes, *Bull. Chem. Soc. Jpn.* **78** (8), 1401–1410 (2005).

159 Sabaté, C.M., Delalu, H., Forquet, V. and Jeanneau, E., Salts of 1-(chloromethyl)-1,1-dimethylhydrazine and ionic liquids, *RSC Adv.* **2** (9), 3691–3699 (2012).

160 Kvick, Å., Joksson, P.-G. and Olovsson, I., Hydrogen-bond studies. XXXVI. Crystal structure of the 1:1 addition of acetic acid with fluorosulfuric acid, *Inorg. Chem.* **8**, 2775–2780 (1969).

161 Burford, N., Macdonald, C.L.B., Parks, T.M., Wu, G., Borecka, B., Kwiatkowski, W. and Cameron, T.S., Preparation and structure of 2-chloro-1,3-dimethyldiaza-2-arsenane, 1,3-dimethyldiaza-2-arsenanium tetrachlorogallate, and butadiene cycloadducts of diazarsenium cations, *Can. J. Chem.* **74** (11), 2209–2216 (1996).

162 Peppel, T., Roth, C., Fumino, K., Paschek, D., Köckerling, M. and Ludwig, R., The influence of hydrogen-bond defects on the properties of ionic liquids, *Angew. Chem. Int. Ed.* **50** (29), 6661–6665 (2011).

163 Craig, F.J., Kennedy, A.R., Mulvey, R.E. and Spicer, M.D., Structures of the potassium aluminates $[\{K_2(Me_3AlOBu^t)_2 \cdot pmdeta\}_\infty]$ and $[\{K(pmdeta)_2\}^+(AlMe_4)^-]$: How the nature of the alane reagent determines which of these products form from alkoxide-containing reaction mixtures, *Chem. Commun.* (16), 1951–1952 (1996).

164 Mutoh, Y., Murai, T. and Yamago, S., Acyclic telluroiminium salts: Isolation and characterization, *J. Am. Chem. Soc.* **126** (51), 16696–16697 (2004).

165 Gálvez-Ruiz, J.C., Guadarrama-Pérez, C., Nöth, H. and Flores-Parra, A., Group 13 complexes of 5-methyl-1,3,5-dithiazinane, *Eur. J. Inorg. Chem.* (3), 601–611 (2004).

166 Gálvez Ruiz, J.C., Nöth, H. and Warchhold, M., Coordination compounds of alkali metal tetrahydroborates with ethers and amines, *Eur. J. Inorg. Chem.* (2), 251–266 (2008).

167 Abdallah, D.J., Bachman, R.E., Perlstein, J. and Weiss, R.G., Crystal structures of symmetrical tetra-*n*-alkyl ammonium and phosphonium halides. Dissection of competing interactions leading to 'biradial' and 'tetraradial' shapes, *J. Phys. Chem. B* **103** (43), 9269–9278 (1999).

168 Klapötke, T.M. and Stierstorfer, J., Azidoformamidinium and 5-aminotetrazolium dinitramide-two highly energetic isomers with a balanced oxygen content, *Dalton Trans.* (4), 643–653 (2009).

169 Drake, G., Hawkins, T., Tollison, K., Hall, L., Vij, A. and Sobaski, S., (1*R*)-4-Amino-1,2,4-triazolium Salts: New Families of Ionic Liquids, in *Ionic Liquids IIIB: Fundamentals, Progress, Challenges, and Opportunities - Transformations and Processes*, Eds. R.D. Rogers and K.R. Seddon, ACS Symposium Series, Vol. **902** (American Chemical Society, Washington, 2005), pp. 259–302.

170 Raabe, I., Wagner, K., Guttsche, K., Wang, M.K., Grätzel, M., Santiso-Quiñones, G. and Krossing, I., Tetraalkylammonium salts of weakly coordinating aluminates: Ionic liquids, materials for electrochemical applications and useful compounds for anion investigation, *Chem. Eur. J.* **15** (8), 1966–1976 (2009).

171 Bulut, S., Klose, P. and Krossing, I., Na[B(hfip)$_4$] (hfip = OC(H)(CF$_3$)$_2$): A weakly coordinating anion salt and its first application to prepare ionic liquids, *Dalton Trans.* **40** (32), 8114–8124 (2011).

172 Henschel, D., Hamann, T., Moers, O., Blaschette, A. and Jones, P.G., Polysulfonylamines, C.L., Hydrogen bonding in crystalline onium dimesylamides: A robust eight-membered ring synthon in the structures of methyluronium and 1,1-dimethyluronium dimesylamide, *Z. Naturforsch. B* **57** (1), 113–121 (2002).

173 Hough-Troutman, W.L., Smiglak, M., Griffin, S., Matthew Reichert, W., Mirska, I., Jodynis-Liebert, J., Adamska, T., Nawrot, J., Stasiewicz, M., Rogers, R.D. and Pernak, J., Ionic liquids with dual biological function: Sweet and anti-microbial, hydrophobic quaternary ammonium-based salts, *New J. Chem.* **33** (1), 26–33 (2009).

174 Nockemann, P., Thijs, B., Pittois, S., Thoen, J., Glorieux, C., Van Hecke, K., Van Meervelt, L., Kirchner, B. and Binnemans, K., Task-specific ionic liquid for solubilizing metal oxides, *J. Phys. Chem. B* **110** (42), 20978–20992 (2006).

175 Theis, B., Burschka, C. and Tacke, R., Optically active zwitterionic λ^5Si,λ^5Si′-disilicates: Syntheses, crystal structures, and behavior in aqueous solution, *Chem. Eur. J.* **14** (15), 4618–4630 (2008).

176 Burford, N., Ragogna, P.J., McDonald, R. and Ferguson, M.J., Phosphine coordination complexes of the diphenylphosphenium cation: A versatile synthetic methodology for P—P bond formation, *J. Am. Chem. Soc.* **125** (47), 14404–14410 (2003).

177 Joo, Y.H., Gao, H.X., Zhang, Y.Q. and Shreeve, J.M., Inorganic or organic azide-containing hypergolic ionic liquids, *Inorg. Chem.* **49** (7), 3282–3288 (2010).

178 Klapötke, T.M., Sabaté, C.M., Penger, A., Rusan, M. and Welch, J.M., Energetic salts of low-symmetry methylated 5-aminotetrazoles, *Eur. J. Inorg. Chem.* (7), 880–896 (2009).

179 Stegemann, H., Oprea, A. and Tebbe, K.F., Formation of cyclic amidiniumiodides and amidiniumpolyiodides from 1,3,5-trimethyl-hexahydro-1,3,5-triazine and iodine, *Z. Anorg. Allg. Chem.* **621** (5), 871–876 (1995).

180 Chesman, A.S.R., Yang, M., Spiccia, N.D., Deacon, G.B., Batten, S.R. and Mudring, A.V., Lanthanoid-based ionic liquids incorporating the dicyanonitrosomethanide anion, *Chem. Eur. J.* **18** (31), 9580–9589 (2012).

181 Burns, N.Z. and Baran, P.S., On the origin of the haouamine alkaloids, *Angew. Chem. Int. Ed.* **47** (1), 205–208 (2008).

182 Fuller, J., Carlin, R.T., De Long, H.C. and Haworth, D., Structure of 1-ethyl-3-methylimidazolium hexafluorophosphate - model for room-temperature molten salts, *Chem. Commun.* (3), 299–300 (1994).

183 Larsen, A.S., Holbrey, J.D., Tham, F.S. and Reed, C.A., Designing ionic liquids: Imidazolium melts with inert carborane anions, *J. Am. Chem. Soc.* **122** (30), 7264–7272 (2000).

184 Hammerl, A., Holl, G., Kaiser, M., Klapötke, T.M., Mayer, P., Nöth, H. and Warchhold, M., New hydrazinium azide compounds, *Z. Anorg. Allg. Chem.* **627** (7), 1477–1482 (2001).

185 Schneider, S., Hawkins, T., Rosander, M., Vaghjiani, G., Chambreau, S. and Drake, G., Ionic liquids as hypergolic fuels, *Energy Fuel* **22** (4), 2871–2872 (2008).

186 Chesman, A.S.R., Yang, M., Mallick, B., Ross, T.M., Gass, I.A., Deacon, G.B., Batten, S.R. and Mudring, A.V., Melting point suppression in new lanthanoid(III) ionic liquids by trapping of kinetic polymorphs: An in situ synchrotron powder diffraction study, *Chem. Commun.* **48** (1), 124–126 (2012).

187 Xue, H., Tong, Z.F., Wei, F.Y. and Qing, S.G., Crystal structure of room-temperature ionic liquid 1-butyl-isoquinolinium gallium tetrachloride [(BIQL)$GaCl_4$], *C. R. Chim.* **11** (1–2), 90–94 (2008).

188 Alcock, N.W., Hagger, R.M., Harrison, W.D. and Wallbridge, M.G.H., Piperidinium tetramethoxyborate, *Acta Crystallogr.* **B38**, 676–677 (1982).

189 Ganesan, K., Alias, Y. and Ng, S.W., Imidazolium-based ionic liquid salts: 3,3′-dimethyl-1,1′-(1,4-phenylenedimethylene)diimidazolium bis(tetrafluoroborate) and 3,3′-di-*n*-butyl-1,1′-(1,4-phenylenedimethylene)diimidazolium bis(trifluoro methanesulfonate), *Acta Crystallogr.* **C64**, o478–o480 (2008).

190 Schneider, S., Drake, G., Hall, L., Hawkins, T. and Rosander, M., Alkene- and alkyne-substituted methylimidazolium bromides: Structural effects and physical properties, *Z. Anorg. Allg. Chem.* **633** (10), 1701–1707 (2007).

191 Gordon, C.M., Holbrey, J.D., Kennedy, A.R. and Seddon, K.R., Ionic liquid crystals: Hexafluorophosphate salts, *J. Mater. Chem.* **8** (12), 2627–2636 (1998).

192 He, L., Tao, G.-H., Parrish, D.A. and Shreeve, J.M., Liquid dinitromethanide salts, *Inorg. Chem.* **50** (2), 679–685 (2011).

193 Fox, P.A., Griffin, S.T., Reichert, W.M., Salter, E.A., Smith, A.B., Tickell, M.D., Wicker, B.F., Cioffi, E.A., Davis, J.H., Rogers, R.D. and Wierzbicki, A., Exploiting isolobal relationships to create new ionic liquids: Novel room-temperature ionic liquids based upon (*N*-alkylimidazole)(amine) BH_2^+ ‘boronium’ ions, *Chem. Commun.* (29), 3679–3681 (2005).

194 Tebbe, K.F. and Loukili, R., Studies of polyhalides. XXXV. On the dimethyl(n-propyl)phenylammoniumtriiodide n-$PrMe_2PhNI_3$ and on the series of dimethyl(isopropyl)phenylammoniumpolyiodides i-$PrMe_2PhNI_x$, with x = 3, 5, 7, 8, and 9, *Z. Anorg. Allg. Chem.* **624** (7), 1175–1186 (1998).

195 Mootz, D. and Hocken, J., The system pyridine-hydrogen chloride – formation and structure of crystalline adducts, *Z. Naturforsch. B.* **44** (10), 1239–1246 (1989).

196 Smiglak, M., Hines, C.C., Reichert, W.M., Vincek, A.S., Katritzky, A.R., Thrasher, J.S., Sun, L., McCrary, P.D., Beasley, P.A., Kelley, S.P. and Rogers, R.D., Synthesis, limitations, and thermal properties of energetically-substituted, protonated imidazolium picrate and nitrate salts and further comparison with their methylated analogs, *New J. Chem.* **36** (3), 702–722 (2012).

197 Callear, S.K., Hursthouse, M.B. and Threlfall, T.L., A systematic study of the crystallisation products of a series of dicarboxylic acids with imidazole derivatives, *CrystEngComm* **12** (3), 898–908 (2010).

198 Laus, G., Kahlenberg, V. and Schottenberger, H., Crystal structure of 1,3-di(benzyloxy)imidazolium bromide, $[C_{17}H_{17}N_2O_2]Br$, *Z. Kristallogr. NCS* **225** (4), 759–760 (2010).

199 Zhang, Y., Gao, H., Guo, Y., Joo, Y.-H. and Shreeve, J.M., Hypergolic *N,N*-dimethylhydrazinium ionic liquids, *Chem. Eur. J.* **16** (10), 3114–3120 (2010).

200 Weigand, J.J., Holthausen, M. and Fröhlich, R., Formation of $[Ph_2P_5]^+$, $[Ph_4P_6]^{2+}$, and $[Ph_6P_7]^{3+}$ cationic clusters by consecutive insertions of $[Ph_2P]^+$ into P—P bonds of the P_4 tetrahedron, *Angew. Chem. Int. Ed.* **48** (2), 295–298 (2009).

201 Corr, M.J., Roydhouse, M.D., Gibson, K.F., Zhou, S.-z., Kennedy, A.R. and Murphy, J.A., Amidine dications as superelectrophiles, *J. Am. Chem. Soc.* **131** (49), 17980–17985 (2009).

202 Fakhraian, H. and Gorgi-douze, A., Synthesis and properties of ionic liquids based on 1-methyl-3-(2-phenethyl)imidazolium, *Can. J. Chem.* **87** (11), 1650–1655 (2009).

203 Dean, P.M., Golding, J.J., Pringle, J.M., Forsyth, M., Skelton, B.W., White, A.H. and MacFarlane, D.R., The effect of alkyl chain length in a series of novel *N*-alkyl-3-benzylimidazolium iodide salts, *CrystEngcomm* **11** (11), 2456–2465 (2009).

204 Gupta, O.D., Twamley, B. and Shreeve, J.M., Perfluoro alkyl 1,3-diketonates of cyclic and acyclic secondary amines, *J. Fluor. Chem.* **127** (2), 263–269 (2006).

205 Wang, K., Zhang, Y., Chand, D., Parrish, D.A. and Shreeve, J.M., Boronium-cation-based ionic liquids as hypergolic fluids, *Chem. Eur. J.* **18** (52), 16931–16937 (2012).

206 Dechambenoit, P., Ferlay, S., Kyritsakas, N. and Hosseini, M.W., Amidinium based ionic liquids, *New J. Chem.* **34** (6), 1184–1199 (2010).

207 Harland, J.J., Payne, J.S., Day, R.O. and Holmes, R.R., Pentacoordinated molecules. LXIX. Steric hindrance in pentacoordinated fluorosilicates – synthesis and molecular-structure of the diphenyl-1-naphthyldifluorosilicate anion and the phenylmethyltrifluorosilicate anion, *Inorg. Chem.* **26** (5), 760–765 (1987).

208 Grigoriev, M.S., Moisy, P., Den Auwer, C. and Charushnikova, I.A., Hydrazinium nitrate, *Acta Crystallogr.* **E61**, i216–i217 (2005).

209 Matsumoto, K., Oka, T., Nohira, T. and Hagiwara, R., Polymorphism of alkali bis(fluorosulfonyl)amides ($M[N(SO_2F)_2]$, M = Na, K, and Cs), *Inorg. Chem.* **52** (2), 568–576 (2013).

210 Henderson, W.A., Young, V.G., Jr., Fox, D.M., De Long, H.C. and Trulove, P.C., Alkyl vs. alkoxy chains on ionic liquid cations, *Chem. Commun.* (35), 3708–3710 (2006).

211 Majhi, P.K., Sauerbrey, S., Schnakenburg, G., Arduengo, A.J., III and Streubel, R., Synthesis of backbone P-functionalized imidazol-2-ylidene complexes: En route to novel functional ionic liquids, *Inorg. Chem.* **51** (19), 10408–10416 (2012).

212 Rudd, M.D., Lindeman, S.V. and Husebye, S., Three-centre, four-electron bonding and structural characteristics of two-coordinate iodine(I) complexes with halogen and chalcogen ligands. Synthesis, spectroscopic characterization and x-ray structural studies of (triiodo)[tris(dimethylamino)phosphaneselenide]iodine and bis{(triiodo)[tri(*N*-morpholyl)phosphaneselenide]iodine(I)}/diiodine molecular complex, *Acta Chem. Scand.* **51** (6–7), 689–708 (1997).

213 Wang, X., Heinemann, F.W., Yang, M., Melcher, B.U., Fekete, M., Mudring, A.-V., Wasserscheid, P. and Meyer, K., A new class of double alkyl-substituted, liquid crystalline imidazolium ionic liquids-a unique combination of structural features, viscosity effects, and thermal properties, *Chem. Commun.* (47), 7405–7407 (2009).

214 Yoshida, Y., Fujii, J., Saito, G., Hiramatsu, T. and Sato, N., Dicyanoaurate(I) salts with 1-alkyl-3-methylimidazolium: Luminescent properties and room-temperature liquid forming, *J. Mater. Chem.* **16** (8), 724–727 (2006).

215 Lork, E., Watson, P.G. and Mews, R., $P_3N_3F_5NPF_2NPF_2NPF_5^{2-}$: A cyclophosphazene with a phosphazene side-chain dianion by F^--induced ring-opening of $P_3N_3F_6$, *J. Chem. Soc. Chem. Commun.* (17), 1717–1718 (1995).

216 Mayer, I., Markovits, G. and Kertes, A.S., X-ray study of solid trilaurylamine salt, *J. Inorg. Nucl. Chem.* **29**, 1377–1380 (1967).

217 Arnold, P.L., Rodden, M., Davis, K.M., Scarisbrick, A.C., Blake, A.J. and Wilson, C., Asymmetric lithium(I) and copper(II) alkoxy-*N*-heterocyclic carbene complexes; crystallographic characterisation and Lewis acid catalysis, *Chem. Commun.* (14), 1612–1613 (2004).

218 Weber, D., Hausner, S.H., Eisengräber-Pabst, A., Yun, S.H., Krause-Bauer, J.A. and Zimmer, H., Unexpected differences in reactivity between tin and lead organyl chlorides – crystal structures of their organylphosphonium salts, *Inorg. Chim. Acta* **357** (1), 125–134 (2004).

219 Aakeröy, C.B., Beffert, K., Desper, J. and Elisabeth, E., Hydrogen-bond directed structural selectivity in asymmetric heterocyclic cations, *Cryst. Growth Des.* **3** (5), 837–846 (2003).

220 Burford, N., Herbert, D.E., Ragogna, P.J., McDonald, R. and Ferguson, M.J., Diphosphine-phosphenium coordination complexes representing monocations with pendant donors and ligand tethered dications, *J. Am. Chem. Soc.* **126** (51), 17067–17073 (2004).

221 Tebbe, K.F. and Loukili, R., Studies on polyhalides. XLI. On ethylmethyldiphenylammoniumpolyiodides $(EtMePh_2N)I_x$ with x=3,5: Preparation and crystal structures of a triiodide $(EtMePh_2N)I_3$ and a pentaiodide $(EtMePh_2N)I_5$, *Z. Anorg. Allg. Chem.* **625** (5), 820–826 (1999).

222 Zhou, Q., Fitzgerald, K., Boyle, P.D. and Henderson, W.A., Phase behavior and crystalline phases of ionic liquid-lithium salt mixtures with 1-alkyl-3-methylimidazolium salts, *Chem. Mater.* **22** (3), 1203–1208 (2010).

223 Tian, X., Pape, T. and Mitzel, N.W., Formamidinium salts of low valent metal halide anions MX_3^-(M = Ge, Sn) and $M_2X_6^{2-}$ (M = Ga, In), *Z. Naturforsch. B* **59** (11–12), 1524–1531 (2004).

224 Schwärtzler, A., Laus, G., Kahlenberg, V., Wurst, K., Gelbrich, T., Kreutz, C., Kopacka, H., Bonn, G. and Schottenberger, H., Quaternary 4-amino-1,2,4-triazolium salts: Crystal structures of ionic liquids and *N*-heterocyclic carbene (NHC) complexes, *Z. Naturforsch. B* **64** (6), 603–616 (2009).

225 Montanari, V., DesMarteau, D.D. and Pennington, W.T., Synthesis and structure of novel perfluorinated iodinanes, *J. Mol. Struct.* **550**, 337–348 (2000).

226 Adamer, V., Wurst, K., Laus, G. and Schottenberger, H., Crystal structure of 1-heptyl-4-(1*H*,1*H*,2*H*,2*H*-perfluorooctyl)-1,2,4-triazolium bis(trifluoromethanesulfonyl)imide, $[C_{17}H_{21}F_{13}N_3][C_2F_6NO_4S_2]$, *Z. Kristallogr. NCS* **226** (2), 233–236 (2011).

227 Storozhuk, T.V., Udovenko, A.A., Mirochnik, A.G., Petrochenkova, N.V. and Karasev, V.E., Antimony(III) chloride complexes with benzylpyridines: Synthesis and luminescence. Crystal structure of bis(2-benzylpyridinium) pentachloroantimonate(III), *Russ. J. Coord. Chem.* **28** (3), 175–182 (2002).

228 Schubert, R.J. and Range, K.J., Tetra-n-butylammonium dicyanoaurate(I), $(n\text{-}C_4H_9)_4NAu(CN)_2$, *Z. Naturforsch. B* **45** (8), 1118–1122 (1990).

229 Brooks, N.R., Schaltin, S., Van Hecke, K., Van Meervelt, L., Binnemans, K. and Fransaer, J., Copper(I)-containing ionic liquids for high-rate electrodeposition, *Chem. Eur. J.* **17** (18), 5054–5059 (2011).

230 Holbrey, J.D., Reichert, W.M., Nieuwenhuyzen, M., Johnston, S., Seddon, K.R. and Rogers, R.D., Crystal polymorphism in 1-butyl-3-methylimidazolium halides: Supporting ionic liquid formation by inhibition of crystallization, *Chem. Commun.* (14), 1636–1637 (2003).

231 Saha, S., Hayashi, S., Kobayashi, A. and Hamaguchi, H., Crystal structure of 1-butyl-3-methylimidazolium chloride. A clue to the elucidation of the ionic liquid structure, *Chem. Lett.* **32** (8), 740–741 (2003).

232 Kärkkäinen, J., Asikkala, J., Laitinen, R.S. and Lajunen, M.K., Effect of temperature on the purity of product in the preparation of 1-butyl-3-methylimidazolium-based ionic liquids, *Z. Naturforsch. B* **59** (7), 763–770 (2004).

233 Golding, J., Hamid, N., MacFarlane, D.R., Forsyth, M., Forsyth, C., Collins, C. and Huang, J., *N*-methyl-*N*-alkylpyrrolidinium hexafluorophosphate salts: Novel molten salts and plastic crystal phases, *Chem. Mater.* **13** (2), 558–564 (2001).

234 Schneider, S., Hawkins, T., Rosander, M., Mills, J., Vaghjiani, G. and Chambreau, S., Liquid azide salts and their reactions with common oxidizers IRFNA and N_2O_4, *Inorg. Chem.* **47** (13), 6082–6089 (2008).

235 Fischer, N., Klapötke, T.M., Stierstorfer, J. and Wiedemann, C., 1-Nitratoethyl-5-nitriminotetrazole derivatives – shaping future high explosives, *Polyhedron* **30** (14), 2374–2386 (2011).

236 Stricker, M., Oelkers, B., Rosenau, C.P. and Sundermeyer, J., Copper(I) and silver(I) bis(trifluoromethanesulfonyl)imide and their interaction with an arene, diverse olefins, and an NTf_2^--based ionic liquid, *Chem. Eur. J.* **19** (3), 1042–1057 (2013).

237 Gooseman, N.E.J., O'Hagan, D., Peach, M.J.G., Slawin, A.M.Z., Tozer, D.J. and Young, R.J., An electrostatic gauche effect in β-fluoro- and β-hydroxy-*N*-ethylpyridinium cations, *Angew. Chem. Int. Ed.* **46** (31), 5904–5908 (2007).

238 Nyamori, V.O., Zulu, S.M. and Omondi, B., 1-(Ferrocen-1-ylmethyl)-3-methylimidazol-3-ium hexafluoridophosphate, *Acta Crystallogr.* **E68** (Pt 4), m353 (2012).

239 Li, X., Zeng, Z., Garg, S., Twamley, B. and Shreeve, J.M., Fluorine-containing ionic liquids from *N*-alkylpyrrolidine and *N*-methylpiperidine and fluorinated acetylacetones: Low melting points and low viscosities, *Eur. J. Inorg. Chem.* (21), 3353–3358 (2008).

240 Peppel, T. and Köckerling, M., Investigations on a series of ionic liquids containing the $[Co(II)Br_3quin]^-$ anion (quin = quinoline), *Cryst. Growth Des.* **11** (12), 5461–5468 (2011).

241 Drab, D.M., Smiglak, M., Shamshina, J.L., Kelley, S.P., Schneider, S., Hawkins, T.W. and Rogers, R.D., Synthesis of *N*-cyanoalkyl-functionalized imidazolium nitrate and dicyanamide ionic liquids with a comparison of their thermal properties for energetic applications, *New J. Chem.* **35** (8), 1701–1717 (2011).

242 Sabaté, C.M. and Delalu, H., Energetic salts of symmetrical dimethylhydrazine (SDMH), *Eur. J. Inorg. Chem.* (5), 866–877 (2012).

243 Vygodskii, Y.S., Lozinskaya, E.I., Shaplov, A.S., Lyssenko, K.A., Antipin, M.Y. and Urman, Y.G., Implementation of ionic liquids as activating media for polycondensation processes, *Polymer* **45** (15), 5031–5045 (2004).

244 Holbrey, J.D., Reichert, W.M., Nieuwenhuyzen, M., Sheppard, O., Hardacre, C. and Rogers, R.D., Liquid clathrate formation in ionic liquid-aromatic mixtures, *Chem. Commun.* (4), 476–477 (2003).

245 Kolomeitsev, A.A., Bissky, G., Barten, J., Kalinovich, N., Lork, E. and Röschenthaler, G.V., Hexaalkylguanidinium and 2-(dialkylamino)-1,3-dimethylimidazolinium trimethyldifluorosiliconates and perfluoroalkoxides. Accidental isolation and molecular structure of $[C(NMe_2)_3]^+F^-{\cdot}6CH_2Cl_2$, *Inorg. Chem.* **41** (23), 6118–6124 (2002).

246 Tang, S., Babai, A. and Mudring, A.-V., Europium-based ionic liquids as luminescent soft materials, *Angew. Chem. Int. Ed.* **47** (40), 7631–7634 (2008).

247 Deleeuw, D.L., Goodrow, M.H., Olmstead, M.M., Musker, W.K. and Doi, J.T., Anchimerically assisted redox reactions. VII. A comparison of the effects of neighboring primary, secondary, and tertiary amine groups on the kinetics and mechanism of thioether oxidation by aqueous iodine, *J. Org. Chem.* **48** (14), 2371–2374 (1983).

248 Yokota, Y., Jacobson, R.A., Logsdon, B.C., Ringrose, S., Setterdahl, A.T. and Verkade, J.G., Synthesis of alkali and alkaline earth metal complexes of open cryptands and their X-ray structures wherein M = Li^+, Na^+ and Ba^{2+}, *Polyhedron* **18** (19), 2519–2532 (1999).

249 Smiglak, M., Hines, C.C., Wilson, T.B., Singh, S., Vincek, A.S., Kirichenko, K., Katritzky, A.R. and Rogers, R.D., Ionic liquids based on azolate anions, *Chem. Eur. J.* **16** (5), 1572–1584 (2010).

250 Ang, H.G., Fraenk, W., Karaghiosoff, K., Klapötke, T.M., Mayer, P., Nöth, H., Sprott, J. and Warchhold, M., Synthesis, characterization, and crystal structures of Cu, Ag, and Pd dinitramide salts, *Z. Anorg. Allg. Chem.* **628** (13), 2894–2900 (2002).

251 Kärnä, M., Lahtinen, M., Hakkarainen, P.-L. and Valkonen, J., Physicochemical properties of new dicationic ether-functionalized low melting point ammonium salts, *Aust. J. Chem.* **63** (7), 1122–1137 (2010).

252 Xue, H., Gao, H., Twamley, B. and Shreeve, J.M., Energetic salts of 3-nitro-1,2,4-triazole-5-one, 5-nitroaminotetrazole, and other nitro-substituted azoles, *Chem. Mater.* **19** (7), 1731–1739 (2007).

253 Mazille, F., Fei, Z.F., Kuang, D.B., Zhao, D.B., Zakeeruddin, S.M., Grätzel, M. and Dyson, P.J., Influence of ionic liquids bearing functional groups in dye-sensitized solar cells, *Inorg. Chem.* **45** (4), 1585–1590 (2006).

254 Saha, S. and Hamaguchi, H.O., Effect of water on the molecular structure and arrangement of nitrile-functionalized ionic liquids, *J. Phys. Chem. B* **110** (6), 2777–2781 (2006).

255 Sun, H.J., Harms, K. and Sundermeyer, J., The crystal structure of a metal-containing ionic liquid: A new octachlorotricuprate(II), *Z. Kristallogr. NCS* **220** (1), 42–44 (2005).

256 Sarkar, A., Pal, S.K., Itkis, M.E., Liao, P.H., Tham, F.S., Donnadieu, B. and Haddon, R.C., Methoxy-substituted phenalenyl-based neutral radical molecular conductor, *Chem. Mater.* **21** (11), 2226–2237 (2009).

257 Mandai, T., Masu, H., Imanari, M. and Nishikawa, K., Comparison between cycloalkyl- and *n*-alkyl-substituted imidazolium-based ionic liquids in physicochemical properties and reorientational dynamics, *J. Phys. Chem. B* **116** (7), 2059–2064 (2012).

258 Wang, K.F., Zhang, L., Zhuang, R.R. and Jian, F.F., An iron(III)-containing ionic liquid: Characterization, magnetic property and electrocatalysis, *Transit. Met. Chem.* **36** (8), 785–791 (2011).

259 Hiemisch, O., Henschel, D., Jones, P.G. and Blaschette, A., Polysulfonyl amines. LXXII. Triphenylcarbenium and triphenylphosphonium di(fluorosulfonyl) amides: Two crystal structures with ordered $(FSO_2)_2N^-$ sites, *Z. Anorg. Allg. Chem.* **622** (5), 829–836 (1996).

260 Ozawa, R., Hayashi, S., Saha, S., Kobayashi, A. and Hamaguchi, H., Rotational isomerism and structure of the 1-butyl-3- methylimidazolium cation in the ionic liquid state, *Chem. Lett.* **32** (10), 948–949 (2003).

261 Sitzmann, M.E., Gilardi, R., Butcher, R.J., Koppes, W.M., Stern, A.G., Thrasher, J.S., Trivedi, N.J. and Yang, Z.Y., Pentafluorosulfanylnitramide salts, *Inorg. Chem.* **39** (4), 843–850 (2000).

262 Chiglien, G., Etienne, J., Jaulmes, S. and Laruelle, P., Crystal structure of azotide of hydrazinium, N_5H_5, *Acta Crystallogr.* **B30**, 2229–2233 (1974).

263 Elaiwi, A., Hitchcock, P.B., Seddon, K.R., Srinivasan, N., Tan, Y.-M., Welton, T. and Zora, J.A., Hydrogen bonding in imidazolium salts and its implications for ambient-temperature halogenoaluminate(III) ionic liquids, *J. Chem. Soc. Dalton Trans.* (21), 3467–3472 (1995).

264 Leclercq, L., Suisse, I., Nowogrocki, G. and Agbossou-Niedercorn, F., Halide-free highly-pure imidazolium triflate ionic liquids: Preparation and use in palladium-catalysed allylic alkylation, *Green Chem.* **9** (10), 1097–1103 (2007).

265 Holbrey, J.D., Visser, A.E., Spear, S.K., Reichert, W.M., Swatloski, R.P., Broker, G.A. and Rogers, R.D., Mercury(II) partitioning from aqueous solutions with a

new, hydrophobic ethylene-glycol functionalized bis-imidazolium ionic liquid, *Green Chem.* **5** (2), 129–135 (2003).

266 Balalaie, S., Kunze, A., Gleiter, R., Rominger, F., Geis, S. and Oeser, T., Inside protonation of 1,8-diazabicyclo[6.6.n]alka-4,11-diynes, *Eur. J. Org. Chem.* (17), 3378–3381 (2003).

267 Moers, O., Wijaya, K., Henschel, D., Blaschette, A. and Jones, P.G., Polysulfonylamines, CXVII 1 – hydrogen bonding in crystalline onium dimesylamides: Six systematically varied sec-ammonium dimesylamides exhibiting six different zero-, one-, or two-dimensional hydrogen bonding patterns, *Z. Naturforsch. B* **54** (11), 1420–1430 (1999).

268 Wang, R.H., Gao, H.X., Ye, C.F., Twamley, B. and Shreeve, J.M., Heterocyclic-based nitrodicyanomethanide and dinitrocyanomethanide salts: A family of new energetic ionic liquids, *Inorg. Chem.* **46** (3), 932–938 (2007).

269 Klapötke, T.M. and Stierstorfer, J., The new energetic compounds 1,5-diaminotetrazolium and 5-amino-1-methyltetrazolium dinitramide – synthesis, characterization and testing, *Eur. J. Inorg. Chem.* (26), 4055–4062 (2008).

270 Klapötke, T.M., Krumm, B., Mayer, P., Piotrowski, H., Polborn, K. and Schwab, I., The first selenonium azides and a selenonium selenocyanate, *Z. Anorg. Allg. Chem.* **628** (8), 1831–1834 (2002).

271 Janikowski, J., Forsyth, C., MacFarlane, D.R. and Pringle, J.M., Novel ionic liquids and plastic crystals utilizing the cyanate anion, *J. Mater. Chem.* **21** (48), 19219–19225 (2011).

272 Klapötke, T.M. and Stierstorfer, J., Synthesis and characterization of the energetic compounds aminoguanidinium, triaminoguanidinium and azidoformamidinium perchlorate, *Cent. Eur. J. Energetic Mater.* **5**, 13–30 (2008).

273 Giuseppetti, G., Tadini, C., Ferloni, P., Zabinska, G. and Torre, S., The crystal-structure of tetraethylammonium tetrafluoroborate, $(C_2H_5)_4NBF_4$, *Z. Kristallogr. NCS* **209** (6), 509–511 (1994).

274 Yoshida, Y. and Saito, G., Ionic liquids based on diethylmethyl(2-methoxyethyl) ammonium cations and bis(perfluoroalkanesulfonyl)amide anions: Influence of anion structure on liquid properties, *Phys. Chem. Chem. Phys.* **13** (45), 20302–20310 (2011).

275 Mehdi, H., Binnemans, K., Van Hecke, K., Van Meervelt, L. and Nockemann, P., Hydrophobic ionic liquids with strongly coordinating anions, *Chem. Commun.* **46** (2), 234–236 (2010).

276 Yoshida, Y., Muroi, K., Otsuka, A., Saito, G., Takahashi, M. and Yoko, T., 1-ethyl-3-methylimidazolium based ionic liquids containing cyano groups: Synthesis, characterization, and crystal structure, *Inorg. Chem.* **43** (4), 1458–1462 (2004).

277 Flemming, A., Hoffmann, M. and Köckerling, M., Tetracyanidoborates with sterically demanding phosphonium cations – thermally resistant ionic liquids, *Z. Anorg. Allg. Chem.* **636** (3–4), 562–568 (2010).

278 Mukai, T. and Nishikawa, K., Syntheses and crystal structures of two ionic liquids with halogen-bonding groups: 4,5-dibromo- and 4,5-diiodo-1-butyl-3-methylimidazolium trifluoromethanesulfonates, *Solid State Sci.* **12** (5), 783–788 (2010).

279 Shklover, V.E., Gridunova, G.V., Struchkov, I.T., Voronkov, M.G., Kryukova, Y.I. and Mirskova, A.N., Crystal and molecular-structure of tris-(2-hydroxyethyl)

ammonium (4-chlorphenylthio)ethanoate, *Proc. Natl. Acad. Sci. USSR* **269** (2), 387–390 (1983).

280 Izsák, D. and Klapötke, T.M., Characterization of the energetic plasticizer methyl-NENA, *Z. Anorg. Allg. Chem.* **637** (14–15), 2135–2141 (2011).

281 Moers, O., Wijaya, K., Hamann, T., Blaschette, A. and Jones, P.G., Polysulfonylamines, CXLIII. Role of weak hydrogen bonds (C—H···O) in the crystal structures of 2,6-dimethylpyridinium, 1-hydroxypyridinium and imidazolium dimesylamide, *Z. Naturforsch. B* **56** (10), 1052–1062 (2001).

282 Neve, F., Crispini, A. and Francescangeli, O., Structural studies on layered alkylpyridinium iodopalladate networks, *Inorg. Chem.* **39** (6), 1187–1194 (2000).

283 Belhocine, T., Forsyth, S.A., Gunaratne, H.Q.N., Nieuwenhuyzen, M., Puga, A.V., Seddon, K.R., Srinivasan, G. and Whiston, K., New ionic liquids from azepane and 3-methylpiperidine exhibiting wide electrochemical windows, *Green Chem.* **13** (1), 59–63 (2011).

284 Dega-Szafran, Z., Gdaniec, M., Grundwald-Wyspianska, M., Kosturkiewicz, Z., Koput, J., Krzyzanowski, P. and Szafran, M., X-ray, FT-IR and PM3 studies of hydrogen bonds in complexes of some pyridines with trifluoroacetic acid, *J. Mol. Struct.* **270**, 99–124 (1992).

285 Sharutin, V.V., Egorova, I.V., Klepikov, N.N., Boyarkina, E.A. and Sharutina, O.K., Bismuth compounds $[Ph_3BuP]^+I^-$, $[Ph_3BuP]^{+2}$ $[Bi_2I_8 \cdot 2Me_2C{=}O]^{2-}$, and $[Ph_3BuP]^{+2}[Bi_2I_8 \cdot 2Me_2S{=}O]^{2-}$: Syntheses and crystal structures, *Russ. J. Coord. Chem.* **35** (3), 186–190 (2009).

286 Maas, G., Rahm, R., Mayer, D. and Baumann, W., Charge dispersal in iminium-substituted alkynes – synthesis, spectral characterization, and crystal structure determination of $CO_2(CO)_6$-complexed propyne iminium salt and structural comparison with its uncomplexed counterpart, *Organometallics* **14** (2), 1061–1066 (1995).

287 Mentzafos, D., Polissiou, M., Grapsas, I., Hountas, K., Georgiadis, M. and Terzis, A., X-ray and NMR studies on the rotational isomerism about the C(5)—C(6) bond of 6-substituted methyl 2,3,4,-tri-O-acetyl-6-X-α-D-glucopyranoside derivatives, *J. Chem. Crystallogr.* **25** (4), 157–164 (1995).

288 Bonnet, L.G. and Kariuki, B.M., Ionic liquids: Synthesis and characterisation of triphenylphosphonium tosylates, *Eur. J. Inorg. Chem.* (2), 437–446 (2006).

289 Zhao, D.B., Fei, Z.F., Scopelliti, R. and Dyson, P.J., Synthesis and characterization of ionic liquids incorporating the nitrile functionality, *Inorg. Chem.* **43** (6), 2197–2205 (2004).

290 Kanatani, T., Matsumoto, K. and Hagiwara, R., Syntheses and physicochemical properties of low-melting salts based on VOF_4^- and $MoOF_5^-$, and the molecular geometries of the dimeric $(VOF_4^-)_2$ and $Mo_2O_4F_6^{2-}$ anions, *Eur. J. Inorg. Chem.* (7), 1049–1055 (2010).

291 Sokka, I., Fischer, A. and Kloo, L., Optimization of the synthesis of non-symmetrical alkyl dimethyl sulfonium halides, *Polyhedron* **26** (17), 4893–4898 (2007).

292 Gonsior, M., Krossing, I. and Mitzel, N., A thallium coated dianion: Trigonal bipyramidal $[F_2Al(OR)_3]^{2-}$ coordinated to three Tl+cations in the ion pair $[Tl_3F_2Al(OR)_3]^+$ $[Al(OR)_4]^-$ R = $CH(CF_3)_2$, *Z. Anorg. Allg. Chem.* **628** (8), 1821–1830 (2002).

293 Fürstner, A., Alcarazo, M., Radkowski, K. and Lehmann, C.W., Carbenes stabilized by ylides: Pushing the limits, *Angew. Chem. Int. Ed.* **47** (43), 8302–8306 (2008).

294 Xie, T., Brockner, W. and Gjikaj, M., New ionic liquid compounds based on tantalum pentachloride $TaCl_5$: Synthesis, structural, and spectroscopic elucidation of the (μ-oxido)-chloridotantalates(V) [BMPy][$TaCl_6$], $[BMPy]_4[(TaCl_6)_2(Ta_2OCl_{10})]$, and $[EMIm]_2[Ta_2OCl_{10}]$, *Z. Anorg. Allg. Chem.* **636** (15), 2633–2640 (2010).

295 De Vreese, P., Brooks, N.R., Van Hecke, K., Van Meervelt, L., Matthijs, E., Binnemans, K. and Van Deun, R., Speciation of copper(II) complexes in an ionic liquid based on choline chloride and in choline chloride/water mixtures, *Inorg. Chem.* **51** (9), 4972–4981 (2012).

296 Ermolaev, V., Miluykov, V., Rizvanov, I., Krivolapov, D., Zvereva, E., Katsyuba, S., Sinyashin, O. and Schmutzler, R., Phosphonium ionic liquids based on bulky phosphines: Synthesis, structure and properties, *Dalton Trans.* **39** (23), 5564–5571 (2010).

297 Henschel, D., Moers, O., Wijaya, K., Wirth, A., Blaschette, A. and Jones, P.G., Polysulfonylamines, CLIII. Weak hydrogen bonding with activated methyl donors: Crystal structures of cholinium, betainium and dimethyl 2-(dimethylamino)ethyl ammonium-dimesylamide, *Z. Naturforsch. B* **57** (5), 534–546 (2002).

298 De Roche, J., Gordon, C.M., Imrie, C.T., Ingram, M.D., Kennedy, A.R., Lo Celso, F. and Triolo, A., Application of complementary experimental techniques to characterization of the phase behavior of $[C_{16}mim][PF_6]$ and $[C_{14}mim][PF_6]$, *Chem. Mater.* **15** (16), 3089–3097 (2003).

299 Lauw, Y., Rüther, T., Horne, M.D., Wallwork, K.S., Skelton, B.W., Madsen, I.C. and Rodopoulos, T., Structural studies on the basic ionic liquid 1-ethyl-1,4-diazabicyclo 2.2.2 octanium bis(trifluoromethylsulfonyl)imide and its bromide precursor, *Cryst. Growth Des.* **12** (6), 2803–2813 (2012).

300 Moers, F.G., Behm, H., Smits, J.M.M. and Beurskens, P.T., Crystal and molecular structure of 1-vinyl-3-methylimidazolium iodide, $C_6H_9N_2I$, *J. Chem. Crystallogr.* **25** (9), 601–603 (1995).

301 Vygodskii, Y.S., Mel'nik, O.A., Lozinskaya, E.I., Shaplov, A.S., Malyshkina, I.A., Gavrilova, N.D., Lyssenko, K.A., Antipin, M.Y., Golovanov, D.G., Korlyukov, A.A., Ignat'ev, N. and Welz-Biermann, U., The influence of ionic liquid's nature on free radical polymerization of vinyl monomers and ionic conductivity of the obtained polymeric materials, *Polym. Adv. Technol.* **18** (1), 50–63 (2007).

302 Kärnä, M., Lahtinen, M., Kujala, A., Hakkarainen, P.-L. and Valkonen, J., Properties of new low melting point quaternary ammonium salts with bis(trifluoromethanesulfonyl)imide anion, *J. Mol. Struct.* **983** (1–3), 82–92 (2010).

303 Marszalek, M., Fei, Z., Zhu, D.-R., Scopelliti, R., Dyson, P.J., Zakeeruddin, S.M. and Grätzel, M., Application of ionic liquids containing tricyanomethanide $[C(CN)_3]^-$ or tetracyanoborate $[B(CN)_4]^-$ anions in dye-sensitized solar cells, *Inorg. Chem.* **50** (22), 11561–11567 (2011).

304 Waşicki, J., Jaskólski, M., Pająk, Z., Szafran, M., Dega-Szafran, Z., Adams, M.A. and Parker, S.F., Crystal structure and molecular motion in pyridine *N*-oxide semiperchlorate, *J. Mol. Struct.* **476** (1–3), 81–95 (1999).

305 Mereiter, K., Mikenda, W. and Steinwender, E., Hydrogen-bonding in 2-hydroxy((di)thio)benzoates. II. X-ray structures of tetraethylammonium salts, *J. Crystallogr. Spectr. Res.* **23** (5), 397–402 (1993).

306 Jarón, T., Wegner, W., Cyrański, M.K., Dobrzycki, Ł. and Grochala, W., Tetrabutylammonium cation in a homoleptic environment of borohydride ligands: $[(n\text{-Bu})_4N][BH_4]$ and $[(n\text{-Bu})_4N][Y(BH_4)_4]$, *J. Solid State Chem.* **191**, 279–282 (2012).

307 Kawahata, M., Endo, T., Seki, H., Nishikawa, K. and Yamaguchi, K., Polymorphic properties of ionic liquid of 1-isopropyl-3-methylimidazolium bromide, *Chem. Lett.* **38** (12), 1136–1137 (2009).

308 Gjikaj, M., Leye, J.-C., Xie, T. and Brockner, W., Structural and spectroscopic elucidation of imidazolium and pyridinium based hexachloridophosphates and niobates, *CrystEngComm* **12** (5), 1474–1480 (2010).

309 Fei, Z., Zhao, D., Pieraccini, D., Ang, W.H., Geldbach, T.J., Scopelliti, R., Chiappe, C. and Dyson, P.J., Development of nitrile-functionalized ionic liquids for C–C coupling reactions: Implication of carbene and nanoparticle catalysts, *Organometallics* **26** (7), 1588–1598 (2007).

310 Smiglak, M., Hines, C.C. and Rogers, R.D., New hydrogen carbonate precursors for efficient and byproduct-free syntheses of ionic liquids based on 1,2,3-trimethylimidazolium and *N,N*-dimethylpyrrolidinium cores, *Green Chem.* **12** (3), 491–501 (2010).

311 Andre, M., Loidl, J., Laus, G., Schottenberger, H., Bentivoglio, G., Wurst, K. and Ongania, K.H., Ionic liquids as advantageous solvents for headspace gas chromatography of compounds with low vapor pressure, *Anal. Chem.* **77** (2), 702–705 (2005).

312 Tebbe, K.F. and Buchem, R., The most iodine-rich polyiodide yet: Fc_3I_{29}, *Angew. Chem. Int. Ed.* **36** (12), 1345–1346 (1997).

313 Svansson, L., Johnston, B.D., Gu, J.H., Patrick, B. and Pinto, B.M., Synthesis and conformational analysis of a sulfonium-ion analogue of the glycosidase inhibitor castanospermine, *J. Am. Chem. Soc.* **122** (44), 10769–10775 (2000).

314 Forsyth, C.M., MacFarlane, D.R., Golding, J.J., Huang, J., Sun, J. and Forsyth, M., Structural characterization of novel ionic materials incorporating the bis(trifluoromethanesulfonyl)amide anion, *Chem. Mater.* **14** (5), 2103–2108 (2002).

315 Lim, S.F., Harris, B.L., Blanc, P. and White, J.M., Orbital interactions in selenomethyl-substituted pyridinium ions and carbenium ions with higher electron demand, *J. Org. Chem.* **76** (6), 1673–1682 (2011).

316 Mutoh, Y.C. and Murai, T., Acyclic selenoiminium salts: Isolation, first structural characterization, and reactions, *Org. Lett.* **5** (8), 1361–1364 (2003).

317 Ronig, B., Pantenburg, I. and Wesemann, L., Meltable stannaborate salts, *Eur. J. Inorg. Chem.* **2002** (2), 319–322 (2002).

318 Klapötke, T.M. and Stierstorfer, J., Triaminoguanidinium dinitramide – calculations, synthesis and characterization of a promising energetic compound, *Phys. Chem. Chem. Phys.* **10** (29), 4340–4346 (2008).

319 Göbel, M. and Klapötke, T.M., Potassium-, ammonium-, hydrazinium-, guanidinium-, aminoguanidinium-, diaminoguanidinium-, triaminoguanidinium- and melaminiumnitroformate – Synthesis, characterization and energetic properties, *Z. Anorg. Allg. Chem.* **633** (7), 1006–1017 (2007).

320 Bracuti, A.J., Crystal and molecular-structure of triethanolammonium nitrate (TEAN), *J. Crystallogr.Spectrosc. Res.* **23** (8), 669–673 (1993).

321 Nadeem, S., Munawar, M.A., Ahmad, S., Smiglak, M., Drab, D.M., Malik, K.I., Amjad, R., Ashraf, C.M. and Rogers, R.D., Solvent-free synthesis of benzothiazole-based quaternary ammonium salts: Precursors to ionic liquids, *Arkivoc*, 19–37 (2010).

322 Vitze, H., Lerner, H.-W. and Bolte, M., Dimethylammonium nitrate, *Acta Crystallogr.* **E63** (12), o4621 (2007).

323 Zhuang, R.R., Jian, F.F. and Wang, K.F., A new binuclear Cd(II)-containing ionic liquid: Preparation and electrocatalytic activities, *J. Organomet. Chem.* **694** (22), 3614–3618 (2009).

324 Gräf-Kavoosian, A., Nafepour, S., Nagel, K. and Tebbe, K.F., Studies on polyhalides, XXXVI. On the octaiodide ion I_8^{2-}: Preparation and crystal structure of [(Crypt-2.2.2)H_2]I_8, of [Ni(phen)$_3$]I_8·2$CHCl_3$ and of the bis(N-alkylurotropinium)octaiodides (UrR)$_2I_8$ with R=methyl and ethyl, *Z. Naturforsch. B* **53** (7), 641–652 (1998).

325 Blake, A.J., McNab, H. and Monahan, L.C., 3-Hydroxypyrroles and 1*H*-pyrrol-3(2*H*)-ones. Part 7. Protonation and *O*-alkylation of simple 1*H*-pyrrol-3(2*H*)-ones: crystal and molecular structure of 3-hydroxy-1-t-butyl-1,2-dihydropyrrolium picrate, *J. Chem. Soc. Perkin Trans. 2* (8), 1463–1468 (1988).

326 Antoniadis, C.D., Hadjikakou, S.K., Hadjiliadis, N., Kubicki, M. and Butler, I.S., Synthesis, X-ray characterisation and studies of the new ionic complex [bis(pyridin-2-yl) disulfide] triiodide, obtained by oxidation of 2-mercaptopyridine with I_2 – Implications in the mechanism of action of antithyroid drugs, *Eur. J. Inorg. Chem.* (21), 4324–4329 (2004).

327 Milewski-Mahrla, B. and Schmidbaur, H., Synthesis and crystal-structures of some tetramethylstibonium hydrogendicarboxylates, *Z. Naturforsch. B* **37** (11), 1393–1401 (1982).

328 Wijaya, K., Moers, O., Blaschette, A. and Jones, P.G., Polysulfonylamines, CXII – a novel solid-state aspect of 2-pyridones: Isolation and x-ray structures of bis(2-pyridone)hydrogen(I) and bis(6-methyl-2-pyridone)hydrogen(I) dimesylamides, *Z. Naturforsch. B* **54** (5), 643–648 (1999).

329 Mazur, L., Pitucha, M. and Rzaczynska, Z., 4-(2-carboxyethyl)morpholin-4-ium chloride, *Acta Crystallogr.* **E63**, o4576 (2007).

330 Abdul-Sada, A.K., Greenway, A.M., Hitchcock, P.B., Mohammed, T.J., Seddon, K.R. and Zora, J.A., Structure of room temperature haloaluminate ionic liquids, *J. Chem. Soc. Chem. Commun.* (24), 1753–1754 (1986).

331 Krieger, B.M., Lee, H.Y., Emge, T.J., Wishart, J.F. and Castner, E.W., Jr., Ionic liquids and solids with paramagnetic anions, *Phys. Chem. Chem. Phys.* **12** (31), 8919–8925 (2010).

332 Katritzky, A.R., Singh, S., Kirichenko, K., Smiglak, M., Holbrey, J.D., Reichert, W.M., Spear, S.K. and Rogers, R.D., In search of ionic liquids incorporating azolate anions, *Chem. Eur. J.* **12** (17), 4630–4641 (2006).

333 Dymek, C.J., Grossie, D.A., Fratini, A.V. and Adams, W.W., Evidence for the presence of hydrogen-bonded ion-ion interactions in the molten-salt precursor, 1-methyl-3-ethylimidazolium chloride, *J. Mol. Struct.* **213**, 25–34 (1989).

334 Ferbinteanu, M., Roesky, H.W., Cimpoesu, F., Atanasov, M., Köpke, S. and Herbst-Irmer, R., New synthetic and structural aspects in the chemistry of alkylaluminum fluorides. The mutual influence of hard and soft ligands and the hybridization as rigorous structural criterion, *Inorg. Chem.* **40** (19), 4947–4955 (2001).

335 Odabaşoğlu, M. and Büyükgüngör, O., Triethanolammonium 2-formylbenzoate, *Acta Crystallogr.* **E63**, o186–o187 (2007).

336 Palmore, G.T.R. and McBride-Wieser, M.T., Pyridinium trifluoroethanoate: Spoked columns of hydrogen-bonded cyclic dimers, *Acta Crystallogr.* **C53**, 1904–1907 (1997).

337 Henschel, D., Moers, O., Wijaya, K., Blaschette, A. and Jones, P.G., Polysulfonylamines, CXVIII – hydrogen bonding in crystalline onium dimesylamides: Three prim.-ammonium dimesylamides exhibiting two- or three-dimensional hydrogen bond patterns, *Z. Naturforsch. B* **54** (11), 1431–1440 (1999).

338 Shattock, T.R., Arora, K.K., Vishweshwar, P. and Zaworotko, M.J., Hierarchy of supramolecular synthons: Persistent carboxylic acid···pyridine hydrogen bonds in cocrystals that also contain a hydroxyl moiety, *Cryst. Growth Des.* **8** (12), 4533–4545 (2008).

339 McKarns, P.J., Heeg, M.J. and Winter, C.H., Synthesis, structure, hydrolysis, and film deposition studies of complexes of the formula $[NbCl_4(S_2R_2)_2][NbCl_6]$, *Inorg. Chem.* **37** (18), 4743–4747 (1998).

340 Laus, G., Schwärtzler, A., Bentivoglio, G., Hummel, M., Kablenberg, V., Wurst, K., Kristeva, E., Schütz, J., Kopacka, H., Kreutz, C., Bonn, G., Andriyko, Y., Nauer, G. and Schottenberger, H., Synthesis and crystal structures of 1-alkoxy-3-alkylimidazolium salts including ionic liquids, 1-alkylimidazole 3-oxides and 1-alkylimidazole perhydrates, *Z. Naturforsch. B* **63** (4), 447–464 (2008).

341 Borodkin, G.I., Nagi, S.M., Bagryanskaya, I.Y. and Gatilov, Y.V., Crystal-structure of 1-phenyl-1,2,3,4,5,6-hexamethylbenzolonium tetrachloroaluminate, *J. Struct. Chem.* **25** (3), 440–445 (1984).

342 Krossing, I., Brands, H., Feuerhake, R. and Koenig, S., New reagents to introduce weakly coordinating anions of type $Al(OR_F)_4^-$: Synthesis, structure and characterization of Cs and trityl salts, *J. Fluor. Chem.* **112** (1), 83–90 (2001).

343 Stewart, J.M., McLaughlin, K.L., Rossiter, J.J., Hurst, J.R., Haas, R.G., Rose, V.J., Ciric, B.E., Murphy, J.A. and Lawton, S.L., Preparation, characterization, and crystallographic data of some substituted pyridinium antimony(III) bromide salts, *Inorg. Chem.* **13** (11), 2767–2769 (1974).

344 Yang, L. and Foxman, B.M., Crystal and molecular structure of ammonium *trans*-2-butenoate, and a preliminary investigation of its solid-state reactivity, *Mol. Cryst. Liq. Cryst.* **456**, 25–33 (2006).

345 Krebs, B., Ahlers, F.P. and Lührs, E., Synthesis, structure, and properties of the novel bromoselenates(II) $[Se_3Br_8]^{2-}$, $[Se_4Br_{14}]^{2-}$ und $[Se_5Br_{12}]^{2-}$ – crystal structures of $[Cu(i\text{-}PropCN)_4]_2[Se_3Br_8]$, $[EtPh_3P]_2[Se_4Br_{14}]$ and $[n\text{-}Prop_4N]_2[Se_5Br_{12}]$, *Z. Anorg. Allg. Chem.* **597** (6), 115–132 (1991).

346 He, L., Tao, G.-H., Parrish, D.A. and Shreeve, J.M., Slightly viscous amino acid ionic liquids: Synthesis, properties, and calculations, *J. Phys. Chem. B* **113** (46), 15162–15169 (2009).

347 Hassel, D. and Hope, H., Structure of the solid compound formed by addition of two molecules of iodine to one molecule of pyridine, *Acta Chem. Scand.* **15**, 407–416 (1961).

348 Leclercq, L., Suisse, I., Roussel, P. and Agbossou-Niedercorn, F., Inclusion of tetrabutylammonium cations in a chiral thiazolium/triflate network: Solid state and solution structural investigation, *J. Mol. Struct.* **1010**, 152–157 (2012).

349 Pernak, J., Smiglak, M., Griffin, S.T., Hough, W.L., Wilson, T.B., Pernak, A., Zabielska-Matejuk, J., Fojutowski, A., Kita, K. and Rogers, R.D., Long alkyl chain quaternary ammonium-based ionic liquids and potential applications, *Green Chem.* **8** (9), 798–806 (2006).

350 Klapötke, T.M., Mayer, P., Schulz, A. and Weigand, J.J., 1,5-Diamino-4-methyltetrazolium dinitramide, *J. Am. Chem. Soc.* **127** (7), 2032–2033 (2005).

351 Gilardi, R., Flippen-Anderson, J., George, C. and Butcher, R.J., A new class of flexible energetic salts: The crystal structures of the ammonium, lithium, potassium, and cesium salts of dinitramide, *J. Am. Chem. Soc.* **119** (40), 9411–9416 (1997).

352 Dubovitskii, F.I., Golovina, N.I., Pavlov, A.N. and Atovmyan, L.O., Structural features of a salts of dinitramid with alkaline metals, *Dokl. Akad. Nauk* **355** (2), 200–202 (1997).

353 Pringle, J.M., Golding, J., Forsyth, C.M., Deacon, G.B., Forsyth, M. and MacFarlane, D.R., Physical trends and structural features in organic salts of the thiocyanate anion, *J. Mater. Chem.* **12** (12), 3475–3480 (2002).

354 Yoshida, Y., Otsuka, A., Saito, G., Natsume, S., Nishibori, E., Takata, M., Sakata, M., Takahashi, M. and Yoko, T., Conducting and magnetic properties of 1-ethyl-3-methylimidazolium (EMI) salts containing paramagnetic irons: Liquids $[EMI][M^{III}Cl_4]$ (M = Fe and $Fe_{0.5}Ga_{0.5}$) and solid $[EMI]_2[Fe^{II}Cl_4]$, *Bull. Chem. Soc. Jpn.* **78** (11), 1921–1928 (2005).

355 Xue, H., Gao, Y., Twamley, B. and Shreeve, J.M., New energetic salts based on nitrogen-containing heterocycles, *Chem. Mater.* **17** (1), 191–198 (2005).

356 Fischer, N., Klapötke, T.M. and Stierstorfer, J., Explosives based on diaminourea, *Propellants, Explos., Pyrotech.* **36** (3), 225–232 (2011).

357 Shanmuganathan, S., Kühl, O., Jones, P.G. and Heinicke, J., Nickel and palladium complexes of enolatefunctionalised *N*-heterocyclic carbenes, *Cent. Eur. J. Chem.* **8** (5), 992–998 (2010).

357a Mandai, T., Masu, H., Seki, H. and Nishikawa, K., Linker-length dependence of crystal structures and thermal properties of bis(imidazolium) salts with tetrafluoroborate anion, *Bull. Chem. Soc. Jpn.* **85** (5), 599–607 (2012).

358 Drake, G., Kaplan, G., Hall, L., Hawkins, T. and Larue, J., A new family of energetic ionic liquids 1-amino-3-alkyl-1,2,3-triazolium nitrates, *J. Chem. Crystallogr.* **37** (1), 15–23 (2007).

359 Klapötke, T.M., Krumm, B., Mayer, P., Piotrowski, H., Schwab, I. and Vogt, M., Synthesis and structures of triorganotelluronium pseudohalides, *Eur. J. Inorg. Chem.* (10), 2701–2709 (2002).

360 Swamy, K.C.K., Kumar, K.P. and Kumar, N.N.B., Further characterization of Mitsunobu-type intermediates in the reaction of dialkyl azodicarboxylates with P(III) compounds, *J. Org. Chem.* **71** (3), 1002–1008 (2006).

361 Cowley, A.H., Stewart, C.A., Whittlesey, B.R. and Wright, T.C., Phosphorus heterocycle synthesis using phosphenium ions and 1,4-dienes, *Tetrahedron Lett.* **25** (8), 815–816 (1984).

362 Ward, D.L., Rhinebarger, R.R. and Popov, A.I., *N-n*-butylpyridinium chloride, *Acta Crystallogr.* **C42**, 1771–1773 (1986).

363 Boyle, P.D., Christie, J., Dyer, T., Godfrey, S.M., Howson, I.R., McArthur, C., Omar, B., Pritchard, R.G. and Williams, G.R., Further structural motifs from the reactions of thioamides with diiodine and the interhalogens iodine monobromide and iodine monochloride: An FT-Raman and crystallographic study, *J. Chem. Soc. Dalton Trans.* (18), 3106–3112 (2000).

364 Hensen, K. and Wagner, P., The crystal-structure of *N*-trimethylsilyl-3,4-dimethyl-pyridinium bromide, *Z. Naturforsch. B* **48** (1), 79–81 (1993).

365 Alder, R.W., Blake, M.E., Bufali, S., Butts, C.P., Orpen, A.G., Schütz, J. and Williams, S.J., Preparation of tetraalkylformamidinium salts and related species as precursors to stable carbenes, *J. Chem. Soc., Perkin Trans. 1* (14), 1586–1593 (2001).

366 Neverov, A.A., Feng, H.X.M., Hamilton, K. and Brown, R.S., Bis (pyridine)-based bromonium ions. Molecular structures of bis(2,4,6-collidine)bromonium perchlorate and bis(pyridine)bromonium triflate and the mechanism of the reactions of 1,2-bis(2′-pyridylethynyl)benzenebrominum triflate and bis(pyridine)bromonium triflate with acceptor olefins, *J. Org. Chem.* **68** (10), 3802–3810 (2003).

367 Foces-Foces, C., Llamas-Saiz, A.L., Lorente, P., Golubev, N.S. and Limbach, H.H., Three 2,4,6-trimethylpyridine-benzoic acid complexes at 150 K, *Acta Crystallogr.* **C55**, 377–381 (1999).

368 Hartmann, F., Mootz, D. and Schwesinger, R., Crystal chemistry of uncharged phosphazene bases.1. Structures of two hydrates and an acetate of tris(dimethylamino)methyliminophosphorane, *Z. Naturforsch. B* **51** (10), 1369–1374 (1996).

369 Nikiforov, V.A., Karavan, V.S., Miltsov, S.A., Selivanov, S.I., Kolehmainen, E., Wegelius, E. and Nissinen, M., Hypervalent iodine compounds derived from *o*-nitroiodobenzene and related compounds: Syntheses and structures, *Arkivoc* (6), 191–200 (2003).

370 Parent, A.R., Landee, C.P. and Turnbull, M.M., Transition metal halide salts of *N*-methylmorpholine: Synthesis, crystal structures and magnetic properties of *N*-methylmorpholinium salts of copper(II), cobalt(II) and manganese(II), *Inorg. Chim. Acta* **360** (6), 1943–1953 (2007).

371 Webb, P.B., Sellin, M.F., Kunene, T.E., Williamson, S., Slawin, A.M.Z. and Cole-Hamilton, D.J., Continuous flow hydroformylation of alkenes in supercritical fluid-ionic liquid biphasic systems, *J. Am. Chem. Soc.* **125** (50), 15577–15588 (2003).

372 Risto, M., Reed, R.W., Robertson, C.M., Oilunkaniemi, R., Laitinen, R.S. and Oakley, R.T., Self-association of the N-methyl benzotellurodiazolylium cation: Implications for the generation of super-heavy atom radicals, *Chem. Commun.* (28), 3278–3280 (2008).

373 Yu, S.F., Lindeman, S. and Tran, C.D., Chiral ionic liquids: Synthesis, properties, and enantiomeric recognition, *J. Org. Chem.* **73** (7), 2576–2591 (2008).

374 Childs, R.F., Varadarajan, A., Lock, C.J.L., Faggiani, R., Fyfe, C.A. and Wasylishen, R.E., Structure of the 2-hydroxyhomotropylium cation – unequivocal evidence for homoaromatic delocalization, *J. Am. Chem. Soc.* **104** (9), 2452–2456 (1982).

375 Borodkin, G.I., Nagi, S.M., Gatilov, Y.V., Shakirov, M.M., Rybalova, T.V. and Shubin, V.G., Effect of crystalline surrounding on the rate of cation rearrangements – degradative rearrangements of 1,1,2,3,4,5,6-heptamethylbenzolonium and 1-phenyl-1,2,3,4,5,6-hexamethylbenzolonium ions, *Russ. J. Org. Chem.* **28** (9), 1806–1826 (1992).

376 Li, Z., Ni, C. and Ma, Y., The reaction of 2′2′3′3′-tetrafluoropropyl-5-iodo-3-oxa-octafluoropentanesulfonate with dimethylsulfoxide and determination of crystal-structure of one of the products, Data from CSD refcode CEXGAN, *Acta Chim. Sin.* **42**, 120–124 (1984).

377 Lozinskaya, E.I., Shaplov, A.S., Kotseruba, M.V., Komarova, L.I., Lyssenko, K.A., Antipin, M.Y., Golovanov, D.G. and Vygodskii, Y.S., 'One-pot' synthesis of aromatic poly(1,3,4-oxadiazole)s in novel solvents – ionic liquids, *J. Polym. Sci. Polym. Chem. Ed.* **44** (1), 380–394 (2006).

378 van den Broeke, J., Stam, M., Lutz, M., Kooijman, H., Spek, A.L., Deelman, B.J. and van Koten, G., Designing ionic liquids: 1-butyl-3-methylimidazolium cations with substituted tetraphenylborate counterions, *Eur. J. Inorg. Chem.* (15), 2798–2811 (2003).

379 Tebbe, K.F. and Kavoosian, A., Studies on polyhalides. XII: Preparation and crystal structure of $[K(crypt\text{-}2.2.2)]_2I_{12}$, *Z. Naturforsch. B* **48** (4), 438–442 (1993).

380 Shah, F.U., Glavatskih, S., Dean, P.M., MacFarlane, D.R., Forsyth, M. and Antzutkin, O.N., Halogen-free chelated orthoborate ionic liquids and organic ionic plastic crystals, *J. Mater. Chem.* **22** (14), 6928–6938 (2012).

381 Khan, M.A. and Tuck, D.G., The structures of tetra-*n*-butylammonium salts of $InCl_4^-$, $InBr_4^-$, $InBrCl_3^-$ and $InBr_3Cl^-$, *Acta Crystallogr.* **B38**, 803–806 (1982).

382 Roesky, H.W., Stasch, A., Hatop, H., Rennekamp, C., Hamilton, D.H., Noltemeyer, M. and Schmidt, H.G., A facile route to group 13 difluorodiorganometalates: $[n\mathrm{Bu}_4\mathrm{N}][\mathrm{R}_2\mathrm{MF}_2]$ (M = Al, Ga, In), *Angew. Chem. Int. Ed.* **39** (1), 171–173 (2000).

383 Karnop, M., duMont, W.W., Jones, P.G. and Jeske, J., Dichlorogermylene-alkyldichlorophosphane reactions revisited: Characterisation of bis(trichlorogermyl) phosphanes, trichlorogermyldiphosphanes, and Ge-P heterocycles, *Chem. Ber. Recl.* **130** (11), 1611–1618 (1997).

384 Oki, M., Yamada, Y. and Murata, S., Sulfur inversion in and molecular-structure of meso-3,4-diethyl-3,4-dimethyl-1-phenylthiolanium tetrafluoroborate, *Bull. Chem. Soc. Jpn.* **61** (3), 707–714 (1988).

385 Deng, D.L., Zhang, Y.H., Dai, C.Y., Zeng, H., Yan, Z.Q., Ye, C.Q. and Ha-Ge, R., Crystal structures of two new monoprotonated products of 1,4,7-trimethyl-1,4,7-triazacyclononane: $[Me_3[9]aneN_3H][SO_3CF_3]$ and $[Me_3[9]aneN_3H]I$, Data from CSD refcode HOZMIS, *Chin. J. Struct. Chem.* **19** (1), 48–52 (2000).

386 Laus, G., Schütz, J., Schuler, N., Kahlenberg, V. and Schottenberger, H., Crystal structure of tetrabutylammonium 2,3,6-tricyano-4-fluoro-5-(trifluoromethyl) phenolate, $[(C_4H_9)_4N][C_{10}F_4N_3O]$, *Z. Kristallogr. NCS* **224** (1), 117–118 (2009).

387 Čížková, M., Kolivoška, V., Císařová, I., Šaman, D., Pospíšil, L. and Teplý, F., Nitrogen heteroaromatic cations by 2 + 2 + 2 cycloaddition, *Org. Biomol. Chem.* **9** (2), 450–462 (2011).

388 Laus, G., Bentivoglio, G., Kahlenberg, V., Griesser, U.J., Schottenberger, H. and Nauer, G., Syntheses, crystal structures, and polymorphism of quaternary pyrrolidinium chlorides, *CrystEngComm* **10** (6), 748–752 (2008).

389 Henderson, W.A., Young, V.G., Passerini, S., Trulove, P.C. and De Long, H.C., Plastic phase transitions in *N*-ethyl-*N*-methylpyrrolidinium bis(trifluoromethanesulfony) imide, *Chem. Mater.* **18** (4), 934–938 (2006).

390 Henderson, W.A., Young Jr., V.G., Pearson, W., Passerini, S., De Long, H.C. and Trulove, P.C., Thermal phase behaviour of *N*-alkyl-*N*-methylpyrrolidinium and piperidinium bis(trifluoromethanesulfonyl)imide salts, *J. Phys. Condens. Matter* **18** (46), 10377–10390 (2006).

391 He, L., Tao, G.-H., Parrish, D.A. and Shreeve, J.M., Nitrocyanamide-based ionic liquids and their potential applications as hypergolic fuels, *Chem. Eur. J.* **16** (19), 5736–5743 (2010).

392 Tebbe, K.F. and Dombrowski, I., Studies of polyhalides. XXXVIII. On potassium-benzo-18-crown-6-polyiodides [K(benzo-18-crown-6)]I_n with n=3, 4, 6: Preparation and crystal structures of a triiodide [K(benzo-18-crown-6)]I_3, an octaiodide $[K(benzo\text{-}18\text{-}crown\text{-}6)]_2I_8 \cdot 1/2C_2H_5OH$ and a dodecaiodide $[K(benzo\text{-}18\text{-}crown\text{-}6)]_2I_{12}$, *Z. Anorg. Allg. Chem.* **625** (1), 167–174 (1999).

393 Borodkin, G.I., Gatilov, Y.V., Nagy, S.M. and Shubin, V.G., Comparative studies of the molecular structure of 1-*R*-1,2,3,4,5,6-hexamethylbenzenonium ions: Molecular and crystal structure of 1-chloromethyl-1,2,3,4,5,6-hexamethylbenzonium tetrachloroaluminate, *J. Struct. Chem.* **37** (3), 464–469 (1996).

394 Zaworotko, M.J., Cameron, T.S., Linden, A. and Sturge, K.C., Structures of the tetrachloroaluminate salts of the *N*-ethylpyridinium, 2-ethylpyridinium, pyridinium and 1-chloromethyl-1,2,3,4,5,6-hexamethylbenzenium cations, *Acta Crystallogr.* **C45**, 996–1002 (1989).

395 Holmes, R.R., Day, R.O., Yoshida, Y. and Holmes, J.M., Hydrogen-bonded phosphate-esters – synthesis and structure of imidazole-containing salts of diphenyl phosphate and (trichloromethyl)phosphonic acid, *J. Am. Chem. Soc.* **114** (5), 1771–1778 (1992).

396 Dixon, D.A., Farnham, W.B., Heilemann, W., Mews, R. and Noltemeyer, M., Structural studies of tris(dialkylamino)sulfonium (TAS) fluorosilicates, *Heteroat. Chem.* **4** (2–3), 287–295 (1993).

397 Castellari, C. and Sabatino, P., Anti-inflammatory drugs. III. Salts of diclofenac with *N*-(2-hydroxyethyl)piperidine, *N*-(2-hydroxyethyl)morpholine and *N*-(2-hydroxyethyl)piperazine, *Acta Crystallogr.* **C52**, 1708–1712 (1996).

398 Schulz, A. and Villinger, A., Binary pnictogen azides-an experimental and theoretical study: $[As(N_3)_4]^-$, $[Sb(N_3)_4]^-$, and $[Bi(N_3)_5(dmso)]^{2-}$, *Chem. Eur. J.* **18** (10), 2902–2911 (2012).

399 Katritzky, A.R., Yang, H.F., Zhang, D.Z., Kirichenko, K., Smiglak, M., Holbrey, J.D., Reichert, W.M. and Rogers, R.D., Strategies toward the design of energetic ionic liquids: Nitro- and nitrile-substituted *N,N′*-dialkylimidazolium salts, *New J. Chem.* **30** (3), 349–358 (2006).

400 Pankratov, V.A., Frenkel, T.M., Shvorak, A.E., Lindeman, S.V. and Struchkov, Y.T., The preparation and X-ray structural study of complexes that catalyze isocyanate cyclotrimerization, *Russ. Chem. Bull.* **42** (1), 81–87 (1993).

401 Hitchcock, P.B., Seddon, K.R. and Welton, T., Hydrogen-bond acceptor abilities of tetrachlorometallate(II) complexes in ionic liquids, *J. Chem. Soc. Dalton Trans.* (17), 2639–2643 (1993).

402 Zarychta, B., Spaleniak, G. and Zaleski, J., 2-Amino-5-butyl-4-methyl-1,3-thiazol-3-ium nitrate, *Acta Crystallogr.* **E59**, o304–o305 (2003).

403 Subramaniam, P., Alias, Y. and Chandrasekaram, K., 1,1′,2,2′-Tetramethyl-3,3′-(p-phenylenedimethylene)diimidazol-1-ium bis(trifluoromethanesulfonate), *Acta Crystallogr.* **E66**, o2455 (2010).

404 Showell, G.A., Gibbons, T.L., Kneen, C.O., Macleod, A.M., Merchant, K., Saunders, J., Freedman, S.B., Patel, S. and Baker, R., Tetrahydropyridyloxadiazoles – semi-rigid muscarinic ligands, *J. Med. Chem.* **34** (3), 1086–1094 (1991).

405 Leclercq, L., Simard, M. and Schmitzer, A.R., 1,3-Dibenzylimidazolium salts: A paradigm of water and anion effect on the supramolecular H-bonds network, *J. Mol. Struct.* **918** (1–3), 101–107 (2009).

406 Amoedo-Portela, A., Carballo, R., Casas, J.S., García-Martínez, E., Gómez-Alonso, C., Sánchez-González, A., Sordo, J. and Vázquez-López, E.M., The coordination chemistry of the versatile ligand bis(2-pyridylthio)methane, *Z. Anorg. Allg. Chem.* **628** (5), 939–950 (2002).

407 Lin, Q.-H., Li, Y.-C., Li, Y.-Y., Wang, Z., Liu, W., Qi, C. and Pang, S.-P., Energetic salts based on 1-amino-1,2,3-triazole and 3-methyl-1-amino-1,2,3-triazole, *J. Mater. Chem.* **22** (2), 666–674 (2012).

408 Joo, Y.-H., Gao, H., Parrish, D.A., Cho, S.G., Goh, E.M. and Shreeve, J.M., Energetic salts based on nitroiminotetrazole-containing acetic acid, *J. Mater. Chem.* **22** (13), 6123–6130 (2012).

409 Linden, A., Petridis, A. and James, B.D., Structural diversity in thallium chemistry. IV. Further examples of solid state bromothallate(III) derivatives obtained by employing certain classes of alkyl diammonium cations, *Inorg. Chim. Acta* **332**, 61–71 (2002).

410 Hassall, K. and White, J., Dimethyl(2-trimethylsilylethyl)[(2-trimethylsilylethyl) dimethylammonio]ammonium tetrakis[3,5-bis(trifluoromethyl)phenyl]borate at 130 K, *Acta Crystallogr.* **E60**, o107–o108 (2004).

411 Xue, H., Arritt, S.W., Twalmley, B. and Shreeve, J.M., Energetic salts from *N*-aminoazoles, *Inorg. Chem.* **43** (25), 7972–7977 (2004).

412 Lukevics, E., Arsenyan, P., Shestakova, I., Domracheva, I., Kanepe, I., Belyakov, S., Popelis, J. and Pudova, O., Synthesis, structure and cytotoxicity of organoammonium selenites, *Appl. Organomet. Chem.* **16** (4), 228–234 (2002).

413 Herler, S., Mayer, P., Schulz, A. and Villinger, A., *N*-methyl-2-pyrrolidone hydrochloride, *Acta Crystallogr.* **E63**, o3991 (2007).

414 Baenziger, N.C. and Nelson, A.D., Crystal structure of the tetrachloroaluminate salt of the heptamethylbenzene cation, *J. Am. Chem. Soc.* **90**, 6602–6607 (1968).

415 Hubig, S.M., Lindeman, S.V. and Kochi, J.K., Charge-transfer bonding in metal-arene coordination, *Coord. Chem. Rev.* **200**, 831–873 (2000).

416 Burford, N., Parks, T.M., Royan, B.W., Borecka, B., Cameron, T.S., Richardson, J.F., Gabe, E.J. and Hynes, R., Aza- and thiaarsolidinium cations: Novel structural features for carbene analogues, *J. Am. Chem. Soc.* **114** (21), 8147–8153 (1992).

417 Ballarin, B., Busetto, L., Cassani, M.C. and Femoni, C., A new gold(III)-aminoethyl imidazolium aurate salt: Synthesis, characterization and reactivity, *Inorg. Chim. Acta* **363** (10), 2055–2064 (2010).

418 Tebbe, K.F., Farida, T., Stegemann, H. and Füllbier, H., Studies on polyhalides. XXIII. Crystal structures of *N*-alkylurotropinium triiodides $UrRI_3$ with R = methyl, ethyl, n-propyl, and n-butyl, *Z. Anorg. Allg. Chem.* **622** (3), 525–533 (1996).

419 Bentzen, S., Nielsen, P.H., Anthoni, U., Christophersen, C. and Gajhede, M., 3-Methoxyazetidinium chloride, *Acta Crystallogr.* **C49**, 932–934 (1993).

420 Houllemare-Druot, S. and Coquerel, G., How far can an unstable racemic compound affect the performances of preferential crystallization? Example with (*R*) and (*S*)-α-methylbenzylamine chloroacetate, *J. Chem. Soc. Perkin Trans. 2* (10), 2211–2220 (1998).

421 Allen, J.L., Boyle, P.D. and Henderson, W.A., Lithium difluoro(oxalato)borate tetramethylene sulfone disolvate, *Acta Crystallogr.* **E67**, m533 (2011).

422 Andersson, S. and Jagner, S., Crystal structure of bis(methyltributylammonium) penta-μ-bromo-di-μ_5-bromo-pentacuprate(I), a compound containing a discrete $[Cu_5Br_7]^{2-}$ aggregate, *J. Crystallogr. Spectr. Res.* **18** (5), 591–600 (1988).

423 Abonia, R., Rengifo, E., Cobo, J., Low, J.N. and Glidewell, C., *N*-Benzylethylammonium nitrate: A three-dimensional hydrogen-bonded framework comprising substructures in zero, one and two dimensions, *Acta Crystallogr.* **C61**, o645–o647 (2005).

424 Geng, H., Zhuang, L.H., Zhang, J., Wang, G.W. and Yuan, A.L., 3,3′-Dimethyl-1,1′-(butane-1,4-diyl)diimidazolium bis(tetrafluoroborate), *Acta Crystallogr.* **E66**, o1267 (2010).

425 Burford, N., Losier, P., Macdonald, C., Kyrimis, V., Bakshi, P.K. and Cameron, T.S., Anionic and steric factors governing coordinative unsaturation at carbenic phosphenium centers, *Inorg. Chem.* **33** (7), 1434–1439 (1994).

426 Ji, J.X., Xu, B., Liu, S. and Wang, J.T., *N-n*-butylpyridinium bromide, *Acta Crystallogr.* **E62**, o516–o517 (2006).

427 Klapötke, T.M., Mayer, P., Sabaté, C.M., Welch, J.M. and Wiegand, N., Simple, nitrogen-rich, energetic salts of 5-nitrotetrazole, *Inorg. Chem.* **47** (13), 6014–6027 (2008).

428 Sangin, J.P. and Brisse, F., Study on the compositions of aliphatic chains. 4. Structure of bis(*n*-decanoate) of piperazinium, $C_4H_{12}N2^{2+} \cdot 2C_{10}H_{19}O_2^-$, *Acta Crystallogr.* **C40**, 2094–2096 (1984).

429 Brisse, F. and Sangin, J.P., Study of compounds with aliphatic chains. 2. The structure of piperazinium bis(*n*-dodecanoate), *Acta Crystallogr.* **B38**, 215–221 (1982).

430 Venkatramani, L. and Craven, B.M., Disordered fatty-acid chains in piperazinium myristate and palmitate, *Acta Crystallogr.* **B47**, 968–975 (1991).

431 Krasan, Y.P., Egorov, Y.P. and Feshchenko, N.G., Synthesis and X-ray diffraction studies of tetraamylphosphonium iodide (C_5H_{11}), Data from CSD refcode QQQDXG, *Zh. Strukt. Khim.* **11** (5), 941 (1970).

432 Brough, P.A., Gallagher, T., Thomas, P., Wonnacott, S., Baker, R., Malik, K.M.A. and Hursthouse, M.B., Synthesis and X-ray crystal structure of 2-acetyl-9-azabicyclo[4.2.1]nonan-3-one. A conformationally locked s-*cis* analogue of anatoxin-a, *J. Chem. Soc., Chem. Commun.* (15), 1087–1089 (1992).

433 Haque, R.A., Washeel, A., Teoh, S.G., Quah, C.K. and Fun, H.-K., 3,5-Bis(3-butylimidazolium-1-ylmethyl)toluene bis(hexafluorophosphate), *Acta Crystallogr.* **E66**, o2797 (2010).

434 Weigand, J.J., Burford, N., Davidson, R.J., Cameron, S. and Seelheim, P., New synthetic procedures to catena-phosphorus cations: Preparation and dissociation of the first cyclo-phosphino-halophosphonium salts, *J. Am. Chem. Soc.* **131** (49), 17943–17953 (2009).

435 Bodrikov, I.V., Bel'skii, V.K., Krasnov, V.L. and Pigin, O.V., Nucleophilic sulfonation of alkenes by sulfur-dioxide-amine-water system – stereochemical principles, data from CSD refcode YEDRIM, *Zh. Org. Khim.* **28** (11), 2228–2238 (1992).

436 Kristl, M., Golič, L. and Volavšek, B., On the synthesis of hydroxylammonium fluoride, *Monatsh. Chem.* **125** (11), 1207–1213 (1994).

437 Roesky, H.W., Noltemeyer, M. and Sheldrick, G.M., Synthesis and structure of trifluoroacetyldicyanomethanide, *Z. Naturforsch. B* **40** (7), 883–885 (1985).

438 Schulze-Matthäi, K., Bendig, J., Neubauer, P. and Ziemer, B., Dimethylphenylsulfonium trifluoromethanesulfonate and methyldiphenylsulfonium trifluoromethanesulfonate, *Acta Crystallogr.* **C56**, e257–e258 (2000).

439 Sangin, J.P. and Brisse, F., Study on the compositions of aliphatic chains. 3. Structure of bis(*n*-heptanoate) of piperazinium, $C_4H_{12}N_2^{2+}.2C_7H_{13}O_2^-$, *Acta Crystallogr.* **C40**, 2091–2093 (1984).

440 Kutuniva, J., Oilunkaniemi, R., Laitinen, R.S., Asikkala, J., Kärkkäinen, J. and Lajunen, M.K., Synthesis and structural characterization of 1-butyl-2,3-dimethylimidazolium bromide and iodide, *Z. Naturforsch. B* **62** (6), 868–870 (2007).

441 Seiler, O., Burschka, C., Penka, M. and Tacke, R., Dianionic tris[oxalato(2-)]-silicate and tris[oxalato(2-)]germanate complexes: Synthesis, properties, and structural characterization in the solid state and in solution, *Z. Anorg. Allg. Chem.* **628** (11), 2427–2434 (2002).

442 Hooper, J.B., Borodin, O. and Schneider, S., Insight into hydrazinium nitrates, azides, dicyanamide, and 5-azidotetrazolate ionic materials from simulations and experiments, *J. Phys. Chem. B* **115** (46), 13578–13592 (2011).

443 Gálvez-Ruiz, J.C., Holl, G., Karaghiosoff, K., Klapötke, T.M., Löhnwitz, K., Mayer, P., Noth, H., Polborn, K., Rohbogner, C.J., Suter, M. and Weigand, J.J., Derivatives of 1,5-diamino-1*H*-tetrazole: A new family of energetic heterocyclic-based salts, *Inorg. Chem.* **44** (12), 4237–4253 (2005).

444 Hady, S.A., Nahringbauer, I. and Olovsson, I., Hydrogen bond studies. 38. The crystal structure of hydrazinium acetate, *Acta Chem. Scand.* **23**, 2764–2772 (1969).

445 Odabaşoğlu, M. and Büyükgüngür, O., 2-Amino-3-methylpyridinium acetate, *Acta Crystallogr.* **E62**, o236–o238 (2006).

446 Decken, A., Jenkins, H.D.B., Nikiforov, G.B. and Passmore, J., The reaction of $Li[Al(OR)_4]$ R = $OC(CF_3)_2Ph$, $OC(CF_3)_3$ with NO/NO_2 giving $NO[Al(OR)_4]$, $Li[NO_3]$ and N_2O. The synthesis of $NO[Al(OR)_4]$ from $Li[Al(OR)_4]$ and $NO[SbF_6]$ in sulfur dioxide solution, *Dalton Trans.* (16), 2496–2504 (2004).

447 Mirzaei, Y.R., Twamley, B. and Shreeve, J.M., Syntheses of 1-alkyl-1,2,4-triazoles and the formation of quaternary 1-alkyl-4-polyfluoroalkyl-1,2,4-triazolium salts leading to ionic liquids, *J. Org. Chem.* **67** (26), 9340–9345 (2002).

448 Tebbe, K.F. and Gilles, T., Studies of polyhalides. XXVII. On tetra(n-propyl) ammonium polyiodides (n-$Pr_4N)I_n$ with n = 3, 5, 7: Preparation and crystal structures of a triiodide (n-$Pr_4N)I_3$, a pentaiodide (n-$Pr_4N)I_5$, and a heptaiodide (n-$Pr_4N)I_7$, *Z. Anorg. Allg. Chem.* **622** (9), 1587–1593 (1996).

449 Bejan, D., Willner, H., Ignatiev, N. and Lehmann, C.W., Synthesis and characterization of bis[bis(pentafluoroethyl)phosphinyl]imides, $M^+N[(C_2F_5)_2P(O)]_2^-$, M = H, Na, K, Cs, Ag, Me_4N, *Inorg. Chem.* **47** (19), 9085–9089 (2008).

450 Aakeröy, C.B., Rajbanshi, A., Li, Z.J. and Desper, J., Mapping out the synthetic landscape for re-crystallization, co-crystallization and salt formation, *CrystEngComm* **12** (12), 4231–4239 (2010).

451 Mostovich, E.A., Mazhukin, D.G., Gatilov, Y.V. and Rybalova, T.V., Coupling of 1,2-bis(alkoxyamino)cyclohexanes with 1,3-dicarbonyl compounds: First synthesis of 1,4-dialkoxy-2,3-dihydro-1,4-diazepinium salts, *Mendeleev Commun.* **17** (1), 48–50 (2007).

452 Fuller, J., The influence of hydrogen bonding on the structure of 1-methylimidazolium D-tartrate, *Acta Crystallogr.* **C51**, 1680–1683 (1995).

453 Fraenk, W., Klapötke, T.M., Krumm, B., Nöth, H., Suter, M., Vogt, M. and Warchhold, M., Pentafluorophenyl and phenyl substituted azidoborates, *Can. J. Chem.* **80** (11), 1444–1450 (2002).

454 Stander-Grobler, E., Strasser, C.E., Schuster, O., Cronje, S. and Raubenheimer, H., Amine-substituted α-N(standard)- and δ-N(remote)-pyridylidene complexes: Synthesis and bonding, *Inorg. Chim. Acta* **376** (1), 87–94 (2011).

455 Hitchcock, P.B., Lewis, R.J. and Welton, T., Vanadyl complexes in ambient-temperature ionic liquids - the 1st X-ray crystal-structure of a tetrachlorooxovanadate(IV) salt, *Polyhedron* **12** (16), 2039–2044 (1993).

456 Gao, H., Joo, Y.-H., Twamley, B., Zhou, Z. and Shreeve, J.M., Hypergolic ionic liquids with the 2,2-dialkyltriazanium cation, *Angew. Chem. Int. Ed.* **48** (15), 2792–2795 (2009).

457 Peppel, T. and Köckerling, M., Salts with the 1,3-dibutyl-2,4,5-trimethyl-imidazolium cation: (DBTMIm)*X* (*X* = Br, PF_6) and $(DBTMIm)_2[MBr_4]$ (M = Co, Ni), *Z. Anorg. Allg. Chem.* **636** (13–14), 2439–2446 (2010).

458 Anders, E., Tropsch, J.G., Irmer, E. and Sheldrick, G.M., Conformational analysis of *N*-benzylpyridinium bromide – a comparison of crystal structure data with results of semiempirical calculations (MINDO/3, MNDO, AM1, and PM3), *Chem. Ber.* **123** (2), 321–325 (1990).

459 Andersson, S., Hakansson, M. and Jagner, S., A far infrared investigation of anionic species present in solution during the crystallization of a polynuclear bromocuprate(I) cluster containing 3-coordinated copper(I), *J. Crystallogr. Spectrosc. Res.* **19** (1), 147–157 (1989).

460 Andersson, S. and Jagner, S., Crystal structure of tetrapropylammonium dichlorocuprate(I) – comparison of anionic configurations in halocuprates(I) crystallizing with symmetrical tetraalkylammonium and related cations, *Acta Chem. Scand. A* **40** (1), 52–57 (1986).

461 Lynch, D.E., Nicholls, L.J., Smith, G., Byriel, K.A. and Kennard, C.H.L., Molecular co-crystals of 2-aminothiazole derivatives, *Acta Crystallogr.* **B55**, 758–766 (1999).

462 Klapötke, T.M. and Stierstorfer, J., The CN_7^- anion, *J. Am. Chem. Soc.* **131** (3), 1122–1134 (2009).

463 Lork, E., Viets, D., Müller, M. and Mews, R., Bis(dimethylamino)trifluoromethylsulfonium salts: $[CF_3S(NMe_2)_2]^+[Me_3SiF_2]^-$, $[CF_3S(NMe_2)_2]^+[HF_2]^-$ and $[CF_3S(NMe_2)_2]^+[CF_3S]^-$, *Z. Anorg. Allg. Chem.* **630** (15), 2692–2696 (2004).

464 Froschauer, C., Wurst, K., Laus, G., Nauer, G. and Schottenberger, H., Crystal structure of bis(1-ethyl-3-methylimidazol-2-ylidene)silver(I) tetracyanoborate, $[(C_6H_{10}N_2)_2Ag][B(CN)_4$, *Z. Kristallogr. NCS* **225** (2), 377–378 (2010).

465 Zhao, D.B., Fei, Z.F., Geldbach, T.J., Scopelliti, R. and Dyson, P.J., Nitrile-functionalized pyridinium ionic liquids: Synthesis, characterization, and their application in carbon – carbon coupling reactions, *J. Am. Chem. Soc.* **126** (48), 15876–15882 (2004).

466 Roesky, H.W., Schmieder, W., Isenberg, W., Sheldrick, W.S. and Sheldrick, G.M., Anions of sulfur with the coordination number 3 – synthesis, structure, and range of existence, *Chem. Ber.* **115** (8), 2714–2727 (1982).

467 Pertlik, F., Mikenda, W. and Steinwender, E., Hydrogen-bonding in 2-hydroxy((di)thio)benzoates. 1. X-ray structures of 2,6-dimethylpiperidinium salts, *J. Crystallogr. Spectrosc. Res.* **23** (5), 389–395 (1993).

468 MacDonald, J.C., Dorrestein, P.C. and Pilley, M.M., Design of supramolecular layers via self-assembly of imidazole and carboxylic acids, *Cryst. Growth. Des.* **1** (1), 29–38 (2001).

469 Tebbe, K.F. and Gilles, T., Studies of polyhalides. XXII [1]. On dimethyldiphenylammoniumpolyiodides $(Me_2Ph_2N)I_n$ with n = 3, 13/3, 6, and 8: Preparation and crystal structures of a triiodide $(Me_2Ph_2N)I_3$, tridecaiodide $(Me_2Ph_2N)_3I_{13}$, dodecaiodide $(Me_2Ph_2N)_2I_{12}$, and hexadecaiodide $(Me_2Ph_2N)_2I_{16}$, *Z. Anorg. Allg. Chem.* **622** (1), 138–148 (1996).

470 Abbott, A.P., Claxton, T.A., Fawcett, J. and Harper, J.C., Tetrakis(decyl)ammonium tetraphenylborate: A novel electrolyte for non-polar media, *J. Chem. Soc. Faraday Trans.* **92** (10), 1747–1749 (1996).

471 Lin, J.C.Y., Huang, C.-J., Lee, Y.-T., Lee, K.-M. and Lin, I.J.B., Carboxylic acid functionalized imidazolium salts: Sequential formation of ionic, zwitterionic, acid-zwitterionic and lithium salt-zwitterionic liquid crystals, *J. Mater. Chem.* **21** (22), 8110–8121 (2011).

472 Sharutin, V.V., Senchurin, V.S., Klepikov, N.N. and Sharutina, O.K., Phosphonium salts $[Ph_3AlkP]_2^+[Hg_2I_6]^{2-}$ and $[Ph_3AlkP]_2^+[Hg_4I_{10}]^{2-}$: Synthesis and structure, *Russ. J. Inorg. Chem.* **54** (2), 232–238 (2009).

473 Chitra, R., Choudhury, R.R., Thiruvenkatam, V., Hosur, M.V. and Row, T.N.G., Molecular interactions in bis(2-aminopyridinium) malonate: A crystal isostructural to bis(2-aminopyridinium) maleate crystal, *J. Mol. Struct.* **1010**, 46–51 (2012).

474 Marchand, A.P., Rajagopal, D., Bott, S.G. and Archibald, T.G., Reactions of 1-aza-3-ethylbicyclo[1.1.0]butane with electrophiles – a facile entry into new, *N*-substituted 3-ethylideneazetidines and 2-azetines, *J. Org. Chem.* **59** (7), 1608–1612 (1994).

475 Schomburg, D., Structural chemistry of penta-coordinated silicon-compounds – crystalline and molecular-structure of tetrapropylammonium phenyltetrafluorosilicate, *J. Organomet. Chem.* **221** (2), 137–141 (1981).

476 Finden, J., Beck, G., Lantz, A., Walsh, R., Zaworotko, M.J. and Singer, R.D., Preparation and characterization of 1-butyl-3-methylimidazolium tetrakis(3,5-

bis(trifluoromethyl)phenyl)borate, [bmim]BARF, *J. Chem. Crystallogr.* **33** (4), 287–295 (2003).

477 Alvanipour, A., Atwood, J.L., Bott, S.G., Junk, P.C., Kynast, U.H. and Prinz, H., Some crown ether chemistry of Ti, Zr and Hf derived from liquid clathrate media, *J. Chem. Soc. Dalton Trans.* (7), 1223–1228 (1998).

478 Lange, I., Henschel, D., Wirth, A., Krahl, J., Blaschette, A. and Jones, P.G., Polysulfonylamine LXII. Dimesylamide as noncoordinating counterion for organotin complex cations – Synthesis and structures of $[R_3Sn(L)(L')]^+[(MeSO_2)_2N]^-$ (L=L′ = HMPA, Ph_3AsO, imidazole, 4-dimethylaminopyridine; L = H_2O und L′=Pyridine-*N*-oxide), *J. Organomet. Chem.* **503** (2), 155–170 (1995).

479 Puranik, V.G., Tavale, S.S., Iyer, V.S., Sehra, J.C. and Sivaram, S., Tetrabutylammonium hydrogen bisbenzoate – crystal structure and study of short hydrogen bonds in hydrogen bisbenzoate anion system, *J. Chem. Soc. Perkin Trans. 2* (8), 1517–1520 (1993).

480 Yang, X., Fei, Z., Geldbach, T.J., Phillips, A.D., Hartinger, C.G., Li, Y. and Dyson, P.J., Suzuki coupling reactions in ether-functionalized ionic liquids: The importance of weakly interacting cations, *Organometallics* **27** (15), 3971–3977 (2008).

481 Aoyagi, N., Shimojo, K., Brooks, N.R., Nagaishi, R., Naganawa, H., Van Hecke, K., Van Meervelt, L., Binnemans, K. and Kimura, T., Thermochromic properties of low-melting ionic uranyl isothiocyanate complexes, *Chem. Commun.* **47** (15), 4490–4492 (2011).

482 Graalmann, O., Hesse, M., Klingebiel, U., Clegg, W., Haase, M. and Sheldrick, G.M., The one-electron oxidation of diazasilacyclopentenes, *Angew. Chem. Int. Ed. Engl.* **22** (8), 621–622 (1983).

483 Childs, R.F., Faggiani, R., Lock, C.J.L. and Mahendran, M., Structure of 1-ethoxyhomotropylium hexachloroantimonate – a nonaromatic homotropylium cation, *J. Am. Chem. Soc.* **108** (13), 3613–3617 (1986).

484 Batsanov, A.S., Howard, J.A.K., Lightfoot, A.P., Twiddle, S.J.R. and Whiting, A., Stereoselective chloro-deboronation reactions induced by substituted pyridine-iodine chloride complexes, *Eur. J. Org. Chem.* (9), 1876–1883 (2005).

485 Evans, P.A. and Clizbe, E.A., Unlocking ylide reactivity in the metal-catalyzed allylic substitution reaction: Stereospecific construction of primary allylic amines with aza-ylides, *J. Am. Chem. Soc.* **131** (25), 8722–8723 (2009).

486 Li, B., Li, Y.-Q., Zheng, W.-J. and Zhou, M.-Y., Synthesis of ionic liquid-supported Schiff bases, *Arkivoc* (11), 165–171 (2009).

487 Henderson, W.A., Herstedt, M., Young, V.G., Passerini, S., De Long, H.C. and Trulove, P.C., New disordering mode for $TFSI^-$ anions: The nonequilibrium, plastic crystalline structure of Et_4NTFSI, *Inorg. Chem.* **45** (4), 1412–1414 (2006).

488 Jensen, B., The Crystal structures of (±)-*cis*- and (−)-(1*S*,3*S*)-*trans*-3-acetoxy-1-methylthiane perchlorates, *Acta Chem. Scand. B* **35** (9), 607–612 (1981).

489 Higashiya, S., Filatov, A.S., Wells, C.C., Rane-Fondacaro, M.V. and Haldar, P., Crystal structures and quantitative structure-property relationships of spirobipyrrolidinium and the oxygen-containing derivatives, *J. Mol. Struct.* **984** (1–3), 300–306 (2010).

490 Currie, M., Estager, J., Licence, P., Men, S., Nockemann, P., Seddon, K.R., Swadźba-Kwaśny, M. and Terrade, C., Chlorostannate(II) ionic liquids: Speciation, Lewis acidity, and oxidative stability, *Inorg. Chem.* **52** (4), 1710–1721 (2013).

491 Bartoszak-Adamska, E., Dega-Szafran, Z., Przedwojska, M. and Jaskolski, M., Crystal and molecular structure of 1 : 1 complex of *N*-methylmorpholine betaine with salicylic acid, *Pol. J. Chem.* **77** (11), 1711–1722 (2003).

492 Allahyari, R., Lee, P.W., Lin, G.H.Y., Wing, R.M. and Fukuto, T.R., Resolution and reactions of the chiral isomers of *O*-ethyl *S*-phenyl ethylphosphonodithioate (fonofos) and its analogs, *J. Agric. Food Chem.* **25** (3), 471–478 (1977).

493 Lynch, D.E., Smith, G., Freney, D., Byriel, K.A. and Kennard, C.H.L., Molecular cocrystals of carboxylic acids. XV. Preparation and characterization of heterocyclic base adducts with a series of carboxylic acids, and the crystal structures of the adducts of 2-aminopyrimidine with 2,6-dihydroxybenzoic acid, 4-aminobenzoic acid, phenoxyacetic acid, (2,4-dichlorophenoxy)acetic acid, (3,4-dichlorophenoxy)-acetic acid and salicylic acid, and 2-aminopyridine with 2,6-dihydroxybenzoic acid, *Aust. J. Chem.* **47** (6), 1097–1115 (1994).

494 Katritzky, A.R., Singh, S., Kirichenko, K., Holbrey, J.D., Smiglak, M., Reichert, W.M. and Rogers, R.D., 1-butyl-3-methylimidazolium 3,5-dinitro-1,2,4-triazolate: A novel ionic liquid containing a rigid, planar energetic anion, *Chem. Commun.* (7), 868–870 (2005).

495 Haiges, R., Boatz, J.A., Vij, A., Vij, V., Gerken, M., Schneider, S., Schroer, T., Yousufuddin, M. and Christe, K.O., Polyazide chemistry: Preparation and characterization of $As(N_3)_5$, $Sb(N_3)_5$, and $[P(C_6H_5)_4][Sb(N_3)_6)]$, *Angew. Chem. Int. Ed.* **43** (48), 6676–6680 (2004).

496 Belabassi, Y., Gushwa, A.F., Richards, A.F. and Montchamp, J.-L., Structural analogues of bioactive phosphonic acids: First crystal structure characterization of phosphonothioic and boranophosphonic acids, *Phosphorus Sulfur Silicon Relat. Elem.* **183** (9), 2214–2228 (2008).

497 Abate, A., Brischetto, M., Cavallo, G., Lahtinen, M., Metrangolo, P., Pilati, T., Radice, S., Resnati, G., Rissanen, K. and Terraneo, G., Dimensional encapsulation of $I^-\cdots I_2\cdots I^-$ in an organic salt crystal matrix, *Chem. Commun.* **46** (16), 2724–2726 (2010).

498 Tang, S. and Mudring, A.-V., Two cyano-functionalized, cadmium-containing ionic liquids, *Eur. J. Inorg. Chem.* (9), 1145–1148 (2009).

499 Gänswein, B. and Behm, H., Problem of the C-S vibration in trifluoromethane sulfonates – structural investigations of $H_3NOH{+}CF_3SO_3$, *Z. Anorg. Allg. Chem.* **428** (1), 248–253 (1977).

500 Fei, Z.F., Zhao, D.B., Geldbach, T.J., Scopelliti, R. and Dyson, P.J., Brønsted acidic ionic liquids and their zwitterions: Synthesis, characterization and pK_a determination, *Chem. Eur. J.* **10** (19), 4886–4893 (2004).

501 Elessawi, M., Elkhalik, S.A., Berthold, H.J. and Wartchow, R., Synthesis and crystal structure of $[P(C_6H_5)_3CH_3]I_3$, *Z. Naturforsch. B* **46** (6), 703–708 (1991).

502 Fischer, N., Goddard-Borger, E.D., Greiner, R., Klapötke, T.M., Skelton, B.W. and Stierstorfer, J., Sensitivities of some imidazole-1-sulfonyl azide salts, *J. Org. Chem.* **77** (4), 1760–1764 (2012).

503 Moers, O., Wijaya, K., Jones, P.G. and Blaschette, A., Polysulfonylamines. CIX. 1,2,4-triazolium di(methanesulfonyl)amidate, *Acta Crystallogr.* **C55**, 754–756 (1999).

504 Gao, R., Yan, B., Mai, T., Hu, Y. and Guan, Y.-L., 3,3-Dinitroazetidinium 2-hydroxybenzoate, *Acta Crystallogr.* **E66**, o3036 (2010).

505 Yamaguchi, S., Akiyama, S. and Tamao, K., Effect of countercation inclusion by [2.2.2]cryptand upon stabilization of potassium organofluorosilicates, *Organometallics* **18** (15), 2851–2854 (1999).

506 Goswami, S., Jana, S., Hazra, A., Fun, H.-K. and Chantraprommа, S., Non-covalent synthesis of ionic and molecular complexes of benzoic acid and substituted 2-aminopyrimidines by varying aryl/alkyl substituents and their supramolecular chemistry, *Supramol. Chem.* **20** (5), 495–500 (2008).

507 Chumakov, Y.M., Simonov, Y.A., Grozav, M., Crisan, M., Bocelli, G., Yakovenko, A.A. and Lyubetsky, D., Hydrogen-bonding network in the organic salts of 4-nitrobenzoic acid, *Cent. Eur. J. Chem.* **4** (3), 458–475 (2006).

508 Jin, S., Guo, J., Liu, L. and Wang, D., Five organic salts assembled from carboxylic acids and bis-imidazole derivatives through collective noncovalent interactions, *J. Mol. Struct.* **1004** (1–3), 227–236 (2011).

509 Dega-Szafran, Z., Kania, A., Grundwald-Wyspiańska, M., Szafran, M. and Tykarska, E., Differences between the N·H·O and O·H·O hydrogen bonds in complexes of 2,6-dichloro-4-nitrophenol with pyridines and pyridine *N*-oxides, *J. Mol. Struct.* **381** (1–3), 107–125 (1996).

510 Mudring, A.V., Babai, A., Arenz, S. and Giernoth, R., The 'noncoordinating' anion Tf_2N^- coordinates to Yb^{2+}: A structurally characterized Tf_2N^- complex from the ionic liquid [mppyr][Tf_2N], *Angew. Chem. Int. Ed.* **44** (34), 5485–5488 (2005).

511 Wessel, J., Lork, E. and Mews, R., Alkali metallocene anions- syntheses and structures, *Angew. Chem. Int. Ed. Engl.* **34** (21), 2376–2378 (1995).

512 Odabaşoğlu, M. and Büyükgüngür, O., 3,6-Dioxaoctane-1,8-diammonium bis(trichloroacetate), *Acta Crystallogr.* **E62**, o739–o741 (2006).

513 Martell, J.M. and Zaworotko, M.J., Synthesis and structure of mixed chloride tetrachloroaluminate salts, *J. Chem. Soc. Dalton Trans.* (6), 1495–1498 (1991).

514 Andersson, S. and Jagner, S., Bis(methyltriphenylphosphonium) di-μ-bromo-dibromodicuprate(I), *Acta Crystallogr.* **C43**, 1089–1091 (1987).

515 Procter, D.J. and Rayner, C.M., Selenoxide-sulfonic acid adducts – a new class of stable, selenoxide-based oxidants, *Tetrahedron. Lett.* **35** (9), 1449–1452 (1994).

516 Chen, S., Wang, M., Wang, F., Yang, F., Shi, S. and Wang, N., Preparation and characterization of two Brønsted acid ionic liquids, *J. Chem. Crystallogr.* **41** (7), 1027–1031 (2011).

517 Consorti, C.S., Suarez, P.A.Z., de Souza, R.F., Burrow, R.A., Farrar, D.H., Lough, A.J., Loh, W., da Silva, L.H.M. and Dupont, J., Identification of 1,3-dialkylimidazolium salt supramolecular aggregates in solution, *J. Phys. Chem. B* **109** (10), 4341–4349 (2005).

518 Monkowius, U.V., Nogai, S.D. and Schmidbaur, H., The tetra(vinyl)phosphonium cation $[(CH_2{=}CH)_4P]^+$, *J. Am. Chem. Soc.* **126** (6), 1632–1633 (2004).

519 Dub, P.A., Rodriguez-Zubiri, M., Daran, J.-C., Brunet, J.-J. and Poli, R., Platinum-catalyzed ethylene hydroamination with aniline: Synthesis, characterization, and studies of intermediates, *Organometallics* **28** (16), 4764–4777 (2009).

520 Funke, W., Hornig, K., Möller, M.H. and Würthwein, E.U., Alkoxy-substituted *N*-acyliminium salts – synthesis, structure, reactivity, *Chem. Ber.* **126** (9), 2069–2077 (1993).

521 Reetz, M.T., Hütte, S. and Goddard, R., Tetrabutylammonium salts of 2-nitropropane, cyclopentadiene and 9-ethylfluorene – crystal structures and use in anionic polymerization, *Z. Naturforsch. B* **50** (3), 415–422 (1995).

522 Grützmacher, H. and Pritzkow, H., 3 Independent molecules in the unit-cell of a phosphorus ylide – changes in the molecular geometry on rotation about the phosphorus carbon ylide bond, *Angew. Chem. Int. Ed. Engl.* **31** (1), 99–101 (1992).

523 Sweidan, K., Kuhn, N. and Maichle-Mössmer, C., Crystal structure of 1,3-diisopropyl-4,5-dimethylimidazolium *E*-2-cyano-1-phenylethenolate, $[C_{11}H_{21}N_2][C_9H_6NO]$, *Z. Kristallogr. NCS* **224** (2), 295–296 (2009).

524 Jegier, J.A. and Atwood, D.A., Base effects on the formation of four- and five-coordinate cationic aluminum complexes, *Inorg. Chem.* **36** (10), 2034–2039 (1997).

525 Junk, P.C., Kepert, C.J., Semenova, L.I., Skelton, B.W. and White, A.H., The structural systematics of protonation of some important nitrogen-base ligands. I – some univalent anion salts of doubly protonated 2,2′: 6′,2″-terpyridyl, *Z. Anorg. Allg. Chem.* **632** (7), 1293–1302 (2006).

526 Bozopoulos, A.P. and Rentzeperis, P.J., Structure of [bis(*p*-methoxyphenyl)] (diethylaminocarbodithioato)iodine(III), *Acta Crystallogr.* **C43**, 914–916 (1987).

527 Fuller, J. and Heimer, N.E., Substituted imidazolium phenylphosphonate salts for nonlinear optical applications, *J. Chem. Crystallogr.* **25** (3), 129–136 (1995).

528 Larsson, G. and Nahringbauer, I., Hydrogen bond studies. XXIII. The crystal structure of potassium hydrogen diformate, *Acta Crystallogr.* **B24**, 666–672 (1968).

529 Sangokoya, S.A., Pennington, W.T. and Robinson, G.H., Alkylation of tellurium tetrachloride by trimethylaluminum – synthesis and molecular-structure of $[Te(CH_3)_3][Al(CH_3)_2Cl_2]$ – a novel organotelluronium-aluminum oligomer, *J. Crystallogr. Spectrosc. Res.* **19** (3), 433–438 (1989).

530 Wolf, J., Labande, A., Daran, J.C. and Poli, R., Nickel(II) complexes with bifunctional phosphine-imidazolium ligands and their catalytic activity in the Kumada-Corriu coupling reaction, *J. Organomet. Chem.* **691** (3), 433–443 (2006).

531 Dubois, R.H., Zaworotko, M.J. and White, P.S., Complex hydrogen-bonded cations – X-ray crystal-structure of $[(C_6H_5CH_2)NH_3)_4Cl][AlCl_4]_3$ and its relevance to the structure of basic chloroaluminate room-temperature melts, *Inorg. Chem.* **28** (11), 2019–2020 (1989).

532 Mootz, D. and Boenigk, D., Fluorides and fluoroacids. XI. Poly(hydrogen fluorides) with the tetramethylammonium cation – preparation, stability ranges, crystal-structures, $[H_nF_{n+1}]^-$ anion homology, hydrogen bonding F–H···F, *Z. Anorg. Allg. Chem.* **544** (1), 159–166 (1987).

533 Batsanov, A.S., Lightfoot, A.P., Twiddle, S.J.R. and Whiting, A., Bis(2,6-dimethylpyridyl)iodonium dibromoiodate, *Acta Crystallogr.* **E62**, o901–o902 (2006).

534 Kataeva, O.N., Litvinov, I.A., Strobykina, I.Y. and Kataev, V.E., Molecular structure of *N*-(2-hydroxyethyl)pyridinium bromide, *Russ. Chem. Bull.* **39** (11), 2371–2373 (1990).

535 Hayashi, H., Karasawa, S. and Koga, N., Molecular structure and magnetic properties of 1-ethyl-2-(1-oxy-3-oxo-4,4,5,5-tetramethylimidazolin-2-yl)-3-methylimidazolium arylcarboxylates and other salts, *J. Org. Chem.* **73** (22), 8683–8693 (2008).

536 Luque, A., Sertucha, J., Castillo, O. and Román, P., Crystal packing and physical properties of pyridinium tetrabromocuprate(II) complexes assembled via hydrogen bonds and aromatic stacking interactions, *New J. Chem.* **25** (9), 1208–1214 (2001).

537 Fei, Z., Ang, W.H., Geldbach, T.J., Scopelliti, R. and Dyson, P.J., Ionic solid-state dimers and polymers derived from imidazolium dicarboxylic acids, *Chem. Eur. J.* **12** (15), 4014–4020 (2006).

538 Luo, J.Q., Ruble, J.R., Craven, B.M. and McMullan, R.K., Effects of H/D substitution on thermal vibrations in piperazinium hexanoate-h_{11},d_{11}, *Acta Crystallogr.* **B52**, 357–368 (1996).

539 Mylrajan, M., Srinivasan, T.K.K. and Sreenivasamurthy, G., Crystal structure of monomethylammonium nitrate, *J. Crystallogr. Spectrosc. Res.* **15** (5), 493–500 (1985).

540 Smith, G., Wermuth, U.D. and Young, D.J., 2-(2,4-Dinitrobenzyl)pyridinium 2-hydroxy-3,5-dinitrobenzoate, *Acta Crystallogr.* **E66**, o1895 (2010).

541 Vaday, S. and Foxman, B.M., Cocrystalline salts of alkynoic acids. II. The crystal and molecular structure of 5-bromo-2-aminopyridinium propynoate, *Cryst. Eng.* **2**, 145–151 (1999).

542 Gololobov, Y.G., Galkina, M.A., Dovgan, O.V., Krasnova, I.Y., Petrovskii, P.V., Antipin, M.Y., Voronzov, II, Lyssenko, K.A. and Schmutzler, R., Intramolecular electrophilic rearrangements in saturated acyclic systems. Reactivity of the zwitterion derived from triisopropylphosphine and ethyl 2-cyanoacrylate with respect to different types of electrophiles, *Russ. Chem. Bull.* **50** (2), 279–286 (2001).

543 Glaser, R., Chen, G.S. and Barnes, C.L., Origin of the stabilization of vinyldiazonium Ions by β-substitution; first crystal structure of an aliphatic diazonium ion: β,β-diethoxyethene-diazonium hexachloroantimonate, *Angew. Chem. Int. Ed. Engl.* **31** (6), 740–743 (1992).

544 Li, D.H., Hao, J.S., Tong, H.B., Huang, S.P., Zhang, Y.B. and Xia, C.Z., 3-(4-methoxyphenyl)-1,2-dimethyl-4,5-dihydroimidazolium iodide, *Acta Crystallogr.* **E60**, o855–o856 (2004).

545 Chen, S.-H., Yang, F.-R., Wang, M.-T. and Wang, N.-N., Synthesis, characterization, and crystal structure of several novel acidic ionic liquids based on the corresponding 1-alkylbenzimidazole with tetrafluoroboric acid, *C. R. Chim.* **13** (11), 1391–1396 (2010).

546 Mootz, D. and Wunderlich, H., Crystalline structure of acid hydrates and oxonium salts. IV. Dioxonium-1,2-ethane disulfonate, *Acta Crystallogr.* **B26**, 1820–1825 (1970).

547 Patil, M.L., Rao, C.V.L., Yonezawa, K., Takizawa, S., Onitsuka, K. and Sasai, H., Design and synthesis of novel chiral spiro ionic liquids, *Org. Lett.* **8** (2), 227–230 (2006).

548 Bailly, F., Barthen, P., Frohn, H.J. and Köckerling, M., Pentafluorophenyliodine(III) compounds. 4 [1] Aryl(pentafluorophenyl)iodoniumtetrafluoroborates: General method of synthesis, typical properties, and structural features, *Z. Anorg. Allg. Chem.* **626** (11), 2419–2427 (2000).

549 Hall, S.R., Allen, F.H. and Brown, I.D., The crystallographic information file (cif) – a new standard archive file for crystallography, *Acta Crystallogr.* **A47**, 655–685 (1991).

550 Thalladi, V.R., Weiss, H.C., Blaser, D., Boese, R., Nangia, A. and Desiraju, G.R., C—H···F interactions in the crystal structures of some fluorobenzenes, *J. Am. Chem. Soc.* **120** (34), 8702–8710 (1998).

551 Thalladi, V.R., Boese, R. and Weiss, H.C., The melting point alternation in α,ω-alkanedithiols, *J. Am. Chem. Soc.* **122** (6), 1186–1190 (2000).

552 Thalladi, V.R., Boese, R. and Weiss, H.C., The melting point alternation in α,ω-alkanediols and α,ω-alkanediamines: Interplay between hydrogen bonding and hydrophobic interactions, *Angew. Chem. Int. Ed.* **39** (5), 918–922 (2000).

553 Boese, R., Kirchner, M.T., Billups, W.E. and Norman, L.R., Cocrystallization with acetylene: Molecular complexes with acetone and dimethyl sulfoxide, *Angew. Chem. Int. Ed.* **42** (17), 1961–1963 (2003).

554 Bond, A.D., In situ co-crystallisation as a tool for low-temperature crystal engineering, *Chem. Commun.* (2), 250–251 (2003).

555 Luzzati, P.V., Structure cristalline de HNO_3·$3H_2O$. I. Resolution de la structure; utilisation de la fonction de Patterson *Acta Crystallogr.* **6**, 142–152 (1953).

556 van den Berg, J.A. and Seddon, K.R., Critical evaluation of C—H···X hydrogen bonding in the crystalline state, *Cryst. Growth Des.* **3** (5), 643–661 (2003).

557 Aakeröy, C.B., Evans, T.A., Seddon, K.R. and Pálinkó, I., The C—H···Cl hydrogen bond: Does it exist? *New J. Chem.* **23** (2), 145–152 (1999).

558 Roth, C., Peppel, T., Fumino, K., Köckerling, M. and Ludwig, R., The importance of hydrogen bonds for the structure of ionic liquids: Single-crystal X-ray diffraction and transmission and attenuated total reflection spectroscopy in the terahertz region, *Angew. Chem. Int. Ed.* **49** (52), 10221–10224 (2010).

559 Steiner, T., The hydrogen bond in the solid state, *Angew. Chem. Int. Ed.* **41** (1), 48–76 (2002).

560 Harder, S., Can C—H···C(π) bonding be classified as hydrogen bonding? a systematic investigation of C—H···C(π) bonding to cyclopentadienyl anions, *Chem. Eur. J.* **5** (6), 1852–1861 (1999).

561 Etter, M.C., Encoding and decoding hydrogen-bond patterns of organic compounds, *Acc. Chem. Res.* **23** (4), 120–126 (1990).

562 Bernstein, J., Davis, R.E., Shimoni, L. and Chang, N.L., Patterns in hydrogen bonding – functionality and graph set analysis in crystals, *Angew. Chem. Int. Ed.* **34** (15), 1555–1573 (1995).

563 Timmermans, J., Plastic crystals: A historical review, *J. Phys. Chem. Solids* **18,** 1–8 (1961).

564 Bernstein, J., *Polymorphism in Molecular Crystals* (International Union of Crystallography, Oxford University Press, New York, 2008).

565 Cherukuvada, S. and Nangia, A., Polymorphism in an API ionic liquid: Ethambutol dibenzoate trimorphs, *CrystEngComm* **14** (23), 7840–7843 (2012).

566 Hayashi, S., Ozawa, R. and Hamaguchi, H., Raman spectra, crystal polymorphism, and structure of a prototype ionic-liquid bmim Cl, *Chem. Lett.* **32** (6), 498–499 (2003).

567 Jayaraman, S. and Maginn, E.J., Computing the melting point and thermodynamic stability of the orthorhombic and monoclinic crystalline polymorphs of the ionic liquid 1-*n*-butyl-3-methylimidazolium chloride, *J. Chem. Phys.* **127** (21), 214504 (2007).

568 Endo, T., Morita, T. and Nishikawa, K., 1,3-dimethylimidazolium hexafluorophosphate: Calorimetric and structural studies of two crystal phases having melting points of ~50 K difference, *Chem. Phys. Lett.* **517** (4–6), 162–165 (2011).

569 Endo, T., Murata, H., Imanari, M., Mizushima, N., Seki, H. and Nishikawa, K., NMR study of cation dynamics in three crystalline states of 1-butyl-3-methylimidazolium hexafluorophosphate exhibiting crystal polymorphism, *J. Phys. Chem. B* **116** (12), 3780–3788 (2012).

570 Endo, T., Murata, H., Imanari, M., Mizushima, N., Seki, H., Sen, S. and Nishikawa, K., A comparative study of the rotational dynamics of PF_6^- anions in the crystals and liquid states of 1-butyl-3-methylimidazolium hexafluorophosphate: results from ^{31}P NMR spectroscopy, *J. Phys. Chem. B* **117** (1), 326–332 (2013).

571 Su, L., Li, L.B., Hu, Y., Yuan, C.S., Shao, C.G. and Hong, S.M., Phase transition of $C_n$mim PF_6 under high pressure up to 1.0 GPa, *J. Chem. Phys.* **130** (18), 184503 (2009).

572 Russina, O., Fazio, B., Schmidt, C. and Triolo, A., Structural organization and phase behaviour of 1-butyl-3-methylimidazolium hexafluorophosphate: An high pressure Raman spectroscopy study, *Phys. Chem. Chem. Phys.* **13** (25), 12067–12074 (2011).

573 Zhang, S.-W., Guzei, I.A., de Villiers, M.M., Yu, L. and Krzyzaniak, J.F., Formation enthalpies and polymorphs of nicotinamide-R-mandelic acid co-crystals, *Cryst. Growth Des.* **12** (8), 4090–4097 (2012).

574 Ling, I., Alias, Y., Sobolev, A.N. and Raston, C.L., Hirshfeld surface analysis of phosphonium salts, *CrystEngComm* **12** (12), 4321–4327 (2010).

575 Chow, H., Dean, P.A.W., Craig, D.C., Lucas, N.T., Scudder, M.L. and Dance, I.G., The subtle tetramorphism of $MePh_3P^+I_3^-$, *New J. Chem.* **27** (4), 704–713 (2003).

576 Berg, R.W., Raman spectroscopy and ab-initio model calculations on ionic liquids, *Monatsh. Chem.* **138** (11), 1045–1075 (2007).

577 Laus, G., Hummel, M., Többens, D.M., Gelbrich, T., Kahlenberg, V., Wurst, K., Griesser, U.J. and Schottenberger, H., The 1: 1 and 1: 2 salts of 1,4-diazabicyclo[2.2.2]octane and bis(trifluoromethylsulfonyl)amine: Thermal behaviour and polymorphism, *CrystEngComm* **13** (17), 5439–5446 (2011).

578 Madelung, E., Das elektrische Feld in Systemen von regelmäßig angeordneten Punktladungen, *Phys. Z.* **XIX**, 524–533 (1918).

579 Glasser, L., Solid-state energetics and electrostatics: Madelung constants and Madelung energies, *Inorg. Chem.* **51** (4), 2420–2424 (2012).

580 Izgorodina, E.I., Bernard, U.L., Dean, P.M., Pringle, J.M. and MacFarlane, D.R., The Madelung constant of organic salts, *Cryst. Growth Des.* **9** (11), 4834–4839 (2009).

581 Glasser, L. and Jenkins, H.D.B., Predictive thermodynamics for condensed phases, *Chem. Soc. Rev.* **34** (10), 866–874 (2005).

582 Jenkins, H.D.B., Tudela, D. and Glasser, L., Lattice potential energy estimation for complex ionic salts from density measurements, *Inorg. Chem.* **41** (9), 2364–2367 (2002).

583 Jenkins, H.D.B., Roobottom, H.K., Passmore, J. and Glasser, L., Relationships among ionic lattice energies, molecular (formula unit) volumes, and thermochemical radii, *Inorg. Chem.* **38** (16), 3609–3620 (1999).

584 Glasser, L. and Jenkins, H.D.B., Lattice energies and unit cell volumes of complex ionic solids, *J. Am. Chem. Soc.* **122** (4), 632–638 (2000).

585 Jenkins, H.D.B. and Glasser, L., Standard absolute entropy, $S°_{298}$, values from volume or density. 1. Inorganic materials, *Inorg. Chem.* **42** (26), 8702–8708 (2003).

586 Glasser, L., Lattice and phase transition thermodynamics of ionic liquids, *Thermochim. Acta* **421** (1–2), 87–93 (2004).

587 Mallouk, T.E., Rosenthal, G.L., Muller, G., Brusasco, R. and Bartlett, N., Fluoride-ion affinities of GeF_4 and BF_3 from thermodynamic and structural data for $(SF_3)_2GeF_6$, ClO_2GeF_5, and ClO_2BF_4, *Inorg. Chem.* **23** (20), 3167–3173 (1984).

588 Kapustinskii, A.F., Lattice energy of ionic crystals, *Q. Rev. Chem. Soc.* **10**, 283–294 (1956).

589 Klapötke, T.M., Stierstorfer, J., Jenkins, H.D.B., van Eldik, R. and Schmeisser, M., Calculation of some thermodynamic properties and detonation parameters of 1-ethyl-3-methyl-*H*-imidazolium perchlorate, [emim][ClO_4], on the basis of CBS-4M and CHEETAH computations supplemented by VBT estimates, *Z. Anorg. Allg. Chem.* **637** (10), 1308–1313 (2011).

590 Desiraju, G.R., Designer crystals: Intermolecular interactions, network structures and supramolecular synthons, *Chem. Commun.* (16), 1475–1482 (1997).

591 Spackman, M.A. and Byrom, P.G., A novel definition of a molecule in a crystal, *Chem. Phys. Lett.* **267** (3–4), 215–220 (1997).

592 McKinnon, J.J., Mitchell, A.S. and Spackman, M.A., Hirshfeld surfaces: A new tool for visualising and exploring molecular crystals, *Chem. Eur. J.* **4** (11), 2136–2141 (1998).

593 McKinnon, J.J., Fabbiani, F.P.A. and Spackman, M.A., Comparison of polymorphic molecular crystal structures through Hirshfeld surface analysis, *Cryst. Growth Des.* **7** (4), 755–769 (2007).

594 McKinnon, J.J., Spackman, M.A. and Mitchell, A.S., Novel tools for visualizing and exploring intermolecular interactions in molecular crystals, *Acta Crystallogr.* **B60**, 627–668 (2004).

595 Spackman, M.A. and McKinnon, J.J., Fingerprinting intermolecular interactions in molecular crystals, *CrystEngComm*, **4**, 378–392 (2002).

596 Parkin, A., Barr, G., Dong, W., Gilmore, C.J., Jayatilaka, D., McKinnon, J.J., Spackman, M.A. and Wilson, C.C., Comparing entire crystal structures: Structural genetic fingerprinting, *CrystEngComm* **9** (8), 648–652 (2007).

597 McKinnon, J.J., Jayatilaka, D. and Spackman, M.A., Towards quantitative analysis of intermolecular interactions with Hirshfeld surfaces, *Chem. Commun.* (37), 3814–3816 (2007).

598 Wolff, S.K., Grimwood, D.J., McKinnon, J.J., Jayatilaka, D. and Spackman, M.A., in 'CrystalExplorer 1.6' (2006).

599 Carvalho, S.P., Wang, R., Wang, H., Ball, B. and Lebel, O., The Bronsted-Lowry reaction revisited: Glass-forming properties of salts of 1,5-dimexylbiguanide, *Cryst. Growth. Des.* **10** (6), 2734–2745 (2010).

600 Saha, S., Roy, R.K. and Ayers, P.W., Are the Hirshfeld and Mulliken population analysis schemes consistent with chemical intuition? *Int. J. Quantum Chem.* **109** (9), 1790–1806 (2009).

601 Krossing, I., Slattery, J.M., Daguenet, C., Dyson, P.J., Oleinikova, A. and Weingartner, H., Why are ionic liquids liquid? A simple explanation based on lattice and solvation energies, *J. Am. Chem. Soc.* **128** (41), 13427–13434 (2006).

602 Benrabah, D., Arnaud, R. and Sanchez, J.Y., Comparative ab-initio calculations on several salts, *Electrochim. Acta* **40** (13–14), 2437–2443 (1995).

603 Johansson, P., Gejji, S.P., Tegenfeldt, J. and Lindgren, J., The imide ion: Potential energy surface and geometries, *Electrochim. Acta* **43** (10–11), 1375–1379 (1998).

604 Arnaud, R., Benrabah, D. and Sanchez, J.Y., Theoretical study of CF_3SO_3Li, $(CF_3SO_2)_2NLi$, and $(CF_3SO_2)_2CHLi$ ion pairs, *J. Phys. Chem.* **100** (26), 10882–10891 (1996).

605 Foropoulos, J. and DesMarteau, D.D., Synthesis, properties, and reactions of bis((trifluoromethyl)sulfonyl)imide, $(CF_3SO_2)_2NH$, *Inorg. Chem.* **23** (23), 3720–3723 (1984).

606 Holbrey, J.D. and Seddon, K.R., Ionic liquids, *Clean Prod. Proc.* **1**, 223–236 (1999).

607 Jaenschke, A., Paap, J. and Behrens, U., Structures of polar magnesium organyls: Synthesis and structure of base adducts of bis(cyclopentadienyl)magnesium, *Z. Anorg. Allg. Chem.* **634** (3), 461–469 (2008).

608 Pajerski, A.D., Squiller, E.P., Parvez, M., Whittle, R.R. and Richey, H.G., Preparation and properties of RMg(macrocycle)$^+$ cations with accompanying anions derived from acidic hydrocarbons or a hindered phenol, *Organometallics* **24** (5), 809–814 (2005).

609 Izgorodina, E.I., Forsyth, M. and MacFarlane, D.R., Towards a better understanding of 'delocalized charge' in ionic liquid anions, *Aust. J. Chem.* **60** (1), 15–20 (2007).

610 MacFarlane, D.R. and Seddon, K.R., Ionic liquids – Progress on the fundamental issues, *Aust. J. Chem.* **60** (1), 3–5 (2007).

611 Taratin, N.V., Lorenz, H., Kotelnikova, E.N., Glikin, A.E., Galland, A., Dupray, V., Coquerel, G. and Seidel-Morgenstern, A., Mixed crystals in chiral organic systems: A case study on (*R*)- and (*S*)-ethanolammonium 3-chloromandelate, *Cryst. Growth Des.* **12** (12), 5882–5888 (2012).

612 Childs, S.L., Stahly, G.P. and Park, A., The salt-cocrystal continuum: The influence of crystal structure on ionization state, *Mol. Pharm.* **4** (3), 323–338 (2007).

613 Aakeröy, C.B., Fasulo, M.E. and Desper, J., Cocrystal or salt: Does it really matter? *Mol. Pharm.* **4** (3), 317–322 (2007).

614 Cruz-Cabeza, A.J., Acid-base crystalline complexes and the pK_a rule, *CrystEngComm* **14** (20), 6362–6365 (2012).

615 Brand, H., Schulz, A. and Villinger, A., Modern aspects of pseudohalogen chemistry: News from CN- and PN-chemistry, *Z. Anorg. Allg. Chem.* **633** (1), 22–35 (2007).

616 Patil, K.C., Soundararajan, R. and Verneker, V.R.P., Differential thermal-analysis of hydrazinium derivatives, *Thermochim. Acta* **31** (2), 259–261 (1979).

617 Luk'yanov, O.A., Agevnin, A.R., Leichenko, A.A., Seregina, N.M. and Tartakovsky, V.A., Dinitramide and its salts. 6. Dinitramide salts derived from ammonium bases, *Russ. Chem. Bull.* **44** (1), 108–112 (1995).

618 Robinson, R.J. and McCrone, W.C., Crystallographic data 169. Hydrazine nitrate (1), *Anal. Chem.* **30**, 1014–1015 (1958).

619 Brodalla, D., Mootz, D., Boese, R. and Osswald, W., Programmed crystal-growth on a diffractometer with focused heat radiation, *J. Appl. Crystallogr.* **18** (October), 316–319 (1985).

620 Taniki, R., Matsumoto, K., Hagiwara, R., Hachiya, K., Morinaga, T. and Sato, T., Highly conductive plastic crystals based on fluorohydrogenate anions, *J. Phys. Chem. B* **117** (3), 955–960 (2013).

621 Driver, G.W. and Mutikainen, I., The complex story of a simple Bronsted acid: Unusual speciation of HBr in an ionic liquid medium, *Dalton Trans.* **40** (41), 10801–10803 (2011).

622 Greaves, T.L., Weerawardena, A., Krodkiewska, I. and Drummond, C.J., Protonic ionic liquids: Physicochemical properties and behavior as amphiphile self-assembly solvents, *J. Phys. Chem. B* **112** (3), 896–905 (2008).

623 Janz, G.J., Allen, C.B., Downey Jr., J.R. and Tomkins, R.P.T., Physical properties data compilations relevant to energy storage. I. Molten salts: Eutectic data, Report NSRDS-NBS 61, Part I, National Bureau of Standards, Washington, DC (1978).

624 Berchiesi, G., Leonesi, D. and Cingolan, A., Ternary system K(CNS,HCOO,NO_3), *Z. Naturforsch. A* **25** (11), 1766–1767 (1970).

625 Liu, C.L. and Angell, C.A., Phase-equilibria, high-conductivity ambient-temperature liquids, and glasses in the pseudo-halide systems $AlCl_3$-MSCN (M = Li, Na, K), *Solid State Ion.* **86** (8), 467–473 (1996).

626 Lee, Y.C., Curtiss, L.A., Ratner, M.A. and Shriver, D.F., Ionic conductivity of new ambient temperature alkali metal glasses $AlCl_3/NaN(CN)_2$, *Chem. Mater.* **12** (6), 1634–1637 (2000).

627 Kubota, K., Nohira, T. and Hagiwara, R., Thermal properties of alkali bis(fluorosulfonyl)amides and their binary mixtures, *J. Chem. Eng. Data* **55** (9), 3142–3146 (2010).

628 Kubota, K., Nohira, T. and Hagiwara, R., New inorganic ionic liquids possessing low melting temperatures and wide electrochemical windows: Ternary mixtures of alkali bis(fluorosulfonyl)amides, *Electrochim. Acta* **66**, 320–324 (2012).

629 Rheingold, A.L., Cronin, J.T., Brill, T.B. and Ross, F.K., Structure of hydroxylammonium nitrate (HAN) and the deuterium homolog, *Acta Crystallogr.* **C43**, 402–404 (1987).

630 Glavič, P. and Slivnik, J., On the synthesis and properties of hydrazinium(1+) fluoride, *J. Fluor. Chem.* **17** (2), 187–190 (1981).

631 Golič, L. and Lazarini, F., Crystal-structure of hydrazinium monofluoride, *Monatsh. Chem.* **105** (4), 735–741 (1974).

632 Seward, R.P., The conductance and viscosity of highly concentrated aqueous solutions of hydrazinium chloride and hydrazinium nitrate, *J. Am. Chem. Soc.* **77**, 905–907 (1955).

633 Sakurai, K. and Tomiie, Y., The crystal structure of hydrazinium chloride, N_2H_5Cl, *Acta Crystallogr.* **5**, 293 (1952).

634 Overberger, C.G. and Lapkin, M., Azo compounds. XV. biradical sources. The synthesis and decomposition of large membered ring azo compounds, *J. Am. Chem. Soc.* **77**, 4651–4657 (1955).

635 Sakurai, K. and Tomiie, Y., The crystal structure of hydrazinium bromide, N_2H_5Br, *Acta Crystallogr.* **5**, 289 (1952).

636 Bottaro, J.C., Penwell, P.E. and Schmitt, R.J., 1,1,3,3-tetraoxo-1,2,3-triazapropene anion, a new oxy anion of nitrogen: The dinitramide anion and its salts, *J. Am. Chem. Soc.* **119** (40), 9405–9410 (1997).

637 Venkatachalam, S., Santhosh, G. and Ninan, K.N., An overview on the synthetic routes and properties of ammonium dinitramide (ADN) and other dinitramide salts, *Propellants, Explos., Pyrotech.* **29** (3), 178–187 (2004).

638 Lukyanov, O.A., Anikin, O.V., Gorelik, V.P. and Tartakovsky, V.A., Dinitramide and its salts. 3. Metallic salts of dinitramide, *Russ. Chem. Bull.* **43** (9), 1457–1461 (1994).

639 Ubbelohde, A.R., *The Molten State of Matter: Melting and Crystal Structure*, (John Wiley & Sons, Ltd, Chichester, 1978), p. 57.

640 Runyon, J.W., Steinhof, O., Dias, H.V.R., Calabrese, J.C., Marshall, W.J. and Arduengo III, A.J., Carbene-based Lewis pairs for hydrogen activation, *Aust. J. Chem.* **64** (8), 1165–1172 (2011).

641 Bąk, J.M., Effendy, Grabowsky, S., Lindoy, L.F., Price, J.R., Skelton, B.W. and White, A.H., True and quasi-isomorphism in tetrakis(acetonitrile)coinage metal(I) salts, *CrystEngComm* **15** (6), 1125–1138 (2013).

642 Hasan, M., Kozhevnikov, I.V., Siddiqui, M.R.H., Femoni, C., Steiner, A. and Winterton, N., *N,N′*-Dialkylimidazolium chloroplatinate(II), chloroplatinate(IV), and chloroiridate(IV) salts and an *N*-heterocyclic carbene complex of platinum(II): Synthesis in ionic liquids and crystal structures, *Inorg. Chem.* **40** (4), 795–800 (2001).

643 Garrison, J.C. and Youngs, W.J., Ag(I) *N*-heterocyclic carbene complexes: Synthesis, structure, and application, *Chem. Rev.* **105** (11), 3978–4008 (2005).

644 Lin, J.C.Y., Huang, R.T.W., Lee, C.S., Bhattacharyya, A., Hwang, W.S. and Lin, I.J.B., Coinage metal-*N*-heterocyclic carbene complexes, *Chem. Rev.* **109** (8), 3561–3598 (2009).

645 Lin, I.J.B. and Vasam, C.S., Preparation and application of *N*-heterocyclic carbene complexes of Ag(I), *Coord. Chem. Rev.* **251** (5–6), 642–670 (2007).

646 Hintermair, U., Englert, U. and Leitner, W., Distinct reactivity of mono- and bis-NHC silver complexes: carbene donors versus carbene-halide exchange reagents, *Organometallics* **30** (14), 3726–3731 (2011).

647 Dutton, J.L., Battista, T.L., Sgro, M.J. and Ragogna, P.J., Diazabutadiene complexes of selenium as Se^{2+} transfer reagents, *Chem. Commun.* **46** (7), 1041–1043 (2010).

648 Dutton, J.L., Tuononen, H.M. and Ragogna, P.J., Tellurium (II)-centered dications from the pseudohalide '$Te(OTf)_2$', *Angew. Chem. Int. Ed.* **48** (24), 4409–4413 (2009).

649 Krossing, I. and Raabe, I., Noncoordinating anions – Fact or fiction? A survey of likely candidates, *Angew. Chem. Int. Ed.* **43** (16), 2066–2090 (2004).

650 Bernard, U.L., Izgorodina, E.I. and MacFarlane, D.R., New insights into the relationship between ion-pair binding energy and thermodynamic and transport properties of ionic liquids, *J. Phys. Chem. C* **114** (48), 20472–20478 (2010).

651 Rowland, C.E., Kanatzidis, M.G. and Soderholm, L., Tetraalkylammonium uranyl isothiocyanates, *Inorg. Chem.* **51** (21), 11798–11804 (2012).

652 Serpell, C.J., Cookson, J., Thompson, A.L., Brown, C.M. and Beer, P.D., Haloaurate and halopalladate imidazolium salts: Structures, properties, and use as precursors for catalytic metal nanoparticles, *Dalton Trans.* **42** (5), 1385–1393 (2013).

653 Heilemann, W. and Mews, R., Tris(dimethylamino)sulfonium-sulfur(IV) salts, *Chem. Ber.* **121** (3), 461–463 (1988).

654 Klanberg, F. and Muetterties, E.L., Nuclear magnetic resonance studies on pentacoordinate silicon fluorides, *Inorg. Chem.* **7**, 155–160 (1968).

655 Matsumoto, K., Hagiwara, R. and Ito, Y., Room temperature molten fluorometallates: 1-ethyl-3-methylimidazolium hexafluoroniobate(V) and hexafluorotantalate(V), *J. Fluor. Chem.* **115** (2), 133–135 (2002).

656 Zhang, Y., Gao, H., Joo, Y.-H. and Shreeve, J.M., Ionic liquids as hypergolic fuels, *Angew. Chem. Int. Ed.* **50** (41), 9554–9562 (2011).

657 Smiglak, M., Metlen, A. and Rogers, R.D., The second evolution of ionic liquids: From solvents and separations to advanced materials-energetic examples from the ionic liquid cookbook, *Acc. Chem. Res.* **40** (11), 1182–1192 (2007).

658 Singh, R.P., Verma, R.D., Meshri, D.T. and Shreeve, J.M., Energetic nitrogen-rich salts and ionic liquids, *Angew. Chem. Int. Ed.* **45** (22), 3584–3601 (2006).

659 Klapötke, T.M., Krumm, B. and Scherr, M., Synthesis and structures of triorganochalcogenium (Te, Se, S) dinitramides, *Eur. J. Inorg. Chem.* (28), 4413–4419 (2008).

660 Ding, J. and Armstrong, D.W., Chiral ionic liquids: Synthesis and applications, *Chirality* **17** (5), 281–292 (2005).

661 Baudequin, C., Bregeon, D., Levillain, J., Guillen, F., Plaquevent, J.C. and Gaumont, A.C., Chiral ionic liquids, a renewal for the chemistry of chiral solvents? Design, synthesis and applications for chiral recognition and asymmetric synthesis, *Tetrahedron-Asymmetry* **16** (24), 3921–3945 (2005).

662 Goldberg, I., Coulombic interactions, hydrogen bonding and supramolecular chirality in pyridinium trifluoromethanesulfonate, *Acta Crystallogr.* **C65**, o509–o511 (2009).

663 Bonanni, M., Soldaini, G., Faggi, C., Goti, A. and Cardona, F., Novel *L*-tartaric acid derived pyrrolidinium cations for the synthesis of chiral ionic liquids, *Synlett* (5), 747–750 (2009).

664 Ríos-Lombardía, N., Busto, E., Gotor-Fernández, V., Gotor, V., Porcar, R., García-Verdugo, E., Luis, S.V., Alfonso, I., García-Granda, S. and Menéndez-Velázquez, A., From salts to ionic liquids by systematic structural modifications: A rational approach towards the efficient modular synthesis of enantiopure imidazolium salts, *Chem. Eur. J.* **16** (3), 836–847 (2010).

665 Busto, E., Gotor-Fernández, V., Rios-Lombardia, N., García-Verdugo, E., Alfonso, I., García-Granda, S., Menéndez-Velázquez, A., Burguete, M.I., Luis, S.V. and Gotor, V., Simple and straightforward synthesis of novel enantiopure ionic liquids via efficient enzymatic resolution of (±)-2-(1*H*-imidazol-1-yl)cyclohexanol, *Tetrahedron. Lett.* **48** (30), 5251–5254 (2007).

666 Feder-Kubis, J., Kubicki, M. and Pernak, J., 3-Alkoxymethyl-1-(1*R*,2*S*,5*R*)-(−)-menthoxymethylimidazolium salts-based chiral ionic liquids, *Tetrahedron-Asymmetry* **21** (21–22), 2709–2718 (2010).

667 Busi, S., Lahtinen, M., Karna, M., Valkonen, J., Kolehmainen, E. and Rissanen, K., Synthesis, characterization and thermal properties of nine quaternary dialkyldiaralkylammonium chlorides, *J. Mol. Struct.* **787** (1–3), 18–30 (2006).

668 Ubbelohde, A.R., *The Molten State of Matter: Melting and Crystal Structure*, (John Wiley & Sons, Ltd, Chichester, 1978), pp. 220–232.

669 McMullan, R. and Jeffrey, G.A., Hydrates of the tetra *n*-butyl and tetra *i*-amyl quaternary ammonium salts, *J. Chem. Phys.* **31**, 1231–1234 (1959).

670 Feil, D. and Jeffrey, G.A., The polyhedral clathrate hydrates, part 2. Structure of the hydrate of tetra *iso*-amyl ammonium fluoride, *J. Chem. Phys.* **35**, 1863–1872 (1961).

671 Swatloski, R.P., Holbrey, J.D. and Rogers, R.D., Ionic liquids are not always green: Hydrolysis of 1-butyl-3-methylimidazolium hexafluorophosphate, *Green Chem.* **5** (4), 361–363 (2003).

672 Zhu, Z.-Q., Jiang, M.-Y., Zhang, C.-G. and Xiao, J.-C., Efficient synthesis of 1-alkyl-3-methylimidazolium fluorides and possibility of the existence of hydrogen bonding between fluoride anion and C(sp^3)–H, *J. Fluor. Chem.* **133**, 160–162 (2012).

673 Lux, D., Schwarz, W. and Hess, H., Methylammoniumfluoride CH_3NH_3F, *Cryst. Struct. Commun.* **8** (1), 41–43 (1979).

674 Mahjoub, A.R., Zhang, X.Z. and Seppelt, K., Reactions of the naked fluoride-ion – syntheses and structures of SeF_6^{2-} and BrF_6, *Chem. Eur. J.* **1** (4), 261–265 (1995).

675 Rudman, R., Lippman, R., Gowda, D.S.S. and Eilerman, D., Polymorphism of crystalline poly(hydroxymethyl) compounds .8. Structures of the tris(hydroxymethyl) aminomethane hydrogenhalides, $(HOH_2C)_3CNH_3^+\cdot X^-$ (X = F, Cl, Br, I), *Acta Crystallogr.* **C39** (September), 1267–1271 (1983).

676 Kornath, A., Neumann, F. and Oberhammer, H., Tetramethylphosphonium fluoride: 'Naked' fluoride and phosphorane, *Inorg. Chem.* **42** (9), 2894–2901 (2003).

677 Christe, K.O., Haiges, R., Boatz, J.A., Jenkins, H.D.B., Garner, E.B. and Dixon, D.A., Why Are $[P(C_6H_5)_4]^+N_3^-$ and $[As(C_6H_5)_4]^+N_3^-$ ionic salts and $Sb(C_6H_5)_4N_3$ and $Bi(C_6H_5)_4N_3$ covalent solids? A theoretical study provides an unexpected answer, *Inorg. Chem.* **50** (8), 3752–3756 (2011).

678 Haiges, R., Schroer, T., Yousufuddin, M. and Christe, K.O., The syntheses and structures of Ph_4EN_3 (E = P, As, Sb), an example for the transition from ionic to covalent azides within the same main group, *Z. Anorg. Allg. Chem.* **631** (13–14), 2691–2695 (2005).

679 Crawford, M.J. and Klapötke, T.M., The synthesis of tetraphenylarsonium azide, $[Ph_4As]^+[N_3]^-$, and the attempted preparation of tetraphenylarsonium azidodithiocarbonate, $[Ph_4As]^+[SCSN_3]^-$, *Heteroat. Chem.* **10** (4), 325–329 (1999).

680 Endo, T., Kato, T. and Nishikawa, K., Effects of methylation at the 2 position of the cation ring on phase behaviors and conformational structures of imidazolium-based ionic liquids, *J. Phys. Chem. B* **114** (28), 9201–9208 (2010).

681 Hagfeldt, A., Boschloo, G., Sun, L., Kloo, L. and Pettersson, H., Dye-sensitized solar cells, *Chem. Rev.* **110** (11), 6595–6663 (2010).

682 Prescott, A.B. and Trowbridge, P.F., Periodides of pyridine, *J. Am. Chem. Soc.* **17**, 859–869 (1895).

683 Ang, H.G., Fraenk, W., Karaghiosoff, K., Klapötke, T.M., Nöth, H., Sprott, J., Suter, M., Vogt, M. and Warchhold, M., Synthesis, characterization, and crystal structures of various energetic urotropinium salts with azide, nitrate, dinitramide and azotetrazolate counter ions, *Z. Anorg. Allg. Chem.* **628** (13), 2901–2906 (2002).

684 Xue, H., Gao, H., Twamley, B. and Shreeve, J.M., Energetic nitrate, perchlorate, azide and azolate salts of hexamethylenetetramine, *Eur. J. Inorg. Chem.* (15), 2959–2965 (2006).

685 Kornath, A. and Blecher, O., Crystal structure and thermal phase transition of tetramethylammonium cyanide, $(CH_3)_4N^+CN^-$, *Z. Naturforsh. B* **54** (3), 372–376 (1999).

686 Erhardt, S., Grushin, V.V., Kilpatrick, A.H., Macgregor, S.A., Marshall, W.J. and Roe, D.C., Mechanisms of catalyst poisoning in palladium-catalyzed cyanation of haloarenes. Remarkably facile C–N bond activation in the $[(Ph_3P)_4Pd]/[Bu_4N]^+CN^-$ system, *J. Am. Chem. Soc.* **130** (14), 4828–4845 (2008).

687 Kuhn, N., Eichele, K., Steimann, M., Al-Sheikh, A., Doser, B. and Ochsenfeld, C., Hydrogen bonds with cyanide ions? The structures of 1,3-diisopropyl-4,5-dimethylimidazolium cyanide and 1-isopropyl-3,4,5-trimethylimidazolium cyanide, *Z. Anorg. Allg. Chem.* **632** (14), 2268–2275 (2006).

688 Bernsdorf, A. and Koeckerling, M., Crystal structure of tetraphenylphosphonium cyanate, $[P(C_6H_5)_4]OCN$, *Z. Kristallogr. NCS.* **227** (1), 85–86 (2012).

689 Kobler, H., Munz, R., Algasser, G. and Simchen, M., Eine einfache Synthese von Tetraalkylammoniumsalzen mit funktionellen Anionen, *Liebigs Ann. Chem.* **1978**, 1937–1945 (1978).

690 Dubovitskii, F.I., Golovina, N.I., Pavlov, A.N. and Atovmyan, L.O., Structural features of salts of dinitramide with alkaline metals, *Dokl. Akad. Nauk.* **355** (20), 200–202 (1997).

691 Strehmel, V., Laschewsky, A., Kraudelt, H., Wetzel, H. and Görnitz, E., '*Ionic Liquids in Polymer Systems*', ACS Symposium Series., Vol. **913** (2005), pp. 17–36.

692 Vitz, J., Erdmenger, T., Haensch, C. and Schubert, U.S., Extended dissolution studies of cellulose in imidazolium based ionic liquids, *Green Chem.* **11** (3), 417–424 (2009).

693 Hurley, F.H. and Wier, T.P., The electrodeposition of aluminium from nonaqueous solutions at room temperature, *J. Electrochem. Soc.* **98**, 207–212 (1951).

694 Hurley, F.H. and Wier, T.P., Electrodeposition of metals from fused quaternary ammonium salts, *J. Electrochem. Soc.* **98**, 203–206 (1951).

695 Graenacher, C., *Cellulose solution*, US Pat., 1943176, 4pp (1934).

696 Sanders, J.R., Ward, E.H. and Hussey, C.L., Aluminum bromide-1-methyl-3-ethylimidazolium bromide ionic liquids. 1. Densities, viscosities, electrical conductivities, and phase-transitions, *J. Electrochem. Soc.* **133** (2), 325–330 (1986).

697 Kim, Y.J. and Varma, R.S., Tetrahaloindate(III)-based ionic liquids in the coupling reaction of carbon dioxide and epoxides to generate cyclic carbonates: H-bonding and mechanistic studies, *J. Org. Chem.* **70** (20), 7882–7891 (2005).

698 Williams, S.D., Schoebrechts, J.P., Selkirk, J.C. and Mamantov, G., A new room-temperature molten-salt solvent system – organic cation tetrachloroborates, *J. Am. Chem. Soc.* **109** (7), 2218–2219 (1987).

699 Desiraju, G.R. (ed.), *Crystal Design: Structure and Function*. Perspectives in Supramolecular Chemistry, Vol. **7th** (John Wiley & Sons, Ltd, Chichester, 2004).

700 Smiglak, M., Holbrey, J.D., Griffin, S.T., Reichert, W.M., Swatloski, R.P., Katritzky, A.R., Yang, H., Zhang, D., Kirichenko, K. and Rogers, R.D., Ionic liquids *via* reaction of the zwitterionic 1,3-dimethylimidazolium-2-carboxylate with protonic acids. Overcoming synthetic limitations and establishing new halide free protocols for the formation of ILs, *Green Chem.* **9** (1), 90–98 (2007).

701 Billard, I., Moutiers, G., Labet, A., El Azzi, A., Gaillard, C., Mariet, C. and Lützenkirchen, K., Stability of divalent europium in an ionic liquid: Spectroscopic investigations in 1-methyl-3-butylimidazolium hexafluorophosphate, *Inorg. Chem.* **42** (5), 1726–1733 (2003).

702 Bonhôte, P., Dias, A.P., Papageorgiou, N., Kalyanasundaram, K. and Grätzel, M., Hydrophobic, highly conductive ambient-temperature molten salts, *Inorg. Chem.* **35** (5), 1168–1178 (1996).

703 Reiter, J., Jeremias, S., Paillard, E., Winter, M. and Passerini, S., Fluorosulfonyl-(trifluoromethanesulfonyl)imide ionic liquids with enhanced asymmetry, *Phys. Chem. Chem. Phys.* **15** (7), 2565–2571 (2013).

704 Beran, M., Příhoda, J. and Taraba, J., A new route to the syntheses of *N*-(fluorosulfuryl)sulfonamide salts: Crystal structure of Ph_4P^+ $[CF_3SO_2NSO_2F]^-$, *Polyhedron* **29** (3), 991–994 (2010).

705 Lopes, J.N.C., Shimizu, K., Pádua, A.A.H., Umebayashi, Y., Fukuda, S., Fujii, K. and Ishiguro, S.I., A tale of two ions: The conformational landscapes of bis(trifluoromethanesulfonyl)amide and *N,N*-dialkylpyrrolidinium, *J. Phys. Chem. B* **112** (5), 1465–1472 (2008).

706 Lopes, J.N.C. and Pádua, A.A.H., Molecular force field for ionic liquids composed of triflate or bistriflylimide anions, *J. Phys. Chem. B* **108** (43), 16893–16898 (2004).

707 Nockemann, P., Pellens, M., Van Hecke, K., Van Meervelt, L., Wouters, J., Thijs, B., Vanecht, E., Parac-Vogt, T.N., Mehdi, H., Schaltin, S., Fransaer, J., Zahn, S., Kirchner, B. and Binnemans, K., Cobalt(II) complexes of nitrile-functionalized ionic liquids, *Chem. Eur. J.* **16** (6), 1849–1858 (2010).

708 Weitze, A., Henschel, D., Blaschette, A. and Jones, P.G., Polysulfonyl Amines. LXIX. Novel pnictogen disulfonylamides: Synthesis of bismuth dimesylamides and crystal structures of the twelve-membered cyclodimer $[Ph_2BiN(SO_2Me)_2]$, and of the ionic complex $[H(OAsPh_3)_2]$+$(MeSO_2)_2N^-$, *Z. Anorg. Allg. Chem.* **621** (10), 1746–1754 (1995).

709 Stenzel, O., Raubenheimer, H.G. and Esterhuysen, C., Biphasic hydroformylation in new molten salts-analogies and differences to organic solvents, *J. Chem. Soc. Dalton Trans.* (6), 1132–1138 (2002).

710 Froschauer, C., Wurst, K., Laus, G. and Schottenberger, H., Crystal structure of 1,3-bis(trimethylsilyl)imidazolium trifluoromethanesulfonate, $[C_9H_{21}N_2Si_2]$ $[CF_3SO_3]$, *Z. Kristallogr. NCS.* **226** (4), 545–546 (2011).

711 Belieres, J.P. and Angell, C.A., Protonic ionic liquids: Preparation, characterization, and proton free energy level representation, *J. Phys. Chem. B* **111** (18), 4926–4937 (2007).

712 Choi, C.S., Mapes, J.E. and Prince, E., Structure of ammonium nitrate(IV), *Acta Crystallogr.* **B28**, 1357–1361 (1972).

713 Mylrajan, M. and Srinivasan, T.K.K., Vibrational study of phase-transitions in $(CH_3)_3NHNO_3$, *J. Chem. Phys.* **89** (3), 1634–1641 (1988).

714 Mylrajan, M. and Srinivasan, T.K.K., A vibrational study of phase-transitions in $(CH_3)_4NNO_3$, *J. Phys. C: Solid State Phys.* **21** (8), 1673–1690 (1988).

715 Ilyukhin, A.B. and Petrosyants, S.P., Crystalline structure of $[NMe_4]NO_3$, *Kristallografiya* **38** (5), 209–211 (1993).

716 Salo, K., Westerlund, J., Andersson, P.U., Nielsen, C., D'Anna, B. and Hallquist, M., Thermal characterization of aminium nitrate nanoparticles, *J. Phys. Chem. A* **115** (42), 11671–11677 (2011).

717 Roobottom, H.K., Jenkins, H.D.B., Passmore, J. and Glasser, L., Thermochemical radii of complex ions, *J. Chem. Educ.* **76** (11), 1570–1573 (1999).

718 Ngo, H.L., LeCompte, K., Hargens, L. and McEwen, A.B., Thermal properties of imidazolium ionic liquids, *Thermochim. Acta* **357**, 97–102 (2000).

719 Moe, N.S., Tetraethyl ammonium fluoroborate as a supporting electrolyte in polarography, *Acta Chem. Scand.* **19**, 1023–1024 (1965).

720 'Database of Select Committee on GRAS Substances (SCOGS) Reviews', http://www.fda.gov/Food/FoodIngredientsPackaging/GenerallyRecognizedasSafeGRAS/GRASSubstancesSCOGSDatabase/default.htm (accessed 1 August 2015).

721 Trefonas, L.M. and Towns, R., Crystal and molecular structure of 8,8-dimethyl-8-azoniabicyclo [5.1.0] octane iodide, *J. Heterocycl. Chem.* **1** (1), 19–22 (1965).

722 Trefonas, L.M. and Majeste, R., Crystal and molecular structure of 9,9-dimethyl-9-azonia-bicyclo [6.1.0] nonane iodide *Tetrahedron* **19**, 929–935 (1963).

723 Trefonas, L.M., Towns, R. and Majeste, R., The crystal molecular structure of *cis*-13,13-dimethyl-13-azoniabicyclo [10.1.0] tridecane iodide, *J. Heterocycl. Chem.* **4**, 511–516 (1967).

724 Duriska, M.B., Grondin, J., Servant, L., Birot, M. and Deleuze, H., Synthesis and characterization of novel ionic liquids: *N*-Substituted aziridinium salts, *J. Heterocycl. Chem.* **49** (3), 652–657 (2012).

725 Zhou, Q., Boyle, P.D., Malpezzi, L., Mele, A., Shin, J.H., Passerini, S. and Henderson, W.A., Phase behavior of ionic liquid-LiX mixtures: Pyrrolidinium cations and $TFSI^-$ anions - linking structure to transport properties, *Chem. Mater.* **23** (19), 4331–4337 (2011).

726 Castellari, I. and Sabatino, P., Antiinflammatory drugs – 1-(2-hydroxyethyl)pyrrolidinium salt of diclofenac, *Acta Crystallogr.* **C50**, 1723–1726 (1994).

727 Barczyński, P., Dega-Szafran, Z. and Szafran, M., Spectroscopic differences between molecular (O–H···N) and ionic pair (O^-···H–N^+) hydrogen complexes, *J. Chem. Soc. Perkin Trans. II* 765–771 (1985).

728 Turner, E.A., Pye, C.C. and Singer, R.D., Use of ab initio calculations toward the rational design of room temperature ionic liquids, *J. Phys. Chem. A* **107** (13), 2277–2288 (2003).

729 Endo, T., Kato, T., Tozaki, K. and Nishikawa, K., Phase behaviors of room temperature ionic liquid linked with cation conformational changes: 1-Butyl-3-methylimidazolium hexafluorophosphate, *J. Phys. Chem. B* **114** (1), 407–411 (2010).

730 Blokhin, A.V., Paulechka, Y.U., Strechan, A.A. and Kabo, G.J., Physicochemical properties, structure, and conformations of 1-butyl-3-methylimidazolium bis(trifluoromethanesulfonyl)imide $[C_4mim]NTf_2$ ionic liquid, *J. Phys. Chem. B* **112** (14), 4357–4364 (2008).

731 Fei, Z.F., Zhao, D.B., Scopelliti, R. and Dyson, P.J., Organometallic complexes derived from alkyne-functionalized imidazolium salts, *Organometallics* **23** (7), 1622–1628 (2004).

732 Li, H., Jin, L.Y. and Tao, R.J., 1,3-Diprop-2-ynyl-1*H*-imidazol-3-ium bromide, *Acta Crystallogr.* **E64**, o900 (2008).

733 Froschauer, C., Sixta, H., Weber, H.K., Laus, G., Kahlenberg, V. and Schottenberger, H., A superior new route to methyl phosphonate-based ionic liquids, *Chem. Lett.* **41** (9), 945–946 (2012).

734 Whittell, G.R., Balmond, E.I., Robertson, A.P.M., Patra, S.K., Haddow, M.F. and Manners, I., Reactions of amine- and phosphane-borane adducts with frustrated Lewis pair combinations of group 14 triflates and sterically hindered nitrogen bases, *Eur. J. Inorg. Chem.* **2010** (25), 3967–3975 (2010).

735 Hensen, K., Gebhardt, F. and Bolte, M., Syntheses and crystal structure determination of addition compounds of alkyldimethylbromosilanes and *N*-trimethylsilylimidazole, *Z. Naturforsch. B* **52** (12), 1491–1496 (1997).

736 Freemantle, M., *An Introduction to Ionic Liquids* (RSC Publications, Cambridge, 2010), p. 12.

737 Holmes, J.D. and Pettit, R., The synthesis and properties of homotropone, *J. Am. Chem. Soc.* **85**, 2531–2532 (1963).

738 Kraft, A., Beck, J., Steinfeld, G., Scherer, H., Himmel, D. and Krossing, I., Synthesis and application of strong Brønsted acids generated from the Lewis acid $Al(OR^F)_3$ and an alcohol, *Organometallics* **31** (21), 7485–7491 (2012).

739 Haiges, R., Rahm, M. and Christe, K.O., Unprecedented conformational variability in main group inorganic chemistry: the tetraazidoarsenite and -antimonite salts $A^+[M(N_3)_4]^-$ ($A = NMe_4$, PPh_4, $(Ph_3P)_2N$; M = As, Sb), five similar salts, five different anion structures, *Inorg. Chem.* **52** (1), 402–414 (2013).

740 Caballero, A., Bennett, S., Serpell, C.J. and Beer, P.D., Iodo-imidazolium salts: Halogen bonding in crystals and anion-templated pseudorotaxanes, *CrystEngComm* **15** (16), 3076–3081 (2013).

741 Haller, H., Ellwanger, M., Higelin, A. and Riedel, S., Investigation of polybromide monoanions of the series $[NAlk]_4Br_9$ (Alk = Methyl, Ethyl, Propyl, Butyl), *Z. Anorg. Allg. Chem.* **638** (3–4), 553–558 (2012).

742 Mootz, D. and Steffen, M., Crystal-structures of acid hydrates and oxonium salts. 20. Oxonium tetrafluoroborates H_3OBF_4, $[H_5O_2]BF_4$, and $[H(CH_3OH)_2]BF_4$, *Z. Anorg. Allg. Chem.* **482** (11), 193–200 (1981).

743 Mootz, D. and Steffen, M., Crystal-structures of acid hydrates and oxonium salts.17. Tetrafluoroboric acid methanol (1-2), cyclic molecules by hydrogen-bonds between ions, *Angew. Chem.-Int. Ed. Engl.* **20** (2), 196 (1981).

744 Henderson, W.A., Fylstra, P., De Long, H.C., Trulove, P.C. and Parsons, S., Crystal structure of the ionic liquid $EtNH_3NO_3$-insights into the thermal phase behavior of protonic ionic liquids, *Phys. Chem. Chem. Phys.* **14** (46), 16041–16046 (2012).

745 Belhocine, T., Forsyth, S.A., Gunaratne, H.Q.N., Nieuwenhuyzen, M., Nockemann, P., Puga, A.V., Seddon, K.R., Srinivasan, G. and Whiston, K., Azepanium ionic liquids, *Green Chem.* **13** (11), 3137–3155 (2011).

746 Becker, M., Harloff, J., Jantz, T., Schulz, A. and Villinger, A., Structure and bonding of tetracyanopyrrolides, *Eur. J. Inorg. Chem.* (34), 5658–5667 (2012).

747 Depuydt, D., Brooks, N.R., Schaltin, S., Van Meervelt, L., Fransaer, J. and Binnemans, K., Silver-containing ionic liquids with alkylamine ligands, *ChemPlusChem* **78** (6), 578–588 (2013).

748 Smith, G., Wermuth, U.D. and Healy, P.C., A dianionic 5-sulfonatosalicylate species in the proton-transfer compound bis(benzylaminium) 3-carboxylato-4-hydroxybenzenesulfonate, *Acta Crystallogr.* **E62**, o1863–o1865 (2006).

749 Froschauer, C., Salchner, R., Laus, G., Weber, H.K., Tessadri, R., Griesser, U., Wurst, K., Kahlenberg, V. and Schottenberger, H., 1,3-Di(alkoxy)imidazolium-based ionic liquids: Improved synthesis and crystal structures, *Aust. J. Chem.* **66** (3), 391–395 (2013).

750 Yang, M., Mallick, B. and Mudring, A.V., On the mesophase formation of 1,3-dialkylimidazolium ionic liquids, *Cryst. Growth Des.* **13** (7), 3068–3077 (2013).

751 Katsyuba, S.A., Vener, M.V., Zvereva, E.E., Fei, Z., Scopelliti, R., Laurenczy, G., Yan, N., Paunescu, E. and Dyson, P.J., How strong is hydrogen bonding in ionic liquids? Combined X-ray crystallographic, infrared/Raman spectroscopic, and density functional theory study, *J. Phys. Chem. B* **117**, 9094–9105 (2013).

752 Prodius, D., Macaev, F., Stingaci, E., Pogrebnoi, V., Mereacre, V., Novitchi, G., Kostakis, G.E., Anson, C.E. and Powell, A.K., Catalytic 'triangles': Binding of iron in task-specific ionic liquids, *Chem. Commun.* **49** (19), 1915–1917 (2013).

753 Klapötke, T.M., Piercey, D.G. and Stierstorfer, J., The 1,3-diamino-1,2,3-triazolium cation: A highly energetic moiety, *Eur. J. Inorg. Chem.* (9), 1509–1517 (2013).

754 Delahaye, E., Göbel, R., Löbbicke, R., Guillot, R., Sieber, C. and Taubert, A., Silica ionogels for proton transport, *J. Mater. Chem.* **22** (33), 17140–17146 (2012).

755 Sabaté, C.M., Delalu, H. and Jeanneau, E., Synthesis, characterization, and energetic properties of salts of the 1-cyanomethyl-1,1-dimethylhydrazinium cation, *Chem. Asian. J.* **7** (5), 1085–1095 (2012).

756 Beichel, W., Yu, Y., Dlubek, G., Krause Rehberg, R., Pionteck, J., Pfefferkorn, D., Bulut, S., Bejan, D., Friedrich, C. and Krossing, I., Free volume in ionic liquids: A connection of experimentally accessible observables from positron annihilation lifetime spectroscopy and pressure-volume-temperature experiments with the molecular structure from X-ray diffraction data, *Phys. Chem. Chem. Phys.* **15**, 8821–8830 (2013).

757 Rouch, A., Castellan, T., Fabing, I., Saffon, N., Rodriguez, J., Constantieux, T., Plaquevent, J.-C. and Génisson, Y., Tartrate-based ionic liquids: Unified synthesis and characterization, *RSC Adv.* **3**, 413–426 (2013).

758 Janikowski, J., Razali, M.R., Forsyth, C.M., Nairn, K.M., Batten, S.R., MacFarlane, D.R. and Pringle, J.M., Physical properties and structural characterization of ionic liquids and solid electrolytes utilizing the carbamoylcyano(nitroso)methanide anion, *ChemPlusChem* **78** (6), 486–497 (2013).

759 Izgorodina, E.I., Chesman, A.S.R., Turner, D.R., Deacon, G.B. and Batten, S.R., Theoretical and experimental insights into the mechanism of the nucleophilic addition of water and methanol to dicyanonitrosomethanide, *J. Phys. Chem. B* **114** (49), 16517–16527 (2010).

760 Aggarwal, R., Singh, S. and Hundal, G., Synthesis, characterization, and evaluation of surface properties of cyclohexanoxycarbonylmethylpyridinium and cyclohexanoxycarbonylmethylimidazolium ionic liquids, *Ind. Eng. Chem. Res.* **52** (3), 1179–1189 (2013).

761 Golding, I.R., Starikova, Z.A., Senchenya, N.G., Petrovskii, P.V., Garbuzova, I.A. and Gololobov, Y.G., Triethylamine-assisted reaction between 2,4-dinitrofluorobenzene and malononitrile, *Russ. Chem. Bull., Int. Ed.* **60** (10), 1995–1998 (2012).

762 Funasako, Y., Inagaki, T., Mochida, T., Sakurai, T., Ohta, H., Furukawa, K. and Nakamura, T., Organometallic ionic liquids from alkyloctamethylferrocenium cations: Thermal properties, crystal structures, and magnetic properties, *Dalton Trans.* **42**, 8317–8327 (2013).

763 Froschauer, C., Weber, H.K., Kahlenberg, V., Laus, G. and Schottenberger, H., Iminium salts by Meerwein alkylation of Ehrlich's aldehyde, *Crystals* **3**, 248–256 (2013).

764 Sharutin, V.V., Senchurin, V.S., Sharutina, O.K., Kunkurdonova, B.B. and Platonova, T.P., Synthesis and structure of bismuth complex $[n\text{-}Bu_4N]_2^+[Bi_2I_8 \cdot 2Me_2S{=}O]^{2-}$, *Russ. J. Inorg. Chem.* **56** (8), 1272–1275 (2011).

765 Beckmann, J., Bolsinger, J., Duthie, A., Finke, P., Lork, E., Lüdtke, C., Mallow, O. and Mebs, S., Mesityltellurenyl cations stabilized by triphenylpnictogens $[MesTe(EPh_3)]^+$ (E = P, As, Sb), *Inorg. Chem.* **51** (22), 12395–12406 (2012).

766 Lava, K., Evrard, Y., Van Hecke, K., Van Meervelt, L. and Binnemans, K., Quinolinium and isoquinolinium ionic liquid crystals, *RSC Adv.* **2** (21), 8061–8070 (2012).

767 Farnham, W.B. and Calabrese, J.C., Novel hypervalent (10-I-2) iodine structures, *J. Am. Chem. Soc.* **108** (9), 2449–2451 (1986).

768 Adamová, G., Gardas, R.L., Nieuwenhuyzen, M., Puga, A.V., Rebelo, L.P.N., Robertson, A.J. and Seddon, K.R., Alkyltributylphosphonium chloride ionic liquids: Synthesis, physicochemical properties and crystal structure, *Dalton Trans.* **41**, 8316–8332 (2012).

769 Smiglak, M., Hines, C.C., Reichert, W.M., Shamshina, J.L., Beasley, P.A., McCrary, P.D., Kelley, S.P. and Rogers, R.D., Azolium azolates from reactions of neutral azoles with 1,3-dimethyl-imidazolium-2-carboxylate, 1,2,3-trimethyl-imidazolium hydrogen carbonate, and *N,N*-dimethyl-pyrrolidinium hydrogen carbonate, *New J. Chem.* **37** (5), 1461–1469 (2013).

770 Holthausen, M.H., Hepp, A. and Weigand, J.J., Synthesis of cationic $R_2P_5^+$ cages and subsequent chalcogenation reactions, *Chem. Eur. J.* **19** (30), 9895–9907 (2013).

771 Junge, D.M., Scadova, D.R., Golen, J.A. and Jasinski, J.P., 1-Butyl-3-ethyl-1*H*-benzimidazol-3-ium tetrafluoroborate, *Acta Crystallogr.* **E68** (Pt 10), o2862 (2012).

772 Pratt, H.D., III, Leonard, J.C., Steele, L.A.M., Staiger, C.L. and Anderson, T.M., Copper ionic liquids: Examining the role of the anion in determining physical and electrochemical properties, *Inorg. Chim. Acta* **396**, 78–83 (2013).

773 Tanzer, E.-M., Zimmer, L.E., Schweizer, W.B. and Gilmour, R., Fluorinated organocatalysts for the enantioselective epoxidation of enals: Molecular preorganisation by the fluorine-iminium ion *gauche* effect, *Chem. Eur. J.* **18** (36), 11334–11342 (2012).

774 Loginov, S.V., Abramkin, A.M., Rybakov, V.B., Sheludyakov, V.D. and Storozhenko, P.A., Molecular and crystal structure of tris-(2-hydroxyethyl)ammonium 4-chlorophenoxy acetate, *Crystallogr. Rep.* **57** (4), 521–523 (2012).

775 Knight, F.R., Randall, R.A.M., Arachchige, K.S.A., Wakefield, L., Griffin, J.M., Ashbrook, S.E., Bühl, M., Slawin, A.M.Z. and Woollins, J.D., Noncovalent interactions in peri-substituted chalconium acenaphthene and naphthalene salts: A combined experimental, crystallographic, computational, and solid-state NMR study, *Inorg. Chem.* **51** (20), 11087–11097 (2012).

776 Maaliki, C., Lepetit, C., Duhayon, C., Canac, Y. and Chauvin, R., Carbene-stabilized phosphenium oxides and sulfides, *Chem. Eur. J.* **18** (50), 16153–16160 (2012).

777 Laus, G. and Kahlenberg, V., Crystal structures of 1-hydroxyimidazole and its salts, *Crystals* **2**, 1492–1501 (2012).

778 Mori, S. and Mochida, T., Organometallic ionic liquids from cationic arene-ruthenium complexes, *Organometallics* **32** (3), 780–787 (2013).

779 Lund, H., Harloff, J., Schulz, A. and Villinger, A., Synthesis and structure of ionic liquids containing the $[Al(OC_6H_4CN)_4]$ anion, *Z. Anorg. Allg. Chem.* **639** (5), 754–764 (2013).

780 Knyazeva, I.R., Sokolova, V.I., Burilov, A.R., Pudovik, M.A., Dobrynin, A.B., Kataeva, O.N. and Sinyashin, O.G., New triamidophosphonium acetals and their condensation with resorcinol and its derivatives, *Russ. Chem. Bull., Int. Ed.* **61** (3), 631–637 (2012).

781 Xue, H., Twamley, B. and Shreeve, J.M., The first 1-alkyl-3-perfluoroalkyl-4,5-dimethyl-1,2,4-triazolium salts, *J. Org. Chem.* **69** (4), 1397–1400 (2004).

782 Sharutin, V.V., Senchurin, V.S., Sharutina, O.K., Pakusina, A.P. and Fastovets, O.A., Synthesis and structure of gold and copper complexes: $[Ph_3PhCH_2P]^+[AuCl_4]^-$, $[NH(C_2H_4OH)_3]^+[AuCl_4]^-\cdot H_2O$, and $[Ph_3EtP]^{2+}[Cu_2Cl_6]^{2-}$, *Russ. J. Inorg. Chem.* **55** (9), 1415–1420 (2010).

783 Sharutin, V.V., Senchyrin, V.S., Sharutina, O.K. and Kunkurdonova, B.B., Crystal structures of triphenylamylphosphonium iodide $[Ph_3AmP]I$ and triphenylamylphosphonium tetraiodide $[Ph_3AmP]I_4$, *Russ. J. Inorg. Chem.* **57** (1), 57–61 (2012).

784 Jensen, B., The structures of acetylcholine hydrogen tartrates, *Acta Crystallogr.* **B38**, 1185–1192 (1982).

785 Mirjafari, A., Pham, L.N., McCabe, J.R., Mobarrez, N., Salter, E.A., Wierzbicki, A., West, K.N., Sykora, R.E. and Davis, J.H., Building a bridge between aprotic and protic ionic liquids, *RSC Adv.* **3** (2), 337–340 (2013).

786 Ilić, D., Damljanović, I., Vukićević, M., Kahlenberg, V., Laus, G., Radulović, N.S. and Vukićević, R.D., Dimethyl(2-oxo-2-ferrocenylethyl)sulfonium iodide-a useful synthetic equivalent of ferrocenoylcarbene in the synthesis of ferrocene-containing cyclopropanes, *Tetrahedron. Lett.* **53** (45), 6018–6021 (2012).

787 Ibad, M.F., Langer, P., Reiß, F., Schulz, A. and Villinger, A., Catalytic trimerization of bis-silylated diazomethane, *J. Am. Chem. Soc.* **134** (42), 17757–17768 (2012).

788 Chang, J.-C., Ho, W.-Y., Sun, I.W., Chou, Y.-K., Hsieh, H.-H., Wu, T.-Y. and Liang, S.-S., Synthesis and properties of new (μ-oxo)bis[trichloroferrate(III)] dianion salts incorporated with dicationic moiety, *Polyhedron* **29** (15), 2976–2984 (2010).

789 Zagumennov, V.A., Sizova, N.A., Lodochnikova, O.A. and Litvinov, I.A., Crystal structure of (2,2-dimethyl-3-methylidenenorbonane)triethyl-phosphonium perchlorate, *J. Struct. Chem.* **53** (1), 206–208 (2012).

790 Crawford, M.J., Karaghiosoff, K., Klapötke, T.M. and Martin, F.A., Synthesis and characterization of 4,5-dicyano-2*H*-1,2,3-triazole and its sodium, ammonium, and guanidinium salts, *Inorg. Chem.* **48** (4), 1731–1743 (2009).

791 Funasako, Y., Nosho, M. and Mochida, T., Ionic liquids from copper(II) complexes with alkylimidazole-containing tripodal ligands, *Dalton Trans.* **42**, 10138–10143 (2013).

792 Peppel, T., Hinz, A. and Köckerling, M., Salts with the $[NiBr_3(L)]^-$ complex anion (L = 1-methylimidazole, 1-methylbenzimidazole, quinoline, and triphenylphosphane) and low melting points: A comparative study, *Polyhedron* **52**, 482–490 (2013).

793 Bucher, N., Szabo, J., Oppel, I.M. and Maas, G., Derivatives of the triaminoguanidinium Ion, 1. Synthesis, crystal and molecular structures of 1,2,3-tris(benzylamino) guanidinium salts, *Z. Naturforsch. B* **67** (6), 631–642 (2012).

794 Holthausen, M.H. and Weigand, J.J., Preparation of cationic $[(R_2N)P_5Cl]^+$-cage compounds from $[(R_2N)PCl]^+$ and P_4, *Z. Anorg. Allg. Chem.* **638** (7–8), 1103–1108 (2012).

795 Sobral, A., Lopes, S.H., Gonsalves, A., Silva, M.R., Beja, A.M., Paixao, J.A. and da Veiga, L.A., Synthesis and crystal structure of new phase-transfer catalysts based on 1,8-diazabicyclo[5.4.0]undec-7-ene and 1,5-diazabicyclo[4.3.0]non-5-ene, in *Trends in Colloid and Interface Science XVI*, Eds. M.G. Miguel and H.D. Burrows, Progress in Colloid and Polymer Science, Vol. **123** (Springer-Verlag, Berlin, 2004), pp. 28–30.

796 Zhang, B., Köberl, M., Pöthig, A., Cokoja, M., Herrmann, W.A. and Kühn, F.E., Synthesis and characterization of imidazolium salts with the weakly coordinating $[B(C_6F_5)_4]^-$ anion, *Z. Naturforsch. B* **67** (10), 1030–1036 (2012).

797 Xuan, X.P., Wang, N. and Xue, Z.K., Synthesis, crystal structure, vibrational spectra and theoretical calculation of 1-carboxymethyl-3-methylimidazolium chloride, *Spectrochim. Acta A* **96**, 436–443 (2012).

798 Xu, F., Matsumoto, K. and Hagiwara, R., The first crystallographic example of a face-sharing fluoroaluminate anion $Al_2F_9^{3-}$, *Dalton Trans.* **42** (6), 1965–1968 (2013).

799 Burba, C.M., Rocher, N.M., Frech, R. and Powell, D.R., Cation-anion interactions in 1-ethyl-3-methylimidazolium trifluoromethanesulfonate-based ionic liquid electrolytes, *J. Phys. Chem. B* **112** (10), 2991–2995 (2008).

800 Ji, S.P., Tang, M., He, L. and Tao, G.H., Water-free rare-earth-metal ionic liquids/ionic liquid crystals based on hexanitratolanthanate(III) anion, *Chem. Eur. J.* **19** (14), 4452–4461 (2013).

801 Coleman, F., Feng, G., Murphy, R.W., Nockemann, P., Seddon, K.R. and Swadźba-Kwaśny, M., Lead(II) chloride ionic liquids and organic/inorganic hybrid materials – a study of chloroplumbate(ii) speciation, *Dalton Trans.* **42**, 5025–5035 (2013).

802 Zhu, S.Z., Arenediazonium bis(trifluoromethanesulphonyl)methides – synthesis and X-ray crystal-structure of p-$CH_3C_6H_4N_2^+(CF_3SO_2)_2CH^-$, *J. Fluor. Chem.* **62** (1), 31–37 (1993).

803 Zhu, S.Z., A new synthetic route to aryl bis(perfluoroalkanesulfonyl)methanes – structures of tolyldiazonium bis(trifluoromethanesulfonyl)methide and 4-nitrophenyl-hydrazono bis(trifluoromethanesulfonyl)methane, *J. Fluor. Chem.* **64** (1–2), 47–60 (1993).

804 Kozma, Á., Petuskova, J., Lehmann, C.W. and Alcarazo, M., Synthesis, structure and reactivity of cyclopropenyl-1-ylidene stabilized S(II), Se(II) and Te(II) mono- and dications, *Chem. Commun.* **49** (39), 4145–4147 (2013).

805 Kozma, Á., Gopakumar, G., Farès, C., Thiel, W. and Alcarazo, M., Synthesis and structure of carbene-stabilized N-centered cations $[L_2N]^+$, $[L_2NR]^{2+}$, $[LNR_3]^{2+}$, and $[L_3N]^{3+}$, *Chem. Eur. J.* **19** (11), 3542–3546 (2013).

806 Mochida, T., Funasako, Y., Inagaki, T., Li, M.J., Asahara, K. and Kuwahara, D., Crystal structures and phase-transition dynamics of cobaltocenium salts with bis(perfluoroalkylsulfonyl)amide anions: Remarkable odd-even effect of the fluorocarbon chains in the anion, *Chem. Eur. J.* **19** (20), 6257–6264 (2013).

807 Inagaki, T., Mochida, T., Takahashi, M., Kanadani, C., Saito, T. and Kuwahara, D., Ionic liquids of cationic sandwich complexes, *Chem. Eur. J.* **18** (22), 6795–6804 (2012).

808 Scholz, F., Himmel, D., Heinemann, F.W., Schleyer, P.V., Meyer, K. and Krossing, I., Crystal structure determination of the nonclassical 2-norbornyl cation, *Science* **341** (6141), 62–64 (2013).

809 Preiss, U.P., Steinfeld, G., Scherer, H., Erle, A.M.T., Benkmil, B., Kraft, A. and Krossing, I., Fluorinated weakly coordinating anions $[M(hfip)_6]$ (M=Nb, Ta): Syntheses, structural characterizations and computations, *Z. Anorg. Allg. Chem.* **639** (5), 714–721 (2013).

810 Briggs, N.E.B., Kennedy, A.R. and Morrison, C.A., 42 salt forms of tyramine: Structural comparison and the occurrence of hydrate formation, *Acta Crystallogr.* **B68**, 453–464 (2012).

811 Saouane, S., Norman, S.E., Hardacre, C. and Fabbiani, F.P.A., Pinning down the solid-state polymorphism of the ionic liquid $[bmim][PF_6]$, *Chem. Sci.* **4** (3), 1270–1280 (2013).

812 Shigemi, M., Takekiyo, T., Abe, H. and Yoshimura, Y., Pressure-induced crystallization of 1-butyl-3-methylimidazolium hexafluorophosphate, *High Pressure Res.* **33** (1), 229–233 (2013).

813 Yoshimura, Y., Abe, H., Imai, Y., Takekiyo, T. and Hamaya, N., Decompression-induced crystal polymorphism in a room-temperature ionic liquid, *N,N*-diethyl-*N*-methyl-*N*-(2-methoxyethyl) ammonium tetrafluoroborate, *J. Phys. Chem. B* **117** (11), 3264–3269 (2013).

814 Rana, U.A., Vijayaraghavan, R., Doherty, C.M., Chandra, A., Efthimiadis, J., Hill, A.J., MacFarlane, D.R. and Forsyth, M., Role of defects in the high ionic conductivity of choline triflate plastic crystal and its acid-containing compositions, *J. Phys. Chem. C* **117** (11), 5532–5543 (2013).

815 Rana, U.A., Vijayaraghavan, R., MacFarlane, D.R. and Forsyth, M., An organic ionic plastic crystal electrolyte based on the triflate anion exhibiting high proton transport, *Chem. Commun.* **47** (22), 6401–6403 (2011).

816 Wally, H., Kratky, C., Weissensteiner, W., Widhalm, M. and Schlogl, K., Ferrocene derivatives.71. Stereochemistry of metallocenes. 56. Synthesis and structure of optically-active ferrocenylaminoalcohols, *J. Organomet. Chem.* **450** (1–2), 185–192 (1993).

817 Porcar, R., Ríos-Lombardía, N., Busto, E., Gotor-Fernández, V., Montejo-Bernardo, J., García-Granda, S., Luis, S.V., Gotor, V., Alfonso, I. and García-

Verdugo, E., Chiral triazolium salts and ionic liquids: From the molecular design vectors to their physical properties through specific supramolecular interactions, *Chem. Eur. J.* **19** (3), 892–904 (2013).

818 Foreiter, M.B., Gunaratne, H.Q.N., Nockemann, P., Seddon, K.R., Stevenson, P.J. and Wassell, D.F., Chiral thiouronium salts: Synthesis, characterisation and application in NMR enantio-discrimination of chiral oxoanions, *New J. Chem.* **37** (2), 515–533 (2013).

819 MacFarlane, D.R., Forsyth, M., Izgorodina, E.I., Abbott, A.P., Annat, G. and Fraser, K., On the concept of ionicity in ionic liquids, *Phys. Chem. Chem. Phys.* **11** (25), 4962–4967 (2009).

820 Ueno, K., Tokuda, H. and Watanabe, M., Ionicity in ionic liquids: Correlation with ionic structure and physicochemical properties, *Phys. Chem. Chem. Phys.* **12** (8), 1649–1658 (2010).

821 Kelley, S.P., Narita, A., Holbrey, J.D., Green, K.D., Reichert, W.M. and Rogers, R.D., Understanding the effects of ionicity in salts, solvates, co-crystals, ionic co-crystals, and ionic liquids, rather than nomenclature, is critical to understanding their behavior, *Cryst. Growth Des.* **13** (3), 965–975 (2013).

822 Andreev, R.V., Borodkin, G.I., Vorob'ev, A.Y., Gatilov, Y.V. and Shubin, V.G., Molecular and crystal structure of 1-amino-X-pyrazinium mesitylenesulfonates, *Russ. J. Org. Chem.* **44** (2), 292–301 (2008).

823 Fumino, K., Fossog, V., Wittler, K., Hempelmann, R. and Ludwig, R., Dissecting anion–cation interaction energies in protic ionic liquids, *Angew. Chem. Int. Ed.* **52**, 2368–2372 (2013).

824 Cojocaru, O.A., Kelley, S.P., Gurau, G. and Rogers, R.D., Procainium acetate versus procainium acetate dihydrate: Irreversible crystallization of a room-temperature active pharmaceutical-ingredient ionic liquid upon hydration, *Cryst. Growth Des.* **13** (8), 3290–3293 (2013).

825 Fischer, N., Fischer, D., Klapötke, T.M., Piercey, D.G. and Stierstorfer, J., Pushing the limits of energetic materials – the synthesis and characterization of dihydroxylammonium 5,5′-bistetrazole-1,1′-diolate, *J. Mater. Chem.* **22** (38), 20418–20422 (2012).

826 Klapötke, T.M., Piercey, D.G. and Stierstorfer, J., The 1,4,5-triaminotetrazolium cation ($CN_7H_6^+$): A highly nitrogen-rich moiety, *Eur. J. Inorg. Chem.* **2012** (34), 5694–5700 (2012).

827 Boneberg, F., Kirchner, A., Klapötke, T.M., Piercey, D.G., Poller, M.J. and Stierstorfer, J., A study of cyanotetrazole oxides and derivatives thereof, *Chem. Asian. J.* **8** (1), 148–159 (2013).

828 Fischer, N., Klapötke, T.M. and Stierstorfer, J., 1-Amino-3-nitroguanidine (ANQ) in high-performance ionic energetic materials, *Z. Naturforsch. B* **67** (6), 573–588 (2012).

829 Farahani, N., Gharagheizi, F., Mirkhani, S.A. and Tumba, K., Ionic liquids: Prediction of melting point by molecular-based model, *Thermochim. Acta* **549**, 17–34 (2012).

830 Kireeva, N., Kuznetsov, S.L. and Tsivadze, A.Y., Toward navigating chemical space of ionic liquids: Prediction of melting points using generative topographic maps, *Ind. Eng. Chem. Res.* **51** (44), 14337–14343 (2012).

831 Yan, F.Y., Xia, S.Q., Wang, Q., Yang, Z. and Ma, P.S., Predicting the melting points of ionic liquids by the quantitative structure property relationship method using a topological index, *J. Chem. Thermodyn.* **62**, 196–200 (2013).

832 Zhang, Y. and Maginn, E.J., Toward fully in silico melting point prediction using molecular simulations, *J. Chem. Theory Comput.* **9**, 1592–1599 (2013).

833 Shimizu, Y., Ohte, Y., Yamamura, Y. and Saito, K., Is the liquid or the solid phase responsible for the low melting points of ionic liquids? Alkyl-chain-length dependence of thermodynamic properties of [$C_n$mim][Tf_2N], *Chem. Phys. Lett.* **470** (4–6), 295–299 (2009).

834 Kick, M., Keil, P. and König, A., Solid-liquid phase diagram of the two ionic liquids EMIMCl and BMIMCl, *Fluid Phase Equilib.* **338**, 172–178 (2013).

INDEX

Note: Page numbers in *italics* refer to Figures; those in **bold** to Tables.

Ionic Liquids Completely UnCOILed: Critical Expert Overviews, First Edition.
Edited by Natalia V. Plechkova and Kenneth R. Seddon.